W9-DHO-042

Earth Science

7TH EDITION

Earth Science

Edward J. Tarbuck
Frederick K. Lutgens

Illinois Central College

Macmillan College Publishing Company
New York

Maxwell Macmillan Canada
Toronto

Maxwell Macmillan International
New York   Oxford   Singapore   Sydney

Cover/photo: The Mittens, Monument Valley Tribal Park, Arizona (Photo by Tom Till)

Title page photo: Southern Alps, New Zealand (Photo by Tom Till)

Editor: Robert McConnin
Production Editor: Rex Davidson
Art Coordinator: Lorraine Woost
Photo Editor: Eloise Marion
Text Designer: Anne Flanagan
Cover Designer: Russ Maselli
Production Buyer: Pamela D. Bennett
Illustrations by Dennis Tasa, Tasa Graphic Arts, Inc.

This book was set in Zapf International by York Graphic Services, Inc. and was printed and bound by R.R. Donnelley & Sons Company. The cover was printed by Lehigh Press, Inc.

Macmillan College Publishing Company
866 Third Avenue
New York, New York 10022

Macmillan College Publishing Company is part of the Maxwell Communication Group of Companies.

Maxwell Macmillan Canada, Inc.
1200 Eglinton Avenue East, Suite 200
Don Mills, Ontario M3C 3N1

Library of Congress Cataloging-in-Publication Data

Tarbuck, Edward J.
  Earth science / Edward J. Tarbuck, Frederick K.
    Lutgens. — 7th ed.
      p.   cm.
  Includes index.
  ISBN 0-02-419025-X
  1. Earth sciences.   I. Lutgens, Frederick K.
  II. Title.
QE26.2.T38   1994
550—dc20   93-33007
                                                          CIP

Printing: 1 2 3 4 5 6 7 8 9   Year: 4 5 6 7 8

Preface

In recent years, media reports have made us increasingly aware of the natural forces at work in our physical environment. News stories graphically portray the violent force of a hurricane, the devastation created by a strong earthquake, and the large numbers left homeless by mudflows and flooding. Such events, and many others as well, are destructive to life and property, and we must be better able to understand and deal with them. However, our natural environment has an even greater importance, for the earth is our home. The earth provides not only the mineral resources so basic to modern society, but it is also the source of most of the ingredients necessary to support life. Therefore, as many members of society as possible should acquire a basic understanding of how the earth works.

With this in mind, we have written a text to help people increase their understanding of our physical environment. We hope this knowledge will encourage some to actively participate in the preservation of the environment, while others may be sufficiently stimulated to pursue a career in the earth sciences. Equally important, however, is our belief that a basic understanding of earth will greatly enhance appreciation of our planet and thereby enrich the reader's life.

Earth Science is a broad and nonquantitative survey at the introductory level of topics in geology, oceanography, meteorology, and astronomy. We have attempted to write a text that is not only informative and timely, but one that is highly usable as well. The language is straightforward and written to be understood by a student with little or no college-level science experience. We have deliberately refrained from using excessive jargon and when new terms are introduced, they are placed in boldface and defined. A list of key terms with page references is found at the end of each chapter, and a glossary is included at the conclusion of the text for easy reference to important terms. Further, review questions conclude each chapter to help the student prepare for exams and quizzes.

In previous editions of this text special attention was given to the quality of photographs and artwork because the earth sciences are highly visual. This emphasis has been maintained in the Seventh Edition. More than 150 new color photographs appear in this revision. The photographs and images were carefully selected to add realism to the subject and to heighten the interest of the reader. Moreover, the already excellent art program of the earlier editions has been strengthened in the Seventh Edition. Because we believe that carefully planned and executed line art will significantly aid student understanding by making difficult concepts less abstract, more than 180 new and redrawn figures appear in the Seventh Edition. Once again, the text has been benefitted greatly from the talents and imaginative production of Dennis Tasa of Tasa Graphic Arts, Inc.

The Seventh Edition of *Earth Science* represents a thorough revision. Extensive rewriting has made many discussions more timely and more readable. Examples include the sections on minerals and metamorphic rocks in Chapter 1, the discussion of mountain building in Chapter 8, and the sections pertaining to changes of state and humidity in Chapter 13. The geology unit includes many other new or revised discussions, including those that deal with glacial deposits (Chapter 4), plate tectonics (Chapter 6), and intrusive igneous features (Chapter 7). Chapter 11 in the oceanography unit includes a revised and expanded section on tidal currents and an updated look at shoreline erosion problems. Many basic discussions in the atmosphere unit have been improved, including those on mechanisms of heat transfer (Chapter 12), atmospheric stability (Chapter 13), and hurricanes (Chapter 15). The text's four-chapter astronomy unit has undergone a great deal of change. Chapter 17, "The Earth's Place in the Universe," has been completely rewritten to improve organization and clarity. Chapters 18 and 19 in the Sixth Edition have been reversed in the Seventh Edition, with the material on lunar geology being placed with the discussion of the solar system. The section on Venus has been updated to reflect information gathered by the *Magellan* spacecraft. Chapter 20, the final chapter in the unit, has also experienced significant rewriting and reorganization, especially the section on stellar evolution.

Because knowledge about our planet and how it works is necessary to our survival and well being, the treatment of environmental and resource topics has always been an important part of *Earth Science*. With each new edition this focus has been given ever-greater emphasis. This is certainly the case with the Seventh Edition. The text integrates a great deal of information about the relationship between people and the physical environment and explores applications of the earth sciences to understanding and solving problems that arise from these interactions. A broad array of hazardous processes are treated, including such phenomena as volcanoes, earthquakes, landslides, floods, tornadoes, and hurricanes. Resources represent another important relationship between earth science and the environment. These include water and soil, a great variety of metallic and nonmetallic minerals, and energy. These substances are the very foundation of modern society. Finally, the planet's rapidly growing human population dramatically influences the magnitude and frequency of many processes that are a natural part of the physical environment. Topics as diverse as river flooding, coastal erosion, and global warming help demonstrate that natural systems do not always adjust to artificial changes in ways that we can anticipate.

A comparison of the Seventh Edition with earlier versions will reveal that a major change to the text is the addition of 62 special interest boxes. This feature was added so that applications, case studies, and interesting facts and related science principles could be explored without significantly disrupting the flow of the basic text discussions. We believe that the boxes provide the reader with a diversity of information that will make the study of earth science more interesting and meaningful.

Although many new topical issues are treated in *Earth Science*, Seventh Edition, it should be emphasized that the main focus of this new edition remains the same as its predecessors—to foster student understanding of basic earth science principles. Whereas student use of the text is a primary concern, the book's adaptability to the needs and desires of the instructor is equally important. Realizing the broad diversity of earth science courses in both content and approach, we have continued to use a relatively non-integrated format to allow maximum flexibility for the instructor. Each of the four major units stands alone; hence, they can be taught in any order. A unit can be omitted entirely without appreciable loss of continuity, and portions of some chapters may be interchanged or excluded at the instructor's discretion.

The authors wish to express their thanks to the many individuals, institutions, government agencies, and businesses that provided photographs and illustrations for use in the text. Further, we would like to acknowledge the aid of our students. Their comments continue to help us maintain our focus on readability and understanding. A special

debt of gratitude goes to those colleagues who prepared indepth prerevision reviews of the Sixth Edition of *Earth Science*. Their critical comments and thoughtful input helped guide our revision and strengthen the text. We wish to thank James R. Albanese, SUNY College, Oneonta; Fred Birmelin, University of Scranton; Robert F. Champlin, Fitchburg State College; Miriam Helen Hill, Indiana University, Southeast; Donald Kampwerth, Okaloosa-Walton Community College; Edward J. Kveton, College of DuPage; Douglas R. Levin, Bryant College; Michael E. Lewis, University of North Carolina, Greensboro; John A. Madson, University of Delaware; Paul D. Nelson, St. Louis Community College at Forest Park; Joel Quam, Augustana College; Gerald E. Schultz, West Texas State University; Douglas Sherman, College of Lake County; Morris L. Sotonoff, Chicago State University; Randall S. Spencer, Old Dominion University; and James D. Stewart, Vincennes University. Our thanks also go to Professor Ken Pinzke, Belleville Area College, and Professor George S. Clark, The University of Manitoba, for their extensive and detailed examinations of the manuscript for the Seventh Edition. Their corrections and suggestions were very helpful. The authors, of course, accept full responsibility for any remaining errors. As always, we appreciate the professional and efficient style of our senior editor, Bob McConnin. Rex Davidson, Anne Flanagan, Russ Maselli, and the rest of the Macmillan production team continue to do a remarkable job. They are true professionals with whom we feel fortunate to be associated.

Supplementary Materials

Available for the first time to accompany *Earth Science*, 7th Edition by Edward J. Tarbuck and Frederick K. Lutgens.

STUDENT STUDY GUIDE by Edward J. Tarbuck, Frederick K. Lutgens, and Gregory J. Ballinger (0–02–419024–1). Includes comprehensive learning and practice exams for reviewing and studying, including a complete answer section.

APPLICATIONS AND INVESTIGATIONS IN EARTH SCIENCE by Edward J. Tarbuck, Frederick K. Lutgens, and Kenneth G. Pinzke (0–02–418011–x). Twenty-two four-color exercises covering: physical geology, meteorology, oceanography, and astronomy, emphasizing elementary principles/concepts.

THE MACMILLAN GEODISC (0–02–419027–6). Combines full-color animations, motion video, and 1500 still images drawn from a variety of geologic sources. This powerful new teaching/learning tool explores such topics as global warming, earthquakes, hurricanes, volcanoes, and more.

Contents

PART 3
The Atmosphere   413

12

13

Introduction

A VIEW OF THE EARTH

A view of the earth from space gives us a unique perspective of our planet (Figure I.1). At first, it may strike us that the earth is a fragile-appearing sphere surrounded by the blackness of space. In fact, it is just a speck of matter in an infinite universe. As we look more closely at our planet from space, it becomes apparent that the earth is much more than rock and soil. Indeed, the most conspicuous features are not the continents but the swirling clouds suspended above the surface and the vast global ocean. From such a vantage point we can appreciate why the earth's physical environment is traditionally divided into three major parts: the solid earth, or lithosphere; the water portion of our planet, the hydrosphere; and the earth's gaseous envelope, the atmos-

Figure I.1
View of the earth from Apollo 17. (Courtesy of NASA)

phere. It should be emphasized that our environment is highly integrated and is not dominated by rock, water, or air alone. Rather, it is characterized by continuous interactions as air comes in contact with rock, rock with water, and water with air. Moreover, the biosphere, the totality of life-forms on our planet, extends into each of the three physical realms and is an equally integral part of the earth.

The interplay and interactions among the spheres of the earth's environment are uncountable. Figure I.2 provides us with one easy-to-visualize example. The shoreline is an obvious meeting place for rock, water, and air. In this scene, ocean waves that were created by the drag of air moving across the water are breaking against the rocky shore. The force of the water can be powerful and the erosional work that is accomplished can be great.

The **hydrosphere** is a dynamic mass of liquid that is continually on the move, from the oceans to the atmosphere, to the land, and back again. The global ocean is certainly the most prominent feature of the hydrosphere, blanketing nearly 71 percent of the earth's surface and accounting for about 97 percent of the earth's water. However, the hydrosphere also includes the fresh water found in streams, lakes, and glaciers, as well as that found underground. Although these latter sources constitute just a tiny fraction of the total, they are much more important than their meager percentage indicates. In addition to providing the fresh water that is so vital to life on the continents, streams, glaciers, and groundwater are responsible for sculpturing and creating many of our planet's varied landforms.

The earth is surrounded by a life-giving gaseous envelope called the **atmosphere**. This thin blanket of air is an integral part of the planet. It not only provides the air that we breathe but also acts to protect us from the sun's intense heat and dangerous radiation. The energy exchanges that continually occur between the atmosphere and the earth's surface and between the atmosphere and space produce the effects we call weather. If, like the moon, the earth had no atmosphere, our planet would not only be lifeless but also many of the processes and interactions that make the surface such a dynamic place could not operate. Without weathering and erosion, the face of our planet might more closely resemble the lunar surface, which has not changed appreciably in nearly 3 billion years.

The **biosphere** includes all life on earth and consists of the parts of the solid earth, hydrosphere, and atmosphere in which living organisms can be found. Plants and animals depend on the physical environment for the basics of life. However, it is important to note that organisms do more than just respond to their physical environment. Indeed, through countless interactions, life-forms help maintain and alter their physical environment. Without life, the makeup and nature of the lithosphere, hydrosphere, and atmosphere would be very different.

Lying beneath the atmosphere and the ocean is the solid earth. It is divided into three principal units: the dense **core**; the less dense **mantle**; and the **crust**,

Figure I.2
The shoreline is one obvious meeting place for rock, water, and air. In this scene, ocean waves that were created by the force of moving air break against the rocky shore. The force of the water can be powerful and the erosional work that is accomplished can be great. (Photo by Pete Saloutos/The Stock Market)

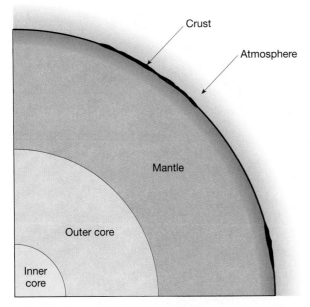

Figure I.3
Cross-sectional view of the earth's layered structure. The inner core, outer core, and mantle are drawn to scale, but the crust's thickness is exaggerated by about three times.

which is the light and very thin outer skin of the earth (Figure I.3). The term **lithosphere** denotes a rigid outer layer of the earth, which includes the crust and part of the upper mantle. The crust is not a layer of uniform thickness; rather, it is characterized by many irregularities. It is thinnest beneath the oceans and thickest where continents exist. Although the crust may seem insignificant when compared with the other units of the solid earth, it was created by the same general processes that were responsible for the earth's present structure. Thus, the crust is important in understanding the history and nature of our planet.

Much of our study of the solid earth focuses on the more accessible surface features. Fortunately, these features represent the outward expressions of the dynamic behavior of the subsurface materials. By examining the most prominent surface features and their global extent, we can obtain clues to the dynamic processes which shape our planet.

The two principal divisions of the earth's surface are the continents and the ocean basins. It is important to realize that the present shoreline is not the boundary between these distinct regions. Rather, along most coasts a gently sloping platform of continental material, called the *continental shelf*, extends seaward from the shore. A glance at Figure I.4 shows

that there can be considerable variation in the extent of the continental shelf from one region to another. For example, the shelf is broad along the East and Gulf coasts of the United States, but relatively narrow along the Pacific margin of the continent. The extent of the continental shelf also varies greatly from one time to another. For instance, during the most recent ice age, when more of the world's water was stored on land in the form of glacial ice, the level of the sea was about 150 meters (500 feet) lower than it is today. Consequently, during this period, more of the earth's surface was dry land. The boundary between the continents and the deep-ocean basins is perhaps best placed about halfway down the *continental slopes*, which are steep dropoffs that lead from the edge of the continental shelves to the deep-ocean basins. Using this as the dividing line, we find that about 60 percent of the earth's surface is represented by the ocean basins, whereas the remaining 40 percent exists as continental masses.

The most obvious difference between the continents and the ocean basins is their relative levels. The average elevation of the continents above sea level is about 840 meters (2750 feet), whereas the average depth of the oceans is about 3800 meters (12,500 feet). Thus, the continents stand on the average about 4.6 kilometers above the level of the ocean floor.

Within these two diverse provinces, great variations in elevation exist. The most prominent features of the continents are linear mountain belts (see Figure I.4). At one time the ocean floor was thought to be a nearly featureless region with little more than an occasional volcanic structure emerging from the otherwise flat, sediment-mantled depths. This perception of the ocean floor was incorrect. Indeed, the ocean basins are now known to contain the most prominent mountain range on earth, the *mid-ocean ridge system*. In Figure I.4 the Mid-Atlantic Ridge and the East Pacific Rise are parts of this system. This broad elevated feature forms a continuous belt that winds for nearly 65,000 kilometers (40,000 miles) around the globe in a manner similar to the seam of a baseball. Rather than consisting of highly deformed rock, such as most of the mountains found on the continents, the oceanic ridge system consists of layer upon layer of once molten rock which has been fractured and uplifted.

The ocean floor also contains extremely deep grooves that are occasionally more than 11,000 meters (36,000 feet) deep. Although these deep-ocean *trenches* are relatively narrow and represent only a

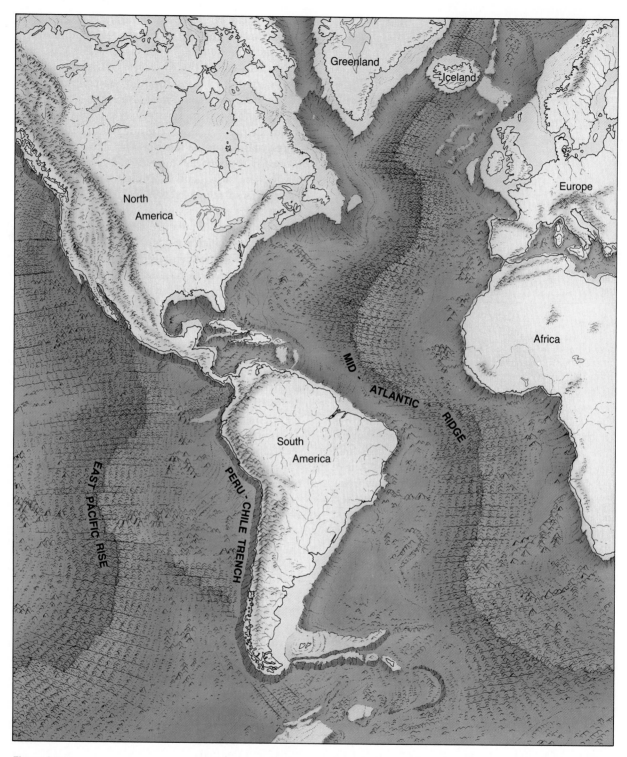

Figure I.4
*Major physical features of the continents and ocean basins. Clearly the diversity of
features on the ocean floor is as varied as on the continents.*

small portion of the ocean floor, they are neverthe-less very important features that mark zones of sig-nificant geological activity. Some trenches are located adjacent to young mountains which flank the conti-nents. For example, in Figure I.4 the Peru-Chile trench off the west coast of South America parallels the An-des Mountains. Other trenches parallel linear island chains called volcanic island arcs.

To summarize this brief view of the earth, we have seen that the physical environment consists of the hy-drosphere, atmosphere, and lithosphere. These three parts are highly integrated and are characterized by countless interactions. In addition, the biosphere ex-tends into each of the spheres of the physical envi-ronment, and must also be considered an integral part of our planet. Finally, we have seen that the ocean basins are not featureless plains, but, like the continents, are characterized by a broad diversity of features.

THE EARTH SCIENCES

Earth science is the name for all the sciences that col-lectively seek to understand the earth and its neigh-bors in space. It includes geology, oceanography, me-teorology, and astronomy.

The focus of Part One of this text, *The Solid Earth*, is the science of **geology**, a word that literally means "study of the earth." Geology is traditionally divided into two broad areas—physical and historical. *Phys-ical geology* examines the materials composing the earth and seeks to understand the many processes that operate beneath and upon its surface. The earth is a dynamic, ever-changing planet. Forces within the earth create earthquakes, build mountains, and pro-duce volcanic structures. At the surface, external processes break rock apart and sculpture a broad ar-ray of landforms. The erosional effects of water, wind, and ice result in a great diversity of landscapes. Since rocks and minerals form in response to the earth's internal and external processes, their inter-pretation is basic to an understanding of our planet. In contrast to physical geology, the aim of *historical geology* is to understand the origin of the earth and the development of the planet through its 4.6-billion-year history. It strives to establish an orderly chrono-logical arrangement of the multitude of physical and biological changes that have occurred in the geologic past (Figure I.5).

Part Two of the text, *The Oceans*, is devoted to the study of **oceanography**. Oceanography is actually not a separate and distinct science. Rather, it involves the application of all sciences in a comprehensive and interrelated study of the oceans in all their aspects and relationships. Oceanography integrates chem-istry, physics, geology, and biology. It includes the study of the composition and movements of seawa-ter, as well as coastal processes, seafloor topography, and marine life.

Part Three, *The Atmosphere*, examines the mixture of gases that is held to the planet by gravity and thins rapidly with altitude. Acted on by the combined ef-fects of the earth's motions and energy from the sun, the formless and invisible atmosphere reacts by pro-ducing an infinite variety of weather, which, in turn, creates the basic pattern of global climates. **Meteo-rology** is the study of the atmosphere and the processes that produce weather and climate. Like oceanography, meteorology involves the application of other sciences in an integrated study of the ocean of air that surrounds the earth.

As Part Four, *Astronomy*, demonstrates, the study of the earth is not confined to investigations of the lithosphere, hydrosphere, and atmosphere and to their many interactions and interrelationships. The earth sciences also attempt to relate our planet to the larger universe. Because the earth is related to all of the other objects in space, the science of **astron-omy**—the study of the universe—is very useful in probing the origins of our own environment. Since we are so closely acquainted with the planet on which we live, it is easy to forget that the earth is just a tiny object in a vast universe. Indeed, the earth is subject to the same physical laws that govern the great many other objects that populate the infinite expanses of space. Thus, to understand explanations of the earth's origin, it is useful to learn something about the other members of our solar system. Moreover, it is helpful to view the solar system as a part of the great as-semblage of stars that comprise our galaxy, which, in turn, is but one of many galaxies (see Box I.1).

POPULATION, RESOURCES, AND ENVIRONMENTAL ISSUES

Environment refers to everything that surrounds and influences an organism. Some of these conditions are biological and social, others are abiotic or nonliving.

BOX I.1

*The Earth's Place in the Cosmos**

*F*or centuries, those who have gazed at the night sky have wondered about the nature of the universe, earth's place within it, and whether or not we are "alone." Today, many exciting discoveries in astronomy are beginning to provide some answers concerning the origin of the universe, the formation and evolution of stars, and how the earth and its materials came into existence.

The realization that the universe is immense and orderly began in the early 1900s when Edwin Hubble and other dedicated scientists demonstrated that the Milky Way Galaxy is one of hundreds of billions of galaxies, each of which contains billions of stars. Today we understand that the earth, its materials, and indeed all living things are the consequence of a sequence

---

* This box feature was prepared by Kenneth Pinzke, Belleville Area College.

of events, governed by natural laws, that occurred within the universe during the past twenty billion years.

Astronomical evidence supports the theory, called the Big Bang, that the universe began perhaps as long as twenty billion years ago when a dense, hot, supermassive concentration of material exploded with cataclysmic force (Figure I.A). Within about one second, the temperature of the expanding universe cooled to approximately 10 billion degrees and fundamental atomic particles called protons and neutrons began to appear. After a few minutes, atoms of the least complex elements—hydrogen and helium—had formed and the initial conversion of energy to matter in the young universe was over.

During the first billion years or so, matter (essentially hydrogen and to a lesser extent helium) in the expanding universe became clumpy and fragmented into enormous clouds and groups of clouds that eventually collapsed to become galaxies and clusters of galaxies. Inside these collapsing clouds, smaller concentrations of matter formed into stars. One of the billions of galaxies to form was the one we call the Milky Way.

The Milky Way is a collection of several hundred billion stars, the oldest of which is about 10 billion years. It is one of a cluster of approximately 28 galaxies, called the Local Group, that exist in our region of the universe. Initially, the oldest stars in the Milky Way formed from nearly pure hydrogen. Later, succeeding generations of younger stars, including our sun, would have heavier, more complex atoms available for their formation.

After forming, all stars go through a series of changes before finally dying out. During the life of most stars, energy is produced as hydrogen nuclei (protons) fuse with other hydrogen nuclei to form helium. During this process, called nuclear fusion, matter is converted to energy. Stars begin to die when their nuclear fuel becomes exhausted. Massive stars often have indescribably explosive deaths. During these violent events, called supernovas, nuclear fusion advances beyond helium and produces heavier, more complex atoms such as oxygen, carbon, and iron. In turn, these products of stellar death may become the materials that compose future generations of stars. It was from the debris

---

The factors in this latter category are collectively referred to as our *physical environment*. The physical environment encompasses water, air, soil, and rock, as well as conditions such as temperature, humidity, and sunlight. The phenomena and processes studied by the earth sciences are basic to an understanding of the physical environment. In this sense, most of earth science may be characterized as environmental science. However, when the term *environmental*

is applied to earth science today, it is usually reserved for those aspects that focus on the relationships between people and the physical environment. Application of the earth sciences is necessary to understand and solve problems that arise from these interactions. Although the primary focus of this book is to gain an understanding of basic earth science principles, many important environmental issues will be explored along the way.

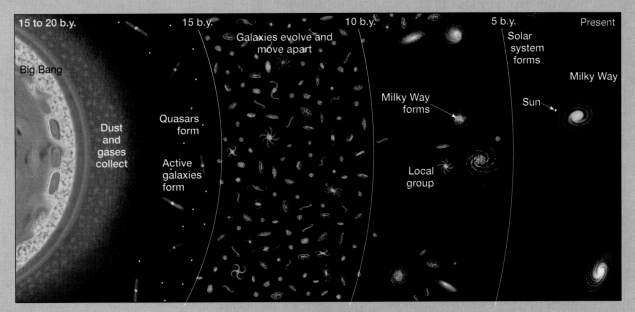

Figure I.A
About 20 billion years ago, an incomprehensibly large explosion sent all the matter of the universe flying outward at incredible speeds. After a few billion years, the material cooled and condensed into the first stars and galaxies. Because the universe is expanding, the evolving galaxies continue to move away from one another. About 5 billion years ago, our solar system began forming in one of these galaxies, the Milky Way.

scattered during the death of a pre-existing star, or stars, that our sun, as well as the rest of the solar system, formed.

Our star, the sun, is at the very least a second-generation star. It, along with the planets and other members of the solar system, began forming nearly 5 billion years ago, some 15 billion years after the Big Bang, from a large, interstellar cloud, called a nebula, which consisted of dust particles and gases enriched in heavy elements from supernova explosions. Gravitational energy caused the nebula to contract and in so doing, it began to rotate and flatten. Inside, smaller concentrations of matter began condensing to form the planets. At the center of the nebula there was sufficient pressure and heat to initiate hydrogen nuclear fusion, and our sun was born.

It has been said that all life on earth is related to the stars. This is true because the very atoms in our bodies, as well as the atoms composing every other thing on earth, owe their origin to a supernova event that occurred billions of years ago, trillions of kilometers away.

Resources are one important area of environmental concern. Resources range from water and soil to metallic and nonmetallic minerals and energy. These materials are the very basis of modern civilization. The mineral and energy resources that are extracted from the crust are the raw materials from which the products used by society are made. Few people who live in highly industrialized nations realize the quantity of resources needed to maintain their present standard of living. For example, the annual per capita consumption of metallic and nonmetallic mineral resources for the United States is nearly 10,000 kilograms (11 tons). This is each person's prorated share of the materials required by industry to provide the vast array of products modern society demands. Figures for other highly industrialized countries are comparable.

Resources are commonly divided into two broad categories. Some are classified as **renewable,** which

GEOLOGIC TIME SCALE

Time Units of the Geologic Time Scale				Development of Plants and Animals
Eon	Era	Period	Epoch	
Phanerozoic	Cenozoic	Quarternary	Holocene ——0.01—— Pleistocene ——1.6——	Humans develop
		Tertiary	Pliocene ——5.3—— Miocene ——23.7—— Oligocene ——36.6—— Eocene ——57.8—— Paleocene ——66.4——	"Age of Mammals"
	Mesozoic	Cretaceous ——144—— Jurassic ——208—— Triassic ——245——	"Age of Reptiles"	Extinction of dinosaurs and many other species First flowering plants First birds Dinosaurs dominant
	Paleozoic	Permian ——286—— Carboniferous: Pennsylvanian ——320—— Mississippian ——360——	"Age of Amphibians"	Extinction of trilobites and many other marine animals First reptiles Large coal swamps Amphibians abundant
		Devonian ——408—— Silurian ——438——	"Age of Fishes"	First insect fossils Fishes dominant First land plants
		Ordovician ——505—— Cambrian ——570——	"Age of Invertebrates"	First fishes Trilobites dominant First organisms with shells
Proterozoic ——2500—— Archean ——3800—— Hadean ——4600——		Collectively called Precambrian, comprises about 87% of the geologic time scale		First multicelled organisms First one-celled organisms Age of oldest rocks Origin of the earth

Figure I.5
The geologic time scale. Numbers on the time scale represent time in millions of years before the present. These dates were added long after the time scale had been established using relative dating techniques. The Precambrian accounts for more than 85 percent of geologic time. (Data from Geological Society of America)

mined or pumped from the ground, there will be no more. Although some nonrenewable resources, such as aluminum, can be used over and over again, others, such as oil, cannot be recycled.

A glance at Figure I.7 shows that the population of our planet is growing rapidly. Although it took until the beginning of the nineteenth century for the number to reach 1 billion, just 130 years were needed for the population to double to 2 billion. Between 1930 and 1975 the figure doubled again to 4 billion

Figure I.6
As world population grows, demand for mineral and energy resources climbs. Geology must deal with the search for additional supplies of traditional and alternative resources as well as the environmental impact of their extraction and use. (Courtesy of Shell Oil Company)

means that they can be replenished over relatively short time spans. Plants and animals for food, natural fibers for clothing, and forest products for lumber and paper are all common examples. Energy from flowing water, wind, and the sun are also considered renewable. By contrast, many other basic resources are classified as **nonrenewable.** Important metals such as iron, aluminum, and copper fall into this category, as do our most important fuels: oil, natural gas, and coal (Figure I.6). Although these and other resources continue to be formed in the earth, the processes that create them are so slow that significant deposits take millions of years to accumulate. In essence, the earth contains fixed quantities of these substances. When the present supplies are

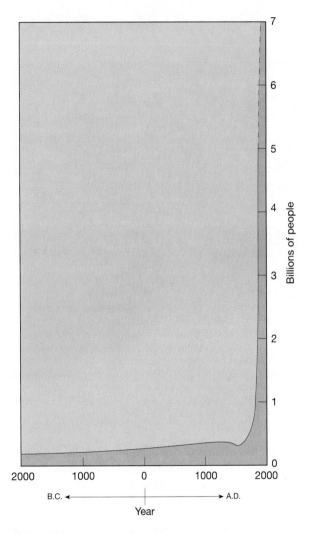

Figure I.7
Growth of world population. It took until 1800 for the number to reach 1 billion. By the year 2000 an estimated 7 billion people will inhabit the planet. The demand for basic resources is growing faster than the rate of the population increase. (Data from the Population Reference Bureau)

A.

B.

C.

Figure I.8
*There are many earth processes and phenomena that are hazardous to people includ-
ing volcanoes, earthquakes, landslides, floods, hurricanes, and tornadoes. **A.** Hurricane
Andrew devastated South Florida in August 1992. (Photo by Allan Tannenbaum/
SYGMA) **B.** One of many structures damaged or destroyed by an earthquake that
struck Mexico City in September 1985. An estimated 7000 lives were lost. (Photo by
James L. Beck) **C.** In June 1991, Mount Pinatubo in the Philippines erupted explosively.
It is one of the many active volcanoes that surround the Pacific Ocean. (Photo by R.
Hoblitt, U.S. Geological Survey)*

and by the year 2000 an estimated 7 billion people
will inhabit the planet. Clearly, as population grows,
the demand for resources expands as well. However,
the rate of mineral and energy resource usage has
climbed more rapidly than the overall growth of population. This fact results from an increasing standard
of living. This is certainly true in the United States,
where about 6 percent of the world's population uses
approximately 30 percent of the annual production
of mineral and energy resources.

How long will the remaining supplies of basic resources be adequate to sustain the rising standard of
living in today's industrialized countries and still provide for the growing needs of developing regions?
How much environmental deterioration are we willing to accept in pursuit of basic resources? Can alternatives be found? If we are to cope with an increasing per capita demand and a growing world
population, it is important that we have some understanding of our present and potential resources.

In addition to the quest for adequate mineral and
energy resources, the earth sciences must also deal
with a broad array of other environmental problems.
Some are local, some are regional, and still others are

global in extent. Serious difficulties face developed and developing nations alike. Urban air pollution, acid rain, ozone depletion, and global warming are just a few that pose significant threats. Other problems involve the loss of fertile soils to erosion, the disposal of toxic wastes, and the contamination and depletion of water resources. The list contines to grow.

In addition to human-induced and human-accentuated problems, people must also cope with the many natural hazards posed by the physical environment (Figure I.8). Earthquakes, landslides, floods, and hurricanes are just four of the many risks. Others such as drought, although not as spectacular, are nevertheless equally important environmental concerns. In many cases, the threat of natural hazards is aggravated by increases in population as more people crowd into places where an impending danger exists or attempt to cultivate marginal lands that should not be farmed.

It is clear that as world population continues its rapid growth, pressures on the environment will increase as well. Therefore, an understanding of the earth is not only essential for the location and recovery of basic resources, but also for dealing with the human impact on the environment and minimizing the effects of natural hazards. Knowledge about our planet and how it works is necessary to our survival and well-being. The earth is the only suitable habitat we have, and its resources are limited.

THE NATURE OF SCIENTIFIC INQUIRY

As members of a modern society, we are constantly reminded of the significant benefits derived from scientific investigations. What exactly is the nature of this inquiry?

All science is based on the assumption that the natural world behaves in a consistent and predictable manner. This implies that the physical laws which govern the smallest atomic particles also operate in the largest, most distant galaxies. Evidence for the existence of these underlying patterns can be found in the physical world as well as the biological world. For example, the same biochemical processes and the same genetic codes that are found in bacterial cells are also found in human cells. The overall goal of science is to discover the underlying patterns in the natural world and then to use this knowledge to make predictions about what should or should not be expected to happen given certain facts or circumstances.

The development of new scientific knowledge involves some basic, logical processes that are universally accepted. To determine what is occurring in the natural world, scientists collect scientific *facts* through observation and measurement (Figure I.9). These data are essential to science and serve as the springboard for the development of scientific theories and laws.

Figure I.9
Scientists who were part of a severe-storm intercept program study a tornado with a portable Doppler radar unit near Hodges, Oklahoma, in May 1989. (Photo by Howard B. Bluestein)

BOX I.2

Do Glaciers Move? An Application of the Scientific Method

Today we know that glaciers can do extraordinary erosional work and that in the past glacial ice has affected vast areas that are now ice-free. We understand much about how glaciers form and move, as well as how they erode and deposit. This and other knowledge about glaciers has been acquired gradually over the past 200 years, yet scientists still have much more to learn about these moving masses of ice.

The study of glaciers provides an early application of the scientific method. High in the Alps of Switzerland and France, small glaciers exist in the upper portions of some valleys. In the late eighteenth and early ninteenth centuries, people who farmed and herded animals in these valleys suggested that glaciers in the upper reaches of the valleys had previously been much larger and had occupied downvalley areas. They based their belief on the fact that the valley floors were littered with angular boulders and other rock debris that seemed identical to the materials that they could see in and near the glaciers at the heads of the valleys.

Although the explanation of these observations seemed logical, others did not accept the notion that masses of ice hundreds of meters thick were capable of movement. The disagreement was settled after a simple experiment was designed and carried out to test the hypothesis that glacial ice can move.

Markers were placed in a straight line completely across an alpine glacier. The position of the line was marked on the valley walls so that if the ice moved, the change in position could be detected. After a year or two the results were clear: the markers on the glacier had advanced down the valley, proving that glacial ice indeed moves. In addition, the experiment demonstrated that ice within a glacier does not move at a uniform rate, because the markers in the center advanced farther than did those along the margins. Although most glaciers move too slowly for direct visual detection, the experiment succeeded in demonstrating that movement nevertheless occurs. In the years that followed, this experiment was repeated many times with greater accuracy using more modern surveying techniques. Each time, the basic relationships established by earlier attempts were verified.

The experiment illustrated in Figure I.B was carried out at

Once a set of scientific facts (or principles) that describe a natural phenomenon are gathered, investigators try to explain how or why things happen in the manner observed. They can do this by constructing a tentative (or untested) explanation, which we call a scientific **hypothesis**. Often several hypotheses are advanced to explain the same facts and observations. For example, there are currently five major hypotheses that have been proposed to explain the origin of the earth's moon. Until recently, the most widely held hypothesis argued that the moon and the earth formed simultaneously from the same cloud of nebular dust and gases. A newer hypothesis suggests that a Mars-sized body impacted the earth. The collision ejected a huge quantity of material into earth's orbit that eventually accumulated into the moon.

Both hypotheses will be submitted to rigorous testing that will undoubtedly result in the modification or rejection of one, or perhaps both, of the proposals. The history of science is littered with discarded hypotheses. One of the best known is the idea that the earth was at the center of the universe, a proposal that was supported by the apparent daily motion of the sun, moon, and stars around the earth.

When a hypothesis has survived extensive scrutiny, and when competing hypotheses have been eliminated, a hypothesis may be elevated to the status of a scientific **theory**. A scientific theory is a well-tested and widely accepted view that scientists agree best explains certain observable facts. However, scientific theories, like scientific hypotheses, are accepted only provisionally. It is always possible that a theory which

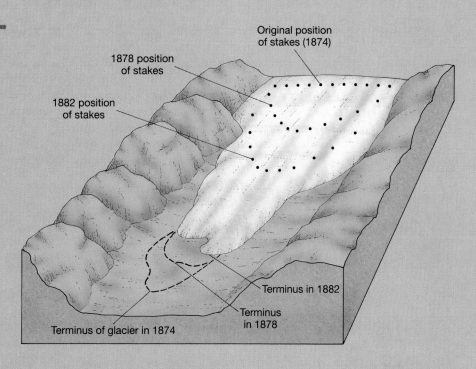

Figure I.B
Ice movement and changes in the terminus at Rhone Glacier, Switzerland. In this classic study of a valley glacier, the movement of stakes clearly showed that ice along the sides of the glacier moves slowest. Also notice that even though the ice front was retreating, the ice within the glacier was advancing.

Original position of stakes (1874)

1878 position of stakes

1882 position of stakes

Terminus in 1882

Terminus in 1878

Terminus of glacier in 1874

Switzerland's Rhone Glacier later in the nineteenth century. It not only traced the movement of markers within the ice, but also mapped the position of the glacier's terminus. Notice that even though the ice within the glacier was advancing, the ice front was retreating. As often occurs in science, experiments and observations designed to test one hypothesis yield new information that requires further analysis and explanation.

has withstood previous testing may eventually be disproven. As theories survive more testing, they are regarded with higher levels of confidence. Theories that have withstood extensive testing, such as the theory of plate tectonics or the theory of evolution, are held with a very high degree of confidence.

Some concepts in science are formulated into scientific laws. A scientific **law** is a generalization about the behavior of nature from which there has been no known deviation after numerous observations or experiments. Scientific laws are generally narrower in scope than theories and can usually be expressed mathematically. Examples include Newton's laws of motion and the laws of thermodynamics. Scientific laws describe what happens in nature, but they do not explain how or why things happen this way. Also, unlike hypotheses, which are inventions of the mind to explain scientific facts, scientific laws are based on observations and measurements and thus are rarely overthrown or seriously modified. Nevertheless, even scientific laws are not necessarily "perpetual truths." As the mathematician Jacob Bronowski so ably stated, "Science is a great many things, but in the end they all return to this: Science is the acceptance of what works and the rejection of what does not."

The processes just described, in which scientists gather facts through observations and formulate scientific hypotheses, theories, and laws, is called the *scientific method*. Contrary to popular belief, the scientific method is not a standard recipe that scientists apply in a routine manner to unravel the secrets of our natural world. Neither is this process a haphazard one.

Some scientific knowledge is gained through the following steps: (1) the collection of scientific facts (data) through observation and measurement (Figure I.10). (2) the development of a working hypothesis to explain these facts; (3) construction of experiments to validate the hypothesis; and (4) the acceptance, modification, or rejection of the hypothesis on the basis of extensive testing. Other scientific discoveries represent purely theoretical ideas, which stood up to extensive ex-amination. Still other scientific advancements have been made when a totally unexpected happening occurred during an experiment. These so-called serendipitous discoveries are more than pure luck, for as Louis Pasteur said, "In the field of observation, chance favors only the prepared mind." Since scientific knowledge is acquired through several avenues, it might be best to describe the nature of scientific inquiry as the *methods* of science, rather than *the* scientific method.

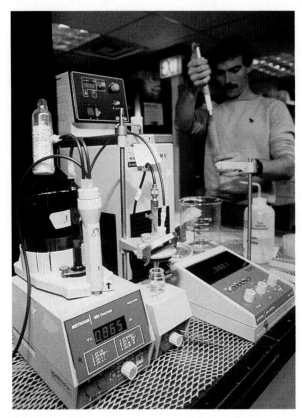

A.

B.

Figure I.10
A. Scientist analyzing sea-floor samples collected by the drilling ship JOIDES Resolu-tion. (Courtesy of Ocean Drilling Program) **B.** Temperature probe of an active lava flow, Kilauea Caldera, Hawaii. Although much can be accomplished in laboratories, a great deal of earth science data must be gathered in the field. (Photo by Norman Banks, U.S. Geological Survey)

1. List and briefly define the three "spheres" that constitute the physical environment.
2. Name the three principal units of the lithosphere. Which one is thinnest?
3. The oceans cover nearly _____ percent of the earth's surface and contain about _____ percent of the planet's total water supply.
4. Is the present shoreline the boundary between the continents and ocean basins? Briefly explain.
5. Name the specific earth science described by each of the following statements.
    (a) The science that deals with the dynamics of the oceans.
    (b) This word literally means "the study of the earth."
    (c) An understanding of the atmosphere is the primary focus of this science.
    (d) This science helps us understand the earth's place in the universe.
6. Contrast renewable and nonrenewable resources. Give one or more examples of each.
7. World population is estimated to reach nearly _____ billion by the year 2000. How does this compare to the world population near the beginning of the nineteenth century?
8. How is a scientific hypothesis different from a scientific theory?
9. A _____ is a generalization about the behavior of nature from which there has been no known deviation.

Review your understanding of important terms in this chapter by defining and explaining the importance of each of the following. Terms are listed in alphabetical order.

astronomy (p. 5)

atmosphere (p. 2)

biosphere (p. 2)

core (p. 2)

crust (p. 2)

geology (p. 5)

hydrosphere (p. 2)

hypothesis (p. 12)

law (p. 13)

lithosphere (p. 3)

mantle (p. 2)

meteorology (p. 5)

nonrenewable resource (p. 9)

oceanography (p. 5)

renewable resource (p. 7)

theory (p. 12)

PART 1

The Solid Earth

Cape Royal's cliffs in Grand Canyon National Park, Arizona. (Photo by Jack W. Dykinga)

1

Minerals and Rocks

Opposite page: Sedimentary rocks exposed in Coconino National Forest near Sedona, Arizona. (Photo by Tom Bean)

Left: West Mitten, Monument Valley, Arizona. (Photo by James E. Patterson)

Above: Sedimentary rocks exposed near Canyonlands National Park, Utah. (Photo by Stephen Trimble)

The study of geology appropriately begins with the study of rocks and minerals, because a knowledge of the materials that make up the earth is basic to our understanding of much of the rest of geology. In the pages that follow we shall examine atoms and elements, the building blocks of minerals, and then investigate the three major groups of rocks. Our inquiry shall attempt to answer some important questions: How do rocks form? How are minerals and rocks identified? What characteristics differentiate one group of rocks from another? What can the study of rocks and minerals tell us about the nature and history of our planet?

MINERALS

The earth's crust and oceans are the source of a wide variety of useful and essential minerals. In fact, practically every manufactured product contains materials obtained from minerals. Most people are familiar with the common uses of many basic metals, including aluminum in beverage cans, copper in electrical wiring, and gold and silver in jewelry. But some people are not aware that pencil lead contains the greasy-feeling mineral graphite and that baby powder comes from a metamorphic rock made of the mineral talc. As the material requirements of modern society grow, the need to locate additional supplies of useful minerals also grows, and becomes more challenging as well.

In addition to the economic uses of rocks and minerals, all of the processes studied by geologists are in some way dependent upon the properties of these basic earth materials. Events such as volcanic eruptions, mountain building, weathering and erosion, and even earthquakes involve rocks and minerals. Consequently, a basic knowledge of earth materials is essential to the understanding of all geologic phenomena.

Many people consider rocks to be rather nondescript objects that are hard and often dirty. Some people may think minerals are merely dietary supplements, or possibly rare ores or precious gems that are mined for their economic value. However, these common perceptions are inadequate.

A **rock** can be defined simply as an aggregate of one or more minerals. Here, the term *aggregate* implies that the minerals are found together as a *mixture* in which the properties of the individual minerals are retained (Figure 1.1). Although most rocks are composed of more than one mineral, certain minerals, along with impurities, are commonly found by themselves in large quantities. In these instances they are considered to be rocks. A common example is the mineral calcite, which frequently is the dominant constituent in large rock units, where it is given the name *limestone*.

By contrast, a **mineral** is defined as a naturally occurring inorganic solid that possesses a definite chemical structure, which gives it a unique set of physical properties. Thus, for any earth material to be considered a mineral, it must exhibit the following characteristics:

1. It must be naturally occurring.
2. It must be inorganic.

Figure 1.1
*Rocks are aggregates of one
or more minerals.*

Granite
(Rock)

Quartz
(Mineral)

Hornblende
(Mineral)

Feldspar
(Mineral)

3. It must be a solid.
4. It must possess a definite chemical structure.

When the term *mineral* is used by geologists, only those substances that fulfill these precise conditions are considered minerals. Consequently, synthetic diamonds, although chemically the same as natural diamonds, are not considered minerals. Further, oil and natural gas, which are not inorganic solids, and coal, which is produced from decayed organic matter, are also not classified as minerals by geologists.

Structure of Minerals

Minerals, like all matter, are made of **elements.** At present, over 100 elements are known, a dozen and a half of which have been produced only in the laboratory. Some minerals such as gold and sulfur are made entirely of one element, but most are a combination of two or more elements joined to form a chemically stable compound. In order to better un-

derstand how elements combine to form compounds, we must first consider the **atom,** the smallest part of matter that still retains the characteristics of an element, because it is this extremely small particle that does the combining.

By the first part of this century, a series of important experiments had shown that atoms are composed of even smaller particles. A simplified model of the structure of an atom is shown in Figure 1.2. Each atom has a central region, called the **nucleus,** which contains very dense, positively charged **protons** and equally dense, neutral particles called **neutrons.** Orbiting the nucleus are negatively charged particles known as **electrons.** Unlike the orderly orbiting of the planets around the sun, electrons move so rapidly that their positions cannot be pinpointed. Hence, a more realistic picture of the positions of electrons can be obtained by envisioning a cloud of negatively charged electrons surrounding the nucleus.

The number of protons in the nucleus determines the **atomic number** and name of the element. For

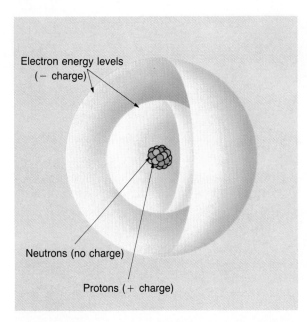

Figure 1.2
Simplified model of an atom. An atom consists of a central nucleus composed of protons and neutrons that is encircled by electrons.

example, all atoms with six protons are carbon atoms, all those with eight protons are oxygen atoms, and so forth. Since uncombined atoms have the same number of electrons as protons, the atomic number also equals the number of electrons surrounding the nucleus. Moreover, since neutrons have no charge, the positive charge of the protons is exactly balanced by the negative charge of the electrons. Consequently, uncombined atoms as a whole are electrically neutral and have no overall electrical charge. Elements

can be considered to be a large collection of electrically neutral atoms, all having the same atomic number.

Elements combine with each other to form a wide variety of more complex substances called compounds. A **compound** is a substance composed of two or more elements bonded together in definite proportions. When the elements separate, the bonds are broken and the compound is destroyed. Through experimentation it has been learned that the forces that bond the atoms are electrical in nature. Further, it is known that chemical bonding results in a change in the electronic structures of the bonded atoms.

When an atom combines chemically, it either gains, loses, or shares electrons with another atom. An atom that gains electrons becomes negatively charged because it has more electrons than protons, whereas atoms that lose electrons become positively charged. Atoms that have an electrical charge because of a gain or loss of electrons are called **ions.** Simply stated, oppositely charged ions attract one another and produce a neutral chemical compound.

An example of chemical bonding involving sodium (Na) and chlorine (Cl) to produce sodium chloride (NaCl, common table salt) is shown in Figure 1.3. When the sodium atom loses one electron it becomes a positive ion, and when the chlorine atom gains one electron it becomes a negative ion. These opposite charges act as a bond, holding the atoms together. Here it is interesting to note that chlorine is a green, poisonous gas and sodium is a silvery metal, which, if held in your hand, gives a severe chemical burn. Together, however, these atoms produce the compound sodium chloride, which looks nothing like either and is a requirement for human life. This

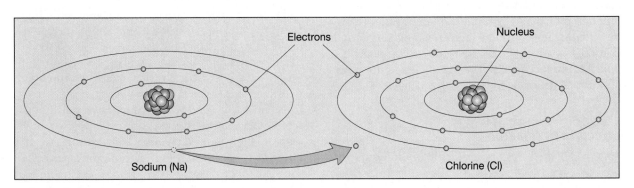

Figure 1.3
Chemical bonding of sodium and chlorine to produce sodium chloride. Through the transfer of one electron from sodium to chlorine, sodium becomes a positive ion and chlorine a negative ion.

example also illustrates a difference between a rock and a mineral. The mineral sodium chloride (called halite) is a chemical compound and has properties unique to it and very different from the elements which make it up. Rocks, on the other hand, are *mixtures* of minerals, with each mineral retaining its own characteristics.

Although electrons play an active role in chemical reactions, they do not contribute significantly to the weight (mass) of an atom. It follows then that the weight of an atom is centered in the nucleus. The **mass number** of an atom is obtained by totaling the number of neutrons and protons in the nucleus. It is not uncommon, however, for atoms of the same element to have varying numbers of neutrons, and therefore, to have different mass numbers. Such atoms are called **isotopes** of that element. For example, carbon has two well-known isotopes, one having a mass number of 12 (carbon-12), the other a mass number of 14 (carbon-14). Recall that all atoms of the same element must have the same number of protons (atomic number) and that carbon always has six protons. Hence, carbon-12 must have six neutrons to give it a mass number of 12, whereas carbon-14 must have eight neutrons to give it a mass number of 14. The term commonly used to express the average of the atomic masses of isotopes for a given element is **atomic weight.*** The atomic weight of carbon is much closer to 12 than 14, because carbon-12 is the more common isotope. Note that in a chemical sense all isotopes of the same element are nearly identical. To distinguish among them would be like trying to differentiate individual members from a group of similar objects, all having the same shape, size, and color, with only some being slightly heavier.

Although the vast majority of atoms are stable, many elements do have isotopes that are unstable. Unstable isotopes such as carbon-14 go through a process of natural disintegration called **radioactivity,** which occurs when the forces that bind the nucleus are not strong enough. The rate at which the unstable nuclei break apart (decay) is measurable and makes such isotopes useful "clocks" in dating the events of earth history. A discussion of radioactivity

and its application in dating events of the geologic past can be found in Chapter 9.

Physical Properties of Minerals

Minerals are naturally occurring solids formed by inorganic processes. Each mineral has an orderly arrangement of atoms (crystalline structure) and a definite chemical composition that give it a unique set of physical properties. Since the internal structure and chemical composition of a mineral are difficult to determine without the aid of sophisticated tests and apparatus, the more easily recognized physical properties are used in identification. A discussion of some diagnostic physical properties follows.

Crystal Form.   Most people think of a crystal as a rare commodity, when in fact most inorganic solid objects are composed of crystals. The reason for this misconception is that most crystals do not exhibit their crystal form. The **crystal form** is the external expression of a mineral that reflects the orderly internal arrangement of atoms. Figure 1.4A illustrates the characteristic form of the iron-bearing mineral pyrite. Generally, when a mineral forms without space restrictions, individual crystals with well-formed crystal faces will develop. Some crystals, such as those of the mineral quartz, have a very distinctive crystal form that can be helpful in identification (Figure 1.4B). However, most of the time crystal growth is interrupted because of competition for space, resulting in an intergrown mass of crystals, none of which exhibits its crystal form.

Color.   Although color is an obvious feature of a mineral, it is often an unreliable diagnostic property. Slight impurities in the common mineral quartz, for example, give it a variety of colors, including pink, purple (amethyst), white, and even black. When a mineral, such as quartz, exhibits a variety of colors, it is said to possess *exotic coloration*. Exotic coloration is usually caused by the inclusion of impurities, such as foreign ions, in the crystalline structure. Other minerals, for example, sulfur, which is yellow, and malachite, which is bright green, are said to have *inherent coloration* because their color is a consequence of their chemical makeup and does not vary significantly.

Streak.   **Streak** is the color of a mineral in its powdered form and is obtained by rubbing the mineral across a piece of unglazed porcelain termed a *streak*

---

* The term *weight* as used here is a misnomer that has been sanctioned by long usage. The correct term is atomic *mass.*

A.

B.

Figure 1.4
Crystal form is the external expression of a mineral's orderly internal structure. **A.**
Pyrite, commonly known as "fool's gold," often forms cubic crystals that may contain
parallel lines called striations. **B.** *Quartz sample that exhibits well-developed hexagonal*
crystals with pyramidal-shaped ends.

plate. Although the color of a mineral may vary from sample to sample, the streak usually does not and is therefore the more reliable property. Streak can also be an aid in distinguishing minerals with metallic lusters from those having nonmetallic lusters. Metallic minerals generally have a dense, dark streak, whereas minerals with nonmetallic lusters do not.

Luster.   **Luster** is the appearance or quality of light reflected from the surface of a mineral. Minerals that have the appearance of metals, regardless of color, are said to have a *metallic luster*. Minerals with a *nonmetallic luster* are described by various adjectives, including vitreous (glassy), pearly, silky, resinous, and earthy (dull).

Hardness.   One of the most useful diagnostic properties of a mineral is **hardness,** a measure of the resistance of a mineral to abrasion or scratching. This property is determined by rubbing a mineral of unknown hardness against one of known hardness, or vice versa. A numerical value can be obtained by using the **Mohs scale** of hardness, which consists of ten minerals arranged in order from 1 (softest) to 10 (hardest) as follows:

Hardness	Mineral
1	Talc
2	Gypsum
3	Calcite
4	Fluorite
5	Apatite
6	Orthoclase
7	Quartz
8	Topaz
9	Corudum
10	Diamond

Any mineral of unknown hardness can be compared to these or to other objects of known hardness. For example, a fingernail has a hardness of 2.5, a copper penny 3, and a piece of glass 5.5. The mineral gypsum, which has a hardness of 2, can easily be scratched with your fingernail. On the other hand, the mineral calcite, which has a hardness of 3, will scratch your fingernail but will not scratch glass. Quartz, the hardest of the common minerals, will scratch a glass plate.

Cleavage.   **Cleavage** is the tendency of a mineral to break along planes of weak bonding that exist between atoms in the crystalline structure. Minerals

Figure 1.5
Sheet-type cleavage common to the micas.

Figure 1.6
Smooth surfaces produced when a mineral with cleavage is broken. The sample on the left (fluorite) exhibits four planes of cleavage (eight sides), whereas the other two samples exhibit three planes of cleavage (six sides). Also notice that the mineral in the center (halite) has cleavage planes that meet at 90-degree angles, whereas the mineral on the right (calcite) has cleavage planes that meet at 75-degree angles.

that possess cleavage are identified by the smooth, flat surfaces produced when the mineral is broken. The simplest type of cleavage is exhibited by the micas (Figure 1.5). Because the micas have excellent cleavage in one direction, they break to form thin, flat sheets. Some minerals have several cleavage planes that produce smooth surfaces when broken, while others exhibit poor cleavage, and still others have no cleavage at all. When minerals break evenly in more than one direction, cleavage is described by the *number of different sets of planes* exhibited and the *angles at which the planes meet* (Figure 1.6).

Cleavage should not be confused with crystal form. When a mineral exhibits cleavage, it will break into pieces that have the same shape as the original sample. By contrast, the quartz crystals shown in Figure 1.4B do not have cleavage, and if broken, would shatter into shapes that do not resemble each other or the original crystals.

Fracture. Minerals that do not exhibit cleavage are said to **fracture.** Those that break into smooth curved surfaces resembling broken glass have a *conchoidal fracture.* Others break into splinters or fibers, but most fracture irregularly.

Specific Gravity. **Specific gravity** is a number that represents the ratio of the weight of a mineral to the weight of an equal volume of water. For example, if a mineral weighs three times as much as an equal volume of water, its specific gravity is 3. With a little practice, you can estimate the specific gravity for minerals by hefting them in your hand. The av-

erage specific gravity for minerals is around 2.7, but some metallic minerals have a specific gravity two or three times that of common rock-forming minerals. For example, galena, which is an ore of lead, has a specific gravity of roughly 7.5, while the specific gravity of 24-carat gold is approximately 20.

Other Properties

In addition to the properties already discussed, some minerals can be recognized by other distinctive properties. For example, halite has a salty taste, thin sheets of mica will bend and elastically snap back, and gold is malleable and can be easily shaped. Talc and graphite both have distinctive feels—talc's is soapy and graphite's is greasy. A few minerals, such as magnetite, have a high iron content and can be picked up with a magnet, while some varieties (lodestone) are natural magnets and will pick up light, metal objects such as pins and paper clips. Some minerals exhibit special optical properties. For example, when a transparent piece of calcite is placed over printed material, the letters appear twice. This optical property is known as *double refraction*. In addition, the streak of many sulfur-bearing minerals smells like rotten eggs.

One very simple chemical test involves placing a small drop of dilute hydrochloric acid on a fresh surface of a mineral. Certain minerals, including some carbonates, will effervesce (fizz) with hydrochloric acid. This test is useful in identifying the mineral cal-

Table 1.1
Relative abundance of the most common elements in the earth's crust.

Element	Approximate Percentage by Weight
Oxygen (O)	46.6
Silicon (Si)	27.7
Aluminum (Al)	8.1
Iron (Fe)	5.0
Calcium (Ca)	3.6
Sodium (Na)	2.8
Potassium (K)	2.6
Magnesium (Mg)	2.1
All others	1.5
Total	100.0

Source: Data from Brian Mason.

cite, which is a common carbonate mineral frequently found in sedimentary and metamorphic rocks.

In summary, a number of special physical and chemical properties are useful in identifying certain minerals. These include taste, smell, elasticity, maleability, feel, magnetism, double refraction, and chemical reaction to hydrochloric acid.

The Silicates

Nearly four thousand minerals are presently known and about 40 to 50 new ones are being identified each year. Fortunately for students beginning to study minerals, no more than a few dozen are abundant. Collectively, these few make up most of the rocks of the earth's crust and as such are classified as the *rock-forming minerals*. It is also interesting to note that only eight elements compose the bulk of these minerals and represent over 98 percent (by weight) of the continental crust (Table 1.1). The two most abundant elements are silicon and oxygen, which combine to form the framework of the most common mineral group, the **silicates.** Every silicate mineral contains oxygen and silicon. Moreover, except for a few minerals such as quartz, every silicate mineral contains one or more additional elements that are needed to produce electrical neutrality.

All silicate minerals have the same basic building block, the **silicon-oxygen tetrahedron.** This structure is composed of four oxygen atoms with a smaller silicon atom positioned in the space between them

(Figure 1.7). In some minerals, the tetrahedra are joined into chains, sheets, or three-dimensional networks by sharing oxygen atoms (Figure 1.8). These larger silicate structures are then connected to one another by other elements. The primary elements that join silicate structures are iron (Fe), magnesium (Mg), potassium (K), sodium (Na), and calcium (Ca). Because iron and magnesium are nearly the same size, they can readily substitute for each other without changing the structure of the mineral. This also holds true for calcium and sodium, which can occupy the same site in a crystal structure. Aluminum (Al) is unique as a substance for silicon in the tetrahedron.

The main groups of silicate minerals and common examples of each are given in Figure 1.9. The feldspars are by far the most abundant group, comprising more than 50 percent of the crust. Quartz, the second most common mineral in the continental crust, is the only one made completely of silicon and oxygen. Notice in Figure 1.9 that each mineral group has a particular silicate structure. A relationship ex-

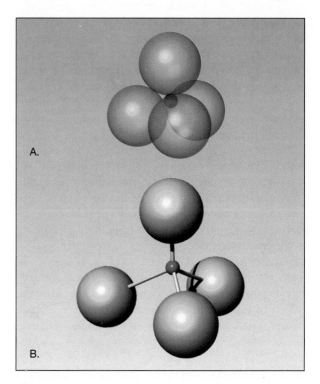

Figure 1.7
Two representations of the silicon-oxygen tetrahedron. ***A.*** *The four large spheres represent oxygen atoms and the blue sphere represents a silicon atom. The spheres are drawn in proportion to the radii of the atoms.* ***B.*** *An expanded view of the tetrahedron using rods to depict the bonds that connect the atoms.*

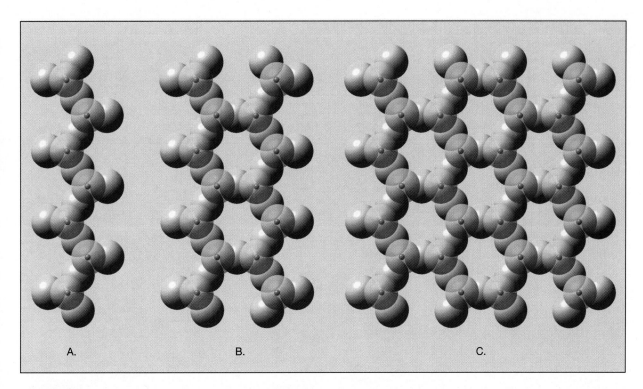

Figure 1.8
Three types of silicate structures. **A.** *Single chains.* **B.** *Double chains.* **C.** *Sheet structures.*

ists between the internal structure of a mineral and the cleavage it exhibits. Because the silicon-oxygen bonds are strong, silicate minerals tend to cleave between the silicon-oxygen structures rather than across them. For example, the micas have a sheet structure and tend to cleave into flat plates (see Figure 1.5). Quartz, which has equally strong silicon-oxygen bonds in all directions, has no cleavage.

Most silicate minerals form when molten rock cools and crystallizes. This process can occur at or near the earth's surface, or at great depths where temperatures and pressures are very high. The environment during crystallization and the chemical composition of the molten rock to a large degree determine the minerals that are produced. In addition, some silicate minerals are stable at the earth's surface and represent the weathered products of pre-existing silicate minerals. Still other silicate minerals are formed under the extreme pressures associated with metamorphism.

The various silicate minerals can be divided into two groups based on their chemical makeup. The so-called dark silicates contain ions of iron and/or magnesium. Examples include hornblende and biotite, which, like other dark silicate minerals, are dark in color and have a specific gravity greater than 3. By contrast, quartz and feldspar are light colored and have an average specific gravity of 2.7. These observed differences are mainly attributable to the presence or absence of iron.

Nonsilicate Minerals

Although many mineral groups are important economically, they are considered scarce compared to the silicates. Table 1.2 lists examples of oxides, sulfides, sulfates, halides, carbonates, and native elements of economic value. This group includes ores of metals such as hematite (iron), sphalerite (zinc), and galena (lead); the native elements gold, silver, and carbon (diamonds); and a host of others such as fluorite, corundum, and malachite.

An important rock-forming group is the *carbonates*, which includes the mineral calcite ($CaCO_3$). This mineral is the major constituent of limestone and marble. Limestone is used commercially for road aggregate, building stone, and as the main ingredient in portland cement.

Two other nonsilicate minerals frequently found in rocks are *halite* and *gypsum*. Both minerals are commonly found in thick layers, where they accumulated when the waters of a saline lake or an an-

Mineral		Idealized Formula	Cleavage	Silicate Structure
Olivine		$(Mg,Fe)_2SiO_4$	None	Single tetrahedron 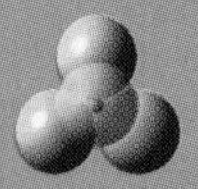
Pyroxene group (Augite)		$(Mg,Fe)SiO_3$	Two planes at right angles	Single chains 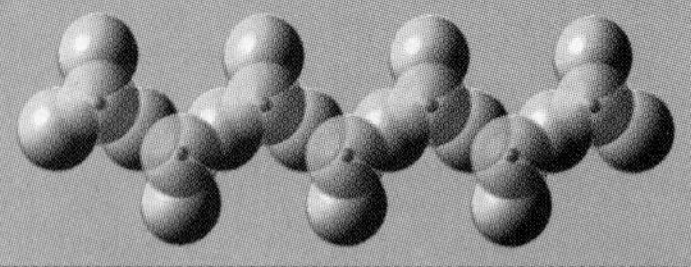
Amphibole group (Hornblende)		$(Ca_2Mg_5)Si_8O_{22}(OH)_2$	Two planes at 60° and 120°	Double chains 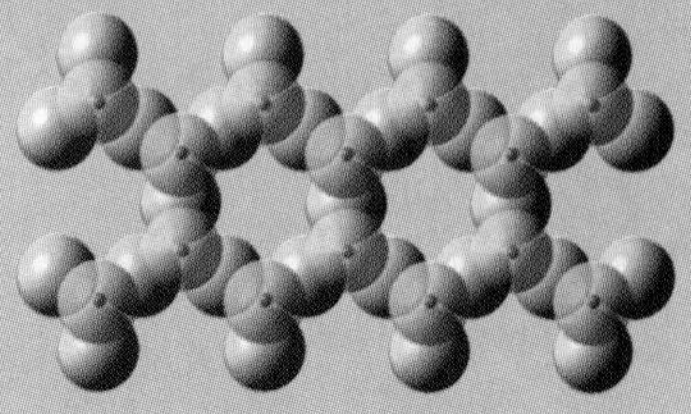
Micas	Biotite	$K(Mg,Fe)_3AlSi_3O_{10}(OH)_2$	One plane	Sheets 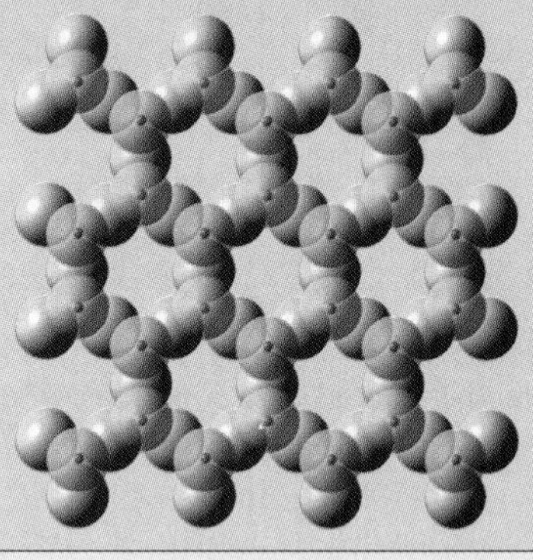
	Muscovite	$KAl_2(AlSi_3O_{10})(OH)_2$		
Feld-spars	Orthoclase	$KAlSi_3O_8$	Two planes at 90°	Three-dimensional networks Very complex structure
	Plagioclase	$(Ca,Na)AlSi_3O_8$		
Quartz		SiO_2	None	

Figure 1.9
Common silicate minerals. Note that the complexity of the silicate structure increases down the chart.

cient sea evaporated. Halite is the mineral name for common table salt. Gypsum is the mineral raw material from which plaster and other similar building materials are made.

THE ROCK CYCLE

The **rock cycle** (Figure 1.10) is one means of viewing many of the interrelationships of geology. By studying the rock cycle we may ascertain the origin of the three basic rock types and gain some insight into the role of various geologic processes in transforming one rock type into another. The concept of the rock cycle, which may be considered as a basic

outline of physical geology, was initially proposed in the late eighteenth century by James Hutton, a founding father of modern geology. This rock cycle, shown in Figure 1.10, uses arrows to indicate chemical and physical processes, and boxes to represent earth materials.

The first rock type, **igneous rock,** originates when molten material called **magma** cools and solidifies. This process, called **crystallization,** may occur either beneath the earth's surface or, following a volcanic eruption, at the surface. Initially, or shortly after forming, the earth's outer shell is believed to have been molten. As this molten material gradually cooled and crystallized, it generated a primitive crust that consisted entirely of igneous rocks.

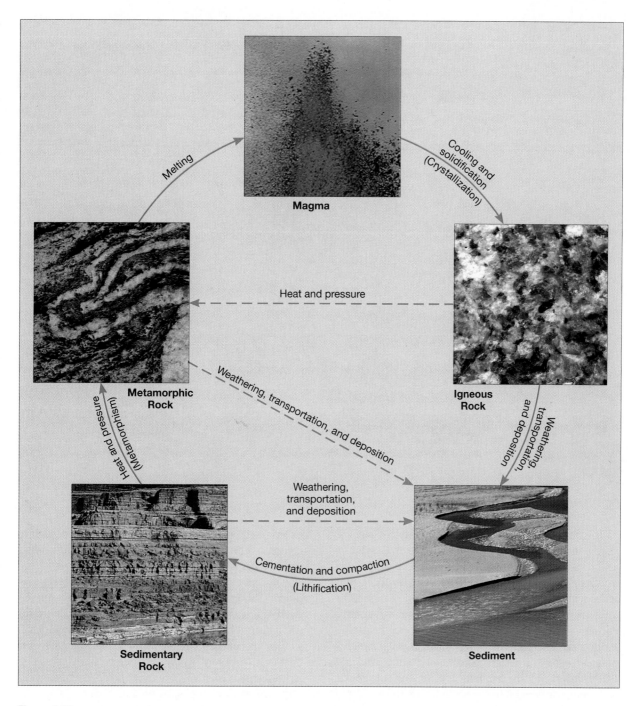

Figure 1.10
The rock cycle. Originally proposed by James Hutton, the rock cycle illustrates the role of the various geologic processes which act to transform one rock type into another.

Table 1.2
Common nonsilicate
mineral groups.

Group	Member	Formula	Economic Use
Oxides	Hematite	Fe_2O_3	Ore of iron
	Magnetite	Fe_3O_4	Ore of iron
	Corundum	Al_2O_3	Gemstone, abrasive
	Ice	H_2O	Solid form of water
Sulfides	Galena	PbS	Ore of lead
	Sphalerite	ZnS	Ore of zinc
	Pyrite	FeS_2	Sulfuric acid production
	Chalcopyrite	$CuFeS_2$	Ore of copper
Sulfates	Gypsum	$CaSO_4 \cdot 2H_2O$	Plaster
	Anhydrite	$CaSO_4$	Plaster
Native elements	Gold	Au	Trade, jewelry
	Copper	Cu	Electrical conductor
	Diamond	C	Gemstone, abrasive
	Sulfur	S	Sulfa drugs, chemicals
	Graphite	C	Pencil lead, dry lubricant
Halides	Halite	NaCl	Common salt
	Fluorite	CaF_2	Steel making, chemicals
Carbonates	Calcite	$CaCO_3$	Portland cement, agricultural lime
	Dolomite	$CaMg(CO_3)_2$	Portland cement, agricultural lime
	Malachite	$Cu_2(OH)_2CO_3$	Ore of copper

If igneous rocks are exposed at the earth's surface, they will undergo **weathering,** in which the day-in and day-out influences of the atmosphere slowly disintegrate and decompose the rock. The resulting material will be picked up, transported, and deposited by any of a number of erosional agents—gravity, running water, glaciers, wind, or waves. Once this material, called **sediment,** is deposited, usually as horizontal beds in the ocean, it will undergo **lithification,** a term meaning "conversion into rock." Sediment is lithified when compacted by the weight of overlying layers or when cemented as percolating groundwater fills the pores with mineral matter. If the resulting **sedimentary rock** is buried deep within the earth or involved in the dynamics of mountain building, it will be subjected to great pressures and heat. The sedimentary rock will react to the changing environment and turn into the third rock type, **metamorphic rock.** When metamorphic rock is sub-jected to still greater heat and pressure, it will melt to create magma, which will eventually solidify as igneous rock.

The full cycle just described does not always take place. "Shortcuts" in the cycle are indicated by dashed lines in Figure 1.10. Igneous rock, for example, rather than being exposed to weathering and erosion at the earth's surface, may be subjected to the heat and pressure found far below and change to metamorphic rock. On the other hand, metamorphic and sedimentary rocks, as well as sediment, may be exposed at the surface and turned into new raw materials for sedimentary rock.

As you study the remaining pages of this chapter, which discuss details of each of the three basic rock types, remember the rock cycle. Although rocks may seem to be unchanging masses, the rock cycle shows that they are not. The changes, however, take time—great amounts of time.

BOX 1.1
Carrot, Karat, or Carat

Like many other words in the English language, the words *carrot, karat,* and *carat* all have the same sound but are different in meaning (Figure 1.A). Such words are called *homonyms.* We all know that a *carrot* is a crunchy, orange vegetable that is supposed to be good for your eyesight. But what about the words *karat* and *carat*?

The term *karat* is used to indicate the purity of gold. Pure or fine gold is 24 karats. Gold that is less than 24 karats is actually an alloy (mixture) of gold and another metal, usually copper or silver. For example, 14-karat gold contains 14 parts of gold (by weight) mixed with 10 parts of other metals. Obviously, gold with a higher karat number is more expensive. Because 24-karat gold is much softer and more malleable than a gold-silver alloy, the alloy forms are more durable.

The third of our homonyms, *carat*, is a unit of weight used for precious gems such as diamonds, emeralds, and rubies. Throughout history, the size of a carat has varied somewhat. However, early in the twentieth century, a carat weight was established at 200 milligrams (or 0.2 gram). To put this in everyday terms, a 1-ounce diamond would be roughly 142 carats. The terms *karat* and *carat* are both derived from the Greek word *keration*, meaning "carob bean." The ancient Greeks used carob beans as a standard of weight.

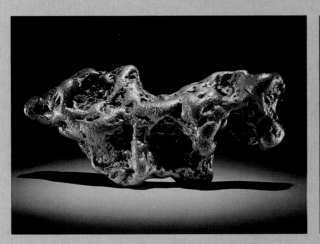

Figure 1.A
The purity of gold, such as this 82-ounce gold nugget, is measured in units called karats. Pure gold is 24 karats. By comparison, the weight of gemstones is measured in carat weight. Each carat equals 0.2 gram. This diamond ring is 29.3 carats. (Courtesy of Smithsonian Institution)

IGNEOUS ROCKS

In our discussion of the rock cycle, it was pointed out that igneous rocks form when magma cools and crystallizes. This molten rock, which originates at depths as great as 200 kilometers within the earth, consists primarily of the elements found in silicate minerals, along with some gases, particularly water vapor, which are confined within the magma by the surrounding rocks. Because the magma body is less dense than the surrounding rocks, it works its way toward the surface, and on occasion breaks through, producing a volcanic eruption.

The spectacular explosions that sometimes accompany an eruption are produced by the gases (volatiles) escaping as the confining pressure lessens near the surface. Sometimes blockage of the vent coupled with surface-water seepage into the magma chamber can produce catastrophic explosions. Along with ejected rock fragments, a volcanic eruption often generates extensive lava flows. **Lava** is similar to magma, except that most of the gaseous component has escaped (Figure 1.11). The rocks that result when lava solidifies are classified as **extrusive,** or **volcanic.** The magma not able to reach the surface eventually crystallizes at depth. Igneous rocks produced in this manner are termed **intrusive,** or **plutonic,** and would never be observed if not for the processes of erosion stripping away the overlying rock.

Crystallization of Magma

Magma is a hot fluid that may contain suspended crystals and a gaseous component. The liquid portion of a magma body is composed of ions that move about freely. However, as magma cools, the random movements of the ions slow and the ions begin to arrange themselves into orderly patterns. This process is called *crystallization.* Usually all of the molten material does not solidify at the same time. Rather, as it cools, numerous small crystals develop. In a systematic fashion, ions are added to these centers of crystal growth. When the crystals grow large enough for their edges to meet, their growth ceases and crystallization continues elsewhere. Eventually, all of the liquid is transformed into a solid mass of interlocking crystals.

The rate of cooling strongly influences the crystallization process, in particular the size of the crystals. When a magma cools very slowly, relatively few centers of crystal growth develop. Slow cooling also allows ions to migrate over relatively great distances. Consequently, slow cooling results in the formation of rather large crystals. On the other hand, when cooling occurs quite rapidly, the ions quickly lose their motion and readily combine. This results in the development of large numbers of tiny crystals that all compete for the available ions. Therefore, the outcome of rapid cooling is the formation of a solid mass composed of very small intergrown crystals.

When the molten material is quenched instantly, there is not sufficient time for the ions to arrange themselves into a crystalline network. Therefore, the solids produced in this manner consist of randomly distributed ions. Rocks that consist of unordered atoms are referred to as *glass* and are quite similar to ordinary manufactured glass.

Figure 1.11
Lava of basaltic composition flowing from a lava tube, Hawaii Volcanoes National Park. (Photo by J. D. Griggs, U.S. Geological Survey)

Texture and Mineral Composition

A large variety of igneous rocks exist and are differentiated on the basis of their texture and mineral content. The term **texture,** when applied to an igneous rock, is used to describe the overall appearance of the rock based on the size and arrangement of its interlocking crystals. Texture is a very important characteristic since it reveals a great deal about the

A.

B.

Extrusive
A
B Intrusive

Figure 1.12
*A. Rhyolite exhibits a fine-grained texture. **B**. Granite is a common igneous rock that has a coarse-grained texture. (Photos by E. J. Tarbuck)*

ure 1.12A). A common feature in many fine-grained igneous rocks are the voids left by gas bubbles trapped as the magma solidifies. These spherical or elongated openings are called *vesicles* and are limited to the upper portion of lava flows. It is in the upper zone of a lava flow that cooling occurs rapidly enough to "freeze" the lava and preserve the openings produced by the expanding gas bubbles. The name *scoria* is commonly applied to rocks that form in this manner (Figure 1.13).

When large masses of magma solidify far below the surface, they form igneous rocks that exhibit a **coarse-grained texture.** These coarse-grained rocks have the appearance of a mass of intergrown crystals, which are roughly equal in size and large enough that the individual minerals can be identified with the unaided eye (Figure 1.12B).

A large mass of magma located at depth may require tens of thousands, even millions, of years to solidify. Since all minerals within a magma do not crystallize at the same rate or at the same time during cooling, it is possible for some to become quite large before others even start to form. If magma containing some large crystals were to change environments by erupting at the surface, for example, the molten

Figure 1.13
Scoria is a volcanic rock that exhibits a vesicular texture. Vesicles form as gas bubbles escape near the top of a lava flow.

environment in which the rock formed. From our discussion of crystallization, we learned that rapid cooling produces small crystals, whereas very slow cooling results in the formation of much larger crystals. As we might expect, the rate of cooling is quite slow in magma chambers lying deep within the crust, whereas a thin layer of lava extruded upon the earth's surface may chill in a matter of hours, and small molten blobs ejected into the air during a violent eruption can solidify almost instantly.

Igneous rocks that form at the earth's surface or as small masses within the upper crust have a very **fine-grained texture,** with the individual crystals being too small to be seen with the unaided eye (Fig-

Figure 1.14
Andesite porphyry. Notice the two distinctively different sizes of crystals. (Photo by E. J. Tarbuck)

portion of the lava would cool quickly. The resulting rock, which has large crystals embedded in a matrix of smaller crystals, is said to have a **porphyritic texture** (Figure 1.14).

During some volcanic eruptions, molten rock is ejected into the atmosphere, where it is quenched very quickly. Rapid cooling of this type may generate rock with a **glassy texture.** As was indicated, glass

results when the ions do not have sufficient time to unite into an orderly crystalline structure. *Obsidian*, a common type of natural glass, is similar in appearance to a dark chunk of manufactured glass (Figure 1.15A). Another volcanic rock that often exhibits a glassy texture is *pumice*. Usually found with obsidian, pumice forms when large amounts of gas escape through lava to generate a gray, frothy mass (Figure 1.15B). In some samples, the voids are quite noticeable, while in others, the pumice resembles fine shards of intertwined glass. Because of the large percentage of voids, many samples of pumice will float when placed in water.

The mineral makeup of an igneous rock is ultimately determined by the chemical composition of the magma from which it crystallized. Such a large variety of igneous rocks exists that it is logical to assume an equally large variety of magmas must also exist. However, geologists have found that various eruptive stages of the same volcano often extrude lavas exhibiting somewhat different mineral compositions, particularly if an extensive time period separated the eruptions. Evidence of this type led geologists to examine the possibility that a single magma might produce rocks of varying mineral content.

A pioneering investigation into the crystallization of magma was carried out by N. L. Bowen in the first quarter of this century. Bowen discovered that as

A.

B.

Figure 1.15
Igneous rocks that exhibit a glassy texture. **A.** *Obsidian, a glassy volcanic rock.*
B. *Pumice, a glassy rock containing numerous tiny voids.*

magma cools in the laboratory, certain minerals crystallize first. At successively lower temperatures, other minerals begin to crystallize, as shown in Figure 1.16).

Bowen also demonstrated that if a mineral remains in the melt after crystallization, it will react with the remaining melt to produce the next mineral in the sequence shown in Figure 1.16. For this reason, this arrangement of minerals became known as *Bowen's reaction series.* On the upper left branch of this reaction series, olivine, the first mineral to form, will react with the remaining melt to become pyroxene. This reaction will continue until the last mineral in the series, biotite, is formed. The right branch of the reaction series is a continuum in which the earliest-formed calcium-rich feldspar crystals react with the sodium ions contained in the melt to become progressively more sodium rich. Ordinarily, these reactions are not complete, so that various amounts of several of these minerals may exist at any given time.

During the last stage of crystallization, after most of the magma has solidified, the minerals muscovite and potassium feldspar form. Finally, the remaining melt (if any) will have a high silica content, which eventually crystallizes as quartz. Although these latter-formed minerals crystallize in the order shown,

they do not react with the melt to produce other minerals.

Bowen's reaction series illustrates the sequence in which minerals crystallize from a magma. One of the consequences of this crystallization scheme is that minerals that form in the same general temperature regime are found together in the same igneous rock. For example, in Figure 1.16, notice that the minerals potassium feldspar, muscovite mica, and quartz are found in the same region of the diagram and are the major constituents of the igneous rock granite.

Bowen demonstrated that minerals crystallize from magma in a systematic fashion. But how does Bowen's reaction series account for the great diversity of igneous rocks? It has been shown that at one or more stages in the crystallization process, a separation of the solid and liquid components of a magma can occur. This happens, for example, if the earlier formed minerals are denser (heavier) than the liquid portion and settle to the bottom of the magma chamber, as shown in Figure 1.17A. This process, called *crystal settling,* is thought to occur frequently with the dark silicates, such as olivine and pyroxene. When the remaining melt solidifies, either in place or in a new location if it migrates out of the magma cham-

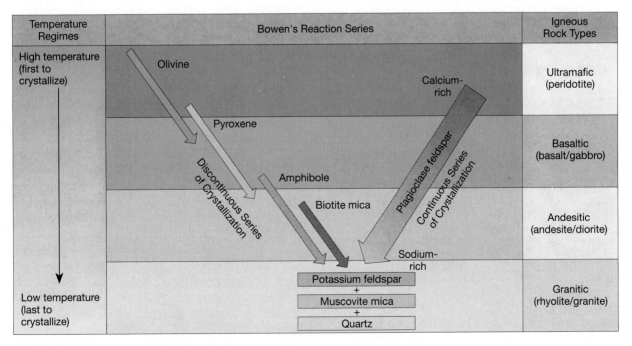

Figure 1.16
Bowen's reaction series shows the sequence in which minerals crystallize from a magma. Compare this figure to the mineral composition of the rock groups in Table 1.3. Note that each rock group consists of minerals that crystallize at the same time.

ber, it will form a rock with a chemical composition much different from the parent magma (Figure 1.17B). The process of developing more than one rock type from a common magma is called **magmatic differentiation.**

At any stage in the evolution of a magma, it can be separated into two or more chemically distinct components. Further, each of these components may undergo further segregation by a variety of processes. Consequently, magmatic differentiation can produce a variety of igneous rocks and a wide range of compositions.

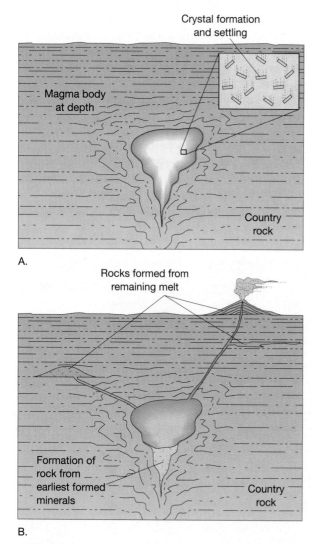

A.

B.

Figure 1.17
*Separation of minerals by crystal settling. **A.** Illustration of how the earliest-formed minerals can be separated from a magma by settling. **B.** The remaining melt could migrate to a number of different locations and, upon further crystallization, generate rocks having a composition much different from that of the parent magma.*

Naming Igneous Rocks

As was stated previously, igneous rocks are most often classified, or grouped, on the basis of their texture and mineral composition. The various igneous textures result from different cooling histories, while the mineral composition of an igneous rock is the consequence of the chemical makeup of the parent magma and the environment of crystallization. As we might expect from the results of Bowen's work, minerals that crystallize under similar conditions are most often found together composing the same igneous rock. A general classification scheme based on texture and mineral composition is provided in Table 1.3.

The rocks on the right side of Table 1.3 consist of the first minerals to crystallize and are higher in iron (Fe) and magnesium (Mg) and lower in silica (SiO_2) than later ones. Their iron content makes them darker in color and slightly more dense than the other rocks. The term **mafic,** derived from **ma**gnesium and **F**e, the chemical symbol for iron, is used for rocks of this composition. Geologists also employ the term *basaltic* (after basalt, a common rock of this composition) to indicate any igneous rock having this composition. *Basalt* is a dark green to black, fine-grained rock composed primarily of pyroxene and calcium-rich feldspar. Basalt is the most common extrusive igneous rock. Many volcanic islands, such as the Hawaiian Islands and Iceland, are composed mainly of basalt. Further, the upper layers of the oceanic crust consist primarily of basalt.

The rocks on the left side of Table 1.3 contain the last minerals to crystallize and consist mainly of **fel**dspar and **si**lica (quartz); consequently, the term **felsic** is applied to them. The felsic rocks are also commonly referred to as having a *granitic* composition. *Granite* is perhaps the best-known igneous rock (see Figure 1.12B). This is partly because of its natural beauty, which is enhanced when it is polished, and partly because of its abundance. Slabs of polished granite are commonly used for tombstones, monuments, and as building stones. Granite is often produced by the processes that generate mountains. Because granite is a by-product of mountain building and is very resistant to weathering and erosion, it frequently forms the core of eroded mountains. For example, Pikes Peak in the Rockies; Mount Rushmore

Table 1.3
Classification of igneous rocks.

	Felsic (Granitic) *light*	Intermediate (Andesitic) *medium*	Mafic (Basaltic) *dark*	Ultramafic
Coarse-grained (intrusive)	Granite	Diorite	Gabbro	Peridotite
Fine-grained (extrusive)	Rhyolite	Andesite	Basalt	Komatite
Mineral Composition	Quartz Potassium feldspar Sodium feldspar	Hornblende Sodium feldspar Calcium feldspar	Calcium feldspar Pyroxene	Olivine Pyroxene
Minor Mineral Constituents	Muscovite Biotite Amphibole	Pyroxene	Olivine Hornblende	Calcium feldspar
Rock color Based on % dark (mafic) minerals	Light-colored Less than 15% dark minerals	Medium-colored 15–40% dark minerals	Dark-gray to black More than 40% dark minerals	Dark-green to black Nearly 100% dark minerals

in the Black Hills; the White Mountains of New Hampshire; Stone Mountain, Georgia; and Yosemite National Park in the Sierra Nevada are all areas where large quantities of granite are exposed at the surface (Figure 1.18).

It should be pointed out that two rocks may have the same mineral constituents but have different tex-tures and hence different names. For example, the coarse-grained intrusive rock granite has a fine-grained volcanic equivalent called *rhyolite* (see Figure 1.12A). Although these rocks are mineralogically the same, they have different textures and do not look at all alike. Further, in contrast to granite, rhyolite is rather uncommon. Yellowstone National Park is one

Figure 1.18
El Capitan, a massive granitic structure in Yosemite National Park, California. (Photo by Galen Rowell)

well-known exception. Here rhyolitic lava flows and ash deposits of similar composition are widespread.

Thus far this discussion has dealt with only two mineral compositions, yet it is important to note that gradations between these types also exist (Table 1.3). For example, an abundant extrusive igneous rock called *andesite* has a mineral composition between that of rocks with a granitic composition and those with a basaltic composition. Andesite is a fine-grained rock of volcanic origin, whose name derives from the Andes Mountains, where numerous volcanoes are composed of this rock type. In addition to the Andes volcanoes, many volcanic structures encircling the Pacific Ocean exhibit this andesitic composition.

Another important igneous rock called *peridotite* contains mostly olivine and pyroxene and thus falls near the very beginning of Bowen's reaction series. Since peridotite is composed almost entirely of ferro-magnesian minerals, its chemical composition is often referred to as *ultramafic*. Although ultramafic rocks are rarely observed on the earth's surface, peridotite is believed to be a major constituent of the upper mantle.

SEDIMENTARY ROCKS

Earlier the rock cycle provided a brief look at the origin of sedimentary rocks. Recall that weathering begins the process. Next, gravity or erosional agents, such as running water, wind, waves, and glacial ice, remove the products of weathering and carry them to a new location where they are deposited. Usually the particles are broken down further during the transport phase. Following deposition, this material, which is now called *sediment*, is lithified. Commonly, compaction and cementation transform the sediment into solid sedimentary rock.

The products of weathering constitute the raw materials for sedimentary rocks. The word *sedimentary* indicates the nature of these rocks, for it is derived from the Latin *sedimentum*, which means "settling," a reference to solid material settling out of a fluid. Most, but not all, sediment is deposited in this fashion. Weathered debris is constantly being swept from bedrock and carried away by water and ice. Eventually the material is deposited in lakes, river valleys, seas, and countless other places. The particles in a desert sand dune, the mud on the floor of a swamp, the gravels in a stream bed, and even household dust

are examples of sediment produced by this never-ending process. Since the weathering of bed-rock and the transport and deposition of the weathering products are continuous, sediment is found almost everywhere. As piles of sediment accumulate, the materials near the bottom are compacted by the weight of the overlying layers. Over long periods, these sediments are cemented together by mineral matter deposited in the spaces between particles to form solid rock.

Geologists estimate that sedimentary rocks account for only about 5 percent (by volume) of the earth's outer 16 kilometers (10 miles). However, the importance of this group of rocks is far greater than this percentage would imply. If we were to sample the rocks exposed at the earth's surface, we would find that the great majority are sedimentary (Figure 1.19). Indeed, about 75 percent of all rock outcrops on the continents are sedimentary. Therefore, we can think of sedimentary rocks as comprising a relatively thin and somewhat discontinuous layer in the uppermost portion of the crust. This fact is readily understood when we consider that sediment accumulates at the surface of the earth.

It is from sedimentary rocks that geologists reconstruct many of the details of earth's history. Because sediments are deposited in a variety of different settings at the earth's surface, the rock layers that they eventually form hold many clues to past surface environments. They may also exhibit characteristics that allow geologists to decipher information about the method and length of sediment transport. Furthermore, it is sedimentary rocks that contain fossils, which are vital tools in the study of the geologic past.

Finally, it should be mentioned that many sedimentary rocks are very important economically. Coal, for example, is classified as a sedimentary rock, whereas our other major energy resources, petroleum and natural gas, are found in association with sedimentary rocks. Still others represent major sources of iron, aluminum, manganese, and fertilizer, as well as numerous materials essential to the construction industry.

Classification

Materials accumulating as sediment have two principal sources. First, sediments may be accumulations of materials that originate and are transported as solid particles derived from weathering. Deposits of this type are termed *detrital* and the sedimentary

Figure 1.19
Sedimentary rocks exposed near Canyonlands National Park, Utah. About 75 percent
of all rock outcrops on the continents are sedimentary. (Photo by Stephen Trimble)

rocks that they form are called **detrital sedimentary rocks.** The second major source of sediment is soluble material produced largely by chemical weathering. When these dissolved substances are precipitated by either inorganic or organic processes, the material is known as chemical sediment and the rocks formed from it are called **chemical sedimentary rocks.**

Detrital Sedimentary Rocks. Though a wide variety of minerals and rock fragments may be found in detrital rocks, clay minerals and quartz are the chief constituents of most sedimentary rocks in this category. As we shall see in Chapter 2, clay minerals are the most abundant product of the chemical weathering of silicate minerals, especially the feldspars. Quartz, on the other hand, is abundant because it is extremely durable and very resistant to chemical weathering. Thus, when igneous rocks such as granite are attacked by weathering processes, individual quartz grains are set free.

Particle size is the primary basis for distinguishing among various detrital sedimentary rocks. Table 1.4 presents the size categories for particles making up detrital rocks. When gravel-sized particles predominate, the rock is called *conglomerate* if the sed-

Table 1.4
Particle size classification for detrital rocks.

Sediment Name	Size Range (mm)	Detrital Rock Name
Gravel	> 2	Conglomerate or breccia
Sand	1/16–2	Sandstone
Silt	1/256–1/16	Siltstone
Clay	< 1/256	Shale

iment is rounded (Figure 1.20A) and *breccia* if the pieces are angular (Figure 1.20B). Angular fragments indicate that the particles were not transported very far from their source prior to deposition. *Sandstone* is the name given rocks when sand-sized grains prevail (Figure 1.20C), whereas *shale*, the most common sedimentary rock, is made of very fine grained sediment (Figure 1.20D). *Siltstone*, another rather fine grained rock, is sometimes difficult to differentiate from rocks such as shale that are composed of even smaller clay-sized sediment.

Particle size is not only a convenient method of dividing detrital rocks; the sizes of the component grains also provide useful information about environments of deposition. Currents of water or air sort the particles by size; the stronger the current, the larger the particle size carried. Gravels, for example, are moved by swiftly flowing rivers as well as by rockslides and glaciers. Less energy is required to transport sand; thus it is common to such features as windblown dunes, as well as some river deposits and beaches. Since silts and clays settle very slowly, accumulations of these materials are generally associated with the quiet waters of a lake, lagoon, swamp, or marine environment.

Although detrital sedimentary rocks are classified primarily by the size of their particles, in certain cases the mineral composition is also part of naming a rock. For example, most sandstones are predominantly quartz-rich, but when appreciable quantities of feldspar are present, the rock is called *arkose*. In addition, rocks consisting of detrital sediments are rarely composed of grains of just one size. Consequently, a rock containing quantities of both sand

and silt can be correctly classified as sandy siltstone or silty sandstone, depending upon which particle size dominates.

Chemical Sedimentary Rocks. In contrast to detrital rocks, which form from the solid products of weathering, chemical sediments derive from material that is carried in solution to lakes and seas. This material does not remain dissolved in the water indefinitely. Rather, some of it precipitates to form chemical sediments. This precipitation may occur directly as the result of inorganic processes or indirectly as the result of the life processes of water-dwelling organisms. Sediment formed in this second way is said to have a *biochemical* origin.

An example of a deposit resulting from inorganic chemical processes is the salt left behind as a body of salt water evaporates. In contrast, many water-dwelling animals and plants extract dissolved mineral matter to form shells and other hard parts. After the organisms die, their skeletons may accumulate on the floor of a lake or ocean.

Limestone is the most abundant chemical sedimentary rock. It is composed chiefly of the mineral calcite ($CaCO_3$) and forms by either inorganic means or as the result of biochemical processes. Limestones having a biochemical origin are by far the most common. As much as 90 percent of the world's limestone may have originated as accumulations of biochemical sediment.

Although most limestone is the product of biological processes, this origin is not always evident because shells and skeletons may undergo considerable change before being converted to rock. However, one easily identified biochemical limestone is *coquina*, a coarse rock composed of poorly cemented shells and shell fragments (Figure 1.21). Another less obvious, but nevertheless familiar, example is *chalk*, a soft, porous rock made up almost entirely of the hard parts of microscopic organisms that are no larger than the head of a pin (Figure 1.22).

Limestones having an inorganic origin form when chemical changes or high water temperatures increase the concentration of calcium carbonate to the point that it precipitates. *Travertine*, the type of limestone commonly seen decorating caverns, is one example. Groundwater is the source of travertine that is deposited in caves. As water drops reach the air in a cavern, some of the carbon dioxide dissolved in the water escapes, causing calcium carbonate to precipitate.

Figure 1.20
Common detrital sedimentary rocks. **A.** *Conglomerate.* **B.** *Breccia.* **C.** *Sandstone.*
D. *Shale. (Photos by E. J. Tarbuck)*

Dissolved silica (SiO_2) precipitates to form varieties of microcrystalline quartz. These include chert (light color), flint (dark), jasper (red), and agate (banded). These chemical sedimentary rocks may have either an inorganic or biochemical origin, but the mode of origin is usually difficult to determine.

Very often, evaporation is the mechanism triggering deposition of chemical precipitates. Minerals commonly precipitated in this fashion include halite, the chief component of *rock salt*, and gypsum, the main ingredient of *rock gypsum*. Both materials have significant commercial importance. Halite is familiar to everyone as the common salt used in cooking and seasoning foods. Of course, it has many other uses

and has been considered important enough that people have sought, traded, and fought over it for much of human history. Gypsum is the basic ingredient of plaster of Paris. This material is used most extensively in the construction industry for wallboard and plaster for interior use.

In the geologic past, many areas that are now dry land were covered by shallow arms of the sea that had only narrow connections to the open ocean. Under these conditions, water continually moved into the bay to replace water lost by evaporation. Eventually the waters of the bay became saturated and salt deposition began. Such deposits are called **evaporites.**

Figure 1.21
This rock, called coquina, consists of shell fragments; therefore, it has a biochemical origin.

On a smaller scale, evaporite deposits may be seen in such places as Death Valley, California. Here, following rains or periods of snowmelt in the mountains, streams flow from surrounding mountains into an enclosed basin. As the water evaporates, *salt flats* form from dissolved materials left behind as a white crust on the ground (Figure 1.23).

Coal is quite different from other sedimentary rocks. Nevertheless, it is often grouped with biochemical sedimentary rocks. However, unlike other rocks in this category, which are calcite- or silica-rich, coal is made of organic matter. Close examination of a piece of coal under a microscope or magnifying glass often reveals the presence of various plant structures such as leaves, bark, and wood that have been chemically altered but are nevertheless still identifiable. This supports the conclusion that coal is the end product of the burial of large amounts of plant material over extended periods.

Along with oil and natural gas, coal is commonly called a **fossil fuel.** Such a designation is certainly appropriate since each time we burn coal we are using energy from the sun that was stored by plants many millions of years ago. We are indeed burning a "fossil."

The initial stage in coal formation is the accumulation of large quantities of plant remains. However,

special conditions are required for such accumulations, because dead plants readily decompose when exposed to the atmosphere or other oxygen-rich environments. One important environment that allows for the buildup of plant material is a swamp. Since stagnant swamp water is oxygen deficient, complete decay (oxidation) of the plant material is not possible. At various times during earth history such environments have been relatively common. With each successive stage in coal formation, higher temperatures and pressures drive off impurities and volatiles as follows:

$$\text{PEAT} \xrightarrow{\text{burial}} \text{LIGNITE} \xrightarrow{\text{greater burial}}$$
(partially   (soft, brown coal)
altered plant
material)

$$\text{BITUMINOUS} \xrightarrow{\text{metamorphism}} \text{ANTHRACITE}$$
(soft, black coal)   (hard, black coal)

Figure 1.22
The White Chalk Cliffs of Dover. (Photo by Jerome Wyckoff)

Figure 1.23
These salt flats in Death Valley, California, are examples of evaporite deposits and are common in basins located in the arid Southwest. A temporary lake is created when water drains from the adjacent mountains into the basin following periods of rain or snowmelt. When the water evaporates, salts are left behind. (Photo by Michael Collier)

Lignite and bituminous coals are sedimentary rocks, but anthracite is a metamorphic rock. Anthracite forms when sedimentary layers are subjected to the folding and deformation associated with mountain building. Although anthracite is a clean-burning fuel, only a relatively small amount is mined. This stems from the fact that anthracite is not widespread and is more difficult and expensive to extract than the flat-lying layers of bituminous coal.

Coal has been an important fuel for centuries. In the nineteenth and early twentieth centuries, cheap and plentiful coal powered the industrial revolution. By 1900 coal was providing 90 percent of the energy produced in the United States. Although still important, coal currently provides less than 20 percent of the United States' energy needs and ranks third as a source of fuel behind oil and natural gas.

According to the U.S. Department of Energy, about 53 percent of the nation's electricity is generated by burning coal. Nearly 80 percent of the approximately 1 billion tons of coal used in the United States in 1991 was used for this purpose. Coal is likely to continue playing an important role in the nation's economy well into the next century.

In summary, the foregoing classification divides sedimentary rocks into two major groups: detrital and chemical. We have seen that the main criterion for subdividing the detrital rocks is particle size, whereas the primary basis for distinguishing among different rocks in the chemical group is their mineral composition. As is true with many (perhaps most) classifications of natural phenomena, the categories presented here are more rigid than is the actual state of nature. As was pointed out, many detrital sedimentary rocks are a mixture of more than one particle size. Furthermore, many of the sedimentary rocks classified as chemical also contain at least small quantities of detrital sediment. Many limestones, for example, contain varying amounts of mud or sand, giving them a "sandy" or "shaly" quality. On the other hand, since practically all detrital rocks are cemented with material that was originally dissolved in water, they too are far from being "pure."

Lithification

Lithification refers to the processes by which unconsolidated sediments are transformed into solid sedimentary rocks. One of the most common processes affecting sediments is *compaction*. As sediments accumulate through time, the weight of overlying material compresses the deeper sediments. As the grains are pressed closer and closer, there is a considerable reduction in pore space. For example, when clays are buried beneath several thousand meters of material, the volume of the clay may be reduced by as much as 40 percent. Since sands and other coarse sediments

A.

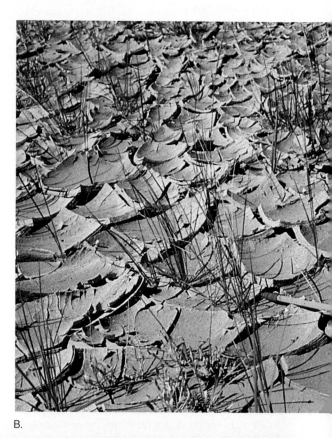

B.

C.

Figure 1.24
A. Ripple marks may indicate a beach or stream channel environment. (Photo by Stephen Trimble) **B.** Mud cracks form when wet mud or clay dries and shrinks, perhaps signifying a tidal flat or desert basin. (Photo by Garrett Deckert) **C.** The cross-bedding in this sandstone indicates that it was once a sand dune. (Photograph used by permission of Dennis Tasa)

are only slightly compressible, compaction is most significant as a lithification process in fine-grained sedimentary rocks such as shale.

Cementation is another important means by which sediments are converted to sedimentary rocks. The cementing materials are carried in solution by water percolating through the open spaces between particles. Through time, the cement precipitates onto the sediment grains, fills the open spaces, and joins the particles. Calcite, silica, and iron oxide are the most common cements. The identification of the cementing material is a relatively simple matter. Calcite cement will effervesce with dilute hydrochloric acid. Silica is the hardest cement and thus produces the hardest sedimentary rocks. When a sedimentary rock has an orange or red color, this usually means iron oxide is present.

Although most sedimentary rocks are lithified by compaction, cementation, or a combination of both, some are made of interlocking crystals. This type of lithification is confined largely to certain chemical sedimentary rocks.

Features of Sedimentary Rocks

Sedimentary rocks are particularly important in the interpretation of earth history. These rocks form at the earth's surface, and as layer upon layer of sediment accumulates, each records the nature of the environment at the time the sediment was deposited. These layers, called **strata,** or **beds,** are the single most characteristic feature of sedimentary rocks (see Figure 1.19). The thickness of beds ranges from microscopically thin to tens of meters thick. Separating the strata are *bedding planes*, flat surfaces along which rocks tend to separate or break. Generally each bedding plane marks the end of one episode of sedimentation and the beginning of another.

As geologists examine sedimentary rocks, much can be deduced. A conglomerate, for example, may indicate a high-energy environment, such as a rushing stream, where only the coarse materials can settle out. If the rock is arkose, it may signify a dry climate, where little chemical alteration of feldspar is possible. Carbonaceous shale is a sign of a low-energy, organic-rich environment such as a swamp or lagoon. Other features found in some sedimentary rocks also give clues to past environments (Figure 1.24).

Fossils, the traces or remains of prehistoric life, are perhaps the most important inclusions found in some

Figure 1.25
Fossil of a trilobite, an ancient marine creature. Fossils are important tools used to determine past environmental conditions and are important time indicators.

sedimentary rock (Figure 1.25). Knowing the nature of the life-forms that existed at a particular time may help to answer many questions about the environment. Was it land or ocean? A lake or swamp? Was the climate hot or cold, rainy or dry? Was the ocean water shallow or deep, turbid or clear? Furthermore, fossils are important time indicators and play a key role in matching up rocks from different places that are the same age. Fossils are important tools used in interpreting the geologic past and will be examined in some detail in Chapter 9.

METAMORPHIC ROCKS

Extensive areas of metamorphic rocks are exposed on every continent in the relatively flat regions known as shields. Moreover, metamorphic rocks are an important component of many mountain belts, where they make up a large portion of a mountain's crystalline core. Even the stable continental interiors, which are generally covered by sedimentary rocks, are underlain by metamorphic basement rocks. In all of these settings the metamorphic rocks are usually highly deformed and intruded by igneous masses. Indeed, significant parts of the earth's continental crust are composed of metamorphic and associated igneous rocks.

Metamorphism involves the transformation of pre-existing rock. Metamorphic rocks can form from igneous, sedimentary, or even from other metamorphic rocks. The term for this process is very appropriate because it literally means to "change form." The agents of change include heat, pressure, and chemically active fluids, while the changes that occur are textural as well as mineralogical.

In some instances rocks are only slightly changed. For example, under low-grade metamorphism the common sedimentary rock shale becomes the metamorphic rock called *slate*. Although slate is more compact than shale, hand samples of these rocks are sometimes difficult to distinguish. In other cases the transformation is so complete that the identity of the original rock cannot be determined. In high-grade metamorphism, such features as bedding planes, fossils, and vesicles that may have existed in the parent rock are completely destroyed. Further, when rocks at depth are subjected to directional pressure, they flow and bend into intricate folds (Figure 1.26). In the most extreme metamorphic environments, the temperatures approach those at which rocks melt. However, during metamorphism the deformed material must remain solid, for once melting occurs, we have entered the realm of igneous activity.

The process of metamorphism takes place when rock is subjected to conditions unlike those in which it originally formed; the rock becomes unstable and gradually changes until a state of equilibrium with the new environment is reached. The changes occur at the temperatures and pressures in the region extending from a few kilometers below the earth's surface to the crust-mantle boundary. Since the formation of metamorphic rocks is completely hidden from view (which is not the case for many sedimentary and some igneous rocks), metamorphism is undoubtedly one of the most difficult processes for geologists to study.

Metamorphism most often occurs in one of two settings. First, during mountain building great quantities of rock are subjected to the intense stresses and high temperatures associated with large-scale deformation. The end result may be extensive areas of metamorphic rocks that are said to have undergone **regional metamorphism.** The greatest volume of metamorphic rock is produced in this fashion. Second, when rock is in contact with, or near, a mass of magma, **contact metamorphism** takes place. In this circumstance the changes are caused primarily

Figure 1.26
Deformed metamorphic rocks exposed in a road cut in the Eastern Highland of Connecticut. (Photo by Phil Dombrowski)

by the high temperatures of the molten material, which in effect "bake" the surrounding rock.

Agents of Metamorphism

As stated earlier, the agents of metamorphism include *heat*, *pressure*, and *chemically active fluids*. During metamorphism, rocks are often subjected to all three metamorphic agents simultaneously. However, the degree of metamorphism and the contribution of each agent vary greatly from one environment to another. In low-grade metamorphism, rocks are subjected to temperatures and pressures only slightly greater than those associated with the lithification of sediments. High-grade metamorphism, on the other hand, involves extreme conditions closer to those at which rocks melt. In addition, the mineralogy of the parent rock determines, to a large extent, the degree to which various metamorphic agents will cause change. For example, when an intruding igneous mass enters a rock unit, hot, ion-rich fluids circulate through the host rock. If the host rock is quartz sandstone, very little alteration may take place. On the other hand, if the host rock is limestone, the impact of these fluids can be dramatic and the effects of metamorphism may extend for several kilometers from the igneous mass.

Perhaps the most important agent of metamorphism is heat, because it provides the energy to drive chemical reactions that result in the recrystallization of minerals. Rocks formed near the earth's surface may be subjected to intense heat when they are intruded by molten material rising from below. Since temperature increases with depth, rocks that originate in a surface environment may also be subjected to extreme temperatures if they are subsequently buried deep within the earth. When buried to a depth of only a few kilometers, certain minerals such as clay become unstable and begin to recrystallize into different minerals that are stable in this environment. Other minerals, particularly those in crystalline igneous rocks, are stable at relatively high temperatures and pressures and therefore require burial to 20 kilometers or more before metamorphism will occur.

Pressure, like temperature, also increases with depth. Buried rocks are subjected to the force exerted by the load above. This confining pressure is analogous to air pressure where the force is applied equally in all directions. In addition to the pressure exerted by the load of material above, rocks are also subjected to *stress* during the process of mountain building. In this situation the applied force is directional, and the material is squeezed as if it had been placed in a vise. Rock located at great depth is quite warm and behaves plastically during deformation. This accounts for its ability to flow and bend into intricate folds.

Chemically active fluids, most commonly water containing ions in solution, also influence the metamorphic process. Some water is contained in the pore spaces of virtually every rock. Water that surrounds the crystals acts as a catalyst by aiding the migration of ions. In some instances, the minerals recrystallize into more stable configurations. In other cases, ion exchange among minerals results in the formation of completely new minerals.

TEXTURAL AND MINERALOGICAL CHANGES

The degree of metamorphism is reflected in the rock's texture and mineralogy. When rocks are subjected to low-grade metamorphism, they become more compact and thus more dense. A common example is the metamorphic rock slate, which forms when shale is subjected to temperatures and pressures only slightly greater than those associated with sedimentary processes. In this case, directed pressure causes the microscopic clay minerals in shale to align into the more compact arrangement found in slate. Under more extreme conditions, pressure causes certain minerals to recrystallize. As described earlier, water is believed to play a very important role in the recrystallization process by aiding the migration of ions. In general, recrystallization encourages the growth of larger crystals. Consequently, many metamorphic rocks consist of visible crystals, much like coarse-grained igneous rocks. The crystals of some minerals, such as micas, which have a sheet structure, and hornblende, which has an elongated structure, will recrystallize with a preferred orientation. The new orientation will be essentially perpendicular to the direction of the compressional force. The resulting mineral alignment usually gives the rock a layered or banded appearance termed **foliated** texture (Figure 1.27). Simply, foliation results whenever the minerals of a rock are brought into parallel alignment.

Not all metamorphic rocks have a foliated texture. Such rocks are said to exhibit a **nonfoliated** texture. Metamorphic rocks composed of only one mineral that forms equidimensional crystals are, as a rule, not visibly foliated. For example, when a fine-grained limestone is metamorphosed, the small calcite crystals combine to form relatively large interlocking crystals. The resulting rock has an appearance similar to a coarse-grained igneous rock. This metamorphic equivalent of limestone is called *marble*.

In some environments, new materials are actually introduced during the metamorphic process. For example, host rock adjacent to a large magma body would be altered by ion-rich *hydrothermal* (hot water) *solutions* released during the latter stages of crystallization. Many metallic ore deposits are formed by the precipitation of minerals from hydrothermal solutions.

Common Metamorphic Rocks

Metamorphic processes cause many changes in preexisting rocks, including increased density, growth of larger crystals, reorientation of the mineral grains into a layered or banded appearance known as foliation, and the transformation of low-temperature minerals into high-temperature minerals. Further, the introduction of ions generates new minerals, some of which are economically important. A description of some of the most common products of these metamorphic processes follows.

Foliated Rocks. *Slate* is a very fine grained foliated rock composed of minute mica flakes (Figure 1.28A). The most noteworthy characteristic of slate is its excellent rock cleavage. This property has made slate a most useful rock for roof and floor tile, blackboards, and billiard tables. Slate is most often generated by the low-grade metamorphism of shale, although less frequently it forms from the metamorphism of volcanic ash. Slate can be almost any color depending on its mineral constituents. Black (carbonaceous) slate contains organic material; red slate gets its color from iron oxide; and green slate is usually composed of chlorite, a micalike mineral formed by the metamorphism of iron-rich silicates.

Schists are strongly foliated rocks, formed by regional metamorphism, that can be readily split into thin flakes or slabs. By definition, schists contain more than 50 percent platy and elongated minerals that commonly include muscovite, biotite, and amphibole. Like slate, the parent material from which many schists originate is shale, but in the case of schist, the metamorphism is more intense.

The term *schist* describes the texture of a rock regardless of composition, unless the composition is specifically included in the rock name. For example, schists composed primarily of muscovite and biotite are called *mica schists* (Figure 1.28B).

Gneiss is the term applied to banded metamorphic

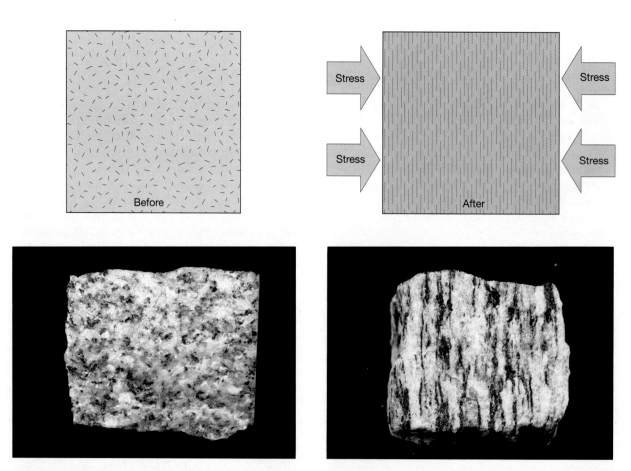

Figure 1.27
Under the pressures of metamorphism, some mineral grains become reoriented and aligned at right angles to the stress. The resulting planar orientation of mineral grains gives the rock a foliated texture. If the coarse-grained igneous rock (granite) on the left underwent intense metamorphism, it could end up closely resembling the metamorphic rock on the right (gneiss). (Photo by E. J. Tarbuck)

A.

B.

Figure 1.28
*A. Slate. **B***. Mica schist. ***C***. Pink
marble. (Photos by E. J.
Tarbuck)*

C.

rocks that contain mostly elongated and granular, as opposed to platy, minerals (see Figure 1.27, right). The most common minerals found in gneisses are quartz and feldspar, with lesser amounts of muscovite, biotite, and hornblende. The segregation of light and dark silicates is developed in gneisses, giving them a characteristic banded texture. While in a plastic state, these banded gneisses are often deformed into rather intricate folds.

Nonfoliated Rocks. *Marble* is a coarse, crystalline rock whose parent rock is limestone (Figure 1.28C). Marble is composed of large interlocking calcite crystals, which form from the recrystallization of smaller grains in the parent rock. Pure marble is white.

Because of its color and relative softness (hardness of 3), marble is a popular building stone. White marble is particularly prized as a stone from which to

Precious stones have been prized by humankind since antiquity. This tradition survives today. Gems are not just prized keepsakes of the rich but are possessed by people of modest means as well. Misinformation abounds about the nature of gems and the minerals of which they are composed. Part of the misinformation stems from the ancient practice of grouping precious stones by color rather than mineral makeup. For example, the more common red spinels were often passed off to royalty as rubies, which are more valuable gems. Even today, when modern techniques of mineral identification are commonplace, yellow quartz is frequently sold as topaz.

Compounding the confusion is the fact that many gems have names that are different from their mineral names. For example, diamond is composed of the mineral of the same name, whereas sapphire is a form of corundum, an aluminum oxide–rich mineral. Although pure aluminum oxide is colorless, a minute amount of foreign element can produce a vividly colored gemstone. Hence, depending on the impurity, sapphires of nearly every color are known to exist. Pure aluminum oxide with trace amounts of titanium and iron produce the most prized blue sapphires. When the mineral corundum contains sufficient quantities of chromium, it exhibits a brilliant red color and the gem is called *ruby*. Large, gem-quality rubies are much rarer than diamonds and thus can command a very high price.

To summarize, when a gem-quality sample of corundum exhibits a red hue, it is called *ruby*, but if it exhibits any other color, the gem is called *sapphire*. If the specimen is not suitable as a

Table 1.A
Important gemstones.

Gem	Mineral Source	Prized Hues
Precious		
Diamond	Diamond	Colorless, yellows
Emerald	Beryl	Greens
Opal	Opal	Brilliant hues
Ruby	Corundum	Reds
Sapphire	Corundum	Blues
Semiprecious		
Alexandrite	Chrysoberyl	Variable
Amethyst	Quartz	Purples
Cat's-eye	Chrysoberyl	Yellows
Chalcedony	Quartz (agate)	Banded
Citrine	Quartz	Yellows
Garnet	Garnet	Reds, greens
Jade	Jadeite or nephrite	Greens
Moonstone	Feldspar	Transparent blues
Peridot	Olivine	Olive greens
Smoky quartz	Quartz	Browns
Spinel	Spinel	Reds
Topaz	Topaz	Purples, reds
Tourmaline	Tourmaline	Reds, blue-greens
Turquois	Turquois	Blues
Zircon	Zircon	Reds

Figure 1.B
Australian sapphires depicting variation in cuts and colors. (Photo by Fred Ward, Black Star)

gem, it simply goes by the mineral name *corundum*. Although common corundum is not a gemstone, it does have value as an abrasive material. Whereas two gems, rubies and sapphires, are composed of the mineral corundum, quartz is the parent mineral of more than a dozen gems. Table 1.A list some well-known gemstones and their mineral names.

What constitutes a gemstone? In essence, certain mineral specimens when cut and polished possess beauty of such quality that they can command a price that makes the process of producing the gem profitable. Gemstones can be divided into two categories: precious and semiprecious. A *precious* gem has beauty, durability, size, and rarity, whereas a *semiprecious* gem generally has only one or two of these qualities. The gems that have traditionally enjoyed the highest esteem are diamonds, rubies, sapphires, emeralds, and some varieties of opal (Table 1.A). All other gemstones are classified as semiprecious. It should be noted, however, that large, high-quality specimens of

so-called semiprecious stones can often command a very high price.

Obviously, beauty is the most important quality that a gem can possess. Today, we prefer translucent stones with evenly tinted colors. The most favored hues appear to be red, blue, green, purple, rose, and yellow. The most prized stones are pigeon-blood rubies, blue sapphires, grass-green emeralds, and canary-yellow diamonds. Colorless gems are generally less desirable except in the case of diamonds that display "flashes of color" known as *brilliance*. Note that gemstones in the "rough" are dull and would be passed over by most laypersons as "just another rock." Gems must be cut and polished by experienced craftsmen before their true beauty can be displayed (Figure 1.B).

The durability of a gem depends upon its hardness; that is, its resistance to abrasion by objects normally encountered in everyday living. For good durability, gems should be as hard as or harder than quartz as defined by the Mohs scale of hardness,

which consists of ten minerals arranged in order from 1 (softest) to 10 (hardest). One notable exception is opal, which is comparatively soft (hardness 5 to 6.5) and brittle. Opal's esteem comes from its "fire," which is a display of a variety of brilliant colors including greens, blues, and reds.

It seems to be human nature to treasure that which is rare. In the case of gemstones, large, high-quality specimens are much rarer than smaller stones. Thus, large rubies, diamonds, and emeralds, which are rare in addition to being beautiful and durable, command the very highest prices.

BOX 1.3

Bingham Canyon, Utah: The Largest Open-Pit Mine

The method used to extract resources from the earth depends upon the nature and location of the material being sought. Many resources are acquired through mining. For large numbers of people, the image of a mine is that of an underground shaft or tunnel. Nevertheless, surface mining accounts for about two-thirds of the world's production of solid mineral resources. *Quarries*, in which gravel or building stone is extracted, are one relatively common type of surface mine. Others are called *strip mines* and *open-pit mines*. Strip mining is employed when a resource, such as coal, occurs near the surface in flat-lying layers. The overlying vegetation, soil, and rock are stripped off, and then the coal or other desired material is removed. When mining is completed, the land can be reclaimed by filling the void with the overburden and then restoring the topsoil and replanting vegetation.

In contrast to strip mines, open-pit mines are the preferred method of extraction where there is a large, three-dimensional body of ore near the surface. As the removal of great tonnages progresses, the excavation enlarges. With time a conical pit is formed with terraced benches spiraling down to the bottom.

The world's largest open-pit mine is the mammoth Bingham Canyon copper mine located about 40 kilometers (25 miles) southwest of Salt Lake City, Utah (Figure 1.C). A mountain once stood where this huge open-pit mine is now. At the top, the rim is nearly 4 kilometers (2.5 miles) across and covers almost 8 square kilometers (3 square miles). The excavation reaches a depth of 900 meters (3000 feet). If a steel tower were erected at the bottom, it would have to be five times taller than the Eiffel Tower to reach the top of the pit.

Underground mining operations that were seeking veins of silver and lead first began at Bingham Canyon in the late nineteenth century. Later, copper was discovered. The ore is known as *porphyry copper*. Similar deposits occur at several locations in the American Southwest and at a number of other sites in a belt that stretches from southern Alaska to northern Chile. As in other places in this belt, the ore at Bingham Canyon is disseminated throughout a porphyritic plutonic rock. The deposit formed after magma was intruded to shallow depths. Following this, shattering created an extensive system of fractures that were penetrated by hydrothermal solutions from which the ore minerals precipitated. Although the percentage of copper in the rock is small, the total volume of copper is huge. Since open-pit operations started in 1906, 5 billion tons of material have been removed, yielding more than 12 million tons of copper. Significant amounts of gold, silver, and molybdenum have also been recovered. The ore body is far from exhausted. Over the next 25 years plans call for an additional 3 billion tons of material to be removed and processed. This largest of artificial excavations has generated most of Utah's mineral production for more than 80 years and has been called the "richest hole on earth."

Like many mines that have been operating for an extended period, the Bingham pit was unregulated during most of its history. Development occurred prior to the present-day appreciation and awareness of the environmental impacts of mining and prior to effective environmental legislation. Today, problems and issues related to groundwater and surface water contamination, air pollution, solid and hazardous wastes, and land reclamation are receiving more serious and long overdue attention.

carve monuments and statues, such as the famous statue of David by Michelangelo. Often the limestone from which marble forms contains impurities that color the marble. Thus, marble can be pink, gray, green, or even black.

Quartzite is a very hard metamorphic rock most often formed from quartz sandstone. Under moderate-to-high-grade metamorphism, the quartz grains in sandstone fuse. Quartzite is typically white, but iron oxide may produce reddish or pinkish stains and dark minerals may impart a gray color.

RESOURCES FROM ROCKS AND MINERALS

The outer layer of the earth, which we call the crust, is only as thick when compared to the remainder of the earth as a peach skin is to a peach, yet it is of supreme importance to us. We depend on it for fossil fuels and as a source of such diverse minerals as the talc for baby powder, salt to flavor food, and gold for world trade. In fact, on occasion, the availability or absence of certain earth materials has altered the course of history. As the material requirements of modern society grow, the need to locate additional supplies of useful minerals also grows, and becomes more challenging as well.

Mineral resources are the endowment of useful minerals ultimately available commercially. Resources include already-identified deposits from which minerals can be extracted profitably, called **reserves,** as well as deposits that are not yet economically or technologically recoverable. In addition, the term **ore** is used to denote those useful metallic minerals that can be mined at a profit. In common usage, the term *ore* is also applied to some nonmetallic minerals such as fluorite and sulfur. However, materials used for such purposes as building stone, road aggregate, abrasives, ceramics, and fertilizers are not usually called ores; rather, they are classified as industrial rocks and minerals.

Recall that more than 98 percent of the earth's crust is composed of only eight elements and, except for oxygen and silicon, all other elements make up a relatively small fraction of common crustal rocks (see Table 1.1). Indeed, the natural concentrations of many elements are exceedingly small. A deposit containing the average percentage of a valuable element is worthless if the cost of extracting it exceeds the value of the material recovered. To be considered of value, an element must be concentrated above the level of its average crustal abundance. Generally, the lower the crustal abundance, the greater the concentration must be. For example, copper makes up only about 0.0135 percent of the crust. However, for a material to be considered a copper ore, it must contain a concentration that is about 50 times this amount. Aluminum, on the other hand, represents 8.13 percent of the crust and must be concentrated to only about four times its average crustal percentage before it can be extracted profitably.

It is important to realize that a deposit may become profitable to extract or lose its profitability because of economic changes. If demand for a metal increases and prices rise, the status of a previously unprofitable deposit changes, and it becomes an ore. The status of unprofitable deposits may also change if a technological advance allows the useful element to be extracted at a lower cost than before. This situation was illustrated recently at the copper mining operation located at Bingham Canyon, Utah, the largest open-pit mine on earth (see Box 1.3). Mining was halted here in 1985 because outmoded equipment had driven the cost of extracting the copper beyond the current selling price. The owners responded by replacing an antiquated 1000-car railroad with conveyor belts and pipelines for transporting the ore and waste. These devices achieved a cost reduction of nearly 30 percent and returned this mining operation to profitability.

Metallic Mineral Resources

Some of the most important accumulations of metals, such as gold, silver, copper, mercury, lead, platinum, and nickel, are produced by igneous and metamorphic processes (Table 1.5). These mineral resources, like most others, result from processes that concentrate desirable materials to the extent that extraction is economically feasible.

The igneous processes that generate some metal deposits are quite straightforward. For example, as a large magma body cools, the heavy minerals that crystallize early tend to settle to the lower portion of the magma chamber. This type of magmatic segregation is particularly active in large basaltic magmas where chromite (ore of chromium), magnetite, and platinum are occasionally generated. Layers of chromite, interbedded with other heavy minerals, are mined from such deposits in the Bushveld Complex in South Africa, which contains over 70 percent of the world's known reserves of platinum.

Table 1.5
Ores of metallic minerals.

Metal	Principal Ores
Aluminum	Bauxite
Chromium	Chromite
Copper	Chalcopyrite
	Bornite
	Chalcocite
Gold	Native gold
Iron	Hematite
	Magnetite
	Limonite
Lead	Galena
Magnesium	Magnesite
	Dolomite
Manganese	Pyrolusite
Mercury	Cinnabar
Molybdenum	Molybdenite
Nickel	Pentlandite
Platinum	Native platinum
Silver	Native silver
	Argentite
Tin	Cassiterite
Titanium	Ilmenite
	Rutile
Tungsten	Wolframite
	Scheelite
Uranium	Uraninite (pitchblende)
Zinc	Sphalerite

Figure 1.29
Light-colored vein deposits emplaced along a series of fractures in dark-colored igneous rock. (Photo by James E. Patterson)

Table 1.6
Uses of nonmetallic minerals.

Mineral	Uses
Apatite	Phosphorus fertilizers
Asbestos (chrysotile)	Incombustible fibers
Calcite	Aggregate; steelmaking; soil conditioning; chemicals; cement; building stone
Clay minerals (kaolinite)	Ceramics; china
Corundum	Gemstones; abrasives
Diamond	Gemstones; abrasives
Fluorite	Steelmaking; aluminum refining; glass; chemicals
Garnet	Abrasives; gemstones
Graphite	Pencil lead; lubricant; refractories
Gypsum	Plaster of Paris
Halite	Table salt; chemicals; ice control
Muscovite	Insulator in electrical applications
Quartz	Primary ingredient in glass
Sulfur	Chemicals; fertilizer manufacture
Sylvite	Potassium fertilizers
Talc	Powder used in paints, cosmetics, etc.

Among the best-known and most important ore deposits are those generated from **hydrothermal** (hot-water) **solutions.** Included in this group are the gold deposits of the Homestake mine in South Dakota; the lead, zinc, and silver ores near Coeur d'Alene, Idaho; the silver deposits of the Comstock Lode in Nevada; and the copper ores of the Keweenaw Peninsula in Michigan.

The majority of hydrothermal deposits are thought to originate from hot, metal-rich fluids that are associated with cooling magma bodies. During solidification, liquids plus various metallic ions accumulate near the top of the magma chamber. Because these hot fluids are very mobile, they can migrate great distances through the surrounding rock before they are eventually deposited. Some of this fluid moves along fractures or bedding planes, where it cools and precipitates the metallic ions to produce **vein deposits** (Figure 1.29). Many of the most productive deposits of gold, silver, and mercury occur as hydrothermal vein deposits.

Another important type of accumulation generated by hydrothermal activity is called a **disseminated deposit.** Rather than being concentrated in narrow veins and dikes, these ores are distributed as minute masses throughout the entire rock mass. Much of the world's copper is extracted from disseminated deposits, including the huge Bingham Canyon copper mine in Utah. Because these accumulations contain only 0.4 to 0.8 percent copper, between 125 and 250 metric tons of ore must be mined for every ton of metal recovered. The environmental impact of these large excavations, including the problems of waste disposal, is significant.

Nonmetallic Mineral Resources

Mineral resources that are not used as fuels or processed for the metals they contain are referred to as **nonmetallic mineral resources.** These materials are extracted and processed either to make use of the nonmetallic elements they contain or for the physical and chemical properties they possess (Table 1.6). Nonmetallic mineral resources are commonly divided into two broad groups—*building materials* and *industrial minerals*. Since some substances have many different uses, they are found in both categories. Limestone, perhaps the most versatile and widely used rock of all, is the best example. As a building material, it is used not only as crushed rock and building stone, but in the making of cement as well. Moreover, as an industrial mineral, it is an ingredient in the manufacture of steel and is used in agriculture to neutralize acidic soils.

Besides aggregate (sand, gravel, and crushed rock) and cut stone, the other important building materials include gypsum for plaster and wallboard, clay for tile and bricks, and cement, which is made from limestone and shale (Figure 1.30). Cement and aggregate go into the making of concrete, a material that is essential to practically all construction.

There is a wide variety of nonmetallic resources that are classified as industrial minerals. People often do not realize the importance of industrial minerals because they see only the products that resulted from their use and not the minerals themselves. That is, many nonmetallics are used up in the process of creating other products. Examples include fluorite and limestone, which are part of the steelmaking process; corundum and garnet, which are used as abrasives to make a piece of machinery; and sylvite, which is used in the production of the fertilizers used to grow a food crop.

Figure 1.30
The primary raw materials for cement—limestone and shale—are widely distributed. Therefore, cement plants such as this one are built to serve relatively small regions. (Photo by E. J. Tarbuck)

Review Questions

1. List the three main particles of an atom and explain how they differ from one another.

2. If the number of electrons in an atom is 35 and the mass number is 80, calculate the following:
   (a) The number of protons.
   (b) The atomic number.
   (c) The number of neutrons.

3. What occurs in an atom to produce an ion?

4. What is an isotope?

5. Although all minerals have an orderly internal arrangement of atoms (crystalline structure), most mineral samples do not demonstrate their crystal form. Explain.

6. Why might it be difficult to identify a mineral by its color?

7. If you found a glassy-appearing mineral while rock hunting and had hopes that it was a diamond, what simple test might help you make a determination?

8. Explain the economic use of corundum as given in Table 1.2 in terms of Mohs hardness scale.

9. Gold has a specific gravity of about 20. If a 25-liter container of water weighs about 25 kilograms, how much would a 25-liter container of gold weigh?

10. What is the difference between silicon and silicate?

11. Explain the statement "One rock is the raw material for another" using the rock cycle.

12. If a lava flow at the earth's surface had a mafic composition, what rock type would the flow likely be (see Table 1.3)? What igneous rock would form from the same magma if it did not reach the surface but instead crystallized at great depth?

13. What does a porphyritic texture indicate about an igneous rock?

14. How are granite and rhyolite different? The same? (See Table 1.3.)

15. Relate the classification of igneous rocks to Bowen's reaction series.

16. What minerals are most common in detrital sedimentary rocks? Why are these minerals so abundant?

17. What is the primary basis for distinguishing among various detrital sedimentary rocks?

18. Distinguish between the two categories of chemical sedimentary rocks.

19. What are evaporite deposits? Name a rock that is an evaporite.

20. Compaction is an important lithification process with which sediment size?

21. What is probably the single most characteristic feature of sedimentary rocks?

22. What is metamorphism? What are the *agents of change*?

23. Distinguish between regional and contact metamorphism.

24. What feature would easily distinguish schist and gneiss from quartzite and marble?

25. How do metamorphic rocks differ from the igneous and sedimentary rocks from which they formed?

26. Contrast resource and reserve.

27. What might cause a mineral deposit that had not been considered an ore to be reclassified as an ore?

28. List two general types of hydrothermal deposits.

29. Nonmetallic resources are commonly divided into two broad groups. List the two groups and some examples of materials that belong to each.

Key Terms

atom (p. 21)

atomic number (p. 21)

atomic weight (p. 23)

chemical sedimentary rock (p. 39)

cleavage (p. 24)

coarse-grained texture (p. 33)

compound (p. 22)

contact metamorphism (p. 46)

crystal form (p. 23)

crystallization (p. 28)

detrital sedimentary rock (p. 38)

disseminated deposit (p. 55)

electron (p. 21)

element (p. 21)

evaporite deposit (p. 41)

extrusive (volcanic) (p. 32)

felsic (p. 36)

fine-grained texture (p. 33)

foliated texture (p. 47)

fossil fuel (p. 42)

fracture (p. 25)

glassy texture (p. 34)

hardness (p. 24)

hydrothermal solution (p. 55)

igneous rock (p. 28)

intrusive (plutonic) (p. 32)

ion (p. 22)

isotope (p. 23)

lava (p. 32)

lithification (p. 30)

luster (p. 24)

mafic (p. 36)

magma (p. 28)

magmatic differentiation (p. 36)

mass number (p. 23)

metamorphic rock (p. 30)

mineral (p. 20)

mineral resource (p. 53)

Mohs scale (p. 24)

neutron (p. 21)

nonfoliated texture (p. 47)

nonmetallic mineral resource (p. 56)

nucleus (p. 21)

ore (p. 53)

porphyritic texture (p. 34)

proton (p. 21)

radioactivity (p. 23)

regional metamorphism (p. 46)

reserve (p. 53)

rock (p. 20)

rock cycle (p. 28)

sediment (p. 30)

sedimentary rock (p. 30)

silicate (p. 26)

silicon-oxygen tetrahedron (p. 26)

specific gravity (p. 25)

strata (beds) (p. 44)

streak (p. 23)

texture (p. 32)

vein deposit (p. 55)

weathering (p. 30)

Weathering, Soil, and Mass Wasting

Opposite page: Rock pinnacles produced by differential weathering, Bryce Canyon National Park, Utah. (Photo by John M. Roberts)

Left: Weathering accentuates differences in rocks to produce some of our most spectacular scenery. (Photo by E. J. Tarbuck)

Above: Differential weathering in the Badlands of South Dakota. (Photo by James E. Patterson)

The earth's surface is constantly changing. Rock is disintegrated and decomposed, moved to lower elevations by gravity, and carried away by water, wind, or ice. In this manner the earth's physical landscape is sculptured. This chapter focuses on the first two steps of this never-ending process— weathering and mass wasting—probing into how and why rock disintegrates and decomposes and what mechanisms act to move it downslope. Soil, an important product of the weathering process and a vital resource, is also examined.

To the casual observer the face of the earth may appear to be without change, unaffected by time. For that matter, less than 200 years ago most people believed that mountains, lakes, and deserts were permanent features of an earth that was thought to be no more than a few thousand years old. Today, however, we know that mountains eventually succumb to weathering and erosion, lakes fill with sediment and vegetation or are drained by streams, and deserts come and go as relatively minor climatic changes occur.

The earth is indeed a dynamic body. Volcanic and tectonic activities* are elevating parts of the earth's surface, while opposing processes are continually removing materials from higher elevations and moving them to lower elevations (Figure 2.1). The latter processes include:

1. **Weathering**—disintegration and decomposition of rock at or near the surface of the earth.
2. **Mass wasting**—transfer of rock material downslope under the influence of gravity.
3. **Erosion**—incorporation and transportation of material by a mobile agent, usually water, wind, or ice.

We will first turn our attention to the process of weathering and the products generated by this activity. However, weathering cannot be easily separated from the other two processes because as weathering breaks rocks apart, it facilitates the movement of rock debris by erosion and mass wasting. On the other hand, the transport of material by erosion and mass wasting furthers the disintegration and decomposition of rock.

WEATHERING

All materials are susceptible to weathering. Consider, for example, the fabricated product concrete, which closely resembles the sedimentary rock called conglomerate. A newly poured concrete sidewalk has a smooth, fresh look. However, not many years later, the same sidewalk will appear chipped, cracked, and rough, with pebbles exposed at the surface. If a tree is nearby, its roots may heave and buckle the concrete as well. The same natural processes that even-

---

* Tectonic activities are activities that result in the deformation of the earth's crust.

tually break apart a concrete sidewalk also act to disintegrate rock.

Why does rock weather? Simply, weathering is the response of earth materials to a changing environment. For instance, after millions of years of uplift and erosion, the rocks overlying a large intrusive igneous body may be removed, exposing it at the surface. This mass of crystalline rock that formed deep below ground where temperatures and pressures are high is now subjected to a very different and comparatively hostile surface environment. In response, this rock mass will gradually change until it is once again in equilibrium, or balance, with its new environment. This transformation of rock is what we call weathering.

In the following sections we will discuss the various types of mechanical and chemical weathering. Although we will consider these two processes separately, keep in mind that they usually work simultaneously in nature.

Mechanical Weathering

When a rock undergoes **mechanical weathering** it is broken into smaller and smaller pieces, each retaining the characteristics of the original material. The end result is many small pieces from a single large one. Figure 2.2 shows that breaking a rock into smaller pieces increases the surface area available for chemi-

cal attack. An analogous situation occurs when sugar is added to a liquid. In this situation, a cube of sugar will dissolve much more slowly than will an equal volume of granules because of the vast difference in surface area. Hence, by breaking rocks into smaller pieces, mechanical weathering increases the amount of surface area available for chemical weathering.

In nature four important physical processes lead to the fragmentation of rock: frost wedging, expan-

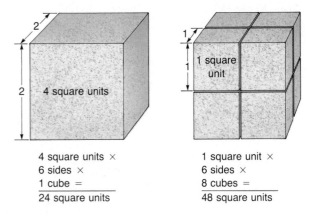

4 square units ×
6 sides ×
1 cube =
_____
24 square units

1 square unit ×
6 sides ×
8 cubes =
_____
48 square units

Figure 2.2
Chemical weathering can occur only to those portions of a rock that are exposed to the elements. Mechanical weathering breaks rock into smaller and smaller pieces, thereby increasing the surface area available for chemical attack.

sion resulting from unloading, thermal expansion, and organic activity.

Frost Wedging.

Alternate freezing and thawing of water is one of the most important processes of mechanical weathering. Water has the unique property of expanding about 9 percent when it freezes. This increase in volume occurs because as ice forms, the water molecules arrange themselves into a very open crystalline structure. As a result, when water freezes, it expands and exerts a tremendous outward force. This can be verified by filling a container with water and freezing it. If sufficient volume does not exist within the container, the formation of ice will cause it to shatter.

In nature, water works its way into cracks or voids in rock and, upon freezing, expands and enlarges the opening. After many freeze-thaw cycles, the rock is broken into pieces. This process is appropriately called **frost wedging** (Figure 2.3). Frost wedging is most pronounced in mountainous regions in the middle latitudes where a daily freeze-thaw cycle often exists. Here sections of rock are wedged loose and may tumble into large piles called **talus slopes** that often form at the base of steep rock outcrops (Figure 2.3).

Unloading.

When large masses of igneous rock, particularly those composed of granite, are exposed by erosion, concentric slabs begin to break loose. The process generating these onionlike layers is called **sheeting** and is thought to occur, at least in part, because of the great reduction in pressure when the overlying rock is stripped away. Accompanying the unloading, the outer layers expand more than the rock below, and thus separate from the rock body. The fractures typically develop parallel to the surface topography and give the exhumed igneous body a domed shape. Continued weathering eventually causes the slabs produced by sheeting to separate and spall off these large structures known as **exfoliation domes**. Excellent examples of exfoliation domes include Stone Mountain, Georgia, and Half Dome (Figure 2.4) and Liberty Cap in Yosemite National Park.

Mine shafts provide us with another example of how rocks behave once the confining pressure is removed. Large rock slabs have been known to explode

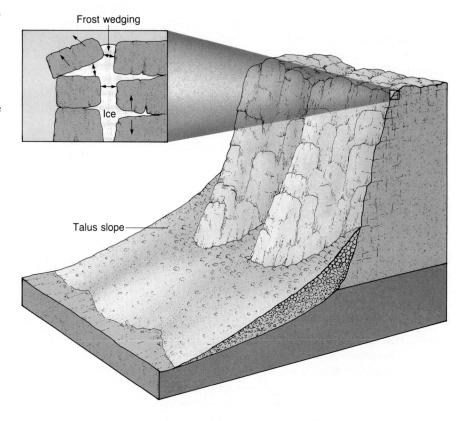

Figure 2.3
Frost wedging. As water freezes it expands, exerting a force great enough to break rock. When frost wedging occurs in a setting such as this, the broken rock fragments fall to the base of the cliff and create a cone-shaped accumulation known as talus.

Frost wedging

Ice

Talus slope

off the walls of newly cut mine shafts because of the reduced pressure. Evidence of this type, plus the fact that fracturing occurs parallel to the floor of a rock quarry when large blocks are removed, strongly supports the process of unloading as the cause of sheeting.

Although many fractures are created by expansion, others are produced by contraction during the crystallization of magma, and still others by tectonic forces during mountain building. Fractures produced by these activities generally form a definite pattern and are called *joints*. Joints are important rock structures that allow water to penetrate to depth and start the process of weathering long before the rock is exposed at the surface.

Thermal Expansion. The daily cycle of temperature change is thought to weaken rocks, particularly in deserts where daily variations may exceed 30°C (54°F). Heating a rock causes it to expand, and cooling causes it to contract. Repeated swelling and shrinking of minerals that have different expansion and contraction rates should logically exert some stress on the rock's outer shell.

Although this process was once thought to be of major importance in the disintegration of rock, laboratory experiments have not substantiated this. In one test, unweathered rocks were heated to temperatures much higher than those normally experienced on the earth's surface and then cooled. This procedure was repeated many times to simulate hundreds of years of weathering, but the rocks showed little apparent change.

Nevertheless, in desert areas pebbles do exhibit unmistakable evidence of shattering from what appears to be temperature changes. A proposed solution to this dilemma suggests that rocks must first be weakened by chemical weathering before they can be broken down by thermal activity. Further, this process may be aided by the rapid cooling of a desert rainstorm. Additional data are needed before a definite answer can be given as to the impact of temperature variation on rock disintegration.

Organic Activity. Weathering is also accomplished by the activities of organisms, including plants, burrowing animals, and humans. Plant roots in search of minerals and water grow into fractures, and as the roots grow they wedge the rock apart (Figure 2.5). Burrowing animals further break down the

Figure 2.4
Summit of Half Dome, an exfoliation dome in Yosemite National Park. (Photo by Stephen Trimble)

rock by moving fresh material to the surface, where physical and chemical processes can more effectively attack it. Decaying organisms also produce acids, which contribute to chemical weathering. Where rock has been blasted in search of minerals or for road construction, the impact is quite noticeable, but on a worldwide scale human activity probably ranks behind burrowing animals in earth-moving accomplishments.

Although usually considered separately from mechanical weathering, the activities of the erosional agents—wind, water, and glaciers—are nonetheless important. For as these mobile agents move rock debris, they relentlessly disintegrate the earth materials they carry.

Chemical Weathering

Chemical weathering involves the complex processes that alter the internal structures of minerals by removing and/or adding elements. During this transformation, the original rock decomposes into substances that are stable in the surface environment. Consequently, the products of chemical weathering will remain essentially unchanged as long as they remain in an environment similar to the one in which they formed.

Figure 2.5
Root wedging widens fractures in rock and aids the process of mechanical weathering. (Photo by E. J. Tarbuck)

Water is by far the most important agent of chemical weathering. Although pure water is nonreactive, a small amount of dissolved material is generally all that is needed to activate it. Oxygen dissolved in water will *oxidize* some materials. For example, when an iron nail is found in the soil, it will have a coating of rust (iron oxide), and if the time of exposure has been long, the nail will be so weak that it can be broken as easily as a toothpick. When rocks containing iron-rich minerals oxidize, a yellow to reddish brown rust will appear on the surface.

Carbon dioxide (CO_2) dissolved in water (H_2O) forms carbonic acid (H_2CO_3), the same weak acid produced when soft drinks are carbonated. Rain dissolves some carbon dioxide as it falls through the atmosphere, and additional amounts released by decaying organic matter are acquired as the water percolates through the soil. Carbonic acid ionizes to form the very reactive hydrogen ion (H^+) and the bicarbonate ion (HCO_3^-).

To illustrate how rock chemically weathers when attacked by carbonic acid, we will consider the weathering of granite, the most abundant continental rock. Recall that granite consists mainly of quartz and potassium feldspar. The weathering of the potassium feldspar component of granite takes place as follows:

$$2KAlSi_3O_8 + 2(H^+ + HCO_3^-) + H_2O \longrightarrow$$

potassium   carbonic acid   water
feldspar

$$Al_2Si_2O_5(OH)_4 + 2KHCO_3 + 4SiO_2$$

clay mineral   potassium   silica
    bicarbonate

In this reaction, the hydrogen ions (H^+) attack and replace potassium ions (K^+) in the feldspar structure, thereby disrupting the crystalline network. Once removed, the potassium is available as a nutrient for plants or becomes the soluble salt potassium bicarbonate ($KHCO_3$), which may be incorporated into other minerals or carried to the ocean in dissolved form by streams. The most abundant products of the chemical breakdown of feldspar are residual clay minerals. Because clay minerals are the end product of weathering, they are very stable under surface conditions. Consequently, clay minerals make up a high percentage of the inorganic material in soils. Moreover, the most abundant sedimentary rock, shale, also contains a high proportion of clay minerals. In addition to the formation of clay minerals during this reaction, some silica is removed from the feldspar structure and is carried away by goundwater (water beneath the earth's surface). This dissolved silica will eventually precipitate to produce nodules of chert or flint, fill in the pore spaces between sediment grains, or be carried to the ocean, where microscopic animals will remove it to build hard silica shells.

To summarize, the weathering of potassium feldspar generates a residual clay mineral, a soluble salt (potassium bicarbonate), and some silica which enters into solution.

Quartz, the other main component of granite, is very resistant to chemical weathering; hence, it remains substantially unaltered when attacked by weakly acidic solutions. As a result, when granite weathers, the feldspar crystals dull and slowly turn to clay, releasing the once-interlocked quartz grains, which still retain their fresh, glassy appearance. Al-

Table 2.1
Products of weathering.

Mineral	Residual Products	Material in Solution
Quartz	Quartz grains	Silica
Feldspars	Clay minerals	Silica K$^+$, Na$^+$, Ca^{2+}
Hornblende	Clay minerals Limonite Hematite	Silica Ca^{2+}, Mg^{2+}
Olivine	Limonite Hematite	Silica Mg^{2+}

though some quartz remains in the soil, much is transported to the sea or to other sites of deposition, where it becomes the main constituent of such features as sandy beaches and sand dunes. In time it may be lithified to form the sedimentary rock sandstone.

Table 2.1 lists the weathered products of some of the most common silicate minerals. Remember that silicate minerals make up most of the earth's crust and that these minerals are essentially composed of only eight elements. When chemically weathered, the silicate minerals yield sodium, calcium, potassium, and magnesium ions that form soluble products which may be removed by groundwater. The element iron combines with oxygen, producing relatively insoluble iron oxides, most notably hematite and linomite, which give soil a reddish-brown or yellowish color. Under most conditions the three remaining elements, aluminum, silicon, and oxygen, join with water to produce residual clay minerals. However, even the highly insoluble clay minerals are very slowly removed by subsurface water.

In addition to altering the internal structure of minerals, chemical weathering causes physical changes as well. For instance, when angular rock masses are attacked by water that enters along joints, the rocks tend to take on a spherical shape. Gradually the corners and edges of the angular blocks become more rounded. The corners are attacked most readily because of the greater surface area for their volume as compared to the edges and faces. This process, called **spheroidal weathering**, gives the weathered rock a more rounded or spherical shape (Figure 2.6).

Sometimes during the formation of spheroidal boulders, successive shells separate from the rock's main body. Eventually the outer shells spall off, al-

Figure 2.6
Spheroidal weathering is evident in this exposure of extensively jointed granite in California's Joshua Tree National Monument. Water moving along the joints gradually enlarges them. Since the rocks are attacked more vigorously on the corners and edges, they take on a spherical shape. (Photo by E. J. Tarbuck)

lowing the chemical weathering activity to penetrate deeper into the boulder. This spherical scaling results because, as the minerals in the rock weather to clay, they increase in size through the addition of water to their structure. This increased bulk exerts an outward force that causes concentric layers of rock to break loose and fall off. Hence, chemical weathering does produce forces great enough to cause mechanical weathering. This type of spheroidal weathering in which shells spall off should not be confused with the phenomenon of sheeting discussed earlier. In sheeting, the fracturing occurs as a result of unloading and the rock layers which separate from the main body are largely unaltered at the time of separation.

Rates of Weathering

The rate at which rock weathers depends on many factors. We have already seen that particle size influences the rate of weathering. So too does the presence or absence of such physical features as joints, which influence the ability of water to penetrate the rock. The mineral makeup of a rock is also a very important factor, which can be demonstrated by comparing headstones carved from different rock types. A headstone made of granite, a rock composed of silicate minerals, is relatively resistant to chemical weathering as we can see by examining the inscriptions on the headstone in Figure 2.7A. This is not true of the marble headstone, which shows signs of extensive chemical alteration over a relatively short pe-

riod (Figure 2.7B). Recall that marble is composed of calcite, which readily dissolves even in a weakly acidic solution.

The silicates, the most abundant mineral group, weather in essentially the same order as their order of crystallization. The minerals that crystallized first formed at much higher temperatures than those that crystallized last. Consequently, these early-formed minerals are not as stable at the earth's surface, where the temperature and pressure are radically different from the environment in which they are formed. By examining Bowen's reaction series (see Figure 1.16), we can see that olivine crystallizes first and is therefore the least resistant to weathering, while quartz, which crystallizes last, is the most resistant.

Climatic factors, particularly temperature and moisture, are of primary significance to the rate of rock weathering. The optimum environment for chemical weathering is a combination of warm temperatures and abundant moisture. In polar regions and at high altitudes, chemical weathering is reduced because frigid temperatures keep the available moisture locked up as ice, whereas in arid regions there is insufficient moisture to foster much chemical weathering. A classic example of how climate affects the rate of weathering was provided when Cleopatra's Needle, a granite obelisk, was moved from Egypt to New York City in the late nineteenth century. After withstanding approximately 35 centuries of exposure in the dry climate of Egypt, the hieroglyphics

Figure 2.7
An examination of headstones reveals the rate of chemical weathering on diverse rock types. The granite headstone **(A.)** *was erected a few years before the marble headstone* **(B.),** *whose inscription date of 1892 is nearly illegible. (Photos by E. J. Tarbuck)*

A.

B.

were almost completely removed from the windward side in less than 75 years in the wet and chemical-laden air of New York City (Figure 2.8).

The sum of these factors determines the type and rate of rock weathering for a given place. However, there is generally enough variation, even within a relatively small area, for the rocks to exhibit some differential weathering. **Differential weathering** simply relates to the fact that rocks exposed at the earth's surface usually do not weather at the same rate. Because of variations in such factors as mineral makeup, degree of jointing, and exposure to the elements, significant differences occur. Consequently, differential weathering and subsequent erosion create many unusual and often spectacular rock formations and landforms. Included are features such as the natural bridges found in Arches National Park (Figure 2.9) and the sculptured rock pinnacles found in Bryce Canyon National Park (see chapter-opening photo).

SOIL

Soil has accurately been called "the bridge between life and the inanimate world." All life owes its existence to a dozen or so elements that must ultimately come from the earth's crust. Once weathering and other processes create soil, plants carry out the intermediary role of assimilating the necessary elements and making them available to animals and people.

With few exceptions, the earth's land surface is covered by **regolith**, the layer of rock and mineral fragments produced by weathering. Some would call this material soil, but soil is more than an accumulation of weathered debris. **Soil** is a combination of mineral and organic matter, water, and air —that portion of the regolith that supports the growth of plants. Although the proportions vary, the major components do not (Figure 2.10). About one-half of the total volume of a good quality surface soil is a mixture of disintegrated and decomposed rock (mineral matter) and **humus**, the decayed remains of animal and plant life (organic matter). The remaining half consists of pore spaces, where air and water circulate.

Although the mineral portion of the soil is usually much greater than the organic portion, humus is a very significant component. In addition to being an important source of plant nutrients, humus enhances the soil's ability to retain water. Since plants require air and water to live and grow, the portion of the soil consisting of pore spaces that allow for the circulation of these fluids is as vital as the solid soil constituents. Soil water is not "pure" water; instead, it is a complex solution containing many soluble nutri-

Figure 2.8
*Chemical weathering of Cleopatra's Needle, a granite obelisk. **A.** Before it was removed from Egypt. (Courtesy of The Metropolitan Museum of Art) **B.** After a span of 75 years in New York City's Central Park. After having survived intact for about 35 centuries in Egypt, the windward side has been almost completely defaced in less than a century. (Courtesy of New York City Parks)*

A.

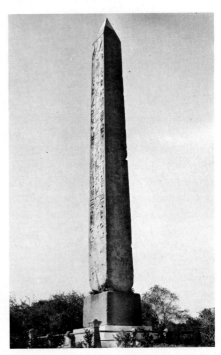

B.

Figure 2.9
Differential weathering is illustrated by Delicate Arch in Arches National Park, Utah. (Photo by Tom Till)

ents. Soil water not only provides the necessary moisture for the chemical reactions that sustain life, it also supplies plants with nutrients in a form they can use. The pore spaces not filled with water contain air. This air is the source of necessary oxygen and carbon dioxide for most microorganisms and plants that live in the soil.

Soil Texture and Structure

Soil texture refers to the relative portions of different particle sizes in a soil. Texture is a very basic soil property because it strongly influences the soil's ability to retain and transmit water and air, both of which are essential to plant growth. Sandy soils may drain too rapidly, whereas the pore spaces of clay-rich soils may be too small for adequate drainage. Moreover, when the clay and silt content is very high, plant roots may have difficulty penetrating the soil.

Soils rarely consist of particles of only one size; therefore, textural categories have been established based upon the varying proportions of clay, slit, and sand. The standard system of classes used by the U.S. Department of Agriculture is shown in Figure 2.11. For example, point *A* on this triangular diagram represents a soil composed of 10 percent silt, 40 percent clay, and 50 percent sand. Such a soil is called a *sandy clay*. The soils called *loam*, which occupy the central portion of the diagram, are those in which no single particle size predominates over the other two. Loam soils are best suited to support plant life because they generally have better moisture characteristics and

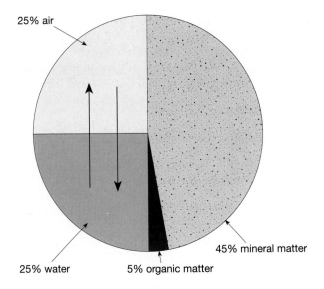

Figure 2.10
Composition (by volume) of a soil in good condition for plant growth. Although the percentages vary, each soil is composed of mineral and organic matter, water, and air.

nutrient storage ability than do soils composed predominantly of clay or coarse sand.

Soil particles are seldom completely independent of one another. Rather, they usually form clumps called *peds* that give soils a particular structure. Four basic soil structures are recognized: platy, prismatic, blocky, and spheroidal. Soil structure is important because it influences the ease of a soil's cultivation as well as the susceptibility of a soil to erosion. In addition, soil structure affects the porosity and permeability of soil (that is, the ease with which water can penetrate). This, in turn, influences the movement of nutrients to plant roots. Prismatic and blocky peds usually allow for moderate water infiltration, whereas platy and spheroidal structures are characterized by slower infiltration rates.

Controls of Soil Formation

Soil is the product of the complex interplay of several factors, including parent material, time, climate, plants and animals, and slope. Although all of these factors are interdependent, it will be helpful to examine their roles separately.

Figure 2.11
The texture of any soil can be represented by a point on this soil texture diagram. Soil texture is one of the most significant factors used to estimate agricultural potential and engineering characteristics. (After U.S. Department of Agriculture)

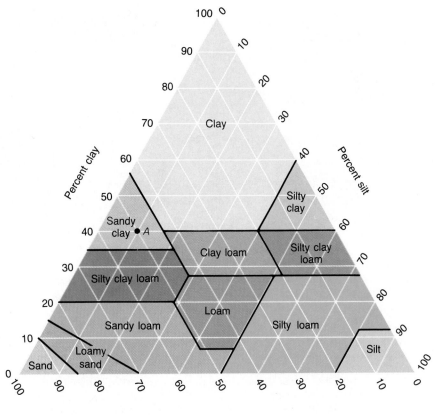

Parent Material. The source of the weathered mineral matter from which soils develop is called the **parent material** and is a major factor influencing a newly forming soil. Gradually the parent material undergoes physical and chemical changes as the processes of soil formation progress. Parent material may either be the underlying bedrock or a layer of unconsolidated deposits. When the parent material is bedrock, the soils are termed *residual soils*. By contrast, those developed on unconsolidated sediment are called *transported soils* (Figure 2.12). It should be pointed out that transported soils form *in place* on parent materials that have been carried from elsewhere and deposited by gravity, water, wind or ice.

The nature of the parent material influences soils in two ways. First, the type of parent material to some degree will affect the rate of weathering and thus the rate of soil formation. For example, the mineral composition of the parent material will influence the rate of chemical weathering. Also, since unconsolidated deposits are already partly weathered, soil development on such materials will likely progress more rapidly than when bedrock is the parent material. Second, the chemical makeup of the parent material will affect the fertility of the soil. If it lacks the necessary elements for plant growth, its usefulness is obviously diminished.

At one time it was thought that the parent material was the primary factor causing differences among soils. Today, soil scientists realize that other factors, especially climate, are more important. In fact, it has been found that similar soils are often produced from different parent materials and that dissimilar soils have developed from the same parent material. Such discoveries reinforce the importance of the other soil-forming factors.

Time. If weathering has been going on for a comparatively short time, the character of the parent material determines to a large extent the characteristics

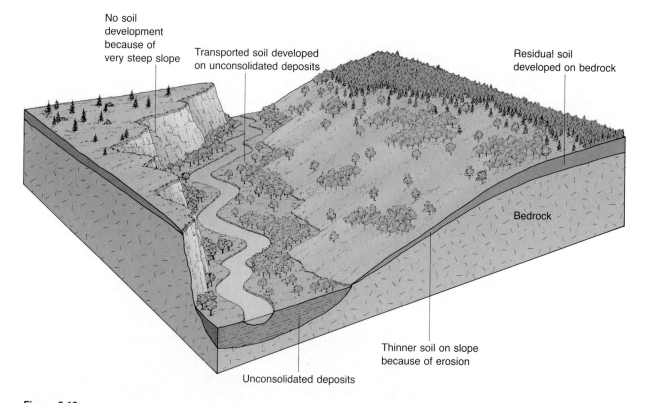

Figure 2.12
The parent material for residual soils is the underlying bedrock. By contrast, transported soils develop on unconsolidated parent materials that have been carried from elsewhere and deposited. Slope is also shown to be a factor. The depths of soils decrease as the steepness of the slopes increase. When the slope exceeds about 45 degrees, weathered material seldom remains for long before moving downslope.

of the soil. As the weathering process continues, the influence of parent material on soil is overshadowed by the other soil-forming factors. Therefore time is an important control of soil formation. It is not possible to list the length of time required for various soils to evolve, because the soil-forming processes act at varying rates under different circumstances. However, it is safe to say that the longer a soil has been forming, the thicker it becomes and the less it resembles the parent material from which it formed.

Climate. Climate is considered to be the most influential control of soil formation. Just as temperature and precipitation are the climatic elements that influence people the most, so too are they the elements that exert the strongest impact on soil formation. Variations in temperature and precipitation determine whether chemical or mechanical weathering will predominate and also greatly influence the rate and depth of weathering. For instance, a hot and wet climate may produce a thick layer of chemically weathered soil in the same amount of time that a cold and less humid climate produces a thin mantle of mechanically weathered debris. Furthermore, the amount of precipitation influences the degree to which various materials are removed from the soil, thereby affecting soil fertility. Finally, climatic conditions are an important control on the type of plant and animal life present.

Plants and Animals. Plants and animals play a vital role in soil formation. The types and abundance of organisms present have a strong influence on the physical and chemical properties of a soil. In fact, for well-developed soils in many regions, the significance of natural vegetation is frequently implied in the descriptions used by soil scientists. Such phrases as prairie soil, forest soil, and tundra soil are common.

The chief function of plants and animals is to supply organic matter to the soil. Certain bog soils are composed almost entirely of organic matter, whereas desert soils may contain only a very small percentage. Although the quantity of organic matter present varies substantially among soils, no soil completely lacks it.

The primary source of organic matter is plants, although animals and the uncountable numbers of microorganisms also contribute. When organic matter is decomposed, it supplies important nutrients for plants, as well as food for the animals and microorganisms living in the soil. Consequently, soil fertility is in part related to the amount of organic matter present. Furthermore, the decay of plant and animal remains causes various organic acids to form. These complex acids hasten the weathering process. Organic matter also has a high water-holding ability and thus aids water retention in a soil.

Microorganisms, including fungi, bacteria, and the single-celled protozoans, play the active role in the decay of plant and animal remains. The end product is humus, a material that no longer resembles the plants and animals from which it formed. In addition, certain microorganisms aid soil fertility because they have the ability to *fix* (change) atmospheric nitrogen into soil nitrogen.

Earthworms and other burrowing animals act to mix the mineral and organic portions of a soil. Earthworms, for example, feed on the organic matter in the soil and thoroughly mix soils in which they live, often moving and enriching many tons per acre each year. Burrows and holes also aid the passage of water and air through the soil.

Slope. The slope of the land can vary greatly over short distances. Such variations, in turn, can lead to the development of a variety of localized soil types. Many of the differences exist because slope significantly affects the amount of erosion and the water content of soil. On steep slopes, soils are often poorly developed. In such situations the quantity of water soaking in is slight, and as a result, the moisture content of the soil may not be sufficient for vigorous plant growth. Further, because of accelerated erosion on steep slopes, the soils there are thin or, in some cases, nonexistent (Figure 2.12). On the other hand, poorly drained and water-logged soils found in bottomlands have a much different character. Such soils are usually very thick and very dark, the dark color resulting from the large quantity of organic matter that accumulates because saturated conditions retard the decay of vegetation. The optimum slope for soil development is a flat-to-undulating upland surface. Here we find good drainage, minimal erosion, and sufficient infiltration of water into the soil.

Slope orientation, the direction the slope faces, is another aspect worthy of mention. In the mid-latitudes of the northern hemisphere, a south-facing slope will receive a great deal more sunlight than will a north-facing slope. In fact, a steep north-facing slope may receive no direct sunlight at all. The difference in the amount of solar radiation received will cause differences in soil temperature and moisture, which in turn may influence the nature of the vegetation and the character of the soil.

Although this section has dealt separately with each of the soil-forming factors, remember that all of them work together to form soil. No single factor is responsible for a soil being as it is, but it is rather the combined influence of parent material, time, climate, plants and animals, and slope.

The Soil Profile

Since soil-forming processes operate from the surface downward, variations in composition, texture, structure, and color gradually evolve at varying depths. These vertical differences, which usually become more pronounced as time passes, divide the soil into zones or layers known as **horizons.** If you were to dig a trench in soil, you would see that its walls are layered. Such a vertical section through all of the soil horizons constitutes the **soil profile** (Figure 2.13).

Four basic horizons are identified and from top to bottom are designated as O, A, B, and C, respectively. The three upper layers may be further divided.

Unlike the layers beneath it, which consist mainly of mineral matter, the O horizon consists largely of organic material. The upper portion of this horizon is primarily plant litter such as loose leaves and other organic debris that are still recognizable. By contrast, the lower portion of the O horizon is made up of partly decomposed organic matter (humus) in which plant structures can no longer be identified.

Underlying the organic-rich O horizon is the A horizon. This zone is largely mineral matter, yet biological activity is high and humus is generally present—up to 30 percent in some instances. As water percolates downward from the surface through the

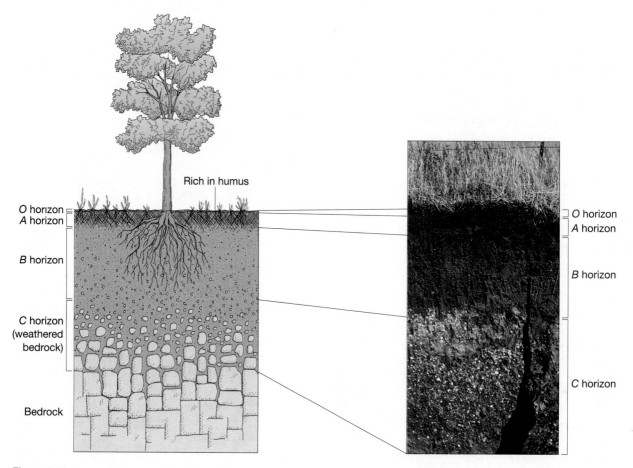

Figure 2.13
Soil profile. Mature soils are characterized by a series of horizontal layers called horizons, which comprise the soil profile.

A horizon, finer particles are carried away with it. This washing out of fine soil components is termed **eluviation.** As a consequence of eluviation, the texture of the A horizon gradually becomes coarser, because a portion of the fine particles is removed. Water percolating downward also dissolves soluble inorganic soil components and carries them to deeper zones. This depletion of soluble materials from the upper soil is termed **leaching.**

Immediately below the A horizon is the B horizon, or *subsoil.* Much of the material removed from the A horizon by eluviation is deposited in the B horizon, which is often referred to as the *zone of accumulation.* The accumulation of fine clay particles derived from the A horizon enhances water retention in the subsoil. However, in extreme cases, clay accumulation can form an extremely dense and impermeable layer called *hardpan.*

Since the B horizon has an intermediate position in the soil profile, it may be considered, at least in part, a transitional zone. For example, living organisms and organic matter are more abundant in the B than in the C horizon, but considerably less so than in the A horizon. The O, A, and B horizons together constitute the **solum,** or "true soil." It is in the solum that the soil-forming processes are active and that living roots and other plant and animal life are largely confined.

Below the solum is the C horizon, a layer characterized by partially altered parent material and little if any organic matter. While the parent material may be so dramatically altered in the solum that its original character is not recognizable, it is easily identifiable in the C horizon. Although this material is undergoing changes that will eventually transform it into soil, it has not yet crossed the threshold that separates regolith from soil.

The characteristics and extent of development can vary greatly among soils in different environments. The boundaries between soil horizons may be very sharp, or the horizons may blend gradually from one to another. Considerable time is needed for soil horizons to develop. Consequently, a well-developed soil profile indicates that environmental conditions have been relatively stable over an extended time span and that the soil is *mature.* By contrast, some soils lack horizons altogether. Such soils are called *immature* because soil building has been going on for only a short time. Immature soils are also characteristic of steep slopes where erosion continually strips away the soil, preventing full development.

Soil Types

In the discussion which follows we shall briefly examine some common soil types. As you read, notice that the characteristics of each of the soil types are primarily manifestations of the prevailing climatic conditions. A summary of the characteristics of the soils discussed in this section is provided in Table 2.2*

The term **pedalfer** gives a clue to the basic characteristic of this soil type. The word is derived from the Greek **ped**on, meaning "soil," and the chemical symbols **Al** (aluminum) and **Fe** (iron). Pedalfers are characterized by an accumulation of iron oxides and aluminum-rich clays in the B horizon. In mid-latitude areas where the annual rainfall exceeds 63 centimeters (25 inches), most of the soluble materials, such as calcium carbonate, are leached from the soil and carried away by underground water. The less soluble iron oxides and clays are carried from the A horizon and deposited in the B horizon, giving it a brown to red-brown color. These soils exist under either grass or forest vegetation where decomposing organic matter provides the acid conditions necessary for leaching. In the United States, pedalfers are found east of a line extending from northwestern Minnesota to south-central Texas.

Pedocal is derived from the Greek **ped**on, meaning "soil," and the first three letters of **cal**cite (calcium carbonate). As the name implies, pedocals are characterized by an accumulation of calcium carbonate. This soil type is found in the drier western United States in association with grassland and brush vegetation. Here rainwater percolating through the soil often evaporates before it can remove soluble materials, chiefly calcium carbonate. The result is a whitish accumulation called *caliche.* In addition, since chemical weathering is less intense in drier areas, pedocals generally contain a smaller percentage of clay minerals than do pedalfers.

In the hot and wet climates of the tropics, soils called **laterites** develop. Since chemical weathering is intense under such climatic conditions, these soils are usually deeper than soils developing over a similar period of time in the mid-latitudes. Not only does leaching remove the soluble materials such as calcite, but the great quantities of percolating water also remove much of the silica, with the result that oxides

---

* For more information on different soil groups and their distribution, see Appendix G.

of iron and aluminum become concentrated in the soil. The iron gives the soil a distinctive red color.

Since bacterial activity is very high in the tropics, laterites contain practically no humus. This fact, coupled with the highly leached nature of these soils, make laterites poor for growing crops. The infertility of these soils has been borne out repeatedly in tropical countries where cultivation has been attempted in such areas (see Box 2.1).

In cold or dry climates, soils are generally very thin and poorly developed. The reasons for this are fairly obvious. Chemical weathering progresses very slowly in such climates, and the scanty plant life yields very little organic matter.

Soil Erosion

Soils represent just a tiny fraction of all earth materials, yet they are a vital resource. Because soils are necessary for the growth of rooted plants, they are the very foundation of the human life-support system. Just as human ingenuity can increase the agri-

cultural productivity of soils through such practices as fertilization and irrigation, soils can be damaged or destroyed by the careless activities of people. Despite their basic role in providing food, fiber, and other basic materials, soils have been one of our most abused resources. Perhaps this neglect and indifference has occurred because a substantial amount of soil seems to remain even in places where soil erosion is serious. Nevertheless, although the loss of fertile topsoil may not be obvious to the untrained eye, it is a growing problem as human activities expand and disturb more and more of the earth's surface.

Soil erosion is a natural process; it is part of the constant recycling of earth materials that we call the rock cycle. Once soil forms, erosional forces, especially water and wind, move soil components from one place to another. Every time it rains, raindrops strike the land with a surprising amount of force (Figure 2.14). Each drop acts like a tiny bomb, blasting movable soil particles out of their positions in the soil mass. Then water flowing across the surface carries away the dislodged soil particles. Because the soil is

Table 2.2
Summary of soil types

Climate	Temperate humid (>63 cm rainfall)	Temperate dry (<63 cm rainfall)	Tropical (heavy rainfall)	Extreme arctic or desert
Vegetation	Forest	Grass and brush	Grass and trees	Almost none, so no humus develops
Typical Area	Eastern U.S.	Western U.S.		
Soil Type	Pedalfer	Pedocal	Laterite	
Topsoil	Sandy; light colored; acid	Commonly enriched in calcite, whitish color	Enriched in iron (and aluminum); brick red color	No real soil forms because there is no organic material. Chemical weathering is very slow
Subsoil	Enriched in aluminum, iron, and clay; brown color	Enriched in calcite; whitish color	All other elements removed by leaching	
Remarks	Extreme development in conifer forests, because abundant humus makes groundwater very acid. Produces light gray soil because of removal of iron	*Caliche* is name applied to the accumulation of calcite	Apparently bacteria destroy humus, so no acid is available to remove iron	

(Zones not developed — between Topsoil and Subsoil of the Tropical column)

BOX 2.1
Laterites and the Rainforest

Laterites are thick red soils that form in the wet tropics and subtropics and are the end product of extreme chemical weathering (Figure 2.A). Because lush tropical rainforests have lateritic soils, it might seem as though they would have great potential for agriculture. However, this is not the case. In fact, just the opposite is true; laterites are among the poorest soils for farming. Why is this the case?

Figure 2.A
Characteristic red color of a well-developed lateritic soil.

Because laterites develop under conditions of high temperatures and heavy rainfall, they are severely leached. Leaching leads to low fertility because plant nutrients are removed by the large volume of downward-percolating water. Therefore, even though the vegetation may be dense and luxuriant, the soil itself contains few available nutrients. Most of the nutrients that support the rainforest are locked up in the trees themselves. The forest is maintained by the decay of plant material. That is, as vegetation dies and decomposes, the roots of the rainforest trees quickly absorb the nutrients before they are leached from the soil. The nutrients are continuously recycled as trees die and decompose. Therefore, when forests are cleared to provide land for farming or to harvest the timber, most of the nutrients are removed as well. What remains is a soil that contains little to nourish the crop growth.

The clearing of rainforests not only removes the supply of plant nutrients, but also leads to accelerated erosion. When vegetation is present, its roots anchor the soil while its leaves and branches provide a canopy that protects the ground from the full force of the frequent heavy rains. The removal of the vegetation also exposes the ground to strong direct sunlight. When baked by the sun, laterites can harden to a bricklike consistency and become practically impenetrable to water and crop roots. In only a few years, lateritic soils in a freshly cleared area may no longer be cultivatable.

The term *laterite* is derived from the Latin word *latere,* meaning "brick," and was first applied to the use of this material for brick making in India and Cambodia. Laborers simply excavated the soil, shaped it, and allowed it to harden in the sun. Ancient but still well-preserved structures built of laterite remain standing today in the wet tropics. Such structures have withstood centuries of weathering because all of the original soluble materials were already removed from the soil by chemical weathering. Laterites are therefore virtually insoluble and thus very stable.

In summary, we have seen that laterites are highly leached soils that are the products of extreme chemical weathering in the warm, wet tropics. Although they may be associated with lush tropical rainforests, these soils are unproductive when vegetation is removed. Moreover, when cleared of plants, laterites are subject to accelerated erosion and can be baked to bricklike hardness by the sun.

moved by thin sheets of water, this process is termed *sheet erosion*. After flowing as a thin, unconfined sheet for a relatively short distance, threads of current typically develop and tiny channels called *rills* begin to form. Still deeper cuts in the soil, known as *gullies*, are created as rills enlarge (Figure 2.15). When cultivation cannot eliminate the channels, we know the rills have grown large enough to be called gullies. Although most dislodged soil particles move only a short distance during each rainfall, substantial quantities eventually leave the fields and make their way downslope to a stream. Once in the stream channel, these soil particles, which can now be called *sediment*, are transported downstream and are eventually deposited.

We know that soil erosion is the ultimate fate of practically all soils. In the past, soil erosion occurred at slower rates than it does today because more of the land surface was covered and protected by trees, shrubs, grasses, and other plants. However, human activities such as farming, logging, and construction, which remove or disrupt the natural vegetation, have greatly accelerated the rate of soil erosion. Without the stabilizing effect of plants, the soil is more easily swept away by the wind or carried downslope by sheet wash.

Natural rates of soil erosion vary greatly from one place to another and depend on soil characteristics as well as such factors as climate, slope, and type of vegetation. Over a broad area, erosion caused by surface runoff may be estimated by determining the sediment loads of the streams that drain the region. When studies of this kind were made on a global scale they indicated that, prior to the appearance of humans, sediment transport by rivers to the ocean amounted to just over 9 billion metric tons per year. By contrast, the amount of material currently transported to the sea by rivers is about 24 billion metric tons per year, or more than two and one-half times the earlier rate.

It is more difficult to measure the loss of soil due to wind erosion. However, the removal of soil by wind is generally much less significant than erosion by flowing water, except during periods of prolonged drought. When dry conditions prevail, strong winds can remove large quantities of soil from unprotected fields. Such was the case in the 1930s in portions of the Great Plains that came to be called the Dust Bowl (see Box 2.2).

In many regions the rate of soil erosion is significantly greater than the rate of soil formation. This means that a potentially renewable resource has become nonrenewable in these places. At present, it is estimated that topsoil is eroding faster than it forms on more than one-third of the world's croplands. The result is lower productivity, poorer crop quality, and reduced agricultural income.

Another problem related to excessive soil erosion involves the deposition of sediment. Each year in the United States hundreds of millions of tons of eroded soil is deposited in lakes, reservoirs, and streams. The detrimental impact of this process can be significant. For example, as more and more sediment is deposited in a reservoir, the capacity of the reservoir is diminished, limiting its usefulness for flood control, water supply, and/or hydroelectric power generation. In addition, sedimentation in streams and other waterways can restrict navigation and lead to costly dredging operations.

Figure 2.14
When it is raining, millions of water drops are falling at velocities approaching 10 meters per second (35 kilometers per hour). When water drops strike an exposed surface, soil particles may splash as high as one meter into the air and land more than a meter away from the point of raindrop impact. Soil dislodged by splash erosion is more easily moved by sheet erosion. (Photo courtesy of U.S. Department of Agriculture)

Figure 2.15
Gully erosion in poorly protected soil. (Photo by James E. Patterson)

In some cases soil particles are contaminated with pesticides used in farming. When these chemicals are introduced into a lake or reservoir, the quality of the water supply is threatened and aquatic organisms may be endangered. In addition to pesticides, nutrients found naturally in soils as well as those added by agricultural fertilizers make their way into streams and lakes, where they stimulate the growth of plants. Over a period of time, excessive nutrients accelerate the process by which plant growth leads to the depletion of oxygen, and an early death of the lake results.

The availability of good soils is critical if the world's rapidly growing population is to be fed. On every continent unnecessary soil loss is occurring because appropriate conservation measures are not being used. Although it is a recognized fact that soil erosion can never be completely eliminated, soil conservation programs can substantially reduce the loss of this basic resource. Windbreaks, terracing, and contour farming techniques are some of the effective measures, as are special tillage practices and crop rotation.

Weathering and Ore Deposits

Weathering creates many important mineral deposits by concentrating minor amounts of metals that are scattered through unweathered rock into economically valuable concentrations. Such a transformation is often termed **secondary enrichment** and takes place in one of two ways. In one situation, chemical weathering coupled with downward-percolating water removes undesirable materials from decomposing rock, leaving the desirable elements enriched in the upper zones of the soil. The second way is basically the reverse of the first. That is, the desirable elements that are found in low concentrations near the surface are removed and carried to lower zones, where they are redeposited and become more concentrated.

The formation of *bauxite*, the principal ore of aluminum, is one important example of an ore created as a result of enrichment by weathering processes. Although aluminum is the third most abundant element in the earth's crust, economically valuable concentrations of this important metal are not common, due to the fact that most aluminum is tied up in silicate minerals, from which it is extremely difficult to extract.

Bauxite forms in rainy tropical climates in association with laterites. In fact, bauxite is sometimes referred to as aluminum laterite. When aluminum-rich source rocks are subjected to the intense and prolonged chemical weathering of the tropics, most of the common elements, including calcium, sodium, and silicon, are removed by leaching. Because aluminum is extremely insoluble in natural solutions, hydrated aluminum oxide (bauxite) becomes concentrated in the soil. Thus, the formation of bauxite depends upon climatic conditions in which chemical weathering and leaching are pronounced, as well as on the presence of an aluminum-rich source rock. Important deposits of nickel and cobalt are also found in laterite soils that develop from igneous rocks rich in other silicate minerals.

Many copper and silver deposits result when weathering processes concentrate metals that are dispersed through a low-grade primary ore. Usually such enrichment occurs in deposits containing pyrite

BOX 2.2
Dust Bowl: Soil Erosion in the Great Plains

During a span of dry years in the 1930s, large dust storms plagued the Great Plains. Because of the size and severity of these storms, the region came to be called the "Dust Bowl," and the time period, the "dirty thirties." The heart of the Dust Bowl consisted of nearly 100 million acres in the panhandles of Texas and Oklahoma, as well as adjacent parts of Colorado, New Mexico, and Kansas (Figure 2.B). To a lesser extent, dust storms were also a problem over much of the Great Plains, from North Dakota to west central Texas.

At times dust storms were so severe that they were called "black blizzards" and "black rollers" because visibility was reduced to only a few feet. Examples of storms that lasted for hours and stripped huge volumes of topsoil from the land are numerous. In the spring of 1934, a wind storm that lasted for a day and a half created a dust cloud that extended for 2000 kilometers (1200 miles). As the sediment moved east, "muddy rains" were experienced in New York, and "black snows" in Vermont. Less than a year later, another storm carried dust more than 3 kilometers (2 miles) into the atmosphere and transported it 3000 kilometers from its source in Colorado to create twilight conditions in the middle of the day in parts of New England and New York.

What caused the Dust Bowl? Clearly, the fact that portions of the Great Plains experience some of North America's strongest winds is important. However, it was the expansion of agriculture that set the stage for the disastrous period of soil erosion.

Mechanization allowed the rapid transformation of the grass-covered prairies of this semiarid region into farms. Between the 1870s and 1930, the area of cultivation in the region expanded nearly tenfold, from about 10 million acres to more than 100 million acres.

As long as precipitation was adequate, the soil remained in place. However, when a prolonged drought struck in the 1930s, the unprotected fields were vulnerable to the wind. The results were severe soil loss, crop failures, and economic hardship.

Beginning in 1939, a return to rainier conditions brought relief. Moreover, farming practices that were designed to reduce soil loss by wind had also been instituted. Although dust storms are less numerous and not as severe as in the "dirty thirties," soil erosion by strong winds still occurs periodically whenever the combination of drought and unprotected soil exists.

Figure 2.B
An abandoned farmstead shows the disastrous effects of wind erosion and deposition during the ``Dust Bowl" period. This photo of a previously prosperous farm was taken in Oklahoma in 1937. (Photo courtesy of Soil Conservation Service, U.S. Department of Agriculture)

(FeS$_2$), the most common and widespread sulfide mineral. Pyrite is important because when it chemically weathers, sulfuric acid forms, which enables percolating waters to dissolve the ore metals. Once dissolved, the metals gradually migrate downward through the primary ore body until they are precipitated. Deposition takes place because of changes that occur in the chemistry of the solution when it reaches the groundwater zone (the zone beneath the surface where all pore spaces are filled with water). In this manner the small percentage of dispersed metal can be removed from a large volume of rock and redeposited as a higher-grade ore in a smaller volume of rock.

MASS WASTING

The earth's surface is never perfectly flat but instead consists of slopes. Some are steep and precipitous; others are moderate or gentle. Some are long and gradual; others are short and abrupt. Some slopes are mantled with soil and covered by vegetation; others consist of barren rock and rubble. Their form and variety are great. Taken together, slopes are the most common elements in our physical landscape. Although most slopes appear to be stable and unchanging, they are not static features because the force of gravity causes material to move downslope. At one extreme, the movement may be gradual and practically imperceptible. At the other extreme, it may consist of a thundering rockfall or avalanche (Figure 2.16).

Landslides are spectacular examples of a common geologic process called mass wasting. Mass wasting refers to the downslope movement of rock, regolith, and soil under the direct influence of gravity. It is distinct from the erosional processes that are examined in subsequent chapters because mass wasting does not require a transporting medium.

Occasionally, news media report the terrifying and often grim details of landslides. On May 31, 1970, one such event occurred when a gigantic rock avalanche buried more than 20,000 People in Yungay and Ranrahirca, Peru. There was little warning of the impending disaster; it began and ended in just a matter of a few minutes. The avalanche started 14 kilometers from Yungay, near the summit of the 6700-meter (22,000-foot) Nevados Huascaran, the loftiest peak in the Peruvian Andes. Triggered by the ground motion from a strong offshore earthquake, a huge mass of rock and ice broke free from the precipitous north face of the mountain. After plunging nearly one kilometer, the material pulverized on impact and immediately began rushing down the mountainside, made fluid by trapped air and melted ice. The initial mass ripped loose additional millions of tons of debris as it roared downhill. Although the material followed a previously eroded gorge, a portion of the debris jumped a 200–300-meter bedrock ridge that had protected Yungay from similar events

Figure 2.16
In 1955 this rockslide above Emerald Bay, Lake Tahoe, Nevada, closed a stretch of highway. (Photo by James E. Patterson)

BOX 2.3
The Vaiont Dam Disaster

A massive rock avalanche in Peru is described at the beginning of the section on mass wasting. As with most occurrences of mass wasting, this tragic episode was triggered by a natural event—in this case, an earthquake. However, disasters also result from the mass movement of surface material triggered by the actions of humans. For example, in 1960 a large dam almost 265 meters tall was built across the Vaiont Canyon in the Italian Alps. Three years later, on the night of October 9, 1963, a violent disaster occurred and the dam was largely responsible.

The bedrock in Vaiont Canyon slanted steeply downward toward the lake impounded behind the dam and was composed of weak, highly fractured limestone strata that contained beds of clay and numerous solution cavities. As the reservoir filled, the lower portions of these rocks became saturated and the clays became swollen and more plastic. The rising water reduced the internal friction that had kept the rock in place. Measurements made shortly after the reservoir was filled hinted at the problem, because they indicated that a portion of the mountain was slowly creeping downhill at the rate of 1 centimeter per week. In September 1963, the rate increased to 1 centimeter per day, then 10–20 centimeters per day, and eventually to as much as 80 centimeters on the day of the disaster. Finally, the mountainside let loose. In just an instant, 240 million cubic meters of rock and rubble slid down the face of Mount Toc and filled nearly 2 kilometers of the gorge to heights of 150 meters above the reservoir level (Figure 2.C). The filling of the reservoir pushed the water completely over the dam in a wave more than 90 meters high. More than 1.5 kilometers downstream, the wall of water was still 70 meters high and everything in its path was completely destroyed. The entire event lasted less than seven minutes, yet it claimed an estimated 2600 lives. This is known as the worst dam disaster in history, but when it was over, the Vaiont Dam was still standing intact. Although the catastrophe was triggered by human interference with the Vaiont River, the slide would have eventually occurred on its own; however, the effects would not have been nearly as tragic.

Figure 2.C
Sketch map of the Vaiont River area showing the limits of the landslide, the portion of the reservoir that was filled with debris, and the extent of flooding downstream. [After G. A. Kiersh, ``Vaiont Reservoir Disaster," Civil Engineering 34 (1964): 32–39]

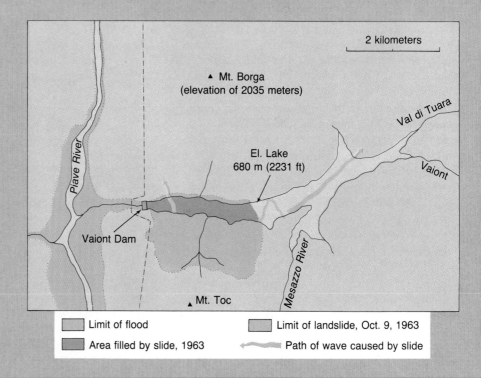

in the past and buried the entire city. After inundating another town in its path, Ranrahirca, the mass of debris finally reached the bottom of the valley where its momentum carried it across the Rio Santa and tens of meters up the valley wall on the opposite side.

Fortunately, landslide disasters that kill hundreds or thousands of people are rare. However, property damages each year can be substantial. In the United States alone, losses related to mass wasting events exceed one billion dollars annually.

As with many geologic hazards, the tragic rock avalanche in Peru was triggered by a natural event—in this case, an earthquake. In fact, most mass wasting events, whether spectacular or subtle, are the result of circumstances that are completely independent of human activities. In places where mass wasting is a recognized threat, steps can often be taken to control downslope movements or limit the damages that such movements can cause. If the potential for mass wasting goes unrecognized or is ignored, the results can be costly and dangerous. It should also be pointed out that, although most downslope movements occur whether people are present or not, many occurrences each year are aggravated or even triggered by human actions.

Controls of Mass Wasting

In the evolution of most landforms, mass wasting is the step that follows weathering. By itself weathering does not produce significant landforms. Rather, landforms develop as the products of weathering are removed from the places where they originate. Once weathering weakens and breaks rock apart, mass wasting transfers the debris downslope, where a stream, acting as a conveyor belt, carries it away. Although there may be many intermediate stops along the way, the sediment is eventually transported to its ultimate destination, the sea. It is the combined effects of mass wasting and running water that produce stream valleys, which are the most common and conspicuous landforms at the earth's surface. If streams alone were responsible for creating the valleys in which they flow, valleys would be very narrow features. However, the fact that most river valleys are much wider than they are deep is a strong indication of the significance of mass wasting processes in supplying material to streams.

Although gravity is the controlling force of mass wasting, other factors play an important part in bringing about the downslope movement of material. Water is one of these factors. When the pores in sediment become filled with water, the cohesion between the particles is destroyed, allowing them to slide past one another with relative ease. For example, when sand is slightly moist, it may stick together quite well. However, if more water is added, filling the openings between the grains, the sand will ooze out in all directions. Thus, saturation reduces the internal resistance of materials, which are then easily set in motion by the force of gravity. When clay is wetted, it becomes very slick—another example of the "lubricating" effect of water. Water also adds considerable weight to a mass of material. The added weight in itself may be enough to cause the material to slide or flow downslope.

Oversteepening of slopes is another cause for many mass movements. Loose, undisturbed particles assume a stable slope called the **angle of repose,** the steepest angle at which material remains stable. Depending upon the size and shape of the particles, the angle varies from 25 to 40 degrees. The larger, more angular particles maintain the steepest slopes. If the angle is increased, the rock debris will adjust by moving downslope. There are many situations in nature where this takes place. A stream undercutting a valley wall and waves pounding against the base of a cliff are but two familiar examples. Furthermore, through their activities, people often create oversteepened and unstable slopes that become prime sites for mass wasting.

Classification of Mass Wasting Processes

There is a broad array of processes that geologists call mass wasting. Four are illustrated in Figure 2.17. Generally the different types are divided and described on the basis of the type of material involved, the kind of motion that is displayed, and the velocity of the movement.

The classification of mass wasting processes based on the material involved in the movement depends upon whether the descending mass began as unconsolidated material or as bedrock. If soil and regolith dominate, terms such as *debris*, *mud*, or *earth* are used in the description. On the other hand, when a mass of bedrock breaks loose and moves downslope, the term *rock* may be part of the description.

In addition to the type of material involved in a mass wasting event, the way in which material moves may also be important. Generally, the kind of motion is described as either a fall, a slide, or a flow.

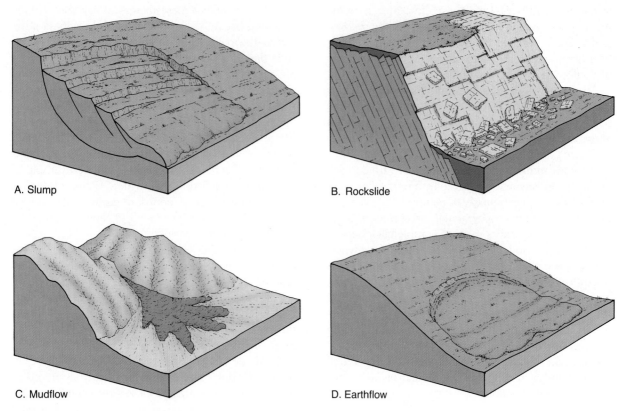

A. Slump

B. Rockslide

C. Mudflow

D. Earthflow

Figure 2.17
*The four processes illustrated here are all considered to be relatively rapid forms of mass wasting. Because materials in slumps **(A.)** and rockslides **(B.)** move along well-defined surfaces, they are said to move by sliding. By contrast, when material moves downslope as a viscous fluid, the movement is described as a flow. Mudflow **(C.)** and earthflow **(D.)** advance downslope in this manner.*

When the movement involves the free-fall of detached individual pieces of any size, it is termed a **fall.** Fall is a common form of movement on slopes that are so steep that loose material cannot remain on the surface. The rock may fall directly to the base of the slope or move in a series of leaps and bounds over other rocks along the way. Many falls result when freeze and thaw cycles or the action of plant roots loosen rock to the point that gravity takes over. Rockfall is the primary way in which talus slopes are built and maintained (Figure 2.18). Sometimes falls may trigger other forms of downslope movement. For example, recall that the Yungay disaster described at the beginning of this section was initiated by a mass of free-falling material that broke from the nearly vertical summit of Nevados Huascaran.

Many mass wasting processes are described as **slides.** Slides occur whenever material remains fairly coherent and moves along a well-defined surface. Sometimes the surface is a joint, a fault, or a bedding plane that is approximately parallel to the slope. However, in the case of the movement called slump, the descending mass moves along a curved surface of rupture. A note of clarification is appropriate at this point. Sometimes the word *slide* is used as a synonym for the word *landslide.* It should be pointed out that, although many people, including geologists, use the term, the world *landslide* has no specific definition in geology. Rather, it should be considered as a popular nontechnical term to describe all perceptible forms of mass wasting, including those in which sliding does not occur.

Figure 2.18
Talus is a slope built of angular rock fragments. Mechanical weathering, especially frost wedging, loosens the pieces of bedrock, which then fall to the base of the cliff. With time, a series of steep, cone-shaped accumulations builds up at the base of the slope. (Photo by John S. Shelton)

The third type of movement common to mass wasting is termed **flow.** Flow occurs when material moves downslope as a viscous fluid. Most flows are saturated with water and typically move as lobes or tongues.

When mass wasting events make the news, a large quantity of material has in all likelihood moved rapidly downslope and has had a disastrous effect upon people and property. Indeed, during events called *rock avalanches*, rock and debris can move downslope at speeds well in excess of 200 kilometers (125 miles) per hour. Many researchers believe that rock avalanches, such as the one that produced the scene in Figure 2.19, must literally "float on air" as they move downslope. That is, high velocities result when air becomes trapped and compressed beneath the falling mass of debris, allowing it to move as a buoyant, flexible sheet across the surface.

Most mass movements, however, do not move with the speed of a rock avalanche. In fact, a great deal of mass wasting is imperceptibly slow. One process that we will examine later, termed *creep*, results in particle movements that are usually measured in millimeters or centimeters per year. Thus, as

you can see, rates of movement can be spectacularly sudden or exceptionally gradual. Although various types of mass wasting are often classified as either rapid or slow, such a distinction is highly subjective because there is a wide range of rates between the two extremes. Even the velocity of a single process at a particular site can vary considerably from one time to another.

Slump

Slump refers to the downward slipping of a mass of rock or unconsolidated material moving as a unit along a curved surface (see Figure 2.17A). Usually the slumped material does not travel spectacularly fast nor very far. This is a common form of mass wasting, especially in thick accumulations of cohesive materials such as clay The surface of rupture beneath the slump block is characteristically spoon-shaped and concave upward or outward. As the movement occurs, a crescent-shaped scarp (cliff) is created at the head and the block's upper surface is sometimes tilted backwards.

Slump commonly occurs because a slope has been oversteepened. The material on the upper portion of a slope is held in place by the material at the bottom of the slope. As this anchoring material at the base is removed, the material above is made unstable and reacts to the pull of gravity. One relatively common example is a valley wall that becomes oversteepened by a meandering river. Another occurs in coastal areas where a cliffed seashore is undercut by wave activity at its base. Slumping may also occur when a slope is overloaded, causing internal stress on the material below. This type of slump often occurs where weak, clay-rich material underlies layers of stronger, more resistant rock such as sandstone. The seepage of water through the upper layers reduces the strength of the clay below and slope failure results.

Rockslide

Rockslides occur when blocks of bedrock break loose and slide downslope (see Figure 2.17B). Such events are among the fastest and most destructive mass movements. Usually rockslides take place in a geologic setting where the rock strata are inclined, or joints and fractures exist, parallel to the slope. If the rock is undercut at the base of the slope, it loses support and the rock eventually gives way. Sometimes

Figure 2.19
Debris deposited atop Sherman Glacier in Alaska by a rock avalanche. The event was triggered by a tremendous earthquake in March 1964. (Photo by Austin Post, U.S. Geological Survey)

the rockslide is triggered when rain or melting snow lubricates the underlying surface to the point where friction is no longer sufficient to hold the rock unit in place. As a result, rockslides tend to be most prevalent during the spring, when heavy rains and melting snow are greatest. Earthquakes are another mechanism that triggers rockslides and other mass movements. The 1811 earthquake at New Madrid, Missouri, for example, caused slides in an area of more than 13,000 square kilometers (5000 square miles) along the Mississippi River valley. In terms of property damage, the most devastating effects of the famous 1964 Alaska earthquake were the slides in Anchorage, some 130 kilometers (80 miles) from the center of the quake.

The Gros Ventre River flows west from the Wind River Range in northwestern Wyoming, through Grand Teton National Park, eventually emptying into the Snake River. On June 23, 1925, a classic rockslide took place in its valley, just east of the small town of Kelly. In the span of only a few minutes a great mass of sandstone, shale, and soil crashed down the south side of the valley, carrying with it a dense pine forest. The volume of debris, estimated at 38 million cubic meters (50 million cubic yards) created a 67–75-meter (220–250 foot)-high dam on the Gros Ventre River (Figure 2.20). Because the river was completely blocked, a lake was created. It filled so quickly that a house that had been 18 meters (60 feet) above the river was floated off its foundation 18 hours after the

Figure 2.20
The scar, 2.4 kilometers (1.5 miles) long and 0.8 kilometer (0.5 mile) wide, left on the side of Sheep Mountain by the Gros Ventre rockslide. (Photo by Stephen Trimble)

slide. In 1927 the lake overflowed the dam, partially draining the lake and resulting in a devastating flood downstream.

Why did the Gros Ventre rockslide take place? Figure 2.21 is a diagrammatic cross-sectional view of the geology of the valley. Notice the following points: (1) the sedimentary strata in this area dip (tilt) 15–21 degrees, (2) underlying the bed of sandstone is a relatively thin layer of clay, and (3) at the bottom of the valley the river had cut through much of the sandstone layer. During the spring of 1925 water from the heavy rains and melting snow seeped through the sandstone, saturating the clay below. Since much of the sandstone layer had been cut through by the Gros Ventre River, the layer had virtually no support at the bottom of the slope. Eventually the sandstone could no longer hold its position on the wetted clay, and gravity pulled the mass down the side of the valley.

The circumstances at this location were such that a rockslide was inevitable.

Mudflow

Mudflow is a relatively rapid type of mass wasting that involves a flow of debris containing a large amount of water (see Figure 2.17C). Mudflows are most characteristic of semiarid mountainous regions and, because of their high water content and the predominance of fine particles, they tend to follow canyons and gullies. Although rains in semiarid regions are infrequent, they are typically heavy. When a cloudburst or rapidly melting mountain snow causes a sudden flood, large quantities of soil and regolith are washed into nearby stream channels because there is usually little or no vegetation to anchor the surface material. This creates a flowing tongue

Figure 2.21
Cross-sectional view of the Gros Ventre rockslide. The slide occurred when the tilted and undercut sandstone bed could no longer maintain its position atop the saturated bed of clay. [After W. C. Alden, ``Landslide and Flood at Gros Ventre, Wyoming,'' Transactions (AIME) 76 (1928): 348]

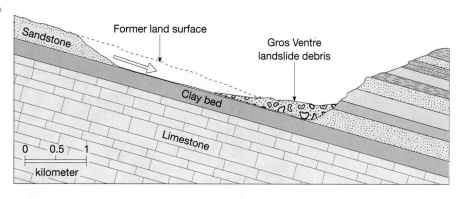

of well-mixed mud, soil, rock, and water. Its consistency may range from that of wet concrete to a soupy mixture not much thicker than muddy water. The rate of flow therefore depends not only on the slope but on the water content as well. When dense, mudflows are capable of carrying or pushing large boulders, trees, and even houses with relative ease. Mudflows periodically are a serious hazard in dry mountainous areas such as portions of southern California. Here, construction of homes on canyon hillsides and the removal of native vegetation by brush fires and other means have increased the frequency of these destructive events (Figure 2.22).

Mudflows are also common on the slopes of some volcanoes, in which case they are termed **lahars.** The word originated in Indonesia, a volcanic region that has experienced many of these often destructive events. Lahars result when highly unstable layers of ash and debris become saturated with water and flow down steep volcanic slopes, generally following existing stream channels. Some are initiated when heavy rainfalls erode volcanic deposits. Others are triggered when large volumes of ice and snow are suddenly melted by heat flowing to the surface from within the volcano or by the near-molten debris emitted during a violent eruption.

In November 1985, Nevado del Ruiz, a large volcano in the Andes Mountains of Colombia, erupted. The eruption melted much of the snow and ice that capped the peak, producing a torrent of hot viscous mud, ash, and debris. The lahar moved outward from the volcano following the valleys and three rain-swollen rivers that radiate from the peak. The flow that moved down the valley of the Lagunilla River was the most destructive, devastating the town of Armero, 48 kilometers (30 miles) from the mountain. Most of the more than 20,000 deaths caused by the event were in this once-thriving agricultural community. Death and property damage also occurred in 13 other villages within the 180-square-kilometer disaster area. Although a great deal of material was explosively ejected from Nevado del Ruiz, it was the lahars triggered by this eruption that made this such a devastating natural disaster.

Earthflow

Unlike mudflows, which are usually confined to channels in semiarid regions, **earthflows** most often form on hillsides in humid areas as the result of excessive rainfall or snowmelt (see Figure 2.17D). When water saturates the soil and regolith on a hillside, the material may break away, leaving a scar on the slope and forming a tongue- or teardrop-shaped mass that flows downslope (Figure 2.23) The materials most commonly involved are rich in clay and silt and contain only small proportions of sand and coarser particles. Earthflows range in size from bodies a few meters long, a few meters wide, and less than one meter deep to masses more than 1 kilometer long, several hundred meters wide, and more than 10 meters deep. Because earthflows are quite viscous, they generally

Figure 2.22
Severe damage resulted when a mudflow buried the lower portion of this house located near the mouth of a canyon in southern California. (Photo by James E. Patterson)

Figure 2.23
This small, tongue-shaped earthflow occurred on a newly formed slope along a recently constructed highway. It formed in clay-rich material following a period of heavy rain. Notice the small slump at the head of the earthflow. (Photo by E. J. Tarbuck)

move at slower rates than the more fluid mudflows described in the preceding section. They are characterized by a slow and persistent movement and may remain active for periods ranging from days to years. Depending upon the steepness of the slope and the material's consistency, measured velocities range from less than 1 millimeter per day up to several meters per day. Over the time span that earthflows are active, movement is typically faster during wet periods than during drier times. In addition to occurring as isolated hillside phenomena, earthflows commonly take place in association with large slumps. In this situation, they may be seen as tongue-like flows at the base of the slump block.

Creep

Movements such as rockslides and rock avalanches are certainly the most spectacular and catastrophic forms of mass wasting. Since these events have been known to kill thousands, they deserve intensive study so that, through more effective prediction, timely warnings and better controls can help save lives. However, because of their large size and spectacular nature they give us a false impression of their importance as a mass wasting process. Indeed, sudden movements are responsible for moving less material than the slow and far more subtle action of creep. Whereas rapid types of mass wasting are characteristic of mountains and steep hillsides, creep can take place on gentle slopes and is thus much more widespread.

Creep is a type of mass wasting that involves the gradual downhill movement of soil and regolith. One of the primary causes of creep is the alternate expansion and contraction of surface material caused by freezing and thawing or wetting and drying. As shown in Figure 2.24, freezing or wetting lifts the soil at right angles to the slope, and thawing or drying allows the particles to fall back to a slightly lower level. Each cycle therefore moves the material a short distance downhill. Creep may also be initiated if the ground becomes saturated with water. Following a heavy rain or snowmelt, a waterlogged soil may lose its internal cohesion, allowing gravity to pull the ma-

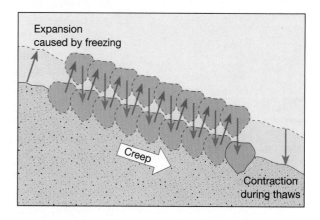

Figure 2.24
The repeated expansion and contraction of the surface material causes a net downslope migration of rock particles—a process called creep.

Figure 2.25
Solifluction lobes northeast of Fairbanks, Alaska. (Photo by James E. Patterson)

terial downslope. Although the movement is imperceptibly slow, its effects are recognizable. Creep causes fences and telephone poles to tilt, and tree trunks will often be bent as a consequence of this movement.

Solifluction is a form of mass wasting that is common in regions underlain by permafrost. **Permafrost** refers to the permanently frozen ground that occurs in association with the earth's harsh tundra and ice cap climates. Solifluction may be regarded as a form of creep in which unconsolidated, water-saturated material gradually moves downslope. Solifluction occurs in a zone above the permafrost which thaws in summer and refreezes in winter. During the summer season, water is unable to percolate into the impervious permafrost layer below. As a result, the layer of unfrozen ground becomes saturated and slowly flows. The process can occur on slopes as gentle as 2–3 degrees. Where there is a well-developed mat of vegetation, a solifluction sheet may move downward in a series of well-defined lobes or as a series of partially overriding folds (Figure 2.25).

Review Questions

1. If two identical rocks were weathered, one mechanically and the other chemically, how would the products of weathering for the two rocks differ?

2. How does mechanical weathering add to the effectiveness of chemical weathering?

3. Describe the formation of an exfoliation dome. Give an example of such a feature.

4. Granite and basalt are exposed at the surface in a hot, wet region.
   (a) Which type of weathering will predominate?
   (b) Which of the rocks will weather most rapidly? Why?

5. Heat speeds up a chemical reaction. Why then does chemical weathering proceed slowly in a hot desert?

6. How is carbonic acid (H_2CO_3) formed in nature? What results when this acid reacts with potassium feldspar?

7. What is the difference between soil and regolith?

8. Using the soil texture diagram (Figure 2.11), name the soil that consists of 60 percent sand, 30 percent silt, and 10 percent clay.

9. What factors might cause different soils to develop from the same parent material, or similar soils to form from different parent materials?

10. Which of the controls of soil formation is most important? Explain.

11. How can slope affect the development of soil? What is meant by the term *slope orientation*?

12. List the characteristics associated with each of the horizons in a well-developed soil profile. Which of the horizons constitute the solum? Under what circumstances do soils lack horizons?

13. Distinguish between pedalfers and pedocals.

14. Soils formed in the humid tropics and the Arctic both contain little organic matter. Do both lack humus for the same reasons?

15. What soil type is associated with tropical rainforests? Since this soil supports the growth of lush natural vegetation, is it also excellent for growing crops? Briefly explain. (See Box 2.1.)

16. Name the primary ore of aluminum and describe its formation.

17. Before soil particles are carried by water moving in tiny channels called rills, they are moved by an unconfined movement of water across the surface termed _____ .

18. In addition to the loss of topsoil from croplands, list three detrimental effects of soil erosion.

19. Briefly describe the conditions that led to the Dust Bowl of the 1930s. (See Box 2.2.)

20. What role does mass wasting play in sculpting the earth's landscape?

21. How did the building of a dam contribute to the Vaiont Canyon disaster? Was the disaster avoidable? (See Box 2.3.)

22. What is the controlling force of mass wasting? What other factors are important?

23. Distinguish among fall, slide, and flow.

24. Why can rock avalanches move at such great speeds?

25. What factors led to the massive rockslide at Gros Ventre, Wyoming (Figure 2.21)?

26. What type of mass wasting event killed thousands of people in the vicinity of the Colombian volcano Nevado del Ruiz in 1985?

27. Compare and contrast mudflow and earthflow.

28. Since creep is an imperceptibly slow process, what evidence might indicate to you that this phenomenon is affecting a slope? Describe the mechanism that creates this slow movement (Figure 2.24).

29. Why is solifluction only a summertime phenomenon?

Key Terms

angle of repose (p. 83)
chemical weathering (p. 65)
creep (p. 89)
differential weathering (p. 69)
earthflow (p. 88)
eluviation (p. 75)
erosion (p. 62)
exfoliation dome (p. 64)
fall (p. 84)
flow (p. 85)
frost wedging (p. 64)
horizon (p. 74)
humus (p. 69)
lahar (p. 88)
laterite (p. 75)
leaching (p. 75)
mass wasting (p. 62)
mechanical weathering (p. 63)
mudflow (p. 87)

parent material (p. 72)
pedalfer (p. 75)
pedocal (p. 75)
permafrost (p. 90)
regolith (p. 69)
rockslide (p. 85)
secondary enrichment (p. 79)
sheeting (p. 64)
slide (p. 84)
slump (p. 85)
soil (p. 69)
soil profile (p. 74)
soil texture (p. 70)
solifluction (p. 90)
solum (p. 75)
spheroidal weathering (p. 67)
talus slope (p. 64)
weathering (p. 62)

3

Running Water and Groundwater

Opposite page: Grand Canyon of the Yellowstone River, Wyoming. An excellent example of a valley in the youthful stage of development. (Photo by Robert Winslow)

Left: Waterfalls in Yosemite National Park, California. (Photo by E. J. Tarbuck)

Above: The smaller American Falls at Niagara Falls. (Photo by James E. Patterson)

93

All the rivers run into the sea; yet the sea is not full; unto the place from whence the rivers come, thither they return again. (Ecclesiastes 1:7)

As the perceptive writer of Ecclesiastes indicated, water is continually on the move, from the ocean to the land and back again. This chapter deals with that part of the water cycle which involves the return of water to the sea, with some water traveling quickly via a rushing stream, and some moving more slowly below the surface. We shall examine the factors that control the movement of water, as well as look at how water sculptures the landscape and at the features that result. To a great extent, the Grand Canyon, Niagara Falls, Old Faithful, and Mammoth Cave all owe their existence to the activities of water on its way to the sea.

THE HYDROLOGIC CYCLE

The amount of water on earth is immense, an estimated 1.36 billion cubic kilometers (326 million cubic miles). Of this total, the vast bulk—97.2 percent—is part of the world ocean. Ice sheets and glaciers account for another 2.15 percent, leaving only 0.65 percent to be divided among lakes, streams, subsurface water, and the atmosphere (Figure 3.1). Although the percentages of the earth's total water found in each of the latter sources is but a small fraction of the total inventory, the absolute quantities are great.

An adequate supply of water is vital to life on earth. With increasing demands on this finite resource, scientists have given a great deal of attention to the exchanges of water among the oceans, the atmosphere, and the continents. This unending circulation of the earth's water supply has come to be called the **hydrologic cycle**. It is a gigantic system powered by energy from the sun in which the atmosphere provides the vital link between the oceans and continents. Water from the oceans, and to a much lesser extent from the continents, is constantly evaporating into the atmosphere. Winds transport the moisture-laden air, often great distances, until the complex processes of cloud formation are set in motion that eventually result in precipitation. The precipitation that falls into the ocean has ended its cycle and is ready to begin another. The water that falls on the continents, however, must still make its way back to the ocean.

What happens to precipitation once it has fallen on land? A portion of the water soaks into the ground, some of it moving downward, then laterally, finally seeping into lakes, streams, or directly into the ocean. When the rate of rainfall is greater than the earth's ability to absorb it, the additional water flows over the surface into lakes and streams. Much of the water which soaks in (**infiltration**) or runs off (**runoff**) eventually finds its way back to the atmosphere because of evaporation from the soil, lakes, and streams. Also, some of the water that infiltrates the ground surface is absorbed by plants, which then release it into the atmosphere. This process is called **transpiration**. Each year a field of crops may transpire an amount of water equivalent to a layer 60 centimeters (2 feet) deep over the entire field, whereas a forest may pump twice this amount into the atmosphere.

Figure 3.1
Lakes and streams represent just a tiny fraction of the earth's water supply. Nevertheless, the absolute quantities of water that flow through this part of the hydrologic cycle are great. South Fork of the Cranberry River in West Virginia. (Photo by David Muench Photography)

When precipitation falls at high elevations or high latitudes, the water may not immediately soak in, run off, or evaporate. Instead, it may become part of a snowfield or a glacier. Glaciers store large quantities of water on land. If present-day glaciers were to melt and release their storage of water, sea level would rise by several tens of meters and submerge many heavily populated coastal areas. As we shall see in Chapter 4, over the past two million years, huge continental ice sheets have formed and melted on several occasions.

A diagram of the earth's water balance, a quantitative view of the hydrologic cycle, is shown in Figure 3.2. Although the amount of water vapor in the air at any one time is just a tiny fraction of the earth's total water supply, the absolute quantities that are cycled through the atmosphere over a one-year period are immense—some 380,000 cubic kilometers—

enough to cover the earth's surface to a depth of about one meter (39 inches). Since the total amount of water vapor in the atmosphere remains about the same, average annual precipitation over the earth must be equal to the quantity of water evaporated. However, for all of the continents taken together, precipitation exceeds evaporation. Conversely, over the oceans, evaporation exceeds precipitation. Since the level of the world ocean is not dropping, runoff from land areas must balance the deficit of precipitation over the oceans. The erosional work accomplished by the 36,000 cubic kilometers of water that annually flows from the land to the ocean is enormous. In fact, this immense volume of moving water represents the single most important agent sculpturing the earth's land surface.

To summarize, the hydrologic cycle represents the continuous movement of water from the oceans to the atmosphere, from the atmosphere to the land, and from the land back to the sea. The wearing down of the earth's land surface is primarily attributable to the last of the steps in this cycle.

RUNNING WATER

Running water is of great importance to people. We depend upon rivers for energy, travel, and irrigation; their fertile floodplains have fostered human progress since the dawn of civilization. Moreover, as the dominant agent of landscape alteration, streams have sculptured much of our physical environment.

Although people have always depended to a great extent on running water, its source eluded them for centuries. It was not until the sixteenth century that they first realized streams were supplied by runoff and underground water, which ultimately had their sources as rain and snow. Runoff initially flows in broad sheets that enter small rills, which carry it to a stream. The word *stream* is used to denote channelized flow of any size, from the smallest brook to the mighty Amazon. Although the terms *river* and *stream* are used synonymously, the term *river* is often preferred when describing a main stream into which several tributaries flow.

Streamflow

Flowing water makes its way to the sea under the influence of gravity. The time required for the journey depends upon the velocity of the stream, which is

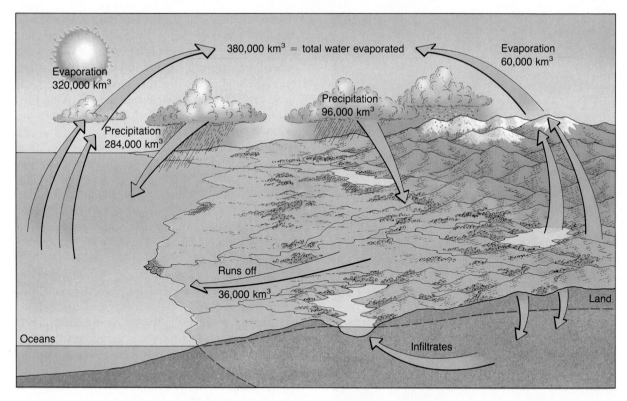

Figure 3.2
The earth's water balance. About 320,000 cubic kilometers of water are evaporated each year from the oceans, while evaporation from the land (including lakes and streams) contributes 60,000 cubic kilometers of water. Of this total of 380,000 cubic kilometers of water, about 284,000 cubic kilometers fall back to the ocean, and the remaining 96,000 cubic kilometers fall on the earth's land surface. Since 60,000 cubic kilometers of water evaporate from the land, 36,000 cubic kilometers of water remain to erode the land during the journey back to the oceans.

measured in terms of the distance the water travels in a given unit of time. Some sluggish streams travel at less than 0.8 kilometer (0.5 mile) per hour, while a few rapid ones reach speeds as high as 32 kilometers (20 miles) per hour. Velocities are determined at gauging stations where measurements are taken at several locations across the channel and then averaged. Along straight stretches the highest velocities are near the center of the channel just below the surface, where friction is lowest. But when a stream curves, its zone of maximum speed shifts toward its outer bank.

The ability of a stream to erode and transport materials is directly related to its velocity; thus, it is a very important characteristic. Even slight variations in velocity can lead to significant changes in the load of sediment transported by the water. Several factors determine the velocity of a stream and therefore control the amount of erosional work a stream may accomplish. These factors include the following: (1) gradient; (2) shape, size, and roughness of the channel; and (3) discharge.

Certainly one of the most obvious factors controlling stream velocity is the **gradient**, or slope, of a stream channel. Gradient is typically expressed as the vertical drop of a stream over a fixed distance. Gradients may vary considerably from one stream to another as well as along the course of a given stream (Figure 3.3). Portions of the lower Mississippi River, for example, have gradients of 10 centimeters per kilometer and less. By way of contrast, some mountain stream channels decrease in elevation at a rate of more than 40 meters per kilometer, 400 times more abruptly than the lower Mississippi. The higher the gradient, the more energy available for streamflow. If two streams were identical in every respect except

Figure 3.3
Rafting the Colorado River's Granite Rapids in the Grand Canyon. Rapids can form where a portion of a stream channel has a steep gradient. (Photo by Robert Winslow)

gradient, the stream with the higher gradient would obviously have the greater velocity.

The cross-sectional shape of a channel determines the amount of water in contact with the channel and hence affects the frictional drag. The most efficient channel is one with the least perimeter for its cross-sectional area. Figure 3.4 compares two types of channels. Although the cross-sectional area of both is identical, the semicircular shape has less water in contact with the channel and therefore less frictional drag. As a result, if all other factors are equal, the water will flow more rapidly in the semicircular channel.

The size and roughness of the channel also affect the amount of friction. An increase in the size of a channel reduces the ratio of perimeter to cross-sectional area and therefore increases the efficiency of flow. The effect of roughness is obvious. A smooth channel promotes a more uniform flow, whereas an irregular channel filled with boulders creates enough turbulence to significantly retard the stream's forward motion.

The **discharge** of a stream is the amount of water flowing past a certain point in a given unit of time. This is usually measured in cubic meters per second or cubic feet per second. Discharge is determined by multiplying a stream's cross-sectional area by its velocity.

The largest river in North America, the Mississippi, discharges an average of 17,300 cubic meters

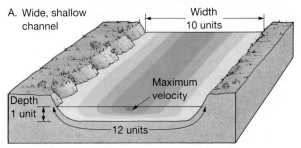

A. Wide, shallow channel

Width 10 units

Maximum velocity

Depth 1 unit

12 units

Cross-sectional area = 10 square units
Perimeter = 12 units

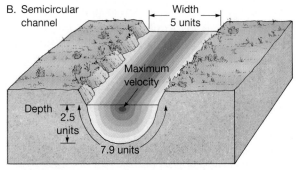

B. Semicircular channel

Width 5 units

Maximum velocity

Depth 2.5 units

7.9 units

Cross-sectional area = 10 square units
Perimeter = 7.9 units

Figure 3.4
Influence of channel shape on velocity. Although the cross-sectional area of these channels is the same, the semicircular channel has less water in contact with the channel, and hence less frictional drag. As a result, the water will flow more rapidly in this channel, all other factors being equal.

(611,000 cubic feet) per second. Although this is a huge quantity of water, it is nevertheless dwarfed by the mighty Amazon, the world's largest river. Draining a rainy region that is nearly three-quarters the size of the conterminous United States, the Amazon discharges 12 times more water than the Mississippi. In fact, the flow of the Amazon accounts for about 15 percent of all the fresh water discharged into the ocean by all of the world's rivers. Just one day's discharge would supply the water needs of New York City for about 9 years!

The discharges of rivers are far from constant. This is true because of such variables as rainfall and snowmelt. If discharge changes, the factors noted earlier must also change. When discharge increases, the width or depth of the channel must increase or the water must flow faster, or some combination of these factors must change. Indeed, measurements show that when the amount of water in a stream increases, the width, depth, and velocity all increase. In order to handle the additional water, the stream will increase the size of its channel by widening and deepening it. As we saw earlier, when the size of the channel increases, proportionally less of the water is in contact with the bed and banks of the channel. This means that friction, which acts to retard the flow, is relatively decreased. The less friction, the more swiftly the water will flow.

Changes Downstream

One useful way of studying a stream is to examine its *longitudinal profile.* Such a profile is simply a cross-sectional view of a stream from its source area (called the *head* or *headwaters*) to its *mouth*, the point downstream where the river empties into another water body. By examining Figure 3.5, you can see that the most obvious feature of a typical longitudinal profile is a constantly decreasing gradient

from the head to the mouth. Although many local irregularities may exist, the overall profile is a smooth concave-upward curve.

The longitudinal profile shows that the gradient decreases downstream. To see how other factors change in a downstream direction, observations and measurements must be made. When data are collected from successive gauging stations along a river, they show that discharge increases toward the mouth. This should come as no surprise since, as we move downstream, more and more tributaries contribute water to the main channel. Furthermore, in most humid regions, additional water is continually being added from the groundwater supply. Since this is the case, the width, depth, and velocity must change in response to the increased volume of water carried by the stream. Indeed, the downstream changes in these variables have been shown to vary in a manner similar to what occurs when discharge increases at one place; that is, width, depth, and velocity all increase systematically.

The observed increase in velocity that occurs downstream contradicts our impressions about wild, rushing mountain streams and wide, placid rivers. The mental picture that we may have of "old man river just rollin' along" is just not so. Although a mountain stream may have the appearance of a raging torrent, its average velocity is often less than that of the river near its mouth. The difference is primarily attributable to the greater efficiency of the larger channel in a downstream direction.

In the headwaters region where the gradient is steep, the water must flow in a relatively small and often boulder-strewn channel. The small channel and rough bed create great drag and inhibit movement by sending water in all directions with almost as much backward motion as forward motion. However, downstream, the material on the bed of the

Figure 3.5
A longitudinal profile is a cross section along the length of a stream. Note the concave-upward curve of the profile, with a steeper gradient upstream and a gentler gradient downstream.

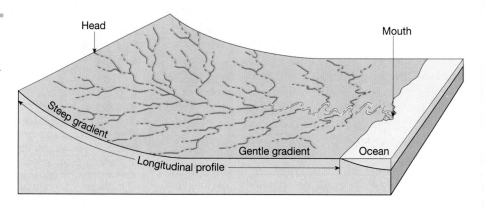

stream becomes much smaller, offering less resistance to flow, and the width and depth of the channel increase to accommodate the greater discharge. These factors, especially the wider and deeper channel, permit the water to flow more freely and hence more rapidly.

In summary, we have seen that there is an inverse relationship between gradient and discharge. Where gradient is high, discharge is small, and where discharge is great, gradient is small. Stated another way, a stream can maintain a higher velocity near its mouth even though it has a lower gradient than upstream because of the greater discharge, larger channel, and smoother bed.

BASE LEVEL AND GRADED STREAMS

An important control over streamflow is base level. **Base level** is the lowest point to which a stream can erode its channel. Two general types of base level exist. Sea level is considered the **ultimate base level**, since it represents the lowest level to which stream erosion could lower the land. **Temporary**, or **local**, **base levels** include lakes, resistant rock, and main streams which act as base levels for their tributaries. All have the capacity to limit a stream at a certain

level. For example, when a stream enters a lake, its velocity quickly approaches zero and its ability to erode decreases. Thus the lake prevents the stream from eroding below its level at any point upstream from the lake. However, since the outlet of the lake can cut downward and drain the lake, the lake is only a temporary hindrance to the stream's ability to downcut its channel. In a similar manner, the layer of resistant rock at the lip of the waterfall in Figure 3.6 acts as a temporary base level. Until the ledge of hard rock is eliminated, it will limit the amount of downcutting upstream.

Any change in base level will cause a corresponding readjustment of stream activities. When a dam is built along a stream course, the reservoir which forms behind it raises the base level of the stream (Figure 3.7). Upstream from the dam the stream gradient is reduced, lowering its velocity and, hence, its sediment-transporting ability. The stream, now unable to transport all of its load, will deposit material, thereby building up its channel. This process continues until the stream again has a gradient sufficient to carry its load. The profile of the new channel would be similar to the old, except that it would be somewhat higher.

If, on the other hand, the base level were lowered, either by an uplift of the land or by a drop in the base

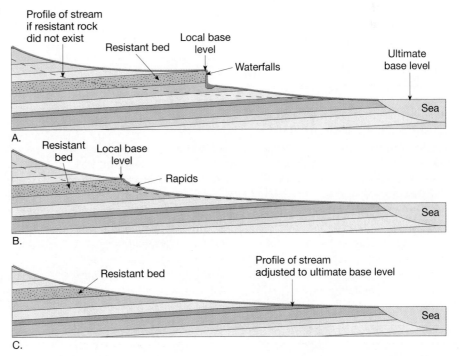

Figure 3.6
A resistant layer of rock can act as a local (temporary) base level. Because the durable layer is eroded more slowly, it limits the amount of downcutting upstream.

BOX 3.1

The Effect of Urbanization on Discharge

When rains occur, stream discharge increases. If the rains are sufficiently heavy, the ability of the channel to contain the discharge is exceeded, and water spills over the banks as a flood. Floods are natural events that should be expected. However, when cities are built, the magnitude and frequency of flooding increases.

The top portion of Figure 3.A is a hypothetical hydrograph that shows the time relationship between a rainstorm and the occurrence of flooding. Notice that the water level in the stream does not rise at the onset of precipitation because time is needed for water to move from the place where it fell to the stream. This time difference is called the *lag time.*

When an area changes from being predominantly rural to largely urban, streamflow is affected. The effect of urbanization on streamflow is illustrated by the bottom hydrograph in Figure 3.A. Notice that after urbanization the peak discharge during a flood is greater, and that the lag time between precipitation and flood peak is shorter than before urbanization. The explanation for this effect is relatively simple. The construction of streets, parking lots, and buildings covers over the ground that once soaked up water. Thus, less wa-

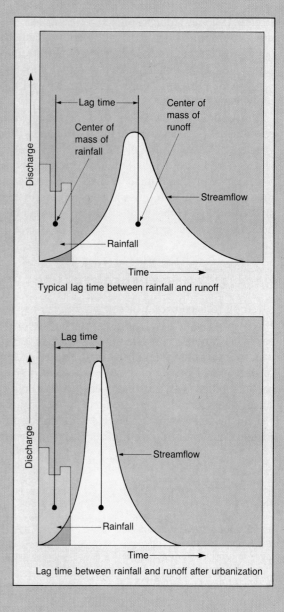

Figure 3.A
When an area changes from rural to urban, the lag time between rainfall and flood peak is shortened. The flood peak is also higher following urbanization. (After L. B. Leopold, U.S. Geological Survey Circular 559, 1968)

Typical lag time between rainfall and runoff

Lag time between rainfall and runoff after urbanization

ter infiltrates the ground, and the rate and amount of runoff increase. Further, since much less water soaks into the ground, the low-water (dry-season) flow in urban streams, which is maintained by the seepage of groundwater into the channel, is greatly reduced. As one might expect, the magnitude of these effects is a function of the percentage of land that is covered by impermeable surfaces.

Urbanization is just one example of human interference with streams. There are many other ways that land use inadvertently influences the flow of streams and the work they carry out. Moreover, there are also many ways by which people intentionally attempt to manipulate and control streams. Some of these are discussed in this chapter.

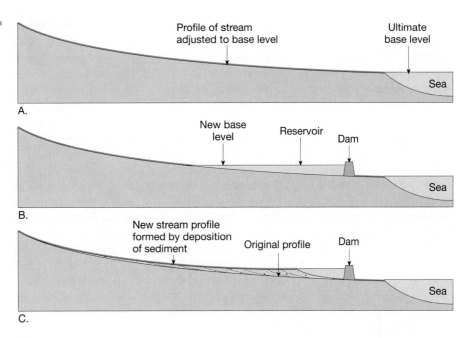

Figure 3.7
When a dam is built and a reservoir forms, the stream's base level is raised. This reduces the stream's velocity and leads to deposition and a reduction of the gradient upstream from the reservoir.

level, the stream would readjust. The stream, now above base level, would have excess energy and would downcut its channel to establish a balance with its new base level. Erosion would first progress near the mouth, then work upstream until the stream profile was adjusted along its full length.

The observation that streams adjust their profile for changes in base level led to the concept of a graded stream. A **graded stream** has the correct slope and other channel characteristics necessary to maintain just the velocity required to transport the material supplied to it. On the average, a graded system is not eroding or depositing material but is simply transporting it. Once a stream has reached this state of equilibrium, it becomes a self-regulating system in which a change in one characteristic causes an adjustment in the others to counteract the effect. Referring again to our example of a stream adjusting to a lowering of its base level, the stream would not be graded while cutting its new channel but would achieve this state after downcutting had ceased.

WORK OF STREAMS

The work of streams includes erosion, transportation, and deposition. These activities go on simultaneously in all stream channels, even though they are presented individually here.

Erosion

Although much of the material carried by streams has been brought in by underground water, overland flow, and mass wasting, streams do contribute to their load by eroding their own channels. If a channel is composed of bedrock, most of the erosion is accomplished by the abrasive action of water armed with sediment, a process analogous to sandblasting. Pebbles caught in eddies serve as cutting tools and bore circular holes called **potholes** into the channel floor. In channels consisting of unconsolidated material, considerable lifting can be accomplished by the impact of water alone.

Transportation

Once streams acquire their load of sediment, they transport it in three ways: (1) in solution (**dissolved load**); (2) in suspension (**suspended load**); and (3) along the bottom (**bed load**).

The dissolved load is brought to the stream by groundwater and to a lesser degree is acquired directly from soluble rock along the stream's course. The quantity of material carried in solution is highly variable and depends upon such factors as climate and the geologic setting. Usually the dissolved load is expressed as parts of dissolved material per million parts of water (parts per million, or ppm). Although some rivers may have a dissolved load of 1000 ppm or more, the average figure for the world's rivers is estimated to be between 115 and 120 ppm. Almost 4 billion metric tons of dissolved mineral matter are supplied to the oceans each year by streams.

Most streams (but not all) carry the largest part of their load in suspension. Indeed, the visible cloud of sediment suspended in the water is the most obvious portion of a stream's load. Usually only fine sand-, silt-, and clay-sized particles can be carried this way, but during floodstage larger particles are carried as well. Also during floodstage, the total quantity of material carried in suspension increases dramatically, as can be verified by persons whose homes have been sites for the deposition of this material.

A portion of a stream's load of solid material consists of sediment that is too large to be carried in suspension. These coarser particles move along the bottom of the stream and constitute the bed load. In terms of the erosional work accomplished by a down-cutting stream, the grinding action of the bed load is of great importance.

The particles composing the bed load move along the bottom by rolling, sliding, and saltation. Sediment moving by **saltation** appears to jump or skip along the stream bed. This occurs as particles are propelled upward by collisions or sucked upward by the current and then carried downstream a short distance until gravity pulls them back to the bed of the stream. Particles that are too large or heavy to move by saltation either roll or slide along the bottom, depending upon their shape.

Unlike the suspended and dissolved loads, which are constantly in motion, the bed load is in motion only intermittently, when the force of the water is sufficient to move the larger particles. Although the bed load may constitute up to 50 percent of the total load of a few streams, it usually does not exceed 10 percent of a stream's total load. For example, consider the distribution of the 750 million tons of material carried to the Gulf of Mexico by the Mississippi River each year. Of this total, it is estimated that approximately 500 million tons are carried in suspension, 200 million tons in solution, and the remaining 50 million tons as bed load. Estimates of a stream's bed load, however, should be viewed cautiously, because this fraction of the load is very difficult to measure accurately. Not only is the bed load more inaccessible than the suspended and dissolved loads, but it moves primarily during periods of flooding when the bottom of a stream channel is most difficult to study.

A stream's ability to carry its load is established using two criteria. First, the **competence** of a stream is a measure of the maximum size of particles it is capable of transporting. The stream's velocity determines its competence. If the velocity of a stream doubles, the impact force of water increases four times; if the velocity triples, the force increases nine times; and so forth. Hence, the large boulders that are often visible during the low-water stage and seem immovable can, in fact, be transported during floodstage because of the stream's increased velocity. Second, the maximum load a stream can carry is termed its **capacity.** The capacity of a stream is directly related to its discharge. The greater the amount of water flowing in the stream, the greater the stream's capacity for hauling sediment.

It should now be clear why the greatest erosion and transportation of sediment occur during floodstage. The increase in discharge not only results in a greater capacity but also in an increased velocity. With rising velocity the water becomes more turbulent, and larger and larger particles are set in motion. In the course of just a few days, or perhaps just

Figure 3.8
The suspended load is clearly visible because it gives this flooding river a brown "muddy" appearance. During floods, both capacity and competency increase. Therefore, the greatest erosion and sediment transport occur during these high-water periods. (Photo by James E. Patterson)

a few hours, a stream in floodstage can erode and transport more sediment than it does during months of normal flow (Figure 3.8).

Deposition

Whenever a stream's velocity decreases, its competence is reduced. As streamflow drops below the critical settling velocity of a certain particle size, sediment in that category begins to settle out. Thus, stream transport provides a mechanism by which solid particles of various sizes are separated. This process, called **sorting**, explains why particles of similar size are deposited together.

The well-sorted material typically deposited by a stream is called **alluvium**, a general term applicable to any stream-deposited sediment. Many different depositional features are composed of alluvium. Some of these features may be found within stream channels, some occur on the valley floor adjacent to the channel, and some exist at the mouth of the stream.

When a stream enters the relatively still waters of an ocean or lake, its forward motion is quickly lost, and the resulting deposits form a **delta.** The finer silts and clays will settle out some distance from the mouth into nearly horizontal layers called *bottomset beds* (Figure 3.9A). Prior to the accumulation of bottomset beds, *foreset beds* begin to form. These beds are composed of coarse sediment, which is dropped almost immediately upon entering a lake or ocean, forming sloping layers. The foreset beds are usually covered by thin, horizontal *topset beds* deposited during floodstage. As the delta grows outward, the gradient of the river is continually lowering, causing the stream to search for a shorter route to base level, a process illustrated in Figure 3.9B. This illustration also shows the main channel dividing into several

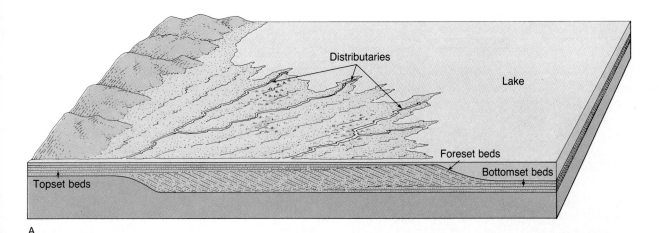

A.

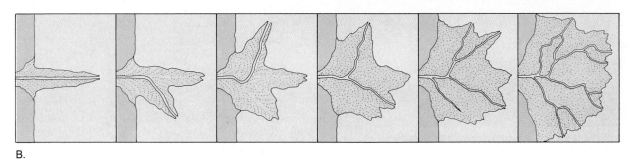

B.

Figure 3.9
A. *Structure of a simple delta that forms in the relatively quiet waters of a lake.*
B. *Growth of a simple delta. Once a stream has extended its channel, the reduced gradient causes it to find a shorter distance to its base level. (After Ward's Natural Science Establishment, Inc., Rochester, N.Y.)*

Figure 3.10
During the past 5000–6000 years, the Mississippi River has built a series of seven coalescing subdeltas. The numbers indicate the order in which the subdeltas were deposited. The present bird-foot delta (number 7) represents the activity of the past 500 years. Without ongoing human efforts, the present course will shift and follow the path of the Atchafalaya River (inset). (After C. R. Kolb and J. R. Van Lopik, Depositional Environments of the Mississippi River Deltaic Plain, *p. 22. Copyright © 1966 by the Houston Geological Society.)*

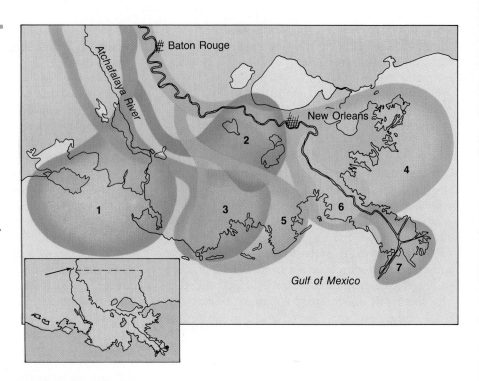

Figure 3.11
Satellite image of the Mississippi delta. For the past 500 years or so, the main flow of the river has been along its present course, extending southeast from New Orleans. During that span, the delta advanced into the Gulf of Mexico at a rate of about 10 kilometers (6 miles) per century. (Photo courtesy of NASA)

smaller ones called **distributaries.** Most deltas are characterized by these shifting channels that act in an opposite way to that of tributaries. Rather than carrying water into the main channel, distributaries carry water away from the main channel in varying paths to base level. After numerous shifts of the channel, the simple delta grows into the idealized triangular shape of the Greek letter delta (Δ), for which it was named. Note, however, that many deltas do not exhibit this idealized shape. Differences in the configurations of shorelines and variations in the nature and strength of wave activity result in many shapes.

Many large rivers have deltas extending over thousands of square kilometers. The delta of the Mississippi River is one example. It resulted from the accumulation of huge quantities of sediment derived from the vast region drained by the river and its tributaries. Today, New Orleans rests where there was ocean less than 5000 years ago. Figure 3.10 shows that portion of the Mississippi delta that has been built over the past 5000–6000 years. As shown, the delta is actually a series of seven coalescing subdeltas. Each formed when the river left its existing channel in favor of a shorter, more direct path to the Gulf of Mexico. The individual subdeltas interfinger and partially cover one another to produce a very complex structure. The present subdelta, called a *bird-foot* delta because of the configuration of its distributaries, has been built by the Mississippi in the last 500 years (Figure 3.11).

Alluvial fans are features similar to deltas which form on land (see Figure 4.27). When mountain streams reach a plain, their gradient is abruptly reduced and they immediately dump much of their load. Usually the coarse material is dropped near the base of the slope, while finer material is carried farther out on the plain.

Rivers that occupy valleys with broad, flat valley floors on occasion build a landform called a **natural levee** that parallels its channel. Natural levees are built by successive floods over many years. When a stream overflows its banks, its velocity immediately diminishes, leaving coarse sediment deposited in strips bordering the channel (Figure 3.12). As the water spreads out over the valley, a lesser amount of fine sediment is deposited over the valley floor. This uneven distribution of material produces the very gen-

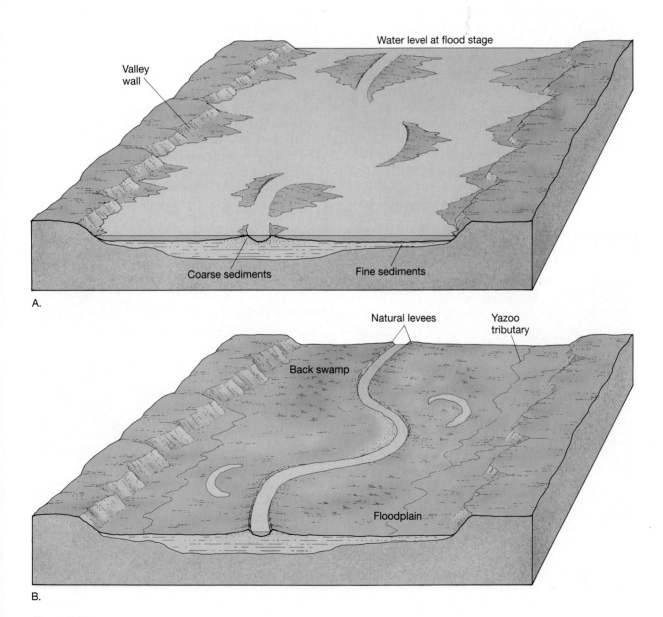

Figure 3.12
Formation of natural levees. After repeated flooding, streams may build very gently sloping levees.

tle slope of the natural levee. The natural levees of the lower Mississippi rise 6 meters (20 feet) above the valley floor. The area behind the levee is characteristically poorly drained for the obvious reason that water cannot flow up the levee and into the river. Marshes called **back swamps** result. A tributary stream that attempts to enter a river with natural levees often has to flow parallel to the main stream until it can breach the levee. Such streams are called **yazoo tributaries** after the Yazoo River, which parallels the Mississippi for over 300 kilometers.

Sometimes artificial levees are built along rivers as a means of flood control. These are usually easy to distinguish from natural levees because their slopes are much steeper. When a river is confined by levees during periods of high water, it deposits material in its channel as the discharge diminishes. This is sediment that otherwise would have been dropped on the floodplain. Thus, each time there is a high flow, deposits are left on the river bed and the bottom of the channel is built up. With the buildup of the bed, less water is required to overflow the original levee. As a result, the height of the levee must be raised to protect the floodplain. For this reason, many levees along the lower Mississippi River have had to be raised to cope with the increasing height of the wa-

ter in the channel. As you can see, artificial levees are not a permanent solution to the problem of flooding. If protection is to be maintained, the structures must be heightened periodically, a process that cannot go on indefinitely.

STREAM VALLEYS

Stream valleys can be divided into two general types. Narrow V-shaped valleys and wide valleys with flat floors exist as the ideal forms, with many gradations between. In some arid regions, where downcutting is rapid and weathering is slow, and in places where rock is particularly resistant, narrow valleys may not be V-shaped but rather may have nearly vertical walls. However, most valleys, even those that are narrow at the base, are much broader at the top than is the width of the channel at the bottom. This would not be the case if the only agent responsible for eroding valleys were the streams flowing through them. The sides of most valleys are shaped by a combination of weathering, sheet flow, and mass wasting.

A narrow V-shaped valley indicates that the primary work of the stream has been downcutting toward base level. The most prominent features of a

Figure 3.13
The American Falls at Niagara Falls. The river plunges over the falls and erodes the shale beneath the more resistant Lockport Dolomite. As a section of dolomite is undercut, it loses support and breaks off. (Photo by James E. Patterson)

narrow valley are **rapids** and **waterfalls.** Both occur where the stream profile drops rapidly, a situation usually caused by variations in the erodibility of the bedrock into which the stream channel is cutting. Resistant beds create rapids by acting as a temporary base level upstream while allowing downcutting to continue downstream. Once erosion has eliminated the resistant rock, the stream profile smooths out again. Waterfalls are places where the stream profile makes a vertical drop. One type of waterfall is exemplified by Niagara Falls (Figure 3.13). Here the falls are supported by a resistant bed of dolomite that is underlain by a less resistant shale. As the water plunges over the lip of the falls it erodes the less resistant shale, undermining a section of dolomite, which eventually breaks off. In this manner the waterfall retains its vertical cliff while slowly but continually retreating upstream. Since its formation Niagara Falls has retreated approximately 11 kilometers (7 miles) upstream.

Once a stream has cut its channel closer to base level it begins to reach a graded condition, and downward erosion becomes less dominant. At this point more of the stream's energy is directed from side to side. The reason for this change is not fully understood, but the reduced gradient probably is an important factor. Nevertheless it does occur, and the result is a widening of the valley as the river cuts away first at one bank and then at the other (Figure 3.14). In this manner the flat valley floor, or **floodplain**, is produced. It is appropriately named because the river is confined to its channel, except during floodstage, when it overflows its banks and inundates the floodplain.

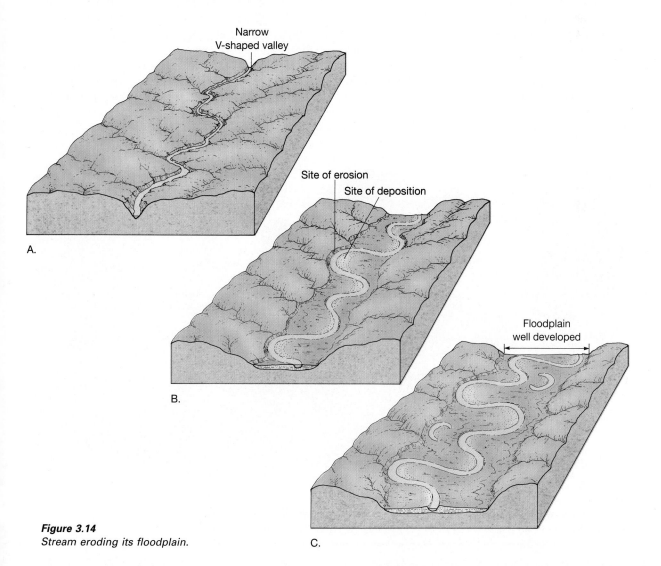

Figure 3.14
Stream eroding its floodplain.

When a river erodes laterally, creating a flood-plain as just described, it is called an *erosional flood-plain*. Floodplains can be depositional in nature as well. *Depositional floodplains* are produced by a major fluctuation in conditions, such as a change in base level. The floodplain in Yosemite Valley is one such feature, and was produced when a glacier gouged the former stream valley about 300 meters (985 feet) deeper than it had been. After the glacial ice melted, the stream readjusted to its former base level by refilling the valley with alluvium.

Streams that flow upon floodplains, whether erosional or depositional, move in sweeping bends called **meanders.** Meanders continually change position by eroding sideways and slightly downstream. The sideways movement occurs because the maximum velocity of the stream shifts toward the outside of the bend, causing erosion of the outer bank (Figure 3.15). At the same time, the reduced current at the inside of the meander results in the deposition of coarse sediment, especially sand. Thus, by eroding its outer bank and depositing material along its inner bank, a stream moves sideways without changing its channel size. Due to the slope of the channel, erosion is more effective on the downstream side of a mean-

der. Therefore, in addition to growing laterally, the bends also gradually migrate down the valley. Sometimes the downstream migration of a meander is slowed when it reaches a more resistant portion of the floodplain. This allows the next meander upstream to "catch up." Gradually the neck of land between the meanders is narrowed. When they get close enough, the river may erode through the narrow neck of land to the next loop (Figure 3.16). The new, shorter channel segment is called a **cutoff** and, because of its shape, the abandoned bend is called an **oxbow lake.**

One method of flood control is to straighten a channel by creating artificial cutoffs. The idea is that by shortening the stream, the gradient, and hence the velocity, are increased. By increasing velocity, the larger discharge associated with flooding can be dispersed more rapidly. Since the early 1930s, the Army Corps of Engineers has created many artificial cutoffs on the Mississippi for the purpose of increasing the channel efficiency and reducing the threat of flooding. In all, the river has been shortened more than 240 kilometers (150 miles). The program has been somewhat successful in reducing the height of the river in flood. However, since the river's tendency

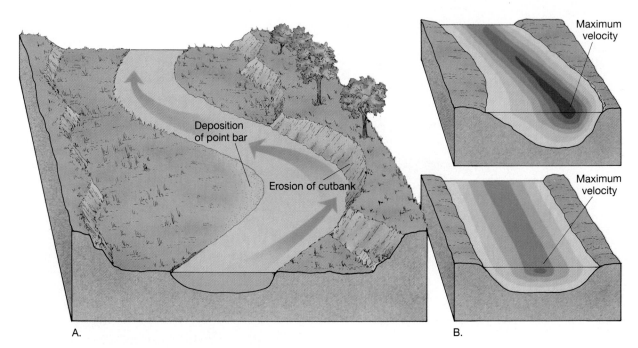

Figure 3.15
Lateral movement of meanders. By eroding its outer bank and depositing material on the inside of the bend, a stream is able to shift its channel.

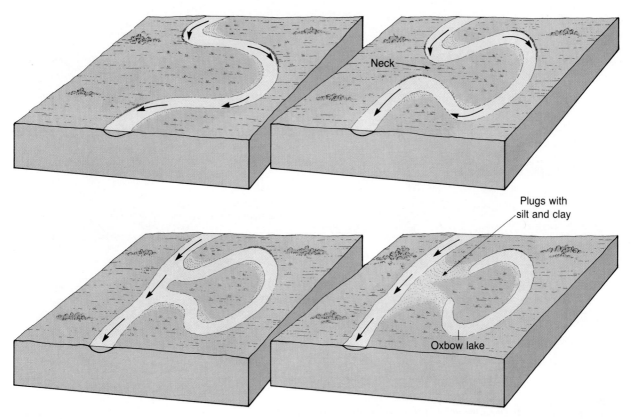

Figure 3.16
Formation of a cutoff and oxbow lake.

to meander still exists, preventing the river from returning to its previous condition has been difficult.

Artificial cutoffs increase the gradient of a stream. This results in an increase in the stream's velocity, which in turn may accelerate erosion of the bed and banks of the channel. A case in point is the Blackwater River in Missouri, whose meandering course was shortened in 1910. Among the many effects of this project was a dramatic increase in the width of the channel caused by the increased velocity of the stream. One bridge collapsed because of bank erosion in 1930. Over the next 17 years the same bridge was replaced on three more occasions, each time with a wider span.

DRAINAGE SYSTEMS AND PATTERNS

A stream is just one small component in a much larger system. Each system consists of a **drainage basin**, the land area that contributes water to the stream. The drainage basin of one stream is separated from another by an imaginary line called a **divide.** Divides range from a ridge separating two small gullies to *continental divides*, which split continents into enormous drainage basins. For example, the continental divide that runs somewhat north-south through the Rocky Mountains separates the drainage which flows west to the Pacific Ocean from that which flows to the Gulf of Mexico. Although divides separate the drainage of two streams, if they are tributaries of the same river, they are both a part of that larger drainage system.

All drainage systems are made up of an interconnected network of streams which together form particular patterns. The nature of a drainage pattern can vary greatly from one type of terrain to another, primarily in response to the kinds of rock on which the streams developed or the structural pattern of faults and folds.

Certainly the most commonly encountered drainage pattern is the **dendritic** pattern (Figure 3.17A).

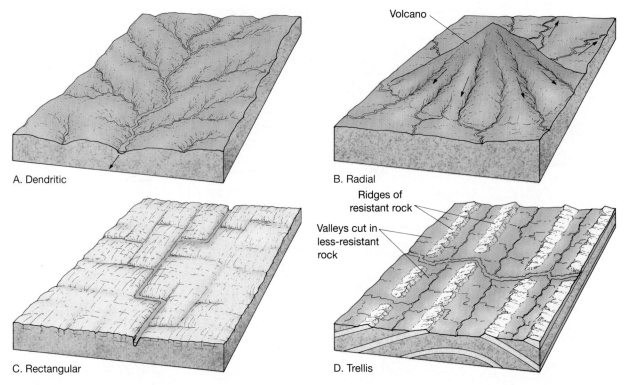

Figure 3.17
Drainage patterns. **A.** *Dendritic.* **B.** *Radial.* **C.** *Rectangular.* **D.** *Trellis.*

This pattern is characterized by an irregular branching of tributary streams that resembles the branching pattern of a deciduous tree. In fact, the word *dendritic* means "treelike." The dendritic pattern forms where the underlying material is relatively uniform. Since the surface material is essentially uniform in its resistance to erosion, it does not control the pattern of streamflow. Rather, the pattern is determined chiefly by the direction of slope of the land.

When streams diverge from a central area like spokes from the hub of a wheel, the pattern is said to be **radial** (Figure 3.17B). This pattern typically develops on isolated volcanic cones and domal uplifts.

Figure 3.17C illustrates a **rectangular** pattern, in which many right-angle bends can be seen. This pattern develops when the bedrock is crisscrossed by a series of joints and/or faults. Since these structures are eroded more easily than unbroken rock, their geometric pattern guides the directions of valleys.

Figure 3.17D illustrates a **trellis** drainage pattern, a rectangular pattern in which tributary streams are nearly parallel to one another and have the appearance of a garden trellis. This pattern forms in areas underlain by alternating bands of resistant and less

resistant rock and is particularly well displayed in the folded Appalachians, where both weak and strong strata outcrop in nearly parallel belts.

STAGES OF VALLEY DEVELOPMENT

Contrary to the popular belief of his day, James Hutton proposed that streams were responsible for cutting the valleys in which they flowed. Later geologic work conducted in stream valleys substantiated Hutton's proposal and further revealed that the development of stream valleys progresses in a somewhat orderly fashion. The evolution of a valley has been arbitrarily divided into three sequential stages: youth, maturity, and old age.

As long as the stream is downcutting to establish a graded condition with its base level, it is considered youthful. Rapids, an occasional waterfall, and a narrow V-shaped valley are all visible signs of the vigorous downcutting that is going on. Other features of youth include a steep gradient, little or no floodplain, and a relatively straight course without meanders (Figure 3.18A). The valley of the Yellowstone River

BOX 3.2
Placer Deposits

A placer deposit is one of the ways in which valuable minerals may be concentrated into economically significant accumulations. Deposits made by streams reflect the sorting action of running water. Usually like-sized grains are deposited together. However, sorting according to the specific gravity of particles also occurs. This latter type of sorting is responsible for the creation of *placers*, which are deposits formed when heavy minerals are mechanically concentrated by currents. Placers associated with streams are among the most common and best known, but the sorting action of waves can also create placers along the shore. Placer deposits usually involve minerals that are not only heavy, but also tough and chemically resistant enough to withstand destruction by weathering processes and transporting currents. Placers form because heavy minerals settle quickly from a current, whereas less dense particles remain suspended and are carried onward. Common sites of accumulation include point bars on the insides of meanders as well as cracks, depressions, and other irregularities on stream beds.

Many economically important placer deposits exist, with accumulations of gold the best known. Indeed, it was the placer deposits discovered in 1848 that led to the famous California gold rush. Years later, similar deposits created a gold rush to Alaska as well. Panning for gold by washing sand and gravel from a flat pan to concentrate the fine "dust" at the bottom was a common method used by early prospectors to recover the precious metal, and is a process similar to that which created the placer in the first place (Figure 3.B). In addition to gold, other heavy and durable minerals form placers. These include platinum, diamonds, and tin.

In some cases, if the source rock for a placer deposit can be located, it too may become an important ore body. By following placer deposits upstream, the original deposit can sometimes be located. This is how the gold-bearing veins of the Mother Lode in California's Sierra Nevada batholith were found, as well as the famous Kimberly diamond mines of South Africa. The placers were discovered first, their source at a later time.

Figure 3.B
Placer deposits form when heavy, durable minerals such as gold are eroded from their original deposits and accumulate in places where currents are relatively weak, such as the inside of a meander. Prospectors who pan for gold exploit such sites. (Photo courtesy of Intermountain Field Operations Center, Denver, Colorado)

---

pictured in the chapter opening photo provides an excellent example of the youthful stage of valley development.

When a stream reaches maturity, downward erosion diminishes and lateral erosion dominates. Thus the mature stream begins to create a floodplain and meander upon it (Figure 3.18B). During the mature stage cutoffs occur, producing oxbows, and a few streams may even produce natural levees (Figure 3.18C). In contrast to the gradient of a youthful stream, the gradient of a mature stream is much lower and the profile is much smoother because all rapids and waterfalls have been eliminated.

A stream enters old age after it has cut its floodplain several times wider than its *meander belt*, which is the width of the meander (Figure 3.18D). When this stage is reached, the stream is rarely near the valley walls; hence it ceases to significantly enlarge the floodplain. Thus the primary work of a river in an old-age valley is the reworking of unconsolidated floodplain deposits. Because this task is easier than cutting bedrock, a stream in an old-age valley shifts more rapidly than a stream in a mature valley. For example, some meanders of the lower Mississippi move 20 meters (66 feet) a year, and its large floodplain is dotted with oxbow lakes and old cut-

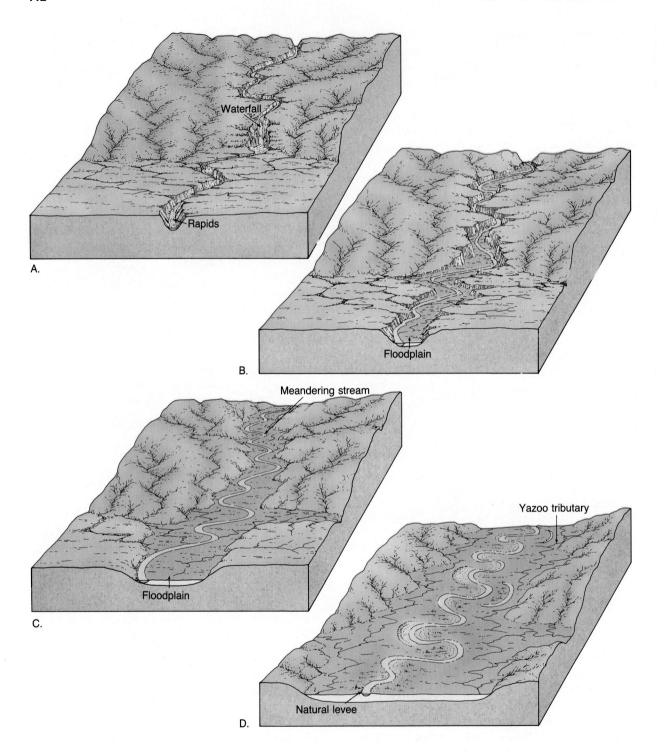

Figure 3.18
*Stages of valley development. **A.** Youth. The youthful stage is characterized by down-cutting and a V-shaped valley. **B.** and **C.** Maturity. Once a stream has sufficiently low-ered its gradient, it begins to erode laterally, producing a wide valley. **D.** Old age. After the valley has been cut several times wider than the width of the meander belt, it has entered old age. (After Ward's Natural Science Establishment, Inc., Rochester, N.Y.)*

offs. Natural levees are also common features of old-age valleys and, when present, are accompanied by back swamps and yazoo tributaries.

Thus far we have assumed that the base level of a stream remains constant as a river progresses from youth to old age. On many occasions, however, the land is uplifted or the base level is lowered. The effect of uplifting on a youthful stream is to increase its gradient and accelerate its rate of downcutting. However, uplifting of a mature stream would cause it to abandon lateral erosion and revert to downcutting. Rivers of this type are said to be **rejuvenated,** and the meanders are known as **entrenched meanders** (Figure 3.19). Mature streams may eventually readjust to uplift by cutting a new floodplain at a level below the old one. The remnants of the old higher floodplain are often present in the form of flat surfaces called *terraces.*

Two additional points concerning valley development should be made. First, the time required for a stream to reach any given stage depends on several factors, including the erosive ability of the stream, the nature of the material through which the stream must cut, and the stream's height above base level. Consequently, a stream which starts out very near base level and has to cut only through unconsolidated sediments may reach maturity in a matter of a few hundred years. On the other hand, the Colorado River, where it is actively cutting the Grand Canyon, has retained its youthful nature for an estimated 5 to 6 million years. Second, individual portions of a stream reach each stage at different times. Often the lower reaches of a stream attain old age while the headwaters are still youthful in character.

CYCLE OF LANDSCAPE EVOLUTION

While streams are cutting their valleys they simultaneously sculpture the land. To describe this unending process, we need a starting point. For this reason only, we will assume the existence of a relatively flat upland area in a humid region. Until a well-established drainage system develops, lakes and ponds will occupy any depressions that exist. As streams form and begin to downcut and erode headward, they will drain the lakes. During the youthful stage the landscape retains its relatively flat surface, interrupted only by narrow stream valleys. As downcutting continues, relief increases, and the flat, youthful landscape is transformed into one consisting of the hills and valleys that characterize the mature stage. Even-

A.

B.

Figure 3.19
Entrenched meanders. **A.** *This high-altitude image shows entrenched meanders of the Delores River in western Colorado. (Courtesy of USDA-ASCS)* **B.** *A close-up view of entrenched meanders of the San Juan River in southern Utah. (Photo by John S. Shelton) In both places, meandering streams began downcutting because of the uplift of the Colorado Plateau.*

tually some of the streams will approach base level, and downcutting will give way to lateral erosion. As the cycle nears the old-age stage, the effects of overland flow and mass wasting, coupled with the lateral erosion by streams, will reduce the land to a *peneplain* ("near plain"), an undulating plain near base level. Although no peneplains are known to exist today, there is evidence that they formed in the past and have since been uplifted. Once a peneplain has formed, uplifting starts the cycle again. Most often, uplifting interrupts the cycles before it reaches old age.

HYDROELECTRIC POWER

Coal, petroleum, and natural gas are the primary fuels of our modern industrial economy. Nearly 90 percent of the energy consumed in the United States today comes from these basic fossil fuels. Although major shortages of oil and gas may not occur in the near term, proven reserves are declining. Moreover, there is a growing concern about the environmental effects of fossil fuel emissions. Among the most serious atmospheric issues linked to the use of these basic fuels are urban air pollution, acid rain, and global (greenhouse) warming.* It seems clear that in the years to come, other sources will have to provide a greater share of our energy needs. **Hydroelectric power,** which is the focus of this section, is one of these alternate energy sources. There are many other possibilities, including geothermal energy, tidal power, solar energy, and energy derived from the wind. These alternatives will be the subjects of later discussions.**

People have used falling water as an energy source for centuries. Through most of history, the mechanical energy produced by water wheels was used to power mills and other machinery. Today the power generated by falling water is used to drive turbines and produce electricity. In the United States, hydroelectric power plants contribute about 4 percent of the country's needs. Most of this energy is produced at large dams, which allow for a controlled flow of water (Figure 3.20). The water impounded in a reservoir is a form of stored energy that can be released at any time to produce electricity.

Although water power is considered a renewable resource, the dams built to provide hydroelectricity have finite lifetimes. All rivers carry suspended sediment that is deposited as soon as the dam is built.

---

* All of these issues are discussed in some detail in Chapter 16.

** The section on geothermal energy is found later in this chapter. Others are as follows: tidal power, Chapter 11; solar energy, Chapter 12; and wind energy, Chapter 14.

Eventually the sediment will completely fill the reservoir. The length of time for this to happen depends on the quantity of suspended material transported by the river. It ranges between about 50 and 300 years. An example is provided by Egypt's huge Aswan High Dam, which was completed in the 1960s. It is now estimated that half the volume of the reservoir will be filled with sediment from the Nile River by the year 2025.

The availability of appropriate sites is an important limiting factor in the development of large-scale

Figure 3.20
Lake Powell is the reservoir that was created when Glen Canyon Dam was built across the Colorado River. As water in the reservoir is released, it drives turbines and produces electricity. Eventually the reservoir will be filled by sediment deposited by the Colorado River. (Photo by Michael Collier)

hydroelectric power plants. Good sites must be able to provide a significant height for the water to fall and a high rate of flow. Hydroelectric dams are found in many parts of the United States, with the greatest concentrations occurring in the Southeast and the Pacific Northwest. Although most of the best sites in the United States have already been developed, the total amount of power produced by hydroelectric sources may still increase in the years to come. However, the relative share provided by this source may decline because alternative energy sources may increase at a faster rate.

In recent years a different type of hydroelectric power production has come into use. Called a *pumped water storage system*, it is actually a type of energy management. During times when demand for electricity is low, unneeded power produced by non-hydroelectric sources is used to pump water from a lower reservoir to a storage area at a higher elevation. Then, when demand for electricity is great, the water stored in the higher reservoir is available to drive turbines and produce electricity to supplement the power supply.

WATER BENEATH THE SURFACE

Groundwater is one of our most important and widely available resources, yet people's perceptions of the subsurface environment from which it comes are often unclear and incorrect. The reason is that the groundwater environment is largely hidden from view except in caves and mines, and the impressions people gain from these subsurface openings are, to a large degree, misleading. Observations on the land surface give an impression that the earth is "solid." This view is not changed very much when we enter a cave and see water flowing in a channel that appears to have been cut into solid rock. Because of such observations many people believe that groundwater occurs only in underground "rivers." In reality, most of the subsurface environment is not "solid" at all but contains a huge volume of openings that exist as spaces between grains of soil and sediment and occur as narrow joints and fractures in bedrock. It is in these tiny openings that groundwater collects and moves.

The importance of groundwater can be demonstrated by comparing its volume with the quantity of water in other parts of the hydrosphere. Of all the world's water, only about six-tenths of one percent is found underground. Nevertheless, the amount of water stored in the rocks and sediments beneath the earth's surface is vast. When the oceans are excluded and only sources of freshwater are considered, the significance of groundwater becomes more apparent. Table 3.1 contains estimates of the distribution of freshwater in the hydrosphere. Clearly the largest volume occurs as glacial ice. Second in rank is groundwater, with slightly more than 14 percent of the total. However, when ice is excluded and just liquid water is considered, more than 94 percent is ground-

Table 3.1
Freshwater of the hydrosphere.

Parts of the Hydrosphere	Volume of Freshwater (km³)	Volume of Freshwater (mi³)	Share of Total Volume of Freshwater (percent)
Ice sheets and glaciers	24,000,000	5,800,000	84.945
Groundwater	4,000,000	960,000	14.158
Lakes and reservoirs	155,000	37,000	0.549
Soil moisture	83,000	20,000	0.294
Water vapor in the atmosphere	14,000	3,400	0.049
River water	1,200	300	0.004
Total	28,253,200	6,820,700	100.000

Source: U.S. Geological Survey Water Supply Paper 2220, 1987.

water. Without question, groundwater represents the largest reservoir of freshwater that is readily available to humans. Its value in terms of economics and human wellbeing is incalculable.

In many parts of the world, wells and springs provide the water needed not only for great numbers of people, but also for crops, livestock, and industry. In the United States, groundwater is the source of about 40 percent of the water used for all purposes exclusive of hydroelectric power generation and power plant cooling. It provides drinking water for more than 50 percent of the population, as well as 40 percent of the water for irrigation and 26 percent of industry's needs. In some areas, however, overuse of this basic resource has caused a number of problems, including streamflow depletion, land subsidence, and increased pumping costs. In addition, groundwater contamination due to human activities is a real and growing threat in many places.

Geologically, groundwater is important as an erosional agent. The dissolving action of groundwater is responsible for producing the surface depressions known as sinkholes as well as creating subterranean caverns (Figure 3.21). Another significant role is as an equalizer of streamflow. Much of the water that flows in rivers is not transmitted directly to the channel after falling as rain. Rather, a large percentage soaks in and then moves slowly underground to stream channels. Groundwater is thus a form of storage that sustains streams during periods when rain does not fall. When we see water flowing in a river during a dry period, it represents rain that fell at some earlier time and was stored underground.

DISTRIBUTION AND MOVEMENT OF GROUNDWATER

When rain falls, it may run off immediately, evaporate, be taken up by plants and transpired, or soak into the ground. This last path is the primary source of practically all subsurface water. The amount of water that takes each of these paths, however, varies considerably both in time and space. The steepness of the slope, the nature of the surface material, the intensity of the rainfall, and the type and amount of vegetation are all influential factors. Heavy rains that fall on steep slopes underlain by impervious materials will obviously result in a high percentage of the water running off. On the other hand, a gentle, steady rain that falls on more gradual slopes composed of materials more easily penetrated by water would result in a much larger percentage of the water soaking into the ground.

Figure 3.21
The cathedral-like beauty of the Big Room at Carlsbad Caverns National Park, New Mexico, is the handiwork of groundwater. (Photo courtesy of the National Park Service)

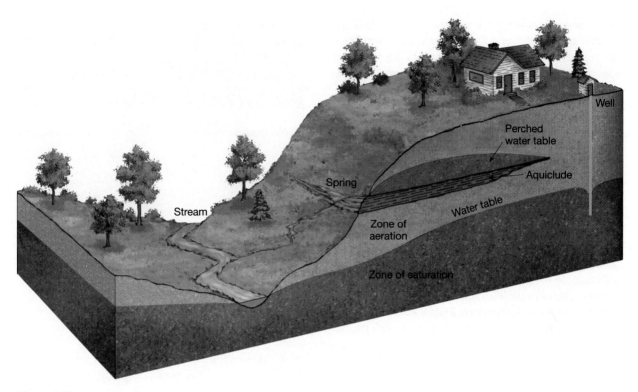

Figure 3.22
This diagram illustrates the relative positions of many features associated with subsurface water.

Some of the water that soaks in does not travel far, because it is held by molecular attraction as a surface film on soil particles. A portion of this moisture evaporates back into the atmosphere, while much of the remainder serves as a source of water for use by plants between rains. Water that is not held near the surface penetrates downward until it reaches a zone where all of the open spaces in sediment and rock are completely filled with water. The water in this **zone of saturation** is called **groundwater.** The upper limit of this zone is known as the **water table.** The area above the water table where the soil, sediment, and rock are not saturated is known as the **zone of aeration** (Figure 3.22). The open spaces here are filled mainly with air.

The water table is rarely level as we might expect a table to be. Instead, its shape is usually a subdued replica of the surface topography, reaching its highest elevations beneath hills and decreasing in height toward valleys (Figure 3.22). Where a swamp is encountered, the water table is right at the surface, whereas lakes and streams generally occupy areas where the land surface is below the water table. Although many factors contribute to the irregular surface of the water table, the most important cause is that groundwater moves very slowly. For this reason, the water tends to "pile up" beneath hills. If rainfall were to cease completely, these water table "hills" would slowly subside and gradually approach the level of the valleys. However, new supplies of rainwater are usually added often enough to prevent this. Nevertheless, in times of extended drought, the water table may drop enough to dry up otherwise productive wells.

Depending upon the nature of the subsurface material, the flow of groundwater and the amount of water that can be stored vary greatly. The quantity of groundwater that can be stored depends upon the **porosity** of the material, that is, the percentage of the total volume of rock or sediment that consists of pore spaces. Although these openings often consist of spaces between particles of sediment or sedimentary rock, such features as vesicles (voids left by gases escaping from lava), joints and faults, and cavities formed by the solution of soluble rocks such as limestone are also common.

Rock or sediment may be very porous and still not allow water to move through it. The **permeability** of a material, its ability to transmit a fluid through interconnected pore spaces, is also an important consideration. In rock or sediment, the smaller the pore spaces, the slower the groundwater moves. If the spaces between particles are too small, the films of water clinging to the grains will come in contact or overlap. As a result, the force of molecular attraction which binds the water to the particles extends across the openings and the water is held quite firmly in place. Clay exemplifies this circumstance. Although the ability of clay to store water is often high, its pore spaces are so small that water is unable to move. Impermeable layers composed of materials such as clay that hinder or prevent water movement are termed **aquicludes**. On the other hand, larger particles, such as sand or gravel, have larger pore spaces. Therefore, the water in the centers of the openings is not bound to the particles by molecular attraction and can move with relative ease. Such permeable rock strata or sediments that transmit groundwater freely are called **aquifers**.

SPRINGS

Springs have aroused the curiosity and wonder of people for thousands of years. The fact that springs were (and to some people still are) rather mysterious phenomena is not difficult to understand, for here is water flowing freely from the ground in all kinds of weather in seemingly inexhaustible supply but with no obvious source. As a result, some interesting but false explanations for the source of springs were proposed that, at least to some extent, have managed to live on to the present. One such erroneous explanation is that springs draw their water from the ocean. How the salt is removed and the water elevated to the great heights it reaches in mountain springs, in defiance of gravity, remain unanswered questions. Another incorrect explanation for springs, one supported by Aristotle, suggested that the water originated in cold subterranean caverns where water vapor condensed from air that had penetrated the earth. Today we know that the source of springs is water from the zone of saturation and that the ultimate source of this water is precipitation.

Whenever the water table intersects the ground surface, a natural flow of groundwater results, which we call a **spring**. There are many circumstances which create springs. Springs such as those in Figure 3.23 form when an aquiclude blocks the downward movement of groundwater and forces it to move lat-

Figure 3.23
Thousand Springs, Gooding County, Idaho. (Photo by James E. Patterson)

erally. Where the permeable bed (aquifer) outcrops in a valley, a spring or series of springs result. In this particular example, the aquifer consists of jointed and vesicular lava flows that form the well-known Thousand Springs along the Snake River in southern Idaho. Another situation that can produce a spring is illustrated in Figure 3.22. Here an aquiclude is situated above the main water table. As water percolates downward, a portion accumulates above the aquiclude to create a localized zone of saturation and a **perched water table**. Springs, however, are not confined to places where a perched water table creates a flow at the surface. Indeed, there are a wide variety of spring types because subsurface conditions vary greatly from place to place.

WELLS

The most common device used by people for removing groundwater is the **well**, an opening bored into the zone of saturation (Figure 3.22). Wells serve as reservoirs into which groundwater moves and from which it can be pumped to the surface. Digging for water dates back many centuries and continues to be an important method of obtaining water. By far the single greatest use of this water in the United States is for irrigation. More than 65 percent of the groundwater used each year is for this purpose. Industrial uses rank a distant second, followed by the amount used in cities and rural areas.

The level of the water table may fluctuate considerably during the course of a year, dropping during dry seasons and rising following periods of rain. Therefore, to ensure a continuous supply of water, a well should penetrate below the water table. Whenever water is withdrawn from a well, the water table in the vicinity of the well is lowered. The extent of this effect, which is termed **drawdown**, decreases with increasing distance from the well. The result is a depression in the water table, roughly conical in shape, known as a **cone of depression** (Figure 3.24). For most small domestic wells, the cone of depression is hardly appreciable. However, when wells are used for irrigation or for industrial purposes, the withdrawal of water can be great enough to create a very wide and steep cone of depression that may substantially lower the water table in an area and cause nearby shallow wells to become dry (Figure 3.24).

ARTESIAN WELLS

The term **artesian** is applied to any situation in which groundwater rises in a well above the level where it was initially encountered. For such a situation to occur, two conditions must exist (Figure 3.25): (1) water must be confined to an aquifer that is inclined so that one end is exposed at the surface, where it can receive water; and (2) impermeable layers, both above and below the aquifer, must be present to prevent the water from escaping. When such a layer is tapped, the pressure created by the weight of the water above will cause the water to rise. If there were no friction the water in the well would rise to the level of the water at the top of the aquifer. However, friction reduces the height of this pressure surface. The greater the distance from the recharge area (area where water enters the inclined aquifer), the greater the friction and the less the rise of water. In Figure 3.25, Well 1 is a *nonflowing artesian well*, because at this location the pressure surface is below ground level. When the pressure surface is above the ground and a well is drilled into the aquifer, a *flowing artesian well* is created (Well 2, Figure 3.25).

Artesian systems act as conduits, transmitting water from remote areas of recharge great distances to the points of discharge. In this manner, water which fell in central Wisconsin years ago is now taken from the ground and used by communities many kilometers away in Illinois. In South Dakota such a system has brought water from the Black Hills in the west, eastward across the state. On a different scale, city water systems may be considered as examples of artificial artesian systems. The water tower, into which water is pumped, may be considered the area of recharge, the pipes the confined aquifer, and the faucets in homes the flowing artesian wells.

ENVIRONMENTAL PROBLEMS ASSOCIATED WITH GROUNDWATER

As with many of our valuable natural resources, groundwater is being exploited at an increasing rate. In some areas, overuse threatens the groundwater supply. In other places, groundwater withdrawal has caused the ground and everything resting upon it to sink. Still other localities are concerned with the possible contamination of their groundwater supply.

Figure 3.24
A cone of depression in the water table often forms around a pumping well. If heavy pumping lowers the water table, some wells may be left dry.

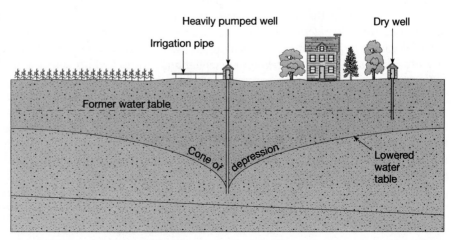

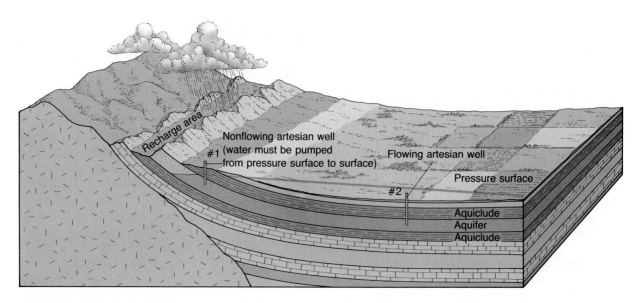

Figure 3.25
Artesian systems occur when an inclined aquifer is surrounded by impermeable beds.

Treating Groundwater as a Nonrenewable Resource

In some regions, groundwater has been and continues to be treated as a nonrenewable resource. Where this occurs, the amount of water available to recharge the aquifer is significantly less than the amount being withdrawn. The High Plains, a relatively dry region that extends from South Dakota to western Texas, provides an example. Here an extensive agricultural economy is largely dependent on irrigation. The widespread occurrence of permeable layers of sand and gravel has permitted the construction of high-yield wells almost anywhere in the region. As a result, there are an estimated 168,000 wells being used to irrigate more than 65,000 square kilometers (16 million acres) of land. In the southern part of this region, which includes the Texas panhandle, the natural recharge of the aquifer is very slow and the problem of declining groundwater levels is acute. In fact, in years of average or below-average precipitation, recharge is negligible because all or nearly all of the meager rainfall is returned to the atmosphere by evaporation and transpiration. Thus, it is only during especially wet years that significant recharge of the aquifer occurs, and this averages only about 5 millimeters per year. Therefore, where intense irrigation has been practiced for an extended period, depletion of groundwater is severe. Declines in groundwater levels at rates as high as 1 meter per year have

led to an overall drop in the water table of between 15 and 60 meters (50 and 200 feet) in some areas. Under these circumstances, it can be said that the groundwater is literally being "mined." Even if pumping were to cease immediately, it could take hundreds or thousands of years for the groundwater to be replenished.

Land Subsidence Caused by Groundwater Withdrawal

As we shall see later in this chapter, surface subsidence can result from natural processes related to groundwater. However, the ground may also sink when water is pumped from wells faster than natural recharge processes can replace it. This effect is particularly pronounced in areas underlain by thick layers of unconsolidated sediments. As water is withdrawn, water pressure drops and the weight of the overburden is transferred to the sediment. The greater pressure packs the sediment grains tightly together and the ground subsides.

Many areas may be used to illustrate land subsidence resulting from the excessive pumping of groundwater from relatively loose sediment. A classic example in the United States occurred in the San Joaquin Valley of California (Figure 3.26). This important agricultural region relies heavily on irrigation. Land subsidence due to groundwater withdrawal began in the valley in the mid-1920s and locally exceeded 8 meters (28 feet) by 1970. At this time, areas within the valley were subsiding at rates in excess of 0.3 meter (1 foot) per year. Then, because of the importation of surface water and a decrease in groundwater pumping, water levels in the aquifer recovered and subsidence ceased. However, during a drought in 1976–1977, heavy groundwater pumping led to a period of renewed subsidence. This time, water levels dropped at a much faster rate than during the previous period because of the reduced storage

capacity caused by earlier compaction of material in the aquifer. In all, more than 13,400 square kilometers (5200 square miles) of irrigable land, one-half the entire valley, were affected by subsidence. A U.S. Geological Survey report referred to the situation as representing ". . . one of the greatest single manmade alterations in the configuration of the Earth's surface. . . ."* Damages to structures, including highways, bridges, water lines, and wells, were extensive. Because subsidence changed the gradients of some streams, flooding was also a costly problem. Many other examples of land subsidence due to groundwater pumping occur in the United States and elsewhere in the world. One well-known example outside the United States occurred in Mexico City. This densely populated urban area rests on a former lake bed. Subsidence occurred as thousands of wells pumped water from the saturated sediments beneath the city. As water was withdrawn, parts of the city subsided by as much as 6 or 7 meters. Some buildings sank to such a point that access from the street is at what used to be the second-floor level!

Groundwater Contamination

The pollution of groundwater is a serious matter, particularly in areas where aquifers supply a large part of the water supply. A very common source of groundwater pollution is sewage which results from an ever-increasing number of septic tanks, as well as inadequate or broken sewer systems, and barnyard wastes.

If water contaminated with bacteria from sewage enters the groundwater system, it may become purified through natural processes. The harmful bacteria may be mechanically filtered out by the sediment through which the water percolates, destroyed by chemical oxidation, and/or assimilated by other organisms. In order for purification to occur, however, the aquifer must be of the correct composition. For example, extremely permeable aquifers such as highly fractured crystalline rock, coarse gravel, or cavernous limestone have such large openings that contaminated groundwater may travel long distances

Figure 3.26
The marks on this utility pole indicate the level of the surrounding land in preceding years. Between 1925 and 1975 this part of the San Joaquin Valley subsided almost 9 meters because of the withdrawal of groundwater and the resulting compaction of sediments. (Photo courtesy of U.S. Geological Survey)

* R. L. Ireland, J. F. Poland, and F. S. Riley, *Land Subsidence in the San Joaquin Valley, California, as of 1980,* U.S. Geological Survey Professional Paper 437-I, Washington, D.C.: U.S. Government Printing Office, 1984, I1.

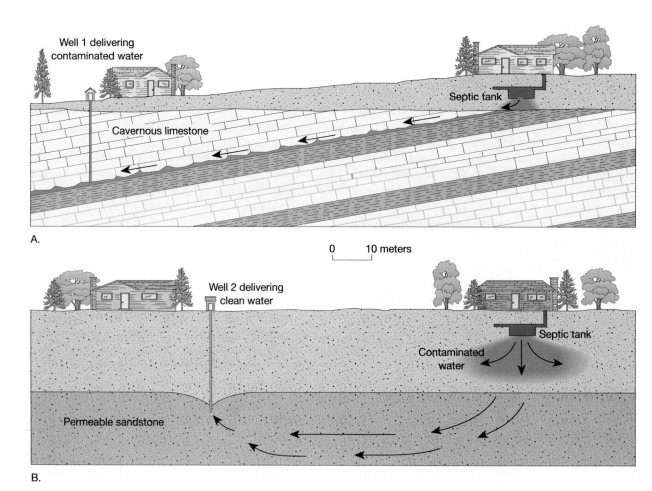

Figure 3.27
A. *Although the contaminated water has traveled more than 100 meters before reaching Well 1, the water moves too rapidly through the cavernous limestone to be purified.* **B.** *As the discharge from the septic tank percolates through the permeable sandstone, it is purified in a relatively short distance.*

without being cleansed. In this case, the water flows too rapidly and is not in contact with the surrounding material long enough for purification to occur. This is the problem at Well 1 in Figure 3.27A. On the other hand, when the aquifer is composed of sand or permeable sandstone, the water can sometimes be purified within distances as short as a few tens of meters. The openings between sand grains are large enough to permit water movement, yet the movement of the water is slow enough to allow ample time for its purification (Well 2, Figure 3.27B).

Since groundwater movement is usually slow, polluted water may go undetected for a considerable time. Most contamination is discovered only after drinking water has been affected. By this time, the volume of polluted water may be very large, and even if the source of contamination is removed immediately, the problem is not solved. Although the sources of groundwater contamination are numerous, the solutions are relatively few. Once the source of the problem has been identified and eliminated, the most common practice in dealing with contaminated aquifers is simply to abandon the water supply and allow the pollutants to be flushed away gradually. This is the least costly and easiest solution, but aquifers must remain unused for many years. To accelerate this process, polluted water is sometimes pumped out and treated. Following removal of the tainted water, the aquifer is allowed to recharge naturally or, in some cases, the treated water or other fresh water is pumped back in. This process, however, is costly and time consuming. It may also be somewhat risky because there is no way to be certain that all of the contamination has been removed. Clearly the most effective solution to groundwater contamination is prevention.

HOT SPRINGS AND GEYSERS

By definition, the water in **hot springs** is 6–9°C (10–15°F) warmer than the mean annual air temperature for the localities where they occur. In the United States alone, there are well over 1000 such springs.

Mineral explorations over the world have shown that temperatures in deep mines and oil wells usually rise with an increase in depth below the surface. Temperatures in such situations increase an average of about 2°C per 100 meters (1°F per 100 feet). Therefore, when groundwater circulates at great depths, it becomes heated, and if it rises to the surface, the water may emerge as a hot spring. The water of some hot springs in the United States, particularly in the East, is heated in this manner. The great majority (over 95 percent) of the hot springs (and geysers) in the United States are found in the West. The reason for such a distribution is that the source of heat for most hot springs is cooling igneous rock, and it is in the West that igneous activity has been most recent.

Geysers are intermittent hot springs or fountains in which columns of water are ejected with great force at various intervals, often rising 30–60 meters (100–200 feet) (Figure 3.28). After the jet of water ceases, a column of steam rushes out, usually with a thundering roar. Perhaps the most famous geyser in the world is Old Faithful in Yellowstone National Park, which erupts about once each hour. Geysers are also found in other parts of the world, including New Zealand and Iceland, where the term *geyser*, meaning "spouter" or "gusher," was coined.

Geysers occur where extensive underground chambers exist within hot igneous rocks. As relatively cool groundwater enters the chambers, it is heated by the surrounding rock. At the bottom of the chamber, the water is under great pressure because of the weight of the overlying water. Consequently, a temperature above 100°C (212°F) is required before it will boil. For example, at the bottom of a 300-meter (1000-foot) chamber, water must attain a temperature of nearly 230°C (450°F) before it will boil. The

Figure 3.28
A wintertime eruption of Old Faithful in Yellowstone National Park. One of the world's most famous geysers, it emits as much as 45,000 liters (12,000 gallons) of hot water and steam about once each hour. (Photo by Robert Winslow)

heating causes the water to expand, with the result that some flows out at the surface. This loss of water reduces the pressure on the remaining water in the chamber. The reduced pressure lowers the boiling point and a small portion of the water deep within the chamber quickly turns to steam and causes the geyser to erupt (Figure 3.29). Following the eruption, cool groundwater again seeps into the chamber and the cycle begins anew.

Groundwater from hot springs and geysers usually contains more material in solution than groundwater from other sources because hot water is a more effective dissolver than cold. When the water contains much dissolved silica, *geyserite* is deposited around the spring. *Travertine*, a form of calcite, is a characteristic deposit at hot springs in limestone regions. Some hot springs contain sulfur. In addition to making the water taste bad, sulfur-rich springs emit an unpleasant odor. Undoubtedly Rotten Egg Spring, Nevada, is such a situation.

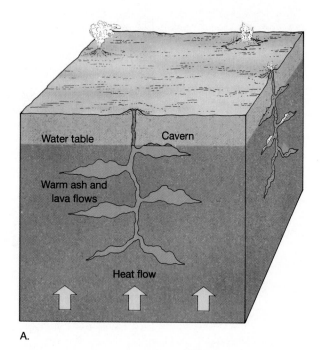

A.

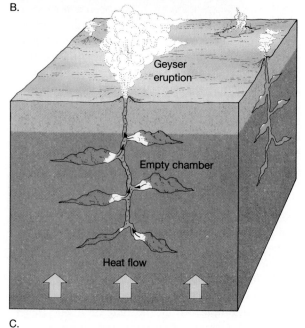

B.

C.

Figure 3.29
*Idealized diagrams of a geyser. A geyser can form if the heat is not distributed by convection. **A.** In this figure, the water near the bottom is heated to near its boiling point. The boiling point is higher there than at the surface because the weight of the water above increases the pressure. **B.** The water higher in the geyser system is also heated; therefore, it expands and flows out at the top, reducing the pressure on the water at the bottom. **C.** When the pressure is reduced at the bottom, boiling occurs. Some of the bottom water flashes into steam, and the expanding steam causes an eruption.*

GEOTHERMAL ENERGY

Geothermal energy is produced by tapping naturally occurring steam and hot water located beneath the surface in regions where subsurface temperatures are high due to relatively recent volcanic activity. Although the development of geothermal power plants has grown quite rapidly in recent years, the idea of using natural steam to generate electricity is not new. As early as 1904, natural steam vents at Larderello, Italy, were used to make power. By 1992, the U.S. Geological Survey reported that nearly 200 separate power units in 17 countries were operating, with a combined capacity of almost 4800 megawatts (million watts). In addition to the United States, countries that lead in the use of geothermal energy to produce electricity are the Philippines, Indonesia, Mexico, Italy, and New Zealand.

The first commercial geothermal power plant in the United States was built in 1960 at The Geysers, north of San Francisco. By 1986, development at this location had grown to almost 1800 megawatts, enough to satisfy the needs of San Francisco and Oakland. However, this peak soon passed and the production of electricity began to decline (see Box 3.3). In addition to The Geysers, geothermal development is occurring elsewhere in the western United States, including Nevada, Utah, and the Imperial Valley in southern California.

Geothermal energy is not used exclusively for generating electricity. In Iceland's capital, Reykjavik, steam and hot water are pumped into buildings throughout the city for space heating and to warm greenhouses, where fruits and vegetables are grown all year. In the United States, localities in several western states use hot water from geothermal sources for space heating.

The most favorable geologic factors for a geothermal reservoir of commercial value include:

1. A potent source of heat, such as a large magma chamber. The chamber should be deep enough to ensure adequate pressure and a slow rate of cooling, and yet not be so deep that the natural circulation of water is inhibited. Magma chambers of this type are most likely to occur in regions of recent volcanic activity;
2. Large and porous reservoirs with channels connected to the heat source, near which water can circulate and then be stored in the reservoir;
3. Capping rocks of low permeability that inhibit the flow of water and heat to the surface. A deep, well-insulated reservoir is likely to contain much more stored energy than an uninsulated, but otherwise similar, reservoir.

It should be emphasized that geothermal power is not an inexhaustible source of energy. When hot fluids are pumped from volcanically heated reservoirs, water cannot be replaced and then heated sufficiently to recharge the reservoir. Experience has shown that the output of steam and hot water from individual wells usually does not last for more than 10 to 15 years. Therefore, more wells must be drilled to maintain power production. Eventually, of course, the field is depleted.

As with other methods of power production, geothermal sources are not expected to provide a high percentage of the world's growing energy needs. Nevertheless, in regions where its potential can be developed, its use will no doubt continue to grow.

THE GEOLOGIC WORK OF GROUNDWATER

The primary erosional work carried out by groundwater is that of dissolving rock. Since soluble rocks, especially limestone, underlie millions of square kilometers of the earth's surface, it is here that groundwater carries on its rather unique and important role as an erosional agent. Although nearly insoluble in pure water, limestone is quite easily dissolved by water containing small quantities of carbonic acid. Most natural water contains this weak acid because rainwater readily dissolves carbon dioxide from the air and from decaying plants. Therefore, when groundwater comes in contact with limestone, the carbonic acid reacts with calcite in the rocks to form calcium bicarbonate, a soluble material that is then carried away in solution.

Caverns

Among the most spectacular results of groundwater's erosional handiwork is the creation of limestone **caverns**. Most are relatively small, yet some have spectacular dimensions. In the United States alone about 17,000 caves have been discovered. Although most are relatively small, some have spectacular dimensions. Carlsbad Caverns in southeastern New Mexico and Mammoth Cave in Kentucky are famous

In 1847, when the man who discovered The Geysers saw the plumes of steam rising from the ground, he thought he had come upon "the gates of hell" (Figure 3.C). A little more than a century later, developers viewed this area as an energy bonanza.

For twenty years (1960–1980), the pace of development of The Geysers was gradual. By 1981, there were 14 generating units dotting the area with a combined output of 943 megawatts. From this point onward, development accelerated, spurred by government incentives, rising oil prices, and the promise of cheap, smog-free energy. By 1985, the outlook for energy production at The Geysers was bright. The area contained 75 percent of the country's installed geothermal electrical-generating capacity, the largest complex of its kind in the world.

Just six years later, in 1991, the promising expansion of The Geysers was over. Although the geothermal field was supplying a full 6 percent of California's electrical power, it was not producing at expected levels. Installed generating capacity had grown to more than 2000 megawatts, but production was only 1500 megawatts. Moreover, steam pressure in the wells was falling rapidly. The problem was straightforward: The Geysers' geothermal field was running out of steam. By the mid-to-late 1990s, electrical output was projected to fall to half of the 1987 level.

When the facts were examined, it was clear that there had never been enough water in the crevices and fractures of the rocks to sustain the expansion of electrical-generating capacity. In most geothermal fields, drilling penetrates zones of superheated water, the volume of which can be measured. However, at The Geysers, wells tap only steam, and steam provides no measure of the volume of water giving rise to it. The reservoir was likened to a boiling teakettle that steams vigorously until it suddenly runs out of water. The amount of water present is not known until it is all gone. One writer observed, "The underlying problem at The Geysers—and a danger for geothermal development everywhere—is overdevelopment of a poorly understood resource. . . ."*

---

* Richard A. Kerr, "Geothermal Tragedy of the Commons," *Science* 253 (12 July 1991): 134. Much of the information in this box was based on this article.

Figure 3.C
The Geysers, a field of steaming fumaroles 115 kilometers north of San Francisco, California. The natural steam beneath these hills was first tapped to power electrical-generating plants in 1960. Today it appears that the field is running out of steam. (Photo courtesy of Pacific Gas and Electric)

examples. One chamber in Carlsbad Caverns has an area equivalent to fourteen football fields and enough height to accommodate the U.S. Capitol Building. At Mammoth Cave, the total length of interconnected caverns extends for more than 540 kilometers (340 miles).

Most caverns are created at or below the water table in the zone of saturation. Here acidic groundwater follows lines of weakness in the rock, such as joints and bedding planes. As time passes, the dissolving process slowly creates cavities and gradually enlarges them into caverns. Material that is dissolved by the groundwater is carried away and discharged into streams.

Certainly the features that arouse the greatest curiosity for most cavern visitors are the stone formations that often exhibit quite bizarre patterns and give some caverns a wonderland appearance. These features are created by the seemingly endless dripping of water over great spans of time. The calcite that is left behind produces the limestone we call travertine. These cave deposits, however, are also commonly called *dripstone*, an obvious reference to their mode of origin.

Although the formation of caverns takes place in the zone of saturation, the deposition of dripstone is not possible until the caverns are above the water table in the zone of aeration. This commonly occurs as nearby streams cut their valleys deeper, lowering the water table as the elevation of the river drops. As soon as the chamber is filled with air, the conditions are right for the decoration phase of cavern building to begin.

Of the various dripstone features found in caverns, perhaps the most familiar are **stalactites**. These iciclelike pendants hang from the ceiling of the cavern and form where water seeps through cracks above. When water reaches air in the cave, some of the dissolved carbon dioxide escapes from the drop and calcite begins to precipitate. Deposition occurs as a ring around the edge of the water drop. As drop after drop follows, each leaves an infinitesimal trace of calcite behind, and a hollow limestone tube is created. Water then moves through the tube, remains suspended momentarily at the end, contributes a tiny ring of calcite, and falls to the cavern floor. The stalactite just described is appropriately called a *soda straw* (Figure 3.30A). Often the hollow tube of the soda straw becomes plugged or its supply of water increases. In either case, the water is forced to flow, and hence deposit, along the outside of the tube. As

A.

B.

Figure 3.30
A. A "live" solitary soda straw stalactite. (Photo by Clifford Stroud, National Park Service). **B.** Stalagmites grow upward from the floor of a cavern. (Photo by E. J. Tarbuck)

deposition continues, the stalactite takes on the more common conical shape.

Formations that develop on the floor of a cavern and reach upward toward the ceiling are called **stalagmites** (Figure 3.30B). The water supplying the calcite for stalagmite growth falls from the ceiling and splatters over the surface. As a result, stalagmites do not have a central tube and are usually more massive in appearance and more rounded on their upper ends than stalactites.

Karst Topography

Many areas of the world have landscapes that, to a large extent, have been shaped by the dissolving power of groundwater. Such areas are said to exhibit **karst topography**. The term is derived from a plateau region located along the northeastern shore of the Adriatic Sea in the border area between Slovenia (formerly a part of Yugoslavia) and Italy where such topography is strikingly developed. The most common geologic setting for karst development is an area where limestone is present at the surface beneath a mantle of soil. In the United States, karst landscapes occur in many areas, including portions of Kentucky, Tennessee, Alabama, southern Indiana, and central and northern Florida. Generally, arid and semiarid areas do not develop karst topography. When solution features exist in such regions, they are likely to be remnants of a time when precipitation was more abundant.

Karst areas characteristically exhibit an irregular terrain punctuated with many depressions called **sinkholes** or, simply, **sinks**. In the limestone areas of Florida, Kentucky, and southern Indiana, there are literally tens of thousands of these depressions varying in depth from just a meter or two to a maximum of more than 50 meters (Figure 3.31).

Sinkholes commonly form in one of two ways. Some develop gradually over many years without any physical disturbance to the rock. In these situations, the limestone immediately below the soil is dissolved by downward-seeping rainwater that is freshly charged with carbon dioxide. These depressions are usually not deep and are characterized by relatively gentle slopes. By contrast, sink-holes can also form suddenly and without warning

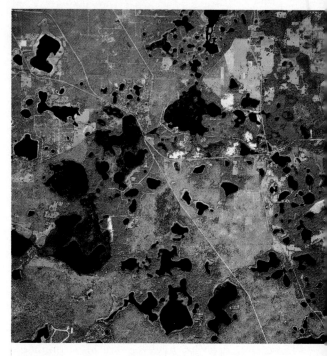

Figure 3.31
This high-altitude infrared image shows an area of karst topography in Florida. The numerous lakes occupy sinkholes. (Courtesy of USDA-ASCS)

when the roof of a cavern collapses under its own weight. Typically the depressions created in this manner are steep-sided and deep. When they form in populous areas, they may represent a serious geologic hazard. Such a case is described in Box 3.4.

In addition to a surface pockmarked by sinkholes, karst regions characteristically show a striking lack of surface drainage. Following a rainfall, runoff is funneled below ground by way of sinks, where it then flows through caverns until it finally reaches the water table. When streams do exist at the surface, their paths are usually short. The names of such streams often give a clue to their fate. In the Mammoth Cave area of Kentucky, for example, there is Sinking Creek, Little Sinking Creek, and Sinking Branch. Other sinkholes become plugged with clay and debris to create small lakes or ponds.

BOX 3.4
The Winter Park Sinkhole

The craterlike sinkhole in Figure 3.D began forming in Winter Park, Florida, on May 8, 1981, just one day before this photograph was taken. Newspaper accounts, such as the one that follows, were front page news and made this sinkhole one of the most publicized ever.

SINKHOLE NIBBLES AWAY AT FLORIDA CITY

WINTER PARK, Fla. (AP)—A giant sinkhole—already several hundred feet wide after swallowing a three-bedroom bungalow, half a swimming pool and six Porsches—nibbled away at a side street yesterday and threatened a main thoroughfare.

"It has slowed down, but it hasn't quit," said Winter Park Fire Capt. Gus LaGarde.

The crater, estimated at between 450 and 600 feet wide and 125 to 170 feet deep, grew by eight to 10 feet yesterday and was filling with water, authorities said.

It developed Friday night and opened rapidly Saturday, when it gulped the wood-frame cottage, cars and part of a foreign car lot and wrecked the city's $150,000 municipal swimming pool.

It was slowly eating its way west yesterday, leaving a group of businesses, their backs lost in Saturday's slide, hanging at the edge of the pit, LaGarde said.

The hole devoured most of a side street yesterday and was about 50 feet from one main thoroughfare and moving closer to several others, he said.

"We're still losing some of the perimeter," he said. "It doesn't appear to get any deeper . . . it continues to eat up the roadway, power poles, anything that gets in the way."

Residents and owners of nearby homes and businesses were warned on Saturday to leave until the sinkhole stopped growing. Some people rented trucks and began moving furniture and other property.*

Sinkhole formation is not uncommon in northern and central Florida. In fact, the Winter Park event was just one of three that occurred in the area over a two-week span. In each case the collapse at the surface was probably triggered by a lowering of the water table brought about by a severe drought. As the water table dropped, the roofs of the underground cavities lost support and fell into the voids below.

---

* Courtesy of Associated Press.

Figure 3.D
Aerial view of a large sinkhole that formed in Winter Park, Florida, in May 1981. (Photo courtesy of George Remaine, Orlando Sentinel Star)

Review Questions

1. Describe the movement of water through the hydrologic cycle. Once precipitation has fallen on land, what paths are available to it?

2. A stream starts out 2000 meters above sea level and travels 250 kilometers to the ocean. What is its average gradient in meters per kilometer?

3. Suppose that the stream mentioned in Question 2 developed extensive meanders so that its course was lengthened to 500 kilometers. Calculate its new gradient. How does meandering affect gradient?

4. Why is the Amazon considered the largest river on earth whereas the Nile is actually longer?

5. When the discharge of a stream increases, what happens to the stream's velocity?

6. Why does the downstream portion of a river have a gentle gradient when compared to the headwater region?

7. Define *base level*. Name the main river in your area. For what streams does it act as base level? What is the base level for the Mississippi River?

8. In what three ways does a stream transport its load?

9. If you collect a jar of water from a stream, what part of its load will settle to the bottom of the jar? What portion will remain in the water?

10. Differentiate between competency and capacity.

11. In what way is a delta similar to an alluvial fan? in what way are they different?

12. Why must the height of many artificial levees be increased periodically?

13. What is the purpose of artificial cutoffs?

14. Briefly describe the way in which minerals accumulate in placers (see Box 3.2).

15. What is a divide?

16. Each of the following statements refer to a particular drainage pattern. Identify the pattern.
    (a) Streams diverging from a central high area such as a dome
    (b) Branching, "treelike" pattern
    (c) A pattern that develops when bedrock is crisscrossed by joints and faults

17. Why is it possible for a youthful valley to be older (in years) than a mature valley?

18. Do mature and old-age valleys make good political boundaries? Explain.

19. Explain why dams built to provide hydroelectricity do not last indefinitely.

20. What percentage of freshwater is groundwater (see Table 3.1)? If glacial ice is excluded and only liquid freshwater is considered, about what percentage is groundwater?

21. Geologically, groundwater is important as an erosional agent. Name another significant geological role of groundwater.

22. Define groundwater and relate it to the water table.

23. How do porosity and permeability differ?

24. Under what circumstances can a material have a high porosity but not be a good aquifer?

25. What problem is associated with the pumping of groundwater for irrigation in the southern part of the High Plains?

26. What is meant by the term *artesian*? Under what circumstances do artesian wells form?

27. Briefly explain what happened in the San Joaquin Valley of California as the result of excessive groundwater withdrawal.

28. Which would be most effective in purifying polluted groundwater: an aquifer composed mainly of coarse gravel, sand, or cavernous limestone?

29. What is the source of heat for most hot springs and geysers? How is this reflected in the distribution of these features?

30. Is geothermal power considered an inexhaustible energy source? Explain.

31. List two conditions required for the development of karst topography.

32. Differentiate between stalactites and stalagmites. How do these features form?

Key Terms

alluvial fan (p. 105)

alluvium (p. 103)

aquiclude (p. 118)

aquifer (p. 118)

artesian well (p. 119)

back swamp (p. 106)

base level (p. 99)

bed load (p. 101)

capacity (p. 102)

cavern (p. 125)

competence (p. 102)

cone of depression (p. 119)

cutoff (p. 108)

delta (p. 103)

dendritic pattern (p. 109)

discharge (p. 97)

dissolved load (p. 101)

distributary (p. 104)

divide (p. 109)

drainage basin (p. 109)

drawdown (p. 119)

entrenched meander (p. 113)

floodplain (p. 107)

geothermal energy (p. 125)

geyser (p. 123)

graded stream (p. 101)

gradient (p. 96)

groundwater (p. 117)

hot spring (p. 123)

hydroelectric power (p. 114)

hydrologic cycle (p. 94)

infiltration (p. 94)

karst topography (p. 127)

meander (p. 108)

natural levee (p. 105)

oxbow lake (p. 108)

perched water table (p. 118)

permeability (p. 117)

porosity (p. 117)

pothole (p. 101)

radial pattern (p. 110)

rapids (p. 107)

rectangular pattern (p. 110)

rejuvenation (p. 113)

runoff (p. 94)

saltation (p. 102)

sinkhole (sink) (p. 128)

sorting (p. 103)

spring (p. 118)

stalactite (p. 127)

stalagmite (p. 128)

suspended load (p. 101)

temporary (local) base level (p. 99)

transpiration (p. 94)

trellis pattern (p. 110)

ultimate base level (p. 99)

waterfall (p. 107)

water table (p. 117)

well (p. 118)

yazoo tributary (p. 106)

zone of aeration (p. 117)

zone of saturation (p. 117)

Glaciers, Deserts, and Wind

Opposite page: *Alpine glaciers in Alaska Range, Denali National Park, Alaska. (Photo by Michael Collier)*

Left: *Picturesque landscape produced by glacial erosion. (Photo by James E. Patterson)*

Above: *Knife-like ridge produced by glacial erosion,
Mont Blanc. (Photo by James E. Patterson)*

Today glaciers cover nearly 10 percent of the earth's land surface; however, in the recent geologic past ice sheets were three times more extensive, covering vast areas with ice thousands of meters thick. Many present-day landscapes still bear the mark of these glaciers. The first part of this chapter examines glaciers and the erosional and depositional features they create. The second part is devoted to dry lands and the geologic work of wind. Since desert and near-desert conditions prevail over an area as large as that affected by the massive glaciers of the Ice Age, the nature of such landscapes is indeed worth investigating.

Many present-day landscapes were modified by the widespread glaciers of the most recent ice age and still strongly reflect the handiwork of ice. The basic character of such diverse places as the Alps, Cape Cod, and Yosemite Valley was fashioned by now vanished masses of glacial ice. Moreover, Long Island, the Great Lakes, and the fiords of Norway and Alaska all owe their existence to glaciers. Glaciers, of course, are not just a phenomenon of the geologic past. As we shall see, they are still sculpturing and depositing in many regions today.

TYPES OF GLACIERS

A **glacier** is a thick ice mass that originates on land from the accumulation, compaction, and recrystallization of snow. Because glaciers are agents of erosion, they must also flow. Indeed, like running water, groundwater, wind, and waves, glaciers are dynamic forces that are capable of accumulating, transporting, and depositing sediment. Although glaciers are found in many parts of the world today, most are located in remote areas.

Literally thousands of relatively small glaciers exist in lofty mountain areas, where they usually follow valleys originally occupied by streams. Unlike the rivers that previously flowed in these valleys, the glaciers advance slowly, perhaps only a few centimeters per day. Because of their setting, these moving ice masses are termed **valley glaciers** or **alpine glaciers** (Figure 4.1). Each is a stream of ice, bounded by precipitous rock walls, that flows downvalley from an accumulation center near its head. Like rivers, valley glaciers can be long or short, wide or narrow, single or with branching tributaries. Generally the widths of alpine glaciers are small compared to their lengths. Some extend for just a fraction of a kilometer, whereas others go on for many tens of kilometers. The west branch of the Hubbard Glacier, for example, runs through 112 kilometers of mountainous terrain in Alaska and the Yukon Territory.

In contrast to valley glaciers, **ice sheets** exist on a much larger scale. These enormous masses flow out in all directions from one or more centers and completely obscure all but the highest areas of underlying terrain. Even sharp variations in the topography beneath the glacier usually appear as relatively subdued undulations on the surface of the ice. Such topographic differences, however, do affect the behavior of the ice sheets, especially near their margins, by

Figure 4.1
Muldrow and Traleika Glaciers are active valley glaciers in Denali National Park, Alaska. (Photo by Michael Collier).

guiding flow in certain directions and creating zones of faster and slower movement. Although many ice sheets have existed in the past, just two achieve this status at present. In the northern hemisphere, Greenland is covered by an imposing ice sheet that occupies 1.7 million square kilometers, or about 80 percent of this large island. Averaging nearly 1500 meters thick, in places the ice extends 3000 meters above the island's bedrock floor. In the south polar realm, the huge Antarctic Ice Sheet attains a maximum thickness of nearly 4300 meters and covers an area of more than 13.9 million square kilometers. Because of the proportions of these huge features, the term *ice sheet* is often preceded by the word *continental*. Indeed the combined areas of present-day continental ice sheets represent almost 10 percent of the earth's land area.

In addition to valley glaciers and ice sheets, other types of glaciers are also identified. Covering some uplands and plateaus are masses of glacial ice called **ice caps**. They resemble ice sheets but are much smaller than the continental-scale features. Ice caps occur in many places, including Iceland and several of the large islands in the Arctic Ocean. Another type, known as **piedmont glaciers**, occupy broad lowlands at the bases of steep mountains and form when one or more valley glaciers emerge from the confining walls of mountain valleys. Here the advancing ice spreads out to form a broad sheet. The size of indi-

vidual piedmont glaciers varies greatly. Among the largest is the broad Malaspina Glacier along the coast of southern Alaska. It covers more than 5000 square kilometers of the flat coastal plain at the foot of the lofty St. Elias range.

GLACIAL MOVEMENT

The movement of glacial ice is generally referred to as *flow*. The fact that glacial movement is described in this way would seem to constitute a paradox—ice is solid, yet it is capable of flow. The way ice flows is complex but is believed to be of two basic types. One mechanism involves internal movement within the ice. Ice behaves as a brittle solid until the pressure or load upon it is equivalent to the weight of about 50 to 60 meters (165 to 200 feet) of ice. Once that load is surpassed, the ice will behave as a plastic material and flow continuously. A second and often equally important mechanism of glacial movement consists of the whole ice mass slipping along the ground. With the exception of some glaciers in polar regions where the ice is probably frozen to the solid bedrock floor, the lowest portions of most glaciers are thought to move by this sliding process.

Figure 4.2 illustrates the effects of these two basic types of glacial motion. This vertical profile through a glacier also shows that all the ice does not flow forward at the same rate. Just as in streams, frictional drag with the bedrock floor results in the lower portions of the glacier moving more slowly.

The uppermost portion of a glacier is often quite appropriately referred to as the *zone of fracture.* Since there is not enough overlying ice to cause plastic flow, this upper part of the glacier consists of brittle ice. Consequently, the ice in this zone is carried along piggyback style by the ice below. When the glacier moves over irregular terrain, the zone of fracture is subjected to tension, with cracks called **crevasses** resulting (Figure 4.3). These gaping cracks, which often make travel across glaciers dangerous, may extend to depths of 50 meters (165 feet). Beyond this depth, plastic flow seals them off.

Unlike streamflow, the movement of glaciers is not readily apparent to the casual observer. If we could watch a glacier in a mountain valley move, we would see that, like the water in a river, all of the ice in the valley does not move downstream at an equal rate. Just as friction with the bedrock floor slows the movement of the ice at the bottom of the glacier, the drag created by the valley walls leads to the flow being greatest in the center of the glacier.

More than 100 years ago, the first measurements of glacial movement were made. In this experiment, stakes were carefully placed in a straight line across the top of a valley glacier. Periodically the positions of the stakes were noted, revealing the type of movement just described (see Figure I.B on page 13).

How rapidly does glacial ice move? Average velocities vary considerably from one glacier to another. Some move so slowly that trees and other vegetation may become well established in the debris that has accumulated on the glacier's surface, whereas others move at rates of up to several meters per day. The advance of some glaciers is characterized by periods of extremely rapid movement followed by periods during which movement is practically nonexistent. For example, Hassanabad Glacier in the Karakoram, a mountain range in Kashmir and northwestern India, advanced 10 kilometers in less than 3 months—a rate of almost 130 meters per day. The precise cause or causes of these sporadic, short-lived advances is not well understood. One proposal is that the base of the glacier may have been frozen to the bedrock and then sudden melting released the ice. Another hypothesis suggests that a block of stagnant ice at the terminus of a glacier in an alpine valley may act as a dam until the buildup of pressure by the flowing ice behind it forces it to give way.

Snow is the raw material from which glacial ice originates; therefore, glaciers form in areas where more snow falls in winter than melts during the

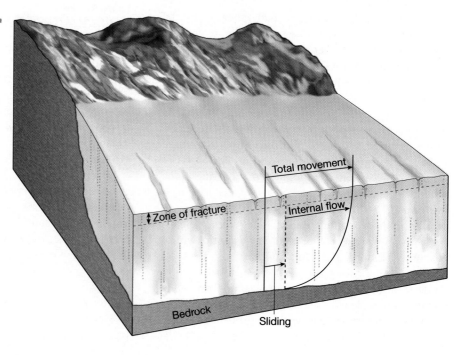

Figure 4.2
Glacial movement is divided into two components. Below approximately 50 meters, ice behaves plastically and flows. In addition, the entire ice mass may slide along the ground. The ice in the zone of fracture is carried along "piggyback" style. Notice that the rate of movement is slowest at the base of the glacier where frictional drag is greatest.

BOX 4.1

Glaciers and the Hydrologic Cycle

Earlier we learned that the earth's water is in constant motion. Time and time again the same water is evaporated from the oceans into the atmosphere, precipitated upon the land, and carried by rivers and underground flow back to the sea. However, when precipitation falls at high elevations or high latitudes, the water may not immediately make its way toward the sea. Instead, it may become part of a glacier. Although the ice will eventually melt and continue its path to the sea, it can be stored as glacial ice for many tens, hundreds, or even thousands of years.

How much water is stored as glacial ice? Estimates by the U.S. Geological Survey indicate that only slightly more than two percent of the world's water is accounted for by glaciers. But this small figure may be misleading when the actual amounts of water are considered. The total volume of all valley glaciers is about 210,000 cubic kilometers, comparable to the combined volume of the world's largest saline and freshwater lakes. Furthermore, 80 percent of the world's ice and nearly two-thirds of the earth's fresh water are represented by Antarctica's ice sheet, which covers an area almost one and one-half times that of the United States. If this ice melted,

sea level would rise an estimated 60 to 70 meters, and the ocean would inundate many densely populated coastal areas (Figure 4.A). The hydrologic importance of the continent and its ice can be illustrated in another way. If Antarctica's ice sheet were melted at a suitable rate it could feed (1) the Mississippi River for more than 50,000 years, (2) all the rivers in the United States for about 17,000 years, (3) the Amazon River for approximately 5000 years, or (4) all the rivers of the world for about 750 years.

As the foregoing discussion illustrates, the quantity of ice on earth today is truly immense. However, present glaciers occupy only slightly more than one-third the area they did in the very recent geologic past.

Figure 4.A
This map of a portion of North America shows the present-day coastline compared to the coastline that existed during the last ice-age maximum (18,000 years ago) and the coastline that would exist if present ice sheets in Greenland and Antarctica melted. (After R. H. Dott, Jr., and R. L. Battan, Evolution of the Earth, New York: McGraw-Hill, 1971. Reprinted by permission of the publisher)

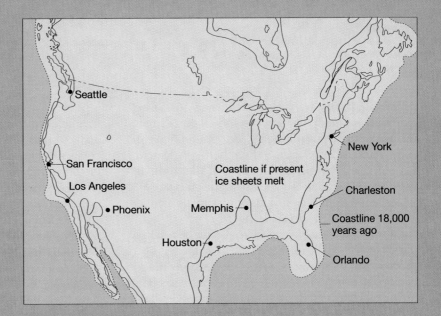

A.

B.

Figure 4.3
A. Crevasses are clearly visible in this aerial view of a valley glacier in Denali National Park, Alaska. (Photo by Michael Collier) B. Close-up view of a crevasse in Antarctica. (Photo by Matt Duvall)

summer. Glaciers are constantly gaining and losing ice. Snow accumulation and ice formation occur in the **zone of accumulation**. Here the addition of snow thickens the glacier and promotes movement. Beyond this area of ice formation is the **zone of wastage**. Here there is a net loss to the glacier as all of the snow from the previous winter melts as does some of the glacial ice (Figure 4.4).

In addition to melting, glaciers also waste as large pieces of ice break off the front of a glacier in a process called *calving*. Calving creates icebergs in places where glaciers reach the sea. Because icebergs are just slightly less dense than seawater, they float very low in the water, with more than 80 percent of their mass submerged. The margins of the Greenland Ice Sheet produce thousands of icebergs each year. Many drift southward and find their way into the North Atlantic, where they can pose a hazard to navigation.

Whether the margin of a glacier is advancing, retreating, or remaining stationary depends upon the budget of the glacier. That is, it depends upon the balance or lack of balance between accumulation on the one hand and wastage (also termed **ablation**) on the other. If ice accumulation exceeds ablation, the glacial front advances until the two factors balance. At this point, the terminus of the glacier is stationary. At a later time when ablation exceeds accumulation, the ice front will retreat until a balance is again reached. Whether the margins of a glacier are advancing, retreating, or stationary, the ice within the glacier continues to flow forward. In the case of a receding glacier, the ice simply does not flow forward rapidly enough to offset ablation. This point is illustrated rather well in Figure I.B on page 13. While the line of stakes within Rhone Glacier continued to move downstream, the terminus of the glacier slowly retreated upstream.

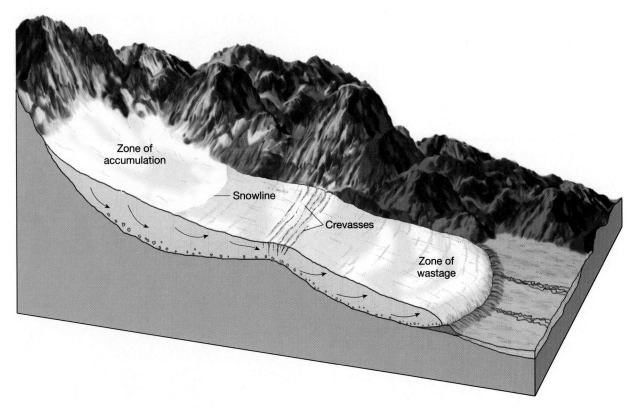

Figure 4.4
The snowline separates the zone of accumulation and the zone of wastage. Above the snowline, more snow falls each winter than melts each summer. Below the snowline, the snow from the previous winter completely melts as does some of the underlying ice. Whether the margin of a glacier advances, retreats, or remains stationary depends upon the balance or lack of balance between accumulation and wastage.

GLACIAL EROSION

Glaciers are capable of carrying on great amounts of erosional work. For anyone who has observed the terminus of an alpine glacier, the evidence of its erosive force is plain. The observer can witness firsthand the melting ice unlocking rock material of all sizes. All signs lead to the conclusion that the ice has scraped, scoured, and torn rock debris from the floor and walls of the valley and carried it downslope. Indeed, as a transporter of sediment, ice has no equal, because once the debris is acquired by the ice, it will not settle out as will the load carried by a stream or by the wind. Consequently, glaciers can carry huge blocks that no other erosional agent could possibly budge. Although glaciers are of limited importance as an erosional agent today, many landscapes that were modified by the widespread glaciers of the recent ice age still reflect to a high degree the work of ice.

Glaciers erode land primarily in two ways. First, as a glacier flows over a fractured bedrock surface, it loosens and lifts blocks of rock, incorporates them into the ice, and carries them off. This process, known as **plucking**, occurs when meltwater penetrates the cracks and joints along the rock floor of the glacier and refreezes. As the water expands, it exerts tremendous leverage that pries the rock loose. In this manner sediment of all sizes becomes part of the glacier's load.

The second major erosional process is **abrasion**. As the ice with its load of rock fragments moves along, it acts as a giant rasp or file and grinds the surface below as well as the rocks within the ice. The pulverized rock produced by the glacial "grist mill" is appropriately called **rock flour**. So much rock flour may be produced that meltwater streams leaving a glacier often have the grayish appearance of skimmed milk—visible evidence of the grinding

Figure 4.5
Results of glacial abrasion. Scratches and grooves in limestone, St. Johns Bay, Newfoundland. (Photo by Peter Kresan)

power of the ice. When the embedded material consists of large fragments, long scratches and grooves called **glacial striations** may be gouged out (Figure 4.5). These linear scratches on the bedrock surface provide clues to the direction of glacial movement. By mapping the striations over large areas, glacial flow patterns can often be reconstructed. On the other hand, not all abrasive action produces striations. When the sediment consists primarily of fine silt-sized particles, the rock surfaces over which the glacier moves may become highly polished.

The erosional effects of valley glaciers and ice sheets are quite different from each other. A visitor to an alpine-glaciated region is likely to see sharp and very angular topography. The reason is that as alpine glaciers move downvalley, they tend to accentuate the irregularities of the mountain landscape by creating steeper canyon walls and making bold peaks even more jagged. By contrast, ice sheets generally override the terrain and hence tend to subdue rather than accentuate the irregularities they encounter.

Although the erosional accomplishments of ice sheets can be tremendous, landforms carved by these

Figure 4.6
Prior to glaciation, a mountain valley is typically narrow and V-shaped. During glaciation, a valley glacier widens, deepens, and straightens the valley, creating a U-shaped glacial trough. This valley is in Glacier National Park, Montana. (Photo by John Montagne)

huge ice masses usually do not inspire the same degree of wonderment and awe as do the erosional features created by valley glaciers. In regions where the erosional effects of ice sheets are significant, glacially scoured surfaces and subdued terrain are the rule. By contrast, in mountainous areas, erosion by valley glaciers produces many truly spectacular features. Much of the rugged mountain scenery so celebrated for its majestic beauty is the product of glacial erosion.

Prior to glaciation, alpine valleys are characteristically V-shaped because streams are well above base level and are therefore downcutting. However, in mountainous regions that have been glaciated, the valleys are no longer narrow. As a glacier moves down a valley once occupied by a stream, the ice modifies it in three ways: The glacier widens, deepens, and straightens the valley, so that what was once a youthful V-shaped valley is transformed into a U-shaped **glacial trough** (Figure 4.6).

Since the amount of the glacial erosion depends in part upon the thickness of the ice, main or trunk glaciers cut their valleys deeper than do smaller tributary glaciers. Thus, after the glaciers have receded, the valleys of feeder glaciers stand above the main trough and are termed **hanging valleys**. Rivers flowing through hanging valleys may produce spectacular waterfalls, such as those in Yosemite National Park, California (Figure 4.7).

At the head of a glacial valley is a characteristic and often imposing feature associated with an alpine glacier—a **cirque**. As Figure 4.8 illustrates, these hollowed-out, bowl-shaped depressions have precipitous walls on three sides but are open on the downvalley side. The cirque represents the focal point of the glacier's source, that is, the area of snow accumulation and ice formation. Although the origin of cirques is still not totally clear, they are believed to begin as irregularities in the mountainside that are subsequently enlarged by frost wedging and plucking along the sides and bottom of the glacier. After the glacier has melted away, the cirque basin is often occupied by a small lake.

Fiords are deep, often spectacular, steep-sided inlets of the sea that exist in many high-latitude areas of the world where mountains are adjacent to the ocean (Figure 4.9). Norway, British Columbia, Greenland, New Zealand, Chile, and Alaska all have coastlines characterized by fiords. They represent glacial troughs that were submerged as the ice left the val-

Figure 4.7
Bridalveil Falls in Yosemite National Park cascades from a hanging valley into the glacial trough below. (Photo by E. J. Tarbuck)

ley and sea level rose following the Ice Age. The depths of fiords may exceed 1000 meters (3300 feet). However, the great depths of these flooded troughs are only partly explained by the post-Ice Age rise in sea level. Unlike the situation governing the downward erosional work of rivers, sea level does not act as base level for glaciers. As a consequence, glaciers

Figure 4.8
*Aerial view of bowl-shaped depressions called cirques in the Uinta Range, Utah.
(Photo by John S. Shelton)*

Figure 4.9
*Like other fiords, Muir Inlet,
Alaska, is a drowned glacial
trough. (Photo by Bruce F.
Molnia, courtesy of
Terraphotographics/BPS)*

are capable of eroding their beds far below the surface of the sea. For example, a valley glacier 300 meters (1000 feet) thick can carve its valley floor more than 250 meters (800 feet) below sea level before downward erosion ceases and the ice begins to float.

The Alps, Northern Rockies, and many other mountain landscapes carved by valley glaciers reveal more than glacial troughs and cirques. In addition, sinuous, sharp-edged ridges called **arêtes** and sharp, pyramid-like peaks called **horns** project above the surroundings. Both features can originate from the same basic process—the enlargement of cirques produced by plucking and frost action. A group of cirques around a single high mountain create the spires of rock called horns. As the cirques enlarge and converge, an isolated horn is produced. The most famous example is the Matterhorn in the Swiss Alps (Figure 4.10). Arêtes can form in a similar manner except that the cirques are not clustered around a point but rather exist on opposite sides of a divide. As the cirques grow, the divide separating them is reduced to a very narrow, knifelike partition. An arête may also be created in another way. When two glaciers occupy parallel valleys, an arête can form when the divide separating the moving tongues of ice is progressively narrowed as the glaciers scour and widen their valleys. The landforms carved by valley glaciers are summarized in Figure 4.11.

GLACIAL DEPOSITS

Glaciers are capable of acquiring and transporting a huge load of debris as they slowly yet steadily advance across the land. Ultimately these materials must be deposited when the ice melts. In regions where glacial sediment is deposited, the sediment can play a truly significant role in forming the physical landscape. For example, in many areas once covered by the ice sheets of the recent Ice Age, the bedrock is rarely exposed because glacial deposits that are tens or even hundreds of meters thick completely mantle the terrain. The general effect of these deposits is to reduce the local relief and thus level the topography. Indeed, much of the familiar country scenery—rocky pastures in New England, wheat fields in the Dakotas, rolling farmland in the Midwest—results directly from glacial deposition.

Long before the theory of an extensive Ice Age was proposed, much of the soil and rock debris covering

Figure 4.10
Horns are sharp, pyramid-like peaks that are fashioned by alpine glaciers. This example is the famous Matterhorn in the Swiss Alps. (Photo by E. J. Tarbuck)

portions of Europe was recognized as coming from elsewhere. At the time, these foreign materials were believed to have been "drifted" into their present positions by floating ice during an ancient flood. As a consequence, the term *drift* was applied to this sediment. Although rooted in a concept that was not correct, this term was so well-established by the time the true glacial origin of the debris became widely recognized that it remained in the glacial vocabulary. Today **drift** is an all-embracing term for sediments of glacial origin, no matter how, where, or in what form they were deposited. One of the features that distinguishes drift from sediments laid down by other erosional agents is that glacial deposits consist primarily of mechanically weathered rock debris that underwent little or no chemical weathering prior to deposition. Thus, minerals that are prone to chemical decomposition are often conspicuous components of glacial sediments.

Glacial drift is divided into two distinct types: (1) materials deposited directly by the glacier, which are known as **till**, and (2) sediments laid down by glacial meltwater, called **stratified drift**. Till is deposited as glacial ice melts and drops its load of rock fragments. Unlike moving water and wind, ice cannot sort the

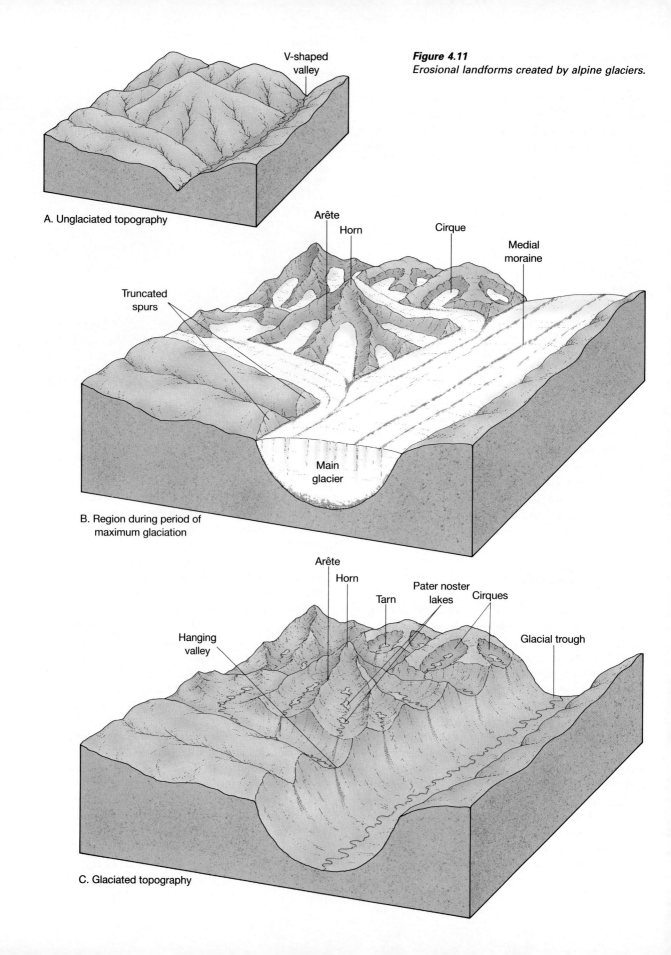

Figure 4.11
Erosional landforms created by alpine glaciers.

V-shaped valley

A. Unglaciated topography

Arête

Horn

Cirque

Medial moraine

Truncated spurs

Main glacier

B. Region during period of maximum glaciation

Arête

Horn

Pater noster lakes

Tarn

Cirques

Hanging valley

Glacial trough

C. Glaciated topography

Figure 4.12
Glacial till is an unsorted mixture of many sediment sizes. (Photo by E. J. Tarbuck)

sediment it carries; therefore, deposits of till are characteristically unsorted mixtures of many particle sizes (Figure 4.12). When boulders are found in the till or lying free on the surface, they are called **glacial erratics** if they are different from the bedrock below. Of course, this means that they must have been derived from a source outside the area where they are found (Figure 4.13). Although the locality of origin for most erratics is unknown, the origin of some can be determined. Therefore, by studying glacial erratics as well as the mineral composition of the till, geologists can sometimes trace the path of a lobe of ice. In portions of New England as well as other areas, erratics may be seen dotting pastures and farm fields. In some places, these rocks were cleared from fields and piled to make fences and walls. Keeping the fields clear, however, was and is an ongoing chore, because each spring the fields have to be cleared of newly exposed erratics that wintertime frost heaving has lifted to the surface.

Stratified drift is sorted according to the size and weight of the fragments. Since ice is not capable of such sorting activity, these sediments are not deposited directly by the glacier as till is, but rather they reflect the sorting action of glacial meltwater. Some deposits of stratified drift are made by streams issuing directly from the glacier which carry debris acquired from in, on, and beneath the ice. Other deposits involve sediment that was originally deposited as till and later picked up, transported, and redeposited by meltwater beyond the margin of the ice. Accumulations of stratified drift often consist largely

of sand and gravel, because the meltwater is not capable of moving larger material and because the finer rock flour remains suspended and is commonly carried far from the glacier. An indication that stratified drift consists primarily of sand and gravel can be seen in many areas where these deposits are actively mined as aggregate for road work and other construction projects.

Moraines, Outwash Plains, and Kettle Holes

Perhaps the most widespread features created by glacial deposition are *moraines*, which are simply layers or ridges of till. Several types of moraines are identified; some are common only to mountain valleys, and others are associated with areas affected by either ice sheets or valley glaciers. Lateral and medial moraines fall in the first category, whereas end moraines and ground moraines are in the second.

The sides of a valley glacier accumulate large quantities of debris from the valley walls. When the glacier wastes away, these materials are left as ridges, called **lateral moraines**, along the sides of the valley (Figure 4.14). **Medial moraines** are formed when two valley glaciers coalesce to form a single ice stream. The till that was once carried along the edges of each glacier joins to form a single dark stripe of

Figure 4.13
When boulders are found in till they are termed glacial erratics. Glaciers can carry huge blocks of rock that no other erosional agent could budge. This erratic boulder is in Yellowstone National Park. (Photo by James E. Patterson)

Figure 4.14
A well-developed lateral moraine deposited by the shrinking Athabaska Glacier in the Canadian Rockies. (Photo by James E. Patterson)

Figure 4.15
Medial moraines form when the lateral moraines of merging valley glaciers join. Since medial moraines could not form if the ice did not advance downvalley, these dark stripes are proof that glacial ice moves. St. Elias Mountains, Yukon Territory. (Photo by Warren Hamilton, U.S. Geological Survey)

debris within the newly enlarged glacier. The creation of these dark stripes within the ice stream is one obvious proof that glacial ice moves, because the medial moraine could not form if the ice did not flow downvalley (Figure 4.15).

As the name implies, **end moraines** form at the terminus of a glacier. Here, while the ice front is stationary, the glacier continues to carry in and deposit large quantities of rock debris, creating a ridge of till tens to hundreds of meters high. The end moraine marking the farthest advance of the glacier is called the *terminal moraine*, and those moraines that formed as the ice front periodically became stationary during retreat are termed *recessional moraines*. As the glacier recedes, a layer of till is laid down, forming a gently undulating surface of **ground moraine**. Ground moraine has a leveling effect, filling in low spots and clogging old stream channels, often leading to a disruption of drainage.

At the same time that an end moraine is forming, water from the melting glacier cascades over the till, sweeping some of it out in front of the growing ridge of unsorted debris. Meltwater generally emerges from the ice in rapidly moving streams that are often choked with suspended material and carry a substantial bed load as well. As the water leaves the glacier, it moves onto the relatively flat surface beyond and rapidly loses velocity. As a consequence, much of its bed load is dropped and the meltwater begins weaving a complex pattern of braided channels. In this way a broad, ramplike surface composed of stratified drift is built adjacent to the downstream edge of most end moraines. When the feature is formed in association with an ice sheet, it is termed an **outwash plain**, and when it is confined to a mountain valley, it is usually referred to as a **valley train**.

Often end moraines, outwash plains, and valley trains are pockmarked with basins or depressions known as **kettles** (Figure 4.16). Kettles form when a block of stagnant ice becomes wholly or partially buried in drift and ultimately melts, leaving a pit in the glacial sediment. Although most kettles do not exceed two kilometers in diameter, some with diameters exceeding 10 kilometers (6 miles) occur in Minnesota. Likewise, the typical depth of most kettles is less than 10 meters (33 feet), although the vertical dimensions of some approach 50 meters. In many cases, water eventually fills the depression and forms a pond or lake. One well-known example is Walden Pond near Concord, Massachusetts. It is here that Henry David Thoreau lived alone for two years in the 1840s and about which he wrote *Walden*, his classic of American literature.

Other Depositional Features

Drumlins are streamlined asymmetrical hills composed of till (Figure 4.17). They range in height from 15 to 60 meters (50 to 200 feet) and average 0.4–0.8 kilometer (0.25–0.50 mile) in length. The steep side of the hill faces the direction from which the ice advanced, while the gentler slope points in the direction the ice moved. Drumlins are not found singly, but rather occur in clusters, called *drumlin fields*. One such cluster, east of Rochester, New York, is estimated to contain about 10,000 drumlins. Although drumlin formation is not fully understood, their streamlined shape would indicate that they were molded in the zone of flow within an active glacier. It is believed that many drumlins originate when glaciers advance over previously deposited drift and reshape the material.

In some areas that were once occupied by glaciers, sinuous ridges composed largely of sand and gravel may be found. These ridges, called **eskers**, are deposits made by streams flowing in tunnels beneath the ice, near the terminus of a glacier. They may be several meters high and extend for many kilometers. In some areas they are mined for sand and gravel, and for this reason, eskers are disappearing in some localities. **Kames** are steep-sided hills that, like eskers, are composed of sand and gravel. Kames are believed to have originated when sediment collected in openings in stagnant ice.

Figure 4.18 depicts a hypothetical area during and following glaciation. It illustrates many of the landforms described in the preceding sections that may be found in regions affected by glaciers.

GLACIERS IN THE GEOLOGIC PAST

At various points in the preceding pages, mention was made of the Ice Age, a time when ice sheets and alpine glaciers were far more extensive than they are today. As was noted earlier, there was a time when the most popular explanation for what we now know

Figure 4.16
These ponds occupy depressions called kettles. A kettle forms when a block of ice that was buried in drift melts and leaves a pit. (Photo by Bruce F. Molnia, courtesy of Terraphotographics/BPS)

to be glacial deposits was that the materials had been drifted in by means of icebergs or perhaps simply swept across the landscape by a catastrophic flood. However, during the nineteenth century, field investigations by many scientists provided convincing proof that an extensive ice age was responsible for these deposits, as well as for many other features.

By the beginning of the twentieth century, geologists had largely determined the areal extent of Ice Age glaciation. Further, during the course of their investigations they discovered that many glaciated regions had not one layer of drift, but several. Moreover, close examination of these older deposits showed well-developed zones of chemical weathering and soil formation as well as the remains of plants that require warm temperatures. The evidence was clear; there had not been just one glacial advance but several, each separated by extended periods when climates were as warm or warmer than at present. The Ice Age had not simply been a time when the ice advanced over the land, lingered for a while, and then receded. Rather, the period was a very complex event characterized by a number of advances and withdrawals of glacial ice.

By early in the twentieth century, a four-fold division of the Ice Age had been established for both North America and Europe. The divisions were based largely on studies of glacial deposits. In North Amer-

ica the four major stages were named for the four midwestern states where deposits of each stage were well exposed and/or were first studied. These are, in order of occurrence, the Nebraskan, Kansan, Illinoian, and Wisconsinan stages. These traditional divisions remained in place until relatively recently, when it was determined that sediment cores from the ocean floor contain a much more complete record of climate change during the Ice Age. Unlike the glacial record on land, which is punctuated by many erosional gaps, seafloor sediments provide an uninterrupted record of climate cycles for this period. Studies of these sea-floor sediments showed that glacial/interglacial cycles had occurred about every 100,000 years. About twenty such cycles of cooling and warming were identified for the span we call the Ice Age.

During the glacial age, ice left its imprint on almost 30 percent of the earth's land area, including about 10 million square kilometers of North America, 5 million square kilometers of Europe, and 4 million square kilometers of Siberia (Figure 4.19). The amount of glacial ice in the Northern Hemisphere was roughly twice that of the Southern Hemisphere. The primary reason is that the southern polar ice could not spread far beyond the margins of Antarctica. By contrast, North America and Eurasia provided great expanses of land for the spread of ice sheets.

Figure 4.17
Drumlins, such as this one in upstate New York, are depositional features associated with continental ice sheets. (Courtesy of Ward's Natural Science Establishment, Inc., Rochester, N.Y.)

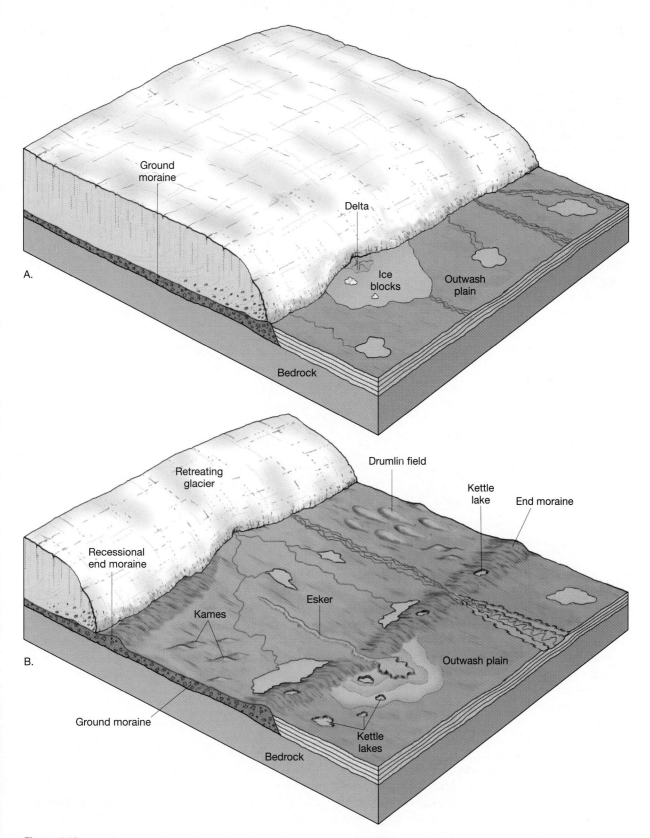

Figure 4.18
This hypothetical area illustrates many common depositional landforms.

Figure 4.19
At their maximum extent, Pleistocene glaciers covered about 10 million square kilometers (4 million square miles) of North America.

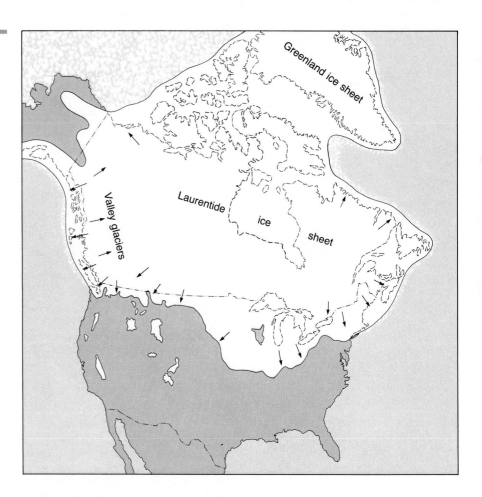

Today we know that the Ice Age began between two and three million years ago. This means that most of the major glacial stages occurred during a division of the geologic time scale called the **Pleistocene epoch**. Although the Pleistocene is commonly used as a synonym for the Ice Age, note that this epoch does not encompass all of the last glacial period. The Antarctic ice sheet, for example, formed at least 14 million years ago, and, in fact, might be much older.

Glaciers have not been ever-present features throughout the earth's long history. In fact, for most of geologic time, glaciers have been absent. Evidence does indicate that in addition to the Pleistocene epoch there were at least three other periods of glacial activity: 2 billion, 600 million, and 250 million years ago. However, the most recent period of glaciation is of greatest interest, because the features of many present-day landscapes are a reflection of the work of Pleistocene glaciers.

SOME INDIRECT EFFECTS OF ICE AGE GLACIERS

In addition to the massive erosional and depositional work carried on by Pleistocene glaciers, the ice sheets had other, sometime profound, effects upon the landscape. For example, as the ice advanced and retreated, animals and plants were forced to migrate. This led to stresses that some organisms could not tolerate. Furthermore, many present-day stream courses bear little resemblance to their preglacial routes. The Missouri River once flowed northward toward Hudson Bay, while the Mississippi River followed a path through central Illinois, and the head of the Ohio River reached only as far as Indiana. Other rivers that today carry only a trickle of water but nevertheless occupy broad channels are testimony to the fact that they once carried torrents of glacial meltwater.

In areas that were centers of ice accumulation, such as Scandinavia and northern Canada, the land has been slowly rising for the past several thousand years. The land is rising because the added weight of the three-kilometer-thick mass of ice downwarped the earth's crust. Following the removal of this immense load, the crust has been adjusting by gradually rebounding upward ever since.

Certainly one of the most interesting and perhaps dramatic effects of the Ice Age was the fall and rise of sea level that accompanied the advance and retreat of the glaciers. In Box 4.1 it was pointed out that sea level would rise by an estimated 60 or 70 meters if the water locked up in the Antarctic Ice Sheet were to melt completely. Such an occurrence would flood many densely populated coastal areas. Since we know that the snow from which glaciers are made ultimately comes from the evaporation of ocean water, the growth of ice sheets must have caused a worldwide drop in sea level. Indeed, estimates suggest that sea level was as much as 100 meters lower than today. Thus, land that is presently flooded by the oceans was dry. The Atlantic Coast of the United States was located more than 100 kilometers (60 miles) to the east of New York City, France and Britain were joined where the English Channel is today, Alaska and Siberia were connected across the Bering Strait, and Southeast Asia was tied by dry land to the islands of Indonesia.

While the formation and growth of ice sheets was an obvious response to significant changes in climate, the existence of the glaciers themselves triggered important climatic changes in the regions beyond their margins. In arid and semiarid areas on all continents, temperatures, and thus evaporation rates, were lower. At the same time, precipitation was moderate. This cooler, wetter climate resulted in the formation of many lakes called **pluvial lakes**, from the Latin term *pluvia* meaning "rain." In North America, the greatest concentration of pluvial lakes occurred in the vast Basin and Range region of Nevada and Utah (Figure 4.20). Although most of the lakes completely disappeared, there are a few small remnants, the Great Salt Lake being the largest and best known.

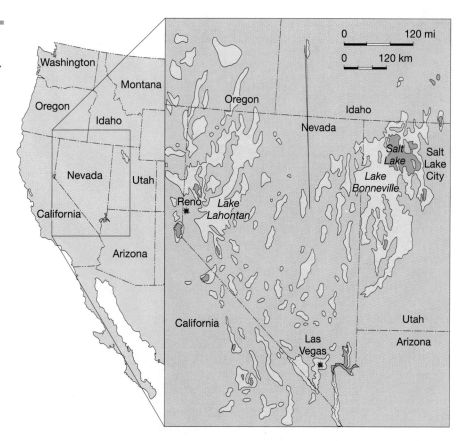

Figure 4.20
Pluvial lakes of the western United States. (After R. F. Flint, Glacial and Quaternary Geology, New York: John Wiley and Sons.)

CAUSES OF GLACIATION

A great deal is known about glaciers and glaciation. Much has been learned about glacier formation and movement, the extent of glaciers past and present, and the features created by glaciers, both erosional and depositional. However, scientists have not yet developed a completely satisfactory explanation for the causes of ice ages.

Any theory that attempts to explain the causes of glacial ages must successfully address two basic questions. First, what causes the onset of glacial conditions? For ice sheets to have formed, average temperatures must have been somewhat lower than at present and perhaps substantially lower than throughout much of geologic time. For that reason, a successful explanation would have to account for the gradual cooling that finally leads to glacial conditions. The second question is: What caused the al-

ternation of glacial and interglacial stages that have been documented for the Pleistocene epoch? Whereas the first question deals with long-term trends in temperature that occur on a scale of millions of years, this second question relates to much shorter-term changes.

Although the literature of science contains a vast array of hypotheses that attempt to explain the possible causes of glacial periods, we will discuss only a few major ideas in an effort to give the current thought on this problem.

Probably the most attractive proposal for explaining why extensive glaciations have occurred only a few times in the geologic past comes from the theory of plate tectonics.* Not only does this theory pro-

---

* A complete discussion of plate tectonics is presented in Chapter 6.

Figure 4.21
A. The supercontinent Pangaea showing the area covered by glacial ice 300 million years ago. B. The continents as they are today. The shading outlines areas where evidence of the old ice sheets exists.

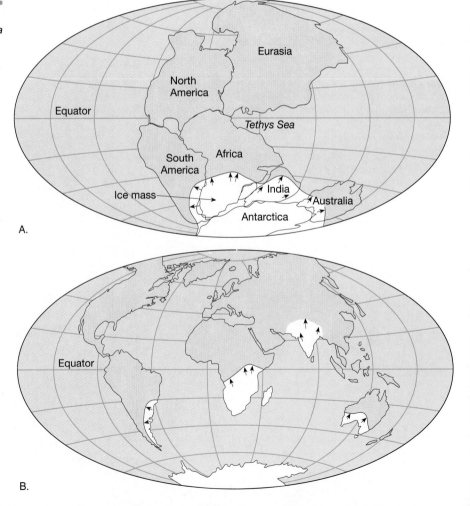

vide geologists with explanations about many previously misunderstood processes and features, it also provides a possible explanation for some previously unexplainable climate changes including the onset of glacial conditions. Since glaciers can form only on land, we know that landmasses must exist somewhere in the higher latitudes before an ice age can commence. Many believe that ice ages have only occurred when the earth's shifting crustal plates carried the continents from tropical latitudes to more poleward positions.

Glacial features in present-day Africa, Australia, South America, and India indicate that these regions experienced an ice age near the end of the Paleozoic era, about 250 million years ago. For many years this puzzled scientists. Was the climate in these relatively tropical latitudes once like it is today in Greenland and Antarctica? Why did glaciers not form in North America and Eurasia? Until the plate tectonics theory was formulated and proven, there was no reasonable explanation. Today scientists realize that the areas containing these ancient glacial features were joined together as a single supercontinent called Pangaea that was located at high latitudes far to the south of their present positions. Later, this landmass broke apart and its pieces, each moving on a different plate, drifted toward their present locations (Figure 4.21). It is now understood that during the geologic past, plate movements accounted for many dramatic climatic changes as landmasses shifted in relation to one another and moved to different latitudinal positions. Changes in oceanic circulation also must have occurred, altering the transport of heat and moisture and, consequently, the climate as well. Since the rate of plate movement is very slow, on the order of a few centimeters per year, appreciable changes in the positions of the continents occur only over great spans of geologic time. Thus, climatic changes brought about by shifting plates are extremely gradual and happen on a scale of millions of years.

Since climatic changes brought about by moving plates are extremely gradual, the plate tectonics theory cannot explain the alternation of glacial and interglacial climates that occurred during the Pleistocene epoch. Therefore we must look to some other triggering mechanism that may cause climatic change on a scale of thousands rather than millions of years. Today many scientists believe, or strongly suspect, that the climatic oscillations that characterized the Pleistocene may be linked to variations in the earth's orbit. This hypothesis was first developed and strongly advocated by the Yugoslavian scientist Milutin Milankovitch and is based on the premise that variations in incoming solar radiation are a principal factor in controlling the earth's climate. Milankovitch formulated a comprehensive mathematical model based on the following elements:

1. Variations in the shape (*eccentricity*) of the earth's orbit around the sun;
2. Changes in *obliquity*; that is, changes in the angle that the earth's axis makes with the plane of the earth's orbit; and
3. The wobbling of the earth's axis, called *precession*.

Using these factors, Milankovitch calculated variations in the receipt of solar energy and the corresponding surface temperature of earth back into time in an attempt to correlate these changes with the climatic fluctuations of the Pleistocene. In explaining climatic changes that result from these three variables, it should be noted that they cause little or no variation in the total amount of solar energy reaching the ground. Instead, their impact is felt because they change the degree of contrast between seasons. Somewhat milder winters in the middle to high latitudes means greater snowfall totals while cooler summers would bring a reduction in snowmelt.

Over the years the astronomical theory of Milankovitch has been widely accepted, then largely rejected, and now, in light of recent investigations, has once again gained significant support. Among the studies that have added credibility and support to the astronomical theory is one in which deep-sea sediments containing certain climatically sensitive microorganisms were analyzed in order to establish a chronology of temperature changes going back nearly one-half million years.[*] This time scale of climatic change was then compared to astronomical calculations of eccentricity, obliquity, and precession to determine if a correlation did indeed exist. Although the study was very involved and mathematically complex, the conclusions were straightforward. The authors found that major variations in climate over the past several hundred thousand years were closely associated with changes in the geometry of the earth's orbit; that is, cycles of climatic change were shown to correspond closely with the periods of obliquity, precession, and orbital eccentricity. More

---

[*] J. D. Hays, John Imbrie, and N. J. Shackleton, "Variations in the Earth's Orbit: Pacemaker of the Ice Ages," *Science* 194 (4270): 1121–32.

specifically, they stated, "It is concluded that changes in the earth's orbital geometry are the fundamental cause of the succession of Quaternary ice ages."*

Let us briefly summarize the ideas that were just described. The theory of plate tectonics provides an explanation for the widely spaced and nonperiodic onset of glacial conditions at various times in the geologic past, whereas the astronomical theory proposed by Milankovitch and recently supported by the work of J. D. Hays and his colleagues furnished an explanation for the alternating glacial and interglacial episodes of the Pleistocene.

In conclusion, it should be emphasized at this point that the proposals just discussed do not represent the only possible explanations for glacial ages. Although interesting and attractive, these ideas are certainly not without critics, nor are they the only proposals currently under study. Other factors may, and in fact probably do, enter into the picture.

DESERTS

The word *desert* literally means "*deserted*" or "*unoccupied.*" For many dry regions this is a very appropriate description. Yet in places where water is available, many people may be found. Nevertheless, the world's dry regions are among the least familiar land areas on earth outside of the polar realm.

Desert landscapes frequently appear stark. Their profiles are not softened by a carpet of soil and abundant plant life. Instead, barren rocky outcrops with steep, angular slopes are common. At some places the rocks are tinted orange and red. At others they are gray and brown and streaked with black. For many visitors desert scenery exhibits a striking beauty; to others, the terrain seems bleak. No matter which feeling is elicited, it is clear that deserts are very different from the more humid places where most people live. As we shall see, arid regions are not dominated by a single geologic process. Rather, the effects of tectonic forces, running water, and wind are all apparent. Because these processes combine in different ways from place to place, the appearance of desert landscapes varies a great deal as well (Figure 4.22).

---

* Ibid., 1131. *Quaternary* refers to the period on the geologic time scale that encompasses the last few million years.

DISTRIBUTION AND CAUSES OF DRY LANDS

The dry regions of the world encompass about 42 million square kilometers, nearly 30 percent of the earth's land surface. No other climatic group covers so large a land area. Within these water–deficient regions, two climatic types are commonly recognized: **desert**, or arid, and **steppe**, or semiarid. The two categories have many features in common; their differences are primarily a matter of degree. The steppe is a marginal and more humid variant of the desert and represents a transition zone that surrounds the desert and separates it from bordering humid climates. The world map showing the distribution of desert and steppe regions reveals that dry lands are concentrated in the subtropics and in the middle latitudes (Figure 4.23).

The heart of the low-latitude dry climates lies in the vicinities of the Tropics of Cancer and Capricorn. A glance at Figure 4.23 shows a virtually unbroken desert environment stretching for more than 9300 kilometers from the Atlantic coast of North Africa to the dry lands of northwestern India. In addition to this single great expanse, the northern hemisphere contains another much smaller area of tropical desert and steppe in northern Mexico and the southwestern United States. In the Southern Hemisphere, dry

Figure 4.22
Rocky outcrops in northern Arizona. The appearance of desert landscapes varies a great deal from place to place. (Photo by Michael Collier)

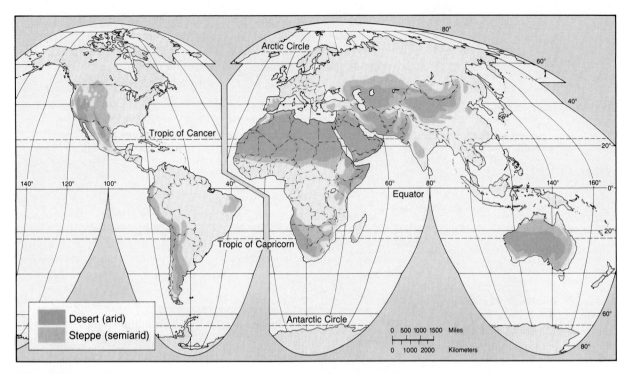

Figure 4.23
Arid and semiarid climates cover about 30 percent of the earth's land area. No other climatic group covers so large an area.

climates dominate Australia. Almost 40 percent of the continent is desert, and much of the remainder is steppe. In addition, arid and semiarid areas are found in southern Africa and make a limited appearance in coastal Chile and Peru. The existence of this dry subtropical realm is primarily the result of the prevailing global distribution of air pressure and winds. Coinciding with dry regions in the lower latitudes are zones of high air pressure known as the *subtropical highs*. These pressure systems are characterized by subsiding air currents. When air sinks, it is compressed and warmed. Such conditions are just the opposite of what is needed to produce clouds and precipitation. Consequently, these regions are known for their clear skies, sunshine, and ongoing drought.

Unlike their low-latitude counterparts, middle-latitude deserts and steppes are not controlled by the subsiding air masses associated with high pressure. Instead, these dry lands exist principally because of their positions in the deep interiors of large landmasses far removed from the oceans, which are the ultimate sources of moisture for cloud formation and precipitation. In addition, the presence of high mountains across the paths of prevailing winds further acts to separate these areas from water-bearing, maritime air masses. As prevailing winds meet mountain barriers, the air is forced to ascent. When air rises it expands and cools, a process that can produce clouds and precipitation. The windward sides of mountains, therefore, are often characterized by relatively high precipitation totals. By contrast, the leeward sides of mountains are usually much drier (Figure 4.24, page 158). This situation exists because air reaching the leeward side has lost much of its moisture and, if the air descends, it is compressed and warmed, making cloud formation even less likely. The dry region that results is often referred to as a **rainshadow desert**. Because most middle-latitude deserts occupy sites on the leeward sides of mountains, they can also be classified as rainshadow deserts. In North America the Coast Ranges, Sierra Nevada, and Cascades are the foremost mountain barriers, whereas in Asia, the great Himalayan chain prevents the summertime monsoon flow of moist Indian Ocean air from reaching into the interior. Because the Southern Hemisphere lacks extensive land areas in the middle latitudes, only a small area of desert and steppe are

BOX 4.2

Common Misconceptions About Deserts

*D*eserts are lifeless, hot, sand-covered landscapes shaped largely by the force of wind. The preceding statement summarizes the image of arid regions that many people hold, especially those living in more humid places. Is it an accurate view? The answer is no. Although there are clearly elements of reality in such an impression, it is a generalization that contains a number of misconceptions (Figure 4.B).

One common fallacy about deserts is that they are lifeless or almost lifeless. Although reduced in amount and different in character, plant and animal life are usually present. Desert plants may differ widely from one part of the world to another, but all have one characteristic in common: they have developed adaptations that make them highly tolerant of drought. Many have waxy leaves, stems, or branches or a thickened cuticle (outermost protective layer) to reduce water loss. Others have very small leaves or no leaves at all. Also, the roots of some species often extend to great depths in order to tap the moisture found there, whereas others produce a shallow but widespread root system that enables them to absorb great amounts of moisture quickly from the infrequent desert downpours. Often the stems of these plants are thickened by a spongy tissue that can store enough water to sustain the plant until the next rainfall comes. Thus, although widely dispersed and providing little ground cover, plants of many kinds flourish in the desert.

A second widely held belief about the world's dry lands is that they are always hot. This fact seems to be reinforced by commonly quoted temperature statistics. The highest accepted temperature record for the United States as well as the entire Western Hemisphere is 57°C (134°F). This long-standing record was set at Death Valley, California, on July 10, 1913. The nearly 59°C (137°F) reading at Azizia, Libya, in North Africa's Sahara Desert on September 13, 1922, is the world record. Despite these remarkably high figures, cold temperatures are also experienced in desert regions. For example, the average daily minimum in January at Phoenix, Arizona, is 1.7°C (35°F), just barely above freezing. At Ulan Bator in Mongolia's Gobi Desert, the average *high* temperature on January days is only −19°C (−2°F)! Dry climates are found from the tropics poleward to the high middle latitudes. Consequently, although tropical deserts lack a cold season, deserts in the middle latitudes do experience seasonal temperature changes.

found in this latitude range, existing primarily near the southern tip of South America in the rainshadow of the towering Andes. In the case of middle-latitude deserts, we have an example of the impact of tectonic processes on climate. Without such mountain-building episodes, wetter climates would prevail where many dry regions exist today.

GEOLOGIC PROCESSES IN ARID CLIMATES

The angular hills, sheer canyon walls, and pebble- or sand-covered surface of the desert contrast sharply with the rounded hills and curving slopes of more humid areas. A desert landscape may at first seem to have been shaped by forces different from those operating in regions where water is more abundant. While the contrasts may be striking, they are not a reflection of different processes but merely the differing effects of the same processes operating under contrasting climatic conditions.

In humid regions, relatively fine textured soils mantle the surface to support an almost continuous cover of vegetation. Here the slopes and rock edges are rounded. Such a landscape reflects the strong influence of chemical weathering in a humid climate. By contrast, much of the weathered debris in deserts consists of unaltered rock and mineral fragments—the result of mechanical weathering processes. In dry lands, rock weathering of any type is greatly reduced

The last two commonly held misconceptions about the world's deserts are more geologic than climatic. One mistaken assumption is that they consist of mile after mile of drifting sand. It is true that sand accumulations do exist in some areas and may be striking features, but they represent only a small percentage of the total desert area. For example, in the Sahara, the world's largest desert, accumulations of sand cover only one-tenth of its area. The sandiest of all deserts is the Arabian, one-third of which consists of sand. The final mistaken assumption is the seemingly logical idea that wind is the most important agent of erosion in deserts. Although wind is relatively more significant in dry areas than anywhere else, most erosional landforms in deserts are created by running water. When the rains come, they usually take the form of thunderstorms. Because the heavy rain associated with these storms cannot all soak in, rapid runoff results. Without a thick vegetative cover to protect the ground, erosion is great.

because of the lack of moisture and the scarcity of organic acids from decaying plants. Chemical weathering, however, is not completely lacking in deserts. Over long spans of time, clays and thin soils do form and many iron-bearing silicate minerals oxidize to produce the rust-colored stain found tinting some desert landscapes.

Although permanent streams are common in humid regions, practically all desert streams are dry most of the time (Figure 4.25A). Desert streams are said to be **ephemeral**, which means that they carry water only in response to specific episodes of rainfall. An average ephemeral stream may flow only a few days or perhaps just a few hours during the year. In some years the channel may carry no water at all.

This fact is obvious even to the casual observer who, while traveling, notices the number of bridges with no streams beneath them or the number of dips in the road where dry channels cross. However, when the rare heavy showers do come, so much rain falls in such a short time that it cannot all soak in. Since the vegetative cover is sparse, runoff is largely unhindered and therefore rapid, often creating flash floods along valley floors (Figure 4.25B). Such floods are quite unlike floods in humid regions. A flood on a river like the Mississippi may take several days to reach its crest and then subside, whereas desert floods arrive suddenly and likewise subside in a short time. Because much of the surface material is not anchored by vegetation, the amount of erosional work

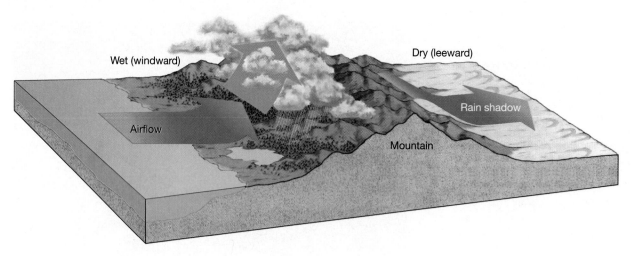

Figure 4.24
Many deserts in the middle latitudes are rainshadow deserts. As moving air meets a mountain barrier, it is forced to rise. Clouds and precipitation on the windward side often result. Air descending the leeward side is much drier. The mountains effectively cut the leeward side off from the source of moisture, producing a rainshadow desert.

that occurs during one of these short-lived events is substantial. In the dry western United States a number of different names are used for ephemeral streams. Two of the most common are *wash* and *arroyo*. In other parts of the world, a dry desert stream may be called a *wadi* (Arabia and North Africa), a *donga* (South America), or a *nullah* (India).

Unlike the drainage in humid regions, stream courses in arid regions are seldom well-integrated. That is, desert streams lack an extensive system of tributaries. In fact, a basic characteristic of deserts is that most streams that originate in them are small and die out before reaching the sea. Because the water table is usually far below the surface, few desert streams can draw upon it. Without a steady supply of water, evaporation and infiltration soon deplete the stream. The few permanent streams that do cross arid regions, such as the Colorado and Nile rivers, originate outside the desert often in mountains where water is more plentiful. Here the supply of water must be great to compensate for the losses that occur as the stream crosses the desert. For example, after the Nile leaves the lakes and mountains of central Africa that are its source, it traverses almost 3000 kilometers (1800 miles) of the Sahara Desert without a single tributary.

It should be emphasized that running water, although an infrequent occurrence, is responsible for

most of the erosional work in deserts. This is contrary to a commonly held belief that the wind is the most important erosional agent sculpturing desert landscapes. Although wind erosion is more significant in dry areas than elsewhere, most desert landforms are nevertheless carved by running water. The main role of wind, as we shall see later in this chapter, is in the transportation and deposition of sediment, creating and shaping the ridges and mounds we call dunes.

THE EVOLUTION OF A DESERT LANDSCAPE

Due to the fact that arid regions typically lack permanent streams, they are characterized as having **interior drainage**, that is, a discontinuous pattern of intermittent streams that do not flow out of the desert to the ocean. In the United States, the dry Basin and Range region provides an excellent example. The region includes southern Oregon, all of Nevada, western Utah, southeastern California, as well as southern Arizona and New Mexico. The name Basin and Range is an apt description for this almost 800,000-square-kilometer region, since it is characterized by more than 200 relatively small mountain ranges which rise 900–1500 meters above the basins that separate them. In this region, as in others like it

Figure 4.25
A. *Most of the time, desert stream channels are dry.* ***B.*** *An ephemeral stream shortly after a heavy shower. Although such floods are short-lived, large amounts of erosion occur. (Photos by James E. Patterson)*

A.

B.

around the world, erosion is carried out for the most part without reference to the ocean (ultimate base level), because drainage is in the form of local interior systems. Even in areas where permanent streams flow to the ocean, few tributaries exist, and thus only a relatively narrow strip of land adjacent to the stream has sea level as the ultimate level of land reduction.

The block models in Figure 4.26 depict the stages of landscape evolution in a mountainous desert such as the Basin and Range region. During and following uplift of the mountains, running water begins carving the elevated mass and depositing large quantities of debris in the basin. In this early stage relief is greatest, for as erosion lowers the mountains and sediment fills the basins, elevation differences diminish.

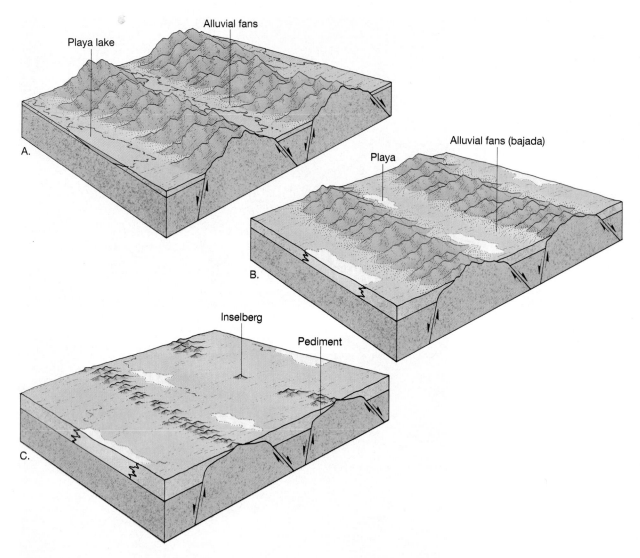

Figure 4.26
*Stages of landscape evolution in a mountainous desert. As erosion of the mountains and deposition in the basins continue, relief diminishes. **A.** Early stage. **B.** Middle stage. **C.** Late stage.*

When the occasional torrents of water produced by sporadic rains move down the mountain canyons, they are heavily loaded with sediment. Emerging from the confines of the canyon, the runoff spreads over the gentler slopes at the base of the mountains and quickly loses velocity. Consequently, most of its load is dumped within a short distance. The result is a cone of debris known as an **alluvial fan** at the mouth of a canyon (Figure 4.27). Since the coarsest material is dropped first, the head of the fan is steepest. Moving down the fan, the size of the sediment

and the steepness of the slope decrease and merge imperceptibly with the basin floor. Over the years, a fan enlarges, eventually coalescing with fans from adjacent canyons to produce an apron of sediment along the mountain front.

On the rare occasions when rainfall is abundant, streams may flow across the alluvial fans to the center of the basin, converting the basin floor into a shallow **playa lake**. Playa lakes are temporary features that last only a few days, or at best a few weeks, before evaporation and infiltration remove

Figure 4.27
Alluvial fans develop where the gradient of a stream changes abruptly from steep to flat. Such a situation exists in Death Valley, California, where streams emerge from the mountains into a flat basin. As a result, Death Valley has many large alluvial fans. (Photo by Michael Collier)

the water. The dry, flat lake bed that remains is termed a *playa.*

Playas are typically composed of fine silts and clays, and occasionally they become encrusted with salts precipitated during evaporation. These precipitated salts may be uncommon. A case in point is the sodium borate (better known as borax) mined from ancient playa lake deposits in Death Valley, California.

With time the mountain front is worn back and a sloping bedrock platform, called a **pediment**, is created adjacent to the steep mountain front. A pediment is an erosional surface, usually covered by a thin veneer of debris, that is formed by the action of running water. Just how the water carves the pediment, however, is unclear and still a matter for debate.

With the ongoing dissection of the mountain mass into an intricate series of valleys and sharp divides as well as the accompanying sedimentation, the local relief continues to diminish. After more time passes,

the steady retreat of the mountain front enlarges the pediment. Eventually this pediment growth results in nearly the entire mountain mass being consumed. Thus, by the late stages of erosion, the mountain areas are reduced to a few large bedrock knobs projecting above the surrounding pediment and sediment-filled basin. These isolated erosional remnants on an old-age desert landscape are called **inselbergs**, a German word meaning "island mountains."

Each of the stages of landscape evolution in an arid climate depicted in Figure 4.26 can be observed in the Basin and Range region. Recently uplifted mountains in an early stage of erosion are found in southern Oregon and northern Nevada. Death Valley, California, and southern Nevada fit into the more advanced middle stage, while the late stage, with its inselbergs and extensive pediments, can be seen in southern Arizona.

WIND EROSION

Although wind erosion is not restricted to arid and semiarid regions, it does its most effective work in these areas. In humid regions moisture binds particles together and vegetation anchors the soil so that wind erosion is negligible. For wind to be effective, dryness and scanty vegetation are important prerequisites. When such circumstances exist, wind may pick up, transport, and deposit great quantities of fine sediment. During the 1930s parts of the Great Plains experienced great dust storms. The plowing under of the natural vegetative cover for farming, followed by severe drought, made the land prime for wind erosion and led to the area being labeled the Dust Bowl.[*]

Moving air, like moving water, is turbulent and able to pick up loose debris and transport it to other locations. Just as in a stream, the velocity of wind increases with height above the surface. Also like a stream, wind transports fine particles in suspension while heavier ones are carried as bed load. However, the transport of sediment by wind differs from that by running water in two significant ways. First, wind has a low density compared to water; thus it is not capable of picking up and transporting coarse materials. Second, because wind is not confined to chan-

---

[*] For more information, see Box 2.2, "Dust Bowl: Soil Erosion in the Great Plains."

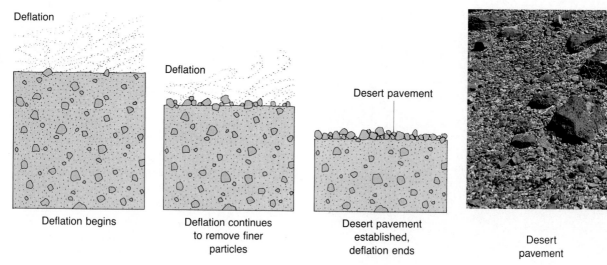

Deflation

Deflation begins

Deflation

Deflation continues
to remove finer
particles

Desert pavement

Desert pavement
established,
deflation ends

Desert
pavement

Figure 4.28
*Formation of desert pavement. Coarse particles gradually become concentrated into a
tightly packed layer as deflation lowers the surface by removing sand and silt. If
left undisturbed, desert pavement will protect the surface from further deflation. (Photo
by Peter Kresan)*

nels, it can spread over large areas, as well as high into the atmosphere.

One way that wind erodes is by **deflation,** the lifting and removal of loose material. Since the competence (ability to transport different-size particles) of moving air is low, it can suspend only fine sediment such as clay and silt. Larger grains of sand are rolled or skipped along the surface (a process called *saltation*) and comprise the bed load. Particles larger than sand are usually not transported by wind. Although the effects of deflation are sometimes difficult to notice because the entire surface is being lowered at the same time, they can be significant. In portions of the 1930s Dust Bowl, the land was lowered by as much as 1 meter in only a few years.

The most noticeable result of deflation in some places are shallow depressions called **blowouts.** In the Great Plains region, from Texas north to Montana, thousands of blowouts can be seen. They range in size from small dimples less than 1 meter deep and 3 meters wide to depressions that are over 45 meters deep and several kilometers across. In the past, several explanations for the depressions were put forth, including one which suggested they were sites of buffalo wallows. Today it is believed that the vast majority were created by deflation.

In portions of many deserts the surface is characterized by a layer of coarse pebbles and cobbles that are too large to be moved by the wind. This stony veneer, called **desert pavement,** is created as deflation lowers the surface by removing sand and silt until eventually only a continuous cover of coarse particles remains (Figure 4.28). Once desert pavement becomes established, a process which may take hundreds of years, the surface is effectively protected from further deflation if left undisturbed. However, since the layer is only one or two stones thick, the passage of vehicles or animals can dislodge the pavement and expose the fine-grained material below. If this happens, the surface is once again subject to deflation.

Like glaciers and streams, wind erodes by abrasion. In dry regions as well as along some beaches, windblown sand cuts and polishes exposed rock surfaces. However, abrasion is often credited for accomplishments beyond its actual capabilities. Such features as balanced rocks that stand high atop narrow pedestals and intricate detailing on tall pinnacles are not the results of abrasion. Since sand seldom travels more than a meter above the surface, the wind's sandblasting effect is obviously quite limited in vertical extent. However, in areas prone to such activity, telephone poles have actually been cut through near their bases. For this reason, collars are often fitted on the poles to protect them from being "sawed" down.

WIND DEPOSITS

Although wind is relatively unimportant as a producer of erosional landforms, wind deposits are significant features in some regions. Accumulations of windblown sediment are particularly conspicuous landscape elements in the world's dry lands and along many sandy coasts. Wind deposits are of two distinctive types: (1) extensive blankets of silt that once were carried in suspension, and (2) mounds and ridges of sand from the wind's bed load.

Loess

In some parts of the world the surface topography is mantled with deposits of windblown silt. Over periods of perhaps thousands of years dust storms deposited this material, which is called **loess.** As can be seen in Figure 4.29, when loess is breached by streams or road cuts, it tends to maintain vertical cliffs and lacks any visible layers. The distribution of loess indicates that there are two primary sources for this sediment: deserts and glacial deposits of stratified drift. The thickest and most extensive loess deposits occur in western and northern China, where accumulations of 30 meters are not uncommon and thicknesses of more than 100 meters have been measured. It is this fine, buff-colored sediment which gives the Yellow River (Hwang Ho) and the adjacent Yellow Sea their names. The sources of China's 800,000 square kilometers of loess are the extensive desert basins of central Asia.

In the United States, deposits of loess are significant in many areas, including South Dakota, Nebraska, Iowa, Missouri, and Illinois as well as portions of the Columbia Plateau in the Pacific Northwest. The correlation between the distribution of loess and important farming regions in the Midwest and eastern Washington is not just a coincidence, because soils derived from this wind-deposited sediment are among the most fertile in the world. Unlike the deposits in China, the loess in the United States, as well as in Europe, is an indirect product of glaciation, for its source was deposits of stratified drift. During the retreat of the glacial ice, many river valleys were choked with sediment provided by meltwater. Strong westerly winds sweeping across the barren floodplains picked up the finer sediment and dropped it as a blanket on the east side of the valleys. Such an origin is confirmed by the fact that loess deposits are thickest and coarsest on the lee side of such major glacier drainage outlets as the Mississippi and Illinois rivers and thin rapidly with increasing distance from the valleys. Furthermore, the angular, mechanically weathered particles composing the loess are essentially the same as the rock flour produced by the grinding action of glaciers.

Sand Deposits

Like running water, wind releases its load of sediment when its velocity falls and the energy available for transport diminishes. Thus sand begins to accumulate wherever an obstruction across the path of the wind slows the movement of the air. Unlike deposits of loess, which form blanketlike layers over large areas, winds commonly deposit sand in mounds or ridges called *dunes*.

As moving air encounters an object, such as a clump of vegetation or a rock, the wind sweeps around and over it, leaving a shadow of more slowly moving air behind the obstacle as well as a smaller zone of quieter air just in front of the obstacle. Some of the saltating sand grains moving with the wind come to rest in these "wind shadows." As the accu-

Figure 4.29
A vertical loess bluff near the Mississippi River in southern Illinois. (Photo by James E. Patterson)

Figure 4.30
*Sand sliding down the steep slip face of a dune. Coral Pink Sand Dunes,
Utah. (Photo by Michael Collier)*

mulation of sand continues, an increasingly efficient wind barrier forms to trap even more sand. If there is a sufficient supply of sand and the wind blows steadily long enough, the mound of sand grows into a dune.

The profile of many dunes shows an asymmetrical shape with the leeward slope being steep and the windward slope more gently inclined. Sand is rolled up the gentle slope on the windward side by the force of the wind. Just beyond the crest of the dune, the wind velocity is reduced and the sand accumulates. As more sand collects, the slope steepens and eventually some of it slides or slumps under the pull of gravity. In this way the leeward slope of the dune, called the **slip face,** maintains an angle of about 34 degrees. Continued sand accumulation, coupled with periodic slides down the slip face, results in the slow

migration of the dune in the direction of air movement (Figure 4.30). As sand is deposited on the slip face, it forms layers inclined in the direction the wind is blowing. These sloping layers are called **cross beds.** When the dunes are eventually buried under layers of sediment and become part of the sedimentary rock record, their asymmetrical shape is destroyed, but the cross beds remain as a testimony to their origin. Nowhere is cross-bedding more prominent than in the sandstone walls of Zion Canyon in Utah (Figure 4.31).

Types of Sand Dunes

Although often complex, dunes are not just random heaps of sediment (Figure 4.32). Rather, they are accumulations that usually assume surprisingly consis-

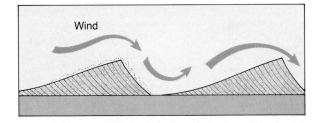

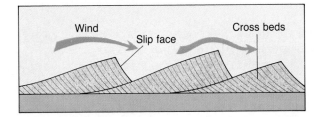

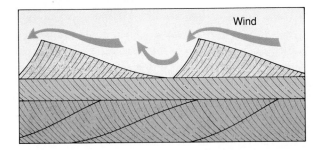

Figure 4.31
Dunes commonly have an asymmetrical shape. The steeper leeward side is called the slip face. Sand grains deposited on the slip face at the angle of repose create the cross bedding of the dunes. A complex pattern develops in response to changes in prevailing winds. The photograph shows the pattern of cross-bedding in the Navajo Sandstone, Zion National Park, Utah. (Photo by E. J. Tarbuck)

Figure 4.32
Barchanoid dunes represent a type that is intermediate between isolated barchans on the one hand and extensive transverse dunes on the other. The gypsum dunes at White Sands National Monument, New Mexico, are an example. (Photo by E. J. Tarbuck)

BOX 4.3

Desertification

On nearly any list of major environmental issues facing the world, one is likely to find reference to the problem of *desertification*. The term by itself simply implies the expansion of desertlike conditions into nondesert areas. Such transformations can result from natural processes that act gradually over extended spans of time. However, the use of the term in recent years has been restricted primarily to situations in which there is a relatively rapid alteration of the land to desertlike conditions as the result of human activities.

Desertification commonly takes place on the margins of deserts. The advancement of desertlike conditions into areas that were previously productive is not a process in which the borders of a desert gradually expand in a uniform manner. Rather, degeneration into desert usually occurs as a patchy transformation of dry but habitable land into dry land that is uninhabitable. It is primarily the product of inappropriate land use and is aided and accelerated by drought. The process may be halted during wet years, only to advance rapidly during succeeding dry years.

On marginal land used for crops, natural vegetation is cleared. During periods of drought, crops fail and the unprotected soil is exposed to the forces of erosion. Gullying of slopes and accumulations of sediment in stream channels are visible signs on the landscape, as

are the clouds of dust created as topsoil is removed by the wind. Where crops are not grown, the raising of livestock leads to degradation of the land. Although the modest vegetation associated with marginal lands may be adequate to maintain local wildlife, it cannot support the intensive grazing of large domesticated herds. Overgrazing reduces or eliminates plant cover. When the vegetative cover is destroyed beyond the minimum required for protection of the soil against erosion, the destruction becomes irreversible.

Desertification first received worldwide attention when drought struck a region in Africa called the *Sahel* in the late 1960s (Figure 4.C). During that period and others since then, the people in this vast expanse lying south of the Sahara Desert have suffered from malnutrition and death by starvation. Livestock

---

tent patterns. A broad assortment of dune forms exists; so to simplify and provide some order, several major types are recognized (Figure 4.33). Of course, there are gradations among different forms as well as irregularly shaped dunes that do not fit easily into any category. Several factors influence the form and size that dunes ultimately assume. These include wind direction and velocity, availability of sand, and the amount of vegetation present.

Barchan Dunes. Solitary sand dunes shaped like crescents and with their tips pointing downwind are called **barchan dunes** (Figure 4.33A). These dunes form where supplies of sand are limited and the surface is relatively flat, hard, and lacking vegetation. They migrate slowly with the wind at a rate of up to 15 meters per year. Their size is usually modest with the largest barchans reaching a height of about 30 meters, while the maximum spread between their horns approaches 300 meters. When the wind direction is nearly constant, the crescent form of these

dunes is nearly symmetrical. However, when the wind direction is not perfectly fixed, one tip becomes longer than the other.

Transverse Dunes. In regions where vegetation is sparse or absent and sand is very plentiful, the dunes form a series of long ridges that are separated by troughs and oriented at right angles to the prevailing wind. Because of this orientation, they are termed **transverse dunes** (Figure 4.33B). Typically, many coastal dunes are of this type. In addition, they are common in arid regions where the extensive surface of wavy sand is sometimes called a sand sea.

There is a relatively common dune form that is intermediate between isolated barchans and extensive waves of transverse dunes. Such dunes, called **barchanoid dunes,** form scalloped rows of sand oriented at right angles to the wind (Figure 4.32C). The rows resemble a series of barchans that have been positioned side by side. Visitors exploring the herds have been decimated, and the loss of productive land has been great. Hundreds of thousands of people have been forced to migrate. As agricultural lands shrink, people must rely on smaller areas for food production. This, in turn, places greater stress on the environment and accelerates the desertification process.

Although human suffering associated with desertification is most serious along the southern margins of the Sahara, the problem is by no means confined to the Sahel. The problem also exists in other parts of Africa as well as on every other continent except Antarctica. Each year millions of acres are lost beyond practical hope for reclamation.

Recurrent droughts may seem to be the obvious reason for desertification, but the stresses placed by people on a tenuous environment with fragile soils is the chief cause.

Figure 4.C
Desertification is most serious along the southern margin of the Sahara in a region known as the Sahel. The lines defining the approximate boundaries of the Sahel represent average annual rainfall in millimeters.

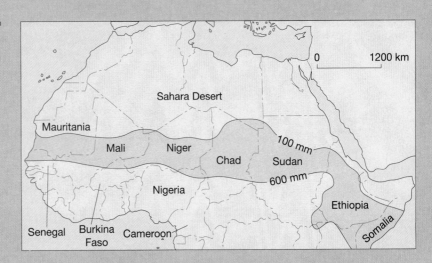

gypsum dunes at White Sands National Monument, New Mexico, will recognize this form (Figure 4.33).

Longitudinal Dunes. **Longitudinal dunes** are long ridges of sand that generally form parallel to the prevailing wind and where sand supplies are limited (Figure 4.32D). Apparently the prevailing wind direction must vary somewhat, but not by more than about 90 degrees. Although the smaller types are only three or four meters high and several tens of meters long, in some large deserts longitudinal dunes can reach great size. For example, in portions of North Africa, Arabia, and central Australia, these dunes may approach a height of 100 meters and extend for distances of more than 100 kilometers.

Parabolic Dunes. Unlike the dunes that have been described thus far, **parabolic dunes** form where vegetation partly covers the sand (Figure 4.32E). The

shape of these dunes resembles the shape of barchans except that their tips point into the wind rather than downwind. Parabolic dunes often form along coasts where there are strong onshore winds and abundant sand. If the sand's sparse vegetative cover is disturbed at some spot, deflation creates a blowout. Sand is then transported out of the depression and deposited as a curved rim which grows higher as deflation enlarges the blowout.

Star Dunes. Confined largely to parts of the Sahara and Arabian deserts, **star dunes** are isolated hills of sand that exhibit a complex form (Figure 4.32F). Their name is derived from the fact that the bases of these dunes resemble multipointed stars. Usually three or four sharp-crested ridges diverge from a central high point that in some cases may approach a height of 90 meters. As their form suggests, star dunes develop where wind directions are variable.

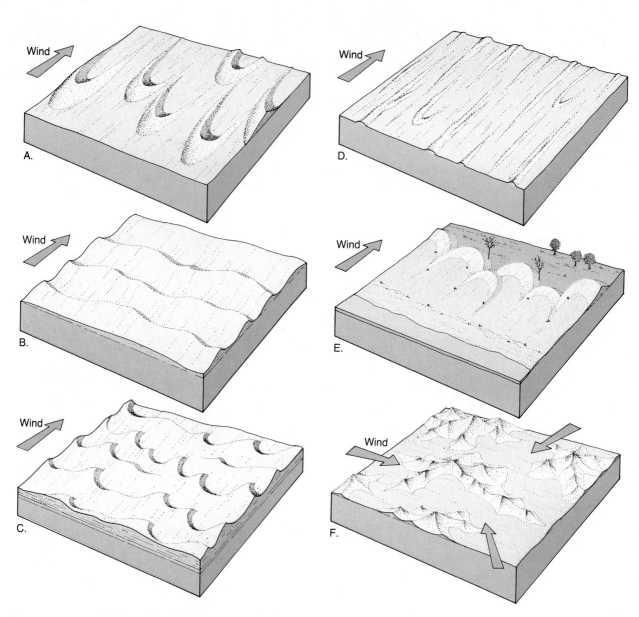

Figure 4.33
Sand dune types. **A.** Barchan dunes. **B.** Transverse dunes. **C.** Barchanoid dunes.
D. Longitudinal dunes. **E.** Parabolic dunes. **F.** Star dunes.

Review Questions

1. What is a glacier? What percentage of the earth's land area do glaciers cover?

2. Each of the following statements refers to a particular type of glacier. Name the type of glacier.
   (a) The term *continental* is often used to describe this type of glacier.
   (b) This type of glacier is also called an *alpine glacier.*
   (c) This is a glacier formed when one or more valley glaciers spreads out at the base of a steep mountain front.
   (d) Greenland is the only example in the Northern Hemisphere.

3. Describe the two components of glacial flow. At what rates do glaciers move? In a valley glacier does all of the ice move at the same rate? Explain.

4. Why do crevasses form in the upper portion of a glacier but not below a depth of about 50 meters?

5. Under what circumstances will the front of a glacier advance? Retreat? Remain stationary?

6. Describe two basic processes of glacial erosion.

7. How does a glaciated mountain valley differ from a mountain valley that was not glaciated?

8. List and describe the erosional features you might expect to see in an area where alpine glaciers exist or have recently existed.

9. What is glacial drift? What is the difference between till and stratified drift? What general effect do glacial deposits have on the landscape?

10. List the four basic moraine types. What do all moraines have in common? Distinguish between terminal and recessional moraines.

11. List and briefly describe four depositional features other than moraines.

12. How does a kettle form?

13. About what percentage of the earth's land surface was covered at some time by Pleistocene glaciers? How does this compare to the area presently covered by ice sheets and glaciers? (Check your answer with Question 1.)

14. List three indirect effects of Ice Age glaciers.

15. How might plate tectonics help us understand the cause of ice ages? Can plate tectonics explain the alternation between glacial and interglacial climates during the Pleistocene?

16. List four common misconceptions about deserts (see Box 4.2).

17. How extensive are the desert and steppe regions of the earth?

18. What is the primary cause of subtropical deserts? Of middle-latitude deserts?

19. Describe the features and characteristics associated with each of the stages in the evolution of a mountainous desert. Where in the United States can these stages be observed?

20. What term refers to the process by which desertlike conditions expand into areas that were previously productive? Is this strictly a natural process? (See Box 4.3.)

21. Why is wind erosion relatively more important in arid regions than in humid areas?

22. List two types of wind erosion and the features which may result from each.

23. Although sand dunes are the best-known wind deposits, accumulations of loess are very significant in some parts of the world. What is loess? Where are such deposits found? What are the origins of this sediment?

24. How do sand dunes migrate?

25. Identify each of the dunes described in the following statements.
   (a) Long ridges of sand oriented parallel to the prevailing wind.
   (b) Solitary, crescent-shaped dunes oriented with their tips pointing downwind.
   (c) Ridges of sand oriented at right angles to the prevailing wind.
   (d) Dunes whose tips point into the wind.
   (e) Scalloped rows of sand oriented at right angles to the wind.
   (f) Isolated dunes consisting of three or four sharp-crested ridges diverging from a central high point.

Key Terms

ablation (p. 138)

abrasion (p. 139)

alluvial fan (p. 160)

alpine glacier (p. 134)

arête (p. 143)

barchan dune (p. 166)

barchanoid dune (p. 166)

blowout (p. 162)

cirque (p. 141)

crevasse (p. 136)

cross beds (p. 164)

deflation (p. 162)

desert (p. 154)

desert pavement (p. 162)

drift (p. 143)

drumlin (p. 147)

end moraine (p. 146)

ephemeral stream (p. 157)

esker (p. 147)

fiord (p. 141)

glacial erratic (p. 145)

glacial striations (p. 140)

glacial trough (p. 141)

glacier (p. 134)

ground moraine (p. 146)

hanging valley (p. 141)

horn (p. 143)

ice cap (p. 135)

ice sheet (p. 134)

inselberg (p. 161)

interior drainage (p. 158)

kame (p. 147)

kettle (p. 147)

lateral moraine (p. 145)

loess (p. 163)

longitudinal dune (p. 167)

medial moraine (p. 145)

outwash plain (p. 147)

parabolic dune (p. 167)

pediment (p. 161)

piedmont glacier (p. 135)

playa lake (p. 160)

Pleistocene epoch (p. 150)

plucking (p. 139)

pluvial lake (p. 151)

rainshadow desert (p. 155)

rock flour (p. 139)

slip face (p. 164)

star dune (p. 167)

steppe (p. 154)

stratified drift (p. 143)

till (p. 143)

transverse dune (p. 166)

valley glacier (p. 134)

valley train (p. 147)

zone of accumulation (p. 138)

zone of wastage (p. 138)

5

Earthquakes and the Earth's Interior

Opposite page: Damage in San Francisco caused by
the Loma Prieta Earthquake, 1989. (Photo by Lawrence
Burr/Gamma-Liaison)

Left: Damage in Marina District, San Francisco. (Photo
by Charles E. Meyers, U.S. Geological Survey)

Above: Earthquake damage to Mexico City, 1985.
(Photo by James L. Beck)

*W*hat is an earthquake? How does a seismograph record the location of a "quake"? Can earthquakes ever be predicted? If we could view the earth's interior, what would it look like? In this chapter we shall try to answer these questions, as well as many others. The study of earthquakes is important, not only because of the devastating effect that some earthquakes have on us, but also because they furnish clues about the structure of the earth's interior.

On October 17, 1989, at 5:04 P.M. Pacific Daylight Time, strong tremors shook the San Francisco Bay area. Millions of Americans and others around the world were just getting ready to watch the third game of the World Series, but instead saw their television sets go black as the shock hit Candlestick Park. Although the Loma Prieta earthquake was centered in a remote section of the Santa Cruz Mountains, about 16 kilometers north of the city of Santa Cruz, major damage occurred in the Marina District of San Francisco 100 kilometers to the north (Figure 5.1). Here, as many as 60 row houses were so badly damaged that they had to be demolished. Although wood frame structures often survive earthquakes with little or no structural damage, many of these homes were built over garages that were supported only by thin wooden columns. Simply, there were no walls on the lower levels to resist the horizontal stresses.

The most tragic result of the violent shaking was the collapse of some double-decked sections of Interstate 880, also known as the Nimitz Freeway (see Figure 5.21). The ground motions caused the upper deck to sway, shattering the concrete support columns along a mile-long section of the freeway. The upper deck then collapsed onto the lower roadway, flattening cars as if they were aluminum beverage cans. Other roadways that were damaged during this earthquake included a 50-foot section of the upper deck of the Bay Bridge, which is a major artery connecting the cities of Oakland and San Francisco. The vibration caused cars on the bridge to bounce up and down vigorously. A motorcyclist on the upper deck described how the roadway bulged and rippled toward him: "It was like bumper cars—only you could die. . .!" Fortunately, only one motorist on the bridge was killed.

The Loma Prieta earthquake lasted just 15 seconds and occurred along the northern segment of the San Andreas fault. This active fault zone and associated faults—including the Hayward fault, which runs through Oakland, and the San Jacinto fault near San Bernadino—extend northward from southern California for over 1000 kilometers. It is along this fault system that two great sections of the earth, the North American plate and the Pacific plate, grind past each other at the rate of a few centimeters per year. Along much of this fault zone the rocks on either side tend to remain locked, resisting the overall motion. In time the stress builds to a point where the strength of the rocks is exceeded and the plates slide past each other, releasing the stored energy in short bursts. The result is an earthquake.

Figure 5.1
*Damage to structures in the Marina District of San Francisco caused by the October 17,
1989, Loma Prieta earthquake. (Photo by Charles E. Meyers, U.S. Geological Survey)*

In spite of the huge economic loss in the Bay Area, it is remarkable that the losses were not even greater. Just a year earlier, a somewhat weaker earthquake killed an estimated 25,000 people in Soviet Armenia. Unquestionably, the efforts to upgrade structures to conform to the building codes of California helped minimize what could have been a catastrophic event.

Although some comfort can be taken from the fact that most buildings in the Bay Area held up well during this event, it is also clear that this was not the long-feared "Big One." Whereas the Loma Prieta earthquake had a magnitude of 7.1 on the Richter scale, the 1985 Mexico City earthquake, with a magnitude of 8.1, involved the release of more than 30 times as much energy. The latter quake was centered along the Pacific Coast nearly 400 kilometers (250 miles) from Mexico City. In less than two minutes the quake battered downtown Mexico City, causing 412 buildings to collapse and killing an estimated 9500

persons. The vibrations from this tremor were felt as far north as Houston, Texas, where skyscrapers swayed. The fact that the central district of Mexico City is built on the soft, moist sediment of an ancient lake bed contributed to the destruction. According to a researcher at the California Institute of Technology, this earthquake caused the lake bed to vibrate "like a bowl of jelly when the seismic waves came through the ground beneath it."

It is estimated that over 30,000 earthquakes, strong enough to be felt, occur worldwide annually. Fortunately, most of these are minor tremors and do very little damage. Generally only about 75 significant earthquakes take place each year, and many of these occur in remote regions. However, occasionally a large earthquake occurs near a large population center. When such an event takes place, it is among the most destructive natural forces on earth. The shaking of the ground coupled with the liquefaction of some soils wreak havoc on buildings and other structures. In addition, when a quake occurs in a populated area, power and gas lines are often ruptured, causing numerous fires. In the 1906 San Francisco earthquake, much of the damage was caused by fires which ran unchecked when broken water mains left firefighters with only trickles of water (Figure 5.2).

WHAT IS AN EARTHQUAKE?

An **earthquake** is the vibration of the earth produced by the rapid release of energy. This energy radiates in all directions from its source, the **focus,** in the form of waves analogous to those produced when a stone is dropped into a calm pond (Figure 5.3). Just as the impact of the stone sets water waves in motion, an earthquake generates seismic waves that ra-

diate throughout the earth. Even though the energy dissipates rapidly with increasing distance from the focus, instruments located throughout the world record the event.

The tremendous energy released by atomic explosions or by volcanic eruptions can produce an earthquake, but these events are usually weak and infrequent. What mechanism does produce a destructive earthquake? Ample evidence exists that the earth is not a static planet. Numerous ancient wave-cut benches can be found many meters above the level of the highest tides, which indicates crustal uplifting. Other regions exhibit evidence of extensive subsidence. In addition to these vertical displacements, offsets in fence lines, roads, and other structures indicate that horizontal movement is also prevalent (Figure 5.4). These movements are usually associated with large fractures in the earth called **faults.** Most of the motion along faults can be satisfactorily ex-

Figure 5.2
San Francisco in flames after the 1906 earthquake. (Reproduced from the collection of the Library of Congress)

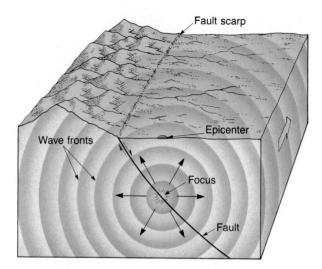

Figure 5.3
The focus of all earthquakes is located at depth. The surface location directly above it is called the epicenter.

plained by the plate tectonics theory. This theory proposes that large slabs of the earth are continually in motion. These mobile plates interact with neighboring plates, straining and deforming the rocks at their edges. It is along faults associated with plate boundaries that most earthquakes occur. Furthermore, earthquakes are repetitive; that is, as soon as one earthquake is over, the continuous motion of the plates begins to add strain to the rocks until they fail again.

The actual mechanism of earthquake generation eluded geologists until H. F. Reid of Johns Hopkins University conducted a study following the great 1906 San Francisco earthquake. The earthquake was accompanied by displacements of several meters along the northern portion of the San Andreas fault, a 1300-kilometer (780-mile) fracture which runs northward through southern California. This large fault zone separates two great sections of the earth, the North American plate and the Pacific plate. Field investigations determined that during this single earthquake the Pacific plate slid as much as 4.7 meters (15 feet) in a northward direction past the adjacent North American plate.

Using land surveys conducted several years apart, Reid discovered that during the 50 years prior to the 1906 earthquake the land at distant points on both sides of the San Andreas fault showed a relative displacement of slightly more than 3 meters (10 feet). The mechanism for earthquake formation which

Reid deduced from this information is illustrated in Figure 5.5. Tectonic forces ever so slowly deform the crustal rocks on both sides of the fault as illustrated by the bent features. Under these conditions, rocks are bending and storing elastic energy, much like a wooden stick would if bent. Eventually, the frictional resistance holding the rocks together is overcome. As slippage occurs at the weakest point (the focus), displacement will exert stress farther along the fault where additional slippage will occur until most of the built-up strain is released. This slippage allows the deformed rock to "snap back." The vibrations we know as an earthquake occur as the rock elastically returns to its original shape. The "springing back" of the rock was termed **elastic rebound** by Reid, since the rock behaves elastically, much like a stretched rubber band does when it is released.

In summary, most earthquakes are produced by the rapid release of elastic energy stored in rock that has been subjected to great differential stress. Once the strength of the rock is exceeded, it suddenly ruptures, which results in the vibrations of an earthquake. Earthquakes also occur along existing faults when the frictional forces on the fault surfaces are overcome.

The intense vibrations of the 1906 San Francisco earthquake lasted about 40 seconds. Although most

Figure 5.4
Slippage along a fault produced an offset in this orange grove located east of Calexico, California. (Photo by John S. Shelton)

BOX 5.1
Severe Earthquakes

Earthquakes are among nature's most destructive forces. Many people in the United States are familiar with the Loma Prieta earthquake that struck the San Francisco Bay Area during the 1989 World Series. Although this event was dramatic and destructive, earthquakes can be much worse. Two of the most damaging earthquakes of the 1980s occurred in Armenia and Mexico.

On December 7, 1988, a violent earthquake jolted northwestern Armenia, devastating several towns and killing an estimated 25,000 persons. Poor construction practices were blamed for the high death toll in this rugged region located near the Caucasus Mountains. To quote a seismologist who was asked about this destructive event, "Earthquakes don't kill people, buildings do." Many of the taller structures in the region were made of precast concrete slabs that were held together by metal hooks. A San Francisco structural engineer studying the destruction stated, "There was very little reinforcing to tie these buildings together. The buildings basically came apart the way they were put together." In addition, many of the village homes had thick floors made of mud and rock, which were deadly when they collapsed. Nearly the entire town of Spitak, which was located near the epicenter, was leveled.

This destructive earthquake registered 6.9 on the Richter scale and was centered near the Armenian-Turkish border. A few minutes after the main shock struck, a large aftershock of magnitude 5.8 collapsed many structures that had been weakened by the main tremor. The cause of this devastating earthquake was the sudden movement along a fault that had not been previously identified. This newly discovered fault is believed to have an origin similar to that of the many other faults that slice through the region. They are all associated with the collision of the Arabian plate with the Eurasian plate.

Although the Armenian earthquake was one of the century's worst in terms of deaths, the magnitude of the shock was much less than that of the earthquake that struck Mexico City in 1985. The latter quake, which registered 8.1 on the Richter scale, was centered along the Pacific Coast nearly 400 kilometers (250 miles) from Mexico City. In less than two minutes, the quake battered downtown Mexico City, causing 412 buildings to collapse and damaging over 7000 others (Figure 5.A). Fortunately, the earthquake struck during the morning rush hour when businesses and schools were not fully occupied. Nevertheless, casualties in Mex-

---

of the displacement along the fracture occurred in this rather short period, additional movements and adjustments in the rocks occurred for several days following the main quake. The adjustments that follow a major earthquake often generate smaller earthquakes called **aftershocks.** Although these aftershocks are usually much weaker than the main earthquake, they can sometimes cause significant destruction to already badly weakened structures. This occurred, for example, during the 1988 Armenian earthquake, when a large aftershock of magnitude 5.8 collapsed many structures that had been weakened by the main tremor. In addition, small earthquakes called **foreshocks** often precede a major earthquake by days or, in some cases, by as much as several years. Monitoring of these foreshocks has been used as a means of predicting forthcoming major earthquakes. We will consider the topic of earthquake prediction in a later section of this chapter.

The tectonic forces responsible for the strain that was eventually released during the 1906 San Francisco earthquake are still active. Currently laser beams* are used to establish the relative motion between the opposite sides of this fault. These measurements have revealed a displacement of from 2 to 5

---

* Laser beams are used in very precise surveying instruments because of their incredibly accurate straight-line qualities.

ico City included more than 9500 deaths, with 30,000 injured and 50,000 left homeless.

The vibrations were felt as far north as Houston, Texas, where skyscrapers swayed. Ironically, coastal towns such as Ixtapa, which were closer to the epicenter than Mexico City, suffered less during the event. The fact that the central district of Mexico City is built on the soft, moist sediments of an ancient lake bed contributed to the destruction.

According to a researcher at the California Institute of Technology, this earthquake caused the lake bed to vibrate "like a bowl of jelly when the seismic waves came through the ground beneath it."

Although the destruction to downtown Mexico City was staggering, the damage was very restricted. Out of a total of more than 800,000 structures located

in the city, only a few hundred buildings collapsed. Further, only one percent of the city was heavily damaged. This is not the first earthquake to batter Mexico City, and it will not be the last. During this century, California has had five earthquakes with Richter magnitudes greater than 7, while Mexico has had over 40, many of which have brought great human suffering.

---

centimeters per year. Although this rate of movement seems slow, it is fast enough to produce substantial movement over millions of years of geologic time. In 30 million years this rate of displacement is sufficient to slide the western portion of California northward so that Los Angeles, on the Pacific plate, would be adjacent to San Francisco on the North American plate. More importantly in the short term, a displacement of just 2 centimeters per year produces 2 meters of offset every 100 years. Consequently, the 4 meters of displacement produced during the 1906 San Francisco earthquake should occur at least every 200 years along this segment of the fault zone.

The San Andreas is undoubtedly the most studied fault system in the world. Over the years investiga-

tions have shown that displacement occurs along discrete segments that are 100 to 200 kilometers long. Further, each fault segment behaves somewhat differently from the others. Some portions of the San Andreas exhibit a slow, gradual displacement known as *creep*, which occurs with little noticeable seismic activity. Other segments regularly slip, producing small earthquakes, while still other segments remain locked and store elastic energy for hundreds of years before rupturing in great earthquakes. The latter process is described as *stick-slip* motion, since the fault exhibits alternating periods of locked behavior followed by sudden slippage. It is estimated that great earthquakes should occur about every 50 to 200 years along those sections of the San Andreas fault

Figure 5.5
Elastic rebound. As rock is deformed it bends, storing elastic energy. Once the rock is strained beyond its breaking point it ruptures, releasing the stored up energy in the form of earthquake waves.

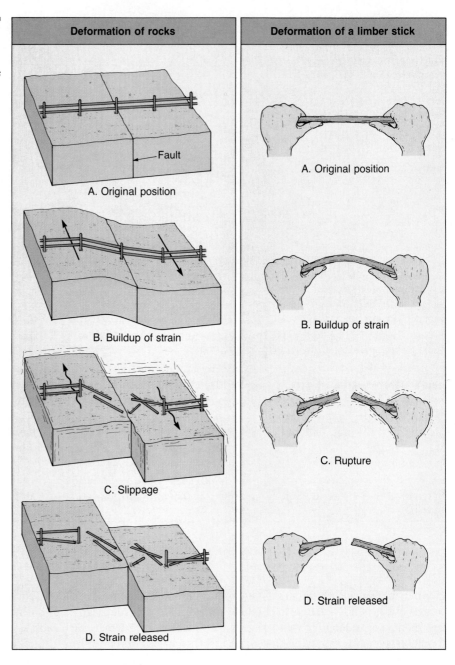

Deformation of rocks	Deformation of a limber stick
A. Original position	A. Original position
B. Buildup of strain	B. Buildup of strain
C. Slippage	C. Rupture
D. Strain released	D. Strain released

that exhibit stick-slip motion. This knowledge is useful when assigning a potential earthquake risk to a given segment of the fault zone.

Not all movement along faults is horizontal. Vertical displacement, in which one side is lifted higher in relation to the other, is also common. Figure 5.6 shows a *fault scarp* (cliff) produced during the 1964 Good Friday earthquake in Alaska. Further, many earthquakes occur at such great depths that no displacement is evident at the surface.

EARTHQUAKE WAVES

The study of earthquake waves, **seismology,** dates back to attempts by the Chinese almost 2000 years ago to determine the direction to the source of each earthquake. The principle used in modern **seismographs,** instruments which record earthquake waves, is rather simple. A weight is freely suspended from a support that is attached to bedrock (Figure 5.7). When waves from a distant earthquake reach the instrument, the inertia* of the weight keeps it stationary, while the earth and the support vibrate. The movement of the earth in relation to the stationary weight is recorded on a rotating drum.

Modern seismographs amplify and record ground motion, producing a trace as shown in Figure 5.8. These records, called **seismograms,** provide a great deal of information about the behavior of seismic waves. Simply stated, seismic waves are elastic energy which radiates outward in all directions from the focus. The propagation (transmission) of this energy can be compared to the shaking of gelatin in a bowl that results when some is spooned out. Whereas the gelatin will have one mode of vibration, seismograms reveal that two main groups of seismic waves are generated by the slippage of a rock mass. One of these wave types travels along the outer layer of the earth. These are called **surface waves.** Others travel through the earth's interior and are called **body waves**. Body waves are further divided into two types

---

* Simply stated, inertia refers to the fact that objects at rest tend to stay at rest and objects in motion tend to remain in motion unless acted upon by an outside force. You probably have experienced this phenomenon when you tried to quickly stop your automobile and your body continued to move forward.

called **primary,** or **P, waves** and **secondary,** or **S, waves.**

Body waves are divided on the basis of their mode of propagation through intervening materials. P waves push (compress) and pull (dilate) rocks in the direction the wave is traveling (Figure 5.9A). This wave motion is analogous to that generated by human vocal cords as they move air to create sound. Solids, liquids, and gases resist a change in volume when compressed and will elastically spring back once the force is removed. Therefore, P waves, which are compressional waves, can travel through all these

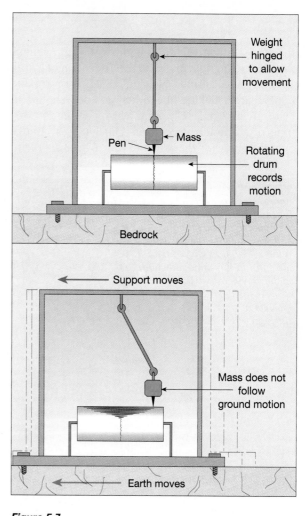

Figure 5.7
Principle of the seismograph. The inertia of the suspended mass tends to keep it motionless, while the recording drum, which is anchored to bedrock, vibrates in response to seismic waves. Thus, the stationary mass provides a reference point from which to measure the amount of displacement occurring as the seismic wave passes through the ground below.

Figure 5.6
Fault scarp produced from vertical movement during the 1964 Alaskan earthquake. (Courtesy of U.S. Geological Survey)

Figure 5.8
Typical seismic record. Note the time interval between the arrival of the first P waves and the arrival of the first S waves.

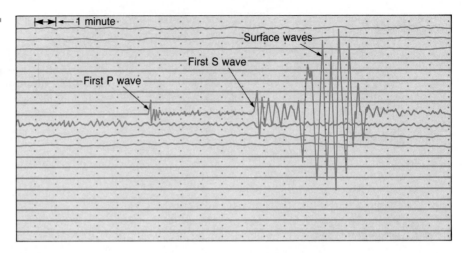

materials. S waves, on the other hand, "shake" the particles at right angles to their direction of travel. This can be illustrated by tying one end of a rope to a post and shaking the other end, as shown in Figure 5.9B. Unlike P waves, which change the volume of the intervening material, S waves change only the shape of the material that transmits them.

Since fluids (gases and liquids) do not resist changes in shape, they will not transmit S waves.

The motion of surface waves is somewhat more complex. As surface waves travel along the ground, they cause the ground and anything resting upon it to move, much like ocean swells toss a ship. In addition to their up-and-down motion, surface waves

Figure 5.9
Types of seismic waves and their characteristic motion.
A. *P waves cause the particles in the material to vibrate back and forth in the same direction as the waves move.* ***B.*** *S waves cause particles to oscillate at right angles to the direction of wave motion.*

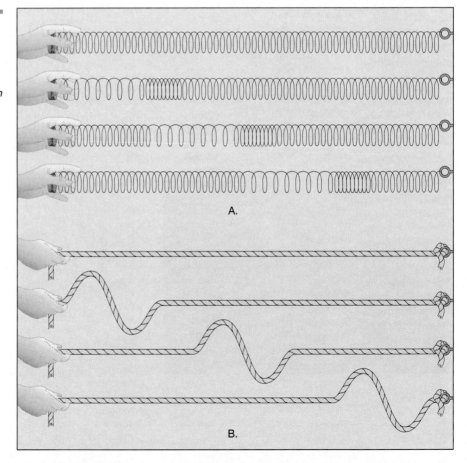

have a side-to-side motion similar to an S wave oriented in a horizontal plane. This latter motion is particularly damaging to the foundations of structures.

By observing a "typical" seismic record, as shown in Figure 5.8, one can see that one of the differences between these seismic waves becomes apparent: P waves arrive at the recording station before S waves, which themselves arrive before the surface waves. This is a consequence of their relative velocities. For purposes of illustration, the velocity of P waves through granite within the crust is about 6 kilometers per second, whereas S waves under the same conditions travel at 3.5 kilometers per second. Differences in density and elastic properties of the transmitting material greatly influence the velocities of these waves. However, in any solid material, P waves travel about 1.7 times faster than S waves, and surface waves can be expected to travel at 90 percent of the velocity of the S waves.

As we shall see, seismic waves allow us to determine the location and magnitude of earthquakes. In addition, seismic waves provide us with a tool for probing the earth's interior.

LOCATION OF EARTHQUAKES

Recall that the focus is the place within the earth where the earthquake waves originate. The **epicenter** is the location on the surface directly above the focus (see Figure 5.3). The difference in velocities of P and S waves provides a method for determining the epicenter. The principle used is analogous to a race between two autos, one faster than the other. The greater the distance of the race, the greater will be the difference in the arrival times at the finish line. Therefore, the greater the interval between the arrival of the first P wave and the first S wave, the greater the distance to the earthquake source.

A system for locating earthquake epicenters was developed through the use of seismograms from earthquakes whose epicenters could be easily pinpointed from physical evidence. From these seismograms, travel-time graphs as shown in Figure 5.10 were constructed. The first travel-time graphs were greatly improved when seismograms became available from nuclear explosions, because the location and time of detonation were well-established.

Using the sample seismogram in Figure 5.8 and the travel-time curves in Figure 5.10, we can determine the distance separating the recording station from the earthquake. The time interval between the

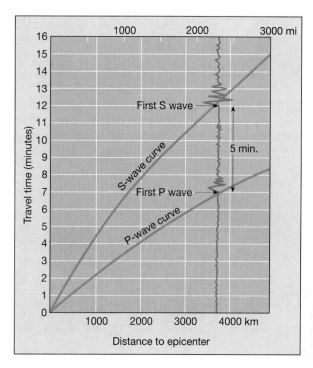

Figure 5.10
A travel-time graph is used to determine the distance to the epicenter. The difference in arrival times of the first P wave and the first S wave in the example is 5 minutes. Thus, the epicenter is roughly 3800 kilometers (2350 miles) away.

arrival of the first P wave and the first S wave is determined, then the place on the travel-time graph which exhibits an equivalent time spread between the P and S wave curves is found. From this information, we can determine that this earthquake occurred 3800 kilometers (2350 miles) from the recording instrument. Although the distance to an earthquake is established in this manner, its location could be in any direction from the observer. As shown in Figure 5.11, the precise location can be found only when the distance is known from three or more different seismic stations. By drawing circles representing the epicenter distance for each of these observatories, an accurate location is established.

About 95 percent of the energy released by earthquakes is concentrated in a few relatively narrow zones that wind around the globe (Figure 5.12). The greatest energy is released along a path near the outer edge of the Pacific Ocean known as the *circum-Pacific belt*. Included in this zone are regions of great seismic activity, such as Japan, the Philippines, Chile, and numerous volcanic island chains, as exemplified by the Aleutian Islands. Another major concentration

Figure 5.11
Earthquake epicenter is located
using the distance obtained
from three seismic stations.

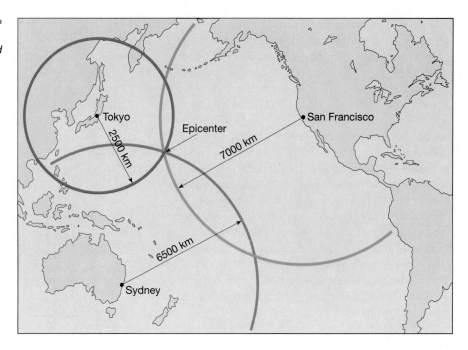

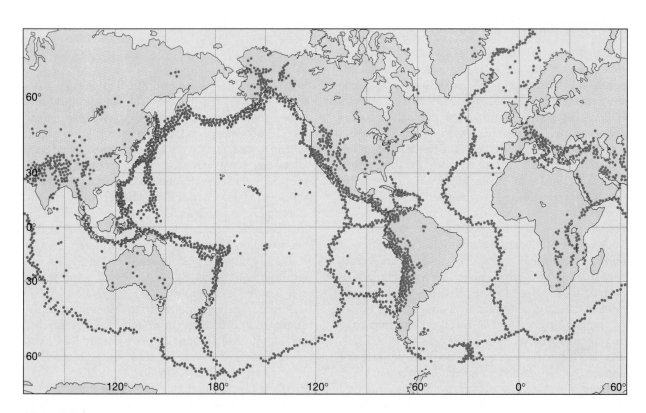

Figure 5.12
World distribution of earthquakes for a nine-year period. (Data from NOAA)

of strong seismic activity runs through the mountainous regions that flank the Mediterranean Sea and continues through Iran and on past the Himalayan complex. Figure 5.12 indicates that yet another continuous belt extends for thousands of kilometers through the world's oceans. This zone coincides with the oceanic ridge system, an area of frequent but low-intensity seismic activity. By comparing this figure with Figure 6.8, we can see a close correlation between the location of earthquake epicenters and plate boundaries, a phenomenon which will be explored in the next chapter.

EARTHQUAKE INTENSITY AND MAGNITUDE

Early attempts to establish the size or strength of earthquakes relied heavily on subjective descriptions.

There was an obvious problem with this method—people's accounts varied widely, making an accurate classification of the quake's intensity difficult. Then, in 1902, a fairly reliable intensity scale based on the amount of damage caused to various types of structures was developed by Giuseppe Mercalli. A modified form of this tool is presently used by the U.S. Coast and Geodetic Survey (Table 5.1).

By definition, earthquake **intensity** is a measure of the effects of a quake at a particular local. It is important to note the earthquake intensity depends not only on the strength of the earthquake, but on other factors as well. These include the distance from the epicenter, the nature of the surface materials, and building design. Recall that the modest 6.9-Richter-magnitude Armenian earthquake of 1988 was very destructive mainly because of inferior construction practices. Moreover, the nature of the surface mate-

Table 5.1
Modified Mercalli intensity scale.

I Not felt except by a very few under especially favorable circumstances.	able in poorly built or badly designed structures.
II Felt only by a few persons at rest, especially on upper floors of buildings.	**VIII** Damage slight in specially designed structures; considerable in ordinary substantial buildings with partial collapse; great in poorly built structures. (Fall of chimneys, factory stacks, columns, monuments, and other vertically oriented features.)
III Felt quite noticeably indoors, especially on upper floors of buildings, but many people do not recognize as an earthquake.	
IV During the day felt indoors by many, outdoors by few. Sensations like heavy truck striking building.	**IX** Damage considerable in specially designed structures. Buildings shifted off foundations. Ground badly cracked.
V Felt by nearly everyone, many awakened. Disturbances of trees, poles, and other tall objects sometimes noticed.	**X** Some well-built wooden structures destroyed. Most masonry and frame structures destroyed with foundations. Ground cracked conspicuously.
VI Felt by all; many frightened and run outdoors. Some heavy furniture moved; few instances of fallen plaster or damaged chimneys. Damage slight.	**XI** Few, if any, (masonry) structures remain standing. Bridges destroyed. Broad fissures in ground.
VII Everybody runs outdoors. Damage negligible in buildings of good design and construction; slight to moderate in well-built ordinary structures; consider-	**XII** Damage total. Waves seen on ground surfaces. Objects thrown upward into the air.

Source: U.S. Coast and Geodetic Survey.

rials contributed significantly to the destruction in central Mexico City during the 1985 quake. Thus, the destruction wrought by earthquakes is not always an adequate means for comparing their relative size. Furthermore, many earthquakes occur in oceanic areas or at great focal depths and either are not felt, or their intensities do not reflect their true strength.

In 1935, Charles Richter of the California Institute of Technology introduced the concept of earthquake **magnitude** when attempting to rank earthquakes of southern California. Today a refined **Richter scale** is used worldwide to describe earthquake magnitude. Using Richter's scale, the magnitude is determined by measuring the amplitude of the largest wave recorded on the seismogram (see Figure 5.8). Although seismographs greatly magnify the ground motion, large-magnitude earthquakes will cause the recording pen to be displaced farther than will small-magnitude earthquakes. In order for seismic stations worldwide to obtain the same magnitude for a given earthquake, adjustments must be made for the weakening of seismic waves as they move from the focus and for the sensitivity of the recording instrument.

The largest earthquakes ever recorded have Richter magnitudes near 8.6. These great shocks released energy roughly equivalent to the detonation of one billion tons of TNT. Conversely, earthquakes with a Richter magnitude of less than 2.0 are usually not felt by humans. Table 5.2 shows how earthquake magnitudes and their effects are related.

As we have seen, earthquakes vary enormously in strength. Furthermore, great earthquakes produce traces having wave amplitudes that are thousands of times larger than those generated by weak tremors. To accommodate this wide variation, Richter used a logarithmic scale to express magnitude. On this scale a tenfold increase in recorded wave amplitude corresponds to an increase of one on the magnitude scale. Thus, the amplitude of the largest surface wave for a 5-magnitude earthquake is 10 times greater than the wave amplitude produced by an earthquake having a magnitude of 4. More importantly, each unit of magnitude increase on the Richter scale equates to roughly a 30-fold increase in the energy released. Thus, an earthquake with a magnitude of 6.5 releases 30 times more energy than one with a magnitude of 5.5, and roughly 900 times (30 × 30) more energy than a 4.5-magnitude quake. A major earthquake with a magnitude of 8.5 releases millions of times more energy than the smallest earthquake felt by humans. This dispels the notion that a moderate earthquake decreases the chances for the occurrence of a major quake in the same region. Thousands of moderate tremors would be needed to release the amount of energy released by one "great" earthquake. Some

Table 5.2
Earthquake magnitudes and expected world incidence.

Richter Magnitudes	Earthquake Effects	Estimated Number per Year
< 2.0	Generally not felt, but recorded.	600,000
2.0–2.9	Potentially perceptible.	300,000
3.0–3.9	Felt by some.	49,000
4.0–4.9	Felt by most.	6200
5.0–5.9	Damaging shocks.	800
6.0–6.9	Destructive in populous regions.	266
7.0–7.9	Major earthquakes. Inflict serious damage.	18
≥ 8.0	Great earthquakes. Produce total destruction to communities near epicenter.	1.4

Source: *Earthquake Information Bulletin* and others.

Figure 5.13
Region most affected by the Good Friday earthquake of 1964. Note the location of the epicenter (red dot). (After U.S. Geological Survey)

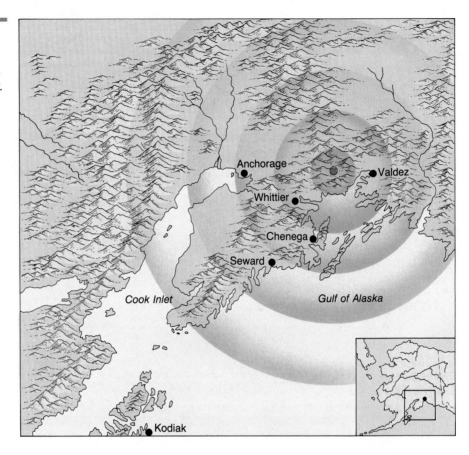

of the world's major earthquakes and their corresponding Richter magnitudes are listed in Table 5.3.

EARTHQUAKE DESTRUCTION

The most violent earthquake to jar North America this century—the Good Friday Alaskan Earthquake—occurred at 5:36 P.M. on March 27, 1964. Felt throughout that state, the earthquake had a magnitude of 8.3–8.4 on the Richter scale and reportedly lasted 3 to 4 minutes. This brief event left 131 persons dead, thousands homeless, and the economy of the state badly disrupted. Had the schools and business districts been open, the toll surely would have been higher. Within 24 hours of the initial shock, 28 aftershocks were recorded, 10 of which exceeded a magnitude of 6 on the Richter scale. The location of the epicenter and the towns that were hardest hit by the quake are shown in Figure 5.13.

Many factors determine the amount of destruction that accompanies an earthquake. The most ob-

vious of these are the magnitude of the earthquake and the proximity of the quake to a populated area. Fortunately, most earthquakes are small and occur in remote regions of the earth. However, about 20 major earthquakes are reported annually, one or two of which are catastrophic.

Destruction Caused by Seismic Vibrations

The 1964 Alaskan earthquake provided geologists with new insights into the role of ground shaking as a destructive force. As the energy released by an earthquake travels along the earth's surface, it causes the ground to vibrate in a complex manner by moving up and down as well as from side to side. The amount of structural damage attributable to the vibrations depends on several factors, including (1) the intensity and duration of the vibrations, (2) the nature of the material upon which the structure rests, and (3) the design of the structure.

All of the multistory structures in Anchorage were damaged by the vibrations; the more-flexible wood-

Table 5.3
Some notable earthquakes.

Year	Location	Deaths (est.)	Magnitude	Comments
1290	Chihli (Hopei), China	100,000		
1556	Shensi, China	830,000		Possibly the greatest natural disaster.
1737	Calcutta, India	300,000		
1755	Lisbon, Portugal	70,000		Tsunami damage extensive.
*1811–12	New Madrid, Missouri	Few		Three major earthquakes.
*1886	Charleston, South Carolina	60		Greatest historical earthquake in the eastern United States.
*1906	San Francisco, California	1500	8.1–8.2	Fires caused extensive damage.
1908	Messina, Italy	120,000		
1920	Kansu, China	180,000		
1923	Tokyo, Japan	143,000	7.9	Fire caused extensive destruction.
1960	Southern Chile	5700	8.5–8.6	Possibly the largest-magnitude earthquake ever recorded.
*1964	Alaska	131	8.4–8.5	
1970	Peru	66,000	7.8	Great rockslide
*1971	San Fernando, California	65	6.5	Damage exceeding one billion dollars.
1975	Liaoning Province, China	Few	7.5	First major earthquake to be predicted.
1976	Tangshan, China	240,000	7.6	Not predicted
1985	Mexico City	9500	8.1	Major damage occurred 400 km from epicenter.
1988	Armenia	25,000	6.9	Poor construction practices contributed to destruction.
*1989	San Francisco Bay Area	62	7.1	Damages exceeded 6 billion dollars.
1990	Northwestern Iran	50,000	7.3	Landslides and poor construction practices caused great damage.

Source: U.S. National Oceanic and Atmospheric Administration.

* U.S. earthquakes.

frame residential buildings fared best. However, many homes were destroyed when the ground failed. A striking example of how construction variations affect earthquake damage is shown in Figure 5.14. We can see in this photo that the steel-frame building on the left withstood the vibrations, whereas the relatively rigid concrete structure was badly damaged.

Most of the large structures in Anchorage were damaged even though they were built to conform to the earthquake provisions of the Uniform Building Code of California. Perhaps some of that destruction can be attributed to the unusually long duration of this earthquake, which was estimated at 3 to 4 minutes. Most earthquakes consist of tremors that last less than one minute. For example, the San Francisco earthquake of 1906 was felt for about 40 seconds, whereas the strong vibrations of the 1989 Loma Prieta earthquake lasted less than 15 seconds.

Figure 5.14
Damage to the five-story J. C. Penney Co. building, Anchorage, Alaska. Very little structural damage was incurred by the adjacent building. (Courtesy of NOAA)

Although the region within 20 to 50 kilometers of an epicenter will experience about the same degree of ground shaking, the destruction will vary considerably within this area. This difference is mainly attributable to the nature of the ground on which the structures are built. Soft sediments, for example, generally amplify the vibration more than solid bedrock. Thus, the buildings in Anchorage, which were situated on unconsolidated sediments, experienced heavy structural damage. By contrast, most of the town of Whittier, although located much nearer to the epicenter than Anchorage, rests on a firm foundation of granite and hence suffered much less damage from the seismic vibrations. However, Whittier was damaged by a seismic sea wave.

The 1985 Mexican earthquake gave seismologists and engineers a vivid reminder of what had been learned following the 1964 Alaskan earthquake. The Pacific coast of Mexico, where the earthquake was centered, experienced unusually mild tremors despite the strength of the quake. As expected, the seismic waves became progressively weaker with increasing distance from the epicenter. However, in the central section of Mexico City, nearly 400 kilometers (250 miles) from the source, the vibrations intensified to five times that experienced in outlying districts. Much of this amplified ground motion can be attributed to soft sediments, remnants of an ancient lake bed, that underlie portions of the city.

To understand what happened in Mexico City, recall that as seismic waves pass through the earth they cause the intervening material to vibrate much as a tuning fork vibrates when struck. Although most objects can be forced to vibrate over a wide range of frequencies, each has a natural period of vibration that is preferred. Various earth materials, like different-length tuning forks, also have different natural periods of vibration.*

Ground motion is amplified when the natural period of ground vibration matches that of the seismic waves. A common example of this phenomenon occurs when a parent pushes a child on a swing. As the parent periodically pushes the child in rhythm with the frequency of the swing, the child moves back and forth in a greater and greater arc. By chance, the column of sediment beneath Mexico City had a natural period of vibration of about two seconds, matching that of the strongest seismic waves. Thus, when the seismic waves began shaking the soft sediments, a resonance developed which greatly increased the amplitude of the vibrations. This amplified motion began throwing the ground back and forth about 40 centimeters (16 inches) every two seconds for a

---

* To demonstrate the natural period of vibration of an object, hold a ruler over the edge of a desk so that most of it is not supported by the desk. Start it vibrating and notice the sound it makes. By changing the length of the unsupported portion of the ruler, the natural period of vibration will change accordingly, as reflected by the change in sound.

Figure 5.15
During the 1985 Mexican earthquake, multistory buildings swayed back and forth as much as one meter. Many, including the hotel shown here, collapsed or were seriously damaged. (Photo by James L. Beck)

period of nearly two minutes. Such shaking was too intense for many of the poorly designed buildings in the city. In addition, buildings of intermediate height (5 to 15 stories) swayed back and forth with a period of about two seconds. Thus, resonance also developed between these buildings and the ground, and most of the building failures were in this height range (Figure 5.15).

In areas where unconsolidated materials are saturated with water, earthquakes can generate a phenomenon known as **liquefaction.** Under these conditions, what had been a stable soil turns into a fluid that is not capable of supporting buildings or other structures (Figure 5.16). As a result, underground objects such as storage tanks and sewer lines may literally float toward the surface, while buildings and

other structures may settle and collapse. In San Francisco's Marina District, foundations failed and geysers of sand and water shot from the ground, indicating that liquefaction occurred during the Loma Prieta earthquake of 1989 (Figure 5.17).

Tsunamis

Most deaths associated with the 1964 Alaskan quake were caused by **seismic sea waves,** or **tsunamis.**[*] These destructive waves have popularly been called

---

[*] Seismic sea waves were given the name *tsunami* by the Japanese, who have suffered a great deal from them. The term *tsunami* is now used worldwide.

Figure 5.16
Effects of liquefaction. This tilted building rests on unconsolidated sediment that imitated quicksand during the 1985 Mexican earthquake. (Photo by James L. Beck)

Figure 5.17
These "mud volcanoes" were produced by the Loma Prieta earthquake of 1989. They formed when geysers of sand and water shot from the ground, an indication that liquefaction occurred. (Photo by Richard Hilton, courtesy of Dennis Fox)

"tidal waves." However, this name is not accurate since these waves are not generated by the tidal effect of the moon or sun.

Most tsunamis result from vertical displacement of the ocean floor during an earthquake as illustrated in Figure 5.18. Once formed, a tsunami resembles the ripples formed when a pebble is dropped into a pond. In contrast to ripples, tsunamis advance at speeds between 500 and 950 kilometers (300 and 600 miles) per hour. Despite this striking characteristic, a tsunami in the open ocean can pass undetected because its height is usually less than one meter and the distance between wave crests ranges from 100 to 700 kilometers. However, upon entering shallower coastal water, these destructive waves are slowed and the water begins to pile up to heights that occasionally exceed 30 meters (100 feet). As the crest of a tsunami approaches shore, it appears as a rapid rise in sea level with a turbulent and chaotic surface.

Usually the first warning of an approaching tsunami is a rather rapid withdrawal of water from beaches. Residents of coastal areas have learned to heed this warning and move to higher ground. About 5 to 30 minutes later the retreat of water is followed by a surge capable of extending hundreds of meters inland. In a successive fashion, each surge is followed by rapid oceanward retreat of the water. These waves, separated by intervals of between 10 and 60 minutes, are able to traverse large stretches of the ocean before their energy is totally dissipated.

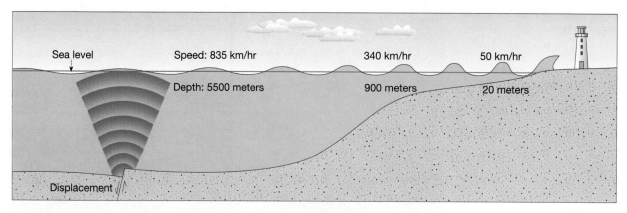

Figure 5.18
Schematic drawing of a tsunami generated by displacement of the ocean floor. The speed of a wave column correlates with ocean depth. As shown, waves moving in deep water advance at speeds in excess of 800 kilometers per hour. Speed gradually slows to 50 kilometers per hour at depths of 20 meters. Decreasing depth slows the movement of the wave column. As waves slow in shallow water, they grow in height until they topple and rush onto shore with tremendous force. The size and spacing of these swells are not to scale.

The tsunami generated in the 1964 Alaskan earthquake inflicted heavy damage to communities in the vicinity of the Gulf of Alaska, completely destroying the town of Chenega. The town of Seward was also heavily damaged as most of its port facilities were demolished by this seismic sea wave (Figure 5.19). The deaths of 107 persons have been attributed to this tsunami. By contrast, only 9 persons died in Anchorage as a direct result of the vibrations. Tsunami damage following the Alaskan earthquake extended along much of the west coast of North America, and in spite of a one-hour warning, 12 persons perished in Crescent City, California, where all of the deaths and most of the destruction were caused by the fifth wave. The first wave crested about 4 meters (13 feet) above low tide and was followed by three progressively smaller waves. Believing that the tsunami had ceased, people returned to the shore, only to be met by the fifth and most devastating wave, which, superimposed upon high tide, crested about 6 meters higher than the level of low tide.

Although most tsunamis are generated by earthquakes, a volcanic eruption in the ocean can generate this destructive phenomenon as well. For example, the 1883 volcanic explosion of Krakatoa generated a tsunami that drowned an estimated 36,000 coastal residents of Java and Sumatra.

Fire

Fire was only a minor consequence in the 1964 Alaskan earthquake, but sometimes it is the most de-

structive result. The 1906 earthquake centered near the city of San Francisco reminds us of the formidable threat of fire. The central city contained mostly large, older wooden structures and brick buildings. Although many of the unreinforced brick buildings were extensively damaged, the greatest destruction was caused by fires which started when gas and electrical lines were severed. The fires raged out of control for three days and devastated over 500 blocks of the city (see Figure 5.2). The problem was compounded by the initial ground shaking, which broke the city's water lines into hundreds of unconnected pieces.

Figure 5.19
The effects of a tsunami at Seward, Alaska. (Courtesy of NOAA)

Figure 5.20
Photo of a small portion of the Turnagain Heights slide. (Photo courtesy of U.S. Geological Survey)

The fire was finally contained when buildings were dynamited along a wide boulevard to provide a fire break. Although only a few deaths were attributed to the fires, that is not always the case. An earthquake which rocked Japan in 1923 triggered an estimated 250 fires, which devastated the city of Yokohama and destroyed more than half the homes in Tokyo. Over 100,000 deaths were attributed to the fires, which were driven by unusually high winds.

Landslides and Ground Subsidence

In the 1964 Alaskan earthquake, it was not ground vibrations directly but landslides and ground subsidence triggered by the vibrations that caused the greatest damage to structures. At Valdez and Seward the violent shaking caused deltaic materials to experience liquefaction; the subsequent slumping carried both waterfronts away. Because of the threat of recurrence, the entire town of Valdez was relocated about 7 kilometers away in a region of stable ground. The destruction at Valdez was compounded by the tragic loss of 31 lives. While waiting for an incoming vessel, the 31 persons died when the dock slid into the sea.

Most of the damage in Anchorage was attributed to landslides caused by the shaking and lurching ground. Many homes were destroyed in Turnagain Heights when a layer of clay lost its strength and over 200 acres of land slid toward the ocean (Figure 5.20). A portion of this landslide was left in its natural condition as a reminder of this destructive event. The site was named "Earthquake Park." Downtown Anchorage was also disrupted as sections of the main

business district dropped by as much as 3 meters (10 feet).

Landslides were also a significant factor in the June 1990, Iranian earthquake. Although building failures were responsible for the loss of many lives, rockslides and rock avalanches also caused much death and destruction. Entire hillside villages in this mountainous region slid downslope or were buried. Equally significant were the transportation difficulties that resulted because roads were blocked by landslide debris. Many who might have been helped did not survive because rescue teams and equipment could not reach isolated villages.

EARTHQUAKE PREDICTION

The vibrations that shook the San Francisco Bay Area during rush hour on October 17, 1989, inflicted 62 deaths and about 6 billion dollars in damages (Figure 5.21), all from an earthquake that lasted less than 15 seconds and had a rating of 7.1 on the Richter scale. Although San Francisco and Oakland suffered considerable damage, seismologists warn that quakes of comparable or greater strength will someday be centered beneath the more heavily populated parts of the San Francisco Bay Area. These earthquakes will be far less kind to the Bay Area unless ways to reduce the risks from earthquakes are forthcoming.

Substantial research programs are underway in Japan, the United States, China, and the former Soviet Union—countries where the risks of earthquakes are high. These projects strive to identify those geologic phenomena that precede major earthquakes. In California, for example, uplift or subsidence of the land, changes in the movements along a fault zone, and a period of seismic quiescence often followed by renewed activity have been found to precede some earthquakes. It therefore seems reasonable that earthquakes may be predicted by continually monitoring ground tile, fault movement, and seismic activity. Perhaps the most ambitious earthquake prediction experiment of this type is underway along a segment of the San Andreas fault near the town of Parkfield in central California. Here, earthquakes of moderate intensity have occurred on a regular basis about once every 22 years since 1857. The most recent rupture was a 5.6-magnitude quake that occurred in 1966. With the next event already overdue, the U.S. Geological Survey has established an elaborate monitoring network. Included are creepmeters,

BOX 5.2
Tsunami Warning System

*T*sunamis are able to traverse large stretches of ocean before their energy is totally dissipated. The tsunami generated by the 1960 Chilean earthquake, in addition to completely destroying villages along an 800-kilometer stretch of coastal South America, traveled 17,000 kilometers across the Pacific to Japan. Here, about 22 hours after the quake, considerable destruction was inflicted upon southern coastal villages of Honshu, the major island of Japan. For several days afterward, tidal gauges located in Hilo, Hawaii, were able to detect these diminishing waves as they bounced about the Pacific.

In 1946, a large tsunami struck the Hawaiian Islands without warning. A wave more than 15 meters high left several coastal villages in shambles. This destruction motivated the United States Coast and Geodetic Survey to establish a tsunami warning system for the coastal areas of the Pacific. From seismic observatories throughout the region, warnings of large earthquakes are reported to the Tsunami Warning Center in Honolulu. Using tidal gauges, a determination is made as to whether a tsunami has been formed. Within an hour a warning is issued. Although tsunamis travel very rapidly, there is sufficient time to evacuate all but the region nearest the epicenter (Figure 5.B). For example, a tsunami generated near the Aleutian islands would take 5 hours to reach Hawaii, and one generated near the coast of Chile would travel 15 hours before reaching Hawaii.

Fortunately, most earthquakes do not generate tsunamis. On the average, only about 1.5 destructive tsunamis are generated worldwide each year. Of these, only about one every ten years is catastrophic.

Figure 5.B
Tsunami travel times to Honolulu, Hawaii, from locations throughout the Pacific. (From NOAA)

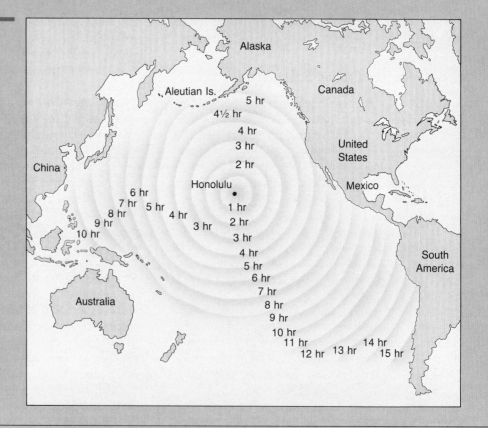

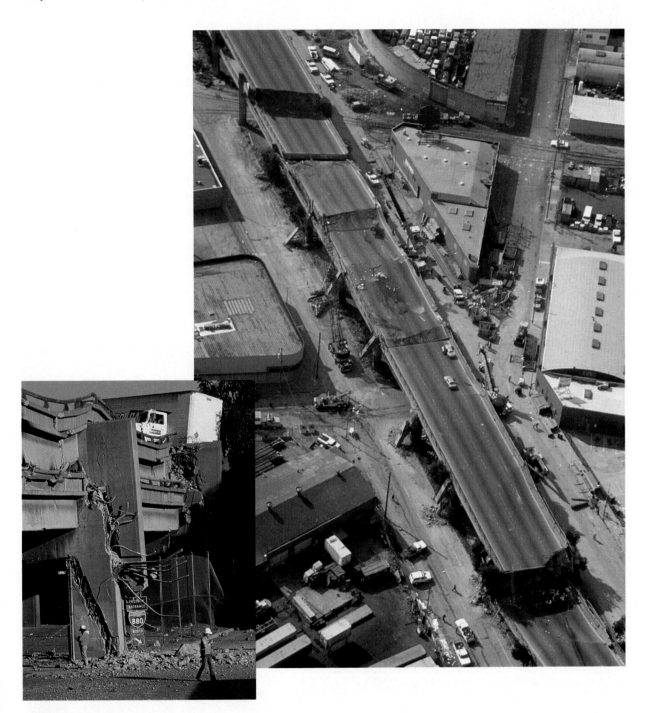

Figure 5.21
Aerial view of collapsed upper deck of Highway 880 caused by the 1989 Loma Prieta Earthquake, California. (Photo by Howard G. Wilshire, U.S. Geological Survey) Inset shows structural damage to support pillars. (Photo by P. M. Duwney/Gamma-Liaison)

tiltmeters, and bore hole strain meters that are used to measure the accumulation and release of strain. Moreover, 70 seismometers of various designs have been installed to record foreshocks as well as the main event. Finally, a network of distance-measuring devices that employ lasers measures movement across the fault. The object is to identify ground displacements that may precede a sizeable rupture.

Although no reliable method of short-range prediction has yet been devised, a few successful predictions have reportedly been made. Most notable was the prediction that foretold the 1975 earthquake in the Lainoning Province of China. Here, for the first and only time, seismologists succeeded in forecasting a large earthquake that was about to destroy a major city. By evacuating an estimated 3 million residents from unreinforced masonry structures, tens of

thousands of lives were spared. Western observers confirmed Chinese reports that almost 90 percent of the structures in the city of Haicheng were heavily damaged. Although this event was predicted months earlier, swarms of small earthquakes which preceded this earthquake undoubtedly aided the prediction and also prompted the citizenry to heed the warning.

Unfortunately, the Chinese were unable to pinpoint the date of the great Tangshan earthquake of 1976. Their long-range warning of a forthcoming earthquake was not sufficiently precise to prevent the loss of an estimated 240,000 lives. The Chinese have also issued false alarms. In a province near Hong Kong, people left their dwellings for over a month, but no earthquake followed. The debate that would precede an order to evacuate a large city in the United

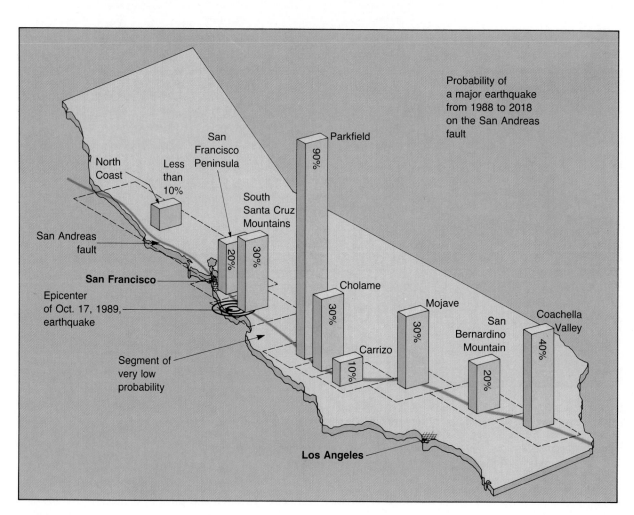

Figure 5.22
Probability of a major earthquake from 1988 to 2018 on the San Andreas fault.

States, such as Los Angeles or San Francisco, would be considerable. The cost of evacuating millions of people, to say nothing of lost work time, and the innumerable other problems associated with an evacuation would have to be weighed against the earthquake's probability.

What constitutes success or failure for earthquake predictions? Most experts agree that a successful earthquake prediction must specify the geographic area involved, the expected magnitude, and the probable time of occurrence. As you might expect, researchers have found that establishing a time for the event is the most elusive factor. This is particularly true if the forecasts attempt to establish the time of a particular quake within the relatively narrow period of a few days, or at most a few weeks. On the other hand, substantial progress has been made in producing long-term earthquake predictions.

Long-term forecasts are based on the premise that earthquakes are repetitive. In other words, as soon as one earthquake is over, the continuing motion of the earth's plates begins to add strain to the rocks until they fail again. This fact has led seismologists to conclude that by studying the history of earthquakes, their occurrences may be predicted. One study conducted by the U.S. Geological Survey gives the probability of a rupture occurring along various segments of the San Andreas fault for the 30 years from 1988 to 2018 (Figure 5.22). From this study, the October 17, 1989, Loma Prieta earthquake in the Santa Cruz Mountains was given a 30 percent probability of producing a 6.5 magnitude earthquake during this time period. The region given the highest probability (90 percent) of generating a quake is the Parkfield section. This area has been called the "Old Faithful" of earthquake zones because activity here has been very regular for many years. One section of the San Andreas between Parkfield and the Santa Cruz Mountains is given a very low probability of generating an earthquake. This area has experienced very little seismic activity in historical times and is thought to exhibit a slow, continual movement known as *creep*.

A significant concern for many Californians is the occurrence of the next "Big One." Although seismologists are not willing to make a formal prediction, most point to southern California as the most likely location. The reason is that southern California has not had a major earthquake since 1857.

In addition to the attention given to California, new concerns are being brought to the remainder of the nation as well. For example, Charleston, South Carolina, was hit by a major earthquake in 1886 and could be a site for renewed activity. In addition, three major earthquakes struck the New Madrid, Missouri, region in 1811–1812. Although this area, near the confluence of the Mississippi and Ohio rivers, was sparsely populated in the early nineteenth century, this is no longer the case. This portion of the Mississippi Valley now has significant agricultural and urban development, with Memphis, Tennessee, being the major city in the region. A 1985 federal study concluded that a magnitude 7.6 earthquake in this area could cause an estimated 2500 deaths, collapse 3000 structures, cause $25 billion in damages, and displace a quarter of a million people in Memphis alone.

THE EARTH'S INTERIOR

The earth's interior lies just below us, yet its accessibility to direct observation is very limited. Most of our knowledge of the earth's interior comes from the study of P and S waves that travel through the earth and emerge at some distant point. Simply stated, the technique involves accurately measuring the time required for seismic waves to travel from the focus of an earthquake or nuclear explosion to a seismographic station. Since the time required for P and S waves to travel through the earth depends upon the properties of the rock materials encountered, seismologists search for variations in travel times that cannot be accounted for simply by differences in the distances traveled. These variations correspond to changes in rock properties. Based upon these seismological data the earth has been divided into four major layers: (1) the **crust**, a very thin outer layer; (2) the **mantle**, a rocky layer located below the crust and having a thickness of 2885 kilometers (1789 miles); (3) the **outer core**, a layer about 2270 kilometers (1407 miles) thick, which exhibits the characteristics of a mobile liquid; and (4) the **inner core**, a solid metallic sphere about 1216 kilometers (754 miles) in radius (Figure 5.23).

In 1909, a pioneering Yugoslavian seismologist, Andrija Mohorovičić, presented the first convincing evidence for layering within the earth. By studying seismic records, he found that the velocity of seismic waves increases abruptly below a depth of 50 kilometers. This boundary that he discovered separates the crust from the underlying mantle and is known as the **Mohorovičić discontinuity** in his honor. For reasons that are obvious, the name for this boundary was quickly shortened to **Moho**.

A few years later another major boundary was discovered by the German seismologist Beno Gutenberg. This discovery was based primarily on the observation that P waves diminish and eventually die out completely about 105 degrees from an earthquake. Then, about 140 degrees away, the P waves reappear, but about two minutes later than would be expected based on the distance traveled. This belt where direct seismic waves are absent is about 35 degrees wide and has been named the **shadow zone** * (Figure 5.24). Gutenberg realized that the shadow zone could be explained if the earth contained a core composed of material unlike the overlying mantle

---

* As more sensitive instruments were developed, weak and delayed P waves that enter this zone via reflection were detected.

New York State alone has experienced over 300 earthquakes large enough to be felt.

Earthquakes in the central and eastern United States occur far less frequently than in California. Yet history indicates that the East is vulnerable. Further, the shocks that have occurred east of the Rockies have generally produced structural damage over a larger area than their counterparts of similar magnitude in California. The reason is that the underlying bedrock in the central and eastern United States is older and more rigid. As a result, seismic waves are able to travel greater distances with less attenuation of intensity than in the western United States. It is estimated that for similar earthquakes, the region of maximum ground motion in the East may be up to ten times larger than in the West. Consequently, the higher rate of earthquake occurrence in the western United States is balanced, at least partially, by the fact that earthquakes occurring in the central and eastern United States are capable of producing damage over a larger area.

Despite the recent geologic history of the surrounding region, Memphis, Tennessee, the largest population center in the area of the New Madrid earthquake, does not have adequate provision regarding earthquakes in its building code. Further, because Memphis is located on unconsolidated floodplain deposits, buildings are more susceptible to damage than similar structures built on bedrock. It has been estimated that if an earthquake the size of the 1811–1812 New Madrid event were to strike in the next decade, it would result in casualties in the thousands and damages in tens of billions of dollars.

Figure 5.C
Damage to Charleston, South Carolina, caused by the August 31, 1886, earthquake. Damage ranged from toppled chimneys and broken plaster to total collapse. (Photo courtesy of U.S. Geological Survey)

and had a radius of 3420 kilometers. The core must somehow hinder the transmission of P waves in a manner similar to the light rays blocked by an opaque object which casts a shadow. However, rather than actually stopping the P waves, the shadow zone is produced by the bending of P waves which enter the core, as shown in Figure 5.24.

It was further learned that S waves could not propagate through the core; therefore, geologists concluded that at least a portion of this region is liquid. This conclusion was further supported by the observation that P-wave velocities suddenly decrease by about 40 percent as they enter the core. Since melting would reduce the elasticity of the rock, all evidence points to the existence of a liquid layer below the rocky mantle.

In 1936, the last major subdivision of the earth's interior was predicted by the discovery of seismic

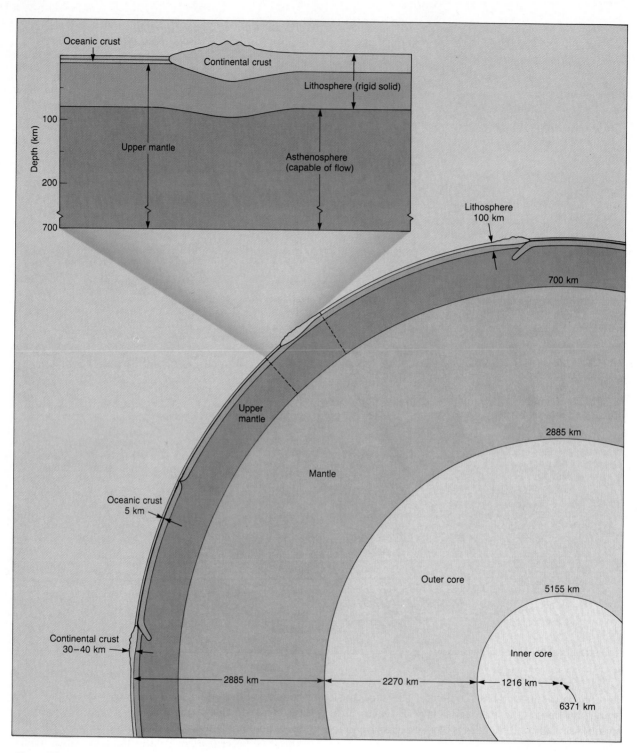

Figure 5.23
Cross-sectional view of the earth showing internal structure.

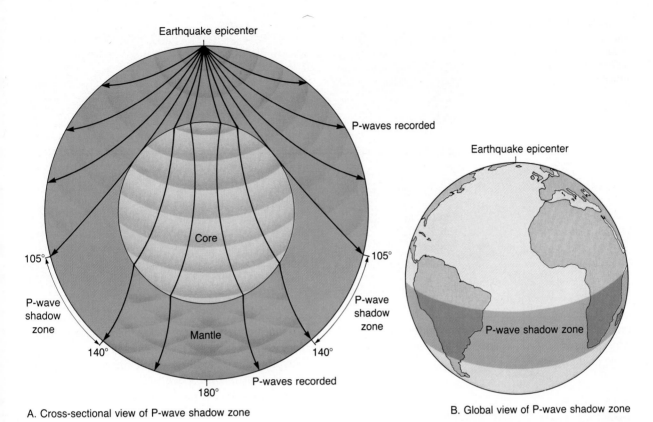

Earthquake epicenter

P-waves recorded

105°

P-wave
shadow
zone

140°

180°

Mantle

Core

P-waves recorded

A. Cross-sectional view of P-wave shadow zone

Earthquake epicenter

P-wave shadow zone

B. Global view of P-wave shadow zone

Figure 5.24
*The abrupt change in physical properties at the mantle-core boundary causes the wave
paths to bend sharply. This abrupt change in wave direction results in a shadow zone
for P waves between about 105 and 140 degrees.*

waves that were reflected from a boundary within
the core. Hence, a core within a core was discovered.
The actual size of the inner core was not accurately
calculated until the early 1960s when underground
nuclear tests were conducted in Nevada. Because the
precise locations and times of the explosions were
known, echoes from seismic waves which bounced
off the inner core provided an accurate means of determining its size. From these data and subsequent
studies, the inner core was found to have a radius of
about 1216 kilometers. Further, P waves passing
through the inner core have appreciably faster travel
times than do those penetrating the outer core exclusively. The apparent increase in elasticity of the inner core material is considered evidence for the solid
nature of the earth's innermost region.

A most important zone exists within the upper
mantle and deserves special mention. This region,

called the **asthenosphere,** is located between the
depths of about 100 kilometers and 350 kilometers,
and may extend down to 700 kilometers (Figure
5.25). In the asthenosphere the velocity of S waves
decreases, indicating to seismologists that this zone
consists of hot, weak rock that is easily deformed. In
addition, as much as 10 percent of the material
within the asthenosphere may be molten.

Situated above the asthenosphere is a cool, rigid
layer called the **lithosphere** (Figure 5.25). Actually
the lithosphere, which is about 100 kilometers thick,
includes the entire crust as well as the uppermost
mantle and is defined as that layer of the earth cool
enough to behave like a rigid solid.

The discovery of the asthenosphere was an important contribution to the theory proposing that the
continents "drift" about. It is thought that the weak,
partially molten nature of the asthenosphere facili-

Figure 5.25
Relative positions of the
asthenosphere and
lithosphere.

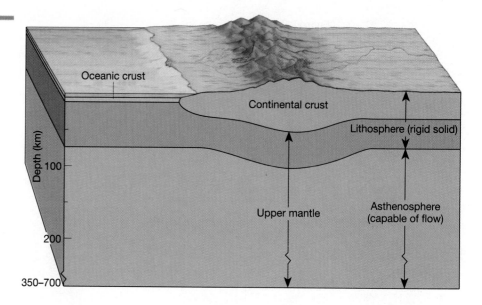

tate the motion of the earth's rigid outer shell. This concept, known as the theory of plate tectonics, is the topic of the next chapter.

COMPOSITION OF THE EARTH

The crust of the earth varies in thickness, being greater than 70 kilometers in some mountainous regions and less than 5 kilometers in some oceanic regions (see Figure 5.23). Early seismic data indicated that the continental crust, which is mostly made of felsic (granite) rocks, is quite different in composition from the oceanic crust. Until recently, however, scientists had only seismic evidence from which to determine the composition of oceanic crust, which lies beneath 3 kilometers of water as well as hundreds of meters of sediment. With the development of the deep-sea drilling ship *Glomar Challenger*, recovery of ocean floor samples became possible. The samples were of basaltic composition—indeed different from the rocks which compose the continents.

Our knowledge of the compositions of the mantle and core is much more speculative. However, we do have some clues. Recall that some of the lava that reaches the earth's surface originates in the partially melted asthenosphere located within the mantle. In the laboratory, experiments have shown that partial melting of a rock called peridotite results in a melt that has a basaltic composition similar to those associated with the volcanic activity of oceanic islands. Ultramafic rocks such as peridotite are thought to

make up the mantle and provide the lava for oceanic eruptions.

Surprisingly, meteorites—"shooting stars"—which fall to the earth from space, are considered evidence of the earth's inner composition. Since meteorites are part of the solar system, they are assumed to be representative samples. Their composition ranges from metallic types made of iron and nickel to stony meteorites composed of ultramafic rock similar to peridotite. Because the earth's crust contains a much smaller percentage of iron than do meteorites, geologists believe that the heavy minerals sank during the early history of the earth. By the same token, the lighter minerals may have floated to the top, creating the crust. Thus, the core of the earth is thought to be mainly iron and nickel, similar to metallic meteorites, while the surrounding mantle is believed to be composed of ultramafic rocks, similar to stony meteorites.

The concept of a molten iron outer core is further supported by the existence of the earth's magnetic field, which acts as a large bar magnet. The most widely accepted mechanism explaining the magnetic field requires that the earth's core be made of a material which conducts electricity, such as iron, and which is mobile enough that circulation can occur. Both of these conditions are met by the model of the earth's core that was established on the basis of seismic data.

Not only does an iron core explain the earth's magnetic field, it also explains the high density of the inner earth, about 13.5 times that of water. Even under the extreme pressure at those depths, average crustal rocks with densities 2.8 times that of water would not have the density calculated for the core.

But iron, which is 3 times more dense than crustal rocks, has the required density.

Review Questions

1. What is an earthquake? Under what circumstances do earthquakes occur?

2. How are faults, foci, and epicenters related?

3. Who was first to explain the actual mechanism by which earthquakes are generated?

4. Explain what is meant by elastic rebound.

5. Faults that are experiencing no active creep may be considered "safe." Rebut or defend this statement.

6. Describe the principle of a seismograph.

7. Using Figure 5.10, determine the distance between an earthquake and a seismic station if the first S wave arrives 3 minutes after the first P wave.

8. List the major differences between P and S waves.

9. Which type of seismic wave causes the greatest destruction to buildings?

10. Most strong earthquakes occur in a zone on the globe known as the _____ .

11. What factor contributed most to the extensive damage that occurred in the central portion of Mexico City during the 1985 earthquake?

12. The 1988 Armenian earthquake had a Richter magnitude of 6.9, far less than the great quakes in Alaska in 1964 and San Francisco in 1906. Nevertheless, the loss of life was far greater in the Armenian event. Why?

13. An earthquake measuring 7 on the Richter scale releases about ___ times more energy than an earthquake with a magnitude of 6.

14. List three factors that affect the amount of destruction caused by seismic vibrations.

15. In addition to the destruction created directly by seismic vibrations, list three other types of destruction associated with earthquakes.

16. Distinguish between the Mercalli scale and the Richter scale.

17. What is a tsunami? How is one generated?

18. Cite some reasons why an earthquake with a moderate magnitude might cause more extensive damage than a quake with a high magnitude.

19. What evidence do we have that the earth's outer core is molten?

20. Contrast the physical makeup of the asthenosphere and the lithosphere.

21. Why are meteorites considered important clues to the composition of the earth's interior?

22. Describe the chemical (mineral) makeup of the following:

    (a) Continental crust.

    (b) Oceanic crust.

    (c) Mantle.

    (d) Core.

Key Terms

aftershock (p. 178)

asthenosphere (p. 201)

body wave (p. 181)

crust (p. 197)

earthquake (p. 176)

elastic rebound (p. 177)

epicenter (p. 183)

fault (p. 176)

focus (p. 176)

foreshock (p. 178)

inner core (p. 197)

intensity (p. 185)

liquefaction (p. 190)

lithosphere (p. 201)

magnitude (p. 186)

mantle (p. 197)

Mohorovičić discontinuity (Moho) (p. 197)

outer core (p. 197)

primary (P) wave (p. 181)

Richter scale (p. 186)

secondary (S) wave (p. 181)

seismic sea wave (tsunami) (p. 190)

seismogram (p. 181)

seismograph (p. 181)

seismology (p. 180)

shadow zone (p. 198)

surface wave (p. 181)

6

Plate Tectonics

Opposite page: The Gulf of Suez (left) and Gulf of Aqaba (right) are major rift zones caused by the separation of the Arabian Peninsula from Africa. (Photo courtesy of Earth Satellite Corporation)

Left: Looking southward down the Red Sea, a rift valley. (Courtesy of NASA)

Above: View of the Gulf of Aden. (Courtesy of NASA)

*W*ill California eventually "slide" into the ocean, as some predict? Have continents really "drifted" apart over the centuries? Answers to these questions and many others that have intrigued geologists for decades are now being provided by a new and exciting theory on large-scale movements taking place within the earth. This theory, called plate tectonics, represents the real frontier of the earth sciences, and its implications are so far-reaching that it can be considered the framework from which most other geologic processes should be viewed.

Early in this century geologic thought was dominated by a belief in the geographic permanency of the ocean basins and continents. During the last few decades, however, vast accumulations of new data have dramatically changed our ideas about the nature and workings of the earth. Earth scientists now realize that the positions of landmasses are not fixed. Rather, the continents gradually migrate over the surface of the globe. The splitting of continental blocks has resulted in the formation of new ocean basins, while older segments of the sea floor are continually being recycled in areas where we find deep-ocean trenches. Further, because of this movement, once-disjointed segments of continental material have collided and formed the earth's great mountain ranges (Figure 6.1). In short, a revolutionary new model of the earth's tectonic* processes has emerged in marked contrast to what was accepted just a few decades ago.

This profound reversal of scientific opinion has been appropriately described as a scientific revolution. Like other scientific revolutions, an appreciable length of time elapsed between the idea's inception and its general acceptance. The revolution began in the early part of the twentieth century as a relatively straightforward proposal that the continents drift about the face of the earth. After many years of heated debate, the idea of drifting continents was rejected by the vast majority of earth scientists as being improbable. However, during the 1950s and 1960s, new evidence began to rekindle interest in this abandoned proposal. By 1968 these new developments led to the unfolding of a far more encompassing theory than continental drift—a theory known as plate tectonics.

CONTINENTAL DRIFT: AN IDEA BEFORE ITS TIME

The idea that continents, particularly South America and Africa, fit together like pieces of a jigsaw puzzle originated with improved world maps. However, little significance was given this idea until 1915, when Alfred Wegener, a German climatologist and geophysicist, published an expanded version of a 1912

---

* Tectonics refers to the deformation of the earth's crust and results in the formation of structural features such as mountains.

Figure 6.1
Teton Range, Grand Teton National Park, Wyoming. (Photo by Tom Till)

lecture in his book *The Origin of Continents and Oceans.* In this monograph, Wegener set forth the basic outline of his radical hypothesis of **continental drift.*** One of his major tenets suggested that a supercontinent he called **Pangaea** (meaning "all land") once existed (Figure 6.2). He further hypothesized that about 200 million years ago this super-

continent began breaking into smaller continents, which then "drifted" to their present positions. Wegener and others who advocated this position collected substantial evidence to support these claims. The fit of South America and Africa, ancient climatic similarities, fossil evidence, and rock structures all seemed to support the idea that these now-separate landmasses were once joined.

Fit of the Continents

Like a few others before him, Wegener first suspected that the continents might have been joined when he noticed the remarkable similarity between the coastlines on opposite sides of the South Atlantic. How-

---

* Wegener's ideas were actually preceded by those of an American geologist, F. B. Taylor, who in 1910 published a paper on continental drift. Taylor's paper provided little supporting evidence for continental drift, which may have been the reason that it had a relatively small impact on the geologic community.

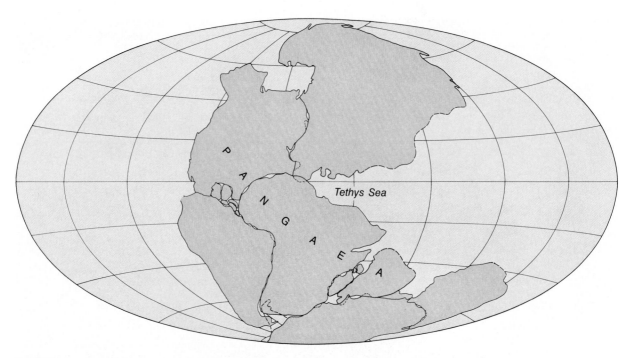

Figure 6.2
*Reconstruction of Pangaea as it is thought to have appeared 200 million years ago.
(After R. S. Dietz and J. C. Holden.* Journal of Geophysical Research *75: 4943. Copyright by American Geophysical Union)*

ever, his use of present-day shorelines to make a fit of the continents was challenged immediately by other earth scientists. These opponents correctly argued that shorelines are continually modified by erosional processes, and even if continental displacement had taken place, a good fit today would be unlikely. Wegener appeared to be aware of this problem, and, in fact, his original jigsaw fit of the continents was only very crude.

A much better approximation of the outer boundary of the continents is the seaward margin of the continental shelf. Today the continental shelf's edge lies several hundred meters below sea level. In the early 1960s, Sir Edward Bullard and two associates produced a map with the aid of computers that attempted to fit the continents at a depth of 900 meters. The remarkable fit that was obtained is shown in Figure 6.3. Although the continents overlap in a few places, these are regions where streams have deposited large quantities of sediment, thus enlarging the continents. The overall fit obtained by Bullard and his associates was better than even the supporters of the continental drift theory suspected it would be.

Fossil Evidence

Although Wegener was intrigued by the remarkable similarities of the shorelines on opposite sides of the Atlantic, he at first thought the idea of a mobile earth improbable. Not until he came across an article citing fossil evidence for the existence of a land bridge connecting South America and Africa did he begin to take his own idea seriously. Through a search of the literature Wegener learned that most paleontologists were in agreement that some type of land connection was needed to explain the existence of identical fossils on the widely separated landmasses.

To add credibility to his argument for the existence of the supercontinent of Pangaea, Wegener cited documented cases of several fossil organisms that had been found on different landmasses but which could not have crossed the vast oceans presently separating the continents. Of particular interest were organisms that were restricted in geographical distribution but which nevertheless appeared in two or more areas that are presently separated by major barriers. The classic example is *Mesosaurus*, a presumably aquatic, snaggle-toothed reptile whose fos-

Figure 6.3
The best fit of South America and Africa along the continental slope at a depth of 500 fathoms (about 900 meters). The area where continental blocks overlap appears in brown. (After A. G. Smith. "Continental Drift." In Understanding the Earth, *edited by I. G. Gass. Courtesy of Artemis Press)*

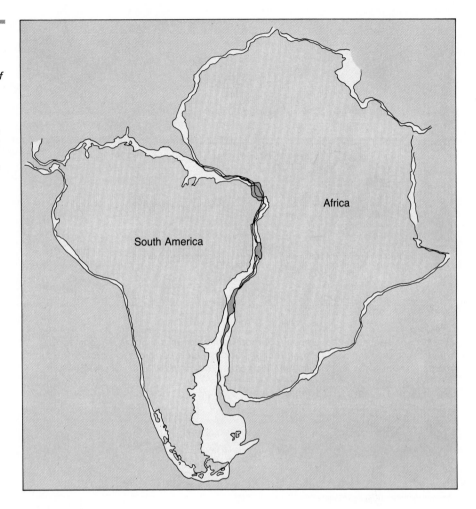

sil remains are known to be limited to eastern South America and southern Africa (Figure 6.4). If *Mesosauraus* had been able to swim well enough to cross a vast ocean, its remains should be widely distributed. Since this was not the case, Wegener argued that South America and Africa must have been joined.

Wegener also cited the distribution of the fossil fern *Glossopteris* as evidence for the existence of Pangaea. This plant, identified by its large seeds that could not be blown very far, was known to be widely dispersed among Africa, Australia, India, and South America during the late Paleozoic era. Later, fossil remains of *Glossopteris* were discovered in Antarctica as well. Wegener knew that these seed ferns and associated flora grew only in a subpolar climate; therefore, he concluded that these landmasses must

have been joined, since they presently include climatic regions that are too diverse to support such flora. For Wegener, fossils proved without question that a supercontinent had existed.

How could these fossil flora and fauna be so similar in places separated by thousands of kilometers of open ocean? The idea of land bridges (isthmian links) was the most widely accepted solution to the problem of migration (Figure 6.5). We know, for example, that during the recent glacial period the lowering of the sea level allowed animals to cross the narrow Bering Straits between Asia and North America.Was it possible then that land bridges once connected Africa and South America? We are now quite certain that land bridges of this magnitude did not exist, for their remnants should still lie below sea level, but are nowhere to be found.

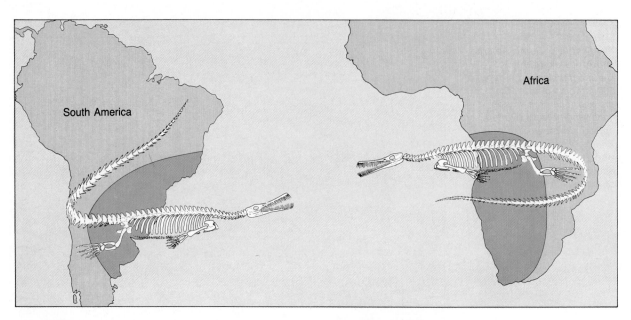

Figure 6.4
Fossils of Mesosaurus have been found on both sides of the South Atlantic and nowhere else in the world. Fossil remains of this and other organisms found on the continents of Africa and South America appear to link these landmasses during the late Paleozoic and early Mesozoic eras.

Figure 6.5
These sketches by John Holden lightheartedly illustrate various explanations that were used to account for the occurrence of similar species on landmasses that are presently separated by vast oceans. (Reprinted with permission of John Holden)

Rock Type and Structural Similarities

Anyone who has worked a picture puzzle knows that in addition to the pieces fitting together, the picture must be continuous as well. The "picture" that must match in the "Continental Drift Puzzle" is represented by the rock types and mountain belts found on the continents. If the continents were once together, the rocks found in a particular region on one continent should closely match in age and type with those found in adjacent positions on the matching continent.

Such evidence has been found in the form of several mountain belts that appear to terminate at one coastline only to reappear on a landmass across the ocean. For instance, the mountain belt that includes the Appalachians trends northeastward through the eastern United States and disappears off the coast of Newfoundland (Figure 6.6A). Mountains of comparable age and structure are found in the British Isles and Scandinavia. When these landmasses are reassembled as in Figure 6.6B, the mountain chains form a nearly continuous belt. Numerous other rock structures exist that appear to have formed at the same time and were subsequently split apart.

Wegener was very satisfied that the similarities in rock structure on both sides of the Atlantic linked these landmasses. In fact, he was too zealous with this evidence and incorrectly suggested that glacial moraines in North America matched up with those of northern Europe. In his own words, "It is just as if we were to refit the torn pieces of a newspaper by matching their edges and then check whether the lines of print run smoothly across. If they do, there is nothing left but to conclude that the pieces were in fact joined in this way."*

Paleoclimatic Evidence

Since Alfred Wegener was a climatologist by training, he was keenly interested in obtaining paleoclimatic (ancient climatic) data in support of continental drift. His efforts were rewarded when he found evidence for apparently dramatic climatic changes. For instance, glacial deposits indicate that near the end of the Paleozoic era (between 220 and 300 million years ago), ice sheets covered extensive areas of the Southern Hemisphere. Layers of glacial till were found in southern Africa and South America, as well as in India and Australia. Below these beds of glacial debris lay striated and grooved bedrock. In some locations the striations and grooves indicated the ice had moved from what is now the sea onto land (Figure 6.7). Much of the land area containing evidence of this late Paleozoic glaciation presently lies within 30 degrees of the equator in a subtropical or tropical climate.

Could the earth have gone through a period sufficiently cold to have generated extensive continental glaciers in what is presently a tropical region? Wegener rejected this explanation because during the late Paleozoic, large swamps existed in the Northern Hemisphere. These swamps with their lush vegetation eventually became the major coal fields of the eastern United States, Europe, and Siberia.

Fossils from these coal fields indicate that the tree ferns which produced the coal deposits had large fronds that are indicative of a tropical setting. Furthermore, the tree trunks lacked growth rings, a characteristic of tropical plants caused by minimal seasonal fluctuations in temperature. Wegener believed that a better explanation for these paleoclimatic regimes is provided when the landmasses are fitted together as a supercontinent with South Africa centered over the South Pole. This would account for the conditions necessary to generate extensive expanses of glacial ice over much of the Southern Hemisphere. At the same time, this geography would place the northern landmasses nearer the tropics and account for their vast coal deposits. Wegener was so convinced that his explanation was correct that he wrote, "This evidence is so compelling that by comparison all other criteria must take a back seat."*

How does a glacier develop in hot, arid Australia? How do land animals migrate across wide expanses of open water? As compelling as this evidence may have been, fifty years passed before most of the scientific community would accept it and the logical conclusions to which it led.

---

* Alfred Wegener, *The Origin of Continents and Oceans.* Translated from the 4th revised German edition of 1929 by J. Birman (London: Methuen, 1966).

---

* Ibid.

Figure 6.6
Matching mountain ranges across the North Atlantic.
A. *The Appalachian Mountains trend along the eastern flank of North America and disappear off the coast of Newfoundland. Mountains of comparable age and structure are found in the British Isles and Scandinavia.* ***B.*** *When these landmasses are placed in their pre-drift locations, these ancient mountain chains form a nearly continuous belt. These folded mountain belts formed roughly 300 million years ago as the landmasses collided during the formation of the supercontinent of Pangaea.*

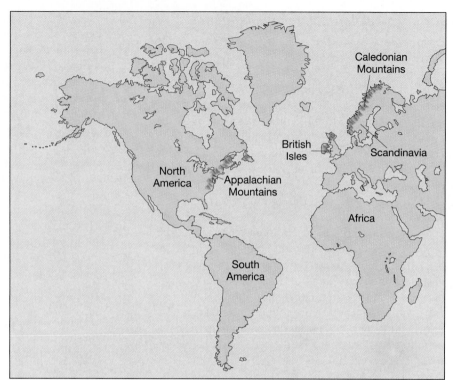

A.

B.

Figure 6.7
A. *Direction of ice movement in the southern supercontinent called Gondwanaland by the founders of the continental drift concept.* **B.** *Glacial striations in the bedrock of Hallet Cove, South Australia, indicate direction of ice movement. (Photo by W. B. Hamilton, U.S. Geological Survey)*

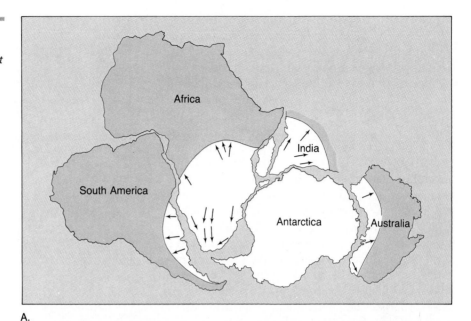

A.

B.

THE GREAT DEBATE

Wegener's proposal did not attract much open criticism until 1924 when his book was translated into English. From this time on, until his death in 1930, his drift hypothesis encountered a great deal of hostile criticism. To quote the respected American geologist T. C. Chamberlin, "Wegener's hypothesis in general is of the foot-loose type, in that it takes

considerable liberty with our globe, and is less bound by restrictions or tied down by awkward, ugly facts than most of its rival theories. Its appeal seems to lie in the fact that it plays a game in which there are few restrictive rules and no sharply drawn code of conduct."

One of the main objections to Wegener's hypothesis stemmed from his inability to provide an acceptable mechanism for continental drift. Wegener proposed two possible energy sources for drift. One of these, the tidal influence of the moon, was presumed by Wegener to be strong enough to give the continents a westward motion. However, the prominent physicist Harold Jeffreys quickly countered with the argument that tidal friction of the magnitude needed to displace the continents would bring the earth's rotation to a halt in a matter of a few years. Further, Wegener proposed that the larger and sturdier continents broke through the oceanic crust, much like ice breakers cut through ice. However, no evidence existed to suggest that the ocean floor was weak enough to permit passage of the continents without themselves being appreciably deformed in the process. By 1929 criticisms of Wegener's ideas were pouring in from all areas of the scientific community. Despite these attacks, Wegener wrote the fourth and final edition of his book, maintaining his basic hypothesis and adding supporting evidence.

Although most of Wegener's contemporaries opposed his views, even to the point of openly ridiculing them, a few considered his ideas plausible. For these few geologists who continued the search, the concept of continents in motion evidently provided enough excitement to hold their interest. Others undoubtedly viewed continental drift as a solution to previously unexplainable observations.

PLATE TECTONICS: A MODERN VERSION OF AN OLD IDEA

During the years that followed Wegener's proposal, great strides in technology permitted mapping of the ocean floor, and extensive data on seismic activity and the earth's magnetic field became available. By 1968 these developments led to the unfolding of a far more encompassing theory than continental drift, known as **plate tectonics.** The implications of plate tectonics are so far–reaching that this theory can be considered the framework from which to view most other geologic processes.

The theory of plate tectonics holds that the outer, rigid lithosphere consists of about twenty rigid segments called **plates** (Figure 6.8). Of these, the largest is the Pacific plate, which is located mostly within the ocean proper, except for a small sliver of North America that includes southwestern California and the Baja Peninsula. Notice from Figure 6.8 that all of the other large plates contain both continental and oceanic crust—a major departure from the conti-

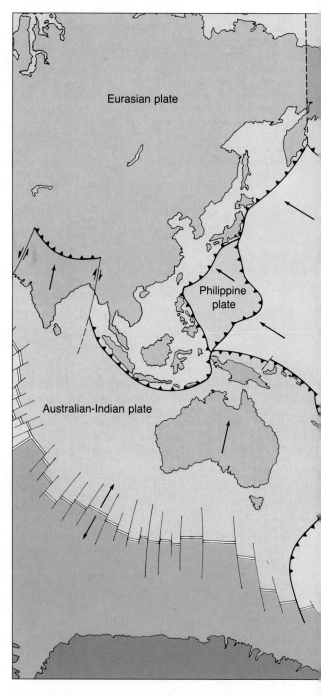

Figure 6.8
Mosaic of rigid plates that constitute the earth's outer shell. **A.** Divergent boundary. **B.** Convergent boundary. **C.** Transform boundary. (After W. B. Hamilton, U.S. Geological Survey)

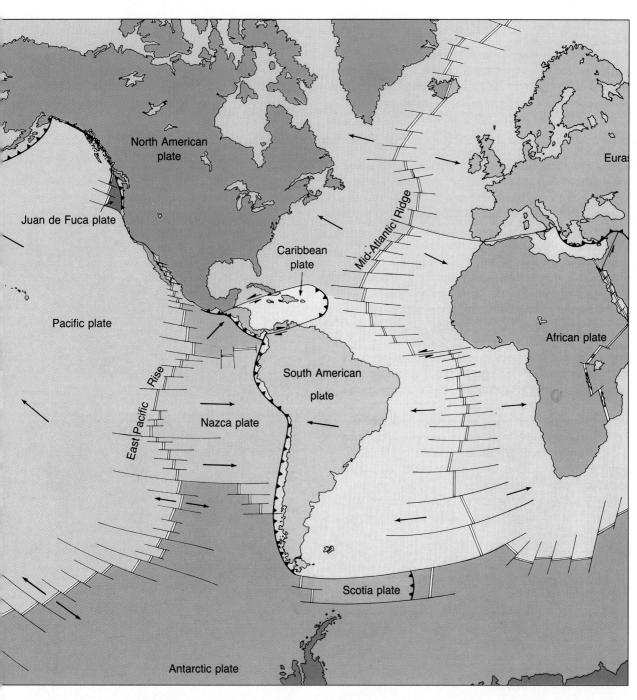

North American
plate

Juan de Fuca plate

Caribbean
plate

Pacific plate

East Pacific Rise

Nazca plate

South American
plate

Mid-Atlantic Ridge

African plate

Euras

Scotia plate

Antarctic plate

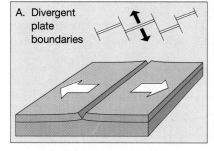

A. Divergent
plate
boundaries

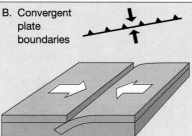

B. Convergent
plate
boundaries

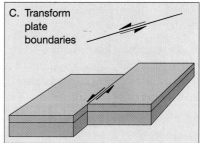

C. Transform
plate
boundaries

nental drift theory, which proposed that the continents moved through, not with, the ocean floor. Many smaller plates, on the other hand, consist exclusively of oceanic material, as, for example, the Nazca plate located off the west coast of South America. Although not clearly defined in Figure 6.8, one small plate that roughly coincides with Turkey is located exclusively within a continent.

The lithosphere is the earth's rigid outer shell. It consists of a number of irregularly shaped plates that vary in thickness. Plates are thinnest in the oceans, where their thicknesses vary from as little as 10 kilometers at the ocean ridges to 100 kilometers in the deep-ocean basins. By contrast, continental blocks are generally 100 to 150 kilometers thick but may extend to 250 kilometers below the stable continental shields. Located below the lithosphere is the hotter and weaker zone known as the *asthenosphere*. The weak nature of the rock within the asthenosphere facilitates motion in the earth's rigid outer shell.

One of the main tenets of the plate tectonics theory is that plates are assumed to be rigid. Therefore, two places on the same plate are not in motion relative to each other. For example, as the plates move, the distance between New York and Denver, located on the same plate, remains unchanged, whereas the distance between New York and London, which are located on different plates, is continually changing. Since each plate moves as a distinct unit, all major interactions between plates occur along plate boundaries. Thus, most of the earth's seismic activity, volcanism, and mountain building occur along these dynamic margins.

PLATE BOUNDARIES

For some time now, tectonic activity has been known to be concentrated along plate boundaries, such as the so-called *Ring of Fire* that encircles the Pacific. Thus, the first approximations of plate margins relied on the distribution of earthquake and volcanic activity. Later work indicated the existence of three distinct types of plate boundaries, each differentiated by the movement it exhibits (Figure 6.9). These are:

1. **Divergent boundaries**—where plates move apart, resulting in upwelling of material from the mantle to create new sea floor.
2. **Convergent boundaries**—where plates move together, causing one of the slabs of lithosphere

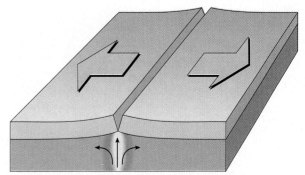

A. Divergent

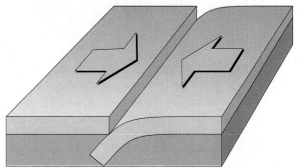

B. Convergent

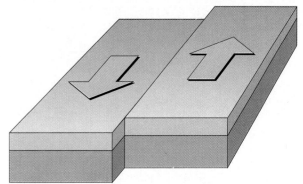

C. Transform

Figure 6.9
*Schematic of plate boundaries. **A.** Divergent boundary.*
***B.** Convergent boundary. **C.** Transform boundary.*

to be consumed into the mantle as it descends beneath an overriding plate.
3. **Transform boundaries**—where plates slide past each other without creating or destroying lithosphere.

Each plate is bounded by a combination of these zones. Movement along any boundary requires that adjustments be made at the others (Figure 6.10).

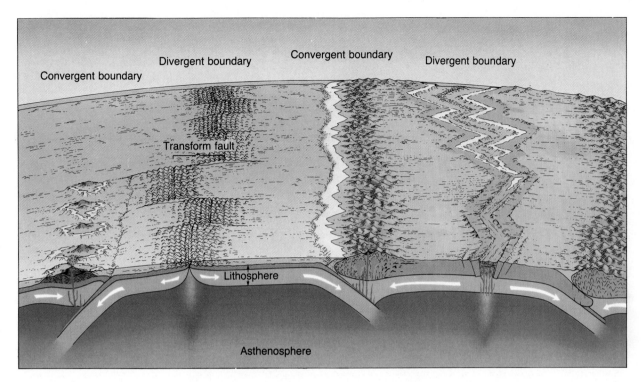

Figure 6.10
View showing the relationship between various types of boundaries.

Divergent Boundaries

Most divergent boundaries, where plate spreading occurs, are situated at the crests of oceanic ridges. Here, as the plates move away from the ridge axis, the fractures created are immediately filled with molten rock that oozes up from the hot asthenosphere. This material cools slowly to produce new slivers of sea floor. In a continuous manner, successive separations and injections of magma add new oceanic crust (lithosphere) between the diverging plates. This mechanism, which has produced the floor of the Atlantic Ocean during the past 165 million years, is called **sea-floor spreading.** The typical rate of spreading at these ridges is estimated to be between 2 and 10 centimeters per year, and averages about 6 centimeters (2 inches) per year. Since new rock is added equally to the trailing edges of both diverging plates, the rate of ocean floor growth is twice the value of the spreading rate. That fact notwithstanding, these seemingly slow rates are rapid enough to have opened and reclosed the Atlantic Ocean more than ten times during the 5-billion-year history of our planet.

Our knowledge of oceanic ridge systems, the sites of sea-floor spreading, comes from soundings taken of the ocean floor, core samples obtained from deep-sea drilling, visual inspection using submersibles, and even first-hand inspection of slices of ocean floor which have been shoved up onto dry land. Ocean ridge systems are characterized by an elevated position and numerous volcanic structures that have grown on the newly formed crust. Because of its accessibility, the Mid-Atlantic Ridge has been studied more thoroughly than other ridge systems. The Mid-Atlantic Ridge is a gigantic submerged mountain range standing 2500–3000 meters (8200–10,000 feet) above the adjacent deep-ocean basins. It extends southward from the Arctic Ocean to beyond the southern tip of Africa. In a few places, such as Iceland, the Mid-Atlantic Ridge has actually grown above sea level. Throughout most of its length, however, this divergent boundary lies 2500 meters below sea level.

Although volcanic structures do contribute to the height of a ridge, the warm, buoyant nature of the intruding magma is the primary reason for its ele-

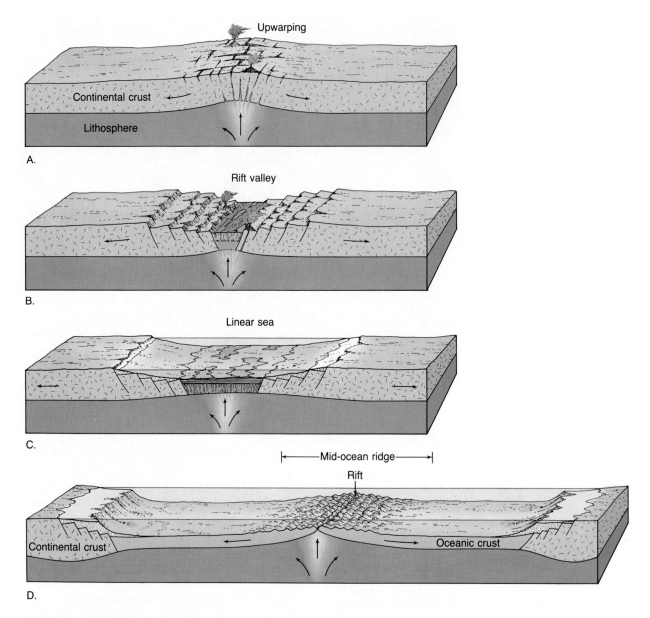

Figure 6.11
A. Rising magma forces the crust upward, causing numerous cracks in the rigid litho-sphere. B. As the crust is pulled apart, large slabs of rock sink, generating a rift zone. C. Further spreading generates a narrow sea. D. Eventually, an expansive ocean basin and ridge system are created.

vated position. As the newly formed lithosphere trav-els away from the spreading center, it gradually cools and contracts. This thermal contraction accounts in part for the greater ocean depths that exist away from the ridge. Almost 100 million years must pass before cooling and contraction cease completely. By this time rock that was once a part of the majestic

ocean mountain system becomes part of the deep-ocean basin.

Not all spreading centers have been in existence as long as the Mid–Atlantic Ridge and not all are found in the middle of large oceans. The Red Sea is believed to be the site of a recently formed divergent bound-ary. Here the Arabian Peninsula separated from

Africa and began to move toward the northeast. Consequently, the Red Sea is providing oceanographers with a view of how the Atlantic Ocean may have looked in its infancy. Another narrow, linear sea produced by sea-floor spreading in the recent geologic past is the Gulf of California.

When a spreading center develops within a continent, the landmass may split into smaller segments as Wegener had proposed for the breakup of Pangaea. The fragmentation of a continent is thought to be associated with the upward movement of hot rock from below. The effect of this activity is to force the crust upward directly above the hot rising plume. Crustal stretching associated with the doming generates numerous tensional cracks as shown in Figure

6.11A. As the plates move from the area of upwelling, the broken slabs are displaced downward, creating downfaulted valleys called **rifts** or **rift valleys** (Figure 6.11B). As the spreading continues, the rift valley will lengthen and deepen, eventually extending out into the ocean. At this point the valley will become a narrow linear sea with an outlet to the ocean similar to the Red Sea today (Figure 6.11C). The zone of rifting will remain the site of igneous activity, continually generating new sea floor in an ever-expanding ocean basin (Figure 6.11D).

The East African rift valleys represent the initial stage in the breakup of a continent as just described (Figure 6.12). The extensive volcanic activity believed to accompany continental rifting is exemplified by

Figure 6.12
East African rift valleys and associated features.

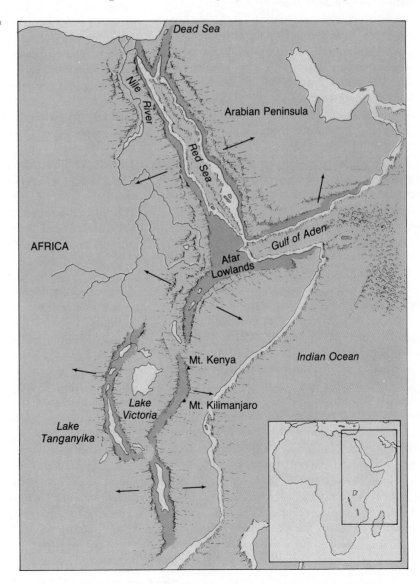

large volcanic mountains such as Kilimanjaro and Mount Kenya. If the rift valleys in Africa remain active, East Africa will eventually part from the mainland in much the same way the Arabian Peninsula did just a few million years ago. However, not all rift valleys develop into full-fledged spreading centers. Running through the central United States is an aborted rift zone extending from Lake Superior to Kansas. The once-active rift valley is filled with rock that was extruded onto the crust more than one billion years ago. Why one rift valley continues to develop while others are abandoned is not yet known.

Convergent Boundaries

At spreading centers new lithosphere is continually being generated; however, since the total surface area of the earth remains essentially constant, lithospere must also be consumed. The zone of plate convergence is the site where lithosphere is reabsorbed into the mantle. When two plates collide, the leading edge of one is bent downward, allowing it to descend beneath the other. The typical angle of descent ranges from 35 to nearly 90 degrees from the surface.

Although all convergent zones are basically similar, the nature of plate collisions is influenced greatly by the type of crustal material involved. Convergence can occur between one oceanic and one continental plate, between two oceanic plates, or between two continental plates, as shown in Figure 6.13.

Oceanic-Continental Convergence.

Whenever the leading edge of a plate capped with continental crust converges with oceanic crust, the less dense continental material apparently remains "floating," while the more dense oceanic slab sinks into the asthenosphere. The region where an oceanic plate descends into the asthenosphere because of convergence is called a **subduction zone.** As the oceanic plate slides beneath the overriding plate, the oceanic plate bends, thereby producing a **deep-ocean trench** adjacent to the zone of subduction (Figure 6.13A). Trenches formed in this manner may be thousands of kilometers long and 8 to 11 kilometers deep.

During a collision between an oceanic slab and a continental block, the oceanic crust is bent, permitting its descent into the asthenosphere (Figure 6.13A). When the descending oceanic plate reaches a depth of about 100 kilometers, partial melting of the water-rich oceanic crust and some of the overlying mantle takes place. The newly formed magma created in

this manner is less dense than the surrounding mantle rocks and, consequently, when sufficient quantities have gathered, the molten rock will slowly rise. Most of the rising magma will be emplaced in the overlying continental crust where it will cool and crystallize at depth. Some of the magma may migrate to the surface where it can give rise to numerous and occasionally explosive volcanic eruptions. The volcanic Andes Mountains are believed to have been produced by such activity when the Nazca plate melted as it plunged beneath the continent of South America (Figure 6.8). The frequent earthquakes that occur within the Andes testify to the activity beyond our view.

Mountains such as the Andes that are believed to be produced in part by volcanic activity associated with the subduction of oceanic lithosphere are called **volcanic arcs.** Two of these volcanic arcs are in the western United States. One, the Cascade Range, is composed of several well-known volcanic mountains, including Mounts Rainier, Shasta, and St. Helens. As the recent eruptions of Mount St. Helens testify, the Cascade Range is still quite active. The magma here arises from the melting of a small remaining segment of the Juan de Fuca plate. The second is the Sierra Nevada, in which Yosemite National Park is located. The Sierra Nevada system is the older of the two and has been inactive for several million years as evidenced by the absence of volcanic cones. Here erosion has stripped away most of the obvious traces of volcanic activity and left exposed the large, crystallized magma chambers that once fed lofty volcanoes.

Oceanic-Oceanic Convergence.

When two oceanic slabs converge, one descends beneath the other, initiating volcanic activity in a manner similar to that which occurs at an oceanic-continental convergent boundary. However, in this case, the volcanoes form on the ocean floor rather than on the continents (Figure 6.13B). If this volcanic activity is sustained, dry land will eventually emerge from the ocean depths. In the early stages of development, this newly formed land consists of a chain of small volcanic islands called an **island arc.** The Aleutian, Mariana, and Tonga islands exemplify such features (Figure 6.14). Island arcs such as these are generally located a few hundred kilometers from an ocean trench where active subduction of the lithosphere is occurring. Adjacent to the island arcs just mentioned are the Aleutian trench, Mariana trench, and the Tonga trench, respectively.

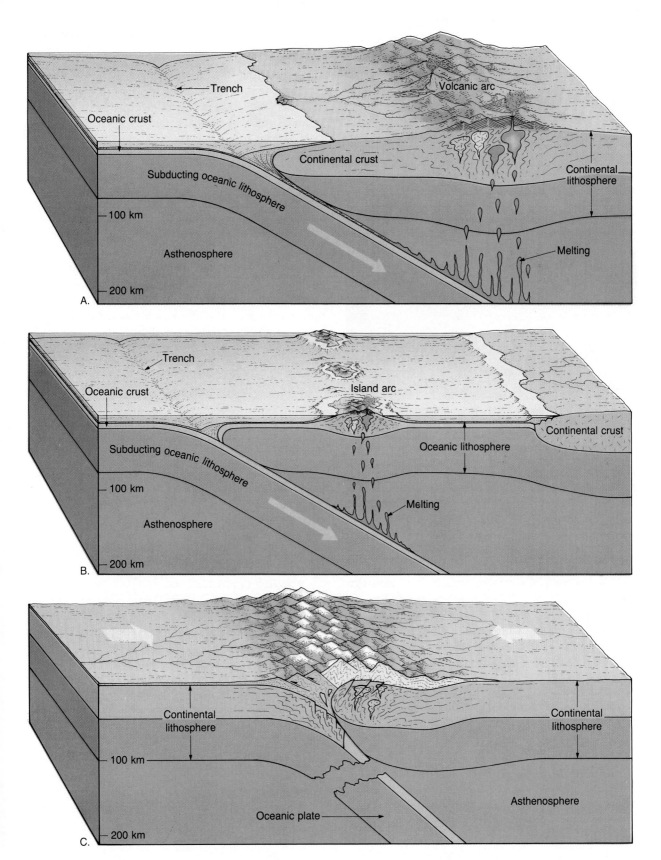

Figure 6.13
Zones of plate convergence. **A.** *Oceanic-continental.* **B.** *Oceanic-oceanic.* **C.** *Continental-continental.*

Figure 6.14
A chain of volcanic islands (island arc) located in the western Pacific. This photo was taken during a space shuttle flight. (Courtesy of NASA)

Over an extended period, numerous episodes of volcanic activity build large volcanic piles on the ocean floor. This volcanic activity, plus the buoyancy of the intrusive igneous rock emplaced within the crust below, gradually increase the size and elevation of the developing arc. This growth, in turn, increases the amount of eroded sediments added to the sea floor. Some of these sediments reach the trench and are deformed and metamorphosed by the compressional forces exerted by the two converging plates. The result of these diverse activities is the development of a mature island arc composed of a complex system of volcanic rocks, folded and metamorphosed sedimentary rocks, and intrusive igneous rocks. Examples of mature island arc systems are the Alaskan Peninsula, the Philippines, and Japan (Figure 6.15).

Continental-Continental Convergence. When two plates carrying continental crust converge, neither plate will subduct beneath the other because of the low density, and thus the buoyant nature, of continental rocks. The result is a collision between the two continental blocks (Figure 6.13C). Such a collision is believed to have happened when the once-separated continent of India "rammed" into Asia and

produced the Himalayas, perhaps the most spectacular mountain range on earth (Figure 6.16). During this collision, the continental crust buckled, fractured, and was generally shortened. In addition to the Himalayas, several other complex mountain systems, including the Alps, Appalachians, and Urals, are thought to have formed in this manner.

Prior to a continental collision, the landmasses involved are separated by an ocean basin. As the continental blocks converge, the intervening sea floor is subducted beneath one of the plates. The partial melting of the descending oceanic slab and mantle rocks generate a volcanic arc. Depending on the location of the subduction zone, the volcanic arc could develop on either of the converging landmasses, or if the subduction zone developed at an appreciable distance into the ocean, an island arc would form. In any case, erosion of the newly formed volcanic arc would add large quantities of sediment to the already sediment-laden continental margins. Eventually, as the intervening sea floor was consumed, these continental masses would collide, thereby squeezing, folding, and generally deforming the sediments as if they were placed in a gigantic vise. The result would be the formation of a new mountain range composed of deformed sedimentary rocks and fragments of the volcanic arc.

Transform Boundaries

The third type of plate boundary is the transform fault, where plates slide past one another without the production of crust, as occurs along oceanic ridges, or without the destruction of crust, as occurs at subduction zones. Transform faults roughly parallel the direction of plate movement and were first identified where they join offset segments of the oceanic ridge system (see Figure 6.15).

The nature of transform faults was discovered in 1965 by J. Tuzo Wilson of the University of Toronto. Wilson suggested that these large fractures connected convergent and divergent plate boundaries into a continuous network that divides the earth's outer shell into several rigid plates. In this role, transform faults provide the means by which the oceanic crust created at the ridge crests can be transported to its site of destruction: the deep-ocean trenches. Figure 6.17 illustrates this activity. Notice that the Juan de Fuca plate moves in a southeasterly direction, eventually being subducted under the west coast of the United States. The southern end of this relatively small plate is bounded by the Mendocino transform

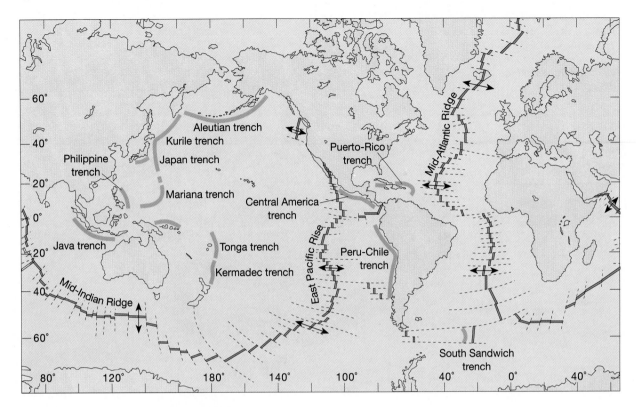

Figure 6.15
Distribution of the world's oceanic trenches, ridge system, and transform faults. Where transform faults offset ridge segments, they permit the ridge to change direction (curve) as can be seen in the Atlantic Ocean.

fault. This transform boundary connects an active spreading center to a subduction zone. Therefore, the fault facilitates the movement of the crustal material created at the ridge crest to its destination beneath the North American continent.

Wilson called these special faults transform faults because the relative motion of the plates can be changed, or transformed, along them. As we saw in the preceding example, divergence occurring at a spreading center can be transformed into convergence at a subduction zone. Since transform faults connect convergent and divergent boundaries in various combinations, other changes in relative plate motion are possible along transform faults.

Most transform faults are located in oceanic crust, but a few, including the San Andreas fault in California, are situated within continents (Figure 6.17). Along the San Andreas fault, the Pacific plate is moving toward the northwest, past the North American plate. If this movement continues for millions of years, that part of California west of the fault zone, including the Baja Peninsula, will become an island off the west coast of the United States

and Canada, and could eventually reach Alaska. However, the more immediate concerns are the earthquakes triggered by movements along this fault system.

TESTING THE MODEL

With the birth of the plate tectonics model, researchers from all of the earth sciences began testing this proposal. Some of the evidence supporting continental drift and sea-floor spreading has already been presented in this chapter. In addition, some of the evidence which was instrumental in solidifying the support for this new concept follows. It should be pointed out that some of this evidence was not new, rather it was a new interpretation of old data that swayed the tide of opinion.

Plate Tectonics and Paleomagnetism

Probably the most persuasive evidence to the geologic community for the acceptance of the plate tectonics theory comes from the study of the earth's magnetic

Figure 6.16
The collision of India and Asia that began about 45 million years ago produced the majestic Himalayas.

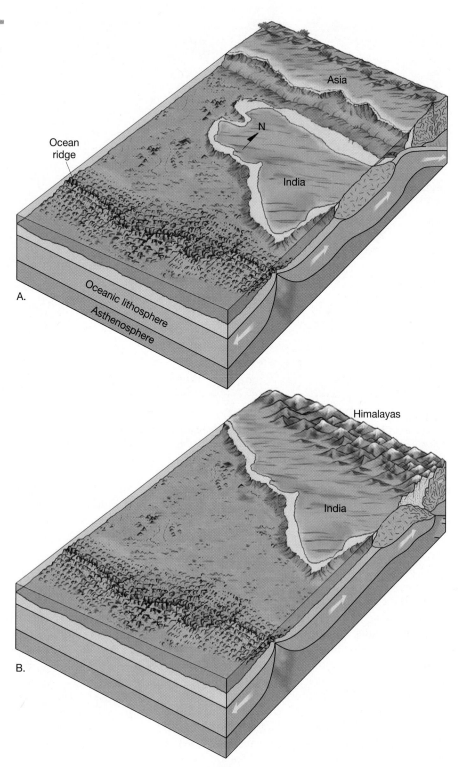

field. Anyone who has used a compass to find direction knows that the earth's magnetic field has a north pole and a south pole. These magnetic poles align closely, but not exactly, with the respective geographic poles. In many respects the earth's magnetic field is very much like that produced by a simple bar magnet. Invisible lines of force pass through the earth and extend from one pole to the other. A compass needle, itself a small magnet free to move about, becomes aligned with these lines of force and thus points toward the magnetic poles.

The technique used to study ancient magnetic fields relies on the fact that certain rocks contain minerals which serve as fossil compasses. These iron-rich minerals such as magnetite are abundant, for example, in lava flows of basaltic composition. When heated above a certain temperature called the **Curie point,** these magnetic minerals lose their magnetism. However, when these iron-rich grains cool below their Curie point (about 580°C) they become magnetized in the direction parallel to the existing magnetic field. Once the minerals solidify, the magnetism they possess will remain "frozen" in this position. In this regard, they behave much like a compass needle inasmuch as they "point" toward the existing magnetic poles. If the rock is moved or the magnetic pole changes position, the rock magnetism will, in most instances, retain its original alignment. Rocks formed thousands or millions of years ago thus "remember" the location of the magnetic poles at the time of their formation and are said to possess fossil magnetism, or **paleomagnetism.**

Polar Wandering. A study of lava flows conducted in Europe in the 1950s led to an interesting discovery. The magnetic alignment in the iron-rich minerals in lava flows of different ages was found to vary widely. A plot of the apparent positions of the magnetic north pole revealed that during the past 500 million years the location of the pole had gradually wandered from a spot near Hawaii northward through eastern Siberia and finally to its present site (Figure 6.18). This was clear evidence that either the magnetic poles had migrated through time, an idea known as **polar wandering,** or the continents had drifted.

Although the magnetic poles are known to move, studies of the magnetic field indicated that the average positions of the magnetic poles correspond closely to the positions of the geographic poles. This

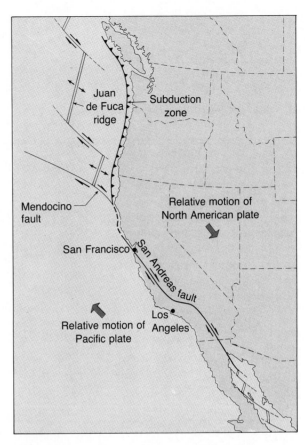

Figure 6.17
The role of transform faults in permitting relative motion between adjacent plates. The Mendocino transform fault permits sea floor generated at the Juan de Fuca ridge to move southeastward past the Pacific plate.

is consistent with our knowledge of the earth's magnetic field, which is generated in part by the rotation of the earth about its axis. If the geographic poles do not wander appreciably, which we believe is true, neither can the magnetic poles. Therefore, a more acceptable explanation for the apparent polar wandering is provided by the plate tectonics theory. If the magnetic poles remain stationary, their apparent movement was produced by the drifting of the continents.

Further evidence for plate tectonics came a few years later when polar wandering curves were constructed for North America and Europe (Figure 6.18A). To nearly everyone's surprise, the curves for North America and Europe had similar paths, except that they were separated by about 30 degrees of longitude. When these rocks solidified, could there have been two magnetic north poles which migrated par-

BOX 6.1

A New Test for Plate Tectonics

Until the late 1980s the evidence supporting the theory of plate tectonics was acquired from the study of geologic phenomena such as volcanoes, earthquakes, and sea-floor sediments. Recently, however, it became possible to test the theory directly. Specifically, scientists are now able to confirm the fact that the plates shift in relation to one another in the way that the plate tectonics theory predicts.

The new evidence comes from two different techniques that allow distances between widely separated points on the earth's surface to be measured with unprecedented accuracy. Called *Satellite Laser Ranging* (SLR) and *Very Long Baseline Interferometry* (VLBI), these methods can detect the motion of any one site with respect to another at a level of better than 1 centimeter per year. Thus, for the first time, scientists can directly measure the relative motions of the earth's plates. Further, because these techniques are quite different, scientists use them to cross-check one against the other by comparing measurements made for the same sites.

The Satellite Laser Ranging system employs ground-based stations that bounce laser pulses off satellites whose orbital positions are well-established. Precise timing of the round-trip travel of these pulses allows scientists to calculate the precise locations of the ground-based stations. By monitoring these stations over time, researchers can establish the relative motions of the sites.

The Very Long Baseline Interferometry system uses large radio telescopes to record signals from very distant quasars (Figure 6.A). Since quasars (quasi-stellar objects) lie billions of light years from the earth, they act as stationary reference points. The millisecond differences in the arrival time of the same signal at different earthbound observational sites provide a means of establishing the distance between receivers.

Confirming data from these two techniques leave little doubt that real plate motion has been detected. Calculations show that Hawaii is moving in a northwesterly direction and approaching Japan at a rate of 8.3 centime-

Figure 6.A
Radio telescopes like these located at Socorro, New Mexico, are used to accurately determine the distance between two distant sites. Data collected by repeated measurements have detected relative plate motions of from 1 to 12 centimeters per year between various sites worldwide. (Photo by Geoff Chester)

ters per year. Moreover, a site located in Maryland is retreating from one in England at a rate of about 1.7 centimeters per year. This rate is roughly equal to the 2.2 centimeters per year of sea-floor spreading that was established from paleomagnetic evidence.

---

allel to each other? This is very unlikely. The differences in these migration paths, however, can be reconciled if the two presently separated continents are placed next to one another, as we now believe they were prior to the opening of the Atlantic Ocean (Figure 6.18B).

Magnetic Reversals. Another discovery in the field of paleomagnetism came when geophysicists learned that the earth's magnetic field periodically reverses polarity; that is, the north magnetic pole becomes the south magnetic pole and vice versa. A rock solidifying during one of the periods of reverse polarity will be magnetized with the polarity opposite that of rocks being formed today. When rocks exhibit the same magnetism as the present magnetic field, they are said to possess **normal polarity,** while those rocks exhibiting the opposite magnetism are said to

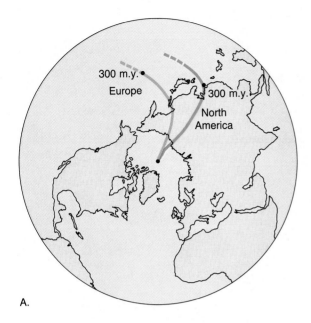

A.

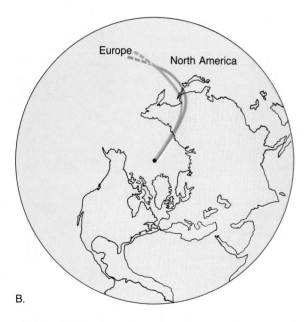

B.

Figure 6.18
Simplified apparent polar wandering paths as established from North American and European paleomagnetic data. **A.** The more westerly path determined from North American data is thought to have been caused by the westward drift of North America by about 24 degrees from Europe. **B.** The positions of the wandering paths when the landmasses are reassembled in their pre-drift locations.

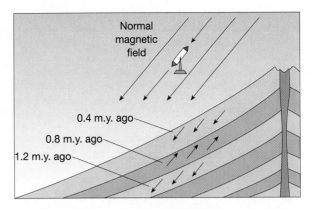

Figure 6.19
Schematic illustration of paleomagnetism preserved in lava flows of various ages. Data such as these from various locales were used to establish the time scale of polarity reversals shown in Figure 6.20A.

have **reverse polarity.** Evidence for magnetic reversals was obtained from lavas and sediments from around the world. Once the concept of magnetic reversals was confirmed, researchers set out to establish a time scale for polarity reversals. There are many areas where volcanic activity has occurred sporadically for periods of millions of years. The task was to measure the directions of paleomagnetism in numerous lava flows of various ages (Figure 6.19). These data were collected from several places and were used to determine the dates when the polarity of the earth's magnetic field changed. Figure 6.20A shows the time scale of the polarity reversals established for the last few million years.

A significant relationship between magnetic reversals and the sea-floor spreading hypothesis was developed from data obtained when very sensitive instruments called *magnetometers* were towed by research vessels across a segment of the ocean floor located off the west coast of the United States. Here workers from the Scripps Institution of Oceanography discovered alternating strips of high- and low-intensity magnetism that trended in roughly a north-south direction. This relatively simple pattern of magnetic variation defied explanation until 1963, when Fred Vine and D. H. Matthews tied the discovery of the high- and low-intensity strips to the concept of sea-floor spreading. Vine and Matthews suggested that the strips of high-intensity magnetism are regions where the paleomagnetism of the ocean crust is of the normal type (Figure 6.20B). Consequently, these positively magnetized rocks enhance the exist-

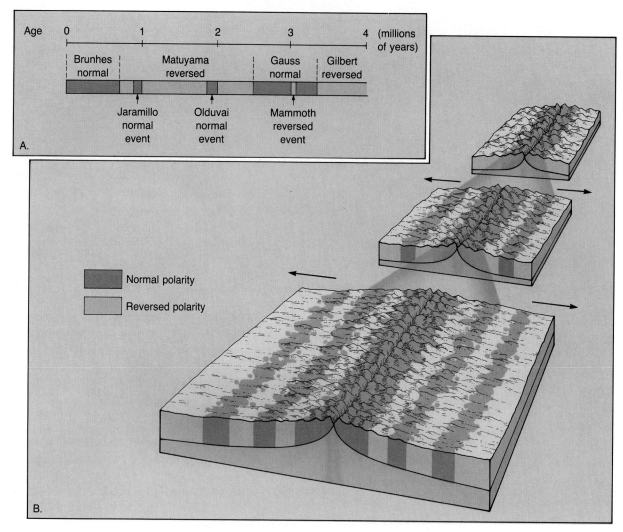

Figure 6.20
*Magnetic evidence in support of the sea-floor spreading hypothesis. **A.** Time scale of the earth's magnetic field in the recent past. This time scale was developed by establishing the magnetic polarity for volcanic lavas of known ages. **B.** New sea floor records the polarity of the magnetic field at the time it formed. Hence it behaves much like a tape recorder, as it records each reversal of the earth's magnetic field. (Data from Allan Cox and G. B. Dalrymple)*

ing magnetic field. Conversely, the low-intensity strips represent regions where the ocean crust is polarized in the reverse direction and, therefore, weaken the existing magnetic field. But how do parallel strips of normally and reversely magnetized rock become distributed across the ocean floor?

Vine and Matthews reasoned that as new basalt was added to the ocean floor at the oceanic ridges, it would be magnetized according to the existing magnetic field. Since new rock is added in approximately equal amounts to the trailing edges of both

plates, we should expect strips of equal size and polarity to parallel both sides of the ocean ridges as shown in Figure 6.20B. This explanation of the alternating strips of normal and reverse polarity, which lay as mirror images across the ocean ridges, was the strongest evidence so far presented in support of the concept of sea-floor spreading.

Now that the dates of the most recent magnetic reversals have been established, the rate at which spreading occurs at the various ridges can be determined accurately. In the Pacific Ocean, for example, the magnetic strips are much wider for corresponding time intervals than those of the Atlantic Ocean. Hence, we conclude that a faster spreading rate exists for the spreading center of the Pacific as compared to the Atlantic. When we apply absolute dates to these magnetic events, we find that the spreading rate for the North Atlantic Ridge is only 1 or 2 centimeters per year.* The rate is somewhat faster for the South Atlantic. The spreading rates for the East Pacific Rise generally range between 3 and 8 centimeters per year, with a maximum rate of about 10 centimeters per year in one segment. Thus, not only had Vine and Matthews discovered a magnetic tape recorder that detailed changes in the earth's magnetic field, but this recorder could also be used to determine the rate of sea-floor spreading.

Plate Tectonics and Earthquakes

By 1968, the basic outline of global tectonics was firmly established. In this same year, three seismolo-

---

* Note that each side of the ridge spreads at this rate.

gists at Lamont-Doherty Geological Observatory published papers demonstrating how successfully the new plate tectonics model accounted for the global distribution of earthquakes (Figure 6.21). In particular, these scientists were able to account for the close association between deep-focus earthquakes and ocean trenches. Furthermore, the absence of deep-focus earthquakes along the oceanic ridge system was also shown to be consistent with the new theory.

The close association between plate boundaries and earthquakes can be seen by comparing the distribution of earthquakes shown in Figure 6.21 with the map of plate boundaries in Figure 6.8. In trench regions where dense slabs of lithosphere plunge into the mantle, this association is especially striking. When the depths of earthquake foci and their locations within the trench systems are plotted, an interesting pattern emerges. Figure 6.22, which shows the distribution of earthquakes in the vicinity of the Japan trench, is an example. Here most shallow-focus earthquakes occur within, or adjacent to, the trench, whereas intermediate- and deep-focus earthquakes occur toward the mainland. A similar distribution

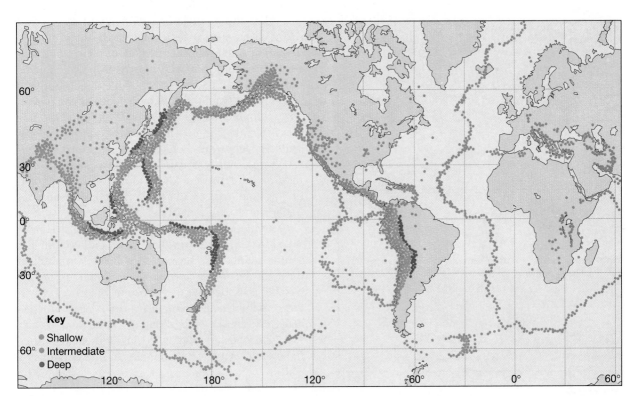

Figure 6.21
Distribution of shallow-, intermediate-, and deep-focus earthquakes. (Data from NOAA)

pattern exists along the western margin of South America where the Nazca plate is being subducted beneath the continent.

In the plate tectonics model, deep-ocean trenches are produced where cold, dense slabs of oceanic lithosphere plunge into the mantle (Figure 6.23). Shallow-focus earthquakes are produced as the descending plate interacts with the overriding lithosphere. As the slab descends further into the asthenosphere, deeper-focus earthquakes are generated. Since the earthquakes occur within the rigid subducting plate rather than in the "plastic" mantle, they provide a method for tracking the plate's descent. Very few earthquakes have been recorded below 700 kilometers (435 miles), possibly because the slab has been heated sufficiently to lose its ridigity.

The plate tectonics model also explains why deep-focus earthquakes are confined to areas adjacent to oceanic trenches (subduction zones), while earthquakes that occur along divergent boundaries and transform fault boundaries have only shallow foci. Notice in Figure 6.21 that deep-focus earthquakes are

confined to the margins of the Pacific basin. The rim of the Pacific basin is also the location of most of the world's deep-ocean trenches. Although the exact cause of deep-focus earthquakes is still debated, their close association with subduction zones is well documented. Furthermore, laboratory studies demonstrate that during subduction the increase in pressure causes certain minerals to go through structural changes that are thought to trigger earthquakes. Since subduction zones are the only regions where crustal rocks are forced to great depth, these should be the only sites of deep-focus earthquakes. Indeed, the absence of deep-focus earthquakes along divergent boundaries and transform faults supports the theory of plate tectonics.

Evidence from Ocean Drilling

Some of the most convincing evidence confirming the plate tectonics theory has come from drilling directly into ocean-floor sediment. From 1968 until 1983, the source of these important data was the Deep Sea Drilling Project, an international program sponsored by several major oceanographic institutions and the National Science Foundation. The primary goal was to gather firsthand information about the age of ocean basins and processes of ocean basin formation. Researchers felt that the predictions concerning seafloor spreading that were based on paleomagnetic data could best be confirmed by the direct sampling of sediments from the floor of the deep-ocean basins. To accomplish this, a new drilling ship, the *Glomar Challenger*, was built. The *Glomar Challenger* represented a significant technological breakthrough, because this ship was capable of lowering drill pipe thousands of meters to the ocean floor and then drilling hundreds of meters into the sediments and underlying basaltic crust.

Operations began in August 1968, and shortly thereafter important evidence was gathered in the South Atlantic. At several sites holes were drilled through the entire thickness of sediments to the basaltic rock below. An important objective was to gather samples of sediment from just above the igneous crust as a means of dating the sea floor at each site.* Since sedimentation begins immediatley after the oceanic crust forms, remains of microorganisms

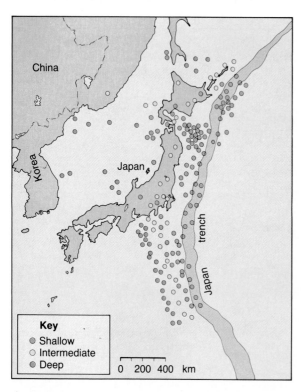

Figure 6.22
Distribution of earthquake foci in the vicinity of the Japan trench. (Data from NOAA)

---

* Radiometric dates of the ocean crust itself are unreliable because of the alteration of basalt by seawater.

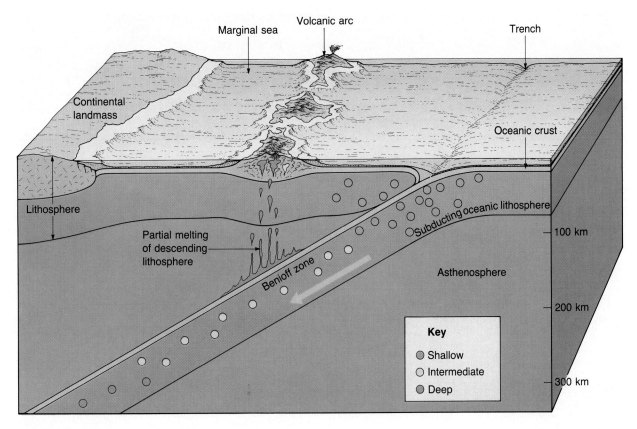

Figure 6.23
Relationship between the descending plate and depth of earthquake foci.

found in the oldest sediments (that is, those resting directly on the basalt) can be used to date the ocean floor at that site. When the oldest sediment from each drill site was plotted against its distance from the ridge crest, it was revealed that the age of the sediment increased with increasing distance from the ridge. This finding was in agreement with the sea-floor spreading hypothesis, which predicted that the youngest oceanic crust is to be found at the ridge crest and that the oldest oceanic crust flanks the continental margins. Further, the rate of sea-floor spreading determined from the ages of sediments was identical to the rate previously estimated from magnetic evidence. Subsequent drilling in the Pacific Ocean verified these findings. These excellent correlations were a striking confirmation of sea-floor spreading.

The data from the Deep Sea Drilling Project also reinforced the idea that the ocean basins are geologically youthful because no sediment with an age in excess of 160 million years was found. By comparison, some continental crust has been dated at 3.9 billion years.

The thickness of ocean-floor sediments provided additional verification of sea-floor spreading. Drill cores from the *Glomar Challenger* revealed that sediments on the ridge crest are almost entirely absent and that the sediment thickens with increasing distance from the ridge. Since the ridge crest is younger than the areas farther from it, this pattern of sediment distribution is exactly what was predicted by the plate tectonics theory.

During its 15 years of operation, the *Glomar Challenger* drilled 1092 holes and obtained more than 96 kilometers (60 miles) of invaluable core samples. Although the Deep Sea Drilling Project ended and the *Glomar Challenger* was retired, the important work of sampling the floors of the world's ocean basins continues. The Ocean Drilling Program has succeeded the Deep Sea Drilling Project and, like its predecessor, it is a major international program. A more technologically advanced drilling ship, the *JOIDES*

Figure 6.24
The JOIDES Resolution, *the drilling ship of the Ocean Drilling Program. This modern drilling ship has replaced the* Glomar Challenger *in the important work of sampling the floors of the world's oceans. (Photo courtesy of Ocean Drilling Program)*

Resolution, now continues the work of the *Glomar Challenger.*[*] The *JOIDES Resolution* can drill in water depths as great as 8100 meters (27,000 feet) and contains onboard laboratories equipped with the largest and most varied array of seagoing scientific research equipment in the world (Figure 6.24). The work of the Ocean Drilling Program is to continue until at least 1995.

Hot Spots

Mapping of seamounts in the Pacific revealed a chain of volcanic structures extending from the Hawaiian Islands to Midway Island and then continuing northward toward the Aleutian trench (Figure 6.25). Radiometric dates of numerous volcanoes in this chain revealed the volcanoes increase in age with an increase in distance from Hawaii. Suiko Seamount, which is located near the Aleutian trench, is 65 million years old, Midway Island is 27 million years old, and the island of Hawaii rose from the sea less than 1 million years ago (Figure 6.25).

Researchers have proposed that a rising plume of mantle material is located below the island of Hawaii. Melting of this hot rock as it enters the low-pressure environment near the surface generates a volcanic area or **hot spot.** Presumably, as the Pacific plate moved over the hot spot, successive volcanic structures emerged. The age of each volcano indicates the time when it was situated over the relatively stationary mantle plume. Kauai is the oldest of the large islands in the Hawaiian chain. Five million years ago, when it was positioned over the hot spot, Kauai was the only Hawaiian Island in existence (Figure 6.25).

---

[*] JOIDES is an acronym for Joint Oceanographic Institutions for Deep Earth Sampling.

Visible evidence of the age of Kauai can be seen by examining the extinct volcanoes that have been eroded into jagged peaks and vast canyons. By contrast, the south slopes of the island of Hawaii consist of fresh lava flows, and two of Hawaii's volcanoes, Mauna Loa and Kilauea, remain active. Recent evidence indicates that a new volcanic pile, named Loihi Seamount, is forming on the ocean floor about 35 kilometers off the southeast coast of Hawaii. Geologically speaking, it should not be long before another tropical island will be added to the Hawaiian chain.

Two island groups parallel the Hawaiian Island–Emperor Seamount chain. One chain consists of the Tuamotu and Line islands, and the other includes the Austral, Gilbert, and Marshall islands. In each case, the most recent volcanic activity has occurred at the southeastern end of the chain, and the islands get progressively older to the northwest. Thus, like the Hawaiian Island–Emperor Seamount chain, these volcanic structures apparently formed by the same motion of the Pacific plate over fixed mantle plumes. This evidence not only supports the fact that the plates do indeed move relative to the earth's interior, but also the hot spot "tracks" provide a frame of reference for tracing the direction of plate motion. Notice, for example, in Figure 6.25 that the Hawaiian Island–Emperor Seamount chain bends. This particular bend in the trace occurred about 40 million years ago when the motion of the Pacific plate changed from nearly due north to a northwesterly path.

Although the existence of mantle plumes is well documented, their exact role in plate tectonics is not altogether clear. Some researchers suggest that mantle plumes originate deep in the mantle, perhaps at the core-mantle boundary. Here a region of abnormally high temperatures produces a rising plume of rock that initiates hot spot volcanism at the earth's surface. Most evidence indicates that hot spots remain relatively stationary. Of the 50 to 120 hot spots believed to exist, about a dozen or so are located near

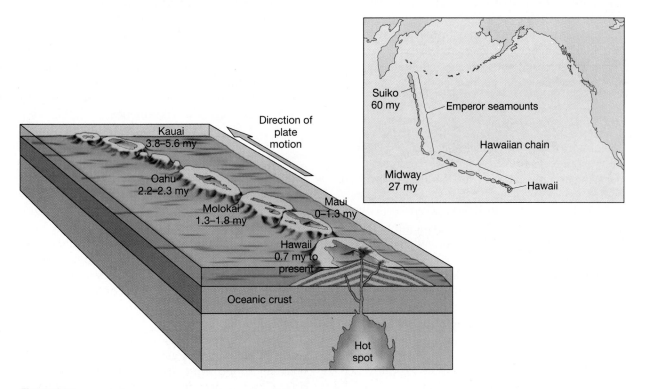

Figure 6.25
The chain of islands and seamounts that extend from Hawaii to the Aleutian trench results from the movement of the Pacific plate over an apparently stationary hot spot. Radiometric dating of the Hawaiian Islands shows that the volcanic activity decreases in age toward the island of Hawaii.

divergent plate boundaries. A hot spot beneath Iceland is thought to be responsible for the unusually large accumulation of lava found in that portion of the Mid-Atlantic Ridge. Another hot spot is believed to be located beneath Yellowstone National Park and may be responsible for the large outpourings of lava and volcanic ash in this area. If the Yellowstone region was indeed modified by hot spot volcanism, there is good reason to expect additional activity in the future.

PANGAEA: BEFORE AND AFTER

Robert Dietz and John Holden have rather precisely projected the gross details of the migrations of individual continents over the past 500 million years. By extrapolating plate motion back in time using such evidence as the orientation of volcanic structures left behind on moving plates, the distribution and movements of transform faults, and paleomagnetism, Dietz and Holden were able to reconstruct Pangaea (Figure 6.26A). The use of radiometric dating helped them establish the time frame for the formation and eventual breakup of Pangaea, and the relatively stationary positions of hot spots through time helped to fix the locations of the continents.

Breakup of Pangaea

The fragmentation of Pangaea began about 200 million years ago. Figure 6.26 illustrates the breakup and subsequent paths taken by the landmasses involved. As we can see in Figure 6.26B, two major rifts initiated the breakup. The rift zone between North America and Africa generated numerous outpourings of Jurassic-age basalts which are presently visible along the eastern seaboard of the United States. Radiometric dating of these basalts indicates that rifting occurred between 200 and 165 million years ago. This date can be used as the birth date of this section of the North Atlantic. The rift that formed in the southern landmass of Gondwanaland developed a "Y"-shaped fracture which sent India on a northward journey and simultaneously separated South America–Africa from Australia-Antarctica.

Figure 6.26C illustrates the position of the continents 135 million years ago, about the time Africa and South America began splitting apart to form the South Atlantic. India can be seen halfway into its journey to Asia, and the southern portion of the North

Atlantic has widened considerably. By the end of the Cretaceous period, about 65 million years ago, Madagascar had separated from Africa, and the South Atlantic had emerged as a full-fledged ocean (Figure 6.26D).

The current map (Figure 6.26E) shows India in contact with Asia, an event that began about 45 million years ago and created the highest mountains on earth, the Himalayas, along with the Tibetan Highlands. By comparing Figures 6.26D and 6.26E, we can see that the separation of Greenland from Eurasia was a recent event in geologic history. Also notice the recent formation of the Baja Peninsula and the Gulf of California. This event is thought to have occurred less than 10 million years ago.

Before Pangaea

Prior to the formation of Pangaea, the landmasses had probably gone through several episodes of fragmentation similar to what we see happening today. Also like today, these ancient continents moved away from each other only to collide again at some other location. During the period between 500 and 225 million years ago, the fragments of an earlier dispersal began collecting to form the continent of Pangaea. Evidence of these earlier continental collisions include the Ural Mountains of the former Soviet Union and the Appalachians of North America.

THE DRIVING MECHANISM

The plate tectonics theory describes plate motion and the effects of this motion. Therefore, acceptance does not rely on a knowledge of the force or forces moving the plates. This is fortunate, since none of the driving mechanisms yet proposed can account for all of the major facets of plate motion. Nevertheless, it is clear that the unequal distribution of heat within the earth is the underlying driving force for plate movement.

One of the first models used to explain the movements of plates was originally proposed by the eminent English geologist Arthur Holmes as a possible driving mechanism for continental drift. Adapted to plate tectonics, this hypothesis suggested that large convection currents within the mantle drive plate motion (Figure 6.27A). The warm, less dense material of the mantle rises very slowly in the regions of oceanic ridges. As the material spreads

BOX 6.2
Plate Tectonics into the Future

Two geologists, Robert Dietz and John Holden, extrapolated present-day plate movements into the future. Figure 6.B illustrates where they envision the earth's landmasses will be 50 million years from now if present plate movements persist for this time span.

In North America we see that the Baja Peninsula and the portion of southern California that lies west of the San Andreas fault will have slid past the North American plate. If this northward migration takes place, Los Angeles and San Francisco will pass each other in about 10 million years, and in about 60 million years, Los Angeles will begin to descend into the Aleutian trench.

Significant changes are seen in Africa, where a new sea emerges as East Africa parts company with the mainland. In addition, Africa will have moved slowly into Europe, perhaps initiating the next major mountain-building stage on our dynamic planet. Meanwhile, the Arabian Peninsula continues to diverge from Africa, allowing the Red Sea to widen and closing the Persian Gulf.

In other parts of the world, Australia is now astride the equator and, along with New Guinea, is on a collision course with Asia. Meanwhile, North and South America are beginning to separate, while the Atlantic and Indian oceans continue to grow at the expense of the Pacific Ocean.

These projections into the future, although interesting, must be viewed with caution because many assumptions must be correct for these events to unfold as just described. Nevertheless, changes in the shapes and positions of continents that are equally profound will undoubtedly occur for millions of years to come.

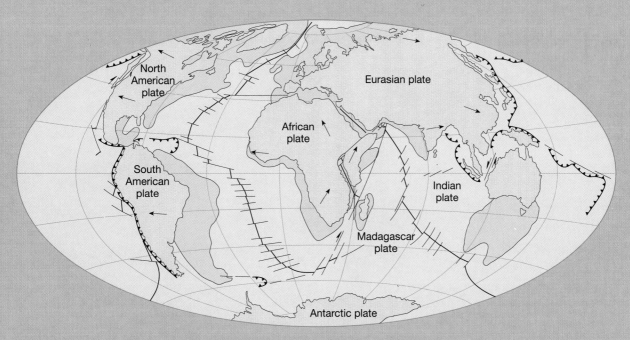

Figure 6.B
The world as it may look 50 million years from now. (From "The Breakup of Pangaea," Robert S. Dietz and John C. Holden. Copyright 1970 by Scientific American, Inc. All rights reserved)

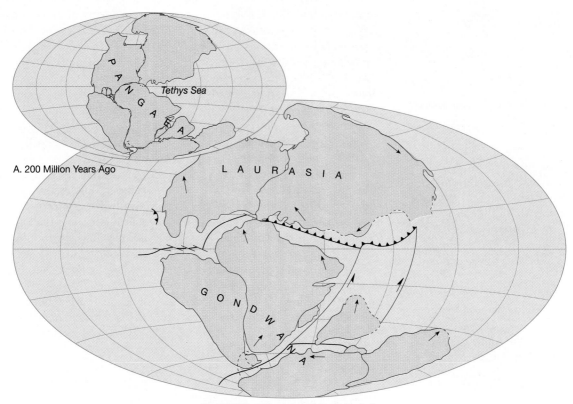

A. 200 Million Years Ago

B. 180 Million Years Ago (Triassic Period)

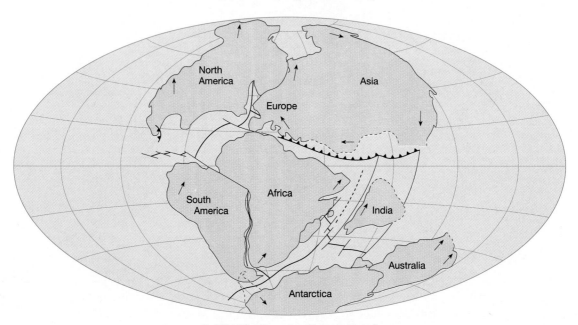

C. 135 Million Years Ago (Jurassic Period)

Figure 6.26
Several views of the breakup of Pangaea over a period of 200 million years according to Dietz and Holden. (After Robert S. Dietz and John C. Holden, Journal of Geophysical Research *75:4939–56. 1970. Copyright by American Geophysical Union)*

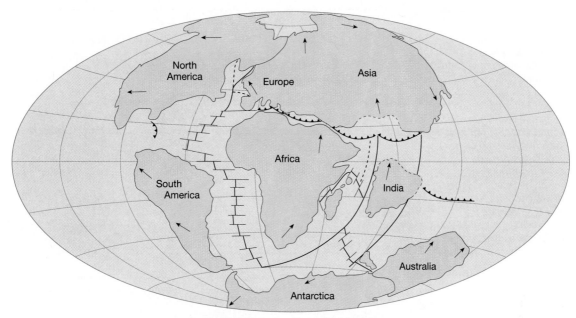

D. 65 Million Years Ago (Cretaceous Period)

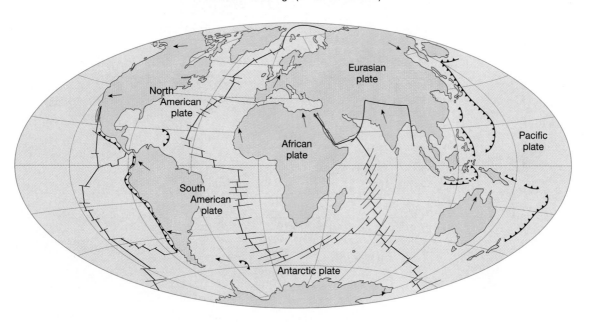

E. Present

laterally, it drags the lithosphere along. Eventually, the material cools and begins to sink back into the mantle, where it is reheated. Partly because of its simplicity, this proposal had wide appeal. However, researchers employing modern research techniques have learned that the flow of material in the mantle is far more complex than that exhibited by simple convection cells. Furthermore, there is considerable debate as to whether mantle circulation is confined to the upper 700 kilometers or involves the whole mantle.

Many other mechanisms that may contribute to or influence plate motion have been suggested. One relies on the fact that as a newly formed slab of oceanic crust moves away from the ridge crest, it gradually cools and becomes denser. Eventually, the cold

Figure 6.27
Proposed models of the driving force for plate tectonics. **A.** *Large convection cells in the mantle carry the lithosphere in a conveyor-belt fashion.* **B.** *Slab-pull results from negative buoyancy of a subducting slab. Slab-push is a form of gravity sliding caused by the elevated position of lithosphere at a ridge crest.* **C.** *The hot plume model suggests that all upward convection is confined to a few narrow plumes, while the downward limbs of these convection cells are the cold, dense subducting oceanic plates.*

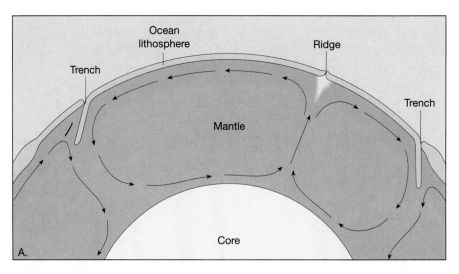

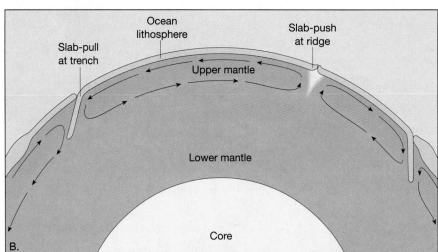

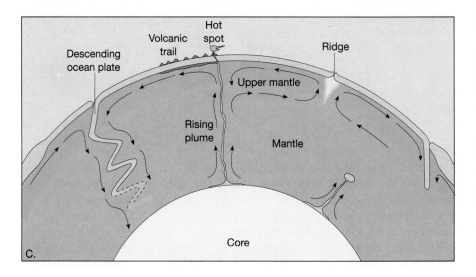

oceanic slab becomes denser than the asthenosphere and begins to descend. When this occurs, the dense sinking slab pulls the trailing lithosphere along. This hypothesis is similar to another model which suggests that the elevated position of an oceanic ridge could cause the lithosphere to slide under the influence of gravity. However, some ridge systems are subdued, which would reduce the effectiveness of the *slab-push* model. Further, some ocean basins, notably the Atlantic, lack subduction zones; thus, the *slab-pull* mechanism cannot adequately explain the spreading occurring at all ridges. Nevertheless, the slab-push and slab-pull phenomena appear to be active in some ridge-trench systems (Figure 6.27B).

One version of the thermal convection model suggests that relatively narrow, hot plumes of rock contribute to plate motion (Figure 6.27C). These hot plumes are presumed to extend upward from the vicinity of the mantle-core boundary. Upon reaching the lithosphere, they spread laterally and facilitate the plate motion away from the zone of upwelling. These mantle plumes reveal themselves as long-lived volcanic areas (hot spots) in such places as Iceland. A dozen or so hot spots have been identified along ridge systems where they may contribute to plate divergence. Recall, however, that many hot spots, including the one which generated the Hawaiian Islands, are not located in ridge areas.

In another version of the hot plume model, all upward convection is confined to a few large cylindrical structures. Embedded in these large zones of upwelling are most of the earth's hot spots. The downward limbs of these convection cells are the cold, dense subducting lithospheric plates. Moreover, advocates of this view suggest that subducting slabs may descend all the way to the core-mantle boundary. However, convincing seismic evidence for the existence of these sheetlike structures below 700 kilometers is lacking.

Although there is still much to be learned about the mechanisms that cause plates to move, some facts are clear. The unequal distribution of heat in the earth generates some type of thermal convection in the mantle which ultimately drives plate motion. Whether the upwelling is mainly in the form of rising limbs of convection currents or cylindrical plumes of various sizes and shapes is yet to be determined. Studies do show, however, that the oceanic ridges are not closely aligned with the active upwelling that originates deep in the mantle. Except for hot spots, upwelling beneath ridges appears to be a shallow feature, responding to the tearing of the lithosphere under the pull of the descending slabs. Furthermore, the descending lithospheric plates are active components of downwelling, and they serve to transport cold material into the mantle.

Review Questions

1. Who is credited with developing the continental drift hypothesis?

2. What was probably the first evidence that led some to suspect the continents were once connected?

3. What was Pangaea?

4. List the evidence that Wegener and his supporters gathered to substantiate the continental drift hypothesis.

5. Explain why the discovery of the fossil remains of *Mesosaurus* in both South America and Africa, but nowhere else, supports the continental drift hypothesis.

6. Early in this century, what was the prevailing view of how land animals migrated across vast expanses of ocean?

7. How did Wegener account for the existence of glaciers in the southern landmasses, while at the same time areas in North America, Europe, and Siberia supported lush tropical swamps?

8. On what basis were plate boundaries first established?

9. What are the three major types of plate boundaries? Describe the relative plate motion at each of these boundaries.

10. What is sea-floor spreading? Where is active sea-floor spreading occurring today?

11. What is a subduction zone? With what type of plate boundary is it associated?

12. Where is lithosphere being consumed? Why must the production and destruction of lithosphere be going on at approximately the same rate?

13. Why is the oceanic portion of a lithospheric plate destroyed, whereas the continental portion is not?

14. Relate the formation of the Andes Mountains to the movement of plates.

15. Discuss the formation of an island arc.

16. In what ways may the origin of the Japanese islands be considered similar to the formation of the Andes Mountains? How do they differ?

17. Briefly describe how the Himalaya Mountains formed.

18. Differentiate between transform faults and the other two types of plate boundaries.

19. Some predict that California will sink into the ocean. Is this idea consistent with the theory of plate tectonics?

20. Define the term *paleomagnetism.*

21. How does the continental drift hypothesis account for polar wandering?

22. Describe the distribution of earthquake epicenters and foci depths as they relate to oceanic trench systems.

23. What is the age of the oldest sediments recovered by deep-ocean drilling? How do the ages of these sediments compare to the ages of the oldest continental rocks?

24. How does sediment thickness vary with increasing distance from the crests of mid-ocean ridges?

25. How do hot spots and the plate tectonics theory account for the fact that the Hawaiian Islands vary in age?

26. With what type of plate boundary are the following places or features associated (be as specific as possible): Himalayas, Aleutian Islands, Red Sea, Andes Mountains, San Andreas fault, Iceland, Japan, Mount St. Helens?

Key Terms

continental drift (p. 209)

convergent boundary (p. 218)

Curie point (p. 227)

deep-ocean trench (p. 222)

divergent boundary (p. 218)

hot spot (p. 234)

island arc (p. 222)

normal polarity (p. 228)

paleomagnetism (p. 227)

Pangaea (p. 209)

plate (p. 216)

plate tectonics (p. 216)

polar wandering (p. 227)

reverse polarity (p. 229)

rift (rift valley) (p. 221)

sea-floor spreading (p. 219)

subduction zone (p. 222)

transform boundary (p. 218)

volcanic arc (p. 222)

7

Igneous Activity

Opposite page: Lava fountains and rivers of lava produced by a recent eruption of Hawaii's Kilauea Volcano. (Photo by Lee Alen Thomas/Douglas Peebles Photography)

Left: Lava flowing from lava tube. (Photo by J. D. Griggs, U.S. Geological Survey)

Above: Lava entering the sea, Hawaii. (Photo by J. D. Griggs, U.S. Geological Survey)

*T*he significance of igneous activity may not be obvious at first glance. However, since volcanoes extrude molten rock that formed at great depth, they provide the only windows we have for direct observation of processes which occur many kilometers below the earth's surface. Furthermore, the gases emitted during volcanic eruptions are thought to be the material from which the atmosphere and oceans evolved. Either of these facts is reason enough for igneous activity to warrant our attention. All volcanic eruptions are spectacular, but why are some destructive and others quiescent? Is the entire island of Hawaii a volcano as high as Mount Everest resting on the ocean floor? Do the large volcanoes of Washington and Oregon present a threat to human lives? This chapter considers these and other questions as we explore the formation and movement of magma.

At 8:32 A.M. on Sunday, May 18, 1980, the largest volcanic eruption to occur in North America in recent times transformed a picturesque volcano into a decapitated remnant (Figure 7.1). On this date in southwestern Washington state, Mount St. Helens erupted with a force hundreds of times greater than that of the atomic bombs dropped on Japan during World War II. The blast blew out the entire north flank of the volcano, leaving a gaping hole. A once prominent volcano that had grown to more than 2900 meters (9500 feet) had in one brief moment been lowered by about 410 meters (1350 feet).

The early morning blast totally devastated a wide swath of timber-rich land on the north side of the mountain (Figure 7.2). Trees within a 400-square-kilometer area lay intertwined and flattened, stripped of their branches and appearing from the air like toothpicks strewn about. The immense force caused trees as far away as 25 kilometers to topple. The gases and ash unleashed from the volcano had temperatures that probably exceeded 800°C (1470°F)! As many as 59 persons were killed by the eruption. Some died from the intense heat and the suffocating cloud of ash and gases. Others perished as they were hurled from the mountain by the force of the blast. Still others were trapped by debris-laden mudflows.

The blast and accompanying mudflows carried ash, trees, and water-saturated rock debris 29 kilometers down the Toutle River. The river quickly became a mud-filled torrent and reached depths of 60 meters in some places. Further, a debris dam was deposited at the outlet of Spirit Lake, causing its level to rise by more than 30 meters (100 feet). For several days the threat of pent-up waters breaching the dam posed another potential hazard.

The eruption of May 18th ejected nearly one cubic kilometer of ash and rock debris (Figure 7.3). Following the devastating explosion, Mount St. Helens continued to emit great quantities of hot gases and ash. Only minutes after the eruption began, a dark plume rose from the volcano. The force of the blast was so strong that some ash was propelled high into the stratosphere, more than 18,000 meters (60,000 feet) above the ground. During the next hours and days, this very fine grained material was carried around the earth by strong upper air winds. Measurable deposits were reported as far away as Oklahoma and Minnesota. Meanwhile, ash fallout in the immediate vicinity accumulated to depths exceeding 2 meters, and the air over Yakima, Washington, 130

Figure 7.1
Before and after photographs show the transformation of Mount St. Helens caused by the May 18, 1980, eruption. ("Before" photo by Stephen Trimble, "after" photo by Jim Hughes, courtesy USDA Forest Service)

Figure 7.2
Forest lands and battered van destroyed by the lateral blast of Mount St. Helens on May 18, 1980. (Photo courtesy of USDA Forest Service)

kilometers to the east, was so filled with ash that residents experienced midnight-like darkness at noon. Crop damage from the volcanic fallout was reported as far away as central Montana.

The events leading to the May 18th eruption began about two months earlier, on March 20th, as a series of minor earth tremors centered beneath the awakening mountain (Figure 7.4A). The first volcanic activity took place on March 27th, when a small amount of ash and steam rose from the summit. Over the next several weeks, sporadic eruptions of varied intensity occurred.

Prior to the main eruption, the primary concern had been the potential hazard of mudflows. These

Figure 7.3
In May 1980, Washington state's Mount St. Helens erupted explosively, sending huge quantities of volcanic ash and debris into the atmosphere. (Photo by R. Hoblitt, U.S. Geological Survey)

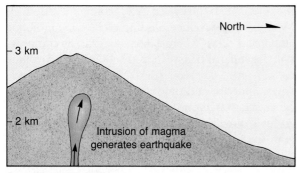

A. March 20, 1980

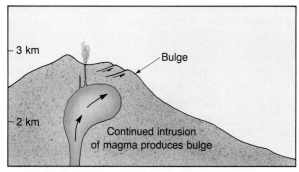

B. April 23, 1980

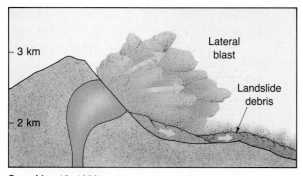

C. May 18, 1980

D. May 18, 1980

Figure 7.4
*Idealized diagrams showing the events in the May 18, 1980, eruption of Mount St. Helens. **A.** First, a sizable earthquake recorded on Mount St. Helens indicates that renewed volcanic activity is possible. **B.** Alarming growth of a bulge on the north flank suggests increasing magma pressure below. **C.** Triggered by an earthquake, a giant landslide relieved the confining pressure on the magma body and initiated an explosive lateral blast. **D.** Within seconds, a large vertical eruption sent a column of volcanic ash to an altitude of about 19 kilometers. This phase of the eruption continued for over nine hours.*

moving lobes of saturated debris were created when ice and snow were melted by heat from the magma within the volcano. The only sign of a potentially hazardous eruption was a bulge on the volcano's north flank (Figure 7.4B). Careful monitoring of this dome-shaped structure indicated a very slow but steady growth rate of a few meters per day. Geologists monitoring the activity suggested that if the growth rate of the bulge changed appreciably, an eruption might quickly follow. Unfortunately, no such variation was detected prior to the explosion. In fact, the seismic activity decreased during the two days preceding the huge blast.

"Vancouver, Vancouver, this is it!" was the only warning to precede the unleashing of tremendous quantities of pent-up gases. The trigger was an earth tremor with a magnitude of 5.1 on the Richter scale. The vibrations sent the north slope of the cone plummeting into the Toutle River, effectively removing the overburden which had trapped the magma below (Figure 7.4 C,D). With the pressure reduced, the water in the magma vaporized and expanded, causing the mountainside to rupture like an overheated steam boiler. Since the eruption originated in the vicinity of the bulge, which was several hundred meters below the summit, the main impact of the eruption was directed laterally rather than vertically. Had the full force of the eruption been upward, far less destruction would have occurred.

Mount St. Helens is only one of the 15 large volcanoes and innumerable smaller ones extending from British Columbia to northern California. Eight of the largest cones have been active in the past few hundred years, while the last eruptive phase of Mount St. Helens came to an end in 1857. Of the remaining seven "active" volcanoes, Mount Baker, Mount Shasta, Lassen Peak, Mount Hood, and Mount Rainier are believed most likely to erupt again. It is hoped that the eruptions of Mount St. Helens will provide geologists with enough data to more effectively evaluate the potential hazards of future volcanic eruptions.

THE NATURE OF VOLCANIC ACTIVITY

Volcanic activity is commonly perceived as a process that produces a picturesque, cone-shaped structure which periodically erupts in a violent manner. However, although some eruptions may be cataclysmic, many are relatively quiescent. The primary factors which determine the nature of volcanic eruptions include the magma's composition, its temperature, and the amount of dissolved gases it contains. These factors affect the magma's **viscosity.** The more viscous the material, the greater its resistance to flow. For example, molasses is more viscous than water. The effect of temperature on viscosity is easily seen. Just as heating molasses makes it more fluid, the mobility of lava is also influenced by temperature changes. As a lava flow cools and begins to congeal, its mobility decreases and eventually the flowing halts.

The chemical composition of magmas was discussed in Chapter 1 with the classification of igneous rocks. One major difference among various igneous rocks and therefore among the magmas from which they originate is their silica (SiO_2) content (Table 7.1). Magmas that produce basaltic rocks contain about 50 percent silica, whereas rocks of granitic composition (granite and its extrusive equivalent, rhyolite) contain over 70 percent silica. The intermediate rock types, andesite and diorite, contain around 60 percent silica. It is important to note that a magma's viscosity is directly related to its silica content. In general, the higher the percentage of silica in magma, the greater its viscosity. It has been shown that the flow of magma is impeded because the silicon-oxygen tetrahedra link together into elongated structures even before crystallization begins. Consequently, because of their high silica content, granitic lavas are very viscous and tend to form comparatively short, thick flows. By contrast, basaltic lavas, which are lower in silica, tend to be quite fluid and have been known to travel distances of 150 kilometers or more before congealing.

The gas content of a magma also affects its mobility. Dissolved gases tend to increase the fluidity of magma. Of far greater consequence is the fact that escaping gases provide enough force to propel molten rock from a volcanic vent. As magma moves into a near-surface environment, such as within a volcano, the confining pressure in the uppermost portion of the magma body is greatly reduced. This reduced confining pressure allows the gases, which had been

Table 7.1
Variations in properties among magmas of differing compositions.

Property	Basaltic	Andesitic	Granitic
Silica content	Least (about 50%)	Intermediate (about 60%)	Most (about 70%)
Viscosity	Least	Intermediate	Highest
Tendency to form lavas	Highest	Intermediate	Least
Tendency to form pyro-clastics	Least	Intermediate	Highest
Melting temperature	Highest	Intermediate	Lowest

dissolved when they were at greater depths, to be released suddenly. At temperatures of 1000°C (1830°F) and low, near-surface pressures, these gases will expand to occupy hundreds of times their original volume. Very fluid basaltic magmas allow the expanding gases to migrate upward and escape from the vent with relative ease. As they escape, the gases will often carry incandescent lava hundreds of meters into the air, producing lava fountains. Although spectacular, such fountains are not generally associated with major explosive events of the type which cause great loss of life and property. Rather, eruptions of fluid basaltic lavas, such as those that occur in Hawaii, are relatively quiescent. At the other extreme, highly viscous magmas impede the upward migration of gases. As a consequence, gases collect as bubbles and pockets that increase in size and pressure until they explosively eject the semi-molten rock from the volcano.

To summarize, we have seen that the quantity of dissolved gases, as well as the ease with which the gases can escape, largely determines the nature of a volcanic eruption. We can now understand why the volcanic eruptions on Hawaii are relatively quiet, whereas the volcanoes bordering the Pacific are explosive and pose the greatest threat to people, because these latter volcanoes generally contain great quantities of gas and emit viscous lavas.

MATERIALS EXTRUDED DURING AN ERUPTION

Many people believe that lava is the primary material extruded from a volcano. This is not always the case. Explosive eruptions that eject huge quantities of broken rock, lava bombs, and fine ash and dust occur nearly as frequently. Moreover, all volcanic eruptions emit large amounts of gas. In this section we will examine each of these materials associated with a volcanic eruption.

Lava Flows

Due to their low silica content, basaltic lavas are usually very fluid and flow in thin, broad sheets or streamlike ribbons. On the island of Hawaii such lavas have been clocked at speeds of 30 kilometers (20 miles) per hour on steep slopes. These velocities are rare, however, and flow rates of 10 to 300 meters per hour are more common. In contrast, the movement of silica-rich lava is on occasion too slow to be perceptible.

When fluid basaltic lavas of the Hawaiian type congeal, they commonly form a relatively smooth skin that wrinkles as the still-molten subsurface lava continues to advance (Figure 7.5A). These are known as **pahoehoe flows** (pronounced *pah-hoy-hoy*) and resemble the twisted braids in ropes. Another common type of basaltic lava has a surface of rough, jagged blocks with dangerously sharp edges and spiny projections (Figure 7.5B). The name **aa** (pronounced *ah ah*) is given to these flows. Active aa

A.

B.

Figure 7.5
A. *Typical pahoehoe (ropy) lava flow, Kilauea, Hawaii.*
B. *Typical slow-moving aa flow. (Photos by J. D. Griggs, U.S. Geological Survey)*

BOX 7.1

The Year Without a Summer

The graph in Figure 7.A allows us to compare the volume of volcanic debris extruded during some well-known eruptions, beginning with Mount Vesuvius in 79 A.D. The eruption of the volcano named *Tambora* is clearly the largest of the modern era. Between April 7 and 12, 1815, this nearly 4000-meter Indonesian structure violently ejected an estimated 30,000 cubic kilometers of volcanic debris. That is 30 times more ash than was emitted during the eruption of Mount St. Helens in May 1980.

Although the Tambora eruption occurred in an isolated part of the world, circulation patterns high in the atmosphere spread its influence far and wide. The impacts of the volcanic dust and gases on climate are believed to have been widespread in the Northern Hemisphere.* According to one researcher who studied the matter, "The extreme cold that prevailed during the spring and summer of 1816 in some regions of the world represents one of the most unusual climatic episodes that has occurred since the advent of instrumental weather observations."*

The effects were especially severe in New England, where 1816 has come to be known as the "year without a summer."

From May through September 1816, an unprecedented series of cold spells affected the northeastern United States and adjacent portions of Canada. The result was a late spring, a cold summer, and an early fall. There were heavy snows in June and frosts in July and August. Across the Northeast, crops were reported killed by the cold. Temperatures in New England averaged up to 3.5°C below normal in June and between 1°C and 2°C below normal in August. Although temperatures this cold had occurred before, there has never been such a protracted span of cold since records have been kept. The reduced averages may seem modest, but they took place in a region where even a small drop in minimum temperatures can mean severe frost.

New England and adjacent Canada were not the only areas to experience a "year without a summer." Patrick Hughes writes, "Although the New England farmer considered it a local tragedy, the abnormal weather was widespread throughout the Northern Hemisphere. In England it was almost as cold as in the United States, and 1816 was a famine year there, as it was in France and Germany."*

The unusual meteorological events of 1816 which followed the massive 1815 eruption of Tambora are regarded by many as a spectacular example of the influence of explosive volcanism on climate. Although the effects were relatively short-lived, this geologic event had a significant atmospheric and human impact.

---

* *American Weather Stories* (Washington, D.C.: National Oceanic and Atmospheric Administration,

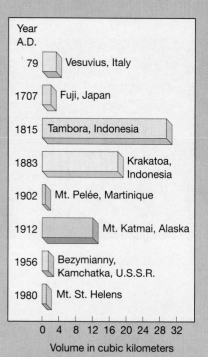

Figure 7.A
Approximate volume of volcanic debris emitted during some well-known eruptions. The 1815 eruption of Tambora, the largest-known eruption in historic time, ejected over 30 times more ash than did Mount St. Helens in 1980.

---

flows are relatively cool and thick and, depending upon the slope, advance at rates from 5 to 50 meters per hour. Further, escaping gases fragment the cool surface and produce numerous voids and sharp spines in the congealing lava. As the molten interior advances, the outer crust is broken further, giving the flow the appearance of an advancing mass of lava rubble.

Gases

Magmas contain varied amounts of dissolved gases held in the molten rock by confining pressure, just as carbon dioxide is held in soft drinks. As with soft drinks, as soon as the pressure is reduced, the gases begin to escape. Obtaining samples from an erupting volcano is often difficult and dangerous, so geologists usually only estimate the amount of gas originally contained within the magma.

The gaseous portion of most magmas is believed to compose from 1 to 5 percent of the total weight, and most of this is in the form of water vapor. Although the percentage may be small, the actual quantity of emitted gas can exceed thousands of tons per day. The composition of the gases is also of interest to scientists, since much evidence points to these as the source of the oceans and the earth's atmosphere. Analysis of samples taken during Hawaiian eruptions indicated that the gases emitted there consist of about 70 percent water vapor, 15 percent carbon dioxide, 5 percent each of nitrogen and sul-

fur compounds, and lesser amounts of chlorine, hydrogen, and argon. Sulfur compounds are easily recognized by their pungent odor and because they readily form sulfuric acid, which when inadvertently inhaled produces a burning sensation.

Pyroclastic Materials

When basaltic lava is extruded, the dissolved gases escape quite freely and continually. As stated earlier, these gases carry incandescent blobs of lava to great heights, thereby producing spectacular lava fountains (see Figure 7.9). Some ejected material may land near the vent and produce a cone structure, while smaller particles will be carried great distances by the wind (Figure 7.6). The gases in highly viscous magmas, on the other hand, are less able to escape and may build up an internal pressure capable of pro-

Figure 7.6
Parícutin Volcano in eruption at night, 1943. (Photo by K. Segerstrom, U.S. Geological Survey)

Figure 7.7
Volcanic bomb. Ejected lava fragments take on a stream-lined shape as they sail through the air.

ducing a violent eruption. Upon release, these superheated gases expand a thousandfold as they blow pulverized rock and lava from the vent. The particles produced by these processes are called **pyroclastics**. These ejected lava fragments range in size from very fine dust and sand-sized volcanic ash, to large volcanic bombs and blocks.

The fine *ash* particles are produced when the extruded lava contains so many gas bubbles that it resembles the froth flowing from a newly opened bottle of champagne. As the hot gases expand explosively, the lava is disseminated into very fine fragments. When the hot ash falls, the glassy shards often fuse to form *welded tuff*. Sheets of this material, as well as ash deposits that consolidate later, cover vast portions of the western United States. Sometimes the frothlike lava is ejected in larger pieces called *pumice*. This material has so many voids that it is often light enough to float in water.

Pyroclastics the size of walnuts, called *lapilli* ("little stones"), and pea-sized particles called *cinders* are also very common. Cinders contain numerous voids and form when ejected lava blobs are pulverized by escaping gases. Particles larger than lapilli are called *blocks* when they are made of hardened lava and *bombs* when they are ejected as incandescent lava. Since volcanic bombs are semimolten upon ejection, they often take on a streamlined shape as shown in Figure 7.7.

VOLCANOES AND VOLCANIC ERUPTIONS

Successive eruptions from a central vent result in a mountainous accumulation of material known as a **volcano**. Located at the summit of many volcanoes is a steep-walled depression called a **crater**, which is connected to a magma chamber via a pipelike conduit, or vent. Some volcanoes have unusually large summit depressions that exceed one kilometer in di-

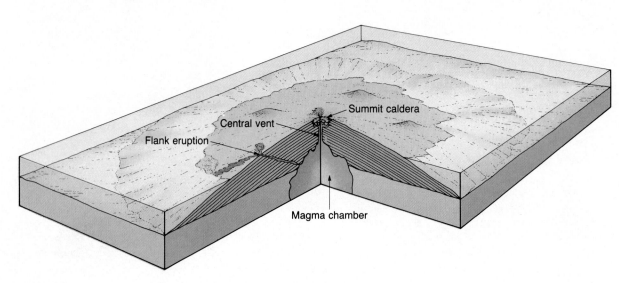

Figure 7.8
Shield volcanoes are built primarily of fluid basaltic lava flows and contain only a small percentage of pyroclastic materials. These broad, slightly domed structures, exemplified by the Hawaiian Islands, are the largest volcanoes on earth.

ameter and are known as **calderas**. When fluid lava leaves a conduit, it is often stored in the crater or caldera until it overflows. On the other hand, lava that is very viscous forms a plug in the pipe which rises slowly or is blown out, often enlarging the crater. However, lava does not always issue from a central crater. Sometimes it is easier for the magma or escaping gases to push through fissures on the volcano's flanks. Continued activity from a flank eruption may build a so-called *parasitic cone*. Mount Etna in Italy, for example, has more than 200 secondary vents. Some of these secondary vents emit only gases and are appropriately called *fumaroles*.

The eruptive history of each volcano is unique; consequently, all volcanoes are somewhat different in form and size. Nevertheless, volcanologists have recognized that volcanoes exhibiting somewhat similar eruptive styles can be grouped. Based on their "typical" eruptive patterns and characteristic form, three groups of volcanoes are generally recognized: shield volcanoes, cinder cones, and composite cones.

Shield Volcanoes

When fluid lava is extruded, the volcano takes the shape of a broad, slightly domed structure called a **shield volcano** (Figure 7.8). Shield volcanoes are built primarily of basaltic lava flows and contain only a small percentage of pyroclastic material. Typically they have a slope of a few degrees at their flanks and generally do not exceed 15 degrees near their summits, as exemplified by the volcanoes of the Hawaiian Islands. Mauna Loa, probably the largest volcano on earth, is one of the five shield volcanoes that together make up the island of Hawaii. Its base rests on the ocean floor 5000 meters below sea level, while its summit reaches a height of 4170 meters (13,680 feet) above the water. Nearly one million years and numerous eruptive cycles were required to build this truly gigantic pile of volcanic rock. Many other volcanic structures, including Midway Island and the Galápagos Islands, have been built in a similar manner from the ocean's depths.

Perhaps the most active and intensively studied shield volcano is Kilauea, located on the island of Hawaii on the southeastern flank of the larger volcano Mauna Loa. Kilauea has erupted more than 50 times in recorded history and is still active today. A testimony to the relatively quiescent nature of these eruptions is the fact that the U.S. Geological Survey's Hawaiian Volcano Observatory is situated on the very rim of the summit caldera. Several months before

each eruptive phase, the summit inflates as magma rises from its source 60 to 100 kilometers below the surface. This molten rock gradually works its way upward and accumulates in smaller reservoirs 3 to 5 kilometers below the summit. For up to 24 hours in advance of each eruption, swarms of earth tremors warn of the impending activity. The 1983 eruption occurred along a 6.5-kilometer (4-mile) fissure located east of the summit caldera (Figure 7.9). Here fountains of fluid lava with heights approaching 100 meters (330 feet) fed a lava flow that extended for 6 kilometers and eventually inundated a few dwellings in the sparsely populated settlement of Royal Gardens. This eruptive phase, like most others, was accompanied by a gradual subsidence of the summit area, nearly equal in volume to the extruded lava.

Cinder Cones

As the name suggests, **cinder cones** are built from ejected lava fragments (see Figure 7.6). Because unconsolidated pyroclastic material maintains a high angle of repose (between 30 and 40 degrees), volcanoes of this type have very steep slopes (Figure 7.10). Cinder cones are rather small, usually less than 300 meters (1000 feet) high, and often form as parasitic cones on or near larger volcanoes. In addition, they frequently occur in groups, where the cones apparently represent the last phase of activity in a region of older basaltic flows. This may result because the contributing magma has cooled and become more viscous.

One of the very few volcanoes whose formation has been observed by geologists from beginning to end is a cinder cone called Parícutin. This volcano's history serves to illustrate the formation and structure of a relatively large cinder cone.

In 1943, about 200 miles west of Mexico City, the volcano Parícutin was born (see Figure 7.6). The eruption site was a cornfield owned by Dionisio Pulido, who with his wife, Paula, witnessed the event as they were preparing the field for planting. For two weeks prior to the first eruption, numerous earth tremors caused apprehension in the village of Parícutin about 3.5 kilometers away. Then around 4:00 P.M. on February 20th, smoke with a sulfurous odor began billowing from a small hole that had been in the cornfield for as long as Señor Pulido could remember. During the night hot, glowing rock fragments thrown into the air from the hole produced a spectacular fireworks display. By the next day the cone had grown to a height of 40 meters and by the

Figure 7.9
Lava fountain along the east rift zone of Kilauea, 1983. (Photo by J. D. Griggs, U.S. Geological Survey)

fifth day it was over 100 meters high. At this time explosive eruptions were throwing hot fragments 1000 meters (3300 feet) above the crater rim. The larger fragments fell near the crater, some remaining incandescent as they rolled down the slope. These fragments built an aesthetically pleasing cone, while finer ash fell over a much larger area, burning and eventually covering the village of Parícutin. Within two years the cone had grown to 400 meters (1310 feet) high and would rise only a few tens of meters more.

The first lava flow came from a fissure that had opened just north of the cone; but after a few months of activity, flows began to emerge from the base of the cone itself. In June of 1944, a clinkery flow 10 meters thick moved over much of the village of San Juan Parangaricutiro, leaving the church steeple exposed

(Figure 7.11). After nine years the activity ceased almost as quickly as it began. Now Parícutin is just another one of the numerous cinder cones dotting the landscape in this region of Mexico. Like the others, it will probably not erupt again.

Composite Cones

The earth's most picturesque volcanoes are termed **composite cones** or **stratovolcanoes**. Most active composite cones are located in a narrow zone that encircles the Pacific Ocean, appropriately named the *Ring of Fire*. Found in this region are Fujiyama in Japan, Mount Mayon in the Philippines, and the picturesque volcanoes of the Cascade Range in the northwestern United States, including Mount St.

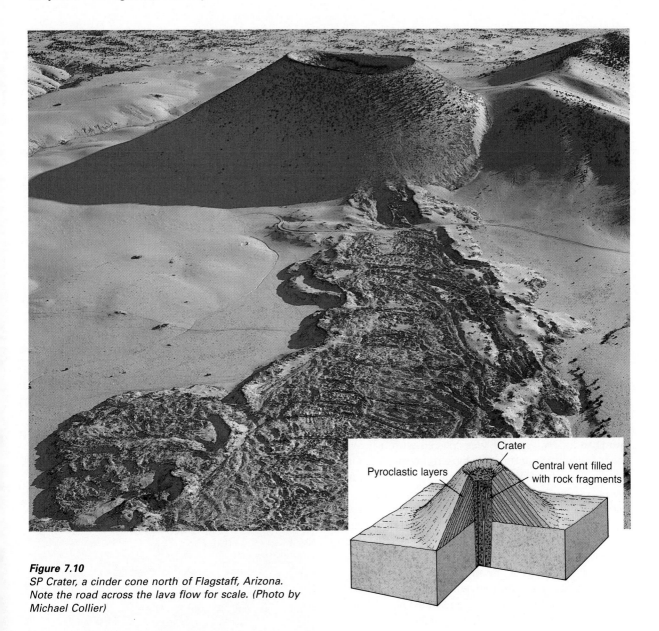

Figure 7.10
SP Crater, a cinder cone north of Flagstaff, Arizona. Note the road across the lava flow for scale. (Photo by Michael Collier)

Crater

Pyroclastic layers

Central vent filled with rock fragments

Helens, Mount Rainier, and Mount Shasta (Figure 7.12).

A composite cone is a large, nearly symmetrical structure built of interbedded lavas and pyroclastic deposits emmitted mainly from a central vent (Figure 7.13). Just as shield volcanoes owe their shape to the fluid nature of the extruded lavas, so too do composite cones reflect the nature of the erupted material. Composite cones are produced when relatively viscous lavas of andesitic composition are extruded. A composite cone may extrude viscous lava for long periods. Then suddenly the eruptive style changes

and the volcano violently ejects pyroclastic material. Most of the ejected pyroclastic material falls near the summit, building a steep-sided mound of cinders. In time this debris will be covered by lava. Occasionally both activities occur simultaneously, and the resulting structure consists of alternating layers of lava and pyroclastics. Two of the most perfect cones, Mount Mayon in the Philippines and Fujiyama in Japan, exhibit the classic form of the stratovolcano with its steep summit area and more gently sloping flanks.

Composite cones represent the most violent type of volcanic activity. Their eruption can be unexpected

Figure 7.11
The village of San Juan Parangaricutiro engulfed by lava from Parícutin, shown in the background. Only the church towers remain. (Photo by Tad Nichols)

In 1902, a nuée ardente from Mount Pelée, a small volcano on the Caribbean island of Martinique, destroyed the port town of St. Pierre. The destruction was instantaneous and so devastating that almost all of St. Pierre's 28,000 inhabitants were killed. Reportedly only a prisoner protected in a dungeon, a shoemaker, and a few people on ships in the harbor were spared (Figure 7.15).

When we compare the destruction of St. Pierre with that of Pompeii, several differences are noticed. Pompeii was totally buried by an event lasting three days, whereas St. Pierre was destroyed in a brief instant and its remains were mantled by only a thin layer of volcanic debris. Also, the structures of Pom-

Figure 7.12
Mount Shasta, California, one of the largest composite cones in the Cascade Range. (Photo by John S. Shelton)

and devastating, as was the A.D. 79 eruption of the Italian volcano we now call Vesuvius. Prior to this eruption, Vesuvius was dormant for centuries. Although minor earthquakes probably warned of the events to follow, Vesuvius was covered with a heavy coat of vegetation and hardly looked threatening. On August 24th, however, the tranquility ended, and in the next three days the city of Pompeii (near Naples) and more than 2000 of its 20,000 residents were buried. They remained so for nearly seventeen centuries, until the city was rediscovered and excavated.

Although the destruction of Pompeii was truly catastrophic, eruptions of a more devastating nature occur when hot gases infused with incandescent ash are ejected, producing a fiery cloud called a **nuée ardente**. Also referred to as *glowing avalanches*, these turbulent steam clouds and companion ash flows travel down steep volcanic slopes at speeds that can approach 200 kilometers (125 miles) per hour (Figure 7.14). The ground-hugging portions of glowing avalanches are rich in particulate matter that is suspended by hot, buoyant gases emitted from the volcanic debris within the moving flow. Thus, these flows, which can include larger rock fragments in addition to ash, travel downslope in a nearly frictionless environment cushioned by expanding volcanic gases. This fact explains why some nuée ardente deposits extend more than 100 kilometers from their source.

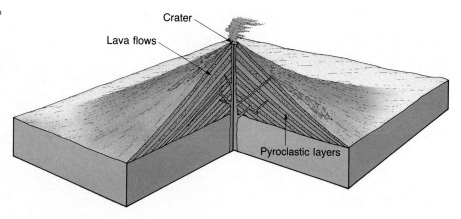

Figure 7.13
Composite cones are built of alternating layers of lava flows and pyroclastics. Generally, summit areas are relatively steep and the flanks have gentler slopes.

peii remained intact except for roofs that collapsed under the weight of the ash. In St. Pierre, masonry walls nearly one meter thick were knocked over like dominoes; large trees were uprooted and cannons were torn from their mounts.

In addition to their violent eruptions, volcanoes are potential hazards in other ways. In particular, large composite cones often generate a type of mudflow known as a *lahar*. These destructive events occur when volcanic ash and debris become saturated with water and flow down steep volcanic slopes, generally following stream valleys. Some lahars are produced when rainfall saturates volcanic deposits, whereas others are triggered as large volumes of ice and snow melt during an eruption.

When Mount St. Helens erupted in May 1980, several lahars formed. The flows and accompanying floods raced down the valleys of the north and south forks of the Toutle River at speeds in excess of 30 kilometers per hour. Water levels reached four meters above flood stage, and nearly all the homes and bridges along the river were destroyed or severely damaged (Figure 7.16). Fortunately, the affected area was not densely populated. This was not the case in November 1985, when Nevado del Ruiz, a volcano in the Andes, erupted and generated a lahar that killed nearly 20,000 people (see the section entitled "Mudflow" in Chapter 2).

Volcanic Pipes and Necks

Most volcanoes are fed by conduits called **pipes** or **vents**, which are believed to be connected to a magma source emplaced near the surface. By contrast, some pipes are thought to extend in a tubelike

Figure 7.14
Nuée ardente races down the slope of Mount St. Helens on August 7, 1980, at speeds in excess of 100 kilometers (60 miles) per hour. (Photo by Peter W. Lipman, U.S. Geological Survey)

Figure 7.15
St. Pierre as it appeared shortly after the eruption of Mount Pelée, 1902. (Reproduced from the collection of the Library of Congress)

fashion directly into the asthenosphere to distances of more than 100 kilometers. Consequently, the materials in these vents are considered to be samples of the asthenosphere that have undergone very little alteration during their ascent. Geologists consider pipes, therefore, to be "windows" into the earth since they allow us to examine earth materials that are normally found at greater depths.

Occasionally a volcanic pipe will reach the surface, but the eruptive phase will cease before any lava is extruded. The upper portions of these pipes often contain a jumble of lava fragments and fragments that were torn from the walls of the vent by violently escaping gases. The best-known of these structures are the diamond-bearing pipes of South Africa. Here the rocks filling the pipes are thought to have origi-

Figure 7.16
A house damaged by debris from lahars that flowed along the Toutle River, west-northwest of Mount St. Helens. The end section of the house was torn free and lodged against trees. (Photo by D. R. Crandell, U.S. Geological Survey)

Figure 7.17
Ship Rock, New Mexico, a volcanic neck. This structure, which stands over 420 meters (1380 feet) high, consists of igneous rock that crystallized in the vent of a volcano that has long since been eroded away. The tabular structure in the foreground is a dike that undoubtedly served to feed lava flows along the flanks of the once active volcano. (Photo by Michael Collier)

nated at depths of about 200 kilometers, where pressures are great enough to generate diamonds and other high-pressure minerals. In these instances the diamonds crystallized at depth and were carried upward by the still-molten portion of the magma.

Volcanoes, like all land areas, are continually being lowered by the forces of weathering and erosion.

Cinder cones are easily eroded, because they are composed of unconsolidated materials. However, all volcanic structures will eventually be worn away. As erosion progresses, the rock occupying the vent is often more resistant and may remain standing above the surrounding terrain long after most of the cone has vanished. Ship Rock, New Mexico, is such a feature, called a **volcanic neck** (Figure 7.17). This structure, higher than many skyscrapers, is but one of many that protrude conspicuously from the red desert landscapes of the southwestern United States.

Formation of Calderas

Earlier it was pointed out that the summits of some volcanoes have nearly circular depressions, known as *calderas*, that exceed one kilometer in diameter. Most calderas are thought to form when the summit of a volcanic structure collapses into the partially emptied magma chamber below (Figure 7.18). Crater Lake in Oregon, 8–10 kilometers wide and 1175 meters deep, is located in such a depression (Figure 7.19). The creation of Crater Lake began about 7000 years ago when the volcano, later to be named Mount Mazama, put forth a violent ash eruption much like that of Vesuvius. However, this ancient eruption was on a much larger scale, extruding an estimated 50–70 cubic kilometers of volcanic material. With the loss of support, 1500 meters of this once prominent 3600-meter cone collapsed. After the collapse, rainwater filled the caldera. Later activity built a small cinder cone called Wizard Island, which today provides a mute reminder of past activity.

Several large calderas are known to exist, with the largest being more than 20 kilometers in diameter. One example, the Valles Caldera located west of Los Alamos, New Mexico, is about 16 kilometers across and 1 kilometer deep. Such features are usually associated with large granitic intrusive bodies located near the surface. Following a violent eruption that may emit tens of cubic kilometers of pyroclastic debris, the roof of the partially emptied magma chamber apparently collapses. Other examples of large calderas believed to have formed in a similar manner are California's Long Valley Caldera and Yellowstone Caldera in Wyoming.

Not all calderas are produced following explosive eruptions. For example, the summits of Hawaii's active shield volcanoes, Mauna Loa and Kilauea, have large calderas that formed as a result of relatively quiet activity. These calderas, which measure 3 to 5

BOX 7.2

The Sizes of Volcanoes

Volcanic cones come in a great variety of shapes and sizes. Nevertheless, volcanologists have recognized that volcanoes exhibiting somewhat similar eruptive styles and shapes can be grouped. Based on their eruptive patterns and characteristic form, three groups of volcanoes are generally recognized: cinder cones, composite cones (stratovolcanoes), and shield volcanoes.

Most cinder cones form over a comparatively short time period. Although the length of these time spans varies, some are known to have formed during a person's lifetime. Parícutin, for example, rose from a Mexican corn field during a period of just nine years. Because their eruption history is short, most cinder cones do not exceed 300 meters (1000 feet) in height (Figure 7.B).

By contrast to cinder cones, composite cones and shield volcanoes have recurring eruptions that can span a period of more than one million years. As shown in Figure 7.B, Mt. Rainier, a composite cone in Washington state, dwarfs Sunset Crater, Arizona, a relatively large cinder cone. Whereas Sunset Crater formed over a period of perhaps a few years between 1064 to 1068 A.D., Mt. Rainier has gradually formed over the past 700,000 years. Further, Mt. Rainier experiences roughly 30 earthquakes each year, making it the most seismically active volcano in the Cascade Range after Mount St. Helens.

Mauna Loa, one of five shield volcanoes that compose the island of Hawaii, is considered the world's largest active volcano. This massive pile of basaltic lava has an estimated volume of 40,000 cubic kilometers that was extruded over a period of approximately one million years. From its base on the floor of the Pacific Ocean to its summit, Mauna Loa approaches 9 kilometers in height. A comparison of Mauna Loa and Mt. Rainier (One of the largest volcanoes in the Cascade Range) is given in Figure 7.B.

Despite its enormous size, Mauna Loa is not the largest known volcano in the solar system. Olympus Mons, a huge shield volcano on Mars, is 25 kilometers high and 600 kilometers wide (see Chapter 19). Moreover, Olympus Mons has an 80-kilometer-wide summit caldera that is nearly large enough to hold the exposed portion of Mauna Loa.

kilometers across and nearly 200 meters deep, formed by collapse as magma was slowly drained from the summit magma chambers during flank eruptions.

FISSURE ERUPTIONS

Although volcanic eruptions from a central vent are the most familiar, by far the largest amounts of volcanic material are extruded from cracks in the crust called **fissures**. Rather than building a cone, these long, narrow cracks distribute volcanic materials over a wide area. An extensive region in the northwestern United States, known as the Columbia Plateau, was formed in this manner (Figure 7.20). Here numerous fissure eruptions extruded very fluid basaltic lava. Successive flows, some 50 meters thick, buried the old landscape as they built a lava plain, which in some places is nearly a mile thick (Figure 7.21). The fluidity is evident, because some lava remained molten long enough to flow 150 kilometers from its source. The term **flood basalts** appropriately describes these waterlike flows.

When silica-rich magma is extruded, **pyroclastic flows** consisting largely of ash and pumice fragments usually result. When these pyroclastic materials are ejected, they move away from the vent at high speeds and may blanket extensive areas before coming to rest. Once deposited, the pyroclastic materials may closely resemble lava flows.

Extensive pyroclastic flow deposits are found in many parts of the world and are most often associated with large calderas. Perhaps the best-known region of pyroclastic flows is the Yellowstone Plateau in northwestern Wyoming. Here a large magma body, rich in silica, still exists a few kilometers below the surface. Several times over the past two million years,

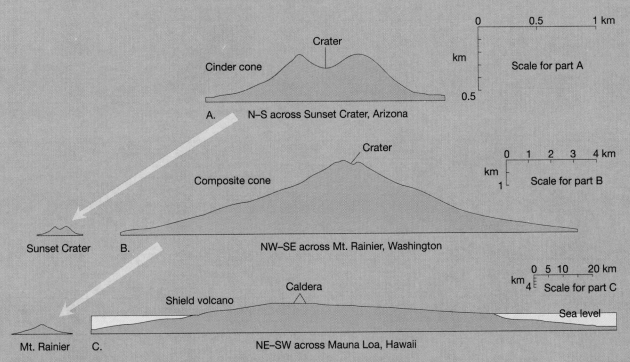

Figure 7.B
Profiles of volcanic landforms. **A.** *Profile of Sunset Crater, Arizona, a typical steep-sided cinder cone.* **B.** *Sunset Crater compared with Mt. Rainier, Washington, a large composite cone of the Cascade Range.* **C.** *Mt. Rainier compared with Mauna Loa, Hawaii, the largest shield volcano in the Hawaiian chain.*

fracturing of the rocks overlying the magma chamber has resulted in huge eruptions accompanied by the formation of calderas. In the northeastern portion of Yellowstone National Park, numerous fossil forests have been discovered, one resting upon another. During periods of inactivity, a forest developed upon the newly formed volcanic surface, only to be covered by ash from the next eruptive phase. Fortunately, no eruption of this type has occurred in modern times.

VOLCANOES AND CLIMATE

The idea that explosive volcanic eruptions change the earth's climate was first proposed many years ago and is still regarded as an explanation for some aspects of climatic variability. Explosive eruptions emit huge quantities of gases and fine-grained debris into the atmosphere. The greatest eruptions are sufficiently powerful to inject material high into the stratosphere, where it spreads around the globe and remains for many months or even years. The basic premise is that this suspended volcanic material (most importantly, droplets of sulfuric acid) will filter out a portion of the incoming solar radiation, which, in turn, will lower air temperatures.

Perhaps the most notable cool period linked to a volcanic event is the "year without a summer" that followed the 1815 eruption of Mount Tambora in Indonesia. The abnormally cold spring and summer of 1816 in many parts of the Northern Hemisphere, including New England, are believed to have been caused by the cloud of volcanic debris and gases ejected from Tambora.

When Mount St. Helens erupted on May 18, 1980, there was almost immediate speculation about the possible effects of this event on our climate. Can an

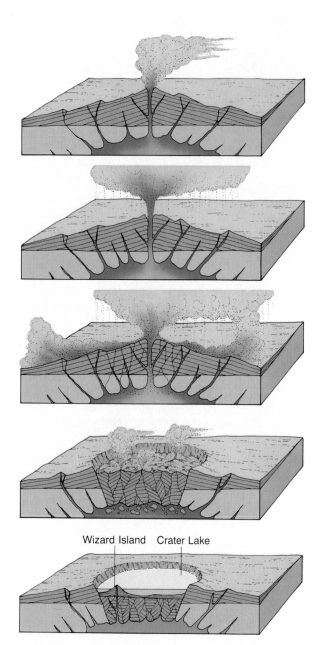

Figure 7.18
Sequence of events that formed Crater Lake, Oregon. About 7000 years ago, the summit of former Mount Mazama collapsed following a violent eruption that partly emptied the magma chamber. Subsequent eruptions produced the cinder cone called Wizard Island. Rainfall and groundwater contributed to form the lake. (After H. Williams. The Ancient Volcanoes of Oregon, p. 47. Courtesy of the University of Oregon)

eruption such as this change our climate? Although spectacular, a single explosive volcanic eruption of the magnitude of Mount St. Helens occurs somewhere in the world every 2 to 3 years. Studies of these events indicate that a very slight cooling of the lower atmosphere does occur. However, it is believed that the cooling is so slight, less than one-tenth of one degree Celsius, as to be inconsequential.

On April 4, 1982, El Chichón, a little-known volcano an Mexico's Yucatan peninsula, erupted. The cloud of debris and sulfur gases lofted into the atmosphere was huge, probably 20 times greater than the cloud from Mount St. Helens. Following the blast, scientists predicted a gradual lowering of temperatures in the Northern Hemisphere, perhaps as great as 0.3–0.5°C. Such a change is large enough to be distinguishable from normal temperature fluctuations, but is probably too small to affect our life-styles. Nevertheless, many scientists agree that such a hemispheric cooling could alter the general pattern of atmospheric circulation for a limited period. Such a change, in turn, could have an effect on the weather in some regions. However, the prediction of specific regional effects still presents a considerable challenge to atmospheric scientists.

In addition to closing Clark Air Force Base and displacing 70,000 Philippinos, the huge eruption of Mount Pinatubo gave scientists another opportunity to test the connection between volcanism and climatic change. In June 1991, this volcano emitted huge quantities of ash and 25 to 30 million tons of sulfur dioxide gas. The gaseous component turned into a haze of tiny sulfuric acid droplets. Because this material was superhot, it penetrated to heights of 20 to 30 kilometers and quickly encircled the entire globe. This sun-blocking haze is probably the most massive since the 1883 eruption of Krakatau. Preliminary data indicated that the average global surface temperature had declined by about 0.6°C (1°F) when compared to the average of the year before. Scientists hope to learn more about the volcano/climate connection over the next few years as Pinatubo's influence fades.

INTRUSIVE IGNEOUS ACTIVITY

Volcanic eruptions can be among the most violent and spectacular events in nature and are therefore worthy of detailed study. Yet most magma is believed to be emplaced at depth. An understanding of intrusive igneous activity is therefore as important to

Figure 7.19
Crater Lake occupies a caldera about 6 miles in diameter.
(Photo from Allstock) Insert shows Wizard Island, a small
cinder cone located at one end of the lake. (Photo by Jeff
Gnass/The Stock Market)

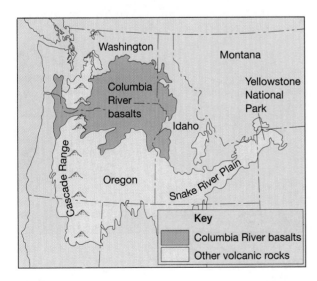

Figure 7.20
Volcanic areas in the northwestern United States. The
Columbia River basalts cover an area of nearly 200,000
square kilometers (80,000 square miles). Activity here
began about 17 million years ago as lava began to pour
out of large fissures, eventually producing a basalt
plateau with an average thickness of more than 1 kilometer. (After U.S. Geological Survey)

Figure 7.21
Basalt flows of the Columbia Plateau. (Photo by E. J.
Tarbuck)

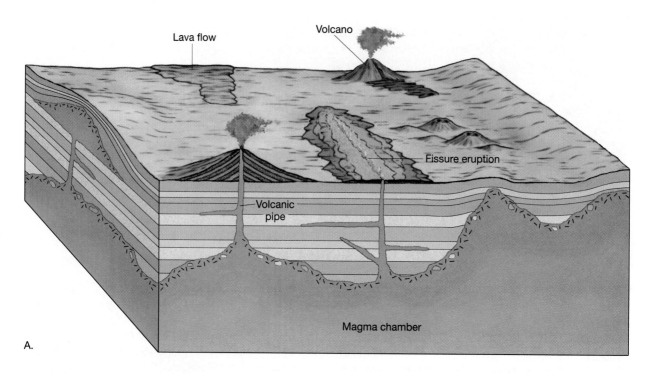

Lava flow

Volcano

Fissure eruption

Volcanic pipe

Magma chamber

A.

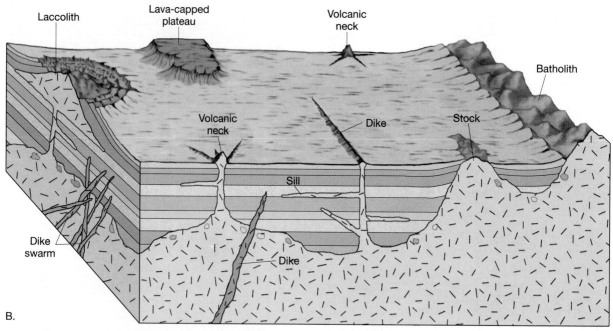

Laccolith

Lava-capped plateau

Volcanic neck

Batholith

Volcanic neck

Dike

Stock

Sill

Dike swarm

Dike

B.

Figure 7.22
Illustrations showing basic igneous structures. **A.** *This cross-sectional view shows the relationship between volcanism and plutonic activity.* **B.** *This view illustrates the basic intrusive igneous structures, some of which have been exposed by erosion long after their formation.*

geologists as is the study of volcanic events. Figure 7.22 shows several types of intrusive igneous bodies that form when magma crystallizes within the earth's crust. Notice that some of these structures have a tabular or sheetlike shape, while others are quite massive. Also observe that some of these bodies cut across existing structures, such as layers of sedimentary rocks, while others form when magma is injected between sedimentary layers. Due to these differences, intrusive igneous bodies are generally classified according to their shape as either *tabular* (sheetlike) or *massive*, and by their orientation with respect to the country (host) rock. Intrusive igneous bodies are said to be *discordant* if they cut across existing sedimentary beds, and *concordant* if they form parallel to the existing sedimentary beds.

Intrusive igneous bodies occur in a great variety of sizes and shapes. **Dikes** are discordant sheetlike bodies produced when magma is injected into fractures that cut across rock layers. Once crystallized, these tabular structures have thicknesses ranging from less than one centimeter to more than one kilometer. The largest have lengths of hundreds of kilometers. Dikes are often oriented vertically and represent pathways followed by molten rock that fed ancient lava flows. Frequently dikes are more resistant to weathering than the surrounding rock. When exposed, these dikes have the appearance of a wall, as shown in Figure 7.17.

Sills are tabular bodies formed when magma is injected along sedimentary bedding surfaces (Figure 7.23). Horizontal sills are the most common, although all orientations, even vertical, are known to exist where the strata have been tilted. Because of their relatively uniform thickness and large extent, sills are believed to form from very fluid magma. As we may expect, sills are most often composed of basaltic magma, which is typically quite fluid. The emplacement of a sill requires that the overlying sedimentary rock be lifted to a height equal to the thickness of the sill. Consequently, sills form only at rather shallow depths where the pressure exerted by the weight of overlying strata is relatively low.

One of the largest and best-known sills in the United States is the Palisades Sill, which is exposed along the west shore of the Hudson River in southeastern New York and northeastern New Jersey. This sill is about 300 meters thick and, due to its resistant nature, has formed an imposing cliff that can be seen easily from the opposite side of the Hudson.

In many respects, sills closely resemble buried lava flows. Both are tabular and often develop shrinkage fractures that produce elongated, pillarlike columns (Figure 7.24). This pattern of cracks, called **columnar joints,** forms as a consequence of contraction that occurs during the cooling of molten rock.

Laccoliths are similar to sills because they form when magma is intruded between sedimentary layers in a near-surface environment. However, the magma that generates laccoliths is more viscous. This less-fluid magma collects as a lens-shaped mass that arches the overlying strata upward. Consequently, a laccolith can occasionally be detected because of the dome-shaped structure it creates at the surface.

By far the largest intrusive igneous bodies are **batholiths.** The Idaho batholith, for example, encompasses an area of more than 40,000 square kilometers. Indirect evidence gathered from gravitational studies indicates that batholiths are also very thick, possibly extending tens of kilometers into the crust. By definition, an intrusive body must have a surface exposure of more than 100 square kilometers (40 square miles) to be considered a batholith. Smaller

Figure 7.23
Salt River Canyon, Arizona. The dark, essentially horizontal band is a sill of basaltic composition that intruded into horizontal layers of sedimentary rock. (Photo by E. J. Tarbuck)

intrusive bodies of this type are termed *stocks*. Many stocks appear to be portions of batholiths that are not yet fully exposed.

Batholiths are usually composed of rock types having chemical compositions near the granitic end of the spectrum, although diorite is relatively common as well. Small batholiths can be rather simple structures composed almost entirely of one rock type. However, studies of large batholiths have shown that they consist of several distinct parts that were intruded over a period of millions of years. The intrusive activity that created the Sierra Nevada batholith, for example, occurred nearly continuously over a 130-million-year time span (Figure 7.25).

Figure 7.25
Half Dome in Yosemite National Park, California. This feature is just a tiny portion of the Sierra Nevada batholith, a huge structure that extends for approximately 400 kilometers. (Photo by Robert MacKinlay/Peter Arnold, Inc.)

Batholiths frequently compose the cores of mountain systems. Here uplift and erosion have removed the surrounding rock to expose the resistant igneous body. Some of the highest mountain peaks, such as Mount Whitney in the Sierra Nevada, are carved from such a granitic mass. Large expanses of granitic rock are also exposed in the stable interiors of the continents, such as the Canadian Shield of North America. These relatively flat outcrops are believed to be the remnants of ancient mountains that erosion has long since leveled.

IGNEOUS ACTIVITY AND PLATE TECTONICS

The origin of magma has been a controversial topic in geology almost from the very beginning of the science. How do magmas of different compositions form? Why do volcanoes located in the deep-ocean basins primarily extrude basaltic lava, whereas those adjacent to oceanic trenches extrude mainly andesitic

Figure 7.24
Devil's Post Pile National Monument, California, exhibits columnar joints and the columns that result. These five-to-seven-sided columns are the consequence of contraction that occurs as a relatively thin layer of rock cools after it has solidified. (Photo by E. J. Tarbuck)

lava? Why does an area of igneous activity, commonly called the *Ring of Fire*, surround the Pacific Ocean? New insights gained from the theory of plate tectonics are providing answers to these questions. We will first examine the origin of magma and then look at the global distribution of volcanic activity as viewed from the model provided by plate tectonics.

Origin of Magma

Based on available scientific evidence, the earth's crust and mantle are composed primarily of solid rock. Further, although the outer core is in a fluid state, this material is very dense and remains deep within the earth. What then is the source of magma that produces the earth's volcanic activity?

Since the molten outer core is not a source of magma, geologists conclude that magma must originate from essentially solid rock located in the crust and mantle. The most obvious way to generate magma from solid rock is to raise its temperature. In a near-surface environment, silica-rich rocks of granitic composition begin to melt at temperatures around 750°C (1400°F), whereas basaltic rocks must reach temperatures above 1000°C (1850°F) before melting commences.

What is the source of heat that melts rock? One source is the heat liberated during the decay of radioactive elements found in the mantle and crust. Workers in underground mines have long recognized that temperatures get higher as they descend to greater depths. Although the rate of temperature increase varies from place to place, it averages between 20°C and 30°C per kilometer in the upper crust.

This gradual increase in temperature with depth is thought to contribute to magma production in two important ways. First, at deep-ocean trenches, slabs of cool oceanic lithosphere descend into the hot mantle. Here heat supplied by the surrounding rocks is thought to be sufficient to melt the subducting oceanic crust and produce basaltic magma. Second, a hot, mantle-derived magma body as just described could migrate to the base of the crust and intrude silica-rich rocks. Because granitic rocks have melting temperatures well below those required to melt basalt, heat derived from the hotter basaltic magma could melt the already warm crustal rocks. The volcanic activity that produced the vast ash flows in Yellowstone National Park is believed to have resulted from such activity. Here basaltic magma from the mantle transported heat to the crust, where the melting of silica-rich rocks generated explosive lavas.

If temperature were the only factor that determined whether or not rock melts, the earth would be a molten ball covered with a thin, solid outer shell. This, of course, is not the case. The reason is that pressure also increases with depth. Since rock must expand when heated, to overcome the effect of confining pressure extra heat is required to melt buried rocks. In general, an increase in the confining pressure causes an increase in the rock's melting temperature, and a reduction in confining pressure causes the rock's melting temperature to decline. Consequently, a drop in confining pressure can lower the melting temperature of rock sufficiently to trigger melting. This occurs whenever rock ascends, thereby moving into zones of lower pressures.

One important difference exists between the melting of a substance that consists of a single compound, such as ice, and the melting of igneous rocks, which are mixtures of several different minerals. Whereas ice melts at a definite temperature, most igneous rocks melt over a temperature range of a few hundred degrees. As a rock is heated, the first liquid to form will contain a higher percentage of the low-melting-temperature minerals than the original rock. Should melting continue, the composition of the melt will steadily approach the overall composition of the rock from which it is derived. Most often, however, melting is not complete. This process, known as **partial melting,** produces most, if not all, magma.

An important consequence of partial melting is the production of a melt with a higher silica content than the parent rock. Recall that basaltic rocks have a relatively low silica content and that granitic rocks have a much higher silica content. Consequently, magmas generated by partial melting are nearer the granitic end of the compositional spectrum than the parent material from which they formed. As we shall see, this idea will help us to understand the global distribution of the various types of volcanic activity.

Distribution of Igneous Activity

Most of the more than 600 active volcanoes that have been identified are located in the vicinity of convergent plate margins. Further, extensive volcanic activity occurs out of view along spreading centers of the oceanic ridge system. In this section we will examine three zones of volcanic activity and relate them to global tectonic activity. These active areas are found along the oceanic ridges, adjacent to ocean trenches, and within the plates themselves (Figure 7.26).

Spreading Center Volcanism.

The greatest volume of volcanic rock is produced along the oceanic ridge system, where sea-floor spreading is active (Figure 7.26). As the rigid lithosphere pulls apart, the pressure on the underlying rocks is lessened. This reduced pressure, in turn, lowers the melting temperature of the mantle rocks. Partial melting of these rocks (primarily peridotite) generates large quantities of basaltic magma that moves upward to fill the newly formed cracks.

Some of the molten basalt reaches the ocean floor, where it produces extensive lava flows or occasionally grows into a volcanic pile. Sometimes this activity produces a volcanic cone that rises above sea level as the island of Surtsey did in 1963 (Figure 7.27). Numerous submerged volcanic cones also dot the flanks of the ridge system and the adjacent deep-ocean floor. Many of these formed along the ridge crests and were moved away as new oceanic crust was created by the process of sea-floor spreading.

Subduction Zone Volcanism.

Recall that ocean trenches are sites where slabs of oceanic crust are bent and move downward into the upper mantle (Figure 7.26). When the descending oceanic plate reaches a depth of about 100 kilometers, partial melting of the water-rich ocean crust and the overlying mantle rocks takes place. The partial melting of these materials is thought to produce basaltic and andesitic magmas.

After a sufficient quantity of magma has accumulated, the melt slowly migrates upward toward the earths surface because it is less dense than the surrounding rock. When subduction volcanism occurs in the ocean, a chain of volcanoes called an *island arc* is produced (Figure 7.26). Examples are numerous in the Pacific and include the Aleutian Islands, the Tonga Islands, and the Mariana Islands.

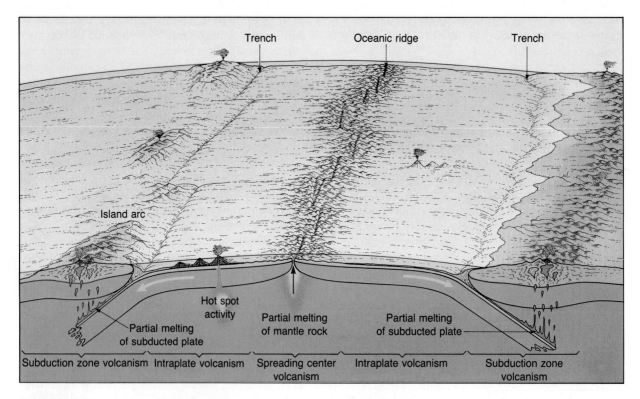

Figure 7.26
Three zones of volcanism. Two of these zones are plate boundaries, and the third includes areas within the plates themselves.

When subduction occurs beneath continental crust, the magma generated is often altered before it solidifies. In particular, magmatic differentiation and the assimilation of crustal fragments into the ascending magma body can lead to a melt exhibiting an andesitic to granitic composition. The volcanoes of the Andes Mountains, from which andesite obtains its name, are examples of this mechanism at work.

Many subduction volcanoes border the Pacific Basin. Because of this pattern, the region has come to be called the *Ring of Fire.* Here volcanism is associated with subduction and partial melting of the Pacific sea floor. As oceanic plates sink, they carry sediments and oceanic crust containing abundant water downward. Since water reduces the melting point of rock, it aids the melting process. Further, the presence of water contributes to the high gas content and explosive nature of volcanoes that make up the *Ring of Fire.* The volcanoes of the Cascade Range in the northwestern United States, including Mount St. Helens, Mount Rainier, and Mount Shasta, are all of this type (Figure 7.28).

Intraplate Volcanism. The processes that actually trigger volcanic activity within a rigid plate are difficult to establish. Activity such as in the Yellowstone region and other nearby areas produced rhyolitic lava, pumice, and ash flows, while extensive basaltic flows cover vast portions of our Northwest. Yet these rocks of greatly varying compositions actually overlie one another in several locations.

Since basaltic extrusions occur on the continents as well as within the ocean basins, the partial melting of mantle rocks is the most probable source for this activity. One proposal suggests that the source of some intraplate basaltic magma comes from rising plumes of hot mantle material. These hot plumes, which may extend to the core-mantle boundary, produce volcanic regions a few hundred kilometers across called **hot spots.** Most that have been identified appear to have persisted for a few tens of millions of years. A hot spot is believed to be located beneath the island of Hawaii and one may have formerly existed beneath the Columbia Plateau.

Generally lavas and ash of granitic composition are extruded from vents located landward of the continental margins. This suggests that remelting of the continental crust may be one mechanism for the formation of these silica-rich magmas. But what mechanism causes large quantities of continental material to melt? One proposal suggests that a thick segment of continental crust occasionally becomes situated over a rising plume of hot mantle material. Rather than producing vast outpourings of basaltic lava, as occurs at oceanic sites such as Hawaii, magma from the rising plume is emplaced at depth.

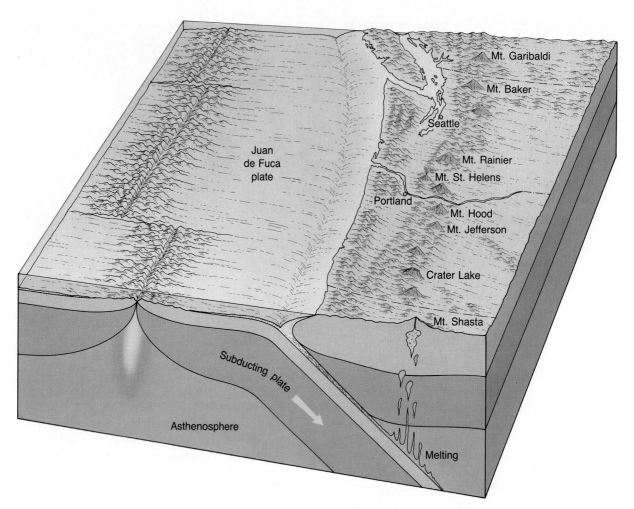

Figure 7.28
Locations of several of the larger composite cones that comprise the Cascade Range.

Here the incorporation and melting of surrounding country rock results in the formation of secondary, silica-rich magma, which slowly migrates upward. Continued hot spot activity supplies heat to the rising mass, thereby aiding its ascent. The activity in the Yellowstone region may have resulted from just this type of activity.

Although the theory of plate tectonics has answered many of the questions that have plagued volcanologists for decades, many new questions have arisen; for example, Why does sea-floor spreading occur in some areas and not others? How do hot spots originate? These are just two questions that are the subject of continuing scientific research.

One of the most popular legends of all time concerns the disappearance of the so-called continent of Atlantis. According to the accounts provided by Plato, an island empire named *Atlantis* disappeared beneath the sea in a single day and night. This event, which reportedly took place between 1500 B.C. and 1400 B.C. and caused the collapse of the Minoan civilization, is thought by many to have been the result of a cataclysmic volcanic eruption.

Research efforts in the eastern Mediterranean have provided evidence apparently linking Atlantis to the volcanic island of Santorin. Although once a majestic volcano, Santorin now consists of five islands located roughly midway between the island of Crete and Greece. Evidence collected from core samples taken around the remnants of Santorin revealed that a violent eruption occurred about 1500 B.C. This eruption generated great volumes of ash and pumice that reached a maximum depth of 60 meters. Many nearby Minoan cities were buried and their ruins preserved beneath the ash.* Even as far away as Crete, enough ash fell to kill crops and possibly livestock grazing on the ash-laden plants. Following the ejection of this large quantity of material, Santorin collapsed, producing a caldera 14 kilometers across. The eruption and collapse of Santorin undoubtedly generated large sea waves that caused widespread destruction to the coastal villages of Crete as well as those on the nearby islands to the north.

The connection between the eruption of Santorin and the disappearance of Atlantis is further supported by the fact that the Minoan civilization also disappeared between 1500 B.C. and 1400 B.C. About 1400 B.C., some of the Minoan traditions began to appear in the culture of Greece.

Most scholars agree that the eruption of Santorin contributed to the collapse of the Minoan civilization. Was this eruption the main cause of the dispersal of this great civilization or only one of many contributing factors? Was Santorin the island continent of Atlantis described by Plato? Whatever the answers to these questions, clearly volcanism can dramatically change the course of human events.

---

* One of these buried cities, Akrotiri, on the island of Thera, has been excavated.

Review Questions

1. What triggered the May 18, 1980, eruption of Mount St. Helens?

2. Name five volcanoes in the western United States that geologists believe will erupt again.

3. What is the difference between magma and lava?

4. What three factors determine the nature of a volcanic eruption? What role does each play?

5. Why is a volcano fed by highly viscous magma likely to be a greater threat than a volcano supplied with very fluid magma?

6. Describe pahoehoe and aa lava.

7. List the main gases released during a volcanic eruption.

8. Analysis of samples taken during Hawaiian eruptions on the island of Hawaii indicate that _____ was the most abundant gas released.

9. Describe each type of pyroclastic material.

10. Compare and contrast the main types of volcanoes (size, shape, eruptive style, and so forth).

11. Name one example of each of the three types of volcanoes.

12. Compare the formation of Hawaii with that of Parícutin.

13. How is a caldera different from a crater?

14. Describe the formation of Crater Lake. Compare it to the caldera formed during the eruption of Kilauea.

15. Briefly describe the mechanism by which explosive volcanic eruptions are thought to influence the earth's climate.

16. What is Ship Rock, New Mexico, and how did it form?

17. Describe each of the four intrusive features discussed in the text.

18. Why might a laccolith be detected at the earth's surface before being exposed by erosion?

19. What is the largest of all intrusive igneous bodies? Is it tabular or massive? Concordant or discordant (see Figure 7.22)?

20. Explain how most magma is thought to originate.

21. Spreading center volcanism is associated with which rock type? What causes rocks to melt in regions of spreading center volcanism?

22. What is the *Ring of Fire?*

23. Are volcanic eruptions in the *Ring of Fire* generally quiet or violent? Name a volcano that would support your answer.

24. The Hawaiian Islands and Yellowstone are thought to be associated with which of the three zones of volcanism?

25. Volcanic islands in the deep ocean are composed primarily of what igneous rock type?

26. Explain why andesitic and granitic rocks are confined to the continents and oceanic margins but are absent from deep-ocean basins.

Key Terms

aa flow (p. 251)

batholith (p. 267)

caldera (p. 255)

cinder cone (p. 255)

columnar joints (p. 267)

composite cone (stratovolcano) (p. 256)

crater (p. 254)

dike (p. 267)

fissure (p. 262)

flood basalt (p. 262)

hot spot (p. 271)

laccolith (p. 267)

nuée ardente (p. 258)

pahoehoe flow (p. 251)

partial melting (p. 269)

pipe (p. 259)

pyroclastic flow (p. 262)

pyroclastics (p. 254)

shield volcano (p. 255)

sill (p. 267)

vent (p. 259)

viscosity (p. 250)

volcanic neck (p. 261)

volcano (p. 254)

8

Mountain Building

Opposite page: *Teton Range, Grand Teton National
Park, Wyoming. (Photo by Tom Till)*

Left: *The Maroon Bells are a part of the Rocky Mountains in Colorado. (Photo by E. J. Tarbuck)*

Above: *The Maroon Bells in autumn. (Photo by
Stephen Trimble)*

277

Mountains provide some of the most spectacular scenery on our planet. This splendor has been captured by poets, painters, and songwriters alike. Geologists believe that at some time all continental regions were mountainous masses and have concluded that the continents grow by the addition of mountains to their flanks. Consequently, by unraveling the secrets of mountain formation, geologists will have taken a major step in understanding the evolution of the earth. If continents do indeed grow by adding mountains to their flanks, how do geologists explain the existence of mountains (the Urals, for example) that are located in the interior of a landmass? In order to answer this and other related questions, this chapter attempts to piece together the sequence of events that is believed to generate these lofty structures.

Mountains often are spectacular features that rise several hundred meters or more above the surrounding terrain (Figure 8.1). Some occur as single isolated masses; the volcanic cone Kilimanjaro, for example, stands almost 6000 meters (20,000 feet) above sea level overlooking the grasslands of East Africa. Other peaks are parts of extensive mountain belts, such as the American Cordillera, which runs continuously from the tip of South America through Alaska. Chains such as the Himalayas consist of youthful, towering peaks that are still rising, whereas others, including the Appalachian Mountains in the eastern United States, are much older and nearly worn down.

The name for the processes that collectively produce a mountain system is **orogenesis,** from the Greek *oros* ("mountain") and *genesis* ("to come into being"). Mountain systems show evidence of enormous forces that have folded, faulted, and generally deformed large sections of the earth's crust. Although the processes of folding and faulting have contributed to the majestic appearance of mountains, much of the credit for their beauty must be given to the work of running water and glacial ice, which sculpture these uplifted masses in an unending effort to lower them to sea level.

The first encompassing explanation of orogenesis came a little over two decades ago as part of the plate tectonics theory. As noted before, the idea of plates colliding has opened many new and exciting avenues to geologists. Before examining mountain building according to the plate tectonics model, however, it will be advantageous first to view the processes of crustal uplifting and rock deformation.

CRUSTAL UPLIFT

The fossilized shells of marine invertebrates are often found in mountain regions, an indication that the sedimentary rock composing the mountain was once below sea level. This is rather convincing evidence that some dramatic changes occurred between the time these animals died and when their fossilized remains were discovered. Evidence for crustal uplift such as this is common in the geologic record. Many examples exist along the coastline of the western United States. When a coastal area remains undisturbed for an extended period, wave action cuts a gently sloping platform. In parts of California, ancient wave-cut platforms can now be found as terraces, hundreds of meters above sea level (Figure 8.2). Each terrace represents a period when that area was at sea level. Unfortunately, the reasons for uplift are not always as easy to determine as the evidence for the movements.

We know that the force of gravity must play an important role in determining the elevation of the land. In particular, the less dense crust is believed to float on top of the denser and more easily deformed rocks of the mantle. The concept of a floating crust in gravitational balance is called **isostasy.** Perhaps the easiest way to grasp the concept of isostasy is to envision a series of wooden blocks of different heights floating in water as shown in Figure 8.3. Note that the thicker wooden blocks float higher in water than the thinner blocks. In a similar manner, mountain belts stand higher above the surface and also extend farther into the supporting material below. In other words, mountainous terrains are supported by light crustal material that extends as "roots" into the denser mantle. Seismological and gravitational studies have confirmed the fact that mountains do indeed consist of unusually thick sections of deformed crustal material. The thickness of continental crust is normally about 35 kilometers, but crustal thicknesses exceeding 70 kilometers have been determined for some mountain chains.

Carrying this idea one step further, the crust beneath the oceans must be thinner than that of the continents because its elevation is lower. Although this is true, oceanic rocks also have a greater density than do continental rocks, another factor contributing to their lower position.

If the concept of isostasy is correct, we should expect that when weight is added to the crust, the crust will respond by subsiding, and that when weight is removed there will be uplifting. (Visualize what happens to a ship as cargo is being loaded and unloaded.) Evidence for this type of movement exists, strongly supporting the concept of *isostatic adjustment.* A classic example is provided by Ice Age ice sheets. When continental glaciers occupied portions of North America during the Pleistocene epoch, the added weight of the 3-kilometer-thick masses of ice caused downwarping of the earth's crust. In the 8000 to 10,000 years since the last ice sheets melted, uplifting of as much as 330 meters has occurred in the Hudson Bay region, where the thickest ice had accumulated.

Figure 8.1
Mount Kangbochen (23,933 feet), Tibet. (Photo by Galen Rowell)

As the foregoing examples illustrate, isostatic adjustment can account for considerable crustal movement. Thus, we can understand why, as erosion lowers the summits of mountains, the crust will rise in response to the reduced load. However, each episode of isostatic adjustment results in somewhat less uplift than the elevation loss due to erosion. The processes of uplifting and erosion will continue until

Figure 8.2
Former wave-cut platforms now exist as a series of elevated terraces on the west side of San Clemente Island off the southern California coast. Once at sea level, the highest terraces are now about 400 meters (1320 feet) above it. (Photo by John S. Shelton)

the mountain block reaches "normal" crustal thickness (Figure 8.4). When this occurs, the mountains will have been eroded to near sea level and the deeply buried portions of the mountains will be exposed at the surface. In addition, as the mountains wear down, the weight of the eroded sediment deposited on the adjacent continental margin will cause the margin to subside.

To summarize, we have learned that the earth's major mountain chains consist of unusually thick sections of deformed crustal material that have been elevated above the surrounding terrain. The support

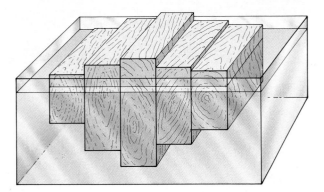

Figure 8.3
This drawing illustrates how wooden blocks of different thicknesses float in water. In a similar manner, thick sections of crustal material float higher than thinner crustal slabs.

for these massive features comes from the buoyancy of deep crustal roots. As erosion lowers the peaks, isostatic adjustment gradually raises the mountains in response. Eventually, the deepest portions of the mountains are brought to the shallower depths of the surrounding crust. The question still to be answered is: How do these thick sections of the earth's crust come into existence?

ROCK DEFORMATION

When rocks are subjected to stresses greater than their own strength, they begin to deform, usually by folding or fracturing (Figure 8.5). It is easy to visualize how individual rocks break, but how are large rock units bent into intricate folds without being appreciably broken during the process? In an attempt to answer this, geologists used the laboratory. Here rocks were subjected to stresses under conditions that simulated those existing at various depths within the crust.

Although each rock type deforms somewhat differently, the general characteristics of rock deformation were determined from these experiments. Geologists discovered that when stress is applied, rocks first respond by deforming elastically. Changes resulting from *elastic deformation* are reversible; that is, like a rubber band, the rock will return to nearly its original size and shape when the stress is removed.

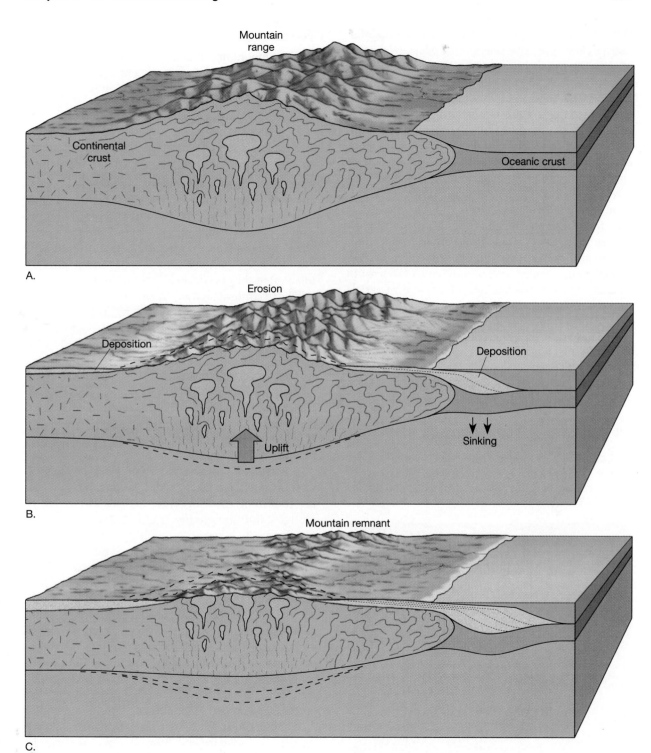

Figure 8.4
This sequence illustrates how the combined effect of erosion and isostatic adjustment results in a thinning of the crust in mountainous regions.

However, once the elastic limit is surpassed, rocks either deform plastically or fracture. *Plastic deformation* results in permanent changes; that is, the size and shape of a rock unit are altered through folding and flowing. Laboratory experiments confirmed the speculation that, at high temperatures and pressures, most rocks deform plastically once their elastic limit is surpassed. Rocks tested under surface conditions also deform elastically, but once they exceed their elastic limit, most behave like a brittle solid and fracture. Recall that the energy for most earthquakes comes from stored elastic energy that is released as rock fractures and snaps back to its original shape.

One factor that researchers cannot duplicate in the laboratory is geologic time. We know that if stress is applied quickly, as with a hammer, rocks tend to fracture. On the other hand, these same materials may deform plastically if stress is applied over an extended period. For example, marble benches have been known to sag under their own weight over a period of a hundred years or so. In nature, small forces applied over long time periods surely play an important role in the deformation of rock.

Figure 8.5
Deformed and metamorphosed rocks in the Panamint Mountains east of Salina Valley, California. (Photo by Michael Collier)

Folds

During mountain building, flat-lying sedimentary and volcanic rocks are often bent into a series of wavelike undulations called **folds** (Figure 8.6). Folds in sedimentary strata are much like those that would form if you were to hold the ends of a sheet of paper and then push them together. In nature, folds come in a wide variety of sizes and configurations. Some folds are broad flextures in which rock units hundreds of meters thick have been slightly warped. Others are very tight, microscopic structures found in metamorphic rocks. Size differences notwithstanding, most folds are the result of compressional stresses that result in the shortening and thickening of the crust. Occasionally, folds are found singly, but most often they occur as a series of undulations.

The two most common types of folds are anticlines and synclines (Figure 8.6). An **anticline** is most commonly formed by the upfolding, or arching, of rock layers. Anticlines are sometimes spectacularly displayed where highways have been cut through deformed strata. Often found in association with anticlines are downfolds, or troughs, called **synclines.** Depending upon their orientation, anticlines and synclines are said to be *symmetrical*, *asymmetrical*, or, if one limb has been tilted beyond the vertical, *overturned*.

Folds do not continue on forever; rather, their ends die out much like the wrinkles in cloth. These folds are said to be *plunging*, since the axis of the fold is plunging into the ground (Figure 8.7). The left side of Figure 8.8 shows an example of a plunging anticline and the pattern produced when erosion removes the upper layers of this structure and exposes its interior. Note that the outcrop pattern of an anticline points in the direction it is plunging, while the opposite is true for synclines. A good example of the kind of topography that results when erosional forces attack folded sedimentary strata is found in the Valley and Ridge Province of the Appalachians. Here resistant sandstone beds remain as imposing ridges separated by valleys cut into more easily eroded shale or limestone beds (see Figure 8.23).

Although most folds are caused by compressional stresses that squeeze and crumble strata, some folds are a consequence of vertical displacement. When unwarping produces a circular or somewhat elongated structure, the feature is called a **dome** (Figure 8.9A). Downwarped structures having a similar shape are termed **basins** (Figure 8.9B). The Black

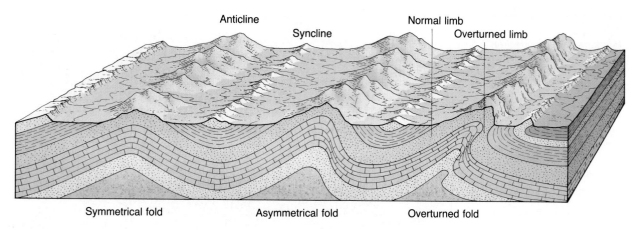

Anticline — Syncline — Normal limb — Overturned limb

Symmetrical fold — Asymmetrical fold — Overturned fold

Figure 8.6
Block diagram of principal types of folded strata. The upfolded or arched structures are anticlines. The downfolds or troughs are synclines. Notice that the limb of an anticline is also the limb of the adjacent syncline.

Hills of western South Dakota are one such domal structure in which erosion has stripped away the upwarped sedimentary beds, exposing older igneous and metamorphic rocks in the center (see Figure 8.18). Several basins also exist in the United States. The basins of Michigan and Illinois have very gently sloping beds similar to saucers. Because large basins contain sedimentary beds sloping at such low angles, they are usually identified by the age of the rocks composing them. The youngest rocks are found near the center and the oldest rocks are at the flanks. This is just the opposite order of a domal structure such as the Black Hills, where the oldest rocks form the core.

Faults

Faults are fractures in the crust along which appreciable displacement has occurred. Occasionally, small faults can be recognized in road cuts where sedimentary beds have been offset a few meters, as shown in Figure 8.10. Faults of this scale usually occur as single discrete breaks. By contrast, large faults, like the San Andreas fault in California, have displacements of hundreds of kilometers and consist of many interconnecting fault surfaces. These so-called fault zones can be several kilometers wide and are often easier to identify from high-altitude images than at ground level.

Faults in which the movement is primarily vertical are called **dip-slip faults** since the displacement is along the inclination, or dip, of the fault plane. Because movement along dip-slip faults can be either up or down the fault plane, two types of dip-slip faults are recognized. In order to distinguish between the two types, it has become practice to call the rock immediately above the fault surface the *hanging wall* and to call the rock below the *foot-wall*. This nomenclature arose from prospectors and miners who ex-

Figure 8.7
Sheep Mountain, a plunging anticline. Note that erosion has cut the flanking sedimentary beds into low ridges that make a "V" pointing in the direction of plunge. (Photo by John S. Shelton)

Figure 8.8
Plunging folds. **A.** *Idealized view.* **B.** *View after extensive erosion.*

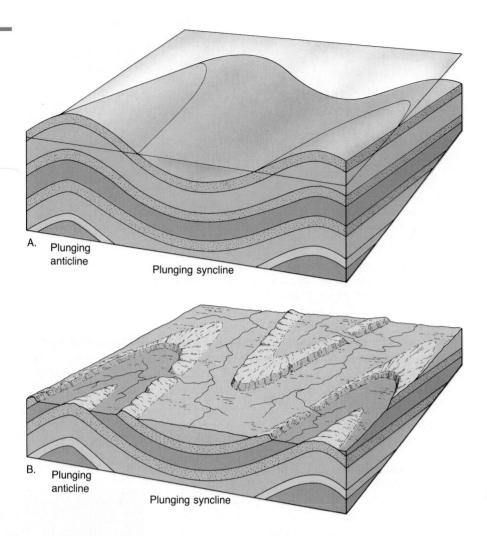

A. Plunging anticline

Plunging syncline

B. Plunging anticline

Plunging syncline

cavated shafts along fault zones because such zones are frequently sites of ore deposition. During these operations, the miners would walk on the rocks below the fault trace (the footwall), and hang their lanterns on the rocks above (the hanging wall). Dip-slip faults are classified as **normal faults** when the hanging wall moves down relative to the footwall (Figure 8.11A). Conversely, **reverse faults** occur when the hanging wall moves up relative to the footwall (Figure 8.11B). Reverse faults having a very low angle are also referred to as **thrust faults** (Figure 8.11C). In mountainous regions such as the Alps and the Appalachians, thrust faults have displaced rock as far as 50 kilometers over adjacent strata. Thrust faults of this type result from strong compressional stresses.

Faults in which the dominant displacement is along the trend, or strike, of the fault are called **strike-slip faults** (Figure 8.11D). Many large strike-slip faults are associated with plate boundaries and are called transform faults. Transform faults have nearly vertical dips and serve to connect large structures, such as segments of an oceanic ridge. The San Andreas fault in California is a well-known transform fault in which the displacement has been on the order of several hundred kilometers (see Box 8.1). When faults have both vertical and horizontal movement, they are called **oblique-slip faults**.

Fault motion provides the geologist with a method of determining the nature of the forces at work within the earth. Normal faults indicate the existence of *tensional stresses* that pull the crust apart. This "pulling apart" can be accomplished either by uplifting that causes the surface to stretch and break, or by horizontal forces that actually rip the crust apart. Normal faulting is known to occur at spreading centers,

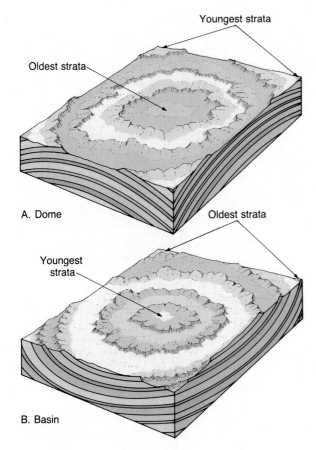

Figure 8.9
Gentle upwarping and downwarping of crustal rocks produce domes and basins. Erosion of these structures results in an outcrop pattern, which is roughly circular or elongated.

where plate divergence is prevalent. Here a central block called a **graben** is bounded by normal faults and drops as the plates separate (Figure 8.12). These grabens produce an elongated valley bounded by upfaulted structures called **horsts.** The Great Rift Valley of East Africa consists of several large grabens, above which tilted horsts produce a linear mountainous topography (see Figure 6.12). This valley, nearly 6000 kilometers (3700 miles) long, contains the excavation sites of some of the earliest human fossils. Other rift valleys included the Rhine Valley in Germany and the valley of the Dead Sea in the Middle East.

Since the blocks involved in reverse and thrust faulting are displaced toward one another, geologists conclude that *compressional forces* are at work. The primary regions of this activity are thought to be the

convergent zones, where plates are colliding. Compressional forces generally produce folds as well as faults, and result in a general thickening and shortening of the material involved.

Among the most common rock structures are fractures called **joints.** Unlike faults, joints are fractures along which no appreciable displacement has occurred. Although some joints have a random orientation, most occur in roughly parallel groups.

We have already considered two types of joints. Earlier we learned that *columnar joints* form when igneous rocks cool and develop shrinkage fractures that produce elongated, pillarlike columns (see Figure 7.24). Also recall that *sheeting* produces a pat-

Figure 8.10
Faulting caused the vertical displacement of these sedimentary beds in southern Nevada. (Photo by E. J. Tarbuck)

BOX 8.1
The San Andreas Fault System

The San Andreas, the best-known and largest fault system in North America, first attracted wide attention after the great 1906 San Francisco earthquake and fire. Following this devastating event, geologic studies demonstrated that a displacement of as much as 5 meters along the fault was responsible for the earthquake. It is now known that this dramatic event is just one of many thousands of earthquakes that have resulted from repeated movements along the San Andreas throughout its 29-million-year history.

Where is the San Andreas fault system located? As shown in Figure 8.A, it trends in a northwesterly direction for nearly 1300 kilometers (780 miles) through much of western California. At its southern end, the San Andreas connects with a spreading center located in the Gulf of California. In the north, the fault enters the Pacific Ocean at Point Arena, where it is thought to continue its northwesterly trend, eventually joining the Mendocino fracture zone. In the central section, the San Andreas is relatively simple and straight. However, at its two extremities, several branches spread from the main trace, so that in some areas the fault zone exceeds 100 kilometers (60 miles) in width.

With the advent of the theory of plate tectonics, geologists have begun to realize the significance of this great fault system. The San Andreas fault is a transform boundary separating two crustal plates that move very slowly over the earth's surface. The Pacific plate, located to the west, moves northwestward relative to the North American plate, causing earthquakes along the fault.

Over much of its extent, a linear trough reveals the presence of the San Andreas fault (Figure 8.A). When the system is viewed from the air, linear scars, offset stream channels, and elongated ponds mark the trace in a most striking manner. On the ground, however, surface expressions of the fault are much more difficult to detect. Some of the most distinctive landforms include long, straight escarpments, narrow ridges, and sag ponds formed by settling of blocks within the fault zone. Further, many stream channels characteristically bend sharply to the right where they cross the fault.

The San Andreas is undoubtedly the most studied of any fault system in the world. Although many questions remain unanswered, geologists have learned that each fault segment exhibits somewhat different behavior. Some portions of the San Andreas exhibit a slow creep with little noticeable seismic activity. Other segments regularly slip, producing small earthquakes, while still other segments seem to store elastic energy for hundreds of years and rupture in great earthquakes. This knowledge is useful when assigning the potential earthquake hazard to a given segment of the fault zone.

Because of the great length and complexity of the San Andreas fault, it is more appropriately referred to as a "fault system." This major fault system consists primarily of the San Andreas fault and several major branches, including the Hayward and Calaveras faults of central California and the San Jacinto and Elsinore faults of southern California. These major segments, plus a vast number of smaller faults that include the Imperial fault, San Fernando fault, and the Santa Monica fault, collectively accommodate the relative motion between the North American and Pacific plates.

Blocks on opposite sides of the San Andreas fault move horizontally in opposite directions, such that if a person stood on one side of the fault, the block on the other side would appear to move to the right when slippage occurred. Since the great San Francisco earthquake of 1906, when as much as 5 meters of displacement took place, geologists have attempted to establish the cumulative displacement along this fault over its 29-million-year history. By matching rock units across the fault, geologists have determined that the total accumulated displacement from earthquakes and creep exceeds 560 kilometers (340 miles)

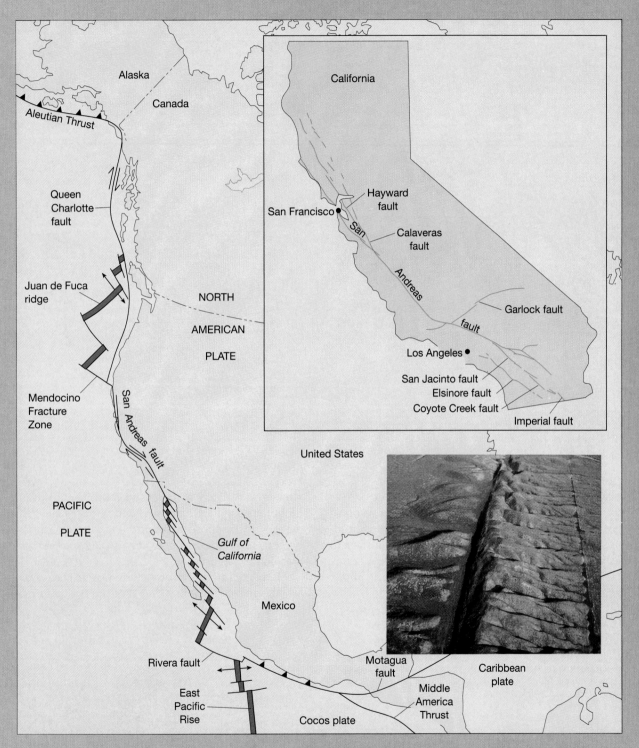

Figure 8.A
Map showing the extent of the San Andreas fault system. Insert map of California shows only a few of the many splinter faults that are part of this great fault system. The aerial photo shows the San Andreas fault as it crosses the Carrizo Plain north of Los Angeles. (Photo by R. E. Wallace, U.S. Geological Survey)

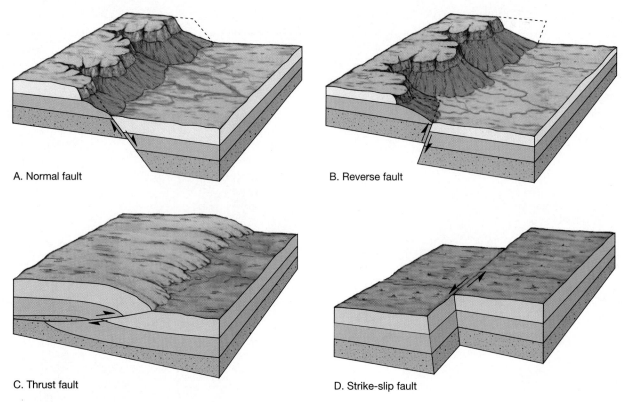

A. Normal fault

B. Reverse fault

C. Thrust fault

D. Strike-slip fault

Figure 8.11
*Block diagrams of four types of faults. **A**. Normal fault. **B**. Reverse fault. **C**. Thrust fault.*
D. *Strike-slip fault.*

tern of gently curved joints that develop more or less parallel to the surface of large exposed igneous bodies such as batholiths. Here the jointing is thought to result from the gradual expansion that occurs when erosion removes the overlying load.

In contrast to the situations just described, most joints are produced when rocks are deformed by the stresses associated with crustal movements during mountain building. Extensive joint patterns can also develop in response to relatively subtle and often barely perceptible regional upwarping and downwarping of the crust.

Many rock units are broken by two or even three sets of intersecting joints that slice the rock into nu-

Figure 8.12
Diagrammatic sketch of downfaulted block (graben) and upfaulted block (horst).

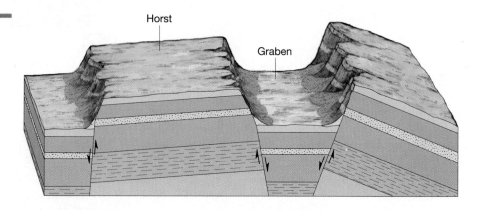

Horst

Graben

Figure 8.13
Chemical weathering is enhanced along joints in granitic rocks near the top of Lembert Dome, Yosemite National Park. (Photo by E. J. Tarbuck)

merous regularly shaped blocks. These joint sets often exert a strong influence on other geologic processes. For example, chemical weathering tends to be concentrated along joints, and in many areas, groundwater movement and the resulting solution activity in soluble rocks is controlled by the joint pattern (Figure 8.13). Moreover, a system of joints can influence the direction that stream courses follow. The rectangular drainage pattern described in Chapter 3 is such a case.

MOUNTAIN TYPES

Even though no two mountain ranges are exactly alike, they can be classified according to their most dominant characteristics. Using this approach, four main categories of mountains result: (1) folded mountains, or complex mountains; (2) volcanic mountains; (3) fault-block mountains; and (4) up-warped mountains. Mountain ranges of the same

type are commonly found in close proximity forming a mountain system. For example, nearly the entire state of Nevada is composed of numerous elongated fault-block mountains that are separated by faulted basins. Further, within any mountainous belt, such as that portion of the American Cordillera in the western United States, mountain ranges representing each of these groups can be found. In addition to these basic varieties, some regions have mountainous topography that was produced without appreciable crustal deformation. For example, plateaus (areas of high-standing rocks that are essentially horizontal) can be deeply dissected into rugged terrains. Although these highlands have the topographical expression of mountains, they lack the structure associated with orogenesis.

In the following sections we will examine three of the four basic mountain types. Volcanic mountains are treated in detail in Chapter 7.

Fault-Block Mountains

Most orogenesis occurs in compressional environments, as evidenced by the predominance of large thrust faults and folded strata in mountainous areas. However, some mountain building is associated with tensional stresses (Figure 8.14). The mountains that form under such circumstances, termed **fault-block mountains**, are bounded on at least one side by high- to moderate-angle normal faults. Some fault-block mountains form in response to broad uplifting, which causes elongation and faulting. Such a situation is exemplified by the fault blocks that rise above the rift valleys of East Africa (see Figure 6.12). Other fault-block mountains appear to develop where isolated blocks are tilted, or lifted nearly vertically, high above adjacent undeformed valleys.

The Basin and Range province, a region that encompasses Nevada and portions of Utah, New Mexico, Arizona, and California, contains excellent examples of fault-block mountains. Here the crust literally has been broken into hundreds of pieces, giving rise to nearly parallel mountain ranges, averaging about 80 kilometers in length, which rise precipitously above the adjacent sediment-laden basins (Figure 8.15).

No single explanation for the development of the Basin and Range is universally accepted. It is clear, however, that important geological events preceded the period of extension and faulting. During this earlier time span, a nearly horizontal oceanic plate was

BOX 8.2

Oil Traps

Unlike the altered organic matter from which they formed, petroleum and natural gas are mobile. These fluids are gradually squeezed from the compacting, mud-rich layers where they originated and may migrate hundreds of kilometers into permeable beds such as sandstone, where openings between sediment grains are larger. The rock layers containing the oil and gas are saturated with water, and, since they are not as dense as water, oil and gas migrate upward through the water-filled pore spaces of the enclosing rocks. Unless something acts to halt this upward migration, the fluids will eventually reach the surface, at which point the volatile components will evaporate.

A geologic environment that allows for economically significant amounts of oil and gas to accumulate is termed an *oil trap.* Although, as we shall see, several geologic structures may act as oil traps, all have two basic conditions in common: a porous, permeable *reservoir rock* that will yield petroleum and natural gas in sufficient quantities to make drilling worthwhile; and a *cap rock,* such as shale, that is virtually impermeable to oil and gas. The cap rock keeps the upwardly mobile oil and gas from escaping at the surface.

Figure 8.B illustrates some common oil and natural gas traps. One of the simplest traps is an *anticline,* an uparched series of sedimentary strata (Part A.). As the strata are folded, the rising oil and gas collect at the apex of the fold. Because of its lower density, the natural gas collects above the oil. Both rest upon the denser water that saturates the reservoir rock. One of the world's largest oil fields, El Nala in Saudi Arabia, is the result of an anticlinal trap, as is the famous Teapot Dome in Wyoming. *Fault traps* form when strata are displaced in such a manner as to bring a dipping reservoir rock opposite an impermeable bed, as shown in Part B. In this case the upward migration of the oil and gas is halted at the fault zone. In the Gulf coastal plain region of the United States important accumulations of oil occur in association with *salt domes.* In such areas, which are characterized by thick accumulations of sedimentary strata, layers of rock salt occurring at great depth have been forced to rise in columns by the pressure of overlying beds. These rising salt columns gradually deform the overlying strata. Since oil and gas migrate to the highest level possible, they accumulate in the upturned sandstone beds adjacent to the salt column (Part C). Yet another important geologic circumstance that may lead to significant accumulations of oil and gas is termed a *stratigraphic trap.* These oil-bearing structures result primarily from the original pattern of sedimentation rather than structural deformation. The stratigraphic trap illustrated in Part D. exists because a sloping bed of sandstone thins to the point of disappearance.

When the lid created by the cap rock is punctured by drilling, the oil and natural gas, which are under pressure, migrate from the pore spaces of the reservoir rock to the drill hole. On rare occasions the fluid pressure may force oil up the drill hole to the surface. Usually, however, a pump is required to lift the oil out.

A drill hole is not the only means by which oil and gas can

subducted eastward under western North America (Figure 8.16A). Compressional stresses in the overlying plate thickened the crust as far inland as the Rocky Mountains of Colorado. Then, about 30 million years ago, the western portion of the North American plate began to override a segment of the spreading center located in the Pacific basin. This event caused the southwestern portion of the United States to become attached to the Pacific plate, which was moving in a northwesterly direction. Today, this movement is most pronounced along the San Andreas fault, where portions of southern California and Mexico's Baja Peninsula have migrated several hundred kilometers toward the northwest. In addition to creating movement along the San Andreas fault system, it has been postulated that this change

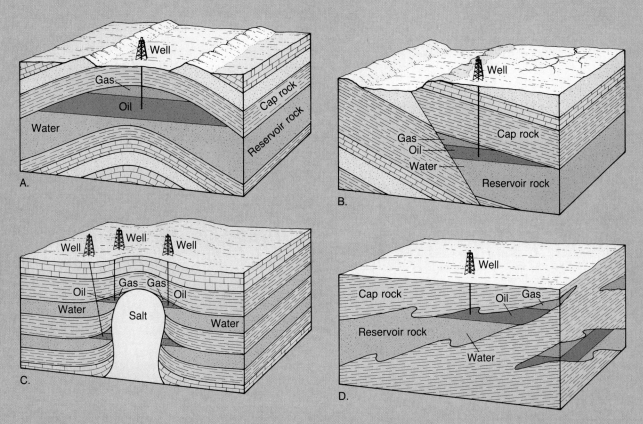

Figure 8.B
*Common oil traps. **A.** Anticline. **B.** Fault trap. **C.** Salt dome. **D.** Stratigraphic (pinch-out) trap.*

escape from a trap. Traps can be broken by natural forces as well. For example, earth movements may create fractures that allow the hydrocarbon fluids to escape, or surface erosion may breach a trap with similar results. The older the rock strata, the greater the chance that deformation or erosion has affected a trap. Indeed, not all ages of rock yield oil and gas in the same proportions. The greatest production comes from the youngest rocks, those of Cenozoic age. Older Mesozoic rocks produce considerably less, followed by even smaller yields from the still older Paleozoic strata. There is virtually no oil produced from the most ancient rocks, those of Precambrian age.

in plate motion also stretched the crustal rocks located farther inland. In particular, this event is believed to have stretched and fractured the crust in the region now occupied by the Basin and Range province, generating the fault-block topography.

Some geologists disagree with part of the preceding scenario. Instead, they suggest that extension in the Basin and Range was triggered when the horizontal oceanic slab that had been thrust under the continent began to sink (Figure 8.16B, bottom). Sinking of the oceanic slab allowed the upwelling of hot material from the asthenosphere. The buoyancy of the warm material caused upwarping and tensional fracturing in the overlying crust. This event was associated with volcanism and east-west extension and fracturing of the crust. Regardless of which, if either,

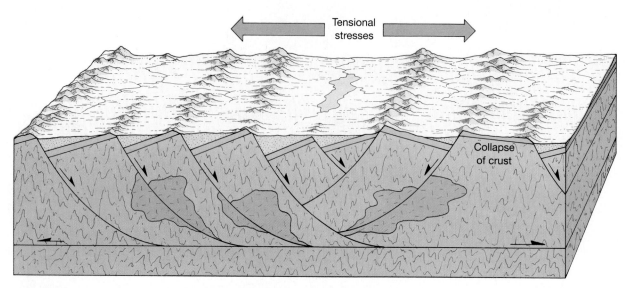

Figure 8.14
Normal faulting in the Basin and Range Province. Here tensional stresses have elongated and fractured the crust into numerous blocks. Movement along these fractures has tilted the blocks, producing parallel mountain ranges called fault-block mountains.

explanation is correct, there is general agreement that the fault-block mountains in this region resulted from tensional forces that stretched the crust by nearly 150 kilometers.

Other fault-block mountains in the United States are the Teton Range of Wyoming and the Sierra Nevada of California. Both are faulted along their eastern flanks, which were uplifted as the blocks tilted downward to the west. Looking west from Jackson Hole, Wyoming, and Owens Valley, California, respectively, the eastern fronts of these ranges rise over 2 kilometers, making them two of the most precipitous mountain fronts in the United States (Figure 8.17).

Figure 8.15
Basin and Range topography north of Tonopah, Navada. These north-south trending mountians are separated by down-faulted basins. Note road in lower left of photo for scale. (Photo by Michael Collier)

Figure 8.16
One proposal for the formation of the Basin and Range Province. **A.** *Nearly horizontal subduction of an oceanic plate produced compressional stresses which generally thickened the crust in the Basin and Range.* **B.** *Sinking of this oceanic slab allowed for the upwelling of hot material from the asthenosphere. The buoyancy of the warm material caused upwarping and tensional fracturing in the crust above. This event was associated with volcanism and east-west extension of the crust by nearly 150 kilometers.*

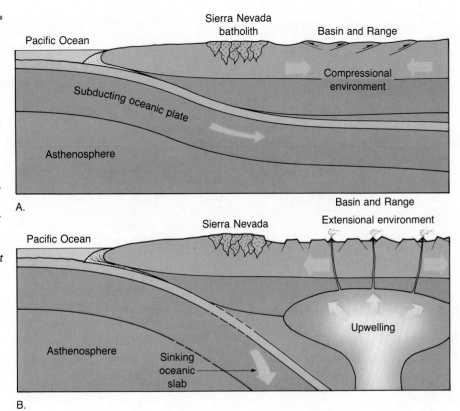

Upwarped Mountains

Upwarped mountains are produced in association with a broad arching of the crust or, in some instances, because of vertical displacement along high-angle faults. Some, such as the Black Hills in western South Dakota and the Adirondack Mountains in upstate New York, consist of older igneous and metamorphic bedrock that was once eroded flat and subsequently mantled with sediment. As the regions were upwarped, erosion removed the veneer of sedimentary strata, leaving a core of igneous and metamorphic rocks standing above the surrounding terrain (Figure 8.18).

Other examples of upwarped mountains are found in the middle and southern Rockies from Montana south through Colorado and into New Mexico. In this group are the Front Range of Colorado, the Sangre de Cristo of New Mexico and Colorado, and the Bighorns of Wyoming. The middle and southern Rockies were pushed almost vertically upward as part of broad upwarping of the crust (see Box 8.3).

Folded Mountains

The largest and most complex mountain systems are the so-called **folded mountains** or **complex mountains.** Although folding is often more conspicuous, faulting, metamorphism, and igneous activity are always present in varying degrees. All major mountain belts, including the Alps, Urals, Himalayas, and Appalachians, are of this type. Since folded mountains represent the world's major mountain systems, the process of mountain building is usually described in terms of their formation. Thus the next section, on mountain building, is devoted to the evolution of these majestic and always complex mountain systems.

MOUNTAIN BUILDING

Mountain building has operated during the recent geologic past in several locations around the world. These relatively young mountainous belts include the American Cordillera, which runs along the western

The portion of the Rocky Mountains that extends from southern Montana to New Mexico was produced by a period of deformation known as the _Laramide Orogeny_. This event, which created some of the most picturesque scenery in the United States, began in the late Cretaceous and peaked about 60 million years ago in early Tertiary time. The mountain ranges generated during the Laramide Orogeny include the Front Range of Colorado, the Sangre de Cristo of New Mexico and Colorado, and the Bighorns of Wyoming. These mountains are structurally much different from the northern Rockies, which include the Canadian Rockies and those portions of the Rock-

ies found in Idaho, western Wyoming and western Montana. Whereas the latter ranges are composed of thick sequences of sedimentary rocks that were deformed by folding and low-angle thrust faulting, the middle and southern Rockies were pushed almost vertically upward as part of broad upwarping of the crust. In general, these mountains consist of Precambrian basement rocks overlain by relatively thin layers of younger strata. Since the time of deformation, much of the sedimentary cover has been eroded from the highest portions of the uplifted blocks, exposing the igneous and metamorphic cores. Examples include a number of granitic outcrops that project as steep summits, such as Pikes Peak and Longs Peak in Colorado's Front Range. In many areas, remnants of the sedimentary strata that once covered this region are visible as prominent angular ridges called _hogbacks_ flanking the crystalline cores of the mountain ranges

(Figure 8.C).

Although the Rockies have been extensively studied for over a century, there is still a good deal of debate regarding the mechanisms that led to uplift. According to the plate tectonics model, most mountain belts are produced along continental margins in association with convergent plate boundaries. Here, crustal buckling forces generate thick sequences of folded, faulted, and metamorphosed strata that are often intruded by massive igneous bodies. However, this model does not fit the middle and southern Rocky Mountains, which consist of elongated, uplifted blocks of Precambrian basement rocks separated by sediment-filled basins.

One hypothesis for the formation of the middle and southern Rocky Mountains has been gaining widespread support. According to this proposal, a subducted plate of oceanic lithosphere remained nearly horizontal as it

margin of the Americas from Cape Horn to Alaska; the Alpine-Himalayan chain, which extends from the Mediterranean through Iran to northern India and into Indochina; and the mountainous terrains of the western Pacific, which include mature island arcs such as Japan, the Philippines, and Sumatra. Most of these young mountain belts have come into existence within the last 100 million years. Some, including the Himalayas, began their growth as recently as 45 million years ago.

In addition to these recently formed complex (folded) mountains, several chains of much older mountains exist on the earth as well. Although these structures are deeply eroded and topographically less prominent, they clearly possess the same structural features found in younger mountains. Typical of this

older group are the Appalachians in the eastern United States and the Urals in the former Soviet Union.

Although major mountains differ from one another in particular details, all possess the same basic structures. Mountain chains generally consist of roughly parallel ridges of folded and faulted sedimentary and volcanic rocks, portions of which have been strongly metamorphosed and intruded by somewhat younger igneous bodies. In most cases, the sedimentary rocks were formed from enormous accumulations of deep-water marine sediments that occasionally exceeded 15 kilometers in thickness, as well as from thinner shallow-water deposits. Moreover, these deformed sedimentary rocks are for the most part older than the mountain building event.

Figure 8.C
Maroon Bells in the Rocky Mountains near Aspen, Colorado. (Photo courtesy of Stockhouse, Inc.)

pushed eastward under North America as far inland as the Black Hills of South Dakota (see Figure 8.16A). As the subducted slab scraped beneath North America, compressional stresses shortened and thickened the rocks in the lower crust. Furthermore, this event locally uplifted blocks of Precambrian basement rocks along high-angle faults to produce the Rockies and intermountain basins. Thus, the Laramide Orogeny may be associated with a special type of convergent plate boundary. Whereas plate subduction along a typical Andean-type plate boundary occurs at a steep angle, the middle and southern Rocky Mountains may have been produced by the nearly horizontal subduction of an oceanic plate.

This fact indicates that a long quiescent period of deposition was followed by an episode of deformation (Figure 8.19).

In order to unravel the events that produce mountains, many studies are conducted in regions that exhibit ancient mountain structures as well as at sites where orogenesis is thought to be in progress. Of particular interest are active subduction zones, where plates are converging. At most modern-day subduction zones, volcanic arcs are forming. This situation is typified by Alaska's Aleutian Islands and by the Andean arc of western South America. Although all volcanic arcs are similar, Aleutian-type subduction zones occur where two oceanic plates converge, whereas Andean-type subduction zones are situated where oceanic crust is being thrust beneath a continental mass. Consequently, the events that generated these particular volcanic arcs followed somewhat different evolutionary paths. Further, although the development of a volcanic arc does result in the formation of mountainous topography, this activity is viewed as just one of the phases in the development of a major mountain belt.

At sites where oceanic crust is being subducted, continental blocks are also being rafted toward one another. Recent studies indicate that the most important cause of orogenesis is the collision of two or more of these crustal fragments. Collisions can occur between a continental block and a variety of landmasses, including island archipelagos such as the Aleutian Islands, or small crustal fragments similar in size to Madagascar, or even other continental-sized

Figure 8.17
*The east face of the Sierra Nevada. Mt. Whitney, the highest point in the United States
outside of Alaska, is shown in the center of the photo. (Photo by James E. Patterson)*

blocks. We shall consider these sites of mountain building in the following sections.

Mountain Building at Subduction Zones

Mountain building along continental margins involves the convergence of an oceanic plate and a plate whose leading edge contains continental crust. Exemplified by the Andes Mountains, this type of convergence generates structures resembling those of a developing volcanic island arc.

The first stage in the development of an Andean-type mountain belt occurs prior to the formation of the subduction zone. During this period the continental margin is *passive*; that is, it is not a plate boundary but a part of the same plate as the adjoining oceanic crust. The East Coast of the United States provides a present-day example of a passive continental margin. Here, as at other passive conti-

nental margins surrounding the Atlantic, deposition of sediment on the continental shelf is producing a thick wedge of shallow-water sandstones, limestones, and shales (Figure 8.20A). Beyond the continental shelf, turbidity currents are depositing sediments on the continental slope and rise.*

At some point the continental margin becomes active; a subduction zone forms and the deformation process begins (Figure 8.20B). A good place to examine an active continental margin is the west coast of South America. Here the Nazca plate is being subducted beneath the South American plate along the Peru-Chile trench (see Figure 6.8). This subduction

---

* A turbidity current is a dense mass of sediment-choked water that moves downslope along the bottom of a lake or the ocean. For more on these currents, see Chapter 10.

Figure 8.18
The Black Hills of South Dakota, an example of a domal structure in which the resistant igneous and metamorphic central core has been exposed by erosion. (After Arthur N. Strahler, Introduction to Physical Geography, 3rd ed., New York: John Wiley & Sons, 1973. Reprinted by permission)

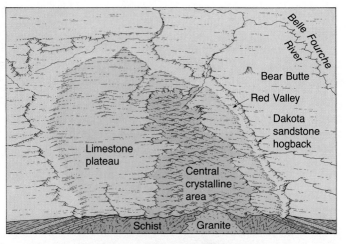

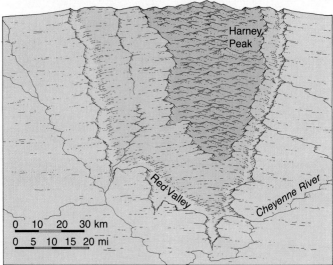

Figure 8.19
Highly deformed sedimentary strata in the Rocky Mountains of British Columbia. (Photo by John Montagne)

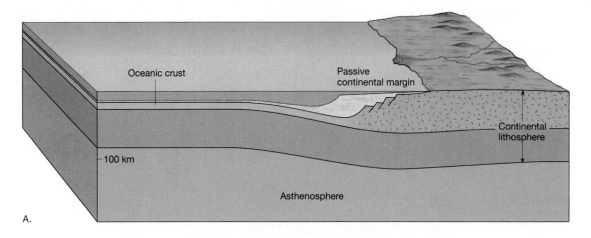

A.

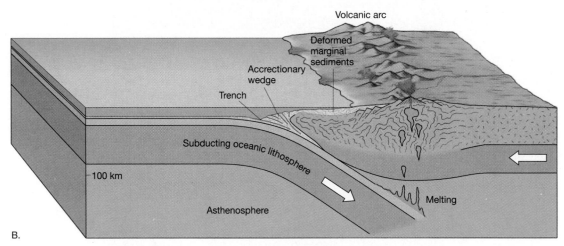

B.

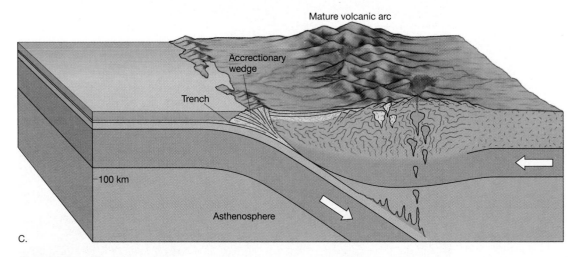

C.

Figure 8.20
Orogenesis along an Andean-type subduction zone. **A.** *Passive continental margin with extensive wedge of sediments.* **B.** *Plate convergence generates a subduction zone, and partial melting produces a developing volcanic arc.* **C.** *Continued convergence and igneous activity further deform and thicken the crust, elevating the mountain belt, while the accretionary wedge grows.*

zone probably formed in conjunction with the breakup of the supercontinent Pangaea. As South America separated from Africa and migrated westward, the oceanic crust adjacent to the west coast of South America was bent and thrust under the continent.

In an idealized Andean-type event, convergence of the continental block and the subducting oceanic plate leads to deformation and metamorphism of the continental margin. Once the oceanic plate descends to about 100 kilometers, partial melting generates magma that migrates upward, intruding and further deforming these strata (Figure 8.20B). During the development of the volcanic arc, sediment derived from the land as well as that scraped from the subducting plate is plastered against the landward side of the trench. This chaotic accumulation of sedimentary and metamorphic rocks with occasional scraps of ocean crust is called an **accretionary wedge** (Figure 8.20B). Prolonged subduction can build an accretionary wedge that is large enough to stand above sea level (Figure 8.20C).

Andean-type mountain belts, like mature island arcs, are composed of two roughly parallel zones. The landward segment is the volcanic arc, which is made up of volcanoes and large intrusive bodies intermixed with high-temperature metamorphic rocks. The belt located seaward of the volcanic arc is the accretionary wedge. It consists of folded, faulted, and metamorphosed sediments and volcanic debris.

One of the best examples of an inactive Andean-type orogenic belt is found in the western United States and includes the Sierra Nevada and the Coast Ranges of California (Figure 8.21). These parallel mountain belts were produced by the subduction of a portion of the Pacific basin under the western edge of the North American plate. The Sierra Nevada batholith is a remnant of a portion of the volcanic arc that was produced by several surges of magma over tens of millions of years. Subsequent uplifting and erosion have removed most evidence of past volcanic activity and exposed a core of crystalline metamorphic and igneous rocks. In the trench region, sediments scraped from the subducting plate, and those provided by the eroding volcanic arc, were intensely folded and faulted into the accretionary wedge which presently constitutes the Franciscan Formation of California's Coast Ranges. Uplifting of the Coast Ranges took place only recently, as evidenced by the unconsolidated sediments still mantling portions of these highlands.

Continental Collisions

So far, we have discussed the formation of mountain belts where the leading edge of just one of the two converging plates contained continental crust. However, it is possible that both of the colliding plates may be carrying continental crust. Because continental lithosphere is apparently too buoyant to undergo any appreciable amount of subduction, a collision between the continental fragments eventually results (Figure 8.22). An example of such a collision occurred about 45 million years ago when India collided with Asia. India, which was once part of Antarctica, split from that continent and moved a few thousand kilometers due north before the collision occurred. The result was the formation of the spectacular Himalaya Mountains and the Tibetan Highlands. Although most of the oceanic crust that separated these landmasses prior to the collision was subducted, some was caught up in the squeeze along with sediment that lay offshore and can now be found elevated high above the sea level. After such a collision, the subducted oceanic plate is believed to decouple from the rigid continental plate and continue its downward path.

The spreading center that propelled India northward is still active; hence, India continues to be thrust into Asia at an estimated rate of a few centimeters per year. However, numerous earthquakes recorded off the southern coast of India may indicate that a new subduction zone is in the making. If formed, it would provide a disposal site for the floor of the Indian Ocean, which is continually being produced at a spreading center located to the southwest. Should this occur, India's northward journey would come to an end and the growth of the Himalayas would cease.

A similar but much older collision is believed to have taken place when the European continent collided with the Asian continent to produce the Ural Mountains, which extend in a north-south direction through the former Soviet Union. Prior to the discovery of plate tectonics, geologists had difficulty explaining the existence of mountain ranges, such as the Urals, that are located deep within the continental interiors. How could thousands of meters of marine sediment be deposited and then become highly deformed while situated in the middle of a large landmass?

Other mountain ranges showing evidence of continental collisions are the Alps and the Appalachians. The Appalachians are thought to have resulted from

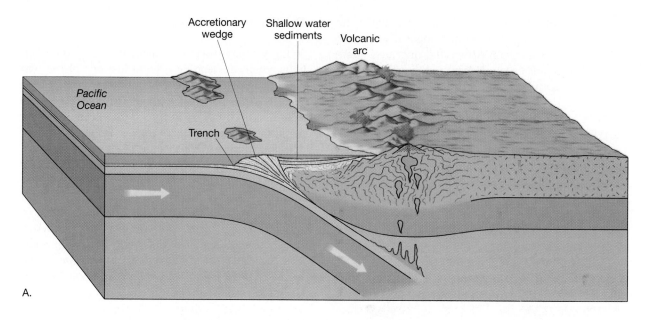

Accretionary wedge · Shallow water sediments · Volcanic arc

Pacific Ocean

Trench

A.

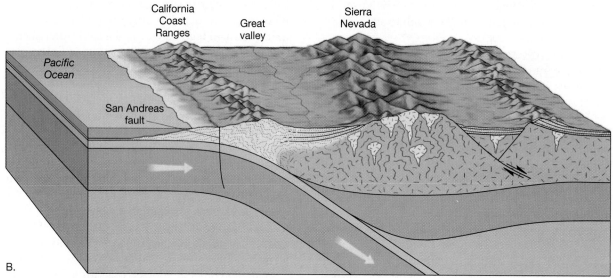

California Coast Ranges · Great valley · Sierra Nevada

Pacific Ocean

San Andreas fault

B.

Figure 8.21
*Simplified diagrams illustrating the formation of the California Coast Ranges and the Sierra Nevada along an Andean-type subduction zone. **A.** Subduction of a portion of the Pacific basin produced a volcanic arc, while sediments scraped from the subducting plate were folded and faulted into an accretionary wedge. **B.** The modern Sierra Nevada batholith is a remnant of the uplifted and eroded volcanic arc, while the California Coast Ranges contain folded and faulted rocks of the accretionary wedge.*

a collision between North America, Europe, and northern Africa. Although they have since separated, these landmasses were juxtaposed as part of the supercontinent Pangaea less than 200 million years ago. Detailed studies in the southern Appalachians indicate that the formation of this mountain belt was more complex than once thought. Rather than form-

ing during a single continental collision, the Appalachians resulted from several distinct episodes of mountain building that occurred over a period of nearly 300 million years. As the continental blocks began to converge, a subduction zone formed seaward of the ancient coastline of North America. The igneous activity associated with the subducting plate

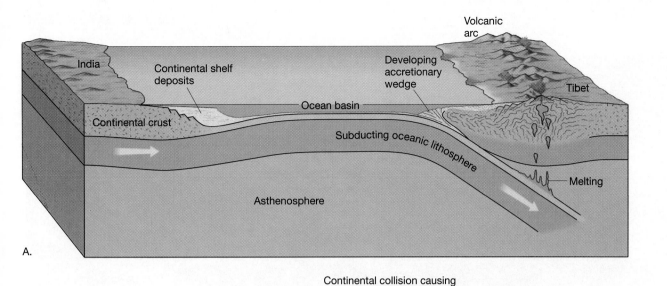

India

Continental shelf
deposits

Continental crust

Ocean basin

Subducting oceanic lithosphere

Asthenosphere

Volcanic
arc

Developing
accretionary
wedge

Tibet

Melting

A.

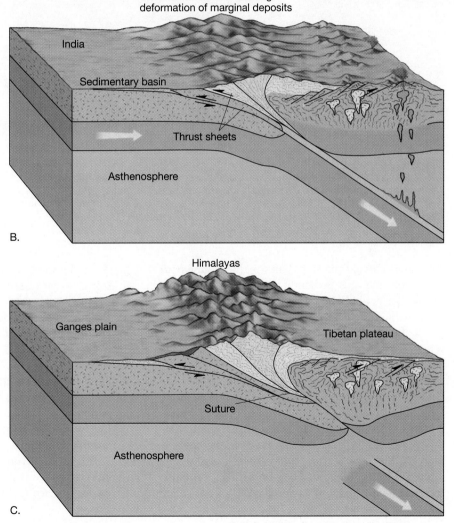

Continental collision causing
deformation of marginal deposits

India

Sedimentary basin

Thrust sheets

Asthenosphere

B.

Himalayas

Ganges plain

Tibetan plateau

Suture

Asthenosphere

C.

Figure 8.22
*Simplified diagrams showing the northward migration and collision of India with the Eurasian plate. **A.** Converging plates generated a subduction zone, while partial melting of the subducting oceanic slab produced a volcanic arc. Sediments scraped from the subducting plate were added to the accretionary wedge. **B.** Eventually the two landmasses collided, deforming and elevating the accretionary wedge and continental shelf deposits. In addition, slices of the Indian crust were thrust up onto the Indian plate. **C.** Continued northward movement further displaced the crustal slices and shelf sediments onto the Indian plate to produce the bulk of the Himalayas. In addition, sediments were thrust onto the Eurasian plate capping the Tibetan Plateau. (Modified after Peter Molnar)*

A.

B.

Figure 8.23
The Valley and Ridge province. **A.** *Aerial view of a small portion of the folded sedimentary strata that were displaced landward during the formation of the Appalachian Mountains. (Photo by John S. Shelton)* **B.** *Satellite view of the Valley and Ridge province. (LANDSAT image courtesy of Phillips Petroleum Company)*

gave rise to a volcanic island arc, perhaps similar to those volcanic arcs presently rimming the western Pacific. Then, about 380 million years ago, the ocean basin behind the volcanic arc began to close.

The final orogeny occurred about 250–300 million years ago when Africa and Europe collided with North America. This event is thought to have displaced the earlier-accreted volcanic arc farther inland along low-angle thrust faults. At some locations the total displacement may have exceeded 250 kilometers (155 miles). This landward displacement further deformed the shallow-water sediments that had flanked North America. Today these folded and faulted sandstones, limestones, and shales compose the essentially unmetamorphosed rocks of the Valley and Ridge Province (Figure 8.23).

In summary, the orogenesis of a complex mountain chain, as typified by the Appalachians, is thought to occur as follows:

1. After the breakup of a continental landmass, a thick wedge of sediments is deposited along the passive continental margins, thereby increasing the size of the newly formed continent.
2. For reasons not yet understood, the ocean basin begins to close and the continents begin to converge.
3. Plate convergence results in the subduction of the intervening oceanic slab and initiates an extended period of igneous activity. This activity results in the formation of a volcanic arc with associated granitic intrusions (see Figure 8.22A).
4. Debris eroded from the volcanic arc and material scraped from the descending plate add to the wedges of sediment along the continental margins.
5. Eventually, the continental blocks collide. This event, which often involves igneous activity, severely deforms and metamorphoses the en-

trapped sediments (see Figure 8.22B). Continental convergence causes these deformed materials, and occasionally slabs of crustal material, to be displaced up onto the colliding plates along thrust faults (see Figure 8.22C). This activity shortens and thickens the crustal rocks, producing an elevated mountain belt.

6. Finally, a change in the plate boundary ends the growth of the mountains. Only at this point do the processes of erosion become the dominant forces in altering the landscape. Prolonged erosion coupled with isostatic adjustments eventually reduce this mountainous landscape to the average thickness of the continents.

This sequence of events is thought to have been duplicated many times during the earth's long history. However, the rate of deformation and the geologic and climatic settings varied in each instance. Thus, the formation of each mountain chain must be regarded as a unique event.

Orogenesis and Continental Accretion

When originally formulated, the plate tectonics theory suggested two mechanisms for orogenesis. First, continental collisions were proposed to explain the formation of such mountainous terrains as the Alps, Himalayas, and Urals. Second, as typified by the Andes, orogenesis associated with the subduction of oceanic lithosphere was thought to be the underlying tectonic process for many circum-Pacific mountain chains. Recent investigations, however, indicate yet another mechanism of orogenesis. This new proposal suggests that relatively small crustal fragments collide and merge with continental margins and that through this process of collision and accretion, many of the mountainous regions rimming the Pacific have been generated (Figure 8.24).

What is the nature of the small crustal fragments and where do they come from? Researchers believe that prior to their accretion to a continental block, some of the fragments may have been microcontinents similar in nature to the present-day island of Madagascar. Many others were island arcs such as Japan, the Philippines, and the Aleutian Islands, which presently rim the Pacific. In addition, others may have been located below sea level and are represented today by submerged platforms rising high above the floor of the western Pacific (see Figure 10.10). Over one hundred of these so-called

oceanic plateaus are known to exist. It is believed that these plateaus originated as submerged continental

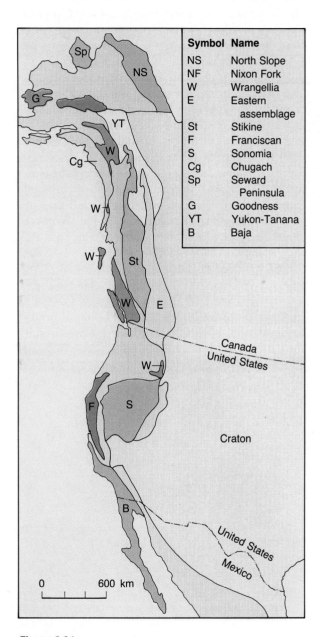

Figure 8.24
Map showing terranes thought to have been added to North America during the past 200 million years. Data from paleomagnetic and fossil evidence indicate that some of these terranes originated thousands of kilometers to the south of their present location. Others such as the Yukon-Tanana terrane are probably displaced parts of the North American continent. (Redrawn after P. Coney and others)

fragments, extinct volcanic island arcs, or submerged volcanic chains associated with hot spot activity.

The widely accepted view today is that as oceanic plates move, they carry the embedded oceanic plateaus or microcontinents to a subduction zone. Here the upper portions of these thickened zones are peeled from the descending plate and thrust in relatively thin sheets upon the adjacent continental block. This newly added material increases the width of the continent and may later be overridden and displaced farther inland by colliding with other fragments. Furthermore, segments of continental crust are continually being displaced along transform faults where they may collide and merge with other crustal fragments. This process is exemplified in western North America where a portion of California and the Baja Peninsula are being displaced toward the northwest along the San Andreas fault. Given enough time, this sliver of continental crust may be accreted to Alaska.

Geologists refer to these accreted crustal blocks as terranes. Simply, the term **terrane** designates any crustal fragment whose geologic history is distinct from that of the adjoining terranes. (Do not confuse the term *terrane* with the word *terrain*, which indicates the lay of the land.) Terranes come in a variety of shapes and sizes; some are no larger than the vol-

canic islands on the floor of the ocean. Others, such as the one composing the entire Indian subcontinent, are quite large.

The idea that orogenesis occurs in association with the accretion of small crustal fragments to a continental mass arose principally from studies conducted in the northern portion of the North American Cordillera. Here it was learned that some mountainous areas, principally those in the orogenic belts of Alaska and British Columbia, contain fossil and magnetic evidence to indicate these strata originated nearer the equator.

It is now believed that many other terranes found in the North American Cordillera were once scattered throughout the eastern Pacific much as we find island arcs and oceanic plateaus distributed in the western Pacific today. Over the last 200 million years, these fragments migrated toward and collided with the west coast of North America (Figure 8.24). Apparently, this activity resulted in the piecemeal addition of fragments to the entire Pacific Coast from the Baja Peninsula to northern Alaska. In a like manner, many modern microcontinents will eventually be accreted to active continental margins, thus resulting in the formation of new orogenic belts.

Review Questions

1. State the three lines of evidence that support the concept of crustal uplift. Can you think of another?

2. What happens to a floating object when weight is added? Subtracted? How do these principles apply to changes in the elevations of mountains? What term is applied to the adjustment that causes crustal uplift of this type?

3. List one line of evidence that supports the idea of a floating crust.

4. What conditions favor rock deformation by folding? By faulting?

5. The San Andreas fault is an excellent example of a _____ fault.

6. Compare the movement of normal and reverse faults. What type of force produces each?

7. At which of the three types of plate boundaries does normal faulting predominate? Reverse faulting? Strike-slip faulting?

8. Describe a horst and a graben. Explain how a graben valley forms and name one.

9. Compare and contrast anticlines and synclines. Domes and basins. Anticlines and domes.

10. Although we classify many mountains as folded, why might this description be misleading?

11. What type of faults are associated with fault-block mountains?

12. During the formation of fault-block mountains, are the forces acting upon the region compressional or tensional?

13. Which of the four types of maintain ranges is exemplified by the following?
    (a) Black Hills
    (b) Basin and Range
    (c) Adirondacks
    (d) Cascades
    (e) Appalachians
    (f) Tetons
    (g) Bighorns
    (h) Himalayas
    (i) Front Range

14. Name the site where sediments are deposited and have a good chance of being squeezed into a mountain range.

15. Which type of plate boundary is most directly associated with mountain building?

16. Describe an accretionary wedge and explain its formation.

17. Would the discovery of a sliver of oceanic crust in the interior of a continent tend to support or refute the theory of plate tectonics? Why?

18. Why might it have been difficult for geologists to conclude that the Appalachian Mountains were formed by plate collision if examples like the Himalayas did not exist?

19. How does the plate tectonics theory explain the existence of fossil marine life on top of the Ural Mountains?

20. In your own words, briefly enumerate the steps involved in the formation of a mountain system according to the plate tectonics model.

Key Terms

Geologic Time and Earth History

Opposite page: The strata exposed in the Grand Canyon contain clues to millions of years of earth history. (Photo by Tom Till)

Left: Fossil trilobite. (Photo by Michael C. Hansen, Ohio Geological Survey)

Above: This extinct coiled cephalopod was a highly developed organism.

In the late eighteenth century James Hutton recognized that the earth is very old. But how old? For many years there was no reliable method to determine the age of the earth or the times of various events in the geologic past. Rather, a geologic time scale was developed that relied on relative dating principles. What are these principles? What part do fossils play? With the discovery of radioactivity and the development and refinement of radiometric dating techniques, geologists now can assign fairly accurate dates to many of the events in earth history. What is radioactivity? Why is it a good "clock" for dating the geologic past? In this chapter we shall attempt to answer the questions raised here, as well as others.

In 1869 John Wesley Powell, who was later to head the U.S. Geological Survey, led a pioneering expedition down the Colorado River and through the Grand Canyon. Writing about the strata that were exposed by the downcutting of the river, Powell said that ". . . the canyons of this region would be a Book of Revelations in the rock-leaved Bible of geology." He was undoubtedly impressed with the millions of years of earth history exposed along the walls of the Grand Canyon (Figure 9.1). Powell realized that the evidence for an ancient earth is concealed in its rocks. Like the pages in a long and complicated history book, rocks record the geological events and changing life forms of the past. The book, however, is not complete. Many pages, especially in the early chapters, are missing. Others are tattered, torn, or smudged. Yet enough of the book remains to allow much of the story to be deciphered. Interpreting earth history is a prime goal of the science of geology. Like a modern-day sleuth, the geologist must interpret clues found preserved in the rocks. By studying rocks, especially sedimentary rocks, and the features they contain, geologists can unravel the complexities of the past.

Geological events by themselves, however, have little meaning until they are put into a time perspective. Studying history, whether it be the Civil War or the Age of Dinosaurs, requires a calendar. Among the major contributions that geology has made to the knowledge of humankind is the geologic time scale and the concept that earth history is exceedingly long. Over many years geologists put together a time scale of earth history in which geologic events can be placed in their proper sequence. Recognizing that earth history has spanned an immense amount of time, geologists worked at finding out just how old the earth is.

SOME HISTORICAL NOTES ABOUT GEOLOGY

The nature of our earth—its materials and processes—has been a focus of study for centuries. Writings about such topics as fossils, gems, earthquakes, and volcanoes date back to the early Greeks, more than 2300 years ago. Certainly the most influential of the Greek philosophers was Aristotle. Unfortunately, Aristotle's explanations about the natural world were not based on scientific methods of inquiry. Instead, they were arbitrary pronouncements. He believed that rocks were created under the "influence" of the stars and that earthquakes occurred when air that was crowded into the ground was heated by central fires and escaped explosively. When confronted with a fossil fish, he explained that "a great many fishes live in the earth motionless and are found when excavations are made." Although Aristotle's explanations may have been adequate for his day, they unfortunately continued to be expounded for many centuries, thus thwarting the acceptance of ideas that were more closely in accord with observations. Frank D. Adams states in *The Birth and Development of the Geological Sciences* (New York: Dover, 1938) that "throughout the Middle Ages Aristotle was regarded as the head and chief of all philosophies; one whose opinion on any subject was authoritative and final."

Catastrophism

During the seventeenth and eighteenth centuries the doctrine of **catastrophism** strongly influenced the formulation of explanations about the dynamics of the earth. Briefly stated, catastrophists believed that the earth's landscape had been developed primarily by great catastrophes. Features such as mountains and canyons, which today we know take great periods of time to form, were explained as having been produced by sudden and often worldwide disasters produced by unknowable causes that no longer operate. This philosophy was an attempt to fit the rate of earth processes to the prevailing ideas on the age of the earth. In the mid-seventeenth century, James Ussher, Anglican Archbishop of Armagh, Primate of all Ireland, published a wok that had immediate and profound influence. A respected scholar of the Bible, Ussher constructed a chronology of human and earth history in which he determined that the earth was only a few thousands of years old, having been created in 4004 B.C. Ussher's treatise earned widespread acceptance among scientific and religious leaders alike, and his chronology was soon printed in the margins of the Bible itself.

The relationship between catastrophism and the age of the earth has been summarized nicely as follows:

> That the earth had been through tremendous adventures and had seen mighty changes during its obscure past was plainly evident to every inquiring eye; but to concentrate these changes into a few brief millenniums required a tailor-made philosophy, a philosophy whose basis was sudden and violent change.*

---

* H. E. Brown, V. E. Monnett, and J. W. Stovall. *Introduction to Geology* (New York: Blaisdell, 1958).

Figure 9.1
Grand Canyon viewed from Yaki Point, Arizona. (Photo by Tom Bean)

The Birth of Modern Geology

The late eighteenth century is generally regarded as the beginning of modern geology, for it was during this time that James Hutton, a Scottish physician and gentleman farmer, published his *Theory of the Earth* in which he put forth a principle that came to be known as the doctrine of **uniformitarianism** (Figure 9.2). Uniformitarianism is a basic part of modern geology. It simply states that the physical, chemical, and biological laws that operate today have also operated in the geologic past. That is to say that the forces and processes that we observe presently shaping our planet have been at work for a very long time. Thus, to understand ancient rocks, we must first understand present-day processes and their results. This idea is commonly stated by saying "the present is the key to the past."

Prior to Hutton's *Theory of the Earth*, no one had effectively demonstrated that geological processes occur over extremely long periods of time. However, Hutton persuasively argued that processes that appear weak and slow-acting could, over long spans of time, produce effects that were just as great as those resulting from sudden catastrophic events. Unlike his predecessors, Hutton cited verifiable observations to support his ideas.

Figure 9.2
James Hutton, the 18th century Scottish geologist who is often called the "father of modern geology." (Photo courtesy of the British Museum)

Since Hutton's literary style was cumbersome and difficult, his work was not widely read nor easily understood. It is the English geologist Sir Charles Lyell who is given the most credit for advancing the basic principles of modern geology. Between 1830 and 1872 Lyell produced eleven editions of his great work, *Principles of Geology*. As was customary, Lyell's book had a rather lengthy subtitle that outlined the main theme of the work: *Being an Attempt to Explain the Former Changes of the Earth's Surface, by Reference to Causes Now in Operation*. In the text, he painstakingly illustrated the concept of the uniformity of nature through time. He was able to show more convincingly than his predecessors that those geologic processes observed today can be assumed to have operated in the past. Although the doctrine of uniformitarianism did not originate with Lyell, he is the person who was most successful in interpreting and publicizing it for society at large.

Today the basic tenets of uniformitarianism are just as viable as in Lyell's day. Indeed, we realize more strongly than ever that the present gives us insight into the past and that the physical, chemical, and bi-

ological laws that govern geological processes remain unchanging through time. However, we also understand that the doctrine should not be taken too literally. To say that geological processes in the past were the same as those occurring today is not to suggest that they always had the same relative importance and operated at precisely the same rate. Although the same processes have prevailed through time, their rates have undoubtedly varied.

The acceptance of the concept of uniformitarianism, however, meant the acceptance of a very long history for the earth, for although processes vary in their intensity, they still take a very long time to create or destroy major landscape features.

For example, rocks containing fossils of organisms that lived in the sea more than 15 million years ago are now part of mountains that stand 3000 meters (9800 feet) above sea level. This means that the mountains were uplifted 3000 meters in about 15 million years, which works out to a rate of only 0.2 millimeter per year! Rates of erosion are equally slow. Estimates indicate that the North American continent is being lowered at a rate of just 3 centimeters per 1000 years. Thus, as you can see, it takes tens of millions of years for nature to build mountains and wear them down again. But even these time spans are relatively short on the time scale of earth history, for the rock record contains evidence that shows the earth has experienced many cycles of mountain building and erosion. Concerning the everchanging nature of the earth through great expanses of geologic time, Hutton stated: "We find no vestige of a beginning, no prospect of an end." A quote from William L. Stokes sums up the significance of Hutton's basic concept:

> In the sense that uniformitarianism implies the operation of timeless, changeless laws or principles, we can say that nothing in our incomplete but extensive knowledge disagrees with it.[*]

It is important to remember that, although many features of our physical landscape may seem to be unchanging in terms of the tens of years we might observe them, they are nevertheless changing, but on time scales of hundreds, thousands, or even many millions of years.

---

[*] William L. Stokes. *Essentials of Earth History* (Englewood Cliffs, New Jersey: Prentice-Hall, 1966), 34.

RELATIVE DATING

During the late nineteenth and early twentieth centuries a number of attempts were made to determine the age of the earth. Although some of the methods appeared promising at the time, none of these early efforts proved to be reliable. What these scientists were seeking was an **absolute date.** Such dates pinpoint the time in history when something took place. Today our understanding of radioactivity allows us to accurately determine absolute dates for rock units that represent important events in the earth's distant past.* However, prior to the discovery of radioactivity, geologists had no accurate and dependable method of absolute dating and had to rely solely on relative dating. **Relative dating** means that rocks are placed in their proper sequence or order. Relative dating will not tell us how long ago something took place, only that it followed one event and preceded another. The relative dating techniques which were developed are still widely used. Absolute dating methods did not replace these techniques; they simply supplemented them. To establish a relative time scale, a few simple principles or rules had to be discovered and applied. Although they may seem obvious to us today, their discovery and acceptance was an important scientific achievement.

Nicolaus Steno, a physician in Florence, Italy, is credited with being the first to recognize a sequence of historical events in an outcrop of sedimentary rock layers. Working in the mountains of western Italy, Steno applied a very simple rule that has come to be the most basic principle of relative dating—the **law of superposition.** The law simply states that in an undeformed sequence of sedimentary rocks, each bed is older than the one above it and younger than the one below. Although it may seem obvious that a layer could not be deposited with nothing beneath it for support, it was not until 1669 that Steno clearly stated the principle. This rule also applies to other surface-deposited materials such as lava flows and beds of ash from volcanic eruptions. Applying the law of superposition to the beds exposed in the upper portion of the Grand Canyon (Figure 9.3), we can easily place the layers in their proper order. Among those that are shown, the sedimentary rocks in the Supai

A.

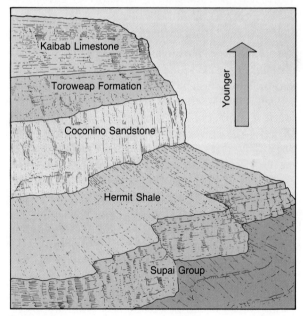

B.

Figure 9.3
Applying the law of superposition to these layers exposed in the upper portion of the Grand Canyon, the Supai Group is oldest and the Kaibab Limestone is youngest. (Photo by E. J. Tarbuck)

---

* Radiometric dating is the subject of a later section in this chapter.

A.

B.

Figure 9.4
A. The principle of original horizontality states that most layers of sediment are deposited in a nearly horizontal position. **B.** When we see folded rock layers such as
these, we can assume they must have been moved into that position by crustal disturbances after their deposition. (Photos by E. J. Tarbuck)

Group are the oldest, followed in order by the Her-
mit Shale, Coconino Sandstone, Toroweap Formation,
and Kaibab Limestone.

Steno is also credited with recognizing the importance of another basic principle, called the **principle
of original horizontality.** Simply stated, most layers of sediment are deposited in a nearly horizontal
position. Thus, if we observe rock layers that are
folded or inclined at a steep angle, they must have
been moved into that position by crustal disturbances
sometime after their deposition (Figure 9.4).

When igneous intrusions or faults cut through
other rocks, they are assumed to be younger than the
structure they cut. For example, when two dikes intersect, the older one must have been opened up in
order to allow the younger one to cut through it. The
younger dike would be continuous, while the older
dike would be interrupted at the point of their intersection. A fault is a fracture in the earth along which
movement occurs. When rocks are cut and offset by
a fault, we know that they must be older than the
fault. Figure 9.5 illustrates this **principle of cross-
cutting relationships.** For example, by applying the
cross-cutting principle we can determine that fault A
occurred after the sandstone layer was deposited but

before the conglomerate was laid down. It is possible to say this because the sandstone is offset by the
fault and the conglomerate is not. We can also state
that dike B and its associated sill are older than dike
A because dike A cuts the sill. In the same manner,
we know that the batholith was emplaced after
movement occurred along fault B but before dike B
was formed. This is true because the batholith cuts
across fault B and dike B cuts across the batholith.

Sometimes inclusions can aid the relative dating
process. **Inclusions** are pieces of one rock unit that
are contained within another. The basic principle is
logical and straightforward. The rock mass adjacent
to the one containing the inclusions must have been
there first in order to provide the rock fragments.
Therefore, the rock mass containing inclusions is the
younger of the two. Figure 9.6 provides an example.
Here the inclusions of granite in the adjacent sedimentary layer indicate the sedimentary layer was
deposited on top of an eroded mass of granite
rather than being intruded from below by a younger
granite.

Layers of rock are said to be **conformable** when
they are found to have been deposited essentially
without interruption. Although particular sites may

exhibit conformable beds representing significant spans of geologic time, there is no place on earth that contains a full set of conformable strata. Throughout earth history, the deposition of sediment has been interrupted over and over again. All such breaks in the rock record are termed unconformities. An **unconformity** represents a period of time during which deposition ceased, erosion removed previously formed rocks, and deposition resumed. In each case uplift and erosion are followed by subsidence and renewed sedimentation. Unconformities are important features because they represent significant geologic events in earth history. Moreover their recognition helps us identify what intervals of time are not represented by strata.

Perhaps the most easily recognized type of unconformity consists of tilted or folded sedimentary rocks that are overlain by younger, more flat-lying strata. These are **angular unconformities** and indicate that during the pause in deposition, a period of deformation (folding or tilting) as well as erosion occurred (Figure 9.7). The angular unconformity pictured in Figure 9.8 is well known to many geologists because it was originally studied and described by James Hutton more than 200 years ago. It was clear to Hutton that the relationship between the nearly vertical lower layers and the gently inclined beds overlying them was evidence for a significant episode of geologic change. Hutton also appreciated the immense time span implied by such relationships.

When contrasted with angular unconformities, **disconformities** are more common, but usually far less conspicuous because the strata on either side are essentially parallel. Many disconformities are difficult

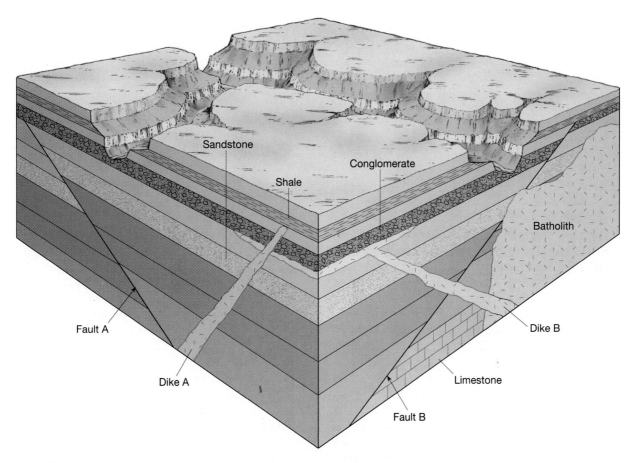

Figure 9.5
Cross-cutting relationships represent one principle used in relative dating. An intrusive rock body is younger than the rocks it intrudes. A fault is younger than the rock layers it cuts.

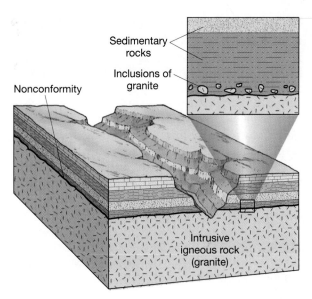

Figure 9.6
Since pieces of granite are contained within the overlying sedimentary bed, we know the granite must be older. When older intrusive igneous rocks are overlain by younger sedimentary layers, a type of unconformity, termed a nonconformity, is said to exist.

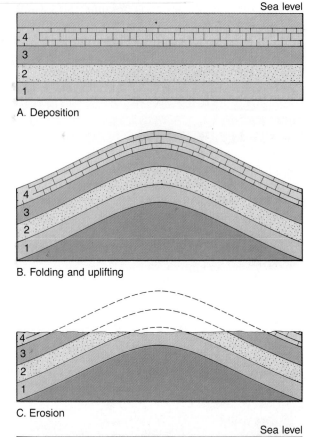

Figure 9.7
Formation of an angular unconformity. An angular unconformity represents a period during which deformation and erosion occurred.

to identify because the rocks above and below are similar and there is little evidence of erosion. Such a break often resembles an ordinary bedding plane. Other disconformities are easier to identify because the ancient erosion surface is cut deeply into the older rocks below.

Nonconformities are the third basic type of unconformity. Here the break separates older metamorphic or intrusive igneous rocks from younger sedimentary strata (see Figure 9.6). Just as angular unconformities and disconformities imply crustal movements, so too do nonconformities. Intrusive igneous masses and metamorphic rocks originate far below the surface. Thus, for a nonconformity to develop, there must be a period of uplift and the erosion of overlying rocks. Once exposed at the surface, the igneous or metamorphic rocks are subject to weathering and erosion prior to subsidence and the renewal of sedimentation.

By applying the principles of relative dating to the hypothetical geologic cross section shown in Figure 9.9, the rocks and the events in earth history they represent can be placed into their proper sequence. The following statements summarize the logic used to interpret the cross section:

1. Applying the law of superposition, beds *A*, *B*, *C*, and *E* were deposited, in that order. Since bed *D* is a sill (a concordant igneous intrusion), it is younger than the rock into which it was intruded. Further evidence that the sill is younger than beds C and E are the inclusions in the sill of fragments from these beds. If the igneous mass contains pieces of surrounding rock, the surrounding rock must have been there first.

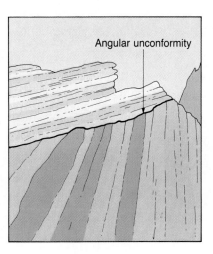

Figure 9.8
This angular unconformity at Siccar Point, Scotland, was first described more than 200 years ago by James Hutton and John Playfair. To produce this feature, the lower layers were deposited, then uplifted, tilted, and eroded. Later, another period of sedimentation produced the flatter-lying beds on top. (Photo by Edward A. Hay)

Figure 9.9
Geologic cross section of a hypothetical region.

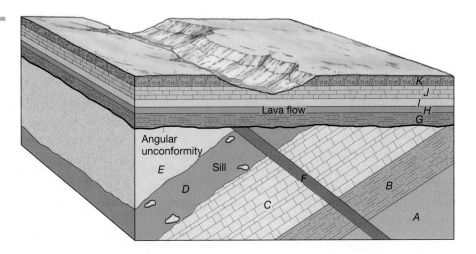

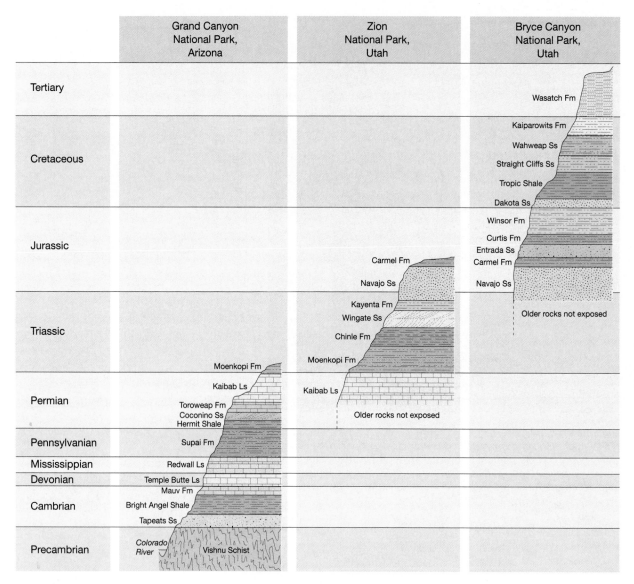

Figure 9.10
Correlation of strata at three locations on the Colorado Plateau reveals the total extent of sedimentary rocks in the region. (After U.S. Geological Survey)

2. Following the intrusion of the sill (*D*), the intrusion of the dike (*F*) occurred. Since the dike cuts through beds *A* through *E*, it must be younger than all of them.

3. Next, the rocks were tilted and then eroded. We know the tilting happened first because the upturned ends of the strata have been eroded. The tilting and erosion, followed by further deposition, produced an angular unconformity.

4. Beds *G, H, I, J,* and *K* were deposited in that order, again using the law of superposition. Although the lava flow (bed *H*) is not a sedimentary rock layer, it is a surface-deposited layer, and thus superposition must be applied.

5. Finally, the irregular surface and the stream valley indicate that another gap in the rock record is being produced by erosion.

In the foregoing example, our goal was to establish a relative time scale for the rocks and events in the area of the cross section. Remember, we do not have any idea how many years of earth history are represented, nor do we know how this area compares to any other.

CORRELATION

In order to develop a geologic time scale that is applicable to the whole earth, rocks of similar age in different regions must be matched up. Such a task is referred to as **correlation.** Within a limited area there are several methods of correlating the rocks of one locality with those of another. A bed or series of beds may be traced simply by walking along the outcrop. However, this may not be possible when the bed is not continuously exposed. Correlation over short distances is often achieved by noting the place of a bed in a sequence of strata, or a bed may be identified in another location if it is composed of very distinctive or uncommon minerals. By correlating the rocks from one place to another, a more comprehensive view of the geologic history of a region is possible. Figure 9.10, for example, shows the correlation of strata at three sites on the Colorado Plateau. No single locale exhibits the entire sequence, but correlation reveals a more complete picture of the sedimentary rock record.

Many geologic studies involve relatively small areas. Although they are important in their own right, their full value is realized only when the rocks are correlated with those of other regions. Although the methods just described may be sufficient to trace a rock formation over relatively short distances, they are not adequate for matching rocks at great distances. When correlation between widely separated areas or between continents is the objective, the geologist must rely upon fossils.

FOSSILS

Paleontology, the branch of geology devoted to the study of ancient life, is based on the study of fossils. Originally the term **fossil** referred to any curious object dug from the ground, but it is now used to mean the remains or traces of organisms preserved from the geologic past. Although geologists studying past life must have a firm background in the biological sciences, they are often quick to point out the differences between the two fields. "If the remains stink they belong to zoology, but if not, to paleontology."

Fossils are of many types. The remains of relatively recent organisms may not have been altered at all. Such objects as teeth, bones, and shells are common examples. Far less common are entire animals, flesh included, that have been preserved because of rather unusual circumstances. Remains of prehistoric elephants called mammoths that were frozen in the Arctic tundra of Siberia and Alaska are examples, as are the mummified remains of sloths preserved in a dry cave in Nevada.

Given enough time, the remains of an organism are likely to be modified. Often fossils become *petrified* (literally, "turned into stone"), meaning that the small internal cavities and pores of the original structure are filled with precipitated mineral matter (Figure 9.11A). In other instances *replacement* may occur. Here the cell walls and other solid material are removed and replaced with mineral matter. Sometimes the microscopic details of the replaced structure are faithfully retained.

Molds and casts constitute another common class of fossils. When a shell or other structure is buried in sediment and then dissolved by underground water, a *mold* is created. The mold faithfully reflects only the shape and surface marking of the organism; it does not reveal any information concerning its internal structure. If these hollow spaces are subsequently filled with mineral matter, *casts* are created (Figure 9.11B).

A type of fossilization called *carbonization* is particularly effective in preserving leaves and delicate animal forms. It occurs when fine sediment encases the remains of an organism. As time passes, pressure squeezes out the liquid and gaseous components and leaves behind a thin residue of carbon (Figure 9.11C). Black shales deposited as organic-rich mud in oxygen-poor environments often contain abundant carbonized remains. If the film of carbon is lost from a fossil preserved in fine-grained sediment, a replica of the surface, called an *impression*, may still show considerable detail (Figure 9.11D).

Delicate organisms, such as insects, are difficult to preserve and consequently are quite rare in the fossil record. Not only must they be protected from decay, they must not be subjected to any pressure that would crush them. One way in which some insects

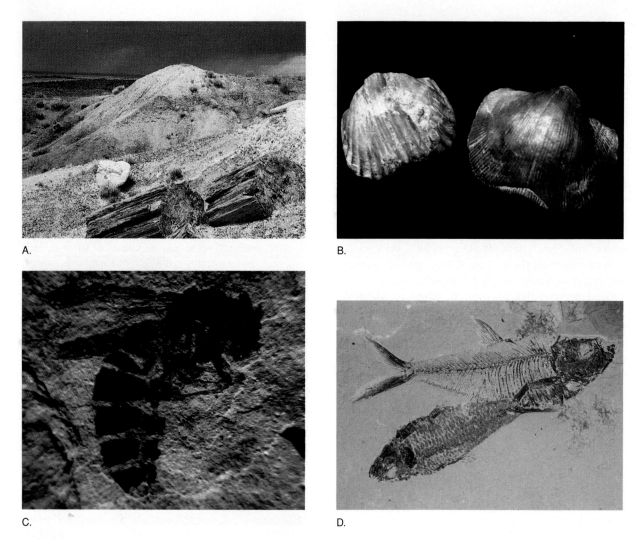

Figure 9.11
*There are many types of fossilization. Six examples are shown here. **A.** Petrified wood in Petrified Forest National Park, Arizona. **B.** Natural casts of shelled invertebrates. **C.** A fossil bee preserved as a thin carbon film. **D.** Impressions are common fossils and often show considerable detail. **E.** Insect in amber. **F.** Dinosaur footprint in fine-grained limestone near Tuba City, Arizona. (Photos A, B, D, and F by E. J. Tarbuck; Photo C courtesy of the National Park Service; Photo E by Breck P. Kent)*

have been preserved is in *amber*, the hardened resin of ancient trees. The fly in Figure 9.11E was preserved after being trapped in a drop of sticky resin. Resin sealed off the insect from the atmosphere and protected the remains from damage by water and air. As the resin hardened, a protective, pressure-resistant case was formed.

In addition to the fossils already mentioned, there are numerous other types, many of them only traces of prehistoric life. Examples of such indirect evidence include:

1. Tracks—animal footprints made in soft sediment that was later lithified (Figure 9.11F).
2. Burrows—tubes in sediment, wood, or rock made by an animal. These holes may later become filled with mineral matter and preserved. Some of the oldest-known fossils are believed to be worm burrows.
3. Coprolites—fossil dung and stomach contents that can provide useful information pertaining to food habits of organisms.
4. Gastroliths—highly polished stomach stones that

E.

F.

were used in the grinding of food by some extinct reptiles.

Conditions Favoring Preservation

Only a tiny fraction of the organisms that have lived during the geologic past have been preserved as fossils. Normally the remains of an animal or plant are totally destroyed. Under what circumstances are they preserved? Two special conditions appear to be necessary: rapid burial and the possession of hard parts.

Usually when an organism perishes, its soft parts are quickly eaten by scavengers or decomposed by bacteria. Occasionally, however, the remains are buried by sediment. When this occurs the remains are protected from the environment where destructive processes operate. Rapid burial therefore is an important condition favoring preservation.

In addition, animals and plants have a much better chance of being preserved as part of the fossil record if they have hard parts. Although traces and imprints of soft-bodied animals such as jellyfish, worms, and insects exist, they are rare. Flesh usually decays so rapidly that preservation is exceedingly unlikely. Hard parts like shells, bones, and teeth predominate in the record of past life.

Since preservation is contingent on special conditions, the record of life in the geologic past is biased. The fossil record of those organisms with hard parts that lived in areas of sedimentation is quite abundant. However, we get only an occasional glimpse of the vast array of other life forms that did not meet the special conditions favoring preservation.

Fossils and Correlation

The existence of fossils had been known for centuries, yet it was not until the late 1700s and early 1800s that their significance as geologic tools was made evident. During this period an English engineer and canal builder, William Smith, discovered that each rock formation in the canals contained fossils unlike those in the beds either above or below. Further, he noted that sedimentary strata in widely separated areas could be identified by their distinctive fossil content. Based upon Smith's classic observations and the findings of many geologists who followed, one of the most important and basic principles in historical geology was formulated: Fossil organisms succeed one another in a definite and determinable order, and therefore any time period can be recognized by its fossil content. This has come to be known as the **principle of fossil succession.** In other words, when fossils are arranged according to their age by using the law of superposition on the rocks in which they are found, they do not present a random or haphazard picture. To the contrary, fossils show progressive changes from simple to complex and docu-

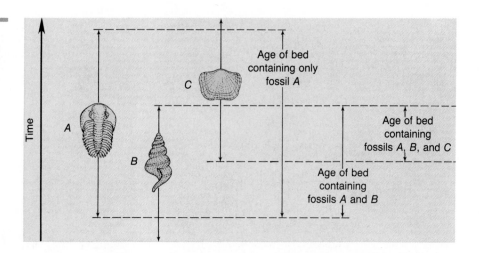

Figure 9.12
Overlapping ranges of fossils
help date rocks more exactly
than using a single fossil.

ment the evolution of life through time. For example, an Age of Trilobites is recognized quite early in the fossil record. Then, in succession, paleontologists recognize an Age of Fishes, an Age of Coal Swamps, an Age of Reptiles, and an Age of Mammals. These "ages" pertain to groups that were especially plentiful and characteristic during particular time periods. Within each of the "ages," there are many subdivisions based, for example, on certain species of trilobites, and certain types of fish, reptiles, and so on. This same succession of dominant organisms, never out of order, is found on every major landmass.

Once fossils were found to be time indicators, they became the most useful means of correlating rocks of similar age in different regions. Geologists pay particular attention to certain fossils called **index fossils.** Due to the fact that these fossils are widespread geographically and are limited to a short span of geologic time, their presence provides an important method of matching rocks of the same age. Rock formations, however, do not always contain a specific index fossil. In such situations, groups of fossils are used to establish the age of the bed. Figure 9.12 illustrates how a group of fossils can be used to date rocks more precisely than could be accomplished by the use of only one of the fossils.

In addition to being important and often essential tools for correlation, fossils are important environmental indicators. Although much can be deduced about past environments by studying the nature and characteristics of sedimentary rocks, a close examination of the fossils present can usually provide a great deal more information. For example, when the remains of certain clam shells are found in limestone,

the geologist feels it is quite reasonable to assume that the region was once covered by a shallow sea. Also, by using what we know of living organisms, we can conclude that fossil animals with thick shells capable of withstanding pounding and surging waves inhabited shorelines. On the other hand, animals with thin, delicate shells probably indicate deep, calm offshore waters. Hence by taking a closer look at the types of fossils, the approximate shoreline may be identified. Further, fossils can be used to indicate the former temperature of the water. Certain kinds of present-day corals must live in warm and shallow tropical seas like those around Florida and the Bahamas. When similar types of coral are found in ancient limestones, they give a good estimate of the marine environment that must have existed when they were alive. The preceding are just a few brief examples of how fossils can help unravel the complex story of earth history.

RADIOACTIVITY AND RADIOMETRIC DATING

In addition to establishing relative dates by using the principles described in the preceding sections, it is also possible to obtain reliable absolute dates for events in the geologic past. For example, we know that the earth is about 4.6 billion years old and that the dinosaurs became extinct about 66 million years ago. Dates that are expressed in millions and billions of years truly stretch our imagination because our personal calendars involve time measured in hours, weeks, and years. Nevertheless the vast expanse of geologic time is a reality and it is radiometric dating that allows us to accurately measure it. In this section we will learn about radioactivity and its application in radiometric dating.

In Chapter 1 we learned that an atom is composed of electrons, protons, and neutrons. Electrons have a negative charge and protons have a positive charge. Since a neutron is actually a proton and electron combined, it has no charge. Protons and neutrons are found in the center, or nucleus, of the atom, and electrons spin around the nucleus in definite paths, called orbits. Practically all (99.9 percent) of an atom's mass is found in the nucleus, indicating electrons have practically no mass at all. By adding together the number of protons and neutrons in the nucleus, the mass number of the atom is determined. The atomic number (the atom's identifying number) is equal to the number of protons. Every element has a different number of protons in the nucleus, and thus a different atomic number. Many elements have atoms with different numbers of neutrons in the nucleus. Called isotopes, these atoms have different mass numbers, but have the same atomic number.

The forces binding protons and neutrons together in the nucleus are strong; however, the nature of these forces is still not fully understood. In some isotopes the nuclei are unstable; that is, the forces binding protons and neutrons together are not strong enough. As a result, the nuclei spontaneously break apart (decay), a process called **radioactivity.** What happens when unstable nuclei break apart? Three common types of radioactive decay are illustrated in Figure 9.13 and are summarized as follows:

1. Alpha particles (α particles) may be emitted from the nucleus. An alpha particle is composed of 2 protons and 2 neutrons. Consequently, the emission of an alpha particle means the mass number of the isotope is reduced by 4 and the atomic number is decreased by 2.

2. When a beta particle (β particle), or electron, is given off from a nucleus, the mass number remains unchanged, because electrons have practically no mass. However, since the electron must have come from a neutron (remember, a neutron is a combination of a proton and an electron), the nucleus contains one more proton than before. Therefore, the atomic number increases by 1.

3. Sometimes an electron is captured by the nucleus. The electron combines with a proton and forms a neutron. As in the last example, the mass number remains unchanged. However, since the nucleus now contains one less proton, the atomic number decreases by 1.

The radioactive isotope is referred to as the *parent*, and the isotopes resulting from the decay of the parent are termed the *daughter products.* Figure 9.14 provides an example of radioactive decay. Here it can be seen that when the radioactive parent, uranium-238 (atomic number 92, mass number 238), decays, it emits 8 alpha particles and 6 beta particles before becoming the stable daughter product lead-206 (atomic number 82, mass number 206).

Certainly among the most important results of the discovery of radioactivity is that it provided a reliable means of calculating the ages of rocks and minerals that contain particular radioactive isotopes, a procedure referred to as **radiometric dating.** Why is radiometric dating reliable? The answer lies in the fact that the rates of decay for many isotopes have been precisely measured and do not vary under the physical conditions that exist in the outer layers of the earth. Therefore, each radioactive isotope used for dating has been decaying at a fixed rate since the formation of the rocks in which it occurs and the products of decay have been accumulating at a corresponding rate. For example, when uranium is incorporated into a mineral that crystallizes from magma, there is no lead (the stable daughter product) from previous decay. As the uranium in this newly formed mineral disintegrates, atoms of the daughter product are trapped and measurable amounts of lead eventually accumulate.

The time required for one-half of the nuclei in a sample to decay, called **half-life,** is a common way of expressing the rate of radioactive disintegration. Figure 9.15 illustrates what occurs when a radioactive parent decays directly into the stable daughter product. When the quantities of parent and daughter are equal (ratio 1:1), we know that one half-life has transpired. When one-quarter of the original parent atoms remain and three-quarters have decayed to the daughter product, the parent/daughter ratio is 1:3 and we know that two half-lives have passed. After three half-lives, the ratio of parent atoms to daughter atoms is 1:7 (one parent atom for every seven daughter atoms). If the half-life of a radioactive isotope is known and the parent/daughter ratio can be determined, the age of the sample can be calculated. For example, assume that the half-life of a hypothetical unstable isotope is 1 million years and the parent/daughter ratio in a sample is 1:15. Such a ratio indicates that four half-lives have passed and that the sample must be 4 million years old.

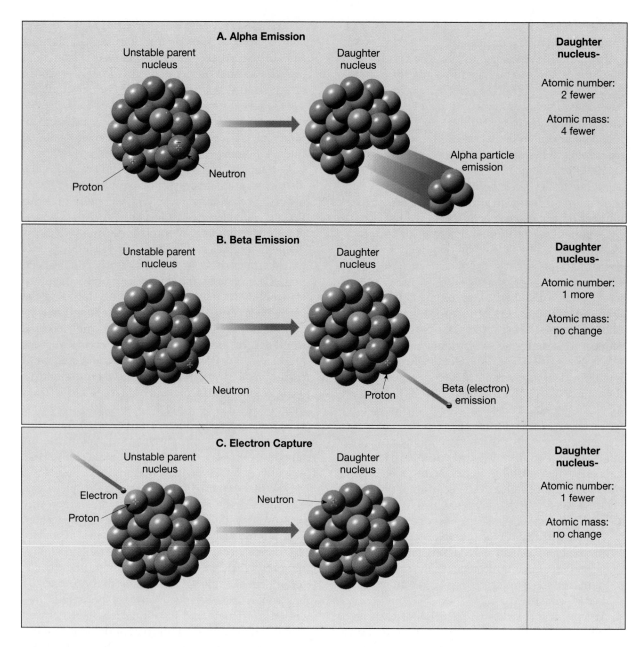

Figure 9.13
Common types of radioactive decay. Notice that in each case the number of protons (atomic number) in the nucleus changes, thus producing a different element.

Notice that the percentage of radioactive atoms that decay during one half-life is always the same. However, the actual number of atoms that decay with the passing of each half-life continually decreases. Thus, as the percentage of radioactive parent atoms declines, the proportion of stable daughter atoms rises, with the increase in daughter atoms just matching the drop in parent atoms. This fact is the key to radiometric dating.

Of the many radioactive isotopes that exist in nature, five have proven particularly important in providing radiometric ages for ancient rocks. Table 9.1 lists these most frequently used isotopes. Rubidium-87 and the two isotopes of uranium are used only for

Figure 9.14
The most common isotope of uranium (U-238) is an example of a radioactive decay series. Before the stable end product (Pb-206) is reached, many different isotopes are produced as intermediate steps.

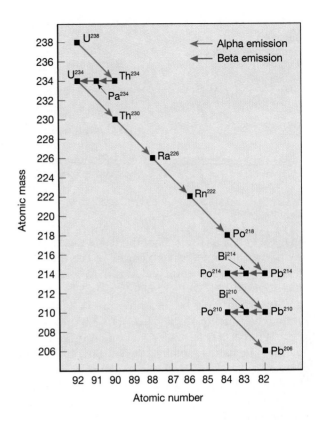

dating rocks that are millions of years old, but potassium-40 is more versatile. Although the half-life of potassium-40 is 1.3 billion years, recent analytical techniques have made possible the detection of tiny amounts of its stable daughter product, argon-40, in some rocks that are younger than 100,000 years. Another important reason for its frequent use is that potassium is an abundant constituent of many common minerals, particularly micas and feldspars. The potassium-argon clock begins when potassium-bearing minerals crystallize from a magma or form within a metamorphic rock. At this point the new minerals will contain K^{40} but will be free of Ar^{40}, because this element is an inert gas that does not chemically combine with other elements. As time passes, the K^{40} steadily decays by electron capture. The Ar^{40} produced by this process remains trapped within the

Figure 9.15
Decay of a radioactive isotope. The parent/daughter ratio changes continually with time. As the proportion of parent decreases, the proportion of daughter rises, with the increase in daughter atoms just matching the decline in parent atoms.

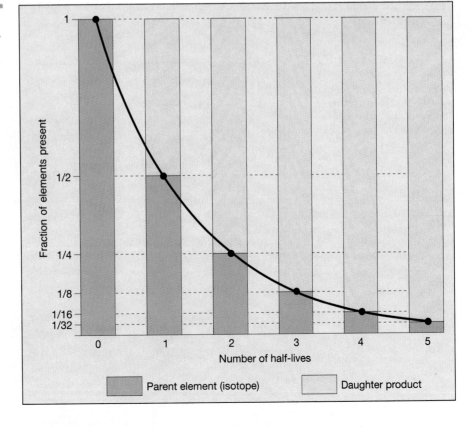

Table 9.1
Radioactive isotopes frequently used in radiometric dating.

Radioactive Parent	Stable Daughter Product	Currently Accepted Half-Life Values
Uranium-238	Lead-206	4.5 billion years
Uranium-235	Lead-207	713 million years
Thorium-232	Lead-208	14.1 billion years
Rubidium-87	Strontium-87	47.0 billion years
Potassium-40	Argon-40	1.3 billion years

mineral's crystal lattice. Since there was no Ar^{40} present when the mineral formed, all of the daughter atoms trapped in the mineral must have come from the decay of K^{40}. To determine a sample's age, the K^{40}/Ar^{40} ratio must be measured precisely and the known half-life for K^{40} applied.

It is important to realize that an accurate radiometric date can be obtained only if the mineral remained a closed system during the entire period since its formation; that is, a correct date is not possible unless there was neither the addition nor loss of parent or daughter isotopes. This is not always the case. In fact, an important limitation of the potassium-

argon method arises from the fact that argon is a gas and may leak from the minerals in which it forms. Indeed, losses can be significant if the rock is subjected to relatively high temperatures. Of course, a reduction in the amount of Ar^{40} leads to an underestimation of the rock's age. Sometimes temperatures are high enough for a sufficiently long period that all argon escapes. When this happens, the potassium-argon clock is reset and dating the sample will give only the time of thermal resetting, not the true age of the rock. For other radiometric clocks, a loss of daughter atoms can occur if the rock has been subjected to weathering or leaching. To avoid such a

Figure 9.16
A. *Production and* **B.** *decay of carbon-14. These sketches represent the nuclei of the respective atoms.*

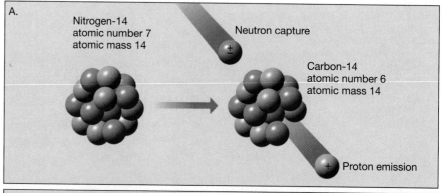

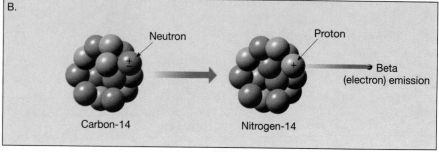

problem, one simple safeguard is to use only fresh, unweathered material and not samples that may have been chemically altered.

If parent/daughter ratios are not always reliable, how can meaningful radiometric dates be obtained? One common precaution against unknown errors is the use of cross checks. Often this simply involves subjecting a sample to two different radiometric methods. If the two dates agree, the likelihood is high that the date is reliable. If, on the other hand, there is an appreciable difference between the two dates, other cross checks must be employed to determine which, if either, is correct.

To date very recent events, carbon-14 (also called *radiocarbon*), the radioactive isotope of carbon, is used. Because the half-life of carbon-14 is only 5730 years, it can be used for dating events from the historic past as well as those from recent geologic history. Until the late 1970s radiocarbon was useful in dating events only as far back as 40,000–50,000 years. However, as was the case with potassium-40, the development of more sophisticated analytical techniques has increased the usefulness of this "clock." In some cases carbon-14 can now be used to date events as far back as 75,000 years. This is a significant accomplishment because geologists can now date many ice-age phenomena that previously could not be dated accurately.

Carbon-14 is continuously produced in the upper atmosphere as a consequence of cosmic ray bombardment, in which cosmic rays (high-energy nuclear particles) shatter the nuclei of gases, releasing neutrons. Some of the neutrons are absorbed by nitrogen (atomic number 7, mass number 14), causing its nucleus to emit a proton. As a result, the atomic number decreases by 1 (to 6), and a different element, carbon-14, is created (Figure 9.16A). This isotope of carbon is quickly incorporated into carbon dioxide, circulates in the atmosphere, and is absorbed by living matter. As a result, all organisms contain a small amount of carbon-14.

While an organism is alive, the decaying radiocarbon is continually replaced, and the proportions of carbon-14 and carbon-12 (the stable and most common isotope of carbon) remain constant. However, when the plant or animal dies, the amount of carbon-14 gradually decreases as it decays to nitrogen-14 by beta emission (Figure 9.16B). By comparing the proportions of carbon-14 and carbon-12 in a sample, radiocarbon dates can be determined. Although carbon-14 is useful in dating only the last small frac-

tion of geologic time, it has become a very valuable tool for anthropologists, archeologists, and historians, as well as for geologists who study very recent earth history. In fact, the development of radiocarbon dating was considered so important that the chemist who discovered this application, Willard F. Libby, received a Nobel prize.

Bear in mind that, although the basic principle of radiometric dating is rather simple, the actual procedure is quite complex, for the analysis which determines the quantities of parent and daughter that are present must be painstakingly precise. In addition, some radioactive materials do not decay directly into the stable daughter product, a fact which may further complicate the analysis. In the case of uranium-238, there are thirteen intermediate unstable daughter products formed before the fourteenth and last daughter product, the stable isotope lead-206, is produced (see Figure 9.14).

Radiometric dating methods have produced literally thousands of dates for events in earth history. Rocks from several localities have been dated at more than 3 billion years, and geologists realize that still-older rocks exist. For example, a granite from South Africa has been dated at 3.2 billion years and contains inclusions of quartzite. Quartzite is a metamorphic rock which originally was the sedimentary rock sandstone. Since sandstone is the product of the lithification of sediments produced by the weathering of pre-existing rocks, we have a positive indication that older rocks existed.

Radiometric dating has vindicated the ideas of Hutton, Darwin, and others who over 150 years ago inferred that geologic time must be immense. Indeed, modern dating methods have proven that there has been enough time for the processes we observe to have accomplished tremendous tasks.

THE GEOLOGIC TIME SCALE

The whole of geologic history has been divided into units of varying magnitude, which together comprise the time scale of earth history (Figure 9.17). The major units of the time scale were delineated during the nineteenth century, principally by workers in western Europe and Great Britain. Since absolute dating was unavailable at that time, the entire time scale was created using methods of relative dating. It has been only in this century that absolute dates have been added.

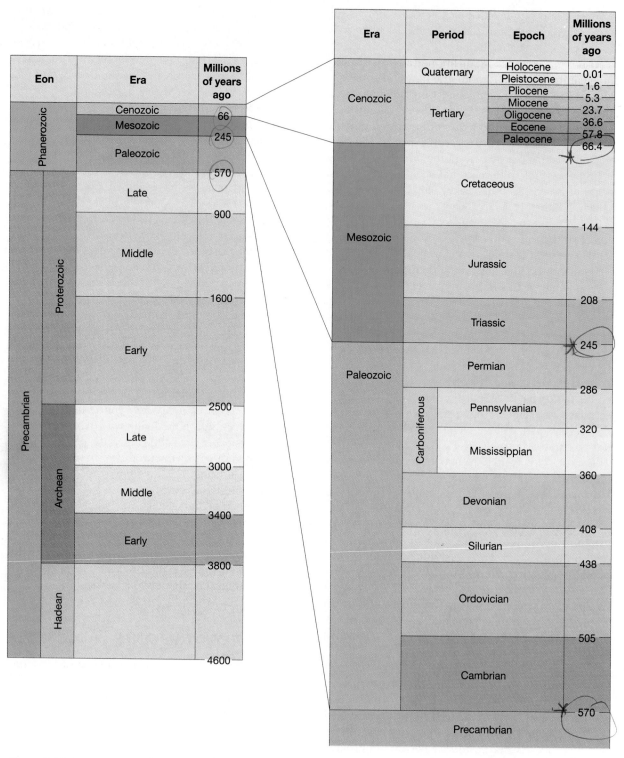

Figure 9.17
The geologic time scale. The absolute dates were added long after the time scale had
been established using relative dating techniques. (Data from Geological Society of
America)

BOX 9.1
The Magnitude of Geologic Time

The concept of geologic time may be a new idea for many. People are accustomed to dealing with increments of time that are measured in hours, days, weeks, and years. Our history books often examine events over spans of centuries, but even a century is difficult to appreciate fully. For most of us, someone or something that is 90 years old is *very old*, and a 1000-year-old artifact is *ancient*.

By contrast, those who study geology must routinely deal with vast time periods—millions or billions (thousands of millions) of years. When viewed in the context of the earth's 4.6-billion-year history, a geologic event that occurred 100 million years ago may be characterized as "recent" by a geologist, and a rock sample that has been dated at 10 million years may be called "young."

An appreciation for the magnitude of geologic time is important in the study of geology because many processes are so gradual that vast spans of time are needed before significant changes occur. Many earth features, which seem to be everlasting and unchanging to us and in fact to generations of people, are indeed slowly changing. Thus, over millions of years, mountains rise and are eroded to hills, and rivers excavate deep canyons.

How long is 4.6 billion years? If you were to begin counting at the rate of one number per second and continued 24 hours a day, 7 days a week, and never stopped, it would take about two lifetimes (150 years) to reach 4.6 billion! Another interesting basis for comparison is as follows:

Compress, for example, the entire 4.5 billion years of geologic time into a single year. On that scale, the oldest rocks we know date from about mid-March. Living things first appeared in the sea in May. Land plants and animals emerged in late November and the widespread swamps that formed the Pennsylvanian coal deposits flourished for about four days in early December. Dinosaurs became dominant in mid-December, but disappeared on the 26th, at about the time the Rocky Mountains were first uplifted. Manlike creatures appeared sometime during the evening of December 31st, and the most recent continental ice sheets began to recede from the Great Lakes area and from northern Europe about 1 minute and 15 seconds before midnight on the 31st. Rome ruled the Western world for 5 seconds from 11:59:45 to 11:59:50. Columbus discovered America 3 seconds before midnight, and the science of geology was born with the writings of James Hutton just slightly more than one second before the end of our eventful year of years.*

The foregoing is just one of many analogies that have been conceived in an attempt to convey the magnitude of geologic time. Although helpful, all of them, no matter how clever, only begin to help us comprehend the vast expanse of earth history.

---

* Don L. Eicher, *Geologic Time,* 2nd ed. (Englewood Cliffs, New Jersey: Prentice-Hall, 1978), pp. 18–19. Reprinted by permission.

The geologic time scale subdivides the 4.6-billion-year history of the earth into many different units and provides a meaningful time frame within which the events of the geologic past are arranged. As shown in Figure 9.17, **eons** represent the greatest expanses of time. The eon that began about 570 million years ago is known as the **Phanerozoic,** a term derived from the Greek words meaning *visible life.* It is an appropriate description because the rocks and deposits of the Phanerozoic eon contain an abundance of fossils that document major evolutionary trends (Figure 9.18). Another glance at the time scale reveals that the Phanerozoic eon is divided into units called **eras.** The eras recognized for this span are the **Paleozoic** ("ancient life"), the **Mesozoic** ("middle life"), and the **Cenozoic** ("recent life"). As the names imply, the eras are bounded by profound worldwide changes in life forms. Each era is subdivided into time units known as **periods.** The Paleozoic has seven, the Mesozoic three, and the Cenozoic two. Each period is characterized by a somewhat less profound change in life forms as compared with the eras. Finally, periods are divided into still smaller units called **epochs.** Seven epochs have been named for the pe-

A.

B.

Figure 9.18
*Unlike Precambrian time, which has a relatively meager fossil record, the Phanerozoic eon is characterized by abundant fossil evidence. **A.** Fossil fish, Eocene epoch, Kemmerer, Wyoming. **B.** Fossil ferns, Pennsylvanian period, St. Clair, Pennsylvania. (Photos by Breck P. Kent)*

riods of the Cenozoic era. The epochs of other periods, however, are not commonly referred to by specific names. Instead, the terms *early, middle,* and *late* are generally applied to the epochs of these earlier periods.

Notice that the detail of the geologic time scale does not begin until about 570 million years ago, the date for the beginning of the first period of the Paleozoic era: the Cambrian period. The more than 4 billion years prior to the Cambrian is divided into three eons, the **Hadean,** the **Archean,** and the **Proterozoic.** It is also common for this vast expanse of time to simply be referred to as the **Precambrian.** Although it represents more than 85 percent of earth history, the Precambrian is not divided into nearly as many smaller time units as the Phanerozoic eon.

Why is the huge expanse of Precambrian time not divided into numerous eras, periods, and epochs? The reason is that Precambrian history is not known in great enough detail. The quantity of information geologists have deciphered about the earth's past is somewhat analogous to the detail of human history. The farther back we go, the less that we know. Certainly more data and information exist about the past ten years than for the first decade of the twentieth century; the events of the nineteenth century have been documented much better than the events of the first century A.D.; and so on. So it is with earth history. The more recent past has the freshest, least disturbed, and most observable record. The farther back in time the geologist goes, the more fragmented the record and clues become.

DIFFICULTIES IN DATING THE GEOLOGIC TIME SCALE

Although reasonably accurate absolute dates have been worked out for the periods of the geologic time scale (see Figure 9.17), the task is not without its difficulties. The primary difficulty in assigning absolute dates to units of time is the fact that not all rocks can be dated radiometrically. Recall that for a radiometric date to be useful, all minerals in the rock must have formed at approximately the same time. For this reason, radioactive isotopes can be used to determine when minerals in an igneous rock crystallized and when pressure and heat created new minerals in a metamorphic rock. However, samples of sedimentary rock can only rarely be dated directly by radiometric means. Although a detrital sedimentary rock may include particles that contain radioactive isotopes, the rock's age cannot be accurately determined because the grains composing the rock are not the same age as the rock in which they occur. Rather, the sediments have been weathered from rocks of diverse ages. Radiometric dates obtained from metamorphic rocks may also be difficult to interpret, since the age of a

BOX 9.2

Unraveling the Precambrian

The Precambrian encompasses an immense amount of geologic time, from the earth's distant beginnings more than 4.6 billion years ago until the start of the Cambrian period about 4 billion years later. In fact, it accounts for about seven-eighths of earth history.

Over large expanses of the continents, outcrops of Precambrian rocks are rare because of an overlying blanket of younger strata. In such regions exposures are confined to the cores of extensively eroded mountains such as the Rockies and to the bottoms of deep canyons such as the Grand Canyon (Figure 9.A). There are, however, extensive regions known as *shields,* where Precambrian rocks are dominant. These geologically complex areas consist largely of deformed metamorphic rocks and were termed shields because of their gently convex profiles, similar to a warrior's shield in cross section. Every continent has one or more of these core regions of ancient Precambrian rocks. In North America, the Canadian shield encompasses 7.2 million square kilometers (2.8 million square miles), when the nearby Greenland shield is included.

The Precambrian is the least understood span of earth history. As a consequence it has not been divided into numerous time units that have gained worldwide acceptance. Indeed, the history of this vast segment of geologic time consists mainly of scattered episodes that are more sketchy and speculative than the geologic history of more recent times. One important reason for this fact is that shields and other Precambrian exposures consist of complex metamorphic rocks that in many cases have been intruded by large igneous bodies. Some rocks have been folded, faulted, intruded, and metamor-

Figure 9.A
Precambrian Vishnu Schist exposed at the bottom of the Grand Canyon is overlain by Paleozoic sedimentary rocks. (Photo by E. J. Tarbuck)

phosed many times. Each event has tended to obscure earlier events, making Precambrian history difficult to decipher.

An additional factor that complicates the study of Precambrian history is the scarcity of fossils. Unfortunately for those attempting to study the Precambrian, abundant fossil evidence does not appear in the geological record until the beginning of the Paleozoic era. Life existed during the Precambrian, but the forms were simple and soft bodied. For that reason, there is only a meager Precambrian fossil record. Although many Precambrian exposures have been studied in great detail, correlation is exceedingly difficult without fossils. This fact helps explain why no widely accepted subdivisions are established for Precambrian time. With the development and application of radiometric dating, a partial solution to the troublesome task of dating and correlating Precambrian rocks now exists. Untangling the complex Precambrian record, however, is still far from being accomplished.

In summary, understanding Precambrian history is difficult because the meager fossil content hinders dating and correlation, the rocks of this age are often highly metamorphosed and deformed, and the record is fragmentary owing to extensive erosion and the presence of younger overlying strata.

Figure 9.19
Absolute dates for sedimentary layers are usually determined by examining their relationship to igneous rocks. (After U.S. Geological Survey)

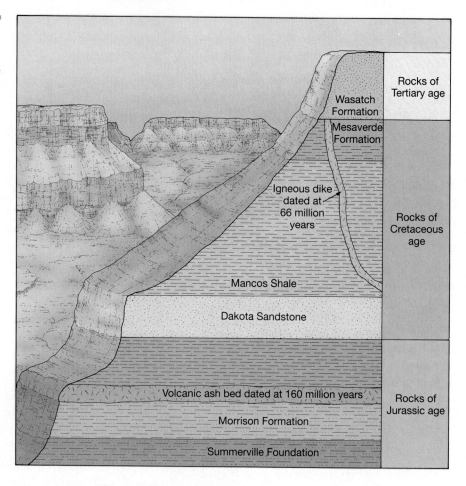

particular mineral in a metamorphic rock does not necessarily represent the time when the rock initially formed. Instead, the date may indicate any one of a number of subsequent metamorphic phases.

If samples of sedimentary rocks rarely yield reliable radiometric ages, how can absolute dates be assigned to sedimentary layers? Usually the geologist must relate them to igneous masses, as in Figure 9.19. In this example, the ages of the volcanic ash bed within the Morrison Formation and the dike cutting the Mancos Shale and Mesaverde Formation are known. The sedimentary beds below the ash are obviously older than the ash, and all the layers above the ash are younger. The dike is younger than the Mancos Shale and the Mesaverde Formation but older than the Wasatch Formation because the dike does not intrude the Tertiary rocks. From this kind of evidence, geologists estimate that a part of the Morrison Formation was deposited about 160 million years ago as indicated by the ash bed. Further, they con-

clude that the Tertiary period began after the intrusion of the dike, 66 million years ago. This is one example of literally thousands that illustrates how dated materials are used to bracket the various episodes in earth history within specific time periods and shows the necessity of combining laboratory dating methods with field observations of rocks.

LIFE OF THE GEOLOGIC PAST

The sequence of occurrence of fossils in rocks of different ages and their transition from one form to another through time represents the documentary record for the evolution of life that has taken place on earth. Due to evolution, different plants and animals have lived at different times in the earth's history.

A century ago the earliest fossils known dated from the beginning of the Cambrian period. An important unanswered problem facing science at that time was the abrupt appearance in the geologic record of complex organisms. Today we know that life evolved during the Precambrian, but the forms were simple and soft bodied. Although fossils from that time are rare, well-preserved remains of minute organisms such as bacteria and blue-green algae (also called cyanobacteria) have been discovered that extend the record of life back nearly 3.5 billion years. The fossil record of animals dates from the late Precambrian and includes distinctive trails and burrows believed to have been created by elongate wormlike animals. Impressions showing completed shapes have also been discovered. Although most, if not all, Precambrian animals were soft bodied and lacked shells, the fossil record is sufficient to demonstrate the existence of a diverse and complex group of multicelled organisms as the Precambrian ended. Thus, the stage was ready for the appearance of even more complex organisms with preservable hard parts as the Cambrian period began.

Paleozoic Life

The geologic time scale shows that the last 570 million years of earth history are divided into three eras: Paleozoic, Mesozoic, and Cenozoic. The Paleozoic era, which encompasses about 325 million years, is by far the longest. Its beginning is marked by the appearance of life forms that possessed hard parts. The most abundant of these early organisms were the trilobites (Figure 9.20). Hard parts greatly enhanced an organism's chance of being preserved as part of the fossil record. Therefore, our knowledge of the diversification of life improves greatly from the Paleozoic onward (Figure 9.21). This abundant fossil evidence has allowed geologists to construct a far more detailed time scale for the last one-eighth of geologic time than for the preceding seven-eighths. Moreover, since every organism is associated with a particular environment, the greatly improved fossil record provided vast and invaluable information for deciphering past geologic environments.

Life in early Paleozoic time was restricted to the seas and consisted of several invertebrate groups (Figure 9.22). Included were trilobites, graptolites, brachiopods, bryozoans, and mollusks. The Cambrian period was the golden age of trilobites. There were more than 600 genera of these mud-burrowing scavengers. By Ordovician times, a group of small marine invertebrates called brachiopods outnumbered the trilobites. Brachiopods are among the most common and abundant Paleozoic fossils. Although the adults lived attached to the sea floor, the young larvae were free-swimming. This mobility accounts for the group's wide geographic distribution. The Ordovician also marked the appearance of abundant cephalopods (Figure 9.22). These highly developed mollusks were very mobile and became the major predators of their time. Cephalopods, whose descendants include the modern squid, octopus, and nautilus, were the first truly large organisms on earth. Whereas the largest trilobites seldom exceeded 30 centimeters (12 inches) in length and the biggest brachiopods were no more than about 20 centimeters (8 inches) across, one species of cephalopod reached a length of nearly 10 meters (30 feet).

Figure 9.20
This 15-cm-long trilobite, Isotelus, *dates from the Ordovician period. (Photo by Michael C. Hansen, Ohio Geological Survey)*

During most of the late Paleozoic, numerous groups of organisms diversified greatly. As the Silurian gave way to the Devonian—some 400 million years ago—plants that had adapted to survival at the water's edge began to move inland. These earliest land plants were leafless, vertical spikes about the size of a person's index finger. However, by the end of the Devonian, the fossil record indicates the existence of forests with trees tens of meters high.

In the oceans, armor-plated fishes that had evolved during the Ordovician made major evolutionary advances. Their armor plates became thin, light-weight scales, which increased their speed and mobility. Other fishes evolved during the Devonian, including primitive sharks and bony fishes, the groups to which virtually all modern fishes belong.

The Devonian period was a time when fishes evolved rapidly; thus, it is often called the "age of fishes."

By late Devonian time, two groups of bony fishes, the lungfish and the lobe-finned fish, adapted to land environments. Like their modern relatives, these fishes had primitive lungs that supplemented the exchange of gases through the gills. Lobe-finned fish are believed to have occupied tidal flats or small ponds. Moreover, in times of drought they may have been able to use their bony fins to "walk" from dried-up pools in search of other ponds. Through time, the lobe-finned fish began to rely more on their lungs and less on their gills. By late Devonian time, true air-breathing amphibians with fishlike heads and tails began to invade the land.

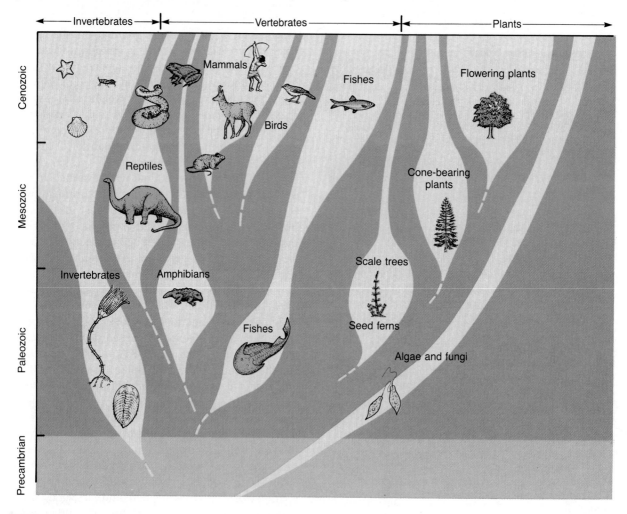

Figure 9.21
This chart indicates the times of appearance and relative abundance of major groups of organisms.

Figure 9.22
The shallow waters of an Ordovician inland sea contained an abundance of marine invertebrates. Shown here are straight-shelled cephalopods, trilobites, brachiopods, snails, and corals. (Courtesy of the Field Museum of Natural History, Chicago)

Although modern amphibians, which include frogs, toads, and salamanders, are small and occupy rather limited biological niches, the conditions during the remainder of the Paleozoic were ideal for these newcomers to the land. Plants and insects, which were their main diet, were very abundant and large. Having only minimal competition from other land dwellers, the amphibians rapidly diversified. Some groups took on roles and forms that were more similar to those of modern reptiles, such as crocodiles, than to those of modern amphibians.

By the Pennsylvanian period, large tropical swamps extended across North America, Europe, and Siberia (Figure 9.23). Trees grew to heights approaching 30 meters (100 feet) with trunks over one meter across. Common trees of the time were the scale trees *Lepidodendron* and *Sigillaria*, the seed ferns, and the scouring rushes called *Calamites*. The coal deposits that fueled the industrial revolution originated in these vast swamps. Further, it was in the lush coal-swamp environment of the late Paleozoic that the amphibians diversified quickly into a variety of species.

The Paleozoic ended with the Permian period, a time when the earth's major landmasses joined to form the supercontinent Pangaea. The redistribution of land and water that resulted from the creation of Pangaea, as well as the changes in elevations of the landmasses, brought about pronounced changes in world climates. Much of the northern continent was elevated above sea level and the climate trended toward a period of greater aridity. This climatic change caused the great scale trees of the coal swamps to become nearly extinct. By the close of the Permian, 75 percent of the amphibian families had disappeared, and plants declined both in number and variety. Although the amphibians declined, their descendants became the most successful and advanced animals on earth. Marine life was not spared during this time of mass extinctions. Many marine invertebrates that had been dominant during the Paleozoic, including all of the remaining trilobites, as well as some types of corals and brachiopods, failed to adapt to the widespread environmental changes at the close of the Permian. These animals, like many others, became extinct.

Mesozoic Life

Spanning nearly 180 million years, the Mesozoic era is divided into three periods: Triassic, Jurassic, and Cretaceous. This episode of geologic history was a time when organisms that had survived the great Permian extinction began to diversify in spectacular ways. On land, dinosaurs became dominant and remained unchallenged for over 100 million years. For

Figure 9.23
Restoration of a Pennsylvanian coal swamp. Shown are scale trees (left), seed ferns (lower left), and scouring rushes (right). Also note the large dragonfly. (Courtesy of the Field Museum of Natural History, Chicago)

this reason the Mesozoic era is often referred to as the "age of dinosaurs."

Early geologists recognized a profound difference between the kinds of fossils in Permian strata and those discovered in younger Triassic rocks. Clearly one-half of the fossil groups in late Paleozoic rocks were missing in rocks of Mesozoic age. On this basis, it was decided to separate the Paleozoic and Mesozoic at the Permian-Triassic boundary.

The life forms at the dawn of the Mesozoic era were those that had survived the great Permian extinctions. These survivors began to diversify to fill the voids left at the close of the Paleozoic. On land, conditions favored those groups that were best able to adapt to drier climates. Among members of the plant kingdom, the gymnosperms were one such group. Unlike the first plants to invade land, the seed-bearing gymnosperms were not dependent on free-standing water for fertilization. Consequently, these plants were not restricted to life near the water's edge.

The gymnosperms quickly replaced the scale trees as the dominant trees of the Mesozoic. Ancient gymnosperms included the cycads, conifers, and ginkgoes. The cycads had large, palmlike leaves and resembled a large pineapple plant, whereas the ginkgoes had fan-shaped leaves much like those of their modern relatives. The largest plants in the Meso-

zoic forests were the conifers, whose modern descendants include the pines, firs, and junipers. The best-known fossil occurrence of these ancient trees is found in northern Arizona's Petrified Forest National Park (see Figure 9.11A). Here, huge petrified logs lie exposed at the surface, having been weathered from rocks of the Triassic Chinle Formation.

Another group that readily adapted to the drier Mesozoic environment was the reptiles, the first true terrestrial life forms. Unlike the amphibians, reptiles have shell-covered eggs that can be laid on land. The elimination of a water-dwelling stage, as, for example, the tadpole stage in the development of frogs, was an important evolutionary step. Of interest is the fact that the watery fluid within the reptilian egg closely resembles seawater in chemical composition. Because the reptile embryo develops in this watery environment, the shelled egg has been characterized as a "private aquarium" in which the embryos of these land vertebrates spend their water-dwelling stage.

With the perfection of the shelled eggs, reptiles quickly became the dominant land animals. They continued this dominance for more than 160 million years. The largest, most awe-inspiring of the Mesozoic reptiles were the dinosaurs. Some of the huge dinosaurs such as *Tyrannosaurus* were carnivorous,

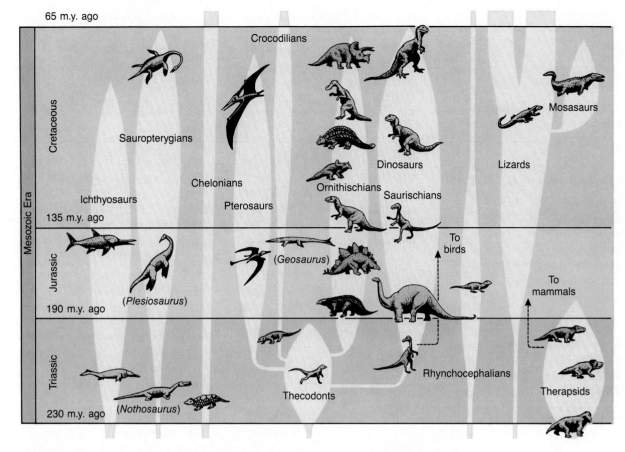

Figure 9.24
Adaptive radiation of reptiles during the Mesozoic era—"the golden age of reptiles."
Dinosaurs, pterosaurs (flying reptiles), and large marine reptiles all became extinct by
the end of the Mesozoic. (Drawing by Robert W. Tope. From Burchfiel et al. Physical
Geology, *Columbus, Ohio: Merrill, 1982)*

whereas others such as the ponderous *Apatosaurus* (formerly *Brontosaurus*) were herbivorous. The extremely long neck of *Apatosaurus* was thought to be an adaptation for feeding on tall conifer trees. However, not all dinosaurs were large. In fact, certain forms closely resembled modern fleet-footed lizards. Further, evidence indicates that some dinosaurs, unlike their present-day relatives, were probably warm blooded.

The reptiles made one of the most spectacular adaptive diversifications in all of earth history (Figure 9.24). One group, the pterosaurs, took to the air. These "dragons of the sky" possessed huge membranous wings that allowed them to exhibit rudimentary flight. Another group of reptiles, exemplified by the fossil *Archaeopteryx*, led to more successful flyers—the birds. Whereas some reptiles took to the skies, others returned to the sea. Included in this lat-

ter group were the fish-eating reptiles: the plesiosaurs and ichthyosaurs. Although these reptiles became proficient swimmers, they retained their reptilian teeth and breathed by means of lungs.

At the close of the Mesozoic many reptile groups became extinct, as had other dominant life forms before them. Of the large number of Mesozoic reptiles, only a few survived to recent times. Among these are the turtles, snakes, crocodiles, and lizards. The huge land-dwelling dinosaurs, the marine plesiosaurs, and the flying pterosaurs all became extinct (Figure 9.24). The demise of the great reptiles is generally attributed to this group's inability to adapt to some radical change in environmental conditions. What could have triggered the extinction of the dinosaurs—the most successful group of land animals ever to have lived? Possible answers are examined in Box. 9.3.

BOX 9.3
The Cretaceous-Tertiary Extinction

The boundaries between divisions on the geologic time scale represent times of significant geological and/or biological change. About 65 million years ago, more than half of all plant and animal species died out. The event marks the end of the Mesozoic era and the beginning of the Cenozoic era. It represents the end of the era in which dinosaurs and other reptiles dominated the landscape and the beginning of the era when mammals became very important (Figure 9.B). Since the Cretaceous (abbreviated K to avoid confusion with other "C" periods) is the last period of the Mesozoic, and the Tertiary (abbreviated T) is the first period of the Cenozoic, the time of this mass extinction is called the Cretaceous-Tertiary or KT boundary.

The dinosaurs met their demise at the KT boundary, along with large numbers of other animal and plant groups, both terrestrial and marine. Equally important, of course, is the fact that many species survived the disaster. Human beings are descended from these survivors. Perhaps this fact explains why an event that occurred 65 million years ago has captured the interest of so many people. The extinction of the great reptiles is generally attributed to this group's inability to adapt to some radical change in environmental conditions. What event could have triggered the sudden extinction of the dinosaurs—the most successful group of land animals ever to have lived?

One modern view proposes that about 65 million years ago, a large asteroid or comet about 10 kilometers in diameter collided with the earth. The impact of such a body would have produced a cloud of dust thousands of times larger than that released during the 1980 eruption of Mount St. Helens. For many months the dust-laden atmosphere would have greatly restricted the amount of sunlight that penetrated to the earth's surface. Without sunlight for photosynthesis, delicate food chains would collapse. It is further hypothesized that large dinosaurs would be affected more adversely by this chain of events than would smaller life forms. In addition, acid rains and global fires may have added to the environmental disaster. It is estimated that when the sunlight returned, more than half of the species on earth, including numerous marine organisms, had become extinct.

What evidence points to this catastrophic collision 65 million years ago? First, a thin layer of sediment nearly 1 centimeter thick has been discovered at the KT boundary. This particular sediment is found worldwide and contains a high level of the element iridium, which is rare in the earth's crust but is found in similar proportions in stony meteorites. Could this layer represent the scattered remains of an asteroid that was responsible for the environmental changes that led to the demise of many reptile groups? Second, this period of mass extinction appears to have affected all land animals larger than dogs. Supporters of this catastrophic-event scenario suggest that small, ratlike mammals could survive a breakdown of food chains lasting perhaps several months. Large animals, they contend, could not survive such an event.

Other scientists disagree with the foregoing hypothesis. They claim that what appears to be a mass extinction over a short period of time did, in fact, occur over a much broader time span. Based upon an examination of the fossil record at the

Cenozoic Life

The Cenozoic era, or "era of recent life," encompasses the past 66 million years of earth history. During this span the physical landscapes and life forms of our modern world came to be. The record for life during the Cenozoic era is more familiar than the records of preceding eras. Mammals replaced the reptiles that had dominated the land during Mesozoic time. More-over, as the Cenozoic era opened, angiosperms (flowering plants with covered seeds) had replaced gymnosperms as the dominant land plants. Marine invertebrates also took on a modern look. Although mollusks, especially pelecypods (clams, oysters) and gastropods (snails), are the most familiar and abundant macroscopic* fossils, microscopic Foraminifera became especially important. Today foraminifers are

Cretaceous-Tertiary boundary, these geologists conclude that the decline of the dinosaurs and numerous other organisms was gradual.

Those who disagree with the impact hypothesis have suggested that volcanism could account for the mass extinction. This competing hypothesis is based on the fact that enormous outpourings of basaltic lavas occurred in India approximately 65 million years ago. Today, these lava flows make up a region known as the Deccan Traps. Although much more extensive, the Deccan Traps resemble the flood basalts of the Columbia Plateau in the Pacific Northwest of the United States. Advocates of the volcanism hypothesis argue that the consequences of an asteroid impact would be quite similar to those associated with extensive volcanism. The first effect would be darkness resulting from large quantities of dust and ash in the atmosphere and causing the food chain to collapse. Furthermore, volcanic eruptions are capable of emitting large quantities of sulfur that could form toxic sulfuric acid rains. To support this scenario, they point to the 1783 eruption at Laki, Iceland. Although small by comparison to the eruptions that produced the Deccan Traps, this volcanic activity killed 75 percent of all livestock and ultimately 24 percent of the inhabitants of Iceland.

Whichever group is correct, the fact remains that one of the more successful groups ever to inhabit the earth died out at the close of the Mesozoic. The decline of the reptiles provided vacancies in habitats for the mammals. Although small and inconspicuous during the Mesozoic, the mammals rose to dominance during the Cenozoic.

Figure 9.B
A composite Mesozoic landscape showing large carnivorous and herbivorous dinosaurs. (Courtesy of the Peabody Museum of Natural History, Yale University)

among the most intensely studied of all fossils, because their widespread occurrence makes them invaluable in correlating Tertiary sediments. These strata are very important to the modern world, for they yield more oil than rocks of any other age.

---

* *Macroscopic* means large enough to be examined with the unaided eye, that is, without instrumentation.

The Cenozoic is often called the "age of mammals," because these animals dominated land life. It could also be called the "age of flowering plants," for the angiosperms enjoyed a similar status in the plant world. Due to advances in seed fertilization and dispersal, angiosperms experienced a period of rapid development and expansion as the Mesozoic drew to a close. Thus, as the Cenozoic era began, angiosperms were already the dominant land plants. Development of the flowering plants, in turn, strongly influenced the evolution of both birds and mammals. Birds that feed on seeds and fruits, for example, evolved rapidly during the Cenozoic in close association with the flowering plants. During the middle Tertiary, grasses developed rapidly and spread over the plains. This fostered the emergence of herbivorous (plant-eating) mammals, which were mainly grazers. The development and spread of grazing mammals in turn established the setting for the evolution of the carnivorous mammals that preyed upon them.

An important evolutionary event was the appearance of primitive mammals in late Triassic time, about the same time as the dinosaurs emerged. Yet throughout the period of dinosaur dominance, mammals remained in the background as small and inconspicuous animals. By the close of the Mesozoic era, dinosaurs and other reptiles no longer dominated the land. It was only after these large reptiles became extinct that mammals came into their own as the dominant land animals. This transition is a major example in the fossil record of the replacement of one large group by another.

Mammals are distinct from reptiles in several respects. Among these differences, mammalian young are born live and mammals maintain a constant body temperature; that is, they are "warm blooded."* This latter adaption allowed mammals to lead more active and diversified lives than reptiles, because they could survive in cold regions and search for food during any season or time of day. Other mammalian adaptations included the development of insulating body hair and more efficient heart and lungs.

Following the reptilian extinctions at the close of the Mesozoic, two groups of mammals, the marsupials and the placentals, evolved and expanded to dominate the Cenozoic. The groups differ principally in their modes of reproduction. Young marsupials are born live but at a very early stage of development. After birth, the tiny and immature young crawl into the mother's external stomach pouch to complete their development. Placental mammals, on the other hand, develop within the mother's body for a much longer period, so that birth occurs after the young are relatively mature and independent.

---

* One minor group of mammals, the monotremes, still lays eggs. The two species in this group, the duck-billed platypus and the spiny anteater, are found only in Australia. Moreover, although modern reptiles are "cold blooded," some paleontologists believe that dinosaurs may have been "warm blooded."

Figure 9.25
Most Pleistocene mammals in this mural, including the mastodon, mammoth, and giant bison in the center, are now extinct. (Peabody Museum of Natural History, Yale University)

Today marsupials are found primarily in Australia, where they went through a separate evolutionary expansion during the Cenozoic, largely isolated from placental mammals. In South America both primitive marsupials and placentals reached the continent before the landmass became completely isolated. Evolution and specialization of both groups continued undisturbed for approximately 40 million years, until the close of the Pliocene epoch when the Central American land bridge was established. Then an invasion of advanced carnivores from North America brought the extinction of many hoofed mammals that had persisted for millions of years. The marsupials, except for opossums, also could not compete and became extinct. Both Australia and South America provide excellent examples of how isolation caused by the separation of continents increased the diversity of animals in the world.

As we have seen, during the Cenozoic era mammals diversified quite rapidly. One tendency was for members of some groups to become very large. For example, by the Oligocene epoch a hornless rhinoceros that stood nearly 5 meters high had evolved. It is the largest land mammal yet known to have existed. As time approached the present, many other types evolved to a large size as well; in fact, more than now exist. Many of these large forms were common as recently as 11,000 years ago. However, a wave of late Pleistocene extinctions rapidly eliminated these animals from the landscape. In North America, huge relatives of the elephant, the mastodon and mammoth, became extinct (Figure 9.25). In addition, saber-toothed cats, giant beavers, large ground sloths, horses, camels, giant bison, and others died out. In Europe, late Pleistocene extinctions included wooly rhinos, large cave bears, and the Irish elk. The reason for this recent wave of large animal extinctions puzzles scientists. Since these animals had survived several major glacial advances and interglacial periods, it is difficult to ascribe these extinctions solely to climatic change. Some scientists believe that early man hastened the decline of these mammals by selectively hunting large forms. Although this hypothesis is advanced more than any other, it is not yet accepted by all.

Review Questions

1. Why did Aristotle have a "bad" influence on the science of geology?

2. Contrast the philosophies of catastrophism and uniformitarianism. How did the proponents of each perceive the age of the earth?

3. Distinguish between absolute and relative dating.

4. What is the law of superposition? How are cross-cutting relationships used in relative dating?

5. When you observe an outcrop of steeply inclined sedimentary layers, what principle allows you to assume that the beds were tilted after they were deposited?

6. Refer to Figure 9.5 and answer the following questions:
   (a) Is fault A older or younger than the sandstone layer?
   (b) Is dike A older or younger than the sandstone layer?
   (c) Was the conglomerate deposited before or after fault A?
   (d) Was the conglomerate deposited before or after fault B?
   (e) Which fault is older, A or B?
   (f) Is dike A older or younger than the batholith?

7. A mass of granite is in contact with a layer of sandstone but does not cut across it. Using a principle described in this chapter, explain how you might determine whether the sandstone was deposited on top of the granite or the granite was intruded from below after the sandstone was deposited.

8. Distinguish among angular unconformity, disconformity, and nonconformity.

9. What is meant by the term *correlation*?

10. Describe several types of fossils. What organisms have the best chance of being preserved as fossils?

11. Describe William Smith's important contribution to the science of geology.

12. Why are fossils such useful tools in correlation?

13. In addition to being important aids in dating and correlating rocks, how else are fossils helpful in geologic investigations?

14. If a radioactive isotope of thorium (atomic number 90, mass number 232) emits 6 alpha particles and 4 beta particles during the course of radioactive decay, what are the atomic number and mass number of the stable daughter product?

15. Why is radiometric dating the most reliable method of dating the geologic past?

16. A hypothetical radioactive isotope has a half-life of 10,000 years. If the ratio of radioactive parent to stable daughter product is 1:3, how old is the rock containing the radioactive material?

17. In order to yield a reliable radiometric date, a mineral must remain a closed system from the time of its formation until the present. Why is this true?

18. Assume that the age of the earth is 5 billion years.
    (a) What fraction of geologic time is represented by recorded history (assume 5000 years for the length of recorded history)?
    (b) The first abundant fossil evidence does not appear until the beginning of the Cambrian period (approximately 600 million years ago). What percent of geologic time is represented by abundant fossil evidence?

19. What subdivisions make up the geologic time scale? What is the primary basis for differentiating the eras?

20. Explain the lack of a detailed time scale for the vast span known as the Precambrian (see Box 9.2).

21. Briefly describe the difficulties in assigning absolute dates to layers of sedimentary rock.

22. Provide the correct period or era for each of the following:
    (a) The era known as the "age of flowering plants."
    (b) By the _____ period large, tropical coal swamps existed.
    (c) The era known as the "age of dinosaurs."
    (d) The _____ period is sometimes called the "golden age of trilobites."
    (e) Mammals became the dominant land animals during the _____ era.
    (f) The period known as the "age of fishes."
    (g) The _____ period was a time of major extinctions, including 75 percent of amphibian families.

23. What is the *KT boundary*? The extinction of what highly successful animal group is associated with this boundary? (See Box 9.3.)

Key Terms

absolute date (p. 313)

angular unconformity (p. 315)

Archean eon (p. 330)

catastrophism (p. 311)

Cenozoic era (p. 329)

conformable (p. 314)

correlation (p. 319)

cross-cutting relationships, principle of (p. 314)

disconformity (p. 315)

eon (p. 329)

epoch (p. 329)

era (p. 329)

fossil (p. 319)

fossil succession, principle of (p. 321)

Hadean eon (p. 330)

half-life (p. 323)

inclusions (p. 314)

index fossil (p. 322)

Mesozoic era (p. 329)

nonconformity (p. 316)

original horizontality, principle of (p. 314)

paleontology (p. 319)

Paleozoic era (p. 329)

period (p. 329)

Phanerozoic eon (p. 329)

Precambrian (p. 330)

Proterozoic eon (p. 330)

radioactivity (p. 323)

radiometric dating (p. 323)

relative dating (p. 313)

superposition, law of (p. 313)

unconformity (p. 315)

uniformitarianism (p. 311)

PART 2

The Oceans

Big Sur Coast, California. (Photo by Larry Ulrich)

10

Ocean Waters and the Ocean Floor

Opposite page: Sunrise over Cape Cod National Seashore. (Photo by David Muench/Allstock)

Left: Sunset over Oregon coast. (Photo by E. J. Tarbuck)

Above: Sunset over Oregon coast. (Photo by E. J. Tarbuck)

*H*ow deep is the ocean? How much of the earth is covered by the global sea? These are basic questions, but they went unanswered for thousands of years. Although people's interest in the oceans undoubtedly dates back to ancient times, it was not until rather recently that these seemingly simple questions began to be answered. The beginning of this chapter deals with these answers. Then, after examining the composition of seawater and the resources extracted from it, we shall take a brief glimpse at the history of oceanography. The remainder of the chapter provides a look at the earth beneath the sea. Suppose that all of the water were drained from the ocean. What would we see? Plains? Mountains? Canyons? Plateaus? Indeed, the ocean conceals all of these features, and more. In fact, the topography of the ocean floor is as varied as that of any continent. And what about the carpet of sediment that covers much of the sea floor? Where did it come from and what can we learn by examining it?

Calling the earth the "water planet" is certainly appropriate, because nearly 71 percent of its surface is covered by the global ocean. Although the ocean comprises a much greater percentage of the earth's surface than the continents, it was only in the relatively recent past that the ocean became an important focus of study. Recently there has been a virtual explosion of data about the oceans, and with it, oceanography has grown dramatically.

Oceanography is actually not a science in itself; rather, it involves the application of all sciences in a comprehensive and interrelated study of the oceans in all of their aspects and relationships. A brief discussion of some of the major subdivisions of oceanography follows. *Physical oceanography* is concerned primarily with energy transmission through ocean water, specifically with such items as wave formation and propagation, currents, tides, energy exchange between ocean and atmosphere, and penetration of light and sound. *Chemical oceanography* is a study of the chemical properties of seawater, of the cause and effect of variation of these properties with time and from place to place, and of the means of measuring these properties. *Biological oceanography* is the study of the interrelationship of marine life with its oceanic environment. The study includes the distribution, life cycles, and population fluctuation of marine organisms. Finally, *geological oceanography* deals with the ocean floor and shore, and embraces such subjects as submarine topography, geological structure, erosion, and sedimentation. The interrelationship of all the sciences is a chief characteristic of oceanography.[*]

EXTENT OF THE OCEANS

A glance at a globe or a look at a photograph of the earth from space reveals a planet dominated by the oceans (Figure 10.1). Indeed, it is for this reason that the earth is often called the *blue planet*. The area of the earth is about 510 million square kilometers (197 million square miles). Of this total, approximately 360 million square kilometers (140 million square

[*] Adapted from H. Dubach and R. Tabor, *Questions About the Oceans*, National Oceanographic Data Center Publication G-13, 67.

Figure 10.1
The earth is sometimes referred to as the "water planet." The global ocean is vast, covering nearly 71 percent of the earth's surface. (Johnny Johnson/Allstock)

miles), or 71 percent, is represented by oceans and marginal seas. The continents comprise the remaining 29 percent, or 150 million square kilometers (58 million square miles).

By studying a globe or world map, it is readily apparent that the continents and oceans are not evenly divided between the Northern and Southern hemispheres. When we compute the percentages of land and water in the Northern Hemisphere, we find that nearly 61 percent of the surface is water, while about 39 percent is land. In the Southern Hemisphere, on the other hand, almost 81 percent of the surface is water, and only 19 percent is land. It is no wonder then that the Northern Hemisphere is called the *land hemisphere*, and the Southern Hemisphere the *water hemisphere*.

Figure 10.2 shows the distribution of land and water for each 5-degree zone in the Northern and Southern hemispheres. Between latitudes 45 degrees north and 70 degrees north there is actually more land than water, whereas between 40 degrees south and 65 degrees south there is almost no land to interrupt the oceanic and atmospheric circulation.

The volume of the ocean basins is many times greater than the volume of the continents above sea level. In fact, the volume of all land above sea level is only $\frac{1}{18}$ that of the ocean. An obvious difference between the continents and the ocean basins is their relative levels. The average elevation of the continents

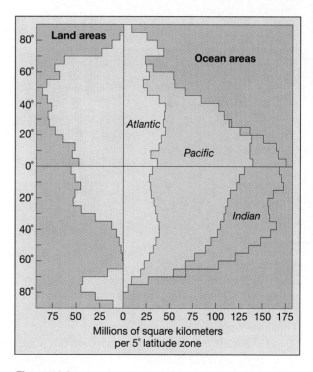

Figure 10.2
Distribution of land and water in each 5-degree latitude belt. (After M. Grant Gross, Oceanography: A View of the Earth, *2nd ed., Englewood Cliffs, N.J.: Prentice-Hall, 1977; and M. Grant Gross,* Oceanography, *4th ed., Columbus, Ohio: Charles E. Merrill, 1980.)*

BOX 10.1

The Seven Seas?

The term *sea* is used two different ways. It is often used interchangeably with *ocean* in referring to bodies of salt water. On the other hand, a sea is also considered to be part of an ocean, or to be a body of water that is substantially smaller than an ocean.

The familiar expression "seven seas" dates back to ancient times and refers to the seas known to the Mohammedans prior to the fifteenth century. These included the Mediterranean Sea, the Red Sea, the East African Sea, the West African Sea, the China Sea, the Persian Gulf, and the Indian Ocean. In more recent times Rudyard Kipling popularized the term "seven seas" by using it as the title for a book of poetry. In fact, since Kipling's time there has been a tendency to divide

the world ocean into seven parts to retain this legendary number. The popular division is the North Atlantic, South Atlantic, North Pacific, South Pacific, Indian, Arctic, and Antarctic (Figure 10.A). Although some speak of an Antarctic Ocean, most (including the International Hydrographic Bureau) regard these waters of the South Polar region merely as extensions of the Atlantic, Pacific, and Indian oceans. Actually, of course, all limits of oceans are arbitrary, as there is only one global ocean.

Figure 10.A
Today the term seven seas *refers to the subdivisions of the global ocean shown here. The Antarctic Ocean is actually just an extension of the adjacent Atlantic, Pacific, and Indian oceans. This "nonocean" is included in order to maintain the traditional seven* seas.

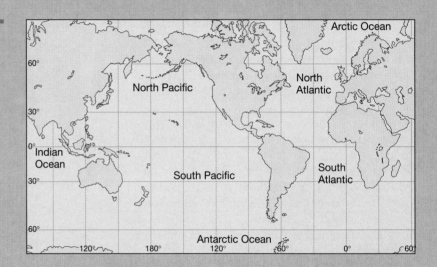

above sea level is about 840 meters (2750 feet), whereas the average depth of the oceans is more than 4.5 times this figure—3800 meters (12,500 feet). Thus, the continents stand on the average about 4.6 kilometers (2.9 miles) above the level of the ocean floor. If the solid earth were perfectly smooth (level) and spherical, the oceans would cover it to a depth of more than 2000 meters.

A comparison of the three major oceans reveals that the Pacific is by far the largest; it is nearly as large as the Atlantic and Indian oceans combined (Figure 10.2). The Pacific contains slightly more than

half of the water in the world ocean, and because it includes few shallow seas along its margins, it has the greatest average depth—3940 meters (12,900 feet).

The Atlantic, bounded by almost parallel continental margins, is a relatively narrow ocean when compared to the Pacific. When the Arctic Ocean is included, the Atlantic has the greatest north-south extent and connects the two polar regions. Because the Atlantic has many shallow adjacent seas, including the Caribbean, Gulf of Mexico, Baltic, and Mediterranean, as well as wide continental shelves

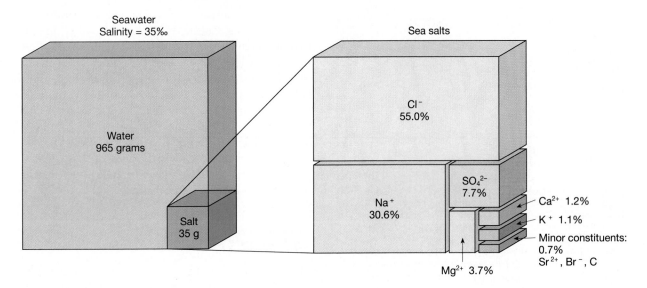

Figure 10.3
Relative proportions of water and dissolved salts in seawater.

along its borders, it is the shallowest of the three oceans, with an average depth of 3310 meters (10,860 feet).

COMPOSITION OF SEAWATER

Seawater is a complex solution of salts, consisting of about 3.5 percent (by weight) dissolved mineral substances. Although the percentage of salts may seem small, the actual quantity is huge. If all of the water were evaporated from the oceans, a layer of salt approaching 60 meters (200 feet) thick would cover the entire ocean floor. The **salinity,** that is, the proportion of dissolved salts to pure water, is commonly expressed in parts per thousand rather than as a percentage (parts per hundred). The symbol used to express parts per thousand is ‰. Thus, the average salinity of the ocean is about 35‰.

The principal elements that contribute to the ocean's salinity are shown in Figure 10.3. If we attempted to make our own seawater, we could come reasonably close by following the recipe shown in Table 10.1. From this table it is evident that most of the salt is sodium chloride (common table salt). Sodium chloride together with the next four most abundant salts comprise 99 percent of the salt in the sea. Although only seven elements make up these five most abundant salts, seawater contains more than 70 of the earth's other naturally occurring elements.

Despite their presence in minute quantities, many of these elements are very important in maintaining the necessary chemical environment for life in the sea.

The relative abundances of the major components in sea salt are essentially constant, no matter where the ocean is sampled. Variations in salinity, therefore, are primarily a consequence of changes in the water content of the solution. As a result, high salinities are found where evaporation is high, as is the case in the dry subtropics. Conversely, where heavy precipitation

Table 10.1
Recipe for artificial seawater.

MIX:	
Sodium chloride (NaCl)	23.48 grams
Magnesium chloride ($MgCl_2$)	4.98
Sodium sulfate (Na_2SO_4)	3.92
Calcium chloride ($CaCl_2$)	1.10
Potassium chloride (KCl)	0.66
Sodium bicarbonate ($NaHCO_3$)	0.192
Potassium bromide (KBr)	0.096
Hydrogen borate (H_3BO_3)	0.026
Strontium chloride ($SrCl_2$)	0.024
Sodium fluoride (NaF)	0.003

ADD:
Water (H_2O) to form 1000 grams of solution.

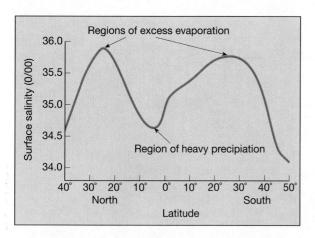

Figure 10.4
The subtropics are dry. On the continents, deserts such as the Sahara occur in this latitude belt. Over the oceans, high evaporation rates remove more water than is replaced by the meager rainfall. Thus, surface salinities are high here. By contrast, near the equator rainfall is heavy, therefore surface salinities are reduced here.

dilutes ocean waters, as in the mid-latitudes and near the equator, lower salinities prevail (Figure 10.4). While salinity variations in the open ocean normally range from 33‰ to 37‰, some seas demonstrate extraordinary extremes. For example, in the restricted waters of the Persian Gulf and the Red Sea, where evaporation far exceeds precipitation, salinity may exceed 42‰. Conversely, very low salinities occur where large quantities of fresh water are supplied by rivers and precipitation. Such is the case for the Baltic Sea, where salinity varies from 2‰ to 7‰.

What are the primary sources for the vast quantities of salts in the ocean? The products of the chemical weathering of rocks on the continents constitute one source. These soluble materials are delivered to the oceans by streams at an estimated rate of more than 2.5 billion tons annually. The second major source of elements found in ocean water is the earth's interior. Through volcanic eruptions, large quantities of water and dissolved gases have been emitted during much of geologic time. This process, called **degassing,** is thought to be the principal source of water in the oceans as well as in the atmosphere. Certain elements, notably chlorine, bromine, sulfur, and boron, are much more abundant in the ocean than in the earth's crust. Since they result from degassing, their abundance in the sea tends to confirm the hypothesis that volcanic action is largely responsible for the present oceans.

Although rivers and volcanic activity continually contribute materials to the oceans, the salinity of seawater is not increasing. In fact, many oceanographers believe that the composition of seawater has been relatively consistent for a large span of geologic time. Why doesn't the sea get saltier? Obviously, material is being removed just as rapidly as it is added. Some elements are withdrawn from seawater by plants and animals as they build hard parts. Others are removed when chemically precipitated as sediment. The net effect is that the overall makeup of seawater remains relatively constant.

RESOURCES FROM SEAWATER

We have seen that the ocean is a great storehouse of dissolved minerals. Some are presently being extracted from seawater while others remain as untapped potential resources. Commercial products obtained from seawater include common salt (sodium chloride), the light-weight metal magnesium, and bromine, which is used principally in gasoline additives and in the manufacture of fireproofing materials. For many years seawater provided a high percentage of the world's magnesium and bromine. In the late 1960s, 70 percent of the world's bromine and 61 percent of its magnesium came from the chemical processing of seawater. Since that time, the production of these two elements has shifted away from seawater for economic reasons. Today the primary sources of magnesium and bromine are water from Utah's Great Salt Lake, and brine wells and evaporite deposits in the United States and other areas. In countries lacking large brine and evaporite deposits, such as Japan and the Scandinavian countries, seawater is still a source of these elements.

Since ancient times the ocean has been an important source of salt for human consumption, and the sea remains a significant supplier. To obtain pure sodium chloride, a series of evaporating ponds is used. The idea is that different salts precipitate from seawater at different salinities. In the first pond, evaporation proceeds until the salt concentration is such that calcium carbonate and calcium sulfate precipitate. Then the remaining brine is transferred to another pond where evaporation continues. Practically pure sodium chloride precipitates there. When most sodium chloride has precipitated, the remaining liquid (now enriched in potassium, bromine, and magnesium) is drained off and the salt is harvested

Figure 10.5
Harvesting salt produced by the evaporation of seawater. (Courtesy of the Leslie Salt Company)

(Figure 10.5). In this manner about 30 percent of the world's salt is produced.

One of the resources derived from seawater is fresh water. The removal of salts and other chemicals from seawater is termed **desalination.** The purpose of this process is to extract low-salinity ("fresh") water suitable for drinking, industry, and agriculture from water that is so saline that it is not suitable for these purposes. Although hundreds of desalination plants are now operating and others are being planned, the cost of desalinated water is still high, and the total production remains relatively small (see Box 10.2).

Although fresh water produced by various desalination technologies is becoming more important as a source of water for human and even industrial use, it is unlikely to be an important supply for agricultural purposes. Since human water requirements are quite low, the cost of supplying areas that do not have adequate natural supplies can often be justified. In addition, the cost of water may represent only a small fraction of the ultimate cost of many industrial goods and hence can be absorbed even in competitive markets. However, the cost of water for irrigation represents a large percentage of the total cost of the crop being produced. For this reason, the cost of desalinated water is still much too high for crop production. Consequently, making the deserts "bloom" by irrigating them with fresh water produced by desalination processes is only a dream and, for economic reasons, is likely to remain so for the foreseeable future.

People have long been intrigued by the prospect of extracting gold from ocean water. For example, following World War I Germany seriously considered extracting gold from seawater to pay its war debt. In fact, a goal of the famous *Meteor* expedition was to investigate the feasibility of gold recovery from the ocean.* Unfortunately, what they found was that the concentration of gold in ocean water is extremely low. To extract a single ounce of gold, 200,000 tons of seawater must be processed. Further, the gold would then have to be separated from the remaining 7000 tons of solids. Thus, even though thousands of kilograms of gold may exist in the waters of the global ocean, no one is now enjoying this wealth because the cost of recovering the gold far exceeds its value.

THE OCEAN'S LAYERED STRUCTURE

By sampling ocean waters, oceanographers have found that temperature and salinity vary with depth. Generally, they recognize a three-layered structure in the open ocean: a shallow surface mixed zone, a transition zone, and a deep zone (Figure 10.6).

Since solar energy is received at the ocean surface, it is here that water temperatures are highest. The mixing of these waters by waves as well as the turbulence from currents creates a rapid vertical heat

---

* This historic expedition is discussed later in the chapter.

BOX 10.2

The Growing Interest in Desalination

During the Persian Gulf war, desalination plants in Saudi Arabia made the news because of fears that oil spills in the gulf would choke intake valves and disrupt the water supply. Meanwhile in California, parched by drought and pressured by population growth, desalination received renewed attention as a possible source of water to supplement dwindling supplies. Indeed, for some regions plagued by chronic water shortages, desalination plants are becoming an important source of fresh water.

In 1990, more than 7500 plants in 120 countries were producing at least 100 cubic meters (26,420 gallons) per day. Just three years earlier, there were fewer than half as many plants. In 1990, desalination plants were capable of producing 13 million cubic meters (3.4 billion gallons) per day, roughly a thirteenfold increase since 1970.

Despite technical advances, converting seawater to fresh water remains costly because it requires large amounts of energy. Fresh water produced by desalination remains three to four times more expensive than water from conventional sources.

Because costs are high, desalination plants are most numerous in areas that can afford the relatively high costs of the fresh water they produce. The arid and oil-rich Persian Gulf region stands out. Expressed as a percentage of world capacity, Saudi Arabia is the world leader with a production of 27 percent, followed by the United States with 12 percent, Kuwait with 11 percent, and the United Arab Emirates with 10 percent.

Most plants in the Middle East use a process known as flash distillation, in which seawater enters a vacuum chamber and boils explosively, producing a salt-free vapor that is captured by condensation. To help contain energy costs, there is increasing interest in building distillation plants adjacent to electric utilities and using the waste heat from power generation to drive the desalination process. Smaller plants typically employ reverse osmosis, which uses high pressure to force salt water through a semipermeable membrane that mechanically filters out both suspended and dissolved solids.

Desalination is not used exclusively to treat seawater. About one-quarter of the plants are devoted to making slightly salty, brackish groundwater usable. Desalination is also used for such purposes as treating polluted groundwater and producing ultrapure water for the electronics industry.

The world's supply of fresh water is unevenly distributed and frequently unreliable. In many regions, conventional sources of fresh water are clearly limited and costs are rising. Moreover, in some areas the pressures of drought are compounded by the addition of more and more people. Although costly, desalination will likely continue to play a significant role in providing reliable supplies of fresh water to coastal communities plagued by chronic shortages.

transfer. Hence, this mixed surface zone is characterized by nearly uniform temperatures. The thickness and temperature of this layer vary, depending upon latitude and season. The zone may attain a thickness of 450 meters (1500 feet) or more, and temperatures ranging between 21°C and 26°C (70°F and 80°F) are common. In equatorial latitudes the temperatures may be even higher. Below the sun-warmed zone of mixing, the temperature falls abruptly with depth. This layer of rapid temperature change, known as the **thermocline,** marks the transition between the warm surface layer and the deep zone of cold water below. Below the thermocline temperatures fall only a few more degrees. At depths greater than about 1500 meters (5000 feet) ocean-water temperatures are consistently less than 4°C (39°F). Since the average ocean depth is approximately 3800 meters (12,500 feet), it is evident that the temperature of most seawater is not much above freezing. In the high polar latitudes, surface waters are cold and temperature changes with depth are slight. Consequently, the three-layered structure is not present.

Salinity variations with depth correspond to the general three-layered system described for temperatures. Generally, in the low and middle latitudes a

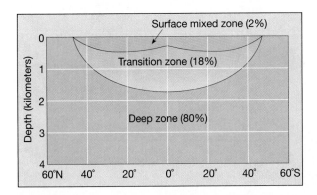

Figure 10.6
Layered structure of the ocean.

surface zone of higher salinity is created when fresh water is removed by evaporation. Below the surface zone salinity decreases rapidly. This layer of rapid change, the **halocline,** corresponds closely to the thermocline. Below the halocline salinity variations are small.

A BRIEF HISTORY OF OCEANOGRAPHY

The modern science of the oceans is often dated back to the time of American naval officer Matthew Fontaine Maury (1806–1873). In 1855 Maury published what many consider the first textbook on oceanography, entitled *The Physical Geography of the*

Sea. The volume includes chapters on the Gulf Stream, the atmosphere, currents, ocean depths, winds, climates, drifts, storms, and more. The book was the end result of a great deal of data gathering and indicated that its author had a profound grasp of the concept that the sea is a single dynamic mechanism. After an injury kept Maury from sailing, he began accumulating and compiling data from ships' logs on currents and winds. In his pursuit of information, Maury enlisted the aid of mariners from all types of ships and of many nationalities. Up to this time there had been little correlated knowledge of wind, weather, tides, and currents; each sailor learned the hard way. Maury's organization of these data revealed previously unknown patterns of ocean currents, and his charts helped reduce sailing times considerably. For example, the time required to sail from the east coast of the United States to Rio de Janeiro, Brazil, was reduced by as much as 10 days, and the trip to California by way of Cape Horn was shortened by 30 days. He also produced the first bathymetric map of the North Atlantic, which showed ocean depths at 1000-fathom intervals (1 fathom equals 1.8 meters, or 6 feet, the approximate distance from fingertip to fingertip of a person with outstretched arms).

Another event of great significance to modern oceanography was the historic 3½-year voyage of the H.M.S. *Challenger* (Figure 10.7). Beginning in December 1872, and ending in May 1876, the *Challenger* expedition made the first, and perhaps most com-

Figure 10.7
H.M.S. Challenger. *(From C. W. Thomson and Sir John Murray,* Report on the Scientific Results of the Voyage of the H.M.S. Challenger, Vol. 1. *Great Britain: Challenger Office, 1895, Plate 1)*

Figure 10.8
Ocean floor profile. **A.** *An echo sounder determines water depth by measuring the time required for a sonic wave to travel from ship to the sea floor and back. The speed of sound in water is 1500 m/sec. Therefore, water depth = ½ (1500 m/sec × echo travel time).* **B.** *Sea-floor profile made by an echo sounder. (Courtesy of Woods Hole Oceanographic Institution)*

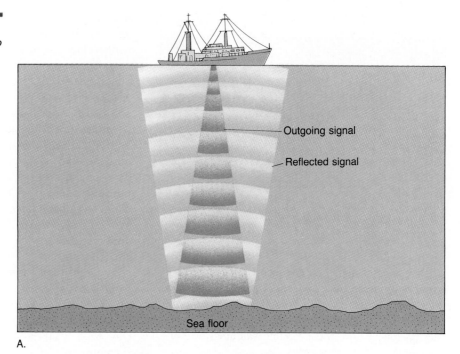

A.

B.

prehensive, study of the global ocean attempted by one agency. The 110,000-kilometer (69,000-mile) trip took the ship and its crew of scientists to every ocean except the Arctic. Periodically they sampled the total depth of water, temperatures at various depths, weather conditions, and the rate and direction of surface and subsurface currents. Samples of water and life were collected at various levels, and the life and sediment of the bottom were sampled with dredges. Over 700 new genera and 4000 new species of animal and plant life were discovered. The data collected clearly indicated that the oceans are teeming with undiscovered life and showed without a question that

life exists at great depths. Previously, many had believed that an *azoic* ("without life") *zone* existed within the ocean beyond a depth of 550 meters (1800 feet). The scientific results of the voyage were published over a 15-year period and filled 50 large volumes. The "*Challenger* Deep-Sea Exploring Expedition," as it was officially called, opened a great descriptive era of oceanography. It is perhaps fitting that from this voyage the term *oceanography* was born.

A remarkable person who contributed to our knowledge of the oceans was the Norwegian adventurer and scientist Fridtjof Nansen. Although best known to oceanographers as a polar explorer, Nansen was also an artist, zoologist, and winner of the Nobel Peace Prize. After studying the waters of the Arctic basin, Nansen developed a hypothesis concerning surface currents in the ice-choked water body. After Nansen convinced his government and the British Royal Geographical Society to help him test his ideas, a special ship, the *Fram*, was built. The *Fram* was designed so that expanding ice would not crush it, but would instead force it up to the surface. For three years, from September 1893 to August 1896, the *Fram* and its crew drifted with the ice. During this pioneering voyage, Nansen not only made important observations about ocean currents but also found that this ice-covered sea reached to true oceanic depths. Further, the drift of the *Fram* disproved the commonly held notion that a continent existed in this northern ocean and showed that the ice covering the polar area throughout the year was not of glacial origin but was a freely moving ice pack that formed directly upon the water surface. Following the voyage, Nansen, realizing the need for more accurate measurements of salinity and temperature, designed an instrument called the *Nansen bottle* that collected water samples from specified depths. For many years the Nansen bottle was a standard oceanographic sampling device.

The voyage during 1925–1927 of the German ship *Meteor* has been considered a sequel to the *Challenger* expedition. This voyage was perhaps the first modern oceanographic trip because it had the advantage of more modern equipment, such as the Nansen bottle, and revolutionary electronic depth-sounding equipment (**echo sounder**). Prior to the echo sounder, ocean depth measurements had to be taken with a weighted line. In deep water the time required for the line to reach bottom can exceed an hour and a half, while the task of hauling in the line is even more time-consuming. Only a limited number of such soundings were practicable; consequently, our knowledge of the sea floor remained very slight. With the development of the echo sounder, depths could be determined in a matter of seconds or minutes, even while the vessel was in motion (Figure 10.8).

The echo sounder apparatus works by transmitting sound waves from the ship toward the ocean bottom. A delicate receiver catches the echo from the bottom, and a highly accurate clock measures the time interval in small fractions of a second. By knowing the velocity of the sound waves in water (about 1500 meters, or 4900 feet, per second), the depth can be calculated very precisely. Since ultrasonic waves are often used because they can be distinguished from the audible sounds during the operation of the ship, the "sound" is not detectable by human ears.

Since World War II, interest in and research concerning the ocean have grown swiftly. Today we live in a period of analytical oceanography, which has replaced the period when purely descriptive studies were the rule. Although sampling and systematic observations are still important, more emphasis is given to theories and hypotheses. Data are gathered that specifically relate to certain ideas, with the goal of proving or disproving these ideas. There are now hundreds of vessels engaged in oceanographic research, some having a special purpose, others equipped to do many tasks. These ships are able to measure the global oceanic system. Although there seems to have been a great emphasis on ocean study over the past century, there is still much more to learn, for every question that is answered about the oceans leads to many others.

In addition, technological breakthroughs are advancing oceanographic study. The research vessel FLIP (*FLoating Instrument Platform*), for example, functions as a very stable free-floating platform for research in the open sea (Figure 10.9A). The drilling ship *JOIDES Resolution*, on the other hand, has added a vast array of new information about the sea floor and its sediments (Figure 10.9B). Moreover, deep-diving manned submersibles are expanding our knowledge of the floor of the ocean by allowing scientists to get a firsthand view of the previously unseen world below.*

THE EARTH BENEATH THE SEA

If all water were removed from the sea basins, what kind of surface would be revealed? It would not be quiet, subdued topography as was once thought, but a surface characterized by great diversity—towering mountain chains, deep canyons, and flat plains. In fact, the scenery would be just as varied as that on the continents (Figure 10.10).

Oceanographers studying the topography of the ocean basins have delineated three major units: continental margins, the ocean basin floor, and mid-

---

* The accomplishments of the *JOIDES Resolution* are detailed in Chapter 6, and deep-diving submersibles are examined more thoroughly later in this chapter.

A.

Figure 10.9
A. *FLIP (**FL**oating **I**nstrument **P**latform) is towed to a research site in the horizontal position. At the site, ballast tanks are flooded with seawater to "flip" it into vertical position. It is outfitted with ingeniously designed hinged equipment that can be used with the vessel in either position.* **B.** *The* JOIDES Resolution *is the drill ship for the Ocean Drilling Program. It represents a significant technological breakthrough because it is capable of lowering drill pipe thousands of meters to the ocean floor and then drilling hundreds of meters into the sediments and underlying crust. It is not just a drilling rig but a floating laboratory with a complete array of scientific equipment and facilities. (Photo A courtesy of Marine Physical Laboratory, Scripps Institution of Oceanography; B courtesy of Ocean Drilling Program)*

B.

ocean ridges. The map in Figure 10.11 (page 362) outlines these provinces for the North Atlantic, and the profile at the bottom of the illustration shows the varied topography. Such profiles usually have their vertical dimension exaggerated many times—40 times in this case—to make topographic features more conspicuous. Because of this, the slopes in the profile of the sea floor in Figure 10.11 appear to be much steeper than they actually are.

Although information about the ocean floor is substantial and growing every day, scientists have nevertheless only scratched the surface. It is worth noting that in many respects we know more about the details of the surface of Venus than of the floor of the ocean. For example, in 1991 and 1992 the *Magellan* spacecraft used special radar to scan and map more than 90 percent of the surface of Venus. By contrast, in the 500 years since Ferdinand Magellan circumnavigated the world's oceans, oceanographers have scanned less than 10 percent of the deep ocean floor at the resolution provided by the *Magellan*

spacecraft. Why have details about the ocean floor come so slowly? To the present day, efforts to accurately map the sea floor are made from ships equipped with echo sounding equipment because radar waves (which are used to map our neighboring planets) do not travel far through seawater. In addition, ships move across the ocean at a mere 18 kilometers per hour rather than soaring through space at 19,500 kilometers per hour.

CONTINENTAL MARGINS

The zones that collectively make up the **continental margin** include the continental shelf, continental slope, and continental rise (Figure 10.12, page 362). The first of these parts, the **continental shelf,** is a gently sloping submerged surface extending from the shoreline toward the deep-ocean basin. Since it is underlain by continental-type crust, it is clearly a flooded extension of the continents. The continental shelf varies greatly in width. Almost nonexistent along some continents, the shelf may extend seaward as far as 1500 kilometers (900 miles) along others. On the average, the continental shelf is about 80 kilometers (50 miles) wide and 130 meters (423 feet) deep at the seaward edge. The average inclination of the continental shelf is less than one-tenth of one degree, a drop of only about 2 meters per kilometer (10 feet per mile). The slope is so slight that it would appear to an observer to be a horizontal surface.

The continental shelves represent 7.5 percent of the total ocean area, which is equivalent to about 18 percent of the earth's total land area. These areas have taken on increased economic and political significance since they have been found to be sites of important mineral deposits, including large reservoirs of petroleum and natural gas, as well as huge sand and gravel deposits. Of course, the waters of the continental shelf contain many important fishing grounds that are significant sources of food.

When compared with many parts of the deep-ocean floor, the surface of the continental shelf is relatively featureless. This is not to say that the shelves are completely smooth. In many regions, long valleys run from the coastline into deeper waters. Many of these shelf valleys are seaward extensions of river valleys on the adjacent landmass. Such valleys were excavated during the Pleistocene epoch (Ice Age). During this time great quantities of water were stored in vast ice sheets on the continent. The result was that the sea level dropped by 100 meters or more, exposing large portions of the continental shelves (see Figure 4.A). Because of the lower sea level, rivers extended their courses, and land-dwelling plants and animals inhabited the newly exposed portions of the continents. Today these areas are again covered by the sea and inhabited by marine organisms. Dredging along the eastern coast of North America has produced the remains of numerous land dwellers, including mammoths, mastodons, and horses. Bottom sampling has also revealed that freshwater peat bogs existed, adding to the evidence that the continental shelves were once above sea level.

Marking the seaward edge of the continental shelf is the **continental slope,** which leads into deep water and has a steep gradient compared to the continental shelf. While the gradient varies from place to place, it has an average drop of about 70 meters per kilometer (370 feet per mile). The continental slope represents the true edge of the continent.

In regions where trenches do not exist, the steep continental slope merges into a more gradual incline known as the **continental rise.** Here the gradient lessens to between 4 and 8 meters per kilometer (20 to 40 feet per mile). Whereas the width of the continental slope averages about 20 kilometers (12 miles), the continental rise may reach for hundreds of kilometers. This feature consists of a thick accumulation of sediment that moved downslope from the continental shelf to the deep-ocean floor. More specifically, the sediments comprising the continental rise are delivered to the base of the continental slope by turbidity currents that follow submarine canyons. When these muddy currents emerge from the mouth of a canyon onto the relatively flat ocean floor, they deposit sediment that forms a **deep-sea fan.** Deep-sea fans have the same basic shape as alluvial fans, which form at the foot of steep mountain slopes on land. As fans from adjacent submarine canyons grow, they coalesce to produce the continuous apron of sediment at the base of the continental slope.

Some continental shelves, such as those along the east coast of the United States, are wide and consist of thick accumulations of shallow-water sediments. These sediments are frequently several kilometers thick and are interbedded with limestones that formed during earlier periods of coral reef building, a process that occurs only in shallow water. The presence of shallow-water sediments buried deep below the surface of the continental shelf led researchers to conclude that these thick accumulations of sediment form along a gradually subsiding continental margin.

This explanation fits very well with our knowledge of how new continental margins are produced during the breakup of a continental landmass. Figure 10.13 on page 363 illustrates the stages in the process. Notice that upwelling of mantle material causes doming, which stretches and fractures the crust. As sea-floor spreading progresses, the stretched and fragmented crust is wedged away from the zone of upwelling where gradual cooling leads to shrinkage and hence subsidence. Sediments carried from adjacent highlands begin to accumulate on the young continental margin. This additional load is believed to contribute to the subsidence.

Passive continental margins of this type surround the Atlantic Ocean. They represent the trailing edges of advancing continents. Over tens to hundreds of

Figure 10.10
The topography of the earth's solid surface is shown on the following two pages. (Copyright © by Marie Tharp)

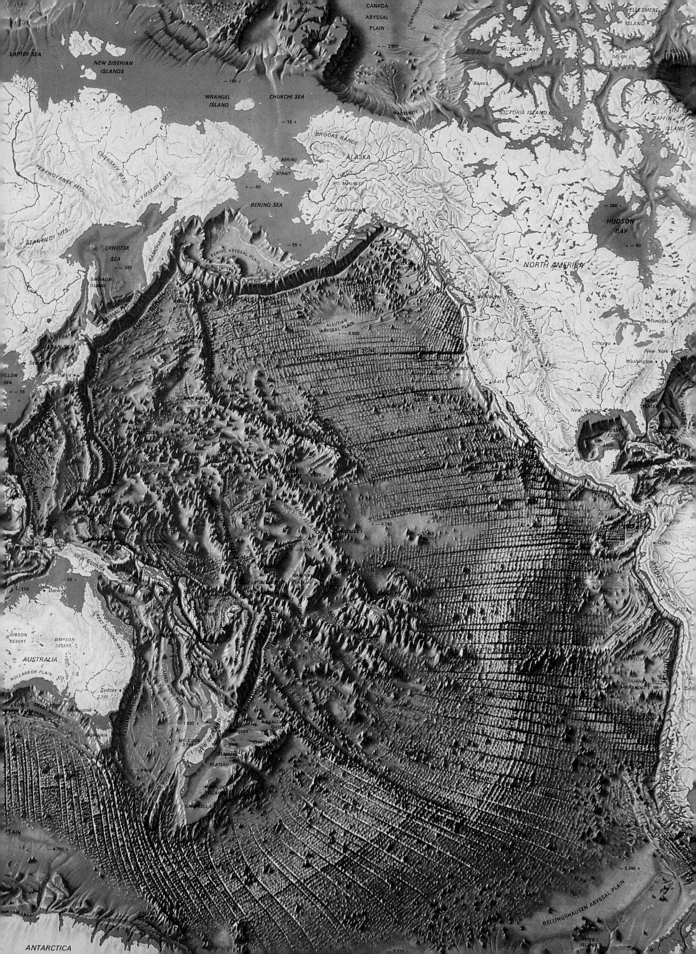

Figure 10.11
This map and profile show the major topographic divisions of the North Atlantic.

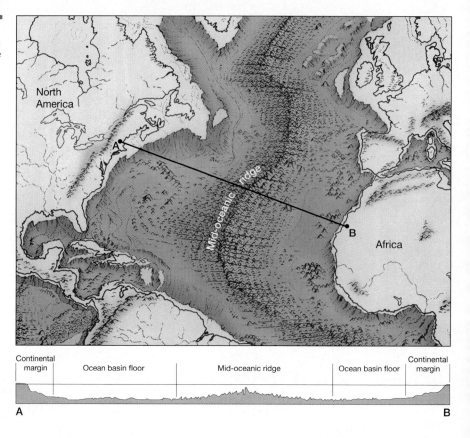

Figure 10.12
Schematic view showing the parts of the continental margin.

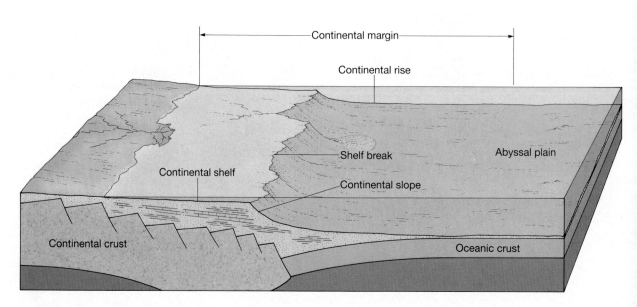

Figure 10.13
*Stages in the breakup of a continental landmass. **A.** Upwelling of mantle rock causes fragmentation of the crust above. **B.** As the landmass moves away from the zone of upwelling, cooling results in the subsidence of the continental margins. **C.** Material derived from the adjacent highlands accumulates into thick wedges of sediment.*

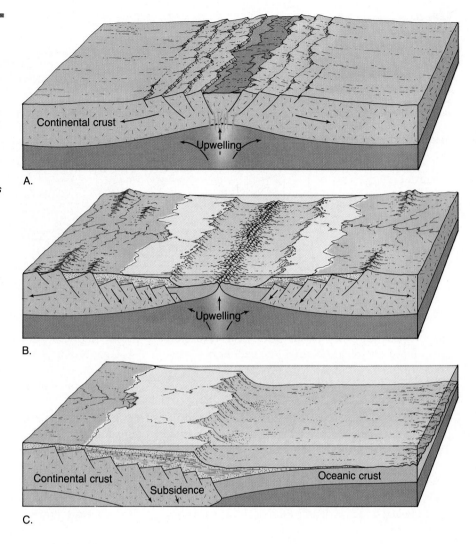

millions of years, the accumulation of material on the passive margins results in a gradual increase in the size of the continent. Moreover, these vast quantities of sediments and sedimentary rocks have an important role in tectonic processes because they are the materials that are deformed during mountain building (see Chapter 8).

Along some mountainous coasts the continental slope descends abruptly into a deep-ocean trench that parallels the landmass. In such cases, the shelf is very narrow or does not exist at all. The side of the trench and the continental slope are essentially the same feature and grade into the adjacent mountains that tower thousands of meters above sea level. These narrow continental margins are primarily located around the Pacific Ocean where plates are converging. The stress between colliding plates results in de-

formation of the continental margin. An example of this activity is found along the west coast of South America. Here the vertical distance from the peaks of the Andes Mountains to the floor of the deep Peru-Chile trench bordering the continent exceeds 12,000 meters.

SUBMARINE CANYONS AND TURBIDITY CURRENTS

Deep, steep-sided valleys known as **submarine canyons** originate on the continental slope and may extend to depths of 3 kilometers (Figure 10.14). Although some of these canyons appear to be the seaward extensions of river valleys, many others do not line up in this manner. Furthermore, since these

Figure 10.14
Submarine canyons are thought to be cut into the continental margins by turbidity currents. These sediment-laden density currents eventually lose momentum and deposit their loads of sediment as deep-sea fans.

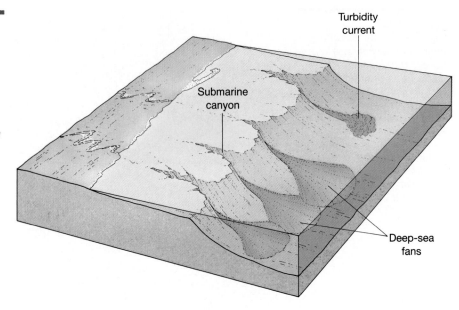

canyons extend to depths far below the maximum lowering of the sea level during the Ice Age, we cannot attribute their formation to stream erosion. These features must be created by some process that operates far below the ocean surface. Most available information seems to favor the view that submarine canyons have been excavated by turbidity currents. **Turbidity currents** are downslope movements of dense, sediment-laden water. They are created when sand and mud on the continental shelf and slope are dislodged and thrown into suspension. Since the mud-choked water is denser than normal sea water,

it flows downslope, eroding and accumulating more sediment as it goes. The erosional work repeatedly carried on by these muddy torrents is thought to be the major force in the excavation of most submarine canyons.

Turbidity currents usually originate along the continental slope and continue across the continental rise, still cutting channels. Eventually they lose momentum and come to rest along the ocean basin floor (Figure 10.15). As these currents slow, suspended sediments begin to settle out. First, the coarser sand is dropped, followed by successively finer accumula-

Figure 10.15
Turbidity currents move downslope, eroding the continental margin to produce submarine canyons. The deposits, called turbidites, are characterized by a decrease in sediment grain size from bottom to top, a phenomenon known as graded bedding.

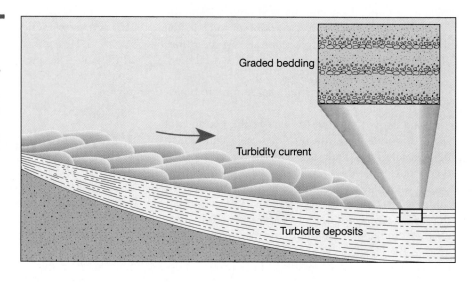

tions of silt and then clay. Consequently, these deposits, called **turbidites,** are characterized by a decrease in sediment grain size from bottom to top, a phenomenon known as **graded bedding** (Figure 10.15).

Although there is still more to be learned about the complex workings of turbidity currents, it has been well established that they are a very important mechanism of sediment transport in the ocean. By the action of turbidity currents, submarine canyons are created and sediments are carried to the deep-ocean floor.

FEATURES OF THE OCEAN BASIN FLOOR

Between the continental margin and the oceanic ridge system lies the ocean basin floor (see Figure 10.11). The size of this region—almost 30 percent of the earth's surface—is roughly comparable to the percentage of the surface that projects above the sea as land. Hence we find ocean trenches, which are dramatically deep grooves in the ocean floor; remarkably flat regions, known as abyssal plains; and steep-sided volcanic peaks, called seamounts.

Deep-Ocean Trenches

Deep-ocean trenches are long, relatively narrow features that represent the deepest parts of the ocean. Several in the western Pacific approach or exceed depths of 10,000 meters (32,800 feet) and at least a portion of one, the Challenger Deep in the Mariana trench, is more than 11,000 meters (36,000 feet) below sea level.

Although deep-ocean trenches represent only a very small portion of the ocean floor area, they are nevertheless significant geological features. Trenches are the sites where moving crustal plates are destroyed as they plunge back into the mantle (see Figure 6.13). In addition to the earthquakes created as one plate descends beneath another, igneous activity is also associated with trench regions. Thus, trenches in the open ocean are often paralleled by volcanic island arcs. Moreover, volcanic mountains, such as those making up a portion of the Andes, are situated adjacent to trenches that parallel the continental margins. As was mentioned in Chapter 6, the melting of a descending plate produces the molten rock that leads to this igneous activity.

Abyssal Plains

Abyssal plains are incredibly flat features; in fact, these regions are likely the most level places on the earth. The abyssal plain found off the coast of Argentina, for example, has less than 3 meters (10 feet) of relief over a distance exceeding 1300 kilometers (800 miles). The monotonous topography of abyssal plains will occasionally be interrupted by the protruding summit of a buried volcanic structure.

By employing seismic profilers, instruments whose signals penetrate far below the ocean floor, researchers have shown that abyssal plains consist of thick accumulations of sediment that were deposited atop the low, rough portions of the ocean floor. The nature of the sediment indicates that these plains consist primarily of sediments transported far out to sea by turbidity currents. The turbidite deposits are interbedded with sediments composed of minute clay-sized particles that continuously settle onto the ocean floor.

Abyssal plains are found as part of the sea floor in all of the oceans. However, they are more widespread where there are no deep-ocean trenches adjacent to the continents. Since the Atlantic Ocean has fewer trenches to act as traps for the sediments carried down the continental slope, it has more extensive abyssal plains than the Pacific.

Seamounts

Dotting the ocean floors are isolated volcanic peaks called **seamounts** that may rise hundreds of meters above the surrounding topography. Although these steep-sided conical peaks are found on the floors of all the oceans, the greatest number have been identified in the Pacific.

Many of these undersea volcanoes begin to rise near oceanic ridges, divergent plate boundaries where the plates of the lithosphere move apart (see Chapter 6). They continue to grow as they ride along on the moving plate. If the volcano rises fast enough, it emerges as an island. Examples in the Atlantic include the Azores, Ascension, Tristan da Cunha, and St. Helena.

During the time they exist as islands, some of these volcanoes are eroded to near sea level by running water and wave action. Over a span of millions of years the islands gradually sink as the moving plate slowly carries them from the oceanic ridge area. These submerged, flat-topped seamounts are called **guyots.** In

BOX 10.3

Evidence for the Existence of Turbidity Currents

For many years the existence of turbidity currents in the ocean was a matter of considerable debate among marine geologists. Not until the 1950s did the speculation begin to subside. Two lines of evidence helped establish turbidity currents as important mechanisms of submarine erosion and sediment transportation. The first important evidence came from records of a rather severe earthquake that took place off the coast of Newfoundland in 1929 and resulted in the breakage of 13 transatlantic telephone and tele-graph cables. At the time, it was presumed that the tremor had caused the multiple breaks. However, when the data were analyzed, it appeared that this was not the case. After plotting the locations of the breaks on a map, researchers saw that all the breaks had occurred along the steep continental slope and the gentler continental rise. Since the time of each break was known from information provided by automatic recorders, a pattern of what had happened could be deduced. The breaks high up on the continental slope took place first, almost concurrently with the earthquake. The other breaks happened in succession, the last occurring 13 hours later, some 720 kilometers (450 miles) from the source of the quake (Figure 10.B). The breaks downslope had obviously taken place too long after the tremor to have been caused by the shock of the earthquake. The existence of a turbidity current, triggered by the quake, thus appeared as a plausible alternative.

As the avalanche of sediment-choked water raced downslope, it snapped the cables in its path. Investigators calculated that the current reached speeds approaching 80 kilometers (50 miles) per hour on the continental slope and about 24 kilometers (15 miles) per hour on the more gently sloping continental rise.

A second compelling line of evidence relating turbidity currents to submarine erosion and transportation of sediment came from the examination of deep-sea sediment samples. These cores show that extensive graded beds of sand, silt, and clay exist in the quiet waters of the deep ocean. Some samples also include fragments of plants and animals that live only in the shallower waters of the continental shelves. No mechanism other than turbidity currents has been identified to satisfactorily explain the existence of these deposits.

More recently, turbidity currents have been measured di-

other instances, guyots may be remnants of eroded volcanic islands that were formed away from the ridge crest, possibly by hot spot activity. Here subsidence occurs after the volcanic activity ceases and the sea floor cools and contracts.

MID-OCEAN RIDGES

Our knowledge of **mid-ocean ridges,** the sites of sea-floor spreading, comes from soundings taken of the ocean floor, core samples obtained from deep-sea drilling, visual inspection using deep-diving submersibles, and even first-hand inspection of slices of ocean floor that have been shoved up onto dry land. Ocean ridge systems are characterized by an elevated position, extensive faulting, and numerous volcanic structures that have developed upon the newly formed crust (Figure 10.10).

Mid-ocean ridges are found in all major oceans and represent more than 20 percent of the earth's surface. They are certainly the most prominent topographic features in the oceans, for they form an almost continuous mountain range, which extends for about 65,000 kilometers (40,000 miles) in a manner similar to the seam on a baseball. Although ocean ridges stand high above the adjacent deep-ocean basins, they are much different from the mountains found on the continents. Rather than containing thick sequences of folded and faulted sedimentary rocks, oceanic ridges consist of layer upon layer of basaltic rocks that have been faulted and uplifted. The term *ridge* may also be misleading since these features are not narrow, but have widths from 500 to 5000 kilo-

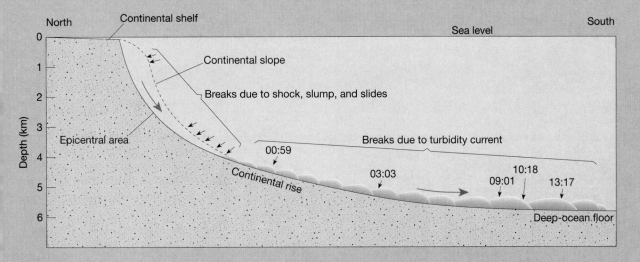

Figure 10.B
Profile of the sea floor showing the events of the November 18, 1929, earthquake off the coast of Newfoundland. The arrows point to cable breaks; the numbers show times of breaks in hours and minutes after the earthquake. Vertical scale is greatly exaggerated. (After B. C. Heezen and M. Ewing, "Turbidity Currents and Submarine Slump and the 1929 Grand Banks Earthquake," American Journal of Science 250: 867)

rectly. In one study, instruments were deployed in Bute Inlet, a deep fiord along the coast of British Columbia in western Canada. The inlet served as a natural laboratory for the study of turbidity current dynamics. The year-long program found turbidity currents to be common events. Although the majority were low-velocity flows that covered limited distances, occasional faster-moving, large-scale currents were also measured. These larger events moved significant amounts of sand down the gently sloping floor of the fiord for distances of 40 to 50 kilometers. The dense currents of sediment in this investigation were linked to submarine landslides in a delta at the head of the fiord.

meters and, in places, may occupy as much as one-half of the total ocean floor area.

Ridges are broad features, yet much of the geologic activity occurs along a relatively narrow region on the ridge crest. This belt, called the **rift zone,** is the site where magma from the asthenosphere moves upward to create new slivers of oceanic crust. Active rift zones are characterized by frequent but generally weak earthquakes and a rate of heat flow that is greater than that of other crustal segments. Here vertical displacement of large slabs of oceanic crust caused by faulting and the growth of volcanic piles contribute to the characteristically rugged topography of the oceanic ridge system. Further, the rocks along the ridge axis appear very fresh and are nearly void of sediment. Away from the ridge axis the topography becomes more subdued and the thickness of the sediments as well as the depth of the water increase. Gradually the ridge system grades into the flat, sediment-laden abyssal plains of the deep-ocean basin.

During sea-floor spreading new material is added about equally to the two diverging plates; hence, we would expect new ocean floor to grow symmetrically about a centrally located ridge. Indeed, the ridge systems of the Atlantic and Indian oceans are located near the middles of these water bodies and as a consequence are named mid-ocean ridges. However, the East Pacific Rise is situated far from the center of the Pacific Ocean. Despite uniform spreading along the East Pacific Rise, much of the Pacific basin that once lay east of this spreading center has been overridden by the westward migration of the American plate.

Figure 10.16
The deep-diving submersible Alvin *is 7.6 meters long, weighs 16 tons, has a cruising speed of 1 knot, and can reach depths as great as 4000 meters. A pilot and two scientific observers are along during a normal 6- to 10-hour dive. (Courtesy of Woods Hole Oceanographic Institution)*

the deep-ocean basin, where it is covered by thick accumulations of sediment.

Partly because of its accessibility to both American and European scientists, the Mid-Atlantic Ridge has been studied more thoroughly than other ridge systems. The Mid-Atlantic Ridge is a gigantic submerged mountain range standing 2500–3000 meters above the adjacent floor of the deep-ocean basins. In a few places, such as Iceland, the ridge actually extends above sea level. Throughout most of its length, however, this divergent plate boundary lies 2500 meters below sea level. Another prominent feature of the Mid-Atlantic Ridge is a deep linear valley extending along the ridge axis. In places, this rift valley is deeper than the Grand Canyon of the Colorado River and two or three times as wide. The name *rift valley* has been applied to this feature because it is strikingly similar to continental rift valleys such as the East African rift valleys. An examination of Figure 10.10 reveals that this central rift is broken into sections that are offset by transform faults.

A CLOSE-UP VIEW OF THE OCEAN FLOOR

Although much has been (and continues to be) learned about the floor of the ocean from echo sounders and other remote sensing equipment, as well as from drilling and sampling by surface ships, oceanographers in the 1970s became aware that direct manned observation was essential to bring about a clearer understanding of many deep-sea phenomena. What was needed was a firsthand view of the previously unseen world below.

Today, the names and accomplishments of deep-diving manned submersibles such as the *Alvin* are common knowledge among oceanographers (Figure 10.16). Manned submersibles are now extending the coverage provided by traditional oceanographic research vessels by allowing scientists to investigate the fine-scale features that previously eluded detection.

One of the pioneering research projects that used deep-diving submersibles was a cooperative venture called Project FAMOUS (French-American Mid-Ocean Undersea Study). In 1974, after three years of preliminary surveying and study by surface ships, two French submersibles and one American vessel made a total of forty-four dives to the floor of the Atlantic. The primary purpose of the project was to examine the structure of the rift valley in the Mid-Atlantic Ridge. The data collected by the three ves-

The primary reason for the elevated position of the ridge system is the fact that newly created oceanic crust is hot and therefore occupies more volume than do cooler rocks of the deep-ocean basin. As the young lithosphere travels away from the spreading center, it gradually cools and contracts. This thermal contraction accounts in part for the greater ocean depths that exist away from the ridge. Almost 100 million years must pass before cooling and contraction cease completely. By this time, rock that was once a part of a majestic oceanic mountain system is located in

Figure 10.17
A photograph taken from the Alvin during Project FAMOUS shows lava extrusions in the rift valley of the Mid-Atlantic Ridge. Large toothpastelike extrusions such as the one in this photograph were common features. A mechanical arm is sampling an adjacent blisterlike extrusion. (Courtesy of Woods Hole Oceanographic Institution)

sels proved invaluable and led to more realistic explanations of how the spreading process works in creating new ocean floor (Figure 10.17).

Later, dives were made by the *Alvin* to a spreading center at 21°N latitude on the East Pacific Rise near the mouth of the Gulf of California. Here, in addition to gathering large quantities of basic data, the scientists aboard the *Alvin* discovered the existence of spectacular geyserlike hot springs. They witnessed two- to five-meter high chimneylike structures spewing dark, mineral-rich, hot (350°C–400°C) water (Figure 10.18). As the heated solutions hit the surrounding 2°C seawater, sulfides of copper, iron, and zinc precipitated immediately, forming mounds of minerals around the steaming vents. In addition to viewing firsthand the formation of massive sulfide deposits, the scientists aboard the *Alvin* found communities of exotic, bottom-dwelling animals living near cooler (20°C) hot springs. The discovery of an animal community thriving more than three kilometers below the surface where no light can reach was totally unexpected. An analysis of the sulfur-rich vent water as well as the stomach contents of some animals revealed that the base of the food chain was sulfur-oxidizing bacteria.

Dives such as the ones briefly highlighted here have demonstrated the value of deep-diving manned submersibles in detecting and studying the fine-scale features of the ocean floor. These vessels now appear to occupy a permanent and important place as tools in oceanographic research.

CORAL REEFS AND ATOLLS

Coral reefs are among the most picturesque features found in the ocean. They are constructed primarily from the calcareous (calcite-rich) skeletal remains and secretions of corals and certain algae. The term *coral reef* is somewhat misleading in that it makes no mention of the skeletons of many small animals and plants found inside the branching framework built by the corals or of the fact that limy secretions of algae help bind the entire structure together.

Coral reefs are confined largely to the warm water of the Pacific and Indian oceans, although a few occur elsewhere. Reef-building corals grow best in waters with an average annual temperature of about 24°C (75°F). They can survive neither sudden temperature changes nor prolonged exposure to temperatures below 18°C (65°F). In addition, these reef-builders require clear sunlit water. For this reason, the limiting depth of active reef growth is about 45 meters (150 feet).

In 1831 the naturalist Charles Darwin set out aboard the British ship H.M.S. *Beagle* on its famous five-year expedition that circumnavigated the globe. One outcome of Darwin's studies during the voyage was the development of a hypothesis on the formation of coral islands, called **atolls.** As Figure 10.19 illustrates, atolls consist of a continuous or broken ring of coral reef surrounding a central lagoon. From the time that Darwin first studied them until shortly af-

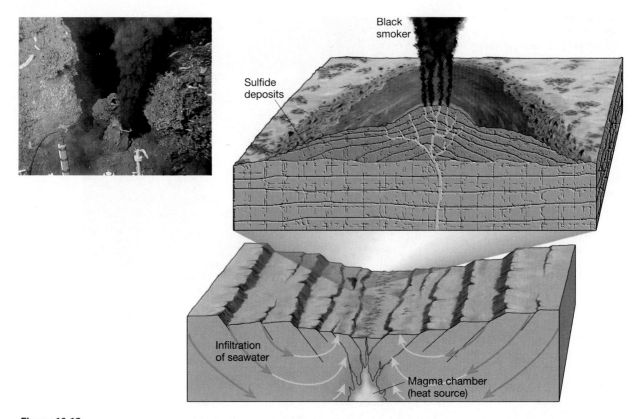

Figure 10.18
Massive sulfide deposits can result from the circulation of seawater through the oceanic crust along active spreading centers. As seawater infiltrates the hot basaltic crust, it leaches sulfur, iron, copper, and other metals. The hot, enriched fluid returns to the sea floor near the ridge axis along faults and fractures. Some metal sulfides may be precipitated in these channels as the rising fluid begins to cool. When the hot liquid emerges from the sea floor and mixes with cold seawater, the sulfides precipitate to form massive deposits. Insert photo shows a close-up view of a black smoker spewing hot, mineral-rich seawater along the East Pacific Rise. (Photo by Dudley Foster, Woods Hole Oceanographic Institution)

ter World War II, their manner of origin challenged people's curiosity.

Darwin's proposal explained what seemed to be a paradox; that is, how can corals, which require warm, shallow, sunlit water no deeper than 45 meters to live, create structures that reach thousands of meters to the floor of the ocean? Commenting on this in *The Voyage of the Beagle*, Darwin stated:

> . . . from the fact of the reef-building corals not living at great depths, it is absolutely certain that throughout these vast areas, wherever there is now an atoll, a foundation must have originally existed within a depth of from 20 to 30 fathoms from the surface.

The essence of Darwin's hypothesis was that coral reefs form on the flanks of sinking volcanic islands.

As an island slowly sinks, the corals continue to build the reef complex upward (Figure 10.20):

> For as mountain after mountain, and island after island slowly sank beneath the water, fresh bases would be successively afforded for the growth of the corals.

Thus atolls, like guyots, are thought to owe their existence to the gradual sinking of oceanic crust. In succeeding years there were numerous challenges to Darwin's proposal. These arguments were not completely put to rest until after World War II when the United States made extensive studies of two atolls (Eniwetok and Bikini) that were going to become sites for testing atomic bombs. Drilling operations at these atolls revealed that volcanic rock did indeed underlie the thick coral reef structure. This finding was a striking confirmation of Darwin's explanation.

A.

B.

Figure 10.19
A. An aerial view of Kayangel Atoll in the Pacific. The light blue waters of the relatively shallow lagoon contrast with the dark blue color of the deep ocean surrounding the atoll. (Photo by Douglas Faulkner/Allstock)
B. View from space of a group of atolls in the Pacific Ocean. (Photo courtesy of NASA)

SEA-FLOOR SEDIMENTS

Except for a few areas, such as near the crests of mid-ocean ridges, the ocean floor is mantled with sediment. Part of this material has been deposited by turbidity currents, and the rest has slowly settled to the bottom from above. The thickness of this carpet of debris varies greatly. In some trenches, which act as traps for sediments originating on the continental margin, accumulations may approach 10 kilometers (6 miles). In general, however, sediment accumula-

tions are considerably less. In the Pacific Ocean, uncompacted sediment measures about 600 meters or less, whereas on the floor of the Atlantic, the thickness varies from 500 to 1000 meters.

Although deposits of sand-sized particles are found on the deep-ocean floor, mud is the most common sediment covering this region. Muds also predominate on the continental shelf and slope, but the sediments in these areas are coarser overall because of greater quantities of sand. Sampling has shown that sands are generally deposited on the continental shelf, forming beaches along the shore. However, in some cases this coarse sediment, which is expected to be found near the shore, occurs in irregular patches at greater depths near the seaward limits of the continental shelves. While some of the sand may have been deposited by local currents that are capable of moving coarse sediment far from shore, much of it appears to result from sand deposition on ancient beaches. Such beaches formed during the Ice Age, when the sea level was much lower than it is today. These patches of sand were then submerged as the sea level rose again.

Types of Sea-Floor Sediments

Sea-floor sediments can be classified according to their origin into three broad categories: (1) lithogenous ("derived from rocks") sediment, (2) biogenous ("derived from organisms") sediment, and (3) hydrogenous ("derived from water") sediment. Although each category is discussed separately, it should be remembered that all sea-floor sediments are mixtures. No body of sediment comes entirely from a single source.

Lithogenous sediment consists primarily of mineral grains that were weathered from continental rocks and transported to the ocean. The sand-sized particles settle near shore. However, since the very smallest particles take years to settle to the ocean floor, they may be carried for thousands of kilometers by ocean currents. As a consequence, virtually every area of the ocean receives some lithogenous sediment. However, the rate at which this sediment accumulates on the deep-ocean floor is indeed very slow. From 5000 to 50,000 years are necessary for a 1-centimeter layer to form. Conversely, on the continental margins near the mouths of large rivers, lithogenous sediment accumulates rapidly. In the Gulf of Mexico, for example, the sediment has reached a depth of many kilometers.

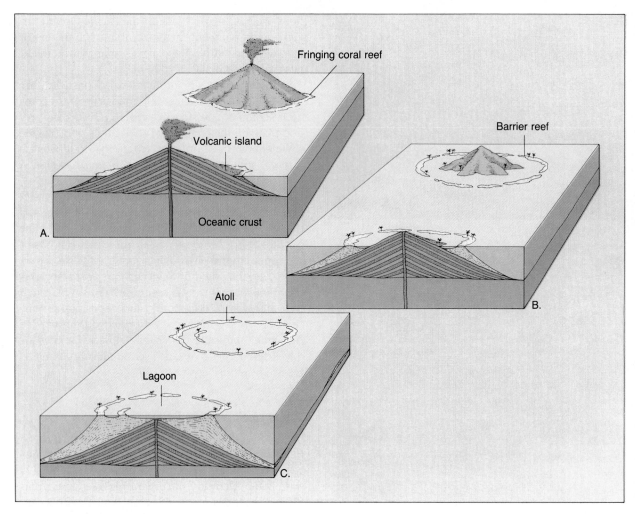

Figure 10.20
Formation of a coral atoll due to the gradual sinking of oceanic crust.

Since fine particles remain suspended in the water for a very long time, there is ample opportunity for chemical reactions to occur. Because of this, the colors of the deep-sea sediments are often red or brown. This results when iron on the particle or in the water reacts with dissolved oxygen in the water and produces a coating of iron oxide (rust).

Biogenous sediment consists of shells and skeletons of marine animals and plants (Figure 10.21). This debris is produced mostly by microscopic organisms living in the sunlit waters near the ocean surface. The remains continually "rain" down upon the sea floor.

The most common biogenous sediments are known as *calcareous* ($CaCO_3$) *oozes*, and as the name implies, they have the consistency of thick mud. These sediments are produced by organisms that inhabit warm surface waters. When calcareous hard parts slowly sink through a cool layer of water, they begin to dissolve. This results because cold seawater contains more carbon dioxide and is thus more acidic than warm water. In seawater deeper than about 4500 meters (15,000 feet), calcareous shells will completely dissolve before they reach bottom. Consequently, calcareous ooze does not accumulate where depths are great.

Other biogenous sediments include *siliceous* (SiO_2) *oozes* and phosphate-rich materials. The former is composed primarily of opaline skeletons of diatoms (single-celled algae) and radiolarians (single-celled

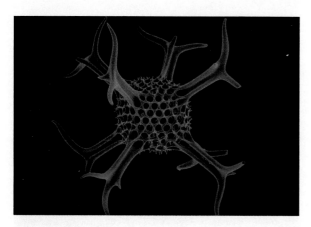

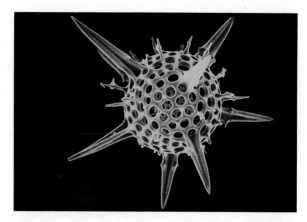

Figure 10.21
Microscopic radiolarian hard parts are examples of biogenous sediments. These photomicrographs have been enlarged hundreds of times. (Photo by Manfred Kage/Peter Arnold, Inc.)

animals), while the latter is derived from the bones, teeth, and scales of fish and other marine organisms.

Hydrogenous sediment consists of minerals that crystallize directly from seawater through various chemical reactions. For example, some limestones are formed when calcium carbonate precipitates directly from the water; however, most limestone is composed of biogenous sediment.

Figure 10.22
Manganese nodules photographed at a depth of 2909 fathoms (5323 meters) beneath the Robert Conrad south of Tahiti. (Courtesy of Lawrence Sullivan, Lamont-Doherty Geological Observatory)

One type of hydrogenous sediment that is significant because of its economic potential is that of **manganese nodules.** These rounded blackish lumps are composed of a complex mixture of minerals that form very slowly on the floor of the ocean basins (Figure 10.22). In fact, their formation rate represents one of the slowest chemical reactions known. By analyzing the radioactive elements continually incorporated into growing nodules, researchers have determined that the growth rates vary from 0.001 to 0.2 millimeter per 1000 years. Some portions of the sea floor are littered with these deposits whereas others lack them altogether. The presence or absence of nodules has been correlated with the sedimentation rate. If sediment accumulates too rapidly (at a rate exceeding about 7 millimeters per 1000 years), newly forming nodules are buried and growth ceases. Since nodule growth is exceedingly slow, why are nodules not buried even where sediment accumulates at less than 7 millimeters per 1000 years? Some scientists suggest that benthic animals living in and on the sea floor are responsible for keeping the nodules at the surface. By stirring the sediment, burrowing organisms are thought to produce a slight lifting effect that, in combination with small surface animals that consume newly arrived sediment from nodule surfaces, keeps nodules from being buried.

Although manganese nodules may contain more than 20 percent manganese, the interest in them as a potential resource lies in the fact that other more valuable metals may be concentrated in them. In addition to manganese, nodules may contain significant

A.   B.

Figure 10.23
A. *The drilling rig of the JOIDES Resolution. As long as the ship is at a particular site, the crew works around the clock retrieving cores from the ocean floor.* **B.** *Scientists examining sea-floor sediments. Analysis occurs onboard as well as at research laboratories around the world. (Photo courtesy of Ocean Drilling Program, Texas A & M University)*

quantities of iron, copper, nickel, and cobalt. All regions containing nodules, however, are not equally good potential sites for mining. Possible mining locations must have abundant nodules (more than 5 kilograms per square meter) and contain the economically optimum mix of cobalt, copper, and nickel. Sites meeting these criteria are relatively limited. Furthermore, before such areas prove to be valuable commercial sources for these metals, the logistics of extracting nodules from the floor of the deep-ocean basins in a cost-effective manner, and the political ramifications of mining the sea bed, must be worked out.

Sea-Floor Sediments and Climatic Change

Due to the fact that instrumental climate records go back only a couple of hundred years (at best), how do scientists find out about climates and climatic changes prior to that time? The obvious answer is that they must reconstruct past climates from indirect evidence; that is, they must examine and analyze phenomena that respond to and reflect changing atmospheric conditions. One of the more interesting

and important techniques for analyzing the earth's climatic history is the study of sediments from the ocean floor.

Although sea-floor sediments are of many types, most contain the remains of organisms that once lived near the sea surface (the ocean-atmosphere interface). When such near-surface organisms die, their shells slowly settle to the ocean floor where they become part of the sedimentary record. One reason that sea-floor sediments are useful recorders of worldwide climatic change is that the numbers and types of organisms living near the sea surface change as the climate changes. This principle is explained by Richard Foster Flint as follows:

> . . . we would expect that in any area of the ocean/atmosphere interface the average annual temperature of the surface water of the ocean would approximate that of the contiguous atmosphere. The temperature equilibrium established between surface seawater and the air above it should mean that . . . changes in climate should be reflected in changes in organisms living near the surface of the deep sea. . . . When we recall that the sea-floor sediments in vast areas of the ocean consist mainly of shells of pelagic foraminifers, and that these

animals are sensitive to variations in water temperature, the connection between such sediments and climatic change becomes obvious.*

Thus, in seeking to understand climatic change as well as other environmental transformations, scientists are tapping the huge reservoir of data in sea-floor sediments. The sediment cores gathered by drilling ships and other research vessels have provided invaluable data that have significantly expanded our knowledge and understanding of past climates (Figure 10.23).

---

* *Glacial and Quaternary Geology* (New York: Wiley, 1971), 718.

Review Questions

1. Define oceanography. Briefly distinguish several subdivisions of this broad discipline.

2. To what does the familiar expression "the Seven Seas" refer today? Did it always have its present-day meaning? (See Box 10.1.)

3. How does the area covered by the oceans compare with that of the continents? Describe the distribution of land and water on earth.

4. Answer the following questions about the Atlantic, Pacific, and Indian oceans:
   (a) Which is the largest in area? Which is smallest?
   (b) Which has the greatest north-south extent?
   (c) Which is shallowest? Explain why its average depth is least.

5. How does the average depth of the ocean compare to the average elevation of the continents?

6. What is meant by *salinity*? What is the average salinity of the ocean?

7. Why do variations in salinity occur? Give some specific examples to illustrate your answer.

8. What is the origin of the minerals dissolved in seawater?

9. List three commercial products that are extracted from seawater.

10. Why is desalination not likely to be a significant source of water for agriculture in the foreseeable future?

11. Briefly describe the contributions to oceanography of the following: Matthew Fontaine Maury, Fridtjof Nansen, the *Challenger* expedition, the *Meteor* expedition, deep-diving submersibles such as the *Alvin*.

12. Assuming that the average speed of sound waves in water is 1500 meters per second, determine the water depth if the signal sent out by an echo sounder requires 6 seconds to strike bottom and return to the recorder (see Figure 10.8A).

13. List the three major subdivisions of the continental margin. Which subdivision is considered a flooded extension of the continent? Which has the steepest slope?

14. How does the continental margin along the west coast of South America differ from the continental margin along the east coast of North America?

15. Defend or rebut the statement "Most submarine canyons found on the continental slope and rise were formed during the Ice Age when rivers extended their valleys seaward."

16. What are turbidites? What is meant by the term *graded bedding*?

17. Discuss the evidence that helped confirm the existence of turbidity currents in the ocean and establish them as significant mechanisms of erosion and sediment transport. (See Box 10.3.)

18. Why are abyssal plains more extensive on the floor of the Atlantic than on the floor of the Pacific?

19. How are mid-ocean ridges and deep-ocean trenches related to sea-floor spreading?

20. What is an atoll? Describe Darwin's proposal on the origin of atolls. Was it ever confirmed?

21. Distinguish among the three basic types of sea-floor sediment.

22. If you were to examine recently deposited biogenous sediment taken from a depth in excess of 4500 meters (15,000 feet), would it more likely be rich in calcareous materials or siliceous materials? Explain.

23. Why are sea-floor sediments useful in studying climates of the past?

Key Terms

11

The Restless Ocean

Opposite page: *Waves crashing at Shore Acres State Park, Oregon. (Photo by Larry Ulrich)*

Left: *Pacific coast, California. (Photo by E. J. Tarbuck)*

Above: *Waves crashing on shore. (Photo by E. J. Tarbuck)*

*T*he restless waters of the ocean are constantly in motion. Winds generate surface currents, the moon and sun produce tides, and density differences create deep-ocean circulation. Further, waves carry the energy from storms to distant shores, where their impact erodes the land. This chapter examines the movements of ocean waters and their importance.

Ocean circulation involves surface currents and movements of deep-water masses. Surface waters are set in motion by the wind (Figure 11.1). Some of these currents are short-lived and affect only small areas. Such water movements are responses to local or seasonal influences. Other surface currents are relatively permanent phenomena that extend over large portions of the oceans. These major horizontal movements of surface waters are closely related to the general atmospheric circulation, which in turn is driven by the unequal heating of the earth by the sun.

In contrast to the largely horizontal movements of surface currents, the deep-ocean circulation has a significant vertical component and accounts for the thorough mixing of deep-water masses. This component of ocean circulation is a response to density differences among water masses and reflects contrasts in temperature and salinity.

SURFACE CURRENTS

> There is a river in the ocean. In the severest droughts it never fails, and in the mightiest floods . . . it never overflows. Its banks and its bottoms are of cold water, while its current is of warm. The Gulf of Mexico is its fountain, and its mouth is in the Arctic Sea. It is the Gulf Stream. There is in the world no other such majestic flow of waters.

This quote from Matthew Fontaine Maury's 1855 book, *The Physical Geography of the Sea*, describes perhaps the best-known and most-studied of the surface ocean currents. However, the Gulf Stream is just a portion of a huge, slowly moving, circular whirl, or **gyre,** that begins near the equator. Similar gyres are found in the South Atlantic, and in the North and South Pacific (Figure 11.2). What creates these large surface current systems? Although our knowledge of the processes that produce and maintain ocean currents is not complete, we do have a general understanding of the principal factors involved.

Where the atmosphere and ocean are in contact, energy is passed from moving air to the water through friction. As a consequence, the drag exerted by winds blowing steadily across the ocean causes the surface layer of water to move. Thus, because winds are the primary driving force of surface currents, there is a relationship between the oceanic circulation and the general atmospheric circulation. A comparison of Figures 11.2 and 14.13 illustrates this. A further clue to the influence of winds on ocean cir-

Figure 11.1
Wind is not only responsible for creating waves, but it also provides the force that drives the ocean's surface circulation. (Photo by Rex Ziak/Allstock)

culation is provided by the currents in the northern Indian Ocean, where there are seasonal wind shifts known as the summer and winter monsoons. When the winds change direction, the surface currents also reverse direction.

Although winds are important in generating surface currents, other factors also influence the movement of ocean waters. The most significant of these is the **Coriolis effect.** Due to the earth's rotation, currents are deflected to the right of their path of motion in the Northern Hemisphere and to the left in the Southern Hemisphere. The Coriolis effect is greater in high latitudes and diminishes toward the equator (for a more complete discussion of the Coriolis effect, see Chapter 14). As a consequence, the direction of surface currents does not coincide with the wind direction. In general, the difference between wind direction and surface-current direction varies from about 15 degrees along shallow coastal areas to a maximum of 45 degrees in the deep oceans.

A closer look at the circulation of one of the major gyres—that of the North Atlantic—will serve to illustrate the general pattern taken by surface currents (Figure 11.2).

North and south of the equator are two westward-moving currents, the North and South Equatorial Currents, which derive their energy principally from the trade winds that blow from the northeast and southeast, respectively, toward the equator. Because of the deflection created by the Coriolis effect, the currents move almost due west. Between these westward-flowing currents there is a weaker, oppositely directed eastward flow, called the Equatorial Countercurrent. The North and South Equatorial Currents pile water up against the eastern coast of South America, and because the trade winds are weak along the equator, some of the piled-up water flows back "downhill" (eastward), creating the countercurrent.

As the North Equatorial Current continues west, the South American continent and the Coriolis effect turn the water to the north. The water then moves northwest through the Caribbean to the mouth of the Gulf of Mexico at the Straits of Yucatan. From here a portion flows parallel to but at some distance from the shore of the Gulf of Mexico, and part curves more sharply toward the east and flows directly toward the north coast of Cuba. These two parts reunite in the Straits of Florida to form the Gulf Stream. A tremendous volume of water flows northward in the Gulf Stream. It can often be distinguished by its deep indigo blue color, which contrasts sharply with the dull green of the surrounding water. When the Gulf Stream encounters the cold waters of the Labrador Current, in the vicinity of the Grand Banks southeast of Newfoundland, there is little mixing of the waters. Rather, the junction is marked by a sharp temperature change. When the warm Gulf Stream water encounters cold air with a low water-holding capacity in this region, the result is a fog that has the appearance of smoke rising from the water.

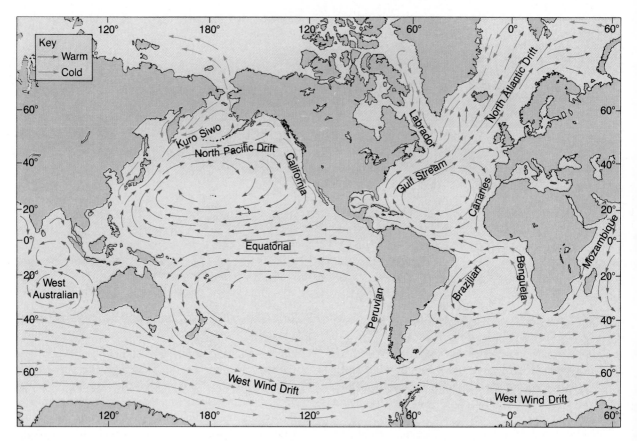

Figure 11.2
Average position and extent of the principal surface ocean currents.

As the Gulf Stream moves along the east coast of the United States, it is strengthened by the prevailing westerly winds and is deflected to the east (to the right) between 35 degrees north and 45 degrees north latitude. As it continues northeastward beyond the Grand Banks, it gradually widens and decreases speed until it becomes a vast, slowly moving current known as the North Atlantic Drift. As the North Atlantic Drift approaches Western Europe, it splits, with part moving northward past Great Britain and Norway. The other part is deflected southward as the cool Canaries Current. As the Canaries Current moves south, it eventually merges into the North Equatorial Current.

The clockwise circulation of the North Atlantic leaves a large central area which has no well-defined currents. This zone of calmer waters is known as the Sargasso Sea, named for the large quantities of *Sargassum,* a type of seaweed encountered there.

In the South Atlantic, surface ocean circulation is very much the same as that in the North Atlantic. The major exception is the circulation of the cold surface waters north of Antarctica. Uncomplicated by large landmasses, the currents in this region move easterly in a globe-circling pattern around the ice-covered continent. They are driven by the prevailing winds from the northwest and deflected to the left by the earth's rotation. The circulation of the Pacific generally parallels that of the Atlantic, and although the monsoons complicate the circulation of the Indian Ocean, the currents there generally coincide with the currents of the South Atlantic.

In addition to producing surface currents, winds may also cause vertical water movements. **Upwelling,** the rising of cold water from deeper layers to replace warmer surface water, is a common wind-induced vertical movement that is most characteristic along the eastern shores of the oceans, most notably along the coasts of California, Peru, and West

Africa. Upwelling occurs in these areas when winds blow toward the equator parallel to the coast. Due to the Coriolis effect, the surface water movement is directed away from the shore. As the surface layer moves away from the coast, it is replaced by water that "upwells" from below the surface. This slow upward flow from depths of 50 to 300 meters (165 to 1000 feet) brings water that is cooler than the original surface water and creates a characteristic zone of lower temperatures near the shore. For swimmers who are accustomed to the waters along the mid-Atlantic shore of the United States, a dip in the Pacific off the coast of central California can be a chilling surprise. In August, when temperatures in the Atlantic are 21°C (70°F) or higher, central California's surf is only about 15°C (60°F). Coastal upwelling also brings to the ocean surface greater concentrations of dissolved nutrients, such as nitrates and phosphates. These nutrient-enriched waters from below promote the growth of plankton, which in turn supports extensive populations of fish.

THE IMPORTANCE OF OCEAN CURRENTS

Since ocean waters are constantly in motion, anyone who navigates the oceans needs to be aware of the horizontal movements we call currents. By understanding the position and strength of ocean currents, sailors soon realize that their voyage times can be reduced. For example, the first good chart of the Gulf Stream, which was printed by Benjamin Franklin, was developed as a result of this need to know (Figure 11.3). P. L. Richardson describes the situation as follows:

> While he was in London as Deputy Postmaster General for the American colonies, Franklin was consulted on the question of why the mail packets took a fortnight longer to sail to America than the merchant ships. In October 1768 Franklin discussed this problem with his cousin Timothy Folger, a Nantucket ship captain then visiting London. Folger told him the packet captains were ignorant of the Gulf Stream and frequently sailed in this current. . . . Folger sketched the Gulf Stream on a chart and added written notes on how to avoid the Gulf Stream and Franklin had the chart printed in 1769 or 1770.*

---

* P. L. Richardson, "Benjamin Franklin and Timothy Folger's First Printed Chart of the Gulf Stream," *Science* 207 (4431): 643.

Even today, the Franklin-Folger chart remains a good summary of the average path of the Gulf Stream.

In addition to being significant considerations in ocean navigation, currents have an important effect on climates. It is known that for the earth as a whole the gains in solar energy equal the losses to space of heat radiated from the earth. When most latitudes are considered individually, however, this is not the case. There is a net gain of energy in the low latitudes and a net loss at higher latitudes. Since the tropics are not becoming progressively warmer, nor the polar regions colder, there must be a large-scale transfer of heat from areas of excess to areas of deficit. This is indeed the case. The transfer of heat by winds and ocean currents equalizes these latitudinal energy imbalances. Ocean water movements account for about a quarter of this total heat transport, and winds the remaining three-quarters (Figure 11.3).

The moderating effect of poleward-moving warm ocean currents is well-known. The North Atlantic Drift, an extension of the warm Gulf Stream, keeps Great Britain and much of northwestern Europe warmer than one would expect for their latitudes. The prevailing westerly winds carry the moderating effects far inland. For example, Berlin (52 degrees north latitude) has an average January temperature similar to that experienced at New York City, which lies 12 degrees farther south, while the January mean at London (51 degrees north latitude) is 4.5°C (8°F) higher than that at New York City.

In contrast to warm ocean currents whose effects are felt most in the middle latitudes in winter, the influence of cold currents is most pronounced in the tropics or during summer months in the middle latitudes. Cool currents, such as the Benguela Current off the west coast of southern Africa, moderate the tropical heat. For example, Walvis Bay (23 degrees south latitude), a town adjacent to the Benguela Current, is 5°C (9°F) cooler in summer than Durban, which is 6 degrees latitude farther poleward but on the eastern side of South Africa, away from the influence of the current.

In addition to influencing temperatures of adjacent land areas, cold currents have other climatic influences. For example, where tropical deserts exist along the west coasts of continents, cold ocean currents have a dramatic impact. The principal west coast deserts are the Atacama in Peru and Chile, and the Namib in southern Africa. The aridity along these coasts is intensified because the lower air is chilled

Figure 11.3
To aid ships crossing the Atlantic, Benjamin Franklin, with the assistance of his cousin, Timothy Folger, produced the first detailed chart of the Gulf Stream about 1769 or 1770. (Courtesy of NOAA) Today satellite images provide views of the Gulf Stream's complexities. Reds and yellows denote warmer waters. The current transports heat from lower latitudes toward the north pole. Meanders of the stream pinch off to form eddies that may move about the ocean for up to two years before dissipating. (Courtesy of O. Brown, R. Evans, and M. Carle/University of Miami Rosenstiel School of Marine and Atmospheric Science)

by cold offshore waters. When this occurs, the air becomes very stable and resists the upward movements necessary to create precipitation-producing clouds. In addition, the presence of cold currents causes temperatures to approach and often reach the dew point, the temperature at which water vapor condenses. As a result, these areas are characterized by high relative humidities and much fog. Thus, not all tropical deserts are hot with low humidities and clear skies. Rather, the presence of cold currents transforms some tropical deserts into relatively cool, damp places that are often shrouded in fog.

DEEP-OCEAN CIRCULATION

Unlike the wind-induced movements of surface and near-surface waters, deep-ocean circulation is governed by gravity and driven by density differences. Two factors—temperature and salinity—are most

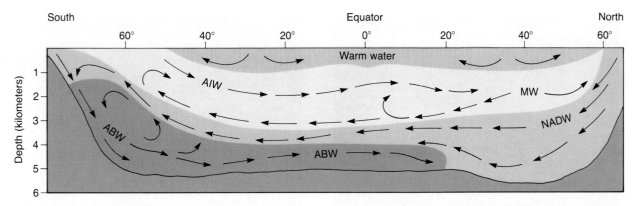

Figure 11.4
Cross section of the deep circulation of the Atlantic Ocean.

significant in creating a dense mass of water. Seawater becomes denser with decreased temperature, increased salinity, or both. Consequently, deep-ocean circulation is called **thermohaline** (*thermo*—"heat," *haline*—"salt") **circulation.** After leaving the surface of the ocean, waters will not reappear at the surface for an average of 500–2000 years.

The basic idea of thermohaline circulation is simple and easily understood. Water at the surface of the ocean is made colder (by heat loss to the atmosphere)

Figure 11.5
Sea ice in the Bering Sea. When seawater freezes, sea salts do not become part of the ice. Consequently, the salt content of the remaining seawater increases, which makes it denser and prone to sink. (Photo by Joel Bennett/Allstock)

or more salty (by removal of water by evaporation or freezing) and becomes more dense. This denser water then sinks toward the ocean bottom and displaces lighter (less dense) water that moves back toward the zone where the denser water formed. In this manner, cold, dense waters flow away from their source near the poles and are replaced by warmer surface waters from lower latitudes.

The deep circulation of the Atlantic Ocean has been the most intensively studied. A simplified cross-sectional view of its deep circulation pattern is seen in Figure 11.4. Arctic and Antarctic waters represent the two major regions where dense water masses are created. Antarctic waters are chilled during the winter. The temperatures here are low enough to form sea ice, and since sea salts are excluded from the ice, the remaining water becomes saltier. The result is the densest water in all of the oceans (Figure 11.5). This cold saline brine slowly sinks to the sea floor, where it becomes Antarctic Bottom Water (ABW in Figure 11.4). Moving northward along the ocean floor, the ABW crosses the equator and reaches as far as 20 degrees north latitude.

North Atlantic Deep Water (NADW in Figure 11.4) is thought to form when warm and highly saline Gulf Stream waters reach the Arctic region near Greenland. The high salinity of the warm surface current results from high evaporation in the low latitudes. In the Arctic these waters are chilled and sink to the bottom of the North Atlantic Basin. From here, the cold, dense water moves south, overriding the denser Antarctic Bottom Waters. North Atlantic Deep Water has been traced almost as far as the Antarctic region, where it becomes obscured by mixing.

Figure 11.6
High tide and low tide on Nova Scotia's Minas Basin in the Bay of Fundy. (Courtesy of Nova Scotia Dept. of Tourism)

Other important subsurface water masses have also been identified. Antarctic Intermediate Water (AIW in Figure 11.4) is formed when the very salty waters of the Brazil Current are chilled as they move south toward the Antarctic. Still another water mass (MW in Figure 11.4) has its source in the Mediterranean Sea. In the warm and dry climate of this region salinity is high because evaporation is great. As this highly saline water is chilled during the winter, its density increases further and it sinks. The water moves along the bottom to the west, passing into the Atlantic at the shallow Strait of Gibraltar. In order to replace the water lost in this outflow, less-dense surface water enters the Mediterranean at Gibraltar. A similar water mass forms in the Red Sea. Since these waters are less dense than the bottom waters, they reach equilibrium after sinking to intermediate depths.

The circulation of deep waters in the Pacific and Indian oceans is different from that of the Atlantic Ocean. Deep water from the Antarctic predominates in these oceans because the shallow barrier at the Bering Strait does not allow Arctic water to flow southward. Consequently, some deep water from the South Polar region may reach as far north as Japan and California.

TIDES

Tides are periodic changes in the elevation of the ocean surface at a specific location. Their rhythmic rise and fall along coastlines have been known since antiquity, and other than waves, they are the easiest ocean movements to observe (Figure 11.6). Although known for centuries, tides were not explained satisfactorily until Sir Isaac Newton applied the law of gravitation to them. Newton showed that there is a mutual attractive force between two bodies, and that since oceans are free to move, they are deformed by this force. Hence tides result from the gravitational attraction exerted upon the earth by the moon, and to a lesser extent by the sun.

To illustrate how tides are produced, we will assume that the earth is a rotating sphere covered to a uniform depth with water (Figure 11.7A). It is easy to see how the moon's gravitational force can cause the water to bulge on the side of the earth nearest the moon. In addition, however, an equally large tidal bulge is produced on the side of the earth directly opposite the moon. Both tidal bulges are caused, as Newton discovered, by the pull of gravity, a force inversely proportional to the square of the distance between two objects. In this case the two objects are the moon and the earth. Because the force of gravity decreases as distance increases, the moon's gravitational pull on the earth is slightly greater on the near side of the earth than on the far side. The result of this differential pulling is to stretch (elongate) the earth. Although the solid earth is stretched by the moon's gravitational pull, the amount of elongation is very slight. However, the world's oceans, which are mobile, are deformed quite dramatically by this effect to produce the two opposing tidal bulges.

Since the position of the moon changes only slightly in a single day, it is the tidal bulges that remain in place while the earth rotates beneath them. For that reason, the earth will carry an observer at any given location alternately into areas of deeper and shallower water. When the person is carried into regions of deeper water, the tide rises, and as the person is carried away, the tide falls. Therefore during one day the observer would experience two high tides and two low tides. In addition to the earth rotating, the tidal bulges also move as the moon revolves around the earth about every 29 days. As a result, the tides, like the time of moonrise, occur about 50 minutes later each day. After about 29 days one cycle is complete and a new one begins.

There may be an inequality between the high tides during a given day. Depending upon the moon's position, the tidal bulges may be inclined to the equator as in Figure 11.7B. This figure illustrates that the

first high tide experienced by an observer in the Northern Hemisphere is considerably lower than the high tide half a day later. On the other hand, a Southern Hemisphere observer would experience the opposite effect.

The sun also influences the tides, but because it is so far away, the effect is considerably less than that of the moon. In fact, the tide-generating potential of the sun is slightly less than half that of the moon. Near the times of new and full moons, the sun and moon are aligned and their forces are added together (Figure 11.8A). Accordingly, the two tide-producing bodies cause higher tidal bulges (high tides) and lower tidal troughs (low tides). These are called the **spring tides.** Spring tides create the largest daily tidal range, that is, the largest variation between high and low tides. Conversely, at about the time of the first and third quarters of the moon, the gravitational forces of the moon and sun on the earth are at right angles, and each partially offsets the influence of the other (Figure 11.8B). As a result, the daily tidal range is less. These are the **neap tides.**

The discussion thus far explains the basic causes and patterns of tides, but keep in mind these theoretical considerations cannot be used to predict either the height or the time of actual tides at a particular place. The shape of coastlines and the configuration of ocean basins greatly influence the tide. Consequently, tides at various locations respond differently to the tide-producing forces. This being the case, the nature of the tide at any place can be determined most accurately by actual observation. The predictions in tidal tables and the tidal data on nautical charts are based upon such observations.

Tides are classified as one of three types according to the tidal pattern occurring at a particular locale. The *semidiurnal* type fits the twice-daily pattern described earlier. There are two high and two low tides each tidal day, with a relatively small difference in the high and low water heights. Tides along the Atlantic coast of the United States are representative of the semidiurnal type, which is illustrated by the tidal curve for Boston in Figure 11.9A. *Diurnal* tides are characterized by a single high and low water height each tidal day (Figure 11.9B). Tides of this type occur along the northern shore of the Gulf of Mexico, among other locations. More common than the diurnal tide, the *mixed* type of tide is characterized by a large inequality in high water heights, low water heights, or both (Figure 11.9C). In this case, there are usually two high and two low waters each day. Such tides are prevalent along the Pacific coast

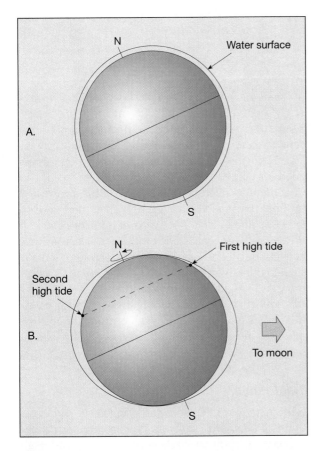

Figure 11.7
Tides on an earth that is covered to a uniform depth with water. Depending upon the moon's position, tidal bulges may be inclined to the equator. In this situation an observer will experience two unequal high tides.

Figure 11.8
*Relationship of the moon and sun to the earth during **A.** spring tides and **B.** neap tides.*

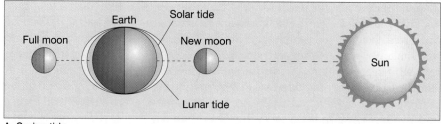

A. Spring tide

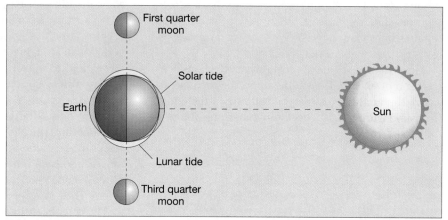

B. Neap tide

of the United States and in many other parts of the world.

Tidal Currents

Tidal current is the term used to denote the horizontal flow of water accompanying the rise and fall of the tide. These water movements induced by tidal forces can be important in some coastal areas. Tidal currents flow in one direction during a portion of the tidal cycle and reverse their flow during the remainder. Tidal currents that advance into the coastal zone as the tide rises are called **flood currents.** Those that move seaward as the level of the tide falls are called **ebb currents.** Periods of little or no current, called *slack water,* separate flood and ebb. The areas affected by these alternating tidal currents are called **tidal flats.** Depending upon the nature of the coastal zone, tidal flats vary in size from narrow strips lying seaward of the beach to zones that may extend for several kilometers.

Although tidal currents are not important in the open sea, they can be rapid in bays, river estuaries, straits, and other narrow places. Off the coast of Brittany, for example, tidal currents that accompany a high tide of 12 meters (40 feet) may attain a speed

of 20 kilometers (12 miles) per hour. While it is generally believed that tidal currents are not major agents of erosion and sediment transport, notable exceptions occur where tides move through narrow inlets. Here they scour the small entrances to many good harbors that would otherwise be blocked.

Sometimes deposits called **tidal deltas** are created by tidal currents (Figure 11.10). They may develop either as *flood deltas* landward of an inlet or as *ebb deltas* on the seaward side of an inlet. Because wave activity and longshore currents are reduced on the sheltered landward side, flood deltas are more common and usually more prominent. They form after the tidal current moves rapidly through an inlet. As it emerges into more open waters from the narrow passage, the current slows and deposits its load of sediment.

Tidal Power

With increased interest in the rising costs and eventual depletion of petroleum, greater attention is being focused upon alternate energy sources. Although several methods of generating electrical energy from the oceans have been proposed, the ocean's energy potential remains largely untapped. The development

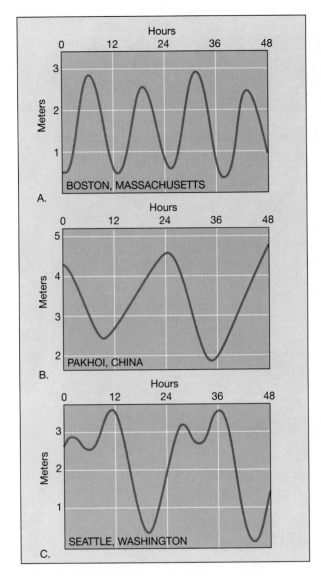

Figure 11.9
Types of tides. **A.** Semidiurnal. **B.** Diurnal. **C.** Mixed.

Tidal power is harnessed by constructing a dam across the mouth of a bay or an estuary in a coastal area having a large tidal range (Figure 11.11). The narrow opening between the bay and the open ocean magnifies the variations in water level that occur as the tides rise and fall. The strong in-and-out flow that results at such a site is then used to drive turbines and electrical generators.

Tidal energy utilization is exemplified by the tidal power plant at the mouth of the Rance River in France (Figure 11.12). By far the largest yet constructed, this plant went into operation in 1966 and produces enough power to satisfy the needs of Brittany and also contribute to the demands of other regions. Much smaller experimental facilities near Murmansk in the former Soviet Union and near Taliang in China are also being used to generate electricity. The United States has not yet tapped its tidal power potential, although a site at Passamaquoddy Bay in Maine, where the tidal range approaches 15 meters (50 feet), has been under review for more than 50 years.

Along most of the world's coasts it is not possible to harness tidal energy. If the tidal range is less than 8 meters (25 feet) or narrow, enclosed bays are absent, tidal power development is uneconomical. For this reason, the tides will never provide a very high proportion of our ever-increasing electrical energy requirements. Nevertheless, the development of tidal

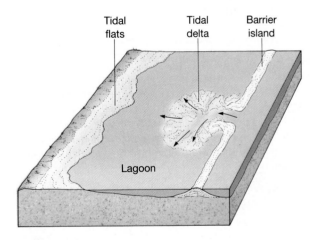

Figure 11.10
Because this tidal delta is forming in the relatively quiet waters on the landward side of a barrier island, it is termed a flood delta. As a rapidly moving tidal current emerges from the inlet, it slows and deposits sediment. The shapes of tidal deltas range from irregular to typical delta-shaped.

of tidal power is the principal example of energy production from the ocean.

Tides have been used as a source of power for centuries. Beginning in the twelfth century, water wheels driven by the tides were used to power gristmills and sawmills. During the seventeenth and eighteenth centuries, much of Boston's flour was produced at a tidal mill. Today, far greater energy demands must be met and more sophisticated ways of using the force created by the perpetual rise and fall of the ocean must be employed.

Figure 11.11
*Simplified diagram showing
the principle of the tidal dam.
(After John J. Fagan, Jr.,* Earth
Environment, *© 1974, Prentice-
Hall, Inc., Englewood Cliffs,
N.J.)*

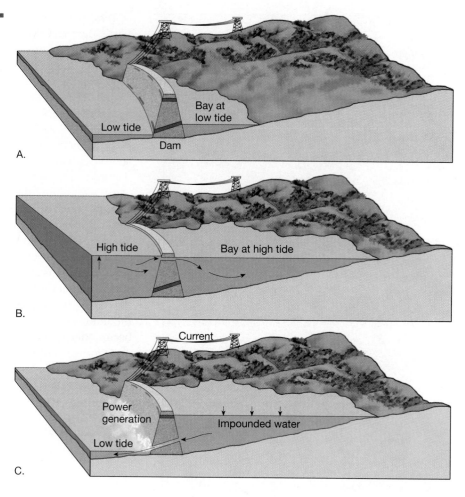

power may be worth pursuing as Paul R. Ryan points
out:

> Although total tidal power potential represents only a
> relatively small proportion of world energy require-
> ments, its realization would nevertheless save a signifi-
> cant amount of fossil fuels. Tidal projects worldwide
> have been estimated to have a potential energy output
> of 635,000 gigawatts, the equivalent of more than a bil-
> lion barrels of oil, a year.*

In addition, electricity produced by the tides consumes no exhaustible fuels and hence creates no noxious wastes.

---

* "Harnessing Power from the Tides: State of the Art,"
Oceanus 22 (1980): 64.

WAVES

The waters of the ocean are constantly in motion. The
restless nature of the water is most noticeable along
the shore—the dynamic interface between land and
sea. Here we can observe the rhythmic rise and fall
of tides and see waves constantly rolling in and
breaking. Sometimes the waves are low and gentle.
At other times, they pound the shore with an awesome fury.

Although it may not be readily apparent to the occasional visitor, the shoreline is constantly being
shaped and modified by the moving ocean waters.
Two well-known areas exhibiting the effects of this
activity are Cape Cod, Massachusetts, and Point
Reyes, California (Figure 11.13). At Cape Cod, waves
striking portions of the eastern shore are causing cliffs
consisting of poorly consolidated glacial sediments to

retreat at rates approaching 1 meter per year. By contrast, the durable bedrock cliffs at Point Reyes are retreating much more slowly. At both locations, wave activity is moving sediment along the shore and building narrow sand bars that protrude into and across some bays. However, the nature of present-day shorelines is not just the result of the relentless attack of the land by the sea. Indeed, the shore is a complex zone whose unique character results from many geologic processes. For example, practically all coastal areas were affected by the worldwide rise in sea level that accompanied the melting of glaciers at the close of the Pleistocene epoch. As the sea edged landward, the shoreline became superimposed upon landscapes that resulted from such diverse processes as stream erosion, glaciation, volcanic activity, and the forces of mountain building.

Wind-generated waves provide most of the energy that shapes and modifies shorelines. Where the land and sea meet, waves that may have traveled unimpeded for hundreds of kilometers suddenly encounter a barrier that will not allow them to advance farther. Stated another way, the shore is where a practically irresistible force confronts an almost immovable object. The conflict that results is never-ending and sometimes dramatic.

The undulations of the water surface, called waves, derive their energy and motion from the wind. If a breeze of less than 3 kilometers (2 miles) per hour starts to blow across still water, small wavelets appear almost instantly. When the breeze dies, the ripples disappear as suddenly as they formed. However, if the wind exceeds 3 kilometers per hour, more stable waves gradually form and progress with the wind.

All waves are described in terms of the characteristics illustrated in Figure 11.14. The tops of the waves are the *crests*, which are separated by *troughs*. The vertical distance between trough and crest is called the **wave height,** and the horizontal distance separating successive crests is the **wavelength.** The **wave period** is the time interval between the passage of successive crests at a stationary point. The height, length, and period that are eventually achieved by a wave depend upon three factors: (1) wind speed; (2) length of time the wind has blown and; (3) **fetch,** the distance that the wind has traveled across the open water. As the quantity of energy transferred from the wind to the water increases, the height and steepness of the waves increase as well. Eventually, a critical point is reached and ocean breakers called *whitecaps* form. For a particular wind speed there is a maximum fetch and duration of wind beyond which waves will no longer increase in size. When the maximum fetch and duration are reached for a given wind velocity, the waves are said to be "fully developed." The reason that waves can grow no further is that they are losing as much energy through the breaking of whitecaps as they are receiving from the wind.

When the wind stops or changes direction, or the waves leave the stormy area where they were created, they continue on without relation to local winds. The waves also undergo a gradual change to *swells* that are lower in height and longer in length and may carry the storm's energy to distant shores. Because many independent wave systems exist at the same time, the sea surface acquires a complex and irregular pattern. Hence the sea waves we watch from the shore are usually a mixture of swells from faraway storms and waves created by local winds.

It is important to realize that in the open sea the motion of the wave is different from the motion of the water particles within it. It is the wave form that

Figure 11.12
The world's first tidal power station to produce electricity was built across the Rance River estuary in 1966. The tide that rushes up this estuary on the northern coast of Brittany is one of the highest in the world, reaching 13.5 meters (44 feet). (Courtesy of Phototeque/Electricité de France)

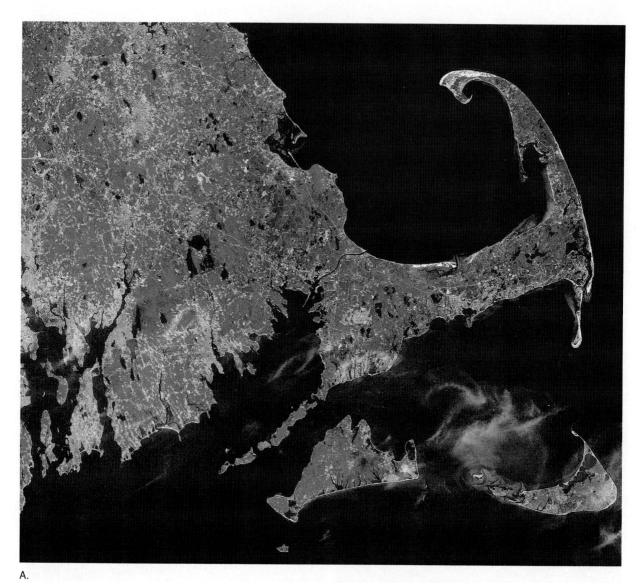

A.

Figure 11.13
A. This satellite image includes the familiar outline of Cape Cod. Boston is just out of the image in the upper left corner. The two large islands off the south shore of Cape Cod are Martha's Vineyard (left) and Nantucket (right). Although the work of waves constantly modifies this coastal landscape, shoreline processes are not responsible for creating it. Rather, the present size and shape of Cape Cod result from the positioning of moraines and other glacial materials deposited during the Pleistocene epoch. (Courtesy of Earth Satellite Corporation) **B.** High-altitude image of the Point Reyes area north of San Francisco, California. The 5.5-kilometer-long south-facing cliffs at Point Reyes (lower left corner) are exposed to the full force of the waves from the Pacific Ocean. Nevertheless, this promontory retreats slowly because the bedrock from which it is formed is very resistant. (Courtesy of USDA-ASCS)

moves forward, not the water itself. Each water particle moves in a nearly circular path during the passage of a wave (Figure 11.14). As a wave passes, a water particle returns almost to its original position. The circular orbits followed by the water particles at the surface have a diameter equal to the wave height. When water is part of the wave crest, in moves in the same direction as the advancing wave form. When it is in the trough, the water moves in the opposite direction. This is demonstrated by observing the behavior of a floating cork as a wave passes. The cork merely seems to bob up and down and sway slightly to and fro without advancing appreciably from its

original position.* For this reason, waves in the open sea are called **waves of oscillation.** The energy contributed by the wind to the water is transmitted not only along the surface of the sea but also downward as well. However, beneath the surface the circular motion rapidly diminishes until at a depth equal to about one-half the wavelength the movement of water particles becomes negligible. This is shown by the

---

* The wind does drag the water forward slightly, causing the surface circulation of the oceans.

B.

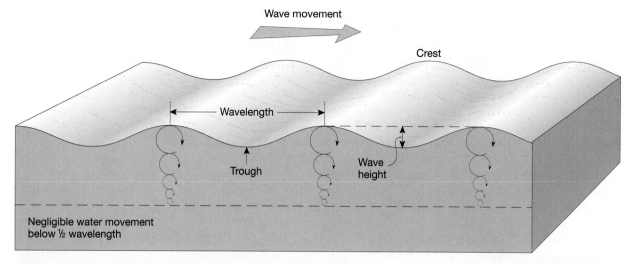

Figure 11.14
This diagram illustrates the basic parts of a wave as well as the movement of water particles with the passage of a wave. Negligible water movement occurs below a depth equal to one-half the wavelength (the level of the dashed line).

rapidly diminishing diameters of water-particle orbits in Figure 11.14.

As long as a wave is in deep water it is unaffected by water depth. However, when a wave approaches the shore the water becomes shallower and influences wave behavior. The wave begins to "feel bottom" at a water depth equal to about one-half its wavelength. Such depths interfere with water movement at the base of the wave and slow its advance.

As the wave continues to advance toward the shore, the slightly faster seaward waves catch up, decreasing the wavelength. As the speed and length of the wave diminish, the wave steadily grows higher. Finally a critical point is reached when the steep wave front is unable to support the wave and it collapses, or *breaks* (Figure 11.15). What had been a wave of oscillation now becomes a **wave of translation** in which the water advances up the shore. The turbu-

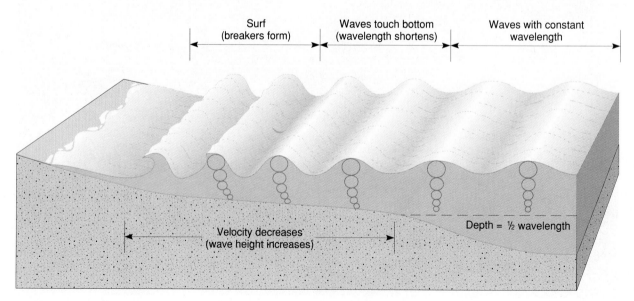

Figure 11.15
Changes that occur when a wave moves onto shore.

lent water created by breaking waves is called **surf.** On the landward margin of the surf zone the turbulent sheet of water from collapsing breakers, called *swash*, moves up the slope of the beach. When the energy of the swash has been expended, the water flows back down the beach toward the surf zone as *backwash*.

WAVE EROSION

During periods of calm weather, wave action is at a minimum. However, just as streams do most of their work during floods, so waves do most of their work during storms. The impact of high, storm-induced waves against the shore can be awesome in its violence. Each breaking wave may hurl thousands of tons of water against the land, sometimes causing the earth to literally tremble. The pressures exerted by Atlantic waves, for example, average nearly 10,000 kilograms per square meter (more than 2000 pounds per square foot) in winter. The force during storms is even greater. During one such storm, for instance, a 1350-ton portion of a steel and concrete breakwater was ripped from the rest of the structure and moved to a useless position toward the shore at Wick Bay, Scotland. Five years later the 2600-ton unit that replaced the first met a similar fate. There are many such stories that demonstrate the great force of breaking waves. It is no wonder then that cracks and crevices are quickly opened in cliffs, seawalls, breakwaters, and anything else that is subjected to these enormous shocks. Water is forced into every opening, causing air in the cracks to become highly compressed by the thrust of crashing waves. When the wave subsides, the air expands rapidly, dislodging rock fragments and enlarging and extending preexisting fractures (Figure 11.16).

In addition to the erosion caused by wave impact and pressure, **abrasion,** the sawing and grinding action of the water armed with rock fragments, is also important. In fact, abrasion is probably more intense in the surf zone than in any other environment. Smooth and rounded stones and pebbles along the shore are obvious reminders of the grinding action of rock against rock in the surf zone. Further, such fragments are used as "tools" by the waves as they cut horizontally into the land.

Along shorelines composed of unconsolidated material rather than hard rock, the rate of erosion by

Figure 11.16
When waves break against the shore, the force of the water can be powerful and the erosional work that is accomplished can be great. (Photo by E. J. Tarbuck)

Figure 11.17
Wave bending around the end of a beach at Stinson Beach, California. (Photo by James E. Patterson)

breaking waves can be extraordinary. In parts of Britain, where waves have the easy task of eroding glacial deposits of sand, gravel, and clay, the coast has been worn back 3–5 kilometers (2–3 miles) since Roman times, sweeping away many villages and ancient landmarks. A similar retreat may be seen along the cliffs of Cape Cod, which in places are retreating at a rate of up to 1 meter (3 feet) per year.

WAVE REFRACTION AND LONGSHORE TRANSPORT

The bending of waves, called **wave refraction,** is an important factor when shoreline processes are considered (Figure 11.17). It is significant because it affects the distribution of energy along the shore, thus strongly influencing where and to what degree erosion, sediment transport, and deposition will take place.

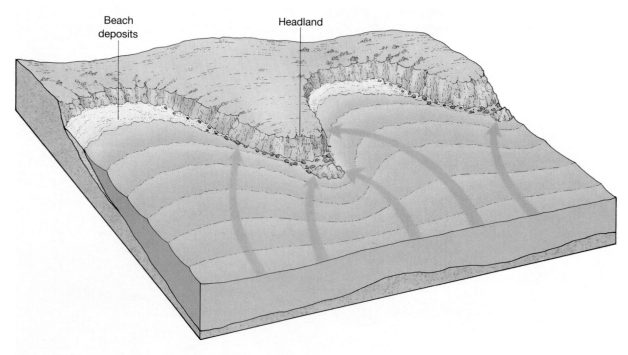

Figure 11.18
Because of wave refraction, the greatest erosional power is concentrated on the headlands. In the bays, the force of the waves is much weaker.

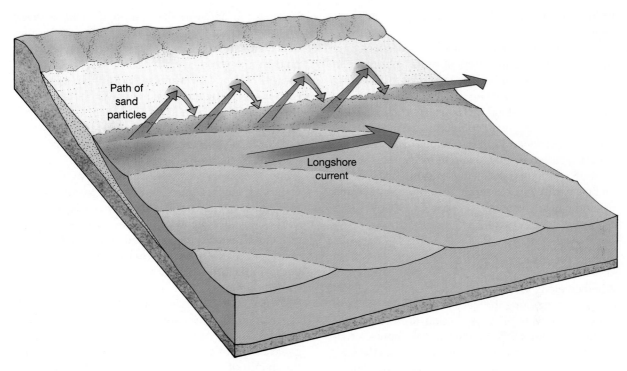

Figure 11.19
Beach drift and longshore currents are created by obliquely breaking waves. These processes transport large quantities of material along the beach and in the surf zone.

Waves seldom approach the shore straight on. Rather, most waves move toward the shore at an angle. When they reach the shallow water of a smoothly sloping bottom, however, they are bent and tend to become parallel to the shore. Such bending occurs because the part of the water nearest the shore touches bottom and slows first, while the end that is still in deep water continues forward at its regular speed. The net result is a wave front that may approach nearly parallel to the shore regardless of the original direction of the wave.

Due to refraction, wave impact is concentrated against the sides and ends of headlands projecting into the water, while wave attack is weakened in bays. This differential wave attack along irregular coastlines is illustrated in Figure 11.18. Since the waves reach the shallow water in front of the headland sooner than they do in adjacent bays, they are bent more nearly parallel to the protruding land and strike it from all three sides. By contrast, refraction in the bays causes waves to diverge and expend less energy. In these zones of weakened wave activity, sediments can accumulate and form sandy beaches. Over a long period, erosion of the headlands and de-

position in the bays will straighten an irregular shoreline.

Although waves are refracted, most still reach the shore at an angle, however slight. Consequently, the uprush of water from each breaking wave (the swash) is oblique. However, the backwash is in the direction of the slope of the beach. The effect of this pattern of water movement is to transport particles of sediment in a zigzag pattern along the beach (Figure 11.19). This movement, called **beach drift,** can transport sand and pebbles hundreds or even thousands of meters each day.

Oblique waves also produce currents within the surf zone that flow parallel to the shore. Since the water here is turbulent, these **longshore currents** easily move the fine suspended sand as well as roll larger sand and gravel along the bottom. When the sediment transported by longshore currents is added to the quantity moved by beach drift, the total amount can be very large. At Sandy Hook, New Jersey, for example, the quantity of sand transported along the shore over a 48-year period averaged almost 750,000 tons per year. For a 10-year period at Oxnard, California, more than

1.5 million tons of sediment moved along the shore each year.

There should be little wonder that beaches have been characterized as "rivers of sand." At any point along a beach there is likely to be more sediment that was derived elsewhere than material eroded from the cliff immediately behind it. It is also worth noting that much of the sediment composing beaches is not wave-eroded debris. Rather, in many areas sediment-laden rivers that discharge into the ocean are the major source of material. For that reason, if it were not for beach drift and longshore currents, many beaches would be nearly sandless.

SHORELINE FEATURES

Whether along the rugged and irregular New England coast or along the steep shorelines of the West Coast, the effects of wave erosion are often easily seen. **Wave-cut cliffs,** as their name implies, originate by the cutting action of the surf against the base of coastal land. As erosion progresses, rocks over-hanging the notch at the base of the cliff crumble into the surf and the cliff retreats. A relatively flat, benchlike surface, the **wave-cut platform,** is left behind by the receding cliff (Figure 11.20). The platform broadens as wave attack continues. Some debris produced by the breaking waves remains along the water's edge as part of the beach, while the remainder is transported farther seaward.

Headlands that extend into the sea are vigorously attacked by the waves because of refraction. The surf erodes the rock selectively, wearing away the softer or more highly fractured rock at the fastest rate. At first, sea caves may form. When two caves on opposite sides of a headland unite, a **sea arch** results (Figure 11.21). Finally the arch falls in, leaving an isolated remnant, or **sea stack,** on the wave-cut platform (Figure 11.21). Eventually it too will be consumed by the action of the waves.

Where beach drift and longshore currents are active, several features related to the movement of sediment along the shore may develop. **Spits** are elongated ridges of sand that project from the land into the mouth of an adjacent bay. Often the end in the

Figure 11.20
Elevated wave-cut platform along the California coast north of San Francisco. A new platform is being created at the base of the cliff. (Photo by John S. Shelton)

Figure 11.21
Sea arch and sea stack along the coast of Iceland. (Photo by Bruce F. Molnia, courtesy of Terraphotographics/BPS)

water hooks landward in response to wave-generated currents. The term **baymouth bar** is applied to a sand bar that completely crosses a bay, sealing it off from the open ocean. Such a feature tends to form across bays where currents are weak, allowing a spit to extend to the other side. A **tombolo,** a ridge of sand that connects an island to the mainland or to another island, forms in much the same manner as does a spit.

The Atlantic and Gulf Coastal Plains are relatively flat and slope gently seaward. The shore zone is characterized by **barrier islands.** These low ridges of sand parallel the coast at distances from 3 to 30 kilometers offshore. From Cape Cod, Massachusetts, to Padre Island, Texas, nearly 300 barrier islands rim the coast (Figure 11.22). Most are from 1 to 5 kilometers wide and between 15 and 30 kilometers long. The highest features are sand dunes, which usually reach heights of 5 to 10 meters; in a few areas unvegetated dunes are more than 30 meters high. The lagoons that separate these narrow islands from the shore represent zones of relatively quiet water that allow small craft traveling between New York and

northern Florida to avoid the rough waters of the North Atlantic.

How barrier islands originate is still not certain. They form possibly in three or more ways. Some are thought to have originated as spits that were subsequently severed from the mainland by wave erosion or by the general rise in sea level following the last episode of glaciation. It is also possible that some barrier islands are created when turbulent waters in the line of breakers heap up sand that has been scoured from the bottom. Since these sand barriers rise above normal sea level, the piling up of sand likely resulted from the work of storm waves at high tides. Finally, some studies suggest that barrier islands may be former sand dune ridges that originated along the shore during the last glacial period, when sea level was lower. As the ice sheets melted, sea level rose and flooded the area behind the beach-dune complex.

There is little question that a shoreline soon undergoes modification regardless of its initial configuration. At first most coastlines are irregular, although the degree of and reason for the irregularity may differ considerably from place to place. Along a coastline that is characterized by varied geology the pounding surf may at first increase its irregularity because the waves will erode the weaker rocks more easily than the stronger ones. Be that as it may, it is

Figure 11.22
This satellite image of New Jersey shows a large spit, known as Sandy Hook, extending north from the coast into the bay. Across the bay is New York City. Further south, barrier islands parallel the coast. Atlantic City is located on one of these islands. (LANDSAT image courtesy of Phillips Petroleum Company, Exploration Projects Section)

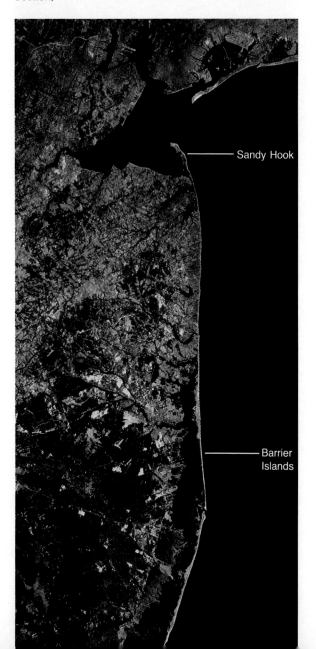

Sandy Hook

Barrier Islands

commonly agreed that if a shoreline remains stable, marine erosion and deposition will eventually produce a more regular coast. Figure 11.23 illustrates the evolution of an initially irregular coast. As waves erode the headlands, creating cliffs and a wave-cut platform, sediment is carried along the shore. Some material is deposited in the bays, while other debris is formed into spits and baymouth bars. At the same time rivers fill the bays with sediment; ultimately a smooth coast results.

SHORELINE EROSION PROBLEMS

Compared with other natural hazards such as earthquakes, volcanic eruptions, and landslides, shoreline erosion is often perceived to be a more continuous and predictable process that appears to cause relatively modest damage to limited areas. This is not always true. The shoreline is a dynamic place that can change rapidly in response to natural forces. Exceptional storms are capable of eroding beaches and cliffs at rates that are far in excess of the long-term average. Such bursts of accelerated erosion not only have a significant impact on the natural evolution of a coast, but can also have a profound impact on people who reside in the coastal zone. Erosion along our coasts has caused, and will continue to cause, significant property damage. Large sums are spent annually not only to repair damage, but also to prevent or control erosion. Already a problem at many sites, shoreline erosion is certain to become an increasingly serious problem as extensive development in coastal areas continues.

Although the same processes cause change along every coast, not all coasts respond in the same way. Interactions among different processes and the relative importance of each process depend upon local factors. These factors include: (1) the proximity of a coast to sediment-laden rivers, (2) the degree of tectonic activity, (3) the topography and composition of the land, (4) prevailing winds and weather patterns, and (5) the configuration of the coastline and nearshore areas.

The shoreline along the Pacific Coast of the United States is strikingly different from that characterizing the Atlantic and Gulf coast regions. Some of the differences are related to plate tectonics. The West Coast represents the leading edge of the North American plate and, because of this, it experiences active uplift and deformation. By contrast, the East Coast is a tec-

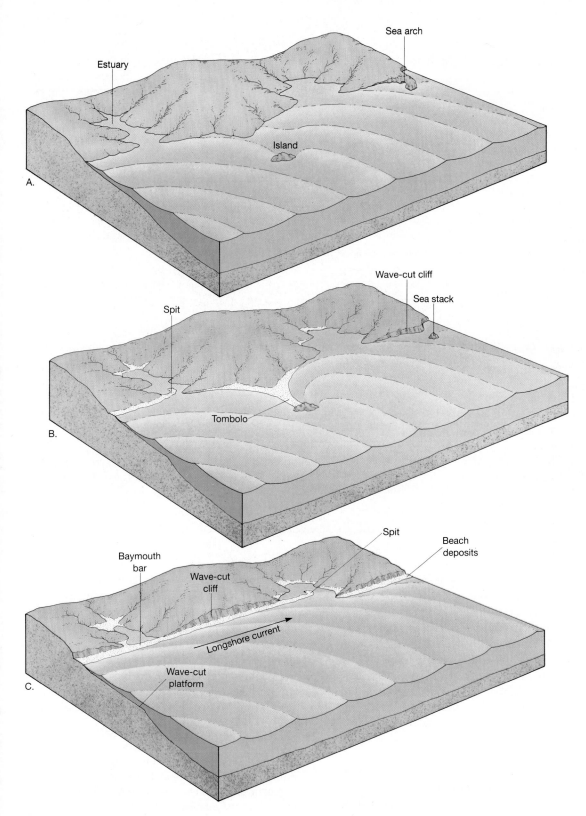

Figure 11.23
These diagrams illustrate the changes that can take place through time along an initially irregular coastline that remains relatively stable. The diagrams also serve to illustrate many of the features described in the section on shoreline features.

BOX 11.1

Ocean City: An Urbanized Barrier Island

Conflicts between human activities and natural systems frequently occur when the two come together in a coastal environment. One striking example of such conflict has occurred at Ocean City, Maryland.*

Ocean City has been a popular beach resort for more than a century. During the 1920s, several large hotels and a boardwalk were built to accommodate visitors. Development was modest but steady until the early 1950s, when a dramatic boom in construction began that lasted nearly 30 years. Early concerns about the coastal environment were raised in the late 1970s and led to federal and state laws to limit dredging and filling of wetlands.

The resort is built on the southern end of Fenwick Island, one of the chain of barrier islands stretching from New York to Florida (Figure 11.A). Ocean City Inlet, which connects the quiet bay on the mainland side of the island with the Atlantic Ocean at the southern end of the island, was opened during the great hurricane of 1933. To maintain the inlet as a navigation channel, two stone jetties were constructed by the U.S. Army Corps of Engineers shortly after the storm. The jetties have stabilized the inlet, but they have drastically altered the sand-transport process near the inlet. The net longshore drift at Ocean City is southerly; it has produced a wide beach at Ocean City north of the jetty, but Assateague Island, south of the inlet, has been starved of sediment. The result is a westerly offset of more than 500 meters in the once-straight barrier island (Figure 11.B).

The most damaging storm to hit Ocean City within historic times was the Ash Wednesday northeaster of early March 1962. It caused severe erosion and flooding along much of the middle Atlantic Coast. For two days, over five high-tide cycles, all of Fenwick Island except the highest dune areas was repeatedly washed over by storm waves superimposed on the 2-meter-high storm surge. Property damage to Ocean City alone was estimated at $7.5 million. Given the dense development of the island over the last 20 years, damage from a similar storm would today be hundreds of millions of dollars.

---

* This discussion is based on one in S. J. Williams et al., *Coasts in Crisis*, U.S. Geological Survey Circular 1075 (Washington, D.C., U.S. Government Printing Office, 1991).

tonically quiet region that is far from any active plate margin. Because of this basic geological difference, the nature of shoreline erosion problems along America's opposite coasts is different.

The discussion that follows will focus first on the problems faced along the East Coast of the United States, especially its vulnerable barrier islands. This will be contrasted with a look at the equally serious erosion problems occurring at many West Coast locations.

Gulf and Atlantic Coasts

During this century growing affluence and increasing demands for recreation have brought unprecedented development to many coastal areas. Much of this development has occurred on barrier islands. Typically, barrier islands consist of a wide beach that is backed by dunes and separated from the mainland by marshy lagoons. The broad expanses of sand and exposure to the ocean have made barrier islands exceedingly attractive sites for development. Unfortunately, development has taken place more rapidly than our understanding of barrier island dynamics.

Because barrier islands face the open ocean, they receive the full force of major storms that strike the coast. When a storm occurs, the barriers absorb the energy of the waves primarily through the movement of sand. Frank Lowenstein describes this process and the dilemma that results:

> Waves may move sand from the beach to offshore areas or, conversely, into the dunes; they may erode the dunes, depositing sand onto the beach or carrying it out to sea; or they may carry sand from the beach and the dunes into the marshes behind the barrier, a process known as overwash. The common factor is movement. Just as a flexible reed may survive a wind that destroys an oak tree, so the barriers survive hurricanes and

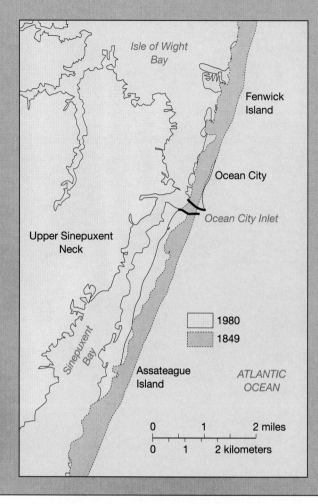

Figure 11.A
Shoreline changes in the Fenwick Island–Assateague Island region between 1849 and 1980 show the dramatic effects of the two large jetties at Ocean City Inlet on the natural sediment-transport processes along the coast. (After U.S. Geological Survey Circular 1075)

Isle of Wight Bay

Fenwick Island

Ocean City

Ocean City Inlet

Upper Sinepuxent Neck

Sinepuxent Bay

Assateague Island

ATLANTIC OCEAN

☐ 1980
☐ 1849

0   1   2 miles
0   1   2 kilometers

Figure 11.B
Oblique view of Ocean City Inlet (about 1980) showing the 500-meter offset between Fenwick Island (background) and Assateague Island (foreground) due to sediment starvation as a result of the construction of jetties in the 1930s. (Photo by S. J. Williams, U.S. Geological Survey)

nor'easters not through unyielding strength but by giving before the storm.

This picture changes when a barrier is developed for homes or a resort. Storm waves that previously rushed harmlessly through gaps between the dunes now encounter buildings and roadways. Moreover, since the dynamic nature of the barriers is readily perceived only during storms, homeowners tend to attribute damage to a particular storm, rather than to the basic mobility of coastal barriers. With their homes or investments at stake, local residents are more likely to seek to hold the sand in place and the waves at bay than to admit that development was improperly placed to begin with.*

Protecting property from storm-induced waves as well as from the ongoing movement of sand by longshore currents has been a significant concern in de-

veloped coastal areas for many years. Attempts at controlling the dynamic beach environment include building such artificial structures as jetties, breakwaters, groins, and seawalls. Such interference can create many new problems and result in unwanted changes that are difficult and expensive to correct.

From relatively early in America's history a principal goal in coastal areas was the development and maintenance of harbors. In many cases, this involved the construction of jetty systems. **Jetties** are usually built in pairs and extend into the ocean at the entrances to rivers and harbors. With the flow of water confined to a narrow zone, the ebb and flow caused by the rise and fall of the tides keep the sand in motion and prevent deposition in the channel. However, as illustrated in Figure 11.24, the jetty may act as a dam against which the longshore current and beach drift deposit sand. At the same time, wave activity removes sand on the other side. Since the other

---

* "Beaches or Bedrooms—The Choice as Sea Level Rises," *Oceanus* 28 (3) (Fall 1985): 22.

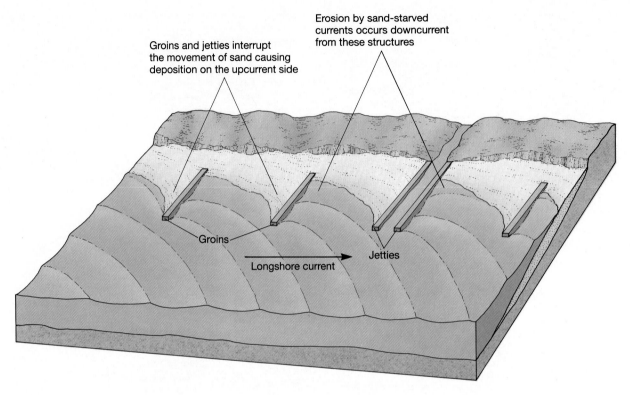

Groins and jetties interrupt the movement of sand causing deposition on the upcurrent side

Erosion by sand-starved currents occurs downcurrent from these structures

Groins

Longshore current

Jetties

Figure 11.24
Jetties and groins interrupt the movement of sand by beach drift and longshore currents. Beach erosion often results downcurrent from the site of the structure.

side is not receiving any new sand, there is soon no beach at all.

To maintain or widen beaches that are losing sand, **groins** are sometimes constructed. A groin is a barrier built at a right angle to the beach for the purpose of trapping sand that is moving parallel to the shore (Figure 11.24). The result is an irregular but wider beach. These structures often do their job so effectively that the longshore current beyond the groin is sand deficient. As a result, the current removes sand from the beach on the leeward side of the groin. Such a situation is illustrated in Figure 11.25. To offset this effect, property owners downcurrent from the structure may erect a groin on their property. In this manner, the number of groins multiplies. An example of such proliferation is the shoreline of New Jersey, where more than three hundred such structures have been built. Since it has been shown that groins often do not provide a satisfactory solution, they are no longer the preferred method of keeping beach erosion in check.

In some coastal areas a **breakwater** may be constructed parallel to the shoreline. The purpose of such a structure is to protect boats from the force of large breaking waves by creating a quiet-water zone near the shore. However, when this is done, the reduced wave activity along the shore behind the structure may allow sand to accumulate. If this happens, the marina will eventually fill with sand while the downstream beach erodes and retreats. At Santa Monica, California, where the building of a breakwater created such a problem, the city had to install a dredge to remove sand from the protected quiet-water zone and deposit it down the beach where longshore currents and beach drift could recirculate the sand (Figure 11.26).

As development has moved ever closer to the beach, seawalls represent yet another structure that is built to defend property from the force of breaking waves. **Seawalls** are simply massive barriers intended to prevent waves from reaching the areas behind the wall. Waves expend much of their energy as

Figure 11.25
A series of groins at Ship Bottom, New Jersey. Since groins trap sand on the upcurrent side, the movement of sand along this coast must be toward the bottom of the photograph. (Photo by John S. Shelton)

they move across an open beach. Seawalls cut this process short by reflecting the force of unspent waves seaward. As a consequence, the beach to the seaward side of the seawall experiences significant erosion and may, in some instances, be eliminated entirely. Once the width of the beach is reduced, the seawall is subjected to even greater pounding by the waves. Eventually this battering will cause the wall to fail and a larger, more expensive wall must be built to take its place.

In recent years the wisdom of building temporary protective structures has been questioned with greater frequency. The feelings of many coastal scientists is expressed in the following excerpt from a position paper that grew out of a conference on America's Eroding Shoreline:

A.

B.

Figure 11.26
A. *The shoreline at Santa Monica pier as it appeared in 1931.* *B.* *The same area in 1949. Construction of the breakwater disrupted longshore transport and caused the seaward growth of the beach. (Photos courtesy of Fairchild Aerial Photography Collection, Whittier College)*

It is now clear that halting the receding shoreline with protective structures benefits only a few and seriously degrades or destroys the natural beach and the value it holds for the majority. Protective structures divert the ocean's energy temporarily from private properties, but usually refocus that energy on the adjacent natural beaches. Many interrupt the natural sand flow in coastal currents, robbing many beaches of vital sand replacement.*

Of the various approaches to stabilizing the sands of barrier islands, **beach nourishment** is now considered the most acceptable method. As the term implies, this practice simply involves the addition of large quantities of sand to the beach system. By building the beaches seaward, beach quality and storm protection are both improved. The source of the sand may be the bottom of a nearby lagoon or inland dunes. In some cases, sand is trucked in and added to the beach. In other instances, the sand is added at an upstream location to be distributed down the coast by wave activity. Beach nourishment, however, is not a permanent solution to the problem of shrinking beaches. A case in point is Virginia Beach, Virginia, one of the largest seaside resort communities on the Atlantic Coast. Each year since 1952, this community has added the equivalent of 30,000 dump trucks' worth of sand along the shore in order to maintain just a thin strip of beach. The price is about $1.5 million per year. When beach nourishment was used to renew 24 kilometers of Miami Beach, the cost was $64 million. Furthermore, in some instances, beach nourishment can lead to unwanted environmental effects. For example, beach replenishment at Waikiki Beach, Hawaii, involved replacing coarse calcareous sand with softer, muddier calcareous sand. Destruction of the soft beach sand by breaking waves increased the water's turbidity and killed offshore coral reefs. At Miami Beach, where quartz sand was replaced by calcareous sand, the increased turbidity damaged local coral communities.

Beach nourishment appears to be an economically viable long-range solution to the beach preservation problem only in areas where there are dense development, large supplies of sand, relatively low wave energy, and reconcilable environmental issues. Unfortunately, few areas possess all of these attributes.

So far two basic responses to shoreline erosion problems have been considered: (1) the building of structures such as seawalls to hold the shoreline in place and (2) the addition of sand to replenish eroding beaches. However, a third option is also available, and that is to relocate buildings away from the beach. For some areas the costs of adding new sand to the beaches and/or building stronger and stronger seawalls may exceed the value of the property to be saved. When this point is reached, abandonment and relocation are the alternatives.

Pacific Coast

In contrast to the broad, gently sloping coastal plains of the East, much of the Pacific Coast is characterized by relatively narrow beaches that are backed by steep cliffs and mountain ranges. Recall that America's western margin is a more rugged and tectonically active region than the East. Since uplift continues, the apparent rise in sea level in the West is negligible or even nonexistent. Nevertheless, like the shoreline erosion problems facing the East's barrier islands, West Coast difficulties also stem largely from the alteration of a natural system by people.

A major problem facing the Pacific shoreline, and especially portions of southern California, is a significant narrowing of many beaches. The bulk of the sand on many of these beaches is supplied by rivers that transport it from the mountainous regions to the coast. Over the years this natural flow of material to the coast has been interrupted by dams built for irrigation and flood control. The reservoirs effectively trap the sand that would otherwise nourish the beach environment. When the beaches were wider, they served to protect the cliffs behind them from the force of storm waves. Now, however, the waves move across the narrowed beaches without losing much energy and cause more rapid erosion of the sea cliffs.

Although the retreat of the cliffs provides materials to replace some of the sand impounded behind dams, it also endangers homes and roads built on the bluffs. In addition, development atop the cliffs aggravates the problem. Urbanization increases runoff, which, if not carefully controlled, can result in serious bluff erosion. Watering lawns and gardens adds significant quantities of water to the slope. This water percolates downward toward the base of the cliff, where it may emerge in small seeps. This action reduces the slope's stability and facilitates mass wasting.

---

* "Strategy for Beach Preservation Proposed," *Geotimes* 30 (12) (December 1985): 15.

BOX 11.2
Sea Level Is Rising

The shifting, dynamic nature of barrier islands and the ineffectiveness of most shoreline protection measures are now relatively well-established facts. Unfortunately, recent research, which indicates that sea level is rising, has compounded this already distressing situation. Studies indicate that sea level has risen between 10 and 15 centimeters over the past century. Furthermore, some investigators predict an accelerated sea-level rise in the years to come—as much as 30 cm or more by the year 2025. Although such a vertical change may seem modest, many coastal geologists believe that any given rise in sea level along the gently sloping Atlantic and Gulf coasts will cause from 10 to 1000 times as much horizontal shoreline retreat (Figure 11.C).

The idea that sea level will continue to rise in the coming decades is linked to the results of climatic studies that predict a global warming trend. Such predictions are based upon the now well-established fact that the carbon dioxide (CO_2) content of the atmosphere has been rising at an accelerating rate for more than a century. The CO_2 is added primarily as a by-product of the combustion of ever-increasing quantities of fossil fuels. If we assume that the use of fossil fuels will continue to rise at pro-

jected rates, current estimates indicate that the atmosphere's CO_2 content will grow by an additional 40 percent or more by some time in the second half of the next century. The importance of CO_2 lies in the fact that it traps a portion of the radiation emitted by the earth and thereby keeps the air near the earth's surface warmer than it would be without CO_2. Because CO_2 is an important heat-absorbing gas; it follows that an increase in the air's CO_2 content should lead to higher atmospheric temperatures. In addition, other trace gases generated by human activities also play a role.*

How is a warmer atmosphere related to a global rise in sea level? First, higher temperatures can cause glacial ice to melt. About one-half of the 10- to 15-centimeter rise in sea level over the past century is attributed to the melting of small glaciers and ice caps. Second, a warmer atmosphere causes an increase in ocean volume through thermal

expansion. That is, higher air temperatures raise the temperature of the upper layers of the ocean, which, in turn, causes the water to expand and sea level to rise. It is also believed that a warmer ocean may spur storm development. Of course, an increase in storm activity would compound an already serious problem in many coastal areas.

Since rising sea level is a gradual phenomenon, it may be overlooked by coastal residents as a significant contributor to shoreline erosion problems. Rather, the blame is assigned to other forces, especially storm activity. Although a given storm may be the immediate cause, the magnitude of its destruction may result from the relatively small sea level rise that allowed the storm's power to cross a much greater land area.

---

* The impact of rising levels of CO_2 and other trace gases on global temperatures is treated in considerable detail in Chapter 16.

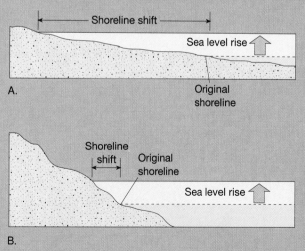

Figure 11.C
The slope of a shoreline is critical to determining the degree to which sea-level changes will affect it. **A.** *When the slope is gentle, small changes in sea level cause a substantial shift.* **B.** *The same sea-level rise along a steep coast results in only a small shoreline shift.*

Figure 11.27
Satellite image of a portion of the East Coast showing Chesapeake Bay, an estuary created when the lower portion of a river valley was submerged by the rise in sea level that followed the end of the Ice Age. (Courtesy of Earth Satellite Corporation)

Shoreline erosion along the Pacific Coast varies considerably from one year to the next, largely because of the sporadic occurrence of storms. As a consequence, when the infrequent but serious episodes of erosion occur, the damage is often blamed on the unusual storms and not on coastal development or the sediment-trapping dams that may be great distances away. If, as predicted, sea level rises at an increasing rate in the years to come, increased shoreline erosion and sea cliff retreat should be expected along many parts of the Pacific Coast.

EMERGENT AND SUBMERGENT COASTS

The great variety of present-day shorelines suggests that they are complex areas. Indeed, to understand the nature of any particular coastal area, many factors must be considered, including rock types, size and direction of waves, number of storms, tidal range, and submarine profile. Moreover, recent tectonic events and changes in sea level must also be taken into account. These many variables make shoreline classification difficult.

One way that many geologists classify coasts is based upon changes that have occurred with respect to sea level. This commonly used, although incomplete, classification divides coasts into two categories: emergent and submergent. **Emergent coasts** develop either because an area has been uplifted or as a result of a drop in sea level. Conversely, **submergent coasts** are created when sea level rises or the land adjacent to the sea subsides.

In some areas the coast is clearly emergent because rising land or a falling water level expose wave-cut cliffs and platforms above sea level. Excellent examples include portions of coastal California where uplift has occurred in the recent geological past. The

elevated wave-cut platform shown in Figure 11.20 illustrates this. In the case of the Palos Verdes Hills, south of Los Angeles, seven different terrace levels exist, indicating seven episodes of uplift. The ever-persistent sea is now cutting a new platform at the base of the cliff. If uplift follows, it too will become an elevated marine terrace.

Other examples of emergent coasts include regions that were once buried beneath great ice sheets. When glaciers were present, their weight depressed the crust and when the ice melted, the crust began to gradually spring back. As a result, prehistoric shoreline features today are found high above sea level. The Hudson Bay region of Canada is one such area, portions of which are still rising at a rate of more than one centimeter per year.

In contrast to the preceding examples, other coastal areas show definite signs of submergence. The shoreline of a coast that has been submerged in the relatively recent past is often highly irregular because the sea typically floods the lower reaches of river valleys flowing into the ocean. The ridges separating the valleys, however, remain above sea level and project into the sea as headlands. These drowned river mouths, which are often called **estuaries,** characterize many coasts today. Along the Atlantic coast, the Chesapeake and Delaware bays are examples of estuaries created by submergence (Figure 11.27). The picturesque coast of Maine, particularly in the vicinity of Acadia National Park, is another excellent example of an area that was flooded by the post-glacial rise in sea level and transformed into a highly irregular submerged coastline.

It should be kept in mind that most coasts have a rather complicated geologic history. With respect to sea level, many have at various times emerged and then submerged again. Each time they retain some of the features created during the previous situation.

Review Questions

1. What is the primary driving force of surface ocean currents? How does the Coriolis effect influence these currents?

2. Briefly describe the North Atlantic gyre. How does the surface circulation of the South Atlantic and Indian oceans differ from that of the North Atlantic?

3. Describe the process of coastal upwelling.

4. How do ocean currents influence climate?

5. What is meant by *thermohaline circulation*?

6. Discuss the origin of ocean tides.

7. Explain why an observer can experience two unequal high tides during a day (see Figure 11.7).

8. How does the sun influence tides?

9. How do semidiurnal, diurnal, and mixed tides differ?

10. Distinguish between flood current and ebb current.

11. What advantages does tidal power production offer? Is it likely that tides will provide a significant proportion of the world's electrical energy needs in the foreseeable future?

12. List three factors that determine the height, length, and period of a wave.

13. Describe the motion of a water particle as a wave passes (see Figure 11.14).

14. Explain what happens when a wave breaks.

15. How do waves cause erosion?

16. What is wave refraction? What is the effect of this process along irregular coastlines?

17. Why are beaches often called "rivers of sand"?

18. Describe the formation of the following features: wave-cut cliff, wave-cut platform, sea stack, spit, baymouth bar, tombolo.

19. List three possible ways that barrier islands form.

20. For what purpose is a groin built? Why might the building of one groin lead to the building of others?

21. How did the building of jetties at Ocean City, Maryland, affect the areas north and south of the inlet? (See Box 11.1).

22. How was the beach at Santa Monica, California, affected when a breakwater was constructed (see Figure 11.26)?

23. How might a seawall increase beach erosion?

24. It is believed that global temperatures will be increasing in the decades to come. How can a warmer atmosphere lead to rise in sea level? (See Box 11.2).

25. Relate the damming of rivers to the shrinking of beaches at some locations along the West Coast of the United States.

26. What observable features would lead you to classify a coastal area as emergent?

27. Are estuaries associated with submergent or emergent coasts? Why?

Key Terms

abrasion (p. 395)
barrier island (p. 399)
baymouth bar (p. 399)
beach drift (p. 397)
beach nourishment (p. 406)
breakwater (p. 404)
Coriolis effect (p. 381)
ebb current (p. 388)
emergent coast (p. 409)
estuary (p. 409)
fetch (p. 391)
flood current (p. 388)
groin (p. 404)
gyre (p. 380)
jetty (p. 403)
longshore current (p. 397)
neap tide (p. 387)
sea arch (p. 398)
sea stack (p. 398)
seawall (p. 404)

spit (p. 398)
spring tide (p. 387)
submergent coast (p. 409)
surf (p. 395)
thermohaline circulation (p. 385)
tidal current (p. 388)
tidal delta (p. 388)
tidal flat (p. 388)
tide (p. 386)
tombolo (p. 399)
upwelling (p. 382)
wave-cut cliff (p. 398)
wave-cut platform (p. 398)
wave height (p. 391)
wavelength (p. 391)
wave of oscillation (p. 393)
wave of translation (p. 394)
wave period (p. 391)
wave refraction (p. 396)

PART 3

The Atmosphere

Thunderstorm over Tucson, Arizona. (Photo by Warren Faidley/International Stock)

12

Composition, Structure, and Temperature

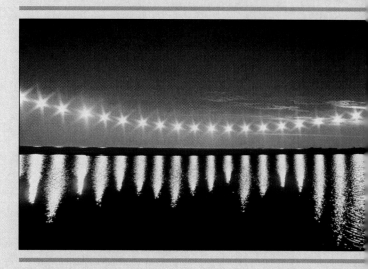

Opposite page: *Winter in Gile State Forest, New Hampshire. (Photo by Tom Till)*

Left: *Midnight sun in Alaska during the summer solstice. (Photo by Brian Stablyk/Allstock)*

Above: *Midnight sun in Alaska during the summer solstice. (Photo by Brian Stablyk/Allstock)*

*T*he study of the atmosphere — the earth's gaseous envelope — goes back thousands of years. Present knowledge suggests that no other planet has the exact mixture of gases or the heat and moisture conditions necessary to sustain life as we know it. The gases that make up the earth's atmosphere and the controls to which they are subject are vital to our existence. In this chapter we begin our examination of the ocean of air in which we all must live. We shall attempt to answer a number of basic questions. What is the composition of the atmosphere? At what point do we leave the atmosphere and enter outer space? What causes the seasons? How is air heated? What factors control temperature variations over the globe?

People's interest in the atmosphere is probably as old as the history of humankind. **Meteorology**, the science of the atmosphere and its weather, goes back many centuries. The term is derived from the title of Aristotle's four-volume treatise, *Meteorologica*, literally "discourse on things above." The study of meteorology has progressed through many phases over the centuries, evolving from the days when weather lore and superstition guided explanations to what it is today—a scientific study using the most modern instruments and equipment.

Certainly there are few other aspects of our physical environment that affect our daily lives more than the phenomena we collectively call the weather (Figure 12.1). The clothes we wear and the activities we engage in are strongly influenced by the weather. Yet in addition to the day-to-day personal decisions we must make about the weather, there have been, and will continue to be, an increasing number of political decisions to make involving the atmosphere. Answers to questions regarding air pollution and its control, the effects of various emissions on global climate and the atmosphere's ozone layer, and the positive or adverse effects of intentional weather modification are all important and, in some cases, perhaps even vital. So there is a need for increased awareness and understanding of our atmosphere and its behavior.

WEATHER AND CLIMATE

If you were to search through the library for books about the atmosphere, many of the titles you would find would contain the term *weather*, while others would have the word *climate*. What is the difference between these two terms? **Weather** is a word used to denote the state of the atmosphere at a particular place for a short period of time. Weather is constantly changing—hourly, daily, and seasonally. **Climate**, on the other hand, might best be described as an aggregate or composite of weather. Stated another way, the climate of a place or region is a generalization of the weather conditions over a long period of time. Therefore, a climatic description is possible only after weather records have been kept for many years.

Although weather and climate are not identical, the nature of both is expressed in terms of the same **elements**, those quantities or properties that are measured regularly. The most important are (1) air temperature, (2) humidity, (3) type and amount of

Figure 12.1
A snowstorm can quickly create slippery, snow-packed highways. Probably no other aspect of our physical environment affects the daily lives of people more than the weather. (Photo by J. L. Atlan/SYGMA)

cloudiness, (4) type and amount of precipitation, (5) air pressure, and (6) the speed and direction of the wind. These elements constitute the variables from which weather patterns and climate types are deciphered. Although we shall study these elements separately at first, keep in mind that they are very much interrelated. A change in any one of the elements will often bring about changes in the others.

COMPOSITION OF THE ATMOSPHERE

In the days of Aristotle all things were thought to be a combination of four fundamental substances—fire, air, earth (soil), and water. Today we know that matter is much more complex. For example, the earth is composed of minerals, which, in turn, are made of

numerous elements and compounds. Furthermore, air, which appears to be a unique substance, is really a mixture of many different gases and tiny solid particles.

The composition of air is not constant; it varies from time to time and from place to place. This variability is readily observable when we drive from the rural countryside into the city and see increases in dust, smoke, and sometimes cloud cover. Be that as it may, if the water vapor, dust, and other variable components were removed from the atmosphere, we would find that its composition is quite uniform. Clean, dry air sampled anywhere on earth is composed almost entirely of two gases—nitrogen and oxygen (Figure 12.2). Nitrogen makes up about 78 percent of the atmosphere, and oxygen represents about 21 percent. Although these gases are the most plentiful components of air and are of great significance to life on earth, they are of minor importance in affecting weather phenomena. The remaining percentage of dry air is mostly the inert gas argon (0.93 percent) plus tiny quantities of a number of other gases. Carbon dioxide, although present in only minute amounts (0.035 percent), is nevertheless an important constituent of air, because it has the ability to absorb heat energy radiated by the earth and thus helps keep the atmosphere warm.

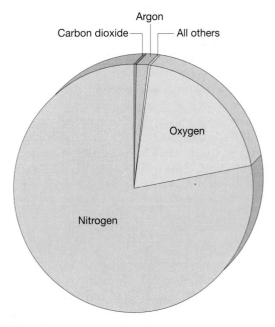

Figure 12.2
Proportional volume of gases composing dry air. Nitrogen and oxygen clearly dominate.

If meteorologists had to choose the most significant component of the atmosphere, they would undoubtedly select water vapor. The amount of water vapor in the air varies considerably, from practically none at all up to about 4 percent by volume. Why is such a small fraction of the atmosphere so significant? Certainly the fact that water vapor is the source of all clouds and precipitation would be enough to justify the choice. However, water vapor has other roles. Like carbon dioxide, it has the ability to absorb heat energy given off by the earth as well as some solar energy. It is therefore an important factor to consider when we examine the heating of the atmosphere. When water changes from one state to another (see Figure 13.2), it absorbs or releases heat energy (termed *latent heat*). As we shall see in later chapters, latent heat plays an important role in the transport of heat from one region to another, and it is the energy source that helps drive many storms.

Most of us probably think of dust as small, barely visible bits of dirt. Nevertheless, from a meteorological standpoint, dust is much more than that, for it includes many microscopic particles that are invisible to the naked eye, among them organic materials like pollen, spores, and seed. Dust particles are most numerous in the lower part of the atmosphere near their primary source, the earth's surface. Still, the upper atmosphere is not free of them, because some dust is carried to great heights by rising currents of air, while other particles are given to the upper atmosphere by meteors that disintegrate as they pass through the earth's envelope of air. Some particles act as surfaces upon which water vapor may condense. This function is very basic to the formation of clouds and fog. In addition, dust may intercept and reflect incoming solar radiation. Thus when the dust content of the atmosphere is high, as it may be following an explosive volcanic eruption, the amount of sunlight reaching the earth's surface can be measurably reduced. Finally, dust in the air contributes to a phenomenon we all have observed—the red and orange colors of sunrise and sunset.

Another important component of the atmosphere is *ozone* (O_3), the triatomic form of oxygen. Ozone is not the same as oxygen we breathe, which has two atoms per molecule (O_2). There is very little of this gas in the atmosphere. If all of the ozone in the atmosphere were brought down to the earth's surface, it would form a layer only about 4 millimeters thick. Furthermore, its distribution is not uniform. In the lowest portion of the atmosphere ozone represents

less than one part in 100 million. It is concentrated well above the surface between 10 and 50 kilometers (6 and 31 miles) with a peak near the altitude of 25 kilometers (Figure 12.3). In this altitude range, oxygen molecules (O_2) are split into single atoms of oxygen after absorbing ultraviolet radiation emitted by the sun. Ozone is then created when an atom of oxygen (O) and a molecule of oxygen (O_2) collide in the presence of a third, neutral molecule that acts as a catalyst by allowing the reaction to take place without itself being consumed in the process. Ozone is concentrated in the 10- to 50-kilometer height range because it is there that a crucial balance exists: The availability of ultraviolet radiation from the sun is in sufficient amounts to produce atomic oxygen, and

Figure 12.3
Giant balloons such as this one are filled with helium and carry instrument packages high into the atmosphere. They provide important data about the region of the atmosphere, known as the stratosphere, in which ozone is concentrated. (Courtesy of National Center for Atmospheric Research/National Science Foundation)

The ozone layer is vital to life on earth as we know it. It filters out most of the biologically damaging ultraviolet rays of the sun that would otherwise reach the earth's surface. Even so, enough of those rays sneak through the layer to cause us

grief if we overexpose ourselves to the midday sun—especially those of us who have pale, sensitive skin. More than just a painful burn, exposure to the sun can lead to an increased risk of skin cancer later in life.

To play it safe, physicians advise us to avoid direct exposure to the sun's rays between the hours of 11 A.M. and 3 P.M., when the rays are most intense. But that involves the nuisance of clock-watching. Besides, depending on your location, the sun may be sufficiently ahead of (or behind) your clock time that a somewhat different range of hours would be more appropriate for sun avoidance.

For an easier way than clock-watching, look at the shadows

around you when you're in the sun. If the shadows are shorter than the heights of the objects that cast them (including your body), then it's wise to seek shade—or to protect yourself with proper clothing or a good sunscreen lotion.

Conversely, when shadows are longer than the heights of the objects that cast them, your skin is relatively safe in the sun. The difference is caused by the ozone layer: when the sun's rays penetrate it at a glancing angle (long shadows on the ground), much less skin-damaging ultraviolet light can pass through the layer than when the sun is more nearly overhead (short shadows on the ground).

there is an atmospheric density that is great enough to bring about the required collisions between atomic and molecular oxygen.

The presence of the ozone layer in our atmosphere is of vital importance to those of us on earth. The reason lies in the capability of ozone to absorb the potentially harmful ultraviolet radiation from the sun. If ozone did not act to filter a great deal of the ultraviolet radiation and if the rays were allowed to reach the surface of the earth, our planet would likely be uninhabitable for most life as we know it. Thus, anything that would act to reduce the amount of ozone in the atmosphere could affect the well-being of life on earth. Just such a concern has been raised.

THE OZONE PROBLEM

Since the early 1970s there has been continuing public concern over the possibility that the amount of ozone in the atmosphere is being reduced as a result of human activities. The greatest human impact on the ozone layer is from a group of chemicals known

as chlorofluorocarbons (abbreviated CFC).* CFCs are used as propellants for aerosol sprays, in the production of certain plastics, as cleaning solvents, and in air conditioning and refrigeration equipment (Figure 12.4). Due to the fact that CFCs are practically inert (that is, not chemically active) in the lower atmosphere, a portion of these gases gradually makes its way upward to the ozone layer, where sunlight separates the chemicals into their constituent atoms. The chlorine atoms released in this way, by a complex series of reactions, have the net effect of destroying ozone molecules. Since ozone filters out most of the damaging ultraviolet radiation in sunlight, a decrease in ozone concentration would permit more of these harmful wavelengths to reach the earth's surface.

Although ozone depletion by CFCs occurs worldwide, measurements have shown that ozone concentrations take an especially sharp drop over

---

* CFCs are also called chlorofluoromethanes (CFMs) or halocarbons, and go by the trade name Freons.

Figure 12.4
Refrigerant leakage from the air conditioning units in these junkyard vehicles is one way that chlorofluorocarbons are released into the atmosphere. (Photo by E. J. Tarbuck)

Antarctica during the Southern Hemisphere spring (September and October). Later, during November and December, the ozone concentration rebounds to more-normal levels. This well-publicized *ozone hole* has intensified and grown larger since its discovery in 1985 (Figure 12.5). The hole is caused in part by the relatively abundant ice particles in the south polar stratosphere. The ice boosts the effectiveness of CFCs in destroying ozone, causing a greater decline than would otherwise occur. The zone of maximum depletion is confined to the Antarctic region by a swirling upper-level wind pattern. When this vortex weakens during the late spring, the ozone-depleted air is no longer restricted and mixes freely with the air from other latitudes where ozone levels are higher. This equatorward movement and mixing contributes to localized decreases in ozone concentrations at lower latitudes.

Each 1-percent decrease in the concentration of stratospheric ozone increases the amount of ultraviolet radiation that reaches the earth's surface by about 2 percent. Therefore, because ultraviolet radiation is known to induce certain types of skin cancer, ozone depletion could seriously affect human

Figure 12.5
This image shows ozone distribution for the Southern Hemisphere for the month of October 1992. The area of greatest depletion, called the ozone ``hole,'' forms over Antarctica during the Southern Hemisphere spring and is shown in shades of pink and dark red. In October 1992, ozone levels reached a minimum of about 120 Dobson units, far below the 220 Dobson units typically seen over Antarctica before the hole forms. The image was produced with data acquired by the Total Ozone Mapping Spectrometer (TOMS), an instrument on board NASA's NIMBUS-7 satellite. (Courtesy of Arlin J. Krueger, NASA/Goddard Space Flight Center)

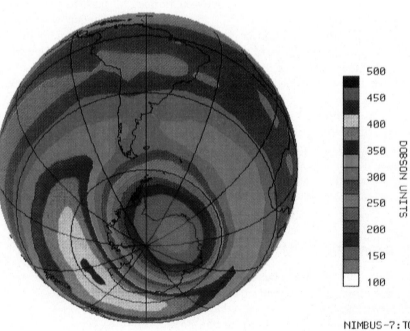

OCTOBER 1992

NIMBUS-7:TOMS
TOTAL OZONE
NASA/GSFC

health, especially among fair-skinned people and those who spend considerable time in the sun. Since up to one-half million cases of these cancers occur in the United States annually, ozone depletion could ultimately lead to many thousands of additional cases each year. The effects of increased ultraviolet radiation on animal and plant life may be as important as the direct effects on human health. Still, too little experimental data exist for scientists to make specific predictions about effects on particular crops or ecosystems. Nevertheless, there is concern that crop yields and quality could be adversely affected. Some scientists also fear that increased ultraviolet radiation in the Antarctic will penetrate the waters surrounding the continent and impair or destroy the microscopic plants, called phytoplankton, that represent the base of the food chain. A decrease in phytoplankton could, in turn, reduce the populations of copepods and krill that sustain fish, whales, penguins, and other marine life in the high latitudes of the Southern Hemisphere.

What can be done to protect the atmospheric ozone layer? The obvious solution is to curtail the production of CFCs. In the 1970s, public concern about ozone depletion prompted the United States, Canada, Norway, and Sweden to ban the use of CFC propellants from most aerosol products. Later, other European countries also limited the CFCs used as propellants. While helpful, these actions were not an adequate solution because not all nations participated, and production for other purposes, such as refrigeration, air conditioning, and solvents, continued to climb. Realizing that the risks of not curbing CFC emissions were difficult to ignore, an international agreement known as the Montreal Protocol was finally concluded under the auspices of the United Nations in late 1987. The treaty specified a 50 percent reduction in CFC production by the end of the century. More than 40 nations eventually endorsed the proposal. The treaty represented a positive international response to the ozone problem. However, subsequent evidence showed that atmospheric ozone levels were dropping more rapidly than had been predicted. Consequently, a stronger response was necessary. In June 1990, a new agreement was reached. The updated protocol calls for a complete phaseout of CFCs by the year 2000, following a 50 percent cut by 1995 and an 85 percent reduction by 1997. Although some nations called for more rapid reductions, this agreement was nevertheless viewed by many as a major step forward in environmental diplomacy.

EXTENT AND STRUCTURE OF THE ATMOSPHERE

To say that the atmosphere begins at the earth's land-sea surface and extends upward is obvious. However, where does the atmosphere end and outer space begin? To help understand the vertical extent of the atmosphere, let us examine the changes in atmospheric pressure with height. Atmospheric pressure is simply the weight of the air above. At sea level, the average pressure is slightly more than 1000 millibars, a pressure that corresponds to a weight of slightly more than 1 kilogram per square centimeter (14.7 pounds per square inch). Obviously the pressure at higher altitudes is less (Figure 12.6). This fact was demonstrated in a dramatic way by some nineteenth-century balloonists who were attempting to study the upper atmosphere. Several lost consciousness upon reaching heights in excess of 6 kilometers; some even perished in the rarefied upper air.

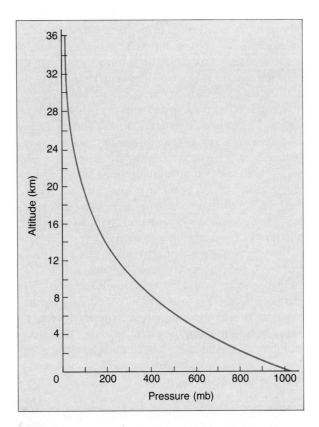

Figure 12.6
Pressure variations with altitude. The rate of pressure decrease with an increase in altitude is not constant. Rather, pressure decreases rapidly near the earth's surface and more gradually at greater heights.

Table 12.1
Percent of sea level pressure encountered at selected altitudes.

Altitude (kilometers)	Altitude (miles)	Percent of Sea Level Pressure
0	0	100
5.6	3.5	50
16.2	10	10
31.2	19.3	1
48.1	29.8	0.1
65.1	40.4	0.01
79.2	49.1	0.001
100	62	0.00003

By examining Table 12.1 you can see that one-half of the atmosphere lies below an altitude of 5.6 kilometers (3.5 miles). At about 16 kilometers (10 miles), 90 percent of the atmosphere has been traversed, and above 100 kilometers (62 miles), only 0.00003 percent of all the gases composing the atmosphere remains. At this latter altitude, the atmosphere is so tenuous that the density of air is less than can be found in the best artificial vacuum at the earth's surface. Even so, traces of our atmosphere extend far beyond this altitude. Thus, to say where the atmosphere ends and outer space begins is quite arbitrary and to a large extent depends upon what phenomena we are studying. There is no definitive, sharp boundary. Certainly the data on vertical pressure changes reveal that the vast bulk of the gases composing the atmosphere are very near the earth's surface and that they gradually merge with the emptiness of space.

By the early part of the twentieth century, much had been learned about the lower atmosphere, and indirect methods had provided some knowledge of the upper air. Data from balloons and kites had revealed that the air temperature dropped with increasing height above the earth's surface. This fact is familiar to anyone who has climbed a high mountain or who has seen pictures of snowcapped mountaintops rising above snow-free lowlands (Figure 12.7).

The atmosphere is divided vertically into four layers on the basis of temperature (Figure 12.8). The bottom layer, where temperature decreases with an increase in altitude, is known as the **troposphere**. The term literally means the region where air "turns over," a reference to the appreciable vertical mixing of the air in this lowermost zone. The temperature decrease in the troposphere is called the **environmental lapse rate**. Although its average value is 6.5°C per kilometer (3.5°F per 1000 feet), a figure known as the *normal lapse rate*, its value is quite variable. Thus, to determine the environmental lapse rate for any particular time and place, as well as gather information about vertical changes in pressure, wind, and humidity, radiosondes are used. The radiosonde is an instrument package that is attached to a balloon and transmits data by radio as it ascends through the atmosphere (Figure 12.9). The thickness of the troposphere is not the same everywhere; it varies with latitude and the season. On the average, the temperature drop continues to a height of about 12 kilometers (7.4 miles).

Figure 12.7
Temperatures drop with an increase in altitude in the troposphere. Therefore it is possible to have snow on a mountaintop and warmer, snow-free lowlands below. (Photo by E. J. Tarbuck)

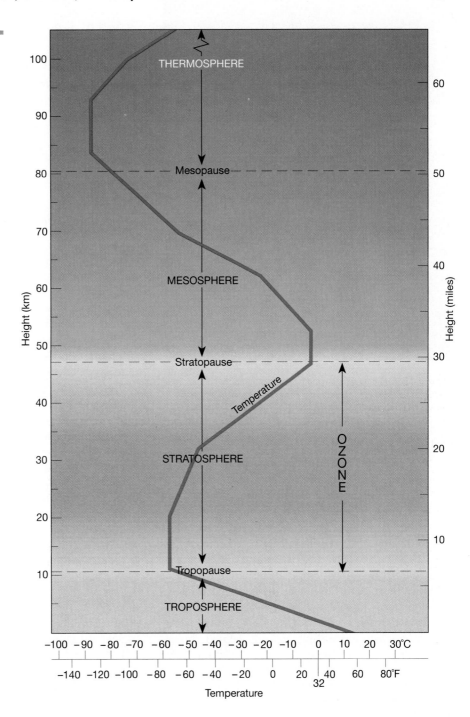

Figure 12.8
Thermal structure of the atmosphere to a height of about 110 kilometers (68 miles).

The troposphere is the chief focus of meteorologists, for it is in this layer that essentially all phenomena that we collectively refer to as weather occur. Almost all clouds and certainly all precipitation, as well as all our violent storms, are born in this lowermost layer of the atmosphere. There should be little wonder why the troposphere is often called the "weather sphere."

Beyond the troposphere lies the **stratosphere**, and the boundary between them is called the **tropopause**. In the stratosphere, the temperature remains constant to a height of about 20 kilometers and then begins a gradual increase that continues until the **stratopause**, at a height of about 50 kilometers above the earth's surface. Below the tropopause, atmospheric properties are readily transferred by

Figure 12.9
Atmospheric soundings using radiosondes supply data on vertical changes in temperature, pressure, and humidity. (Courtesy of NOAA)

large-scale turbulence and mixing; above it, in the stratosphere, they are not. The reason for the increased temperatures in the stratosphere is that the atmosphere's ozone is concentrated in this layer. Recall that ozone absorbs ultraviolet radiation from the sun. As a consequence, the stratosphere is heated.

In the **mesosphere** temperatures again decrease with height until, at the **mesopause**, approximately 80 kilometers (50 miles) above the surface, the temperature approaches −90°C. Extending upward from the mesopause and having no well-defined upper limit is the **thermosphere**, a layer that contains only a minute fraction of the atmosphere's mass. In the extremely rarefied air of this outermost layer, temperatures again increase due to the absorption of very short wave solar energy by atoms of oxygen and nitrogen. While temperatures rise to extremely high values of more than 1000°C, such temperatures are not strictly comparable to those experienced near the earth's surface. Temperature is defined in terms of the average speed at which molecules move. Since the gases of the thermosphere are moving at very high speeds, the temperature is obviously very high. Even

so, the gases are so sparse that only a very few of these fast-moving air molecules collide with a foreign body, and hence only an insignificant quantity of energy is transferred. For this reason, the temperature of a satellite orbiting the earth in the thermosphere is determined chiefly by the amount of solar radiation it absorbs and not by the temperature of the surrounding air. If an astronaut inside were to expose his or her hand, it would not feel hot.

EARTH-SUN RELATIONSHIPS

The earth intercepts only a minute percentage of the energy given off by the sun—less than one two-billionth. This may seem to be an insignificant amount until we realize that it is several hundred thousand times the electrical-generating capacity of the United States. Energy from the sun is undoubtedly the most important control of our weather and climate. Solar radiation represents more than 99.9 percent of the energy that heats the earth. Solar energy is not distributed evenly over the earth's surface, however. It is this unequal heating that drives the ocean's currents and creates the winds, which, in turn, transport heat from the tropics to the poles in an unending attempt to reach an energy balance. If the sun were to be "turned off," global winds would quickly subside. Yet as long as the sun shines, the winds will blow and the phenomena we know as weather will persist. Therefore, in order to have a basic understanding of atmospheric processes, we must understand what causes the time and space variations in the amount of solar energy reaching the earth.

Motions of the Earth

The earth has two principal motions—rotation and revolution. **Rotation** is the spinning of the earth about its axis, which is an imaginary line running through the poles. Our planet rotates once every 24 hours, producing the daily cycle of daylight and darkness. Half the earth is always experiencing daylight, and the other half darkness. The line separating the dark half of the earth from the lighted half is called the **circle of illumination**. **Revolution** refers to the movement of the earth in its orbit around the sun. Hundreds of years ago, most people believed that the earth was stationary in space. We now know that the earth is traveling at more than 107,000 kilometers per hour in an elliptical orbit about the sun.

The distance between the earth and sun averages about 150 million kilometers (93 million miles). Because the earth's orbit is not perfectly circular, the distance varies during the course of a year. Each year, on about January 3, our plant is 147 million kilometers from the sun, closer than at any other time. This position is called **perihelion.** About six months later, on July 4, the earth is 152 million kilometers from the sun, farther away than at any other time. This position is called **aphelion**. Variations in the amount of solar radiation received by the earth, as the result of its slightly elliptical orbit, are nevertheless slight and of little consequence when explaining major seasonal temperature variations. By way of illustration, consider that the earth is closest to the sun during the Northern Hemisphere winter.

The Seasons

We know that it is colder in winter than in summer, but if the variations in solar distance do not cause this, what does? All of us have made adjustments for the continuous change in the length of daylight that occurs throughout the year by planning our outdoor activities accordingly. The gradual but significant change in length of daylight certainly accounts for some of the difference we notice between summer and winter. Further, a gradual change in the **altitude** (angle above the horizon) of the noon sun during the course of a year's time is evident to most people. At mid-summer the sun is seen high above the horizon as it makes its daily journey across the sky. But as summer gives way to autumn, the noon sun appears lower and lower in the sky and sunset occurs earlier each evening.

The seasonal variation in the altitude of the sun affects the amount of energy received at the earth's surface in two ways. First, when the sun is directly overhead (90-degree angle), the solar rays are most concentrated. The lower the angle, the more spread out and less intense is the solar radiation that reaches the surface. This idea is illustrated in Figure 12.10. Second, and of lesser importance, the angle of the sun determines the amount of atmosphere the rays must traverse (Figure 12.11). When the sun is directly overhead, the rays pass through a thickness of only 1 atmosphere, while rays entering a 30-degree angle travel through twice this amount, and 5-degree rays travel through a thickness roughly equal to 11 atmospheres. The longer the path, the greater are the chances for absorption, reflection, and scattering by the atmosphere, which reduce the intensity at the surface. The same effects account for the fact that the midday sun can be literally blinding, while the setting sun can be a sight to behold.

Figure 12.10
Changes in the sun angle cause variations in the amount of solar energy reaching the earth's surface. The higher the angle, the more intense the solar radiation. Notice that one unit of solar energy striking at a 30° angle spreads over twice as much area as that striking at a 90° angle.

Figure 12.11
Rays striking at a low angle must travel through more of the atmosphere than rays striking at a higher angle and thus are subject to greater depletion by reflection and absorption.

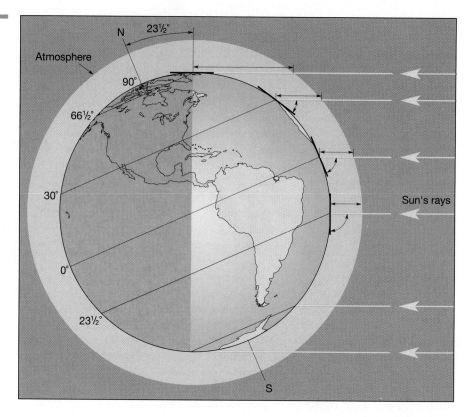

It is important to remember that the earth is spherical. Hence, on any given day only places located at a particular latitude receive vertical (90-degree) rays from the sun. As we move either north or south of this location, the sun's rays strike at an ever-decreasing angle. Thus, the nearer a place is to the latitude receiving vertical rays of the sun, the higher will be its noon sun.

What causes the yearly fluctuations in the sun angle and length of daylight? They occur because the earth's orientation to the sun continually changes. The earth's axis is not perpendicular to the plane of its orbit around the sun, but instead is tilted 23½ degrees from the perpendicular. This is termed the **inclination of the axis**, and as we shall see, if the axis were not inclined, we would have no seasonal changes. In addition, because the axis remains pointed in the same direction (toward the North Star) as the earth journeys around the sun, the orientation of the earth's axis to the sun's rays is constantly changing (Figure 12.12). On one day each year the axis is such that the Northern Hemisphere is "leaning" 23½ degrees toward the sun. Six months later, when the earth has moved to the opposite side of its

orbit, the Northern Hemisphere leans 23½ degrees away from the sun. On days between these extremes, the earth's axis leans at amounts less than 23½ degrees to the rays of the sun. This change in orientation causes the vertical rays of the sun to make a yearly migration from 23½ degrees north of the equator to 23½ degrees south of the equator. This migration in turn causes the altitude angle of the noon sun to vary by as much as 47 degrees (23½ + 23½) during the course of a year at places located poleward of latitude 23½ degrees. For example, a mid-latitude city like New York (about 40 degrees north latitude) has a maximum noon sun angle of 73½ degrees when the sun's vertical rays reach their farthest northward location and a minimum noon sun angle of 26½ degrees six months later.

Historically, four days each year have been given special significance based on the annual migration of the direct rays of the sun and its importance to the yearly weather cycle. On June 21 or 22 the earth is in a position such that the axis in the Northern Hemisphere is tilted 23½ degrees toward the sun (Figure 12.13A). At this time the vertical rays of the sun strike 23½ degrees north latitude (23½ degrees north of the equator), a line of latitude known as the **Tropic of Cancer**. For people in the Northern Hemisphere June 21 or 22 is known as the **summer solstice**. Six months later, on about December 21 or 22, the earth is in the opposite position, with the sun's vertical rays striking at 23½ degrees south latitude (Figure 12.13C). This line is known as the **Tropic of Capricorn**. For those in the Northern Hemisphere, December 21 or 22 is the **winter solstice**. However, at the same time in the Southern Hemisphere, people are experiencing just the opposite—the summer solstice.

The equinoxes occur midway between the solstices. September 22 or 23 is the date of the **autumnal equinox** in the Northern Hemisphere, and March 21 or 22 is the date of the **vernal**, or **spring, equinox**. On these dates, the vertical rays of the sun strike at the equator (0 degrees latitude) because the earth is in such a position in its orbit that the axis is tilted neither toward nor away from the sun (Figure 12.13B).

Further, the length of daylight versus darkness is also determined by the position of the earth in its or-

bit. The length of daylight on June 21, the summer solstice in the Northern Hemisphere, is greater than the length of night. This fact can be established from Figure 12.13A by comparing the fraction of a given latitude that is on the "day" side of the circle of illumination with the fraction on the "night" side. The opposite is true for the winter solstice, when the nights are longer than the days (Table 12.2). Again for comparison let us consider New York City, which has 15 hours of daylight on June 21 and only 9 hours on December 21. Also note from Table 12.2 that on June 21 the farther you are north of the equator the longer is the period of daylight, until the Arctic Circle is reached, where the length of daylight is 24 hours. During an equinox (meaning "equal night"), the length of daylight is 12 hours everywhere on earth, because the circle of illumination passes directly through the poles, dividing the latitudes in half.

As a review of the characteristics of the summer solstice for the Northern Hemisphere, examine Figure 12.13A and Table 12.2 and consider the following facts:

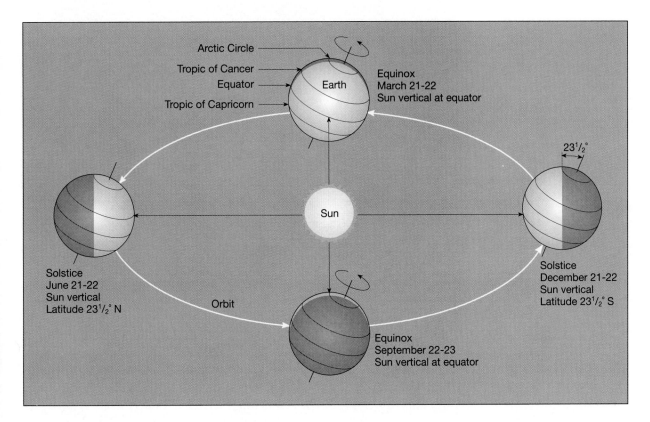

Figure 12.12
Earth-sun relationships.

Figure 12.13
Characteristics of the solstices and equinoxes.

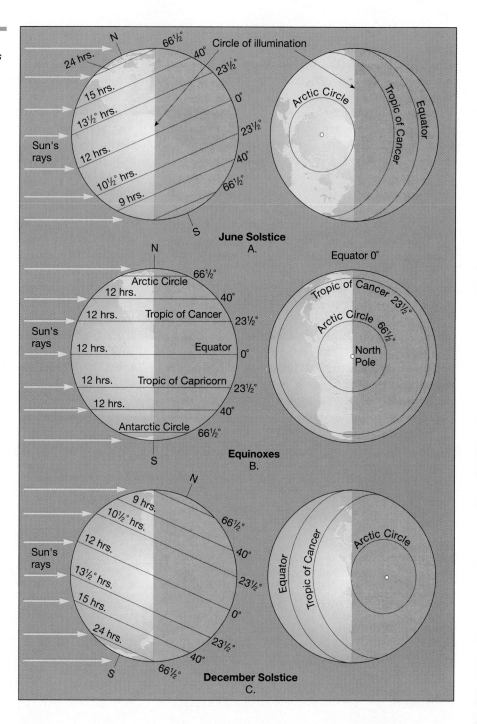

1. The date of occurrence is June 21 or 22.
2. The vertical rays of the sun are striking the Tropic of Cancer (23½ degrees north latitude).
3. Locations in the Northern Hemisphere are experiencing their length of daylight. (Opposite for the Southern Hemisphere.)
4. Locations north of the Tropic of Cancer are expe-

riencing their highest noon sun angles. (Opposite for places south of the Tropic of Capricorn.)

5. The farther you are north of the equator, the longer the period of daylight is, until the Arctic Circle is reached, where the day is 24 hours long. (Opposite for the Southern Hemisphere.)

Table 12.2
Length of daylight.

Latitude (degrees)	Summer Solstice	Winter Solstice	Equinoxes
0	12 h	12 h	12 h
10	12 h 35 min	11 h 25 min	12
20	13 12	10 48	12
30	13 56	10 04	12
40	14 52	9 08	12
50	16 18	7 42	12
60	18 27	5 33	12
70	24 h (for 2 mo)	0 00	12
80	24 h (for 4 mo)	0 00	12
90	24 h (for 6 mo)	0 00	12

The facts about the winter solstice are just the opposite. It should now be apparent why a midlatitude location is warmest in the summer, for it is then that days are longest and the sun's altitude is highest.

In summary, seasonal variations in the amount of solar energy reaching places on the earth's surface are caused by the migrating vertical rays of the sun and the resulting variations in sun angle and length of daylight. All places at the same latitude have identical sun angles and lengths of daylight. If the earth-sun relationships just described were the only controls of temperature, we would expect these places to have identical temperatures as well. Obviously this is not the case. Although the altitude of the sun is the main control of temperature, it is not the only control, as we shall see.

MECHANISMS OF HEAT TRANSFER

All forms of matter, whether solid, liquid, or gas, are composed of atoms or molecules that are in constant motion. Because of this vibratory motion, all matter is said to contain *thermal energy*. Whenever a substance is heated, its atoms move faster and faster. This leads to an increase in thermal energy. It is the average motion of the atoms or molecules in objects that we sense when we determine how hot or cold something is. We commonly describe the "hotness" or "coldness" of an object using the measurement called *temperature*. More specifically, temperature is a measure of the average motion of the atoms or molecules within a substance.

Although closely related, temperature and heat are very different concepts. Whereas temperature relates to the average motion of molecules, heat is the energy that flows because of temperature differences. In all situations, heat is transferred from warmer to cooler objects. Thus, if two objects of dif-ferent temperature are in contact, the warmer object will become cooler and the cooler object will become warmer until they both reach the same temperature.

Three mechanisms of heat transfer are recognized: conduction, convection, and radiation. Since radiation is the only one of these that can travel through the relative emptiness of space, the vast majority of energy coming to and leaving the earth must be in this form. Radiation also plays an important role in transferring heat from the earth's land-sea surface to the atmosphere and vice versa.

Conduction and Convection

Conduction is familiar to most of us. Anyone who has attempted to pick up a metal spoon that was left in a hot pan has discovered that heat was conducted through the spoon. Conduction is the transfer of heat through matter by molecular activity. The energy of molecules is transferred through collisions from one molecule to another, with the heat flowing from the higher temperature to the lower temperature. The

BOX 12.2

When Are the Seasons?

Have you ever been caught in a snowstorm in early December, only to be told by the person doing the TV weather forecast that winter does not begin until December 21? Or perhaps you have endured several consecutive days of 100-degree temperatures only to discover that summer has not yet started. The idea of dividing the year into four seasons clearly origi- nated from the earth-sun rela- tionships discussed in this chap- ter. This astronomical definition of the seasons defines winter (Northern Hemisphere) as the period from the winter solstice (December 21–22) to the spring equinox (March 21–22), and so forth (Table 12.A). This is also the definition used most widely by the news media. Yet it is not unusual for portions of the United States and Canada to have significant snowfalls weeks before the "official" start of win- ter (Figure 12.A).

Because the weather phenom- ena we normally associate with each season do not coincide well with the astronomical seasons, meteorologists prefer to divide the year into four 3-month peri- ods based primarily on tempera- ture. Thus, winter is defined as December, January, and Febru- ary, the three coldest months of the year in the Northern Hemi- sphere (Table 12.A). Summer, on the other hand, is defined as the three warmer months: June, July, and August. Spring and au- tumn are the transition periods between these two seasons. Inasmuch as these four 3-month periods better reflect the temper- atures and weather that we as- sociate with the respective sea- sons, this definition of the seasons is more useful for mete- orological discussions.

Table 12.A
Occurrence of the seasons in the Northern Hemisphere.

Seasons	Astronomical Seasons	Meteorological Seasons
Spring	March 21 or 22 to June 21 or 22	March, April, May
Summer	June 21 or 22 to September 22 or 23	June, July, August
Autumn	September 22 or 23 to December 21 or 22	September, October, November
Winter	December 21 or 22 to March 21 or 22	December, January, February

ability of substances to conduct heat varies consid- erably. Metals are good conductors, as those of us who have touched a hot spoon have quickly learned. Air, on the other hand, is a very poor conductor of heat. Consequently, conduction is important only be- tween the earth's surface and the air directly in con- tact with the surface. As a means of heat transfer for the atmosphere as a whole, conduction is the least significant and can be disregarded when considering most meteorological phenomena.

Heat gained by the lowest layer of the atmosphere from radiation or conduction is most often trans- ferred by convection. **Convection** is the transfer of heat by the movement of a mass or substance from one place to another. It can take place only in liquids and gases. Convective motions in the atmosphere are responsible for the redistribution of heat from equa- torial regions to the poles and from the surface up- ward. The term **advection** is usually reserved for horizontal convective motions such as winds, while *convection* is used to describe vertical motions in the atmosphere.

Figure 12.A
Fall scene in White Mountain National Forest, New Hampshire. (Photo by Tom Till)

Figure 12.14 summarizes the various mechanisms of heat transfer. The heat produced by the campfire passes through the pan by conduction, warming the water at the bottom of the pan. Convection currents carry the warmed water throughout the container, heating the remaining water. Meanwhile, the camper is warmed by radiation from the fire and the pan. Furthermore, since metals are good conductors, the camper's hand is likely to be burned if a pot holder is not used. In most situations involving heat transfer, conduction, convection, and radiation occur simultaneously.

Radiation

From our everyday experience we know that the sun emits light and heat as well as the rays that give us a suntan. Although these forms of energy comprise a major portion of the total energy that radiates from the sun, they are only part of a large array of energy called **radiation** or **electromagnetic radiation**. This array or spectrum of electromagnetic energy is shown in Figure 12.15. All radiation, whether x rays, radio waves, or heat waves, is capable of transmitting energy through the vacuum of space at 300,000 kilometers (186,000 miles) per second and only

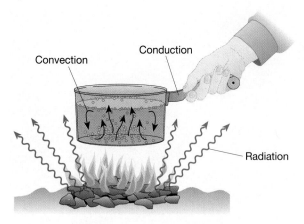

Figure 12.14
Illustration of conduction, convection, and radiation.

slightly slower through our atmosphere. Nineteenth century physicists were so puzzled by the seemingly impossible task of energy traveling without a medium to transmit it that they assumed that a material, which they named ether, existed between the sun and the earth. This medium was thought to transmit radiant energy in much the same way that air transmits sound waves produced by the vibration of one

person's vocal cords to another person's eardrums. Today we know that, like gravity, radiation requires no material to transmit it.

In some respects, the transmission of radiant energy parallels the motion of the gentle swells in the open ocean. Not unlike ocean swells, these electromagnetic waves, as they are called, come in various sizes. For our purpose, the most important difference between electromagnetic waves is their wavelength, or distance from one crest to the next. Radio waves have the longest wavelengths, ranging to tens of kilometers, whereas gamma waves are the very shortest, being less than one billionth of a centimeter long. **Visible light**, as the name implies, is the only portion of the spectrum we can see. We often refer to visible light as white light since it appears "white" in color. However, it is easy to show that white light is really an array of colors, each color corresponding to a particular wavelength (Figure 12.16). Using a prism, we can divide white light into colors of the rainbow, violet having the shortest wavelength—0.4 micrometer (1 micrometer is 0.0001 centimeter)—and red having the longest—0.7 micrometer (see Figure 19.2). Located adjacent to red, and having a longer wavelength, is **infrared** radiation, which we cannot see but which we can detect as heat. The closest invisi-

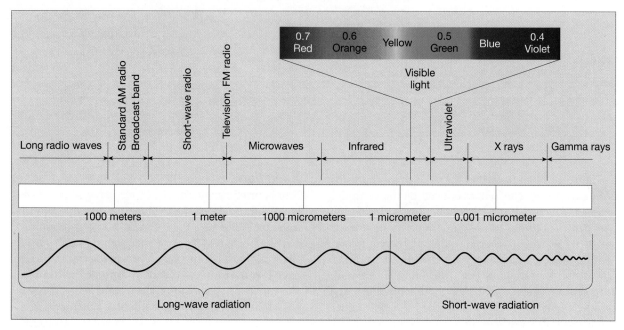

Figure 12.15
The electromagnetic spectrum.

Figure 12.16
Visible light consists of an array of colors we commonly call the "colors of the rainbow." Rainbows are relatively common optical phenomena produced by the bending and reflection of light by drops of water. (Photo by Gordon Garradd/Photo Researchers, Inc.)

ble waves to violet are called **ultraviolet** rays and are responsible for sunburn after an intense exposure to the sun. Although we divide radiant energy into groups based on our ability to perceive them, all forms of radiation are basically the same. When any form of radiant energy is absorbed by an object, the result is an increase in molecular motion, which causes a corresponding increase in temperature.

To better appreciate how the sun's radiant energy interacts with the earth's atmosphere and surface, we must have a general understanding of the basic laws governing radiation. Although the mathematical implications of these laws are beyond the scope of this book, the concepts themselves are worth summarizing.

1. All objects, at whatever temperature, emit radiant energy. Hence, not only hot objects like the sun but also the earth, including its polar ice caps, continually emit energy.

2. Hotter objects radiate more total energy per unit area than do colder objects. The sun, with a surface temperature of 6000 K,* emits hundreds of thousands of times more energy than the earth, which has an average surface temperature of 288 K.

3. The hotter the radiating body, the shorter the wavelength of maximum radiation. For example, a very hot metal rod will emit visible radiation, producing a white glow. Upon cooling, it will emit more of its energy in longer wavelengths and glow a reddish color. Eventually no light will be given off, but if you place your hand near the rod, the still-longer infrared radiation will be detectable as heat. The sun, with a surface temperature of 6000 K, radiates maximum energy at 0.5 micrometer, which is in the visible range. The maximum radiation for the earth occurs at a wavelength of 10 micrometers, well within the infrared (heat) range. Because the maximum earth radiation is roughly 20 times longer than the maximum solar radiation, terrestrial radiation is often called long-wave radiation, and solar radiation is called short-wave radiation.

4. Objects that are good absorbers of radiation are good emitters as well. Technically, a perfect emitter is any object that radiates, for every wavelength, the maximum intensity of radiation possible for that temperature. The earth's surface and the sun approach being perfect radiators since they absorb and radiate with nearly 100 percent efficiency for their respective temperatures. On the other hand, gases are selective absorbers and radiators. Thus the atmosphere, which is nearly transparent to (does not absorb) certain wavelengths of radiation, is nearly opaque (a good absorber) to others. Our experience tells us that the atmosphere is transparent to visible light; hence, it readily reaches the earth's surface. This is not the case for long-wave terrestrial radiation, as we shall see.

---

* To convert kelvins to degrees Celsius, add the degree symbol and subtract 273.

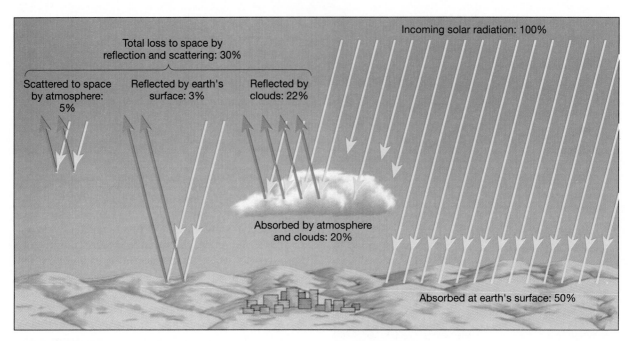

Figure 12.17
Distribution of incoming solar radiation by percentage. More solar energy is absorbed by the earth's surface than by the atmosphere. Consequently, the air is not heated directly by the sun but is heated chiefly by the earth's surface.

HEATING THE ATMOSPHERE

Although the atmosphere is largely transparent to incoming solar radiation, only about 25 percent penetrates directly to the earth's surface without some sort of interference by the atmosphere. The remainder is either *absorbed* by the atmosphere, *scattered* about until it reaches the earth's surface or returns to space, or is *reflected* back to space (Figure 12.17). What determines whether radiation will be absorbed, scattered, or reflected outward? As we shall see, it depends greatly on the wavelength of the energy being transmitted, as well as on the size and nature of the intervening material.

Although solar radiation travels in a straight line, the gases and dust particles in the atmosphere can redirect this energy, a process called *scattering*. Some of the light is back-scattered and lost to space, while the remainder continues downward, where it interacts with other molecules that scatter it further by changing the direction of the light beam, but not its wavelength.

The light that reaches the earth's surface after having its direction changed is called **diffused light**. It is diffused light that keeps a shaded area or room

lighted when direct sunlight is absent. Scattering also produces the blue color of the sky. Air molecules are more effective scatterers of the shorter-wavelength (blue and violet) portion of "white" sunlight than the longer-wavelength (red and orange) portion. Thus, when we look in a region of the sky away from the direct solar rays, we see predominantly blue light, which was more readily scattered. On the other hand, the sun appears to have a yellowish to reddish tint when viewed near the horizon. When the sun is in this position, the solar beam must travel through a great deal of atmosphere before it reaches your eye. For this reason, most of the blues and violets will be scattered out, leaving a beam of light composed mostly of reds and yellows. This phenomenon is particularly pronounced on a day when fine dust or smoke particles are present.

About 30 percent of the solar energy reaching the outer atmosphere is reflected back to space. Included in this figure is the amount sent skyward by back-scattering. This energy is lost to the earth and does not play a role in heating the atmosphere.

The fraction of the total radiation encountered that is reflected by a surface is called **albedo**. Thus the albedo for the earth as a whole (planetary

Table 12.3
Albedo of various surfaces.

Surface	Percent Reflected
Clouds, stratus	
<150 meters thick	25–63
150–300 meters thick	45–75
300–600 meters thick	59–84
Average of all types and thicknesses	50–55
Concrete	17–27
Crops, green	5–25
Forest, green	5–10
Meadows, green	5–25
Ploughed field, moist	14–17
Road, blacktop	5–10
Sand, white	30–60
Snow, fresh-fallen	80–90
Snow, old	45–70
Soil, dark	5–15
Soil, light (or desert)	25–30
Water	8*

* Typical albedo value for a water surface. The albedo of a water surface varies greatly depending upon the sun angle. If the sun angle is greater than 30 degrees, the albedo is less than 5 percent. When the sun is near the horizon (sun angle less than 3 degrees), the albedo is more than 60 percent.

albedo) is 30 percent. However, the albedo from place to place as well as from time to time in the same locale varies considerably, depending upon the amount of cloud cover and particulate matter in the air, as well as upon the angle of the sun's rays and the nature of the surface. A lower angle means that more atmosphere must be penetrated, thus making the "obstacle course" longer, and therefore the loss of solar radiation greater (see Figure 12.11). Table 12.3 gives the albedo for various surfaces and clouds. Note that the angle at which the sun's rays strike a water surface greatly affects it albedo.

As stated earlier, gases are selective absorbers, meaning that they absorb strongly in some wavelengths, moderately in others, and only slightly in still others. When a gas molecule absorbs light waves, this energy is transformed into internal molecular motion, which is detectable as a rise in temperature. Nitrogen, the most abundant constituent in the atmo-

sphere, is a rather poor absorber of all types of incoming radiation. Oxygen (O_2) and ozone (O_3) are efficient absorbers of ultraviolet radiation. Oxygen removes most of the shorter ultraviolet radiation high in the atmosphere, and ozone absorbs longer-wavelength ultraviolet rays in the stratosphere between 10 and 50 kilometers. The absorption of ultraviolet radiation in the stratosphere accounts for the high temperatures experienced there. The only other significant absorber of incoming solar radiation is water vapor, which along with oxygen and ozone accounts for most of the solar radiation absorbed within the atmosphere.

For the atmosphere as a whole, none of the gases are effective absorbers of radiation with wavelengths between 0.3 and 0.7 micrometer. This region of the spectrum corresponds to the visible range, to which a large fraction of solar radiation belongs. This explains why most visible radiation reaches the ground and why we say that the atmosphere is transparent to incoming solar radiation. Thus, direct solar energy is a rather ineffective "heater" of the earth's atmosphere. The fact that the atmosphere does not acquire the bulk of its energy directly from the sun but rather from reradiation from the earth's surface is of the utmost importance to the dynamics of the weather machine.

Approximately 50 percent of the solar energy that strikes the top of the atmosphere reaches the earth's surface directly or indirectly (diffused) and is absorbed. Most of this energy is then reradiated skyward. Since the earth has a much lower surface temperature than the sun, terrestrial radiation is emitted in longer wavelengths than solar radiation. The bulk of terrestrial radiation has wavelengths between 1 and 30 micrometers, placing it well within the infrared range. The atmosphere as a whole is a rather efficient absorber of radiation between 1 and 30 micrometers (terrestrial radiation). Water vapor and carbon dioxide are the principal absorbing gases in that range. Water vapor absorbs roughly five times more terrestrial radiation than do all the other gases combined and accounts for the warm temperatures found in the lower troposphere, where it is most highly concentrated. Because the atmosphere is quite transparent to shorter-wavelength solar radiation and more readily absorbs longer-wavelength terrestrial radiation, the atmosphere is heated from the ground up rather than vice versa. This explains the general drop in temperature with increasing altitude experienced in the troposphere. The farther from the "radiator," the colder it becomes.

Figure 12.18
The heating of the atmosphere. Most of the solar radiation that is not reflected back to space passes through the atmosphere and is absorbed at the earth's surface. The earth's surface, in turn, emits longer wavelength radiation. A portion of this energy is absorbed by certain gases in the atmosphere.

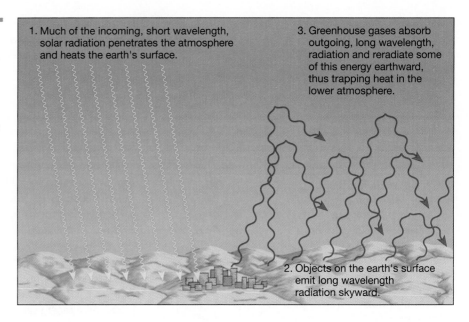

1. Much of the incoming, short wavelength, solar radiation penetrates the atmosphere and heats the earth's surface.

3. Greenhouse gases absorb outgoing, long wavelength, radiation and reradiate some of this energy earthward, thus trapping heat in the lower atmosphere.

2. Objects on the earth's surface emit long wavelength radiation skyward.

When the gases in the atmosphere absorb terrestrial radiation, they warm, but they eventually radiate this energy away. Some travels upward, where it may be reabsorbed by other gas molecules, a possibility less likely with increasing height because the concentration of water vapor decreases with altitude. The remainder travels downward and is again absorbed by the earth. For this reason, the earth's surface is continually being supplied with heat from the atmosphere as well as from the sun. Without these absorptive gases in our atmosphere, the earth would not be a suitable habitat for humans and numerous other life forms.

This very important phenomenon has been termed the **greenhouse effect*** because it was once thought that greenhouses were heated in a similar manner (Figure 12.18). The gases of our atmosphere, especially water vapor and carbon dioxide, act very much like the glass in the greenhouse. They allow shorter-wavelength solar radiation to enter, where it is absorbed by the objects inside. These objects in turn reradiate the heat, but at longer wavelengths, to which glass is nearly opaque. The heat therefore is trapped in the greenhouse. However, a more important factor in keeping a greenhouse warm is the fact

that the greenhouse itself prevents mixing of air inside with cooler air outside. Nevertheless, the term *greenhouse effect* is still used.

The role that these absorbing gases play in keeping the atmosphere warm is well known to those residing in mountainous regions. More radiant energy is received on mountaintops than in the valleys below because there is less atmosphere to hinder its arrival. Yet the decrease in water vapor content with altitude allows much of the heat to escape these lofty peaks. This loss more than compensates for the extra radiation received. As a result, valleys remain warmer than the adjacent mountains even though the valleys receive less solar radiation.

The importance of water vapor in keeping our atmosphere warm is easily demonstrated. Figure 12.19 shows the maximum and minimum temperatures for various cities in the United States on a typical July day. By comparing the temperatures for cities in the arid Southwest (Nevada and southern Arizona, for example) with temperatures in the more humid East, the following can be seen:

1. Maximum temperatures are, for the most part, higher in the arid regions.
2. Minimum temperatures often are lower in the dry Southwest as compared with the humid East.
3. The difference between the maximum and minimum temperatures (a calculation known as the **daily temperature range**) is in almost every case greater in the arid areas.

---

*   The term *greenhouse effect* is frequently seen and heard in the news in connection with a possible global warming that is directly related to human activities. This topic is treated at some length in Chapter 16.

The higher maximum temperatures in arid regions are easily explained, for with little or no cloud cover, a maximum amount of sunshine is received. Further, the records of the stations in the dry regions indicate that these places cool rapidly at night, resulting in lower minimum temperatures. A hot, muggy day in a humid location often means a warm night as well, because the water vapor in the air traps some of the heat being radiated from the earth's surface. In the desert, however, there is very little water vapor in the air. Therefore, the terrestrial radiation escapes, and the air near the surface cools rapidly. From this example, it is quite evident that water vapor plays an important role in absorbing heat, and when it is lacking, the air cools rapidly, resulting in a high daily temperature range.

As we have discussed, variations in the amount of water vapor (an invisible gas) are seen to strongly influence the daily temperature range. It should be pointed out, however, that daily temperature ranges can also be affected by other factors, including cloud cover (Figure 12.20). If a day is overcast, the difference between the maximum and minimum temperatures will be smaller than if the day were clear. When the sky is overcast, clouds reflect incoming solar radiation back to space and reduce daytime heating. At night, clouds act as a "blanket" by absorbing radiation emitted by the earth and reradiating it back to the surface. Therefore, nighttime temperatures are not as low as they otherwise would have been.

SOLAR ENERGY

Solar energy is one of several so-called alternative energy sources that may provide a greater share of our future energy needs. The term *solar energy* generally refers to the direct use of the sun's rays to supply energy for the needs of people. The simplest, and perhaps most widely used, solar collectors are south-facing windows. As sunlight passes through the glass, energy is absorbed by objects in the room. These ob-

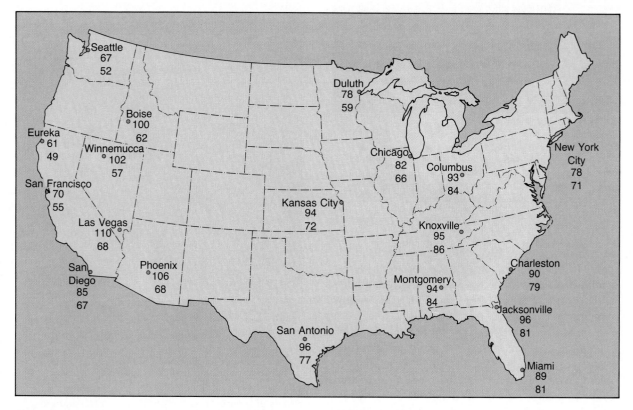

Figure 12.19
Daily maximum and minimum temperatures (°F) for a typical July day.
°C = (°F − 32)/1.8.

A. Cloudy location (day)

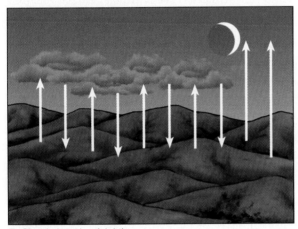

B. Cloudy location (night)

Figure 12.20
*An overcast sky causes the daily temperature range to be lower than it would otherwise be. **A.** By reflecting incoming sunlight, clouds reduce daytime heating.
B. By acting as a "blanket," clouds reduce nighttime cooling.*

heating, this application of solar energy benefits from the fact that hot water is required throughout the year, even in the tropics. In Israel, for example, about 20 percent of all homes are equipped with some type of solar device. Although solar energy is free, the necessary equipment and its installation are not. The initial costs of setting up a system, including a supplemental heating unit, can be substantial. Nevertheless, over the long term, solar energy is economical in many parts of the United States and will become even more cost effective as the prices of other fuels increase.

Research is currently under way to improve the technologies for concentrating solar energy so as to produce high-temperature heat. One method being examined uses an array of mirrors that track the sun and keep its rays focused on a receiving tower. A prototype facility, with 2000 mirrors, has been constructed near Barstow, California. Solar energy focused on the tower heats water in pressurized panels

Figure 12.21
Panels of solar cells at the ARCO solar installation located near Hesperia, California. Electricity provided by the facility is used by customers of Southern California Edison. (Photo by E. J. Tarbuck)

jects, in turn, radiate heat that warms the air in the room. In the United States the use of south-facing windows, along with better-insulated and more airtight construction, is widely practiced, substantially reducing heating costs.

More elaborate systems used for home heating involve an active solar collector. These roof-mounted devices are normally large, blackened boxes that are covered with glass. The heat collected in the system can be transferred by the circulation of air or by fluids that are circulated through a network of tubes. Solar collectors are also used successfully to heat water for domestic and commercial needs. Unlike space to over 500°C. The superheated water is then transferred to turbines, which turn electrical generators. This pilot project has demonstrated that solar collectors have the potential of becoming commercially viable.

Another type of collector uses solar cells that convert the sun's energy directly into electricity. Currently, this technology is used only for special applications such as solar-powered calculators and to provide electrical power for space vehicles. A large experimental facility that uses solar cells is located near Hesperia, California (Figure 12.21). However, the future of this technology is uncertain. In addition to being relatively inefficient, solar cells are presently very expensive to produce and are easily damaged by the elements.

TEMPERATURE MEASUREMENT AND DATA

Changes in air temperature are probably noticed by people more often than changes in any other element of weather. At a weather station, the temperature is read on a regular basis from instruments mounted in an instrument shelter (Figure 12.22). The shelter protects the instruments from direct sunlight and allows a free flow of air. In addition to a standard mercury thermometer, the shelter is likely to contain a thermograph, which is an instrument that makes a continuous record of temperature, and a set of maximum-minimum thermometers. As their name implies, these thermometers record the highest and lowest temperatures during a given day.

The daily maximum and minimum temperatures are the bases for many of the temperature data compiled by meteorologists:

1. By adding the maximum and minimum temperatures and then dividing by two, the **daily mean** is calculated.
2. The daily temperature range is computed by finding the difference between the maximum and minimum temperatures for a given day.
3. The **monthly mean** is calculated by adding together the daily means for each day of the month and dividing by the number of days in the month.
4. The **annual mean** is an average of the twelve monthly means.
5. The **annual temperature range** is computed by finding the difference between the highest and lowest monthly means.

Mean temperatures are particularly useful for making comparisons, whether on a daily, monthly, or annual basis. It is quite common to hear a weather reporter state, "Last month was the hottest July on record," or "Today Chicago was ten degrees warmer than Miami." Temperature ranges are also useful statistics, because they give an indication of extremes.

In order to make weather data more useful to people, many different applications have been devised. One familiar category of applications relates to human perceptions of temperature. People who watch or listen to weather reports on television and radio are probably acquainted with indices that attempt to portray levels of human comfort and discomfort. Such indices are based on the fact that the sensation of temperature that the human body feels is often

Figure 12.22
Standard instrument shelter. A shelter protects instruments from direct sunlight and allows for the free flow of air. (Courtesy of Qualimetrics, Inc.)

BOX 12.3

Windchill: The Cooling Power of Moving Air

M̲ost everyone is familiar with the wintertime cooling power of moving air. When the wind blows on a cold day, we realize that comfort would improve if the wind were to stop. A stiff breeze penetrates ordinary clothing and reduces its capacity to retain body heat while causing exposed parts of the body to chill rapidly. Not only is cooling by evaporation heightened in this situation, but the wind is also acting to carry heat away from the body by constantly replacing warmer air with colder air.

During the late 1930s and early 1940s, pioneering experiments performed in Antarctica by the polar scientist Paul Siple led to the development of the concept of windchill. Today during cold weather it is common to hear the term *windchill* mentioned prominently during weather reports. The figure that is actually being reported is the *windchill equivalent temperature* (WET for short). WET is not an actual temperature, but an expression that relates wind speed to the temperature we perceive. The windchill charts (Tables 12.B and 12.C) illustrate the combined effects of wind and temperature on the cooling rate of the human body by translating the cooling power of the atmosphere with wind to a temperature under nearly calm conditions. (Rather than using an absolute calm, a wind of about 6 kilometers or 4 miles per hour is assumed because it represents the air motion that a person feels while walking briskly in calm air.) Therefore on a day when the temperature is −8°C and the wind speed is 30 kilometers per hour, the sensation of temperature, that is, the windchill equivalent temperature, would be reported as −25°C, or 17°C less than the actual air temperature. It should be emphasized that the temperature of the skin does not drop to −25°C. Skin temperature can fall no lower than the air temperature, which is −8°C in this example. What the WET does indicate is that any exposed skin will lose heat at a rate equal to the rate that occurs when the temperature is −25°C and the air is relatively calm. By examining Tables 12.B and 12.C, it is clear that the cooling power of the wind increases as wind speed rises and as temperature decreases. It should also be pointed out that in contrast to a cold and windy day, a calm and sunny day in winter often feels warmer than the thermometer reading. In this situation, the warm feeling is caused by the absorption of direct solar radiation by the body.

It is important to remember that the windchill equivalent temperature is only an estimate of human discomfort. The degree of discomfort felt by different people will vary because it is influenced by many factors. Even if clothing is assumed to be the same, individuals vary widely in their responses due to such things as age, physical condition, state of health, and level of activity. Nevertheless, as a relative measure, WET is useful because it allows people to make more informed judgments regarding the potential harmful effects of wind and cold.

different from the actual temperature of the air as recorded by a thermometer. The human body is a heat engine that is continually releasing energy, and anything that influences the rate of heat loss from the body also influences the sensation of temperature that the body feels, thereby affecting human comfort. Several factors play a part in controlling the thermal comfort of the human body, and certainly air temperature is a major factor. Other environmental conditions, such as relative humidity, wind, and solar radiation, are also significant, however. Many of us know that a cold and windy winter day can feel much colder than the air temperature would seem to indicate. Box 12.3 focuses on this frequently used wintertime application known as *windchill.*

CONTROLS OF TEMPERATURE

The controls of temperature are those factors that cause variations in temperature from place to place. Earlier in this chapter we examined the single greatest cause for temperature variations—differences in the receipt of solar radiation. Since variations in sun

Table 12.B
Windchill Equivalent Temperature (°C)

Actual Temperature (°C)	Wind Speed (km per hr)								
	6	10	20	30	40	50	60	70	80
12	12	9	5	3	1	0	−0	−1	−1
8	8	5	0	−3	−5	−6	−7	−7	−8
4	4	0	−5	−8	−11	−12	−13	−14	−14
0	0	−4	−10	−14	−17	−18	−19	−20	−21
−4	−4	−8	−15	−20	−23	−25	−26	−27	−27
−8	−8	−13	−21	−25	−29	−31	−32	−33	−34
−12	−12	−17	−26	−31	−35	−37	−39	−40	−40
−16	−16	−22	−31	−37	−41	−43	−45	−46	−47
−20	−20	−26	−36	−43	−47	−49	−51	−52	−53
−24	−24	−31	−42	−48	−53	−56	−58	−59	−60
−28	−28	−35	−47	−54	−59	−62	−64	−65	−66
−32	−32	−40	−52	−60	−65	−68	−70	−72	−73
−36	−36	−44	−57	−65	−71	−74	−77	−78	−79
−40	−40	−49	−63	−71	−77	−80	−83	−85	−86

Source: NOAA, National Weather Service.

Table 12.C
Windchill Equivalent Temperature (°F)

Air Temperature (°F)	Wind Speed (mi per hr)								
	5	10	15	20	25	30	35	40	45
40	37	28	22	18	15	13	11	10	9
35	32	22	16	11	8	5	3	2	1
30	27	16	9	4	0	−2	−4	−6	−7
25	22	10	2	−3	−7	−10	−12	−14	−15
20	16	4	−5	−10	−15	−18	−20	−22	−23
15	11	−3	−11	−17	−22	−25	−28	−29	−31
10	6	−9	−18	−25	−29	−33	−35	−37	−39
5	1	−15	−25	−32	−37	−41	−43	−45	−47
0	−5	−21	−32	−39	−44	−48	−51	−53	−55
−5	−10	−27	−38	−46	−52	−56	−59	−61	−62
−10	−15	−33	−45	−53	−59	−63	−67	−69	−70
−15	−20	−40	−52	−60	−66	−71	−74	−77	−78
−20	−26	−46	−58	−67	−74	−79	−82	−85	−86
−25	−31	−52	−65	−74	−81	−86	−90	−93	−94
−30	−36	−58	−72	−82	−89	−94	−98	−101	−102

Source: NOAA, National Weather Service.

angle and length of daylight are a function of latitude, they are responsible for warm temperatures in the tropics and colder temperatures at more poleward locations. However, latitude is not the only control of temperature; if it were, we would expect that all places along the same parallel would have identical temperatures. This is clearly not the case. For example, Eureka, California, and New York City are both coastal cities at about the same latitude and both have an average annual mean temperature of 11°C (52°F). However, New York City is 9°C (16°F) warmer than Eureka in July and 10°C cooler in January. Quito and Guayaquil, two cities in Ecuador, are within a relatively few kilometers of each other, yet the mean annual temperatures at these two cities differ by 12°C (21°F). To explain these situations and countless others, we must realize that factors other than variations in solar radiation also exert a strong influence upon temperatures. Among the most important of these other factors are the differential heating of land and water, ocean currents,* altitude, and geographic position.

Land and Water

The heating of the earth's surface directly influences the heating of the air above. Therefore, in order to understand variations in air temperature, we must examine the nature of the surface. Different land surfaces absorb varying amounts of incoming solar energy, which in turn cause variations in the temperature of the air above. The largest contrast, however, is not between different land surfaces, but between land and water. Land heats more rapidly and to higher temperatures than water and cools more rapidly and to lower temperatures than water. Temperature variations, therefore, are considerably greater over land than over water.

Among the reasons for the differential heating of land and water are the following:

1. The specific heat (amount of energy needed to raise 1 gram of a substance 1°C) is far greater for water than for land. Thus, water requires a great deal more heat to raise its temperature the same amount as an equal quantity of land.
2. Land surfaces are opaque, so heat is absorbed only at the surface. Water, being more transparent, allows heat to penetrate to a depth of many meters.
3. The water that is heated often mixes with water below, thus distributing the heat through an even larger mass.
4. Evaporation (a cooling process) from water bodies is greater than that from land surfaces.

Monthly temperature data for two cities will demonstrate the moderating influence of a large water body and the extremes associated with land (Figure 12.23). Vancouver, British Columbia, is located along a windward coast, whereas Winnipeg, Manitoba, is in a continental position far from the influence of water. Both cities are at about the same latitude and thus experience similar sun angles and lengths of daylight. Winnipeg, however, has a mean January temperature that is 20°C lower than Vancouver's. Conversely, Winnipeg's July mean is 2.6°C higher than Vancouver's. Although their latitudes are nearly the same, Winnipeg, which has no water influence, experiences must greater temperature extremes than Vancouver, which does.

On a different scale, the moderating influence of water may also be demonstrated when temperature variations in the Northern and Southern hemispheres

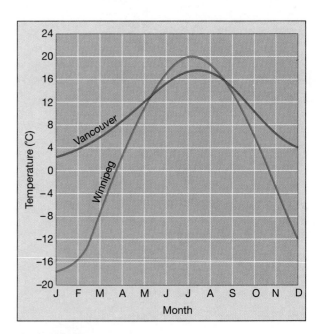

Figure 12.23
Monthly mean temperatures (°C) for Vancouver, British Columbia, and Winnipeg, Manitoba, demonstrate the moderating influence of a large water body and the greater extremes associated with an interior location.

---

* For a discussion of the effects of ocean currents on temperature, see Chapter 11.

Table 12.4
Variation in mean annual temperature range (°C) with latitude.

Latitude	Northern Hemisphere	Southern Hemisphere
0	0	0
15	3	4
30	13	7
45	23	6
60	30	11
75	32	26
90	40	31

are compared. Sixty-one percent of the Northern Hemisphere is covered by water, and land accounts for the remaining 39 percent. However, the figures for the Southern Hemisphere (81 percent water, 19 percent land) reveal why it is correctly called the water hemisphere. Table 12.4 portrays the considerably smaller annual temperature variations in the water-dominated Southern Hemisphere as compared with the Northern Hemisphere.

Altitude

The two cities in Ecuador mentioned earlier—Quito and Guayaquil—demonstrate the influence of altitude upon mean temperatures. Although both cities are near the equator and relatively close to each other, the annual mean at Guayaquil is 25°C (77°F) as compared to Quito's mean of 13°C (55°F). The difference is explained largely by the difference in the cities' elevations. Guayaquil is only 12 meters (40 feet) above sea level, whereas Quito is high in the Andes Mountains at 2800 meters (9200 feet). Recall that temperatures drop an average of 6.5°C per kilometer in the troposphere; thus cooler temperatures are to be expected at greater heights. Even so, the magnitude of the difference is not explained completely by the normal lapse rate. If the normal lapse rate is used, we would expect Quito to be about 18°C cooler than Guayaquil; the difference, however, is only 12°C. Places at high elevations, such as Quito, are warmer than the value calculated using the normal lapse rate to the same altitude in the free atmosphere because of the absorption and reradiation of solar energy by the ground surface.

Geographic Position

The geographic setting may greatly influence the temperatures experienced at a particular locale. For example, a windward coastal location, that is, a place subject to prevailing ocean winds, experiences considerably different temperatures from a coastal location where the prevailing winds are directed from the land toward the ocean. In the first situation, the place will experience the full moderating influence of the ocean—cool summers and mild winters, compared with an inland station at the same latitude. However, a leeward coastal site will have a more continental temperature pattern because the winds do not carry the ocean's influence onshore. Eureka, California, and New York City, the two cities mentioned in the section on temperature controls, illustrate this aspect of geographic position. The annual temperature range at New York City is 19°C (34°F) higher than that at Eureka.

Seattle and Spokane, both in the state of Washington, illustrate a second aspect of geographic position—mountains that act as barriers. Although Spokane is only bout 360 kilometers (220 miles) east of Seattle, the towering Cascade Range separates the cities. Consequently, while Seattle's temperatures show a marked marine influence, Spokane's are more typically continental. Spokane is 7°C (13°F) cooler than Seattle in January and 4°C (7°F) warmer than Seattle in July. The annual range at Spokane is 11°C (20°F) greater than at Seattle. The Cascade Range effectively cuts Spokane off from the moderating influence of the Pacific Ocean.

WORLD DISTRIBUTION OF TEMPERATURE

Temperature distribution is shown on a map by using **isotherms**, which are lines that connect places of equal temperature. On world maps that depict global patterns, temperatures are often corrected to sea level to eliminate the complications caused by altitude variations. January (Figure 12.24) and July (Figure 12.25) are selected most often for analysis because, for most locations, they represent temperature extremes.

On Figures 12.24 and 12.25 the isotherms generally trend east and west and show a decrease in temperatures poleward from the tropics, illustrating one of the most fundamental and best-known aspects of the world distribution of temperature—that the effectiveness of incoming solar radiation in heating the

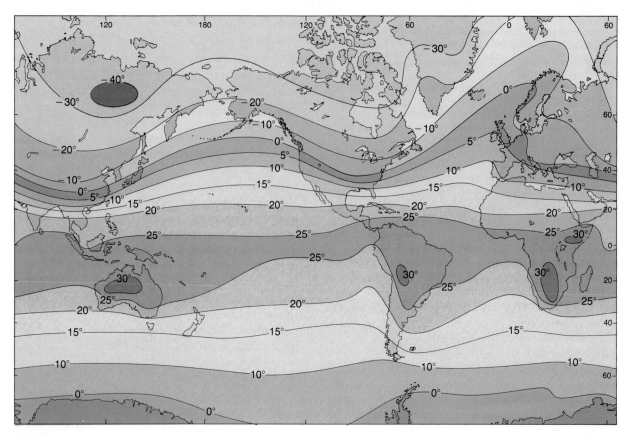

Figure 12.24
World distribution of mean temperatures (°C) for January.

earth's surface and the atmosphere above is largely a function of latitude. Further, there is a latitudinal shifting of temperatures caused by the seasonal migration of the sun's vertical rays.

The added effect of the differential heating of land and water is also reflected on the January and July temperature maps. The warmest and coldest temperatures are found over land. Hence, since temperatures do not fluctuate as much over water as over land, the north-south migration of isotherms is greater over the continents than over the oceans. In addition, it is clear that the isotherms in the Southern Hemisphere, where there is little land and the oceans predominate, are much more regular than in the Northern Hemisphere, where they bend sharply northward in July and southward in January over the continents.

Isotherms also reveal the presence of ocean currents. Warm currents cause isotherms to be deflected toward the pole, whereas cold currents cause an equatorward bending. The horizontal transport of water poleward warms the overlying air and results in air temperatures that are higher than otherwise would be expected for the latitude. Conversely, currents moving toward the equator produce air temperatures cooler than expected.

Since Figures 12.24 and 12.25 show the seasonal extremes of temperature, they can be used to evaluate variations in the annual range of temperature from place to place. A comparison of the two maps shows that a station near the equator will record a very small annual range because it experiences little variation in the length of daylight and always has a relatively high sun angle. By contrast, a station in the middle latitudes experiences much wider variations in sun angle and length of daylight, and thus larger variations in temperature. Therefore, we can state that the annual temperature range increases with an increase in latitude. Moreover, land and water also affect seasonal temperature variations, especially

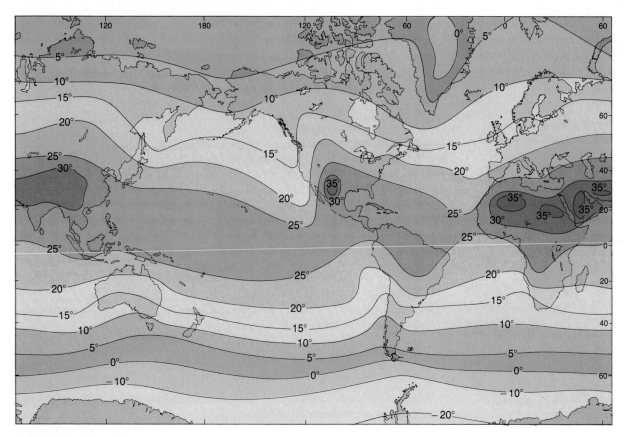

Figure 12.25
World distribution of mean temperatures (°C) for July.

outside the tropics. A continental location must endure hotter summers and colder winters than a coastal location. Consequently, the annual range will increase with an increase in continentality.

A classic example of the effect of latitude and continentality on annual temperature range is Yakutsk, a city in Siberia, approximately 60 degrees north latitude and far from the influence of water. As a result, Yakutsk has an average annual temperature range of 62.2°C (112°F), one of the highest in the world.

Review Questions

1. Distinguish between weather and climate.

2. List the basic elements of weather and climate.

3. What are the major components of clean, dry air (Figure 12.2)? Are they important meteorologically?

4. Why are water vapor and dust important constituents of our atmosphere?

5. **(a)** Why is ozone important to life on earth?
   (b) What are CFCs and what is their connection to the ozone problem?
   (c) If the ozone layer is reduced, what will be the probable impact on human health?

6. What percentage of the atmosphere is found below the following altitudes: 5.6 kilometers, 16.2 kilometers, 31.2 kilometers (see Table 12.1)?

7. Why is the troposphere called the "weather sphere"?

8. Why do temperatures rise in the stratosphere?

9. Are the variations in the amount of solar energy received by the earth because of its elliptical orbit important to understanding seasonal temperature variations? Explain your answer.

10. After examining Table 12.2, write a generalization that relates the season, the latitude, and the length of daylight.

11. Distinguish between the astronomical and the meteorological definitions of winter (Northern Hemisphere) (see Box 12.2).

12. Describe the relationship between the temperature of a radiating body and the wavelengths it emits.

13. Distinguish among the three basic mechanisms of heat transfer.

14. Figure 12.17 illustrates what happens to incoming solar radiation. The percentages shown, however, are only global averages. In particular, the amount of solar radiation reflected (albedo) may vary considerably. What factors might cause variations in albedo?

15. How does the earth's atmosphere act as a "greenhouse"?

16. On a warm summer day, one city had a daily temperature range of 25°C (45°F), while another experienced a range of only 8.3°C (15°F). One of these cities is located in Nevada, and the other in Indiana. Which location likely had the highest daily temperature range? Explain.

17. Briefly describe two methods by which solar energy might be used to produce electricity.

18. How are the following temperature data computed: Daily mean, daily range, monthly mean, annual mean, annual range?

19. Referring to Table 12.B in Box 12.3, determine the windchill equivalent temperature if the wind speed were 10 kilometers per hour and the air temperature were −8°C. What if the air temperature remained unchanged but the wind speed were 30 kilometers per hour?

20. Quito, Ecuador, is located on the equator and is not a coastal city. It has an average annual temperature of only 13°C (55°F). What is the likely cause for this low average temperature?

21. In what ways can geographic position be considered a control of temperature?

22. Yakutsk is located in Siberia at about 60 degrees north latitude. This Russian city has one of the highest averge annual temperature ranges in the world: 62.2°C (112°F). Explain the reasons for the very high annual temperature range.

Key Terms

advection (p. 430)
albedo (p. 434)
altitude (of the sun) (p. 425)
annual mean (p. 439)
annual temperature range (p. 439)
aphelion (p. 425)
autumnal equinox (p. 427)
circle of illumination (p. 424)
climate (p. 416)
conduction (p. 429)
convection (p. 430)
daily mean (p. 439)
daily temperature range (p. 436)
diffused light (p. 434)
electromagnetic radiation (p. 431)
element (of weather and climate) (p. 416)
environmental lapse rate (p. 422)
greenhouse effect (p. 436)
inclination of the axis (p. 426)
infrared (p. 432)
isotherm (p. 443)

mesopause (p. 424)
mesosphere (p. 424)
meteorology (p. 416)
monthly mean (p. 439)
perihelion (p. 425)
radiation (p. 431)
revolution (p. 424)
rotation (p. 424)
spring equinox (p. 427)
stratopause (p. 423)
stratosphere (p. 423)
summer solstice (p. 427)
thermosphere (p. 424)
Tropic of Cancer (p. 427)
Tropic of Capricorn (p. 427)
tropopause (p. 423)
troposphere (p. 422)
ultraviolet (p. 433)
vernal equinox (p. 427)
visible light (p. 432)
weather (p. 416)
winter solstice (p. 427)

13

Moisture

Opposite page: *Lenticular clouds over the eastern Sierra Nevada in California. (Photo by Galen Rowell)*

Left: *Clouds at sunset, Tucson, Arizona. (Photo by E. J. Tarbuck)*

Above: *Clouds at sunset, Tucson, Arizona. (Photo by E. J. Tarbuck)*

A s you observe day-to-day weather changes, many questions may come to mind concerning the role of water in the air. What is humidity, and how is it measured? Why do clouds form on some occasions but not on others? What processes produce clouds and precipitation? Why are some clouds thin, white, and harmless, while others are towering, gray, and ominous? What is the difference between sleet and hail? What is fog? Are all fogs alike? This chapter investigates these and other questions as the topic of moisture in the air is examined.

Water vapor constitutes only a small fraction of the gases in the atmosphere, varying from nearly 0 to about 4 percent by volume. The importance of water in the air is far greater than these small percentages would indicate. Scientists agree that when it comes to understanding atmospheric processes, water vapor is the most important gas in the atmosphere. In Chapter 12 we examined the role of water vapor as an important heat-absorbing gas. Of course, water vapor is also the source of all forms of condensation and precipitation. Clouds and fog, as well as rain, snow, sleet, and hail, are among the most conspicuous and observable of weather phenomena (Figure 13.1). Although clouds of many types are familiar to all of us, their names are not. Moreover, we all expect rain to fall when certain clouds are present, yet we have no idea of the complex processes that must take place in a cloud to produce rain. The formation of a single average raindrop requires water from nearly one million cloud droplets. What mechanisms foster the creation of raindrops? The large variations in the amount of precipitation from place to place as well as local differences from time to time have a significant impact on the nature of the physical landscape and on people's life-styles as well.

CHANGES OF STATE

Water vapor is an odorless, colorless gas that mixes freely with the other gases of the atmosphere. Unlike oxygen and nitrogen, the two most abundant components of the atmosphere, water vapor changes from one state of matter (solid, liquid, or gas) to another at the temperatures and pressures experienced near the surface of the earth. It is because of this occurrence, which allows water to leave the oceans as a gas and return again as a liquid, that the vital hydrologic cycle exists. The processes that involve a change of state require that heat be absorbed or released (Figure 13.2). This heat energy is measured in calories. One **calorie** is the amount of heat required to raise the temperature of 1 gram of water 1°C. Thus when 10 calories of heat are added to 1 gram of water, a 10°C temperature rise occurs.

Under certain conditions, heat may be added to a substance without an accompanying temperature change. This situation occurs during a change in state. When heat is supplied to a glass of ice water (0°C), for example, the temperature remains constant until all the ice has melted. Where has the heat

Figure 13.1
Clouds are among the most observable of weather phenomena. (Photo by Henry Lansford)

release of latent heat aids the formation of the towering clouds often seen on warm summer days, as well as being the major source of energy for violent storms such as hurricanes.

The process of converting a liquid to a gas is termed **evaporation.** It takes approximately 600 calories of energy to convert 1 gram of water to water vapor (Figure 13.2). The energy absorbed by the water molecules during evaporation is used solely to give them the motion needed to escape the surface of the liquid and become a gas. This energy is referred to as *latent heat of vaporization*. During the process of evaporation it is the higher-temperature (faster-moving) molecules that escape the surface. As a result, the average molecular motion (temperature) of the remaining water is reduced—hence the common expression "evaporation is a cooling process." You have undoubtedly experienced this cooling effect when stepping dripping wet from a swimming pool or bathtub.

Condensation denotes the process whereby water vapor changes to the liquid state. In order for condensation to occur, the water molecules must release energy (*latent heat of condensation*) equivalent to what was absorbed during evaporation. This energy plays an important role in producing violent weather and can act to transfer great quantities of heat energy from tropical oceans to more poleward locations. When condensation occurs in the atmosphere, it results in the formation of such phenomena as fog and clouds.

Melting is the process by which a solid is changed to a liquid. It requires absorption of approximately 80 calories of energy per gram of water. **Freezing,** the reverse process, releases these 80 calories per gram as *latent heat of fusion*.

gone? In this case, the energy was used to disrupt the internal crystalline structure of the ice cubes and cause them to melt. Because this heat energy is not associated with a temperature change, it is referred to as **latent** (meaning hidden) **heat.** This energy is not available as heat until the liquid returns to the solid state. As we shall examine later, latent heat plays an important role in many atmospheric processes. In particular, the energy derived from the

Figure 13.2
Changes of state.

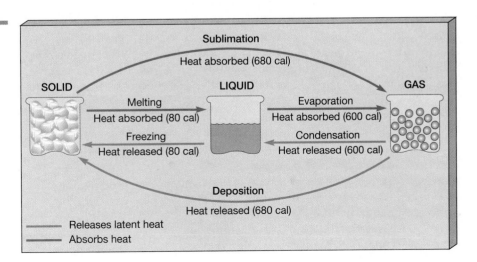

Figure 13.3
Frost on a window pane. (Photo by Bob Firth/International Stock Photo)

The last of the processes illustrated in Figure 13.2 are sublimation and deposition. **Sublimation** is the term used to describe the conversion of a solid directly to a gas without passing through the liquid state. You may have observed this change as you watched the sublimation of dry ice (frozen carbon dioxide). The term **deposition** is used to denote the reverse process, the conversion of a vapor directly to a solid. This change occurs, for example, when water vapor is deposited as ice on solid objects such as grass or windows (Figure 13.3). These deposits are called *white frost* or *hoar frost* and are frequently referred to simply as *frost*. A household example of the process of deposition is the "frost" that accumulates in a freezer compartment. As shown in Figure 13.2, sublimation and deposition involve an amount of energy equal to the total of the other two processes.

HUMIDITY

Humidity is the general term used to describe the amount of water vapor in air. Several methods are used to quantitatively express humidity. Among these are specific humidity and relative humidity.

Before we consider these humidity measures individually, it is important to understand the concept of **saturation.** Imagine a closed jar half full of water and overlain with dry air at the same temperature. As the water begins to evaporate from the water surface, a small increase in pressure can be detected in the air above. This increase is the result of the motion of the water vapor molecules that were added to the air through evaporation. In the open atmosphere this pressure is termed **vapor pressure** and is defined as that part of the total atmospheric pressure that can be attributed to the water vapor content. In the closed container, as more and more molecules escape from the water surface, the steadily increasing vapor pressure in the air above forces more and more of these molecules to return to the liquid. Eventually the number of vapor molecules returning to the surface will balance the number leaving. At that point, the air is said to be saturated, or filled to capacity. However, if we increase the temperature of the water and air in the jar, more water will evaporate before a balance is reached. Consequently, at higher temperatures more moisture is required for saturation. Stated another way, the water vapor capacity of air is temperature dependent, with warm air having a much greater capacity than cold air. The amount of water vapor required for saturation at various temperatures is shown in Table 13.1.

One method used to express humidity, specific humidity, specifies the amount of water vapor contained in a unit of air. **Specific humidity** is expressed as

Table 13.1
Water vapor capacity (at average sea level pressure).

Temperature (°C)	Grams/kg
−40	0.1
−30	0.3
−20	0.75
−10	2
0	3.5
5	5
10	7
15	10
20	14
25	20
30	26.5
35	35
40	47

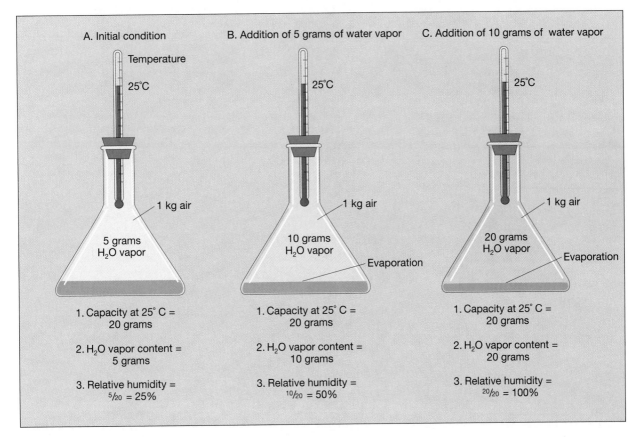

Figure 13.4
When the temperature remains constant, relative humidity will increase as water vapor is added to the air. Here the capacity remains constant at 20 grams per kilogram, while the relative humidity rises from 25 percent to 100 percent as the water vapor content increases.

the weight of water vapor per weight of a chosen mass of air, including the water vapor. Since it is measured in units of weight (usually in grams per kilogram), specific humidity is not affected by changes in pressure or temperature.

The most familiar, and perhaps the most misunderstood, term used to describe the moisture content of air is relative humidity. Stated in an admittedly oversimplified manner, **relative humidity** is the ratio of the air's water vapor content to its water vapor capacity at a given temperature. To illustrate, we see from Table 13.1 that at 25°C air is saturated when it contains 20 grams of water vapor per kilogram of air. Thus, if the air contains 10 grams per kilogram on a 25°C day, the relative humidity is expressed as 10/20 or 50 percent. Further, if air with a temperature of 25°C had a water vapor content of 20 grams per kilogram, the relative humidity would be ex-

pressed as 20/20 or 100 percent. On those occasions when the relative humidity reaches 100 percent, the air is said to be saturated.

Since relative humidity is based on the air's water vapor content as well as on its capacity, relative humidity can be changed in either of two ways. First, if moisture is added to or subtracted from air, its relative humidity will change (Figure 13.4).

The second condition that affects relative humidity is air temperature. Examine Figure 13.5 carefully and note in Part A that when air at 20°C contains 7 grams of water vapor per kilogram, it has a relative humidity of 50 percent. When the air is cooled from 20°C to 10°C as shown in Part B, the relative humidity increases from 50 percent to 100 percent. We can conclude from this that when the water vapor content remains constant, a decrease in temperature results in an increase in relative humidity. There is no

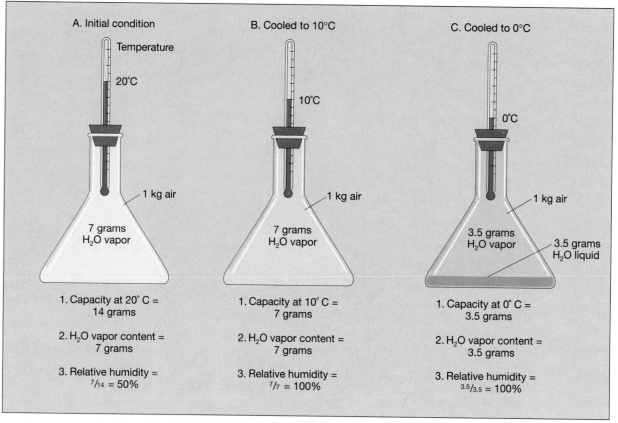

Figure 13.5
*When the water vapor content remains constant, the relative humidity can be
changed by increasing or decreasing the air temperature. In this example, when the
temperature of the air in the flask was lowered from 20°C to 10°C, the relative humidity increased from 50 percent to 100 percent. Further cooling (from 10°C to 0°C)
causes one-half of the water vapor to condense. In nature, cooling of air below its due
point generally causes condensation in the form of clouds, dew, or fog.*

reason, however, to assume that cooling ceases when
the air reaches saturation. What happens when the
air is cooled below the temperature at which saturation occurs? Part C of Figure 13.5 illustrates this situation. Notice from Table 13.1 that when the flask is
cooled to 0°C, the air is saturated at 3.5 grams of water vapor per kilogram of air. Since this flask originally contained 7 grams of water vapor, 3.5 grams of
water vapor will condense to form liquid droplets
that collect on the walls of the container. In the
meantime, the relative humidity of the air inside remains at 100 percent. This brings up an important
concept. When air aloft is cooled below its saturation
point, some of the water vapor condenses to form
clouds. Since clouds are made of liquid droplets, this

moisture is no longer part of the *water vapor* content
of the air.

We can generalize the effects of temperature on
relative humidity as follows. When the water vapor
content of air remains at a constant level, a decrease
in air temperature results in an increase in relative
humidity, and an increase in temperature causes
a decrease in relative humidity. In Figure 13.6 the
variations in temperature and relative humidity during a typical day demonstrate the relationship just
described.

Another important idea related to relative humidity is the dew-point temperature. **Dew point** is the
temperature to which air would have to be cooled in
order to reach saturation. Note that in Figure 13.5

unsaturated air at 20°C is cooled to 10°C before saturation occurs. Therefore, 10°C would be the dew-point temperature for this air. If this same parcel of air were cooled further, the air's capacity would be exceeded, and the excess vapor would condense.

In summary, relative humidity indicates how near the air is to being saturated, and specific humidity denotes the quantity of water vapor contained in that air.

HUMIDITY MEASUREMENT

Specific humidity is difficult to measure directly. Nevertheless it may be readily computed by consulting an appropriate table or graph if the air temperature and relative humidity are known.

Relative humidity is most commonly measured using a psychrometer or a hygrometer. The **psychrometer** consists of two identical thermometers mounted side by side (Figure 13.7). One thermome-

Figure 13.6
Typical daily variations in temperature and relative humidity during a spring day at Washington, D.C.

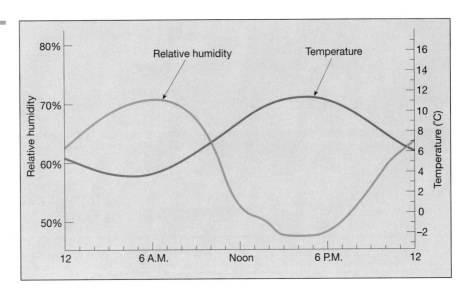

ter, the *dry-bulb*, gives the present air temperature. The other, called the *wet-bulb* thermometer, has a thin muslin wick tied around the end. To use the psychrometer, the cloth sleeve is saturated with water and a continuous current of air is passed over the wick, either by swinging the instrument freely in the air or by fanning air past it. As a consequence, water evaporates from the wick and the temperature of the wet bulb drops. The loss of heat that was required to evaporate water from the wet bulb lowers the thermometer reading.

The amount of cooling that takes place is directly proportional to the dryness of the air. The drier the air, the more the cooling. Therefore, the larger the difference between the thermometers, the lower the relative humidity; the smaller the difference, the higher the relative humidity. If the air is saturated, no evaporation will occur, and the two thermometers will have identical readings.

To determine the precise relative humidity from the thermometer readings, a standard table is used (Table 13.2). With the same information, but using

Table 13.2
Relative humidity (percent). *

Dry bulb (°C)	Dry-Bulb Temperature Minus Wet-Bulb Temperature = Depression of the Wet Bulb																					
	1	2	3	4	5	6	7	8	9	10	11	12	13	14	15	16	17	18	19	20	21	22
−20	28																					
−18	40																					
−16	48	0																				
−14	55	11																				
−12	61	23																				
−10	66	33	0																			
−8	71	41	13																			
−6	73	48	20	0																		
−4	77	54	32	11																		
−2	79	58	37	20	1																	
0	81	63	45	28	11																	
2	83	67	51	36	20	6																
4	85	70	56	42	27	14																
6	86	72	59	46	35	22	10	0														
8	87	74	62	51	39	28	17	6														
10	88	76	65	54	43	33	24	13	4													
12	88	78	67	57	48	38	28	19	10	2												
14	89	79	69	60	50	41	33	25	16	8	1											
16	90	80	71	62	54	45	37	29	21	14	7	1										
18	91	81	72	64	56	48	40	33	26	19	12	6	0									
20	91	82	74	66	58	51	44	36	30	23	17	11	5									
22	92	83	75	68	60	53	46	40	33	27	21	15	10	4	0							
24	92	84	76	69	62	55	49	42	36	30	25	20	14	9	4	0						
26	92	85	77	70	64	57	51	45	39	34	28	23	18	13	9	5						
28	93	86	78	71	65	59	53	47	42	36	31	26	21	17	12	8	4					
30	93	86	79	72	66	61	55	49	44	39	34	29	25	20	16	12	8	4				
32	93	86	80	73	68	62	56	51	46	41	36	32	27	22	19	14	11	8	4			
34	93	86	81	74	69	63	58	52	48	43	38	34	30	26	22	18	14	11	8	5		
36	94	87	81	75	69	64	59	54	50	44	40	36	32	28	24	21	17	13	10	7	4	
38	94	87	82	76	70	66	60	55	51	46	42	38	34	30	26	23	20	16	13	10	7	5
40	94	89	82	76	71	67	61	57	52	48	44	40	36	33	29	25	22	19	16	13	10	7

* To determine the relative humidity and dew point, find the air (dry-bulb) temperature on the vertical axis (far left) and the depression of the wet bulb on the horizontal axis (top). Where the two meet, the relative humidity or dew point is found. For example, use a dry-bulb temperature of 20°C and a wet-bulb temperature of 14°C. From Table 13.2, the relative humidity is 51 percent, and from Table 13.3, the dew point is 10°C.

a different table (Table 13.3), the dew-point temperature may also be calculated.

The second commonly used instrument for measuring relative humidity, the **hygrometer**, can be read directly, without the use of tables. The hair hygrometer operates on the principle that hair or certain synthetic fibers change their length in proportion to changes in the relative humidity, lengthening as relative humidity increases and shrinking as the relative humidity drops. The tension of a bundle of hairs is linked mechanically to an indicator that is calibrated between 0 and 100 percent. Thus we need only glance at the dial to determine the relative humidity. Unfortunately, the hair hygrometer is less accurate than the psychrometer. Furthermore, it requires frequent calibration and is slow in responding to changes in humidity, especially at low temperatures.

A different type of hygrometer is used in remote-sensing instrument packages such as radiosondes that transmit upper-air observations back to ground stations. The electric hygrometer contains an electrical conductor coated with a moisture-absorbing chemical. It works on the principle that the passage of current varies as the relative humidity varies.

Table 13.3
Dew-point temperature (°C).*

Dry bulb (°C)	Dry-Bulb Temperature Minus Wet-Bulb Temperature = Depression of the Wet Bulb																					
	1	2	3	4	5	6	7	8	9	10	11	12	13	14	15	16	17	18	19	20	21	22
−20	−33																					
−18	−28																					
−16	−24																					
−14	−21	−36																				
−12	−18	−28																				
−10	−14	−22																				
−8	−12	−18	−29																			
−6	−10	−14	−22																			
−4	−7	−12	−17	−29																		
−2	−5	−8	−13	−20																		
0	−3	−6	−9	−15	−24																	
2	−1	−3	−6	−11	−17																	
4	1	−1	−4	−7	−11	−19																
6	4	1	−1	−4	−7	−13	−21															
8	6	3	1	−2	−5	−9	−14															
10	8	6	4	1	−2	−5	−9	−14	−28													
12	10	8	6	4	1	−2	−5	−9	−16													
14	12	11	9	6	4	1	−2	−5	−10	−17												
16	14	13	11	9	7	4	1	−1	−6	−10	−17											
18	16	15	13	11	9	7	4	2	−2	−5	−10	−19										
20	19	17	15	14	12	10	7	4	2	−2	−5	−10	−19									
22	21	19	17	16	14	12	10	8	5	3	−1	−5	−10	−19								
24	23	21	20	18	16	14	12	10	8	6	2	−1	−5	−10	−18							
26	25	23	22	20	18	17	15	13	11	9	6	3	0	−4	−9	−18						
28	27	25	24	22	21	19	17	16	14	11	9	7	4	1	−3	−9	−16					
30	29	27	26	24	23	21	19	18	16	14	12	10	8	5	1	−2	−8	−15				
32	31	29	28	27	25	24	22	21	19	17	15	13	11	8	5	2	−2	−7	−14			
34	33	31	30	29	27	26	24	23	21	20	18	16	14	12	9	6	3	−1	−5	−12	−29	
36	35	33	32	31	29	28	27	25	24	22	20	19	17	15	13	10	7	4	0	−4	−10	
38	37	35	34	33	32	30	29	28	26	25	23	21	19	17	15	13	11	8	5	1	−3	−9
40	39	37	36	35	34	32	31	30	28	27	25	24	22	20	18	16	14	12	9	6	2	−2

* See footnote to Table 13.2.

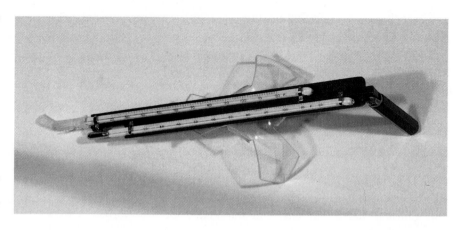

Figure 13.7
Sling psychrometer. This instrument is used to determine relative humidity and dew point. The dry-bulb thermometer gives the current air temperature. The thermometers are spun until the temperature of the wet-bulb thermometer stops declining. Then the thermometers are read and the data used in conjunction with Tables 13.2 and 13.3. (Photo by E. J. Tarbuck)

CONDENSATION ALOFT: ADIABATIC TEMPERATURE CHANGES

Up to this point we considered some basic properties of water vapor as well as how its variability is measured. We are now ready to examine one of the important roles that water vapor plays in weather processes. Recall that condensation occurs when water vapor changes to a liquid. The result of condensation may be the formation of dew, fog, or clouds. Although the forms may be different, all require saturated air in order to develop. As indicated earlier, saturation occurs either when water vapor is added to the air or, more commonly, when the air is cooled to its dew point. Heat near the earth's surface is read-

ily exchanged between the ground and the air above. Thus, radiation cooling of the earth's surface during evening hours accounts for the formation of dew and some types of fog. Yet clouds often form during the warmest part of the day. Consequently, some other mechanism must operate during cloud formation.

The process that is responsible for most cloud formation is easily visualized if you have ever pumped up a bicycle tire and noticed that the pump barrel became quite warm. The heat you felt was the consequence of the work you did on the air to compress it. When energy is used to compress air, an equivalent amount of energy is released as heat. Conversely, air that is allowed to escape from a bicycle tire cools as it expands. This results because the expanding air

Figure 13.8
Rising air cools at the dry adiabatic rate of 10° C per 1000 meters, until the air reaches the dew point and condensation (cloud formation) begins. As air continues to rise, the latent heat released by condensation reduces the rate of cooling. The wet adiabatic rate is therefore always less than the dry adiabatic rate.

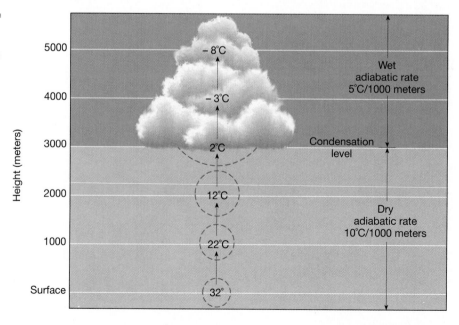

pushes (does work on) the surrounding air and must cool by an amount equivalent to the energy expended. You probably have experienced the cooling effect of an expanding gas while applying a spray deodorant. As the compressed gas propellant in the aerosol can is released, it quickly expands and cools. This drop in temperature occurs even though heat is neither added nor subtracted. Such variations are known as **adiabatic temperature changes** and result when air is compressed or allowed to expand. In summary, when air is allowed to expand, it cools, and when it is compressed, it warms.

Any time air moves upward, it passes through regions of successively lower pressure. As a result, the ascending air expands and cools adiabatically. Unsaturated air cools at the rather constant rate of 10°C for every 1000 meters of ascent (1°C per 100 meters). Conversely, descending air comes under increasingly higher pressures, compresses, and is heated 10°C for every 1000 meters of descent. This rate of cooling or heating applies only to unsaturated air and is known as the **dry adiabatic rate.** If air rises high enough, it will cool sufficiently to cause condensation. From this point on along its ascent, latent heat stored in the water vapor will be liberated. Although the air will continue to cool after condensation begins, the released latent heat works against the adiabatic process, thereby reducing the rate at which the air cools. This slower rate of cooling caused by the addition of latent heat is called the **wet adiabatic rate** of cooling. Since the amount of latent heat released depends upon the quantity of moisture present in the air, the wet adiabatic rate varies from 5°C per 1000 meters for air with a high moisture content to 9°C per 1000 meters for dry air. Figure 13.8 illustrates the role of adiabatic cooling in the formation of clouds. Note that from the surface up to the condensation level the air cools at the dry adiabatic rate. The wet adiabatic rate commences at the condensation level.

STABILITY

As we have learned, if air rises, it will cool and eventually produce clouds. Why does air rise on some occasions, but not on others? Why do the size of clouds and the amount of precipitation vary so much when air does rise? The answers to these questions are closely related to the stability of the air. Imagine, if you will, a large bubble of air with a thin flexible cover that allows it to expand but prevents it from mixing with the surrounding air. If the imaginary bubble were forced to rise, its temperature would decrease because of expansion. By comparing the bubble's temperature to that of the surrounding air, we can determine the stability of the bubble. If the bubble's temperature is lower than that of its environment, it will be heavier, and if allowed to move freely, it would sink to its original position. Air of this type, termed *stable air*, resists vertical displacement. On the other hand, if our imaginary bubble were warmer, and therefore lighter, than the surrounding air, it would continue to rise until it reached an altitude having the same temperature, much as a hot-air balloon rises as long as it is lighter than the surrounding air. This type of air is called *unstable air*.

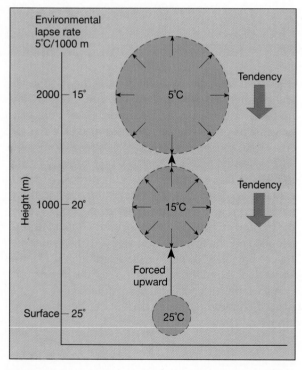

Figure 13.9
In a stable atmosphere, as an unsaturated parcel of air is lifted, it expands and cools at the dry adiabatic rate of 10° C per 1000 meters. Because the temperature of the rising parcel of air is lower than that of the surrounding environment, it will be heavier and, if allowed to do so, will sink to its original position.

Determination of Stability

In an actual situation, the stability of the air is determined by examining the temperature of the atmosphere at various heights. As you recall, this measure is termed the *environmental lapse rate*. It is important not to confuse the environmental lapse rate, which is the temperature of the atmosphere as determined from observations made by radiosondes and airplanes, with adiabatic temperature changes. The latter measure is the change in temperature due to expansion or compression as a parcel of air rises or descends.

For illustration, we will examine a situation where the environmental lapse rate is 5°C per 1000 meters (Figure 13.9). Under this condition, when air at the surface has a temperature of 25°C, the air at 1000 meters will be 5 degrees cooler, or 20°C, the air at 2000 meters will have a temperature of 15°C, and so forth. At first glance it appears that the air at the surface is lighter than the air at 1000 meters since it is 5 degrees warmer. However, if the air near the surface is unsaturated and were to rise to 1000 meters, it would expand and cool at the dry adiabatic rate of 10°C per 1000 meters. Therefore, upon reaching 1000 meters, its temperature would have dropped 10°C. Being 5 degrees cooler than its environment, it would be heavier and tend to sink to its original position. Hence, we say that the air near the surface is potentially cooler than the air aloft and therefore will not rise. Using similar reasoning, if the air at 1000 meters subsided, adiabatic heating would increase its temperature 10 degrees by the time it reached the surface, making it warmer than the surrounding air, so its buoyancy would cause it to return. The air just described is stable and resists vertical movement.

Stated quantitatively, **absolute stability** prevails when the environmental lapse rate is less than the wet adiabatic rate. Figure 13.10 depicts this situation using an environmental lapse rate of 5°C per 1000 meters and a wet adiabatic rate of 6°C per 1000 meters. Note that at 1000 meters the temperature of the surrounding air is 15°C, while the rising parcel of air has cooled to 10°C and is therefore the heavier air. Even if this stable air were to be forced above the condensation level, it would remain cooler and heavier than its environment and would have a tendency to return to the surface.

The most stable conditions occur when the temperature in a layer of air actually increases with altitude. When such a reversal occurs, a *temperature inversion* is said to exist. Temperature inversions fre-

Figure 13.10
Absolute stability prevails when the environmental lapse rate is less than the wet adiabatic rate. The rising parcel of air is therefore always cooler and heavier than the surrounding air. When stable air is forced to rise, it spreads out, producing flat layered clouds.

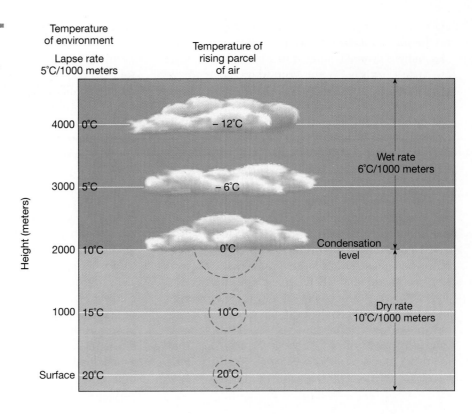

quently occur on clear nights as a result of radiation cooling at the earth's surface. Under these conditions, an inversion is created because the ground and the air immediately above will cool more rapidly than the air aloft. When warm air overlies cooler air, it acts as a lid and prevents appreciable vertical mixing. Because of this, temperature inversions are responsible for trapping pollutants in a narrow zone near the earth's surface. This idea will be explored more fully in Chapter 16.

At the other extreme, air is said to exhibit **absolute instability** when the environmental lapse rate is greater than the dry adiabatic rate. As shown in Figure 13.11, the ascending parcel of air is always

Figure 13.11
Absolute instability illustrated using an environmental lapse rate of 12° C per 1000 meters. The rising air is always warmer and therefore lighter than the surrounding air.

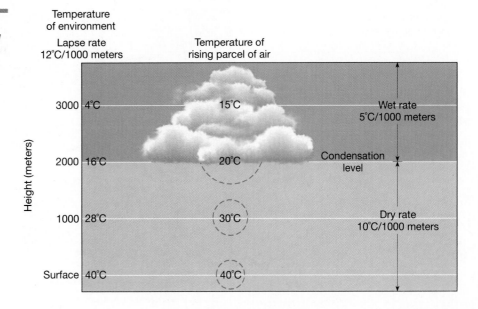

warmer than its environment and will continue to rise because of its own buoyancy. However, absolute instability is generally limited to a narrow zone near the earth's surface. On hot, sunny days the air above some surfaces is heated more than the air over adjacent surfaces. These invisible pockets of more intensely heated air, being less dense than the air aloft, will rise like a hot-air balloon. This phenomenon produces the small fluffy clouds we associate with fair weather. Occasionally, when the surface air is considerably warmer than the air aloft, clouds with great vertical development can form. Clouds of this type are associated with heavy precipitation.

A more common type of atmospheric instability is called **conditional instability.** This situation prevails when moist air has an environmental lapse rate between the dry and wet adiabatic rates (between 5 and 10°C per 1000 meters). Simply, the atmosphere is said to be conditionally unstable when it is *stable* for an *unsaturated* parcel of air and *unstable* for a *saturated* parcel of air. Notice in Figure 13.12 that the rising parcel of air is cooler than the surrounding air for the first 4000 meters. With the addition of

latent heat above the lifting condensation level, the parcel becomes warmer than the surrounding air. From this point along its ascent the parcel will continue to rise without an outside force. Thus, conditional instability depends on whether or not the rising air is saturated. The word *conditional* is used because the air must be forced upward, such as over mountainous terrain, before it becomes unstable and rises because of its own buoyancy.

In summary, the stability of air is determined by examining the temperature of the atmosphere at various heights. In simple terms, a column of air is deemed unstable when the air near the bottom of this layer is significantly warmer (less dense) than the air aloft, indicating a steep environmental lapse rate. Under these conditions the air actually turns over, since the warm air below rises and displaces the colder air aloft. Conversely, the air is considered to be stable when the temperature drops gradually with increasing altitude. The most stable conditions occur during a temperature inversion when the temperature actually increases with height. Under these conditions, there is very little vertical air movement.

Figure 13.12
Conditional instability illustrated using an environmental lapse rate of 8°C per 1000 meters, which lies between the dry adiabatic rate and the wet adiabatic rate. The rising parcel of air is cooler than the surrounding air below 4000 meters and warmer above 4000 meters.

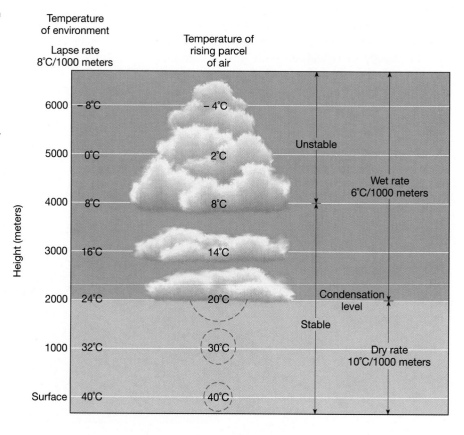

Figure 13.13
Cauliflower-shaped clouds provide evidence of unstable conditions in the atmosphere. (Photo by Henry Lansford)

Stability and Daily Weather

From the previous discussion we can conclude that stable air resists vertical movement, whereas unstable air ascends freely because of its own buoyancy. But how do these facts manifest themselves in our daily weather? Since stable air resists upward movement, we might conclude that clouds will not form when stable conditions prevail in the atmosphere. Although this seems reasonable, processes do exist that force air aloft. These will be discussed in the following section. On occasions when stable air is forced aloft, the clouds that form are widespread and have little vertical thickness when compared to their horizontal dimension, and precipitation, if any, is light to moderate. By contrast, clouds associated with unstable air are towering and are usually accompanied by heavy precipitation (Figure 13.13). For this reason we can conclude that on a dreary, overcast day with light drizzle, stable air is forced aloft. On the other hand, during a day when cauliflower-shaped clouds appear to be growing as if bubbles of hot air are surging upward, we can be fairly certain that the ascending air is unstable.

In summary, the role of stability in determining our daily weather is very important. To a large degree stability determines the type of clouds that develop and whether precipitation will come as a gentle shower or a heavy downpour.

FORCEFUL LIFTING

Earlier we learned that on some summer days, surface heating can be sufficient to cause pockets of air to be warmed more than the surrounding air. If the temperature of this air is sufficiently high and ample moisture is present, these unstable parcels will rise to produce convective clouds and occasionally precipitation. By contrast, air that is stable or conditionally unstable will not rise on its own. Rather it requires some mechanism to initiate the vertical movement. Three such mechanisms are orographic lifting, frontal wedging, and convergence.

Orographic lifting occurs when elevated terrains, such as mountains, act as barriers to the flow of air (Figure 13.14). As air ascends a mountain slope, adiabatic cooling often generates clouds and copious precipitation. In fact, many of the rainiest places in the world are located on windward mountain slopes (see Box 13.4).

In addition to providing the lift to render air unstable, mountains remove additional moisture in other ways. By slowing the horizontal flow of air, they cause convergence and retard the passage of storm systems. Moreover, the irregular topography of mountains enhances differential heating and surface instability. These combined effects account for the generally higher precipitation associated with mountainous regions compared with surrounding lowlands.

Figure 13.14
Orographic lifting.

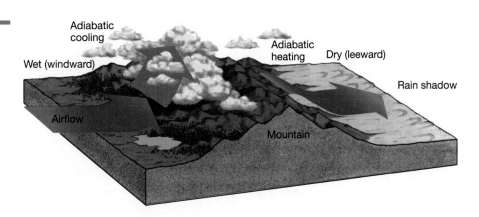

By the time air reaches the leeward side of a mountain, much of the moisture has been lost; and if the air descends, it warms, making condensation and precipitation even less likely. As shown in Figure 13.14, the result can be a **rainshadow desert.** The Great Basin desert of the western United States lies only a few hundred kilometers from the Pacific but is effectively cut off from the ocean's moisture by the imposing Sierra Nevada (Figure 13.15). The Gobi desert of Mongolia, the Takla Makan of China, and the Patagonia desert of Argentina are other examples of deserts found on the leeward sides of mountains.

If orographic lifting were the only mechanism that forced air aloft, the relatively flat central portion of North America would be an expansive desert. Fortunately, this is not the case. **Frontal wedging** occurs when cool air acts as a barrier over which warmer, less dense air rises. Figure 13.16 illustrates frontal wedging of stable and unstable air. As this figure shows, forceful lifting is important in producing clouds. The stability of the air, however, determines to a great extent the type of clouds formed and the amount of precipitation that may be expected. It should be noted that these weather-producing fronts

Figure 13.15
Rainshadow desert. The arid conditions in Death Valley can be partially attributed to the adjacent mountains, which effectively remove the moisture from air originating over the Pacific. (Photo by E. J. Tarbuck)

Figure 13.16
A. *When stable air is lifted, layered clouds usually result.*
B. *When warm, unstable air is forced to rise over cooler air, "towering" clouds develop.*

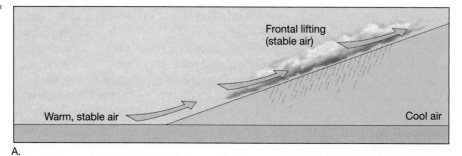

Frontal lifting (stable air)

Warm, stable air

Cool air

A.

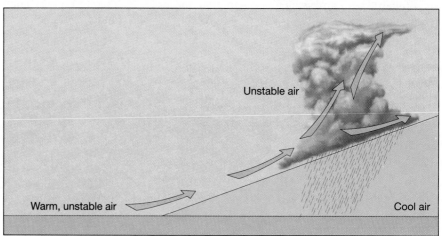

Unstable air

Warm, unstable air

Cool air

B.

are associated with storm systems called *middle-latitude cyclones*. Because these cyclones are responsible for producing a high proportion of the precipitation in the middle latitudes, we will examine them closely in Chapter 15.

Whenever air flows together, **convergence** is said to occur. Such flow results in general upward movement, because when air converges, it occupies a smaller and smaller area, which necessitates that the height of the air column increase. Consequently, air within the column must move upward.

The Florida peninsula provides an excellent example of the role that convergence can play in initiating cloud development and precipitation. On warm days the airflow is from the ocean to the land along both coasts of Florida. This leads to general convergence over the peninsula. Such a pattern of air movement and the uplift that results is aided by intense solar heating. The result is the greatest frequency of midafternoon thunderstorms in the United States. More importantly, convergence as a mechanism of forceful lifting is a major contributor to the stormy weather associated with middle-latitude cyclones and hurricanes.

NECESSARY CONDITIONS FOR CONDENSATION

As we learned earlier in this chapter, condensation occurs when water vapor in the air changes to a liquid. The result of this process may be dew, fog, or clouds. Although these forms of condensation are quite different, they have two things in common. First, for any form of condensation to occur, the air must be saturated. Saturation occurs either when air is cooled to its dew point, which most commonly happens, or when water vapor is added to the air. Second, there generally must be a surface on which the water vapor may condense. When dew occurs, objects at or near the ground serve this purpose. When condensation occurs in the air above the ground, tiny bits of particulate matter, known as **condensation nuclei,** serve as surfaces for water vapor condensation. The importance of these nuclei should be noted, since in their absence a relative humidity well in excess of 100 percent is needed to produce clouds. However, condensation nuclei such as microscopic dust, smoke, and salt particles are profuse in the lower atmosphere. Due to this abundance of particles, relative humidity rarely exceeds 101 percent. Some particles, such as salt from the ocean, are particularly good nuclei because they absorb water. These particles are termed **hygroscopic** ("water-seeking") **nuclei.**

When condensation takes place, the initial growth rate of cloud droplets is rapid but diminishes quickly because the excess water vapor is readily consumed by the numerous competing particles. This results in the formation of a cloud consisting of millions upon millions of tiny water droplets, all so fine that they remain suspended in air. The slow growth of these cloud droplets by additional condensation and the immense size difference between cloud droplets and raindrops suggest that condensation alone is not responsible for the formation of drops large enough to fall as rain. We will first examine clouds and then return to the question of how precipitation forms.

CLOUDS

Clouds are among the most conspicuous and observable aspects of the atmosphere and its weather. **Clouds** are a form of condensation best described as visible aggregates of minute droplets of water or of tiny crystals of ice. In addition to being prominent and sometimes spectacular features in the sky, clouds are of continual interest to meteorologists, because they provide a visible indication of what is going on in the atmosphere. Anyone who observes clouds with the hope of recognizing different types often finds that there is a bewildering variety of these familiar white and gray masses streaming across the sky. Still, once one comes to know the basic classification scheme for clouds, most of the confusion vanishes.

Clouds are classified on the basis of their appearance and height (Figure 13.17). Three basic forms are recognized: cirrus, cumulus, and stratus. **Cirrus** clouds are high, white, and thin. They are separated, or detached, and form delicate veil-like patches or extended wispy fibers that often have a feathery appearance. The **cumulus** form consists of globular individual cloud masses. Normally they exhibit a flat base and have the appearance of rising domes or towers. Such clouds are frequently described as having a cauliflower structure. **Stratus** clouds are best described as sheets or layers that cover much or all of the sky. While there may be minor breaks, there are no distinct individual cloud units. All other clouds reflect one of these three basic forms or are combinations or modifications of them.

Figure 13.17
Classification of clouds according to height and form. (After Ward's Natural Science Establishment, Inc., Rochester, N.Y.)

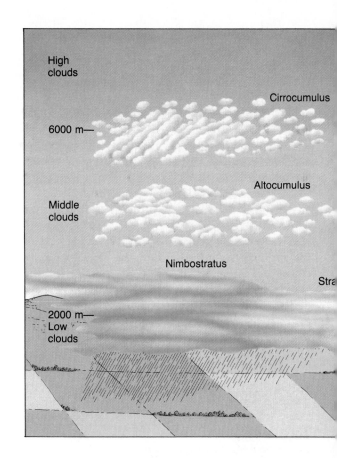

Three levels of cloud heights are recognized: high, middle, and low. **High clouds** normally have bases above 6000 meters, **middle clouds** generally occupy heights from 2000 to 6000 meters, and **low clouds** form below 2000 meters. The altitudes listed for each height category are not hard and fast. There is some seasonal as well as latitudinal variation. For example, at high latitudes or during cold winter months in the mid-latitudes, high clouds are often found at lower altitudes.

Three cloud types make up the family of high clouds (above 6000 meters): *cirrus, cirrostratus,* and *cirrocumulus.* Cirrus clouds are thin and delicate and sometimes appear as hooked filaments called "mares' tails" (Figure 13.18A). As the names suggest, cirrocumulus clouds consist of fluffy masses (Figure 13.18B), whereas cirrostratus clouds are flat layers (Figure 13.18C). Because of the low temperatures and small quantities of water vapor present at high altitudes, all high clouds are thin and white and are made up of ice crystals. Further, these clouds are not considered precipitation makers. However, when cirrus clouds are followed by cirrocumulus clouds and increased sky coverage, they may warn of impending stormy weather.

Clouds that appear in the middle range (2000 to 6000 meters) have the prefix *alto* as part of their name. *Altocumulus* clouds are composed of globular masses that differ from cirrocumulus clouds in that they are larger and denser (Figure 13.18D). *Altostratus* clouds create a uniform white to grayish sheet covering the sky with the sun or moon visible as a bright spot (Figure 13.18E). Although infrequent, precipitation in the form of light snow or drizzle may accompany these clouds.

There are three members in the family of low clouds: *stratus, stratocumulus,* and *nimbostratus.* Stratus are a uniform fog-like layer of clouds that frequently covers much of the sky. On occasions these clouds may produce light amounts of precipitation. When stratus clouds develop a scalloped bottom that appears as long parallel rolls or broken globular patches, they are called stratocumulus clouds. Nimbostratus clouds derive their name from the Latin

nimbus, which means "rainy cloud," and *stratus,* which means "to cover with a layer" (Figure 13.18F). As the name suggests, nimbostratus clouds are one of the chief precipitation producers. Nimbostratus clouds form in association with stable conditions. We might not expect clouds to grow or persist in stable air, yet cloud growth of this type is common when air is forced to rise, as occurs along a front or near the center of a cyclone where converging winds cause air to ascend. Such forced ascent of stable air leads to the formation of a stratified cloud layer that is large horizontally compared to its depth.

Some clouds do not fit into any one of the three height categories mentioned. Such clouds have their bases in the low height range but often extend upward into the middle or high altitudes. Consequently, clouds in this category are called **clouds of vertical development.** They are all related to one another and are associated with unstable air. Although *cumulus* clouds are often connected with "fair" weather, they may grow dramatically under the proper circumstances (Figure 13.18G). Once upward

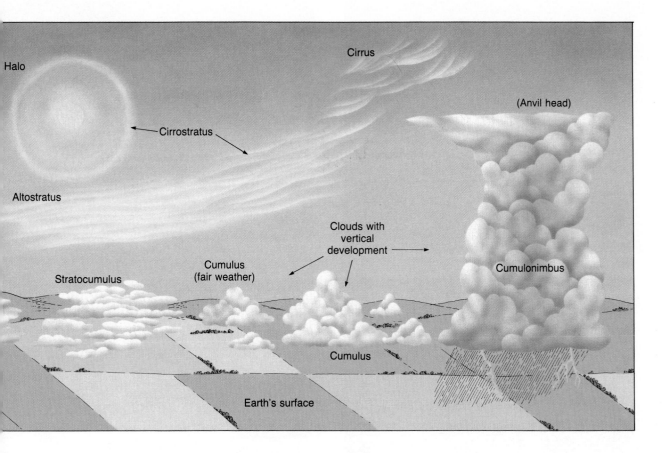

Cirrus

Halo

(Anvil head)

Cirrostratus

Altostratus

Clouds with
vertical
development

Cumulonimbus

Cumulus
(fair weather)

Stratocumulus

Cumulus

Earth's surface

movement is triggered, acceleration is powerful and clouds with great vertical extent form. The end result is often a towering cloud, called a *cumulonimbus*, that may produce rain showers or a thunderstorm (Figure 13.18H).

Definite weather patterns can often be associated with particular clouds or certain combinations of cloud types, so it is important to become familiar with cloud descriptions and characteristics. Table 13.4, on page 471, lists the ten basic cloud types that are recognized internationally and gives some characteristics of each.

FOG

Fog is generally considered to be an atmospheric hazard. When it is light, visibility is reduced to 2 or 3 kilometers. However, when it is dense, visibility may be cut to a few tens of meters or less, making travel by any mode not only difficult but often dangerous as well. Officially, visibility must be reduced to 1 kilo-

meter or less before fog is reported. While this figure is arbitrary, it does permit a more objective criterion for comparing fog frequencies at different locations.

Fog is defined as a cloud with its base at or very near the ground. Physically, there is no basic difference between a fog and a cloud; the appearance and structure of both are the same. The essential difference is the method and place of formation. While clouds result when air rises and cools adiabatically, fogs (with the exception of upslope fogs) are the consequence of radiation cooling or the movement of air over a cold surface. In other circumstances, fogs form when enough water vapor is added to the air to bring about saturation (evaporation fogs).

Fogs Caused by Cooling

When warm, moist air is blown over a cool surface, the result may be a blanket of fog called **advection fog** (Figure 13.19). Examples of such fogs are very common. The foggiest location in the United States, and perhaps in the world, is Cape Disappointment,

A. Cirrus

B. Cirrocumulus

C. Cirrostratus

Figure 13.18
These photos depict common forms of several different cloud types. (Photos A, B, D, E, F, G, and H by E. J. Tarbuck. Photo C courtesy of Ward's Natural Science Establishment, Inc. Photo H by Tom Bean/DRK Photo)

D. Altocumulus

E. Altostratus

F. Nimbostratus

G. Cumulus

H. Cumulonimbus

Figure 13.19
Advection fog rolling into San Francisco Bay. (Photo by Horst Osterwinter/International Stock Photo)

Washington. The name is indeed appropriate, since this station averages 2552 hours of fog each year. The fog experienced at Cape Disappointment, as well as that at other West Coast locations, is produced when warm, moist air from the Pacific Ocean moves over the cold California Current and is then carried onto shore by the prevailing winds. Advection fogs are also quite common in the winter season when warm air from the Gulf of Mexico is blown over cold, often snow-covered surfaces of the Midwest and East.

Radiation fog forms on cool, clear, calm nights, when the earth's surface cools rapidly by radiation. As the night progresses, a thin layer of air in contact with the ground is cooled below its dew point. As the air cools and becomes heavier, it drains into low areas, resulting in "pockets" of fog. The largest "pockets" are often river valleys, where rather thick accumulations may occur (Figure 13.20).

As its name implies, **upslope fog** is created when relatively humid air moves up a gradually sloping plain or, in some cases, up the steep slopes of a mountain (Figure 13.21). Due to the upward movement, air expands and cools adiabatically. If the dew point is reached, an extensive layer of fog may form. In the United States, the Great Plains offers an excellent example. When humid easterly or southeasterly winds move westward from the Mississippi River toward the Rocky Mountains, the air gradually rises, resulting in an adiabatic decrease of about 13°C. When the difference between the air temperature and dew point of westward-moving air is less than 13°C, an extensive fog often results in the western plains.

Evaporation Fogs

When cool air moves over warm water, enough moisture may evaporate from the water surface to produce saturation. As the rising water vapor meets the cold air, it immediately recondenses and rises with the air that is being warmed from below. Since the water has a steaming appearance, the phenomenon is called **steam fog** (Figure 13.22). Steam fog is fairly common over lakes and rivers in the fall and early

Figure 13.20
Radiation fog in downtown Peoria, Illinois. (Photo by David Zalaznik/Peoria Journal Star)

winter, when the water may still be relatively warm and the air is rather crisp. Steam fog is often quite shallow because as the steam rises it reevaporates in the unsaturated air above.

When frontal wedging occurs, warm air is lifted over colder air. If the resulting clouds yield rain, and the cold air below is near the dew point, enough rain will evaporate to produce fog. A fog formed in this manner is called **frontal fog**, or **precipitation fog.** The result is a more or less continuous zone of con-

densed water droplets reaching from the ground up through the clouds.

In summary, both steam fog and frontal fog result from the addition of moisture to a layer of air. As we learned, the air is usually cool or cold and already near saturation. Since air's capacity to hold water vapor at low temperatures is small, only a relatively modest amount of evaporation is necessary to produce saturated conditions and fog.

Table 13.4
Cloud types and characteristics.

Cloud Family and Height	Cloud Type	Characteristics
High clouds—above 6000 meters (20,000 feet)	Cirrus	Thin, delicate, fibrous, ice-crystal clouds. Sometimes appear as hooked filaments called "mares' tails." (Figure 13.18A)
	Cirrocumulus	Thin, white, ice-crystal clouds in the form of ripples, waves, or globular masses all in a row. May produce a "mackerel sky." Least common of the high clouds. (Figure 13.18B)
	Cirrostratus	Thin sheet of white, ice-crystal clouds that may give the sky a milky look. Sometimes produce halos around the sun or moon. (Figure 13.18C)
Middle clouds—2000–6000 meters (6500–20,000 feet)	Altocumulus	White to gray clouds often composed of separate globules; "sheep-back" clouds. Figure 13.18D
	Altostratus	Stratified veil of clouds that are generally thin and may produce very light precipitation. When thin, the sun or moon may be visible as a "bright spot," but no halos are produced. (Figure 13.18E)
Low clouds—below 2000 meters (6500 feet)	Stratocumulus	Soft, gray clouds in globular patches or rolls. Rolls may join together to make a continuous cloud
	Stratus	Low uniform layer resembling fog but not resting on the ground. May produce drizzle.
	Nimbostratus	Amorphous layer of dark gray clouds. One of the chief precipitation-producing clouds. (Figure 13.18F)
Clouds of vertical development—500–18,000 meters (1600–60,000 feet)	Cumulus	Dense, billowy clouds often characterized by flat bases. May occur as isolated clouds or closely packed. (Figure 13.18G)
	Cumulonimbus	Towering cloud sometimes spreading out on top to form an "anvil head." Associated with heavy rainfall, thunder, lightning, hail, and tornadoes. (Figure 13.18H)

Figure 13.21
Patches of upslope fog forming on the sides of a mountain valley. (Courtesy of Ward's Natural Science Establishment, Inc., Rochester, N.Y.)

Figure 13.22
Early-morning steam fog on the Madison River, Montana. (Photo by Robert Winslow)

FORMATION OF PRECIPITATION

Although all clouds contain water, why do some produce precipitation while others drift placidly overhead? This seemingly simple question perplexed meteorologists for many years. First, cloud droplets are very small, averaging less than 10 micrometers in diameter (for comparison, a human hair is about 75 micrometers in diameter). Because of their small size, cloud droplets fall incredibly slowly. Theoretically, an average cloud droplet falling from a cloud base at 1000 meters would require about 48 hours to reach the ground. Of course, it would never complete its journey. Even falling through humid air, a cloud droplet would evaporate before it fell a few meters below the cloud base. In addition, clouds are made up of many billions of these droplets, all competing for the available water vapor; thus, their continued growth via condensation is very slow.

A raindrop large enough to reach the ground without evaporating contains roughly a million times more water than a cloud droplet. Therefore, for precipitation to form, millions of cloud droplets must somehow coalesce (join together) into drops large enough to sustain themselves during their descent. Two mechanisms have been proposed to explain this phenomenon: the Bergeron process and the collision-coalescence process.

The **Bergeron process,** named after its discoverer, relies on two interesting properties of water. First, cloud droplets do not freeze at 0°C as expected. In fact, pure water suspended in air does not freeze until it reaches a temperature of nearly −40°C (−40°F). Water in the liquid state below 0°C is referred to as **supercooled.** Supercooled water will readily freeze if sufficiently agitated. This explains why airplanes collect ice when they pass through a liquid cloud composed of supercooled droplets. In addition, supercooled droplets will freeze upon contact with solid particles that have a crystal form closely resembling that of ice. These materials are termed **freezing nuclei.** The need for freezing nuclei to initiate the freezing process is similar to the requirement for condensation nuclei in the process of condensation. However, in contrast to condensation nuclei, freezing nuclei are very sparse in the atmosphere and do not generally become active until the temperature reaches −10°C or less. Only at temper-

Table 13.5
Relative humidity with respect to ice when relative humidity with respect to water is 100 percent.

Temperature (°C)	Relative Humidity with Respect to:	
	Water (%)	Ice (%)
0	100	100
−5	100	105
−10	100	110
−15	100	116
−20	100	121

atures well below freezing will ice crystals begin to form in clouds, and even at that, they will be few and far between. Once ice crystals form, they are in direct competition with the supercooled droplets for the available water vapor.

This brings us to the second interesting property of water. When air is saturated (100 percent relative humidity) with respect to water, it is supersaturated (relative humidity greater than 100 percent) with respect to ice. Table 13.5 shows that at −10°C, when the relative humidity is 100 percent with respect to water, the relative humidity with respect to ice is nearly 110 percent. Thus, ice crystals cannot coexist with water droplets, because the air always "appears" supersaturated to the ice crystals. So the ice crystals begin to consume the "excess" water vapor, which lowers the relative humidity near the surrounding droplets. In turn, the water droplets evaporate to replenish the diminishing water vapor, thereby providing a continual source of vapor for the growth of the ice crystals (Figure 13.23).

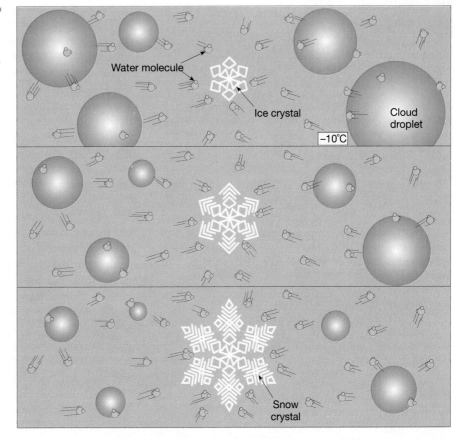

Figure 13.23
The Bergeron process. Ice crystals grow at the expense of cloud droplets until they are large enough to fall. The size of these particles has been greatly exaggerated.

Water molecule

Ice crystal

−10°C

Cloud droplet

Snow crystal

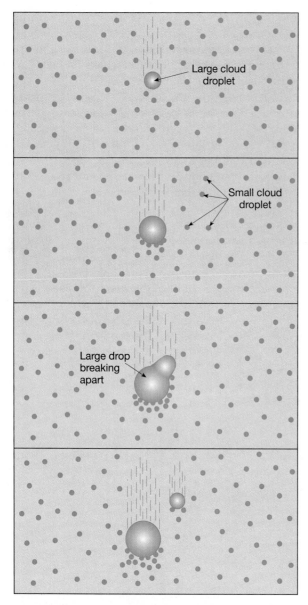

Figure 13.24
The collision-coalescence process. Because large cloud droplets fall more rapidly than smaller droplets, they are able to sweep up the smaller ones in their path and grow. Most cloud droplets are so small that the motion of the air keeps them suspended. Even if these small cloud droplets were to fall, they would evaporate long before reaching the surface.

Because the level of supersaturation with respect to ice can be quite great, the growth of ice crystals is generally rapid enough to generate crystals large enough to fall. During their descent, these ice crystals enlarge as they intercept cloud drops, which freeze upon them. Air movement will sometimes break up these delicate crystals and the fragments will serve as freezing nuclei. A chain reaction develops, producing many ice crystals, which by accretion form into large crystals called snowflakes. When the surface temperature is above 4°C (39°F), snowflakes usually melt before they reach the ground and continue their descent as rain. Even a summer rain may have begun as a snowstorm in the clouds overhead.

Cloud seeding to produce precipitation utilizes the Bergeron process just described. By adding freezing nuclei (commonly silver iodide) to supercooled clouds, the growth of these clouds can be markedly changed. This process is discussed in greater detail in Chapter 16.

A few decades ago, meteorologists believed that the Bergeron process was responsible for the formation of most precipitation. However, it was discovered that copious rainfall can be associated with clouds located well below the freezing level (*warm clouds*), particularly in the tropics. This led to the proposal of a second mechanism thought to produce precipitation—the **collision-coalescence process.**

Research has shown that clouds composed entirely of liquid droplets must contain droplets larger that 20 micrometers if precipitation is to form. These large droplets form when "giant" condensation nuclei are present, or when hygroscopic particles such as sea salt exist. Hygroscopic particles begin to remove water vapor from the air at relative humidities under 100 percent and can grow quite large. Since the rate at which drops fall is size dependent, these "giant" droplets fall most rapidly. As such, they collide with the smaller, slower droplets and coalesce. Becoming larger in the process, they fall even more rapidly (or in an updraft, rise more slowly), increasing their chances of collision and rate of growth (Figure 13.24). After a great many such collisions they are large enough to fall to the surface without completely evaporating. Due to the number of collisions required for growth to raindrop size, droplets in clouds with great vertical thickness and abundant moisture have a better chance of reaching the required size. Updrafts also aid in this process since they allow the droplets to traverse the cloud repeatedly. Raindrops can grow to a maximum size of 5 millimeters when

Most everyone has seen *condensation trails,* or *contrails* as they are popularly called, in the wake of an aircraft flying on a clear day. Most contrails are produced by jet aircraft engines that expel large quantities of hot, moist air. As this moist, hot air mixes with the cooler air aloft, a streamline cloud is produced. Because it often takes a few seconds for sufficient cooling to occur, the visible contrail frequently forms somewhat behind the aircraft. Why do contrails occur on some occasions and not on others? Contrails form under the same conditions as any other cloud, that is, when the air reaches saturation and condensation nuclei exist in sufficient numbers. Most contrails form when the exhaust gases add sufficient moisture to the air to cause saturation. Further, it has been demonstrated that the exhaust gases of aircraft engines donate condensation nuclei that promote the development of contrails.

Although contrails are frequent occurrences, most have very short life spans. Once formed, these streamlined clouds continue to mix with the cold, dry air aloft and ultimately evaporate. However, if the air aloft is near saturation, contrails may exist for long periods of time. Under these conditions the air flow aloft usually spreads the streamlike cloud into more extensive bands of cirrus clouds (Figure 13.A). With the increase in air traffic during the last few decades, an overall increase in cloudiness has been recorded, particularly near major transportation hubs. Research is currently underway to find ways to reduce the formation of contrails. One study suggests that exhaust contrails might be prevented by adding very tiny particles to engine exhaust. These condensation nuclei, being much smaller than the particles found in exhaust gases, would produce cloud droplets that are too small to be visible.

Figure 13.A
Aircraft contrails. Condensation trails produced by jet aircraft often spread out to form broad bands of cirrus clouds. (Photo by E. J. Tarbuck)

Figure 13.25
Glaze forms when supercooled raindrops freeze on contact with objects. (Photo by E. J. Tarbuck)

they fall at the rate of 33 kilometers (20 miles) per hour. At this size and speed the water's surface tension, which holds the drop together, is surpassed by the drag imposed by the air, which in turn succeeds in pulling the drops apart. The resulting breakup of a large raindrop produces numerous smaller drops which begin anew the task of sweeping up cloud droplets. Drops that are less than 0.5 millimeter upon reaching the ground are termed *drizzle* and require about ten minutes to fall from a cloud 1000 meters (3280 feet) overhead.

The collision-coalescence process is not quite as simple as described. First, as the larger droplets descend, they produce an air stream around them similar to that produced by an automobile when it is driven rapidly down the highway. If an automobile is driven at night and we use the bugs that are often out to be analogous to the cloud droplets, it is easy to visualize how most cloud droplets are swept aside. The larger the cloud droplet (or bug), the better chance it will have of colliding with the giant droplet (or car). Second, collision does not guarantee coalescence. Experimentation has indicated that the presence of atmospheric electricity may be the key to what holds these droplets together once they collide. If a droplet with a negative charge should collide with a positively charged droplet, their electrical attraction may bind them together.

FORMS OF PRECIPITATION

Much of the world's precipitation begins as snow crystals or other solid forms such as hail or graupel. Entering the warmer air below the cloud, these ice particles often melt and reach the ground as raindrops. In some parts of the world, particularly the subtropics, precipitation often forms in clouds that are warmer than 0°C (32°F). These rains frequently occur over the ocean where cloud condensation nuclei are not plentiful and those that exist vary in size. Under such conditions, cloud droplets can grow rapidly by the collision-coalescence process to produce copious amounts of rainfall.

Because atmospheric conditions vary greatly from place to place as well as seasonally, several different forms of precipitation are possible. Rain and snow are the most common and familiar, but the other forms of precipitation listed in Table 13.6 are important as well. The occurrence of sleet, glaze, and hail are often associated with important weather events. Although limited in occurrence and sporadic in both time and space, these forms, especially glaze and hail, may on occasion cause considerable damage.

Sleet is a wintertime phenomenon and refers to the fall of small, clear-to-translucent particles of ice. For sleet to be produced, a layer of air with temperatures above freezing must overlie a subfreezing layer near the ground. When the raindrops leave the warmer air and encounter the colder air below, they solidify, reaching the ground as small pellets of ice no larger than the raindrops from which they formed.

On some occasions, when the vertical distribution of temperatures is similar to that associated with the formation of sleet, freezing rain or **glaze** results instead. In such situations, the subfreezing air near the ground is not thick enough to allow the raindrops to freeze. The raindrops, however, do become supercooled as they fall through the cold air and consequently turn to ice upon colliding with solid objects. The result can be a thick coating of ice that is sufficiently heavy to break tree limbs and down power lines as well as make walking or motoring extremely hazardous (Figure 13.25).

Hail is precipitation in the form of hard, rounded pellets or irregular lumps of ice (Figure 13.26). Furthermore, large hailstones often consist of a series of nearly concentric shells of differing densities and degrees of opaqueness. Most frequently, hailstones have a diameter of about 1 centimeter, but they may vary in size from 5 millimeters to more than 10 centimeters in diameter. The largest hailstone on record fell on Coffeyville, Kansas, September 3, 1970 (Figure 13.26). With a 14-centimeter diameter and a circumference of 44 centimeters, this "giant" weighed 766 grams (over 1½ pounds)! The destructive effects of heavy hail are well-known, especially to farmers whose crops can be devastated in a few short minutes and to persons whose windows are shattered.

Hail is produced only in cumulonimbus clouds where updrafts are strong and where there is an abundant supply of supercooled water. First, rain is lifted above the freezing level by the rapidly ascending air. Once frozen, these small ice granules grow by collecting supercooled cloud droplets as they fall through the cloud. If they encounter another strong updraft, they may be carried upward again and begin the downward journey anew. Each trip above the freezing level may be represented by an additional layer of ice.

Hailstones, however, may also form from a single descent through an updraft. In this situation, the lay-

Table 13.6
Forms of precipitation.

Type	Approximate Size	State of Matter	Description
Mist	0.005 to 0.05 mm	Liquid	Droplets large enough to be felt on the face when air is moving 1 meter/second. Associated with stratus clouds.
Drizzle	Less than 0.5 mm	Liquid	Small uniform drops that fall from stratus clouds, generally for several hours.
Rain	0.5 to 5 mm	Liquid	Generally produced by nimbostratus or cumulonimbus clouds. When heavy, it can show high variability from one place to another.
Sleet	0.5 to 5 mm	Solid	Small, spherical to lumpy ice particles that form when raindrops freeze while falling through a layer of subfreezing air. Because the ice particles are small, damage, if any, is generally minor. Sleet can make travel hazardous.
Glaze	Layers 1 mm to 2 cm thick	Solid	Produced when supercooled raindrops freeze on contact with solid objects. Glaze can form a thick coating of ice having sufficient weight to seriously damage trees and power lines.
Rime	Variable accumulations	Solid	Deposits usually consisting of ice feathers that point into the wind. These delicate, frostlike accumulations form as supercooled cloud or fog droplets encounter objects and freeze on contact.
Snow	1 mm to 2 cm	Solid	The crystalline nature of snow allows it to assume many shapes including six-sided crystals, plates, and needles. Produced in supercooled clouds where water vapor is deposited as ice crystals that remain frozen during their descent.
Hail	5 mm to 10 cm or larger	Solid	Precipitation in the form of hard, rounded pellets or irregular lumps of ice. Produced in large convective, cumulonimbus clouds, where frozen ice particles and supercooled water coexist.
Graupel	2 to 5 mm	Solid	Sometimes called soft hail, graupel forms as rime collects on snow crystals to produce irregular masses of "soft" ice. Because these particles are softer than hailstones, they normally flatten out upon impact.

A.

B.

Figure 13.26
*Hail. **A.** Results of a Texas hail storm. (Photo by Warren Faidley/Weatherstock)*
***B.** A cross section of the Coffeyville hailstone. This largest recorded hailstone fell over Kansas in 1970, and weighed 0.75 kilogram (1.67 pounds). (Courtesy of the National Center for Atmospheric Research)*

ered structure is attributed to variations in the rate at which supercooled droplets accumulate and freeze, which, in turn, is related to differences in the amount of supercooled water in different parts of the cumulonimbus tower.

In either case, hailstones grow by the addition of supercooled water upon growing ice pellets. The ultimate size of the hailstone depends primarily upon three factors: the strength of the updrafts, the concentration of supercooled water, and the length of the path through the cloud.

Rime is a deposit of ice crystals formed by the freezing of supercooled fog or cloud droplets on objects whose surface temperature is below freezing. When rime forms on trees, it adorns them with its characteristic ice feathers that can be spectacular to behold (Figure 13.27). In these situations, objects such as pine needles act as freezing nuclei, causing the supercooled droplets to freeze on contact. On occasions when the wind is blowing, only the windward surfaces of objects will accumulated the layer of rime.

A.

B.

Figure 13.27
*Rime consists of delicate ice crystals that form when supercooled fog or cloud droplets freeze upon contact with objects. **A.** Rime on wire fence. **B.** Rime on pine trees.*
(Photos by Henry Lansford)

Figure 13.28
Precipitation measurement.
A. *The standard rain gauge allows for accurate rainfall measurement to the nearest 0.025 centimeter (0.01 inch). Because the cross-sectional area of the measuring tube is only one-tenth as large as the collector, rainfall is magnified 10 times.* ***B.*** *The tipping-bucket rain gauge contains two "buckets" that hold the equivalent of 0.025 centimeter (0.01 inch) of precipitation. When one bucket fills it tips and the other bucket takes its place. Each event is recorded as 0.01 inch of rainfall.*

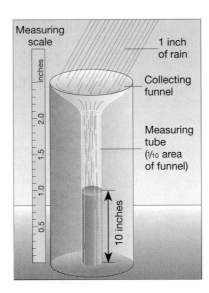

A. Standard rain gauge

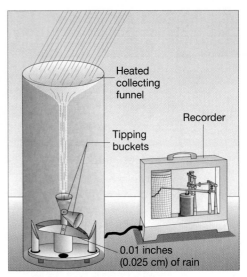

B. Tipping–bucket gauge

PRECIPITATION MEASUREMENT

The most common form of precipitation, rain, is probably the easiest to measure. Any open container that has a consistent cross section throughout can be a rain gauge. In general practice, however, more sophisticated devices are used in order to measure small amounts of rainfall more accurately as well as reduce losses resulting from evaporation. The *standard rain gauge* (Figure 13.28A) has a diameter of about 20 centimeters (8 inches) at the top. Once the water is caught, a funnel conducts the rain through a narrow opening into a cylindrical measuring tube that has a cross-sectional area only one-tenth as large as the receiver. Consequently, rainfall depth is magnified ten times, which allows for accurate measurements to the nearest 0.025 centimeter (0.01 inch); the narrow opening minimizes evaporation. When the amount of rain is less than 0.025 centimeter, it is generally reported as being a *trace* of precipitation.

In addition to the standard rain gauge, several types of recording gauges are routinely used. These instruments record not only the amount of rain but also its time of occurrence and intensity (amount per unit of time). Two of the most common gauges are the tipping-bucket gauge and weighing gauge.

A *tipping-bucket gauge* consists of two compartments, each one capable of holding 0.025 centimeter of rain, situated at the base of a 25-centimeter fun-

nel (Figure 13.28B). When one "bucket" fills, it tips and empties its water. Meanwhile, the other "bucket" takes its place at the mouth of the funnel. Each time a compartment tips, an electrical circuit is closed and the amount of precipitation is automatically recorded on a graph. The *weighing gauge*, as the name would indicate, works on a different principle. The precipitation is caught in a cylinder that rests on a spring balance. As the cylinder fills, the movement is transmitted to a pen that records the data.

No matter which rain gauge is used, proper exposure is critical. Errors arise when the gauge is shielded from obliquely falling rain by buildings, trees, or other high objects. Hence, the instrument should be at least as far away from such obstructions as the objects are high. Another cause for error is the wind. It has been shown that with increasing wind and turbulence, it becomes more difficult to collect a representative quantity of rain. To offset this effect, a windscreen is often placed around the instrument so that rain falls into the gauge and is not carried across it.

When snow records are kept, two measurements are normally taken: depth and water equivalent. Usually the depth of snow is measured with a calibrated stick. The actual measurement is not difficult, but choosing a representative spot often poses a dilemma. Even when winds are light or moderate, snow drifts freely. As a rule, it is best to take several measurements in an open place away from trees and

BOX 13.4

Precipitation Records and Mountainous Terrain

Many of the rainiest places in the world are located on windward mountain slopes. Typically these rainy areas are situated such that the mountains act as a barrier to the general circulation. Thus, the prevailing winds are forced to ascend the sloping terrain, thereby generating clouds and often abundant precipitation. A station at Mt. Waialeale, Hawaii, for example, records the highest average annual rainfall in the world, some 1234 centimeters (486 inches). The station is located on the windward (northeast) coast of the island of Kauai at an elevation of 1569 meters (5148 feet). Incredibly, only 31 kilometers (19 miles) away, lies the popular sunspot Barking Sands, with an annual precipitation that averages less than 50 centimeters (20 inches).

The greatest recorded rainfall for a twelve-month period occurred at Cherrapunji, India, where an astounding 2647 centimeters (1042 inches), over 86 feet, fell. Cherrapunji, which is located at an elevation of 1293 meters (4309 feet), lies just north of the Bay of Bengal in an ideal location to receive the full effect of India's wet, summer monsoon. Most all of this rainfall occurred in the summer, particularly during the month of July, when a record 930 centimeters (366 inches) fell. For comparison, 10 times more rain fell in a single month at Cherrapunji than falls in an average year at Chicago, Illinois.

Because mountains can be sites of abundant precipitation, they are frequently very important sources of water. This is particularly true for arid regions such as the southwestern United States. Here the snow pack that accumulates high in the mountains during the winter is a major source of water for the summer season when precipitation is light and demand is high (Figure 13.B). Reservoirs in the Sierra Nevada, for example, accumulate and store spring runoff, which is then delivered to cities such as Los Angeles by way of an extensive network of canals. The record for greatest annual snowfall in the United States goes to Paradise Ranger Station, located on Mount Rainier, where 2850 centimeters (1122 inches) of snow fell during the winter of 1971–72.

Figure 13.B
Heavy winter snowfall in the Colorado Rockies is visible along Trail Ridge Road. (Photo by Henry Lansford)

obstructions and then average them. To obtain the water equivalent, samples may be melted and then weighed or measured as rain.

The quantity of water in a given volume of snow is not constant. A general ratio of 10 units of snow to 1 unit of water is often used when exact informa- tion is not available, but the actual water content of snow may deviate widely from this figure. It may take as much as 30 centimeters of light and fluffy dry snow or as little as 4 centimeters of wet snow to pro- duce 1 centimeter of water.

Review Questions

1. Summarize the processes by which water changes from one state to another. Indi- cate whether heat energy is absorbed or liberated.

2. After studying Table 13.1, write a generalization relating temperature and the ca- pacity of air to hold water vapor.

3. How do relative and specific humidity differ?

4. Referring to Figure 13.6, answer the following questions:
   (a) During a typical day, when is the relative humidity highest? Lowest?
   (b) At what time of day would dew most likely form?
   Write a generalization relating air temperature and relative humidity.

5. If the temperature remains unchanged and the specific humidity decreases, how will relative humidity change?

6. On a cold winter day when the temperature is −10°C and the relative humidity is 50 percent, what is the specific humidity (refer to Table 13.1)? What is the specific humidity for a day when the temperature is 20°C and the relative humidity is 50 percent?

7. Explain the principle of the sling psychrometer and the hair hygrometer.

8. Using the standard tables (Tables 13.2 and 13.3), determine the relative humidity and dew-point temperature if the dry-bulb thermometer reads 16°C and the wet- bulb thermometer reads 12°C. How would the relative humidity and dew point change if the wet-bulb thermometer read 8°C?

9. On a warm summer day when the relative humidity is high, it may seem even warmer than the thermometer indicates. Why do we feel so uncomfortable on a "muggy" day?

10. Why does air cool when it rises through the atmosphere?

11. Explain the difference between environmental lapse rate and adiabatic cooling.

12. If unsaturated air at 23°C were to rise, what would its temperature be at 500 me- ters? If the dew-point temperature at the condensation level were 13°C, at what altitude would clouds begin to form?

13. Why does the adiabatic rate of cooling change when condensation begins? Why is the wet adiabatic rate not a constant figure?

14. The contents of an aerosol can are under very high pressure. When you push the nozzle on such a can, the spray feels cold. Explain.

15. How do orographic lifting and frontal wedging act to force air to rise?

16. Explain why the Great Basin area of the western United States is so dry. What term is applied to such a situation?

17. How does stable air differ from unstable air? Describe the general nature of the clouds and precipitation expected with each.

18. What is the function of condensation nuclei in cloud formation? The function of the dew point?

19. As you drink an ice-cold beverage on a warm day, the outside of the glass or bot- tle becomes wet. Explain.

20. What is the basis for the classification of clouds?

21. Why are high clouds always thin?

22. Which cloud types are associated with the following characteristics: thunder, halos, precipitation, hail, mackerel sky, lightning, mares' tails?

23. List five types of fog and discuss the details of their formation.

24. What is the difference between precipitation and condensation?

25. List the forms of precipitation and the circumstances of their formation.

26. Sometimes, when rainfall is light, the amount is reported as a trace. When this occurs, how much (or little) rain has fallen?

Key Terms

absolute instability (p. 461)

absolute stability (p. 460)

adiabatic temperature change (p. 459)

advection fog (p. 467)

Bergeron process (p. 472)

calorie (p. 450)

cirrus (p. 465)

cloud (p. 465)

cloud of vertical development (p. 466)

collision-coalescence process (p. 474)

condensation (p. 451)

condensation nuclei (p. 465)

conditional instability (p. 462)

convergence (p. 465)

cumulus (p. 465)

deposition (p. 452)

dew point (p. 454)

dry adiabatic rate (p. 459)

evaporation (p. 451)

fog (p. 467)

freezing (p. 451)

freezing nuclei (p. 472)

frontal fog (p. 471)

frontal wedging (p. 464)

glaze (p. 476)

hail (p. 476)

high cloud (p. 466)

humidity (p. 452)

hygrometer (p. 457)

hygroscopic nuclei (p. 465)

latent heat (p. 451)

low cloud (p. 466)

melting (p. 451)

middle cloud (p. 466)

orographic lifting (p. 463)

precipitation fog (p. 471)

psychrometer (p. 455)

radiation fog (p. 470)

rainshadow desert (p. 464)

relative humidity (p. 453)

rime (p. 478)

saturation (p. 452)

sleet (p. 476)

specific humidity (p. 452)

steam fog (p. 470)

stratus (p. 465)

sublimation (p. 452)

supercooled (p. 472)

upslope fog (p. 470)

vapor pressure (p. 452)

wet adiabatic rate (p. 459)

Pressure and Wind

Opposite page: *A sailboat participating in the 1992
America's Cup. (Photo courtesy of Sygma)*

Left: *Strong winter winds. (Photo by Teresa Tarbuck)*

Above: *Windy conditions. (Photo by E. J. Tarbuck)*

We have already dealt with the two elements of weather and climate that generally are of the greatest interest to people—temperature and moisture. In this chapter we shall investigate the remaining two elements—pressure and wind—that may not seem as important but indeed are. It is the wind that often brings changes in temperature and moisture conditions, and it is pressure differences that drive the wind. Moreover, we will also see that an understanding of atmospheric pressure patterns is a key factor in understanding global and regional precipitation distribution. Among the questions we will try to answer in this chapter are: How can the weight of the air be measured? What are the factors that control the winds? What is a prevailing wind? Why are "highs" and "lows" always shown on the weather map? How can the weather be predicted by checking the barometer?

Of the various elements of weather and climate, changes in air pressure are the least noticeable. When listening to a weather report, we are generally interested in moisture conditions (humidity and precipitation), temperature, and perhaps wind. It is the rare individual, however, who wonders about air pressure. Although the hour-to-hour and day-to-day variations in air pressure are not perceptible to human beings, they are very important in producing changes in our weather. Variations in air pressure from place to place are responsible for the movement of air (wind), as well as being a significant factor in weather forecasting (Figure 14.1). As we shall see, air pressure is tied very closely to the other elements of weather in a cause-and-effect relationship.

PRESSURE MEASUREMENT

In Chapter 12 we saw that air has weight; at sea level, it exerts a pressure of 1 kilogram per square centimeter (14.7 pounds per square inch). The air pressure at a particular place is simply the force exerted by the weight of the air above. With an increase in altitude the weight of the air, and thus the pressure, decreases rapidly at first, then much more slowly (see Figure 12.6 and Table 14.1).

When meteorologists measure atmospheric pressure, they employ a unit called the *millibar*. Standard sea level pressure is expressed as 1013.2 millibars. Although the millibar has been the unit of measure on all United States weather maps since January 1940, you might be better acquainted with the expression "inches of mercury," which is used by the media to describe atmospheric pressure. In the United States the National Weather Service converts millibar values to inches of mercury for public and aviation use.

The use of mercury for measuring air pressure dates from 1643, when Torricelli, a student of the famous Italian scientist Galileo, invented the **mercurial barometer**. Torricelli correctly described the atmosphere as a vast ocean of air that exerts pressure on us and all objects about us. To measure this force he filled a glass tube, which was closed at one end, with mercury. The tube was then inverted into a dish of mercury (Figure 14.2A). Torricelli found that the mercury flowed out of the tube until the weight of the column was balanced by the pressure that the atmosphere exerted on the surface of the mercury in the dish. In other words, the weight of mercury in the column equalled the weight of a similar diame-

Figure 14.1
Wind is simply air that flows horizontally with respect to the earth's surface. Air is set in motion by variations in air pressure from place to place. (Photo by C. J. Walker/Palm Beach Post)

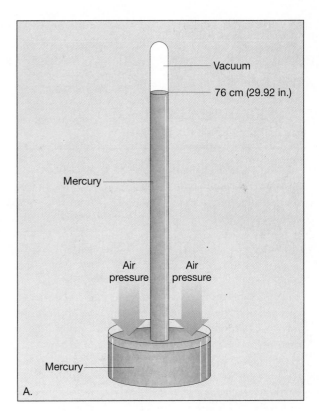

ter column of air that extended from the ground to the top of the atmosphere. Torricelli noted that when air pressure increased, the mercury in the tube rose; conversely, when air pressure decreased so did the height of the column of mercury. The length of the column of mercury, therefore, became the measure of the air pressure. With some refinements the mercurial barometer invented by Torricelli is still the standard pressure-measuring instrument used today.

Table 14.1
Pressure changes with altitude.

Altitude (kilometers)	Altitude (miles)	Pressure (millibars)
0	0	1013
1.0	0.6	899
2.0	1.2	795
3.0	1.9	701
4.0	2.5	617
5.0	3.1	540
10.0	6.2	265
20.0	12.4	55
30.0	18.6	12
40.0	24.8	3

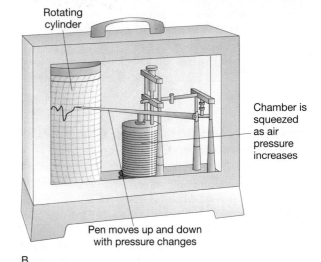

Figure 14.2
A. Simple mercurial barometer. The weight of the column of mercury is balanced by the pressure exerted on the dish of mercury by the air above. If the pressure decreases, the column of mercury falls; if the pressure increases, the column rises. B. An aneroid barograph makes a continuous record of pressure changes. One important advantage of the aneroid barometer is that it is easily adapted to a recording mechanism.

The cockpit of nearly every aircraft contains a *pressure altimeter*, an instrument that allows a pilot to determine the altitude of a plane. A pressure altimeter is essentially an aneroid barometer, and as such, responds to changes in air pressure. Recall that air pressure decreases with an increase in altitude and that the pressure distribution with height is well-established. To make an altimeter, an aneroid barometer is simply marked in meters instead of millibars. For example, we see in Table 14.1 that a pressure of 795 millibars "normally" occurs at a height of 2 kilometers (2,000 meters). Therefore, if such a pressure were experienced, the al-

timeter would indicate an altitude of 2 kilometers.

Because of temperature variations, actual pressure conditions are usually different than those represented as standard. Consequently, the altitude of an aircraft is seldom the same as that shown by its altimeter. On those occasions when the barometric pressure aloft is lower than specified by the Standard Atmosphere, the plane will be flying lower than the height indicated by the altimeter. This could be dangerous, especially if the pilot is flying a small plane during a period of poor visibility through mountainous terrain. To correct for these situations, the pilot makes an altimeter correction before takeoff and landing. In addition, up-to-date altimeter readings are usually supplied by the airport control tower.

Above 5.5 kilometers (18,000 feet) where commercial jets fly and pressure changes are more gradual, corrections cannot be made as precisely as at lower

levels. Consequently, such aircraft have their altimeters set by the Standard Atmosphere and fly paths of constant pressure instead of constant altitude. Stated another way, when an aircraft flies a constant altimeter setting, a pressure variation will result in a change in height. When pressure increases along a flight path, the plane will climb, and when pressure decreases, the plane will descend. There is little risk of midair collisions because all high-flying aircraft adjust their altitude in a similar manner. Large commercial aircraft also use radio altimeters to measure heights above the terrain. The time required for a radio signal to reach the surface and return is used to accurately determine the height of the plane above the ground. This system is not without its drawbacks. Since a radio altimeter provides the elevation above the ground rather than above sea level, a knowledge of the underlying terrain is required.

Standard atmospheric pressure at sea level equals 29.92 inches of mercury.

The need for a smaller and more portable instrument for measuring air pressure led to the development of the **aneroid** ("without liquid") **barometer**. Based on a different principle from the mercurial barometer, this instrument consists of partially evacuated metal chambers that have a spring inside, keeping them from collapsing. The metal chambers, being very sensitive to air pressure variations, change shape, compressing as the pressure increases and expanding as the pressure decreases. Aneroids are often used in making **barographs**, instruments that continuously record pressure changes (Figure 14.2B). Another important adaptation of the aneroid is its use as an **altimeter** in aircraft (see Box 14.1).

FACTORS AFFECTING WIND

We have discussed the upward movement of air and its importance in cloud formation. As important as vertical motion is, far more air is involved in horizontal movement, the phenomenon we call **wind**. Although we know that air will move vertically if it is warmer, and consequently more buoyant, than the surrounding air, what causes air to move horizontally? Simply stated, wind is the result of horizontal differences in air pressure. Air flows from areas of higher pressure to areas of lower pressure. You may have experienced this when opening a vacuum-packed can of coffee. The noise you hear is caused by air rushing from the higher pressure outside the can to the lower pressure inside. Wind is nature's at-

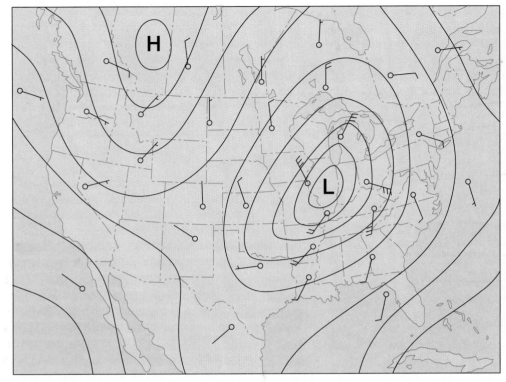

Figure 14.3
Isobars are lines used to connect places of equal barometric pressure. They show the distribution of pressure on weather maps. The lines usually curve and often join where cells of high and low pressure exist. The "arrows" indicate the expected airflow surrounding cells of high and low pressure and are plotted as "flying" with the wind. Wind speed is indicated by flags and feathers as shown along the right hand side of this drawing.

ff	Miles per hour
◎	Calm
	1–2
	3–8
	9–14
	15–20
	21–25
	26–31
	32–37
	38–43
	44–49
	50–54
	55–60
	61–66
	67–71
	72–77
	78–83
	84–89
	119–123

tempt to balance similar inequalities in air pressure. Because unequal heating of the earth's surface generates these pressure differences, solar radiation is the ultimate driving force of wind.

If the earth did not rotate, and if there were no friction, air would flow directly from areas of higher pressure to areas of lower pressure. But because both of these factors exist, wind is controlled by the following combination of forces: (1) the pressure gradient force, (2) Coriolis effect, (3) friction, and (4) the tendency of a moving object to continue moving in a straight line. The last factor is often referred to as centrifugal force. The magnitude of centrifugal force is small compared to the other forces, and thus is of minor importance, except in rapidly rotating storms such as tornadoes and hurricanes. Discussions of the other factors follow.

Pressure Gradient Force

Pressure differences create wind, and the greater these differences, the greater the wind speed. Over the earth's surface, variations in air pressure are determined from barometric readings taken at hundreds of weather stations. These pressure data are shown on a weather map using **isobars**, lines that connect places of equal air pressure (Figure 14.3). The spacing of isobars indicates the amount of pressure change occurring over a given distance and is expressed as the **pressure gradient**.

You might find it easier to visualize a pressure gradient if you think of it as being analogous to the slope of a hill. A steep pressure gradient, like a steep hill, causes greater acceleration of an air parcel than does a weak pressure gradient. Thus, the relationship between wind speed and the pressure gradient is straightforward: Closely spaced isobars indicate a steep pressure gradient and high winds, whereas widely spaced isobars indicate a weak pressure gradient and light winds. Figure 14.3 illustrates the relationship between the spacing of isobars and wind speed. Notice that wind speeds are greater in Ohio and Kentucky, where isobars are more closely spaced, than in the western states where isobars are more widely spaced.

The pressure gradient is the driving force of wind, and it has both magnitude and direction. While its magnitude is determined from the spacing of isobars, the direction of force is always from areas of higher pressure to areas of lower pressure and at right angles to the isobars. Once the air starts to move, the Coriolis effect and friction come into play, but then only to modify the movement, not to produce it.

Coriolis Effect

Figure 14.3 shows the typical air movements associated with high- and low-pressure systems. As expected, the air moves out of the regions of higher pressure and into the regions of lower pressure. However, the wind does not cross the isobars at right angles as the pressure gradient force directs. This deviation is the result of the earth's rotation and has been named the **Coriolis effect** after the French scientist who first expressed its magnitude quantitatively. All free-moving objects, including the wind, are deflected to the *right* of their path of motion in the Northern Hemisphere and to the *left* in the Southern Hemisphere. The reason for this deflection can be illustrated by imagining the path of a rocket launched from the North Pole toward a target located on the equator (Figure 14.4). If the rocket took an hour to reach its target, during its flight the earth would have rotated 15 degrees to the east. To someone standing on the earth, it would look as if the rocket veered off its path and hit the earth 15 degrees west of its target. The true path of the rocket was straight and would appear so to someone out in space looking down at the earth. It was the earth turning under the rocket that gave it its *apparent* deflection. Note that the rocket was deflected to the right of its path of motion because of the counterclockwise rotation of the Northern Hemisphere. Clockwise rotation produces a similar deflection in the Southern Hemisphere, but to the left of the path of motion. The same deflection is experienced by wind regardless of the direction it is moving.

We attribute this apparent shift in wind direction to the Coriolis effect. This deflection: (1) is always directed at right angles to the direction of airflow; (2) affects only wind direction, not wind speed; (3) is affected by wind speed (the stronger the wind, the greater the deflection); and (4) is strongest at the poles and weakens equatorward, where it eventually becomes nonexistent.

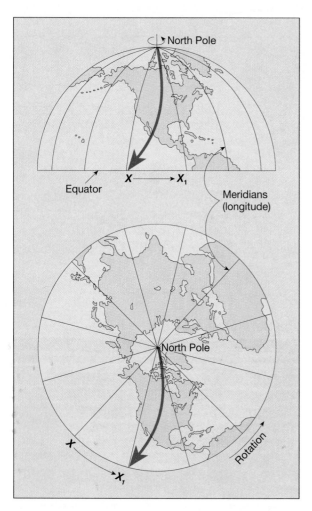

Figure 14.4
The Coriolis effect. During the rocket's flight from the North Pole to point X, the earth rotates eastward, moving point X to point X₁. The rotation gives the rocket's trajectory a curved path when plotted on the earth's surface. This deflection is termed the Coriolis effect.

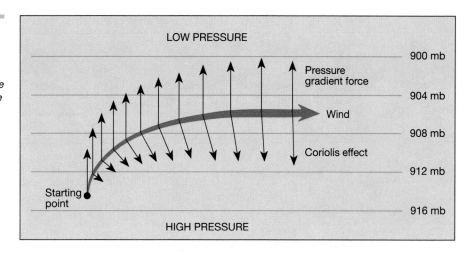

Figure 14.5
The geostrophic wind. Upper-level winds are deflected by the Coriolis effect until the Coriolis effect just balances the pressure gradient force. Above 600 meters (1970 feet), where friction is negligible, these winds will flow nearly parallel to the isobars and are called geostrophic winds.

Friction

Friction as a factor affecting wind is important only within the first few kilometers of the earth's surface. It acts to slow the movement of air and, as a consequence, alters wind direction. To illustrate friction's effect on wind direction, let us look first at a situation in which it has no role. Above the friction layer, the pressure gradient force and the Coriolis effect are directing the flow of air. Under these conditions, the pressure gradient force will cause the air to start moving across the isobars. As soon as the air starts to move, the Coriolis effect will act at right angles to this motion. The faster the wind speed, the greater the deflection. Eventually, the Coriolis effect will balance the pressure gradient force and the wind will blow parallel to the isobars (Figure 14.5). Upper-air winds generally take this path and are called **geostrophic winds**. Due to the lack of friction, geostrophic winds travel at higher speeds than do surface winds (Figure 14.6).

The most prominent features of upper-level flow are the **jet streams**. First encountered by high-flying bombers during World War II, these fast-moving "rivers" of air travel between 120 and 240 kilometers (75 and 150 miles) per hour in a west-to-east direction. One such stream is situated over the polar front, which is the zone separating cool polar air from warm subtropical air.

Below 600 meters (2000 feet), friction complicates the airflow just described. Recall that the Coriolis effect is proportional to wind speed. By lowering the wind speed, friction reduces the Coriolis effect. Since the pressure gradient force is not affected by wind speed, it wins the tug of war shown in Figure 14.7. The result is a movement of air at an angle across the isobars toward the area of lower pressure. The roughness of the terrain determines the angle of airflow across the isobars. Over the smooth ocean surface, friction is low and the angle is small. Over rugged terrain, where friction is higher, the angle that air makes as it flows across the isobars can be as great as 45 degrees. In summary, upper airflow is nearly parallel to the isobars, while the effect of friction causes the surface winds to move more slowly and cross the isobars at an angle.

CYCLONES AND ANTICYCLONES

Among the most common features on any weather map are areas designated as pressure centers. **Cyclones**, or **lows**, are centers of low pressure, and **anticyclones**, or **highs**, are high-pressure centers. As Figure 14.8 illustrates, the pressure decreases from the outside toward the center in a cyclone, whereas in an anticyclone, just the opposite is the case—the pressure increases from the outside toward the center. By knowing just a few basic facts about centers of high and low pressure, you can greatly increase your understanding of current and forthcoming weather.

Cyclonic and Anticyclonic Winds

From the preceding section, we learned that the two most significant factors that affect wind are pressure differences and the Coriolis effect. Winds move from

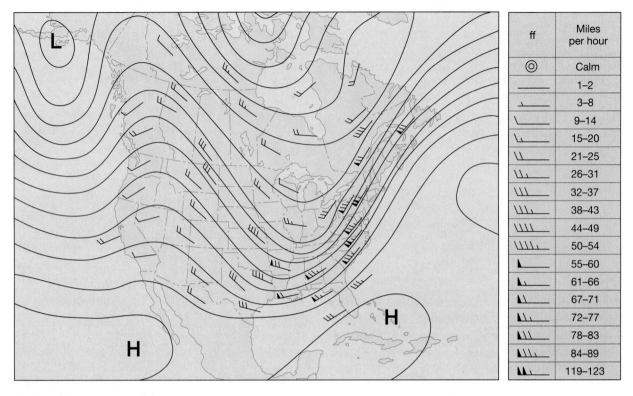

ff	Miles per hour
◎	Calm
——	1–2
⌐	3–8
⌐	9–14
⌐	15–20
⌐	21–25
⌐	26–31
⌐	32–37
⌐	38–43
⌐	44–49
⌐	50–54
▌	55–60
▌	61–66
▌	67–71
▌	72–77
▌	78–83
▌	84–89
▌	119–123

Figure 14.6
Upper-air winds. This map shows the direction and speed of the upper-air wind for a particular day. Note that the airflow is nearly parallel to the contours. These isolines are height contours for the 500-millibar level.

higher pressure to lower pressure and are deflected to the right or left by the earth's rotation. When these controls of airflow are applied to pressure centers in the Northern Hemisphere, the result is that winds blow inward and counterclockwise around a low and outward and clockwise around a high (Figure 14.8). Of course, in the Southern Hemisphere the Coriolis effect deflects the winds to the left, and therefore winds around a low are blowing clockwise, and winds around a high are moving counterclockwise. However, in whatever hemisphere, friction causes a net inflow **(convergence)** around a cyclone and a net outflow **(divergence)** around an anticyclone.

Figure 14.7
Comparison between upper-level winds and surface winds showing the effects of friction on airflow. Friction slows surface wind speed, which weakens the Coriolis effect, causing the winds to cross the isobars and move toward the lower pressure.

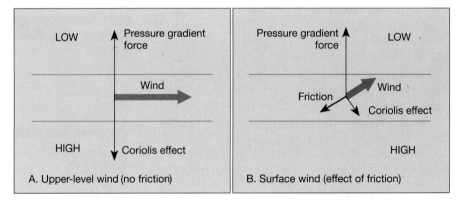

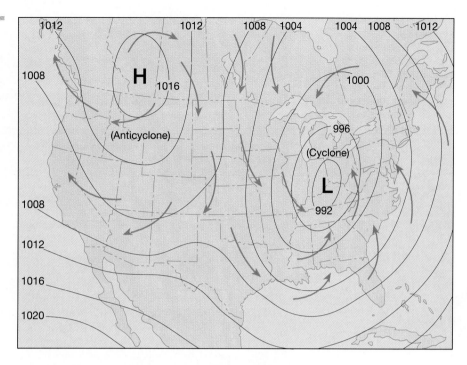

Figure 14.8
Cyclonic and anticyclonic winds in the Northern Hemisphere. Arrows show that winds blow into and counterclockwise around a low. By contrast, around a high winds blow outward and clockwise.

Weather Generalizations About Highs and Lows

Rising air is associated with cloudy conditions and precipitation, whereas subsidence produces adiabatic heating and clearing conditions. In this section we will learn how the movement of air can itself create pressure change and hence generate winds. Upon doing so, we will examine the interrelationship between horizontal and vertical flow, and their effects on the weather.

Let us first consider the situation around a surface low-pressure system where the air is spiraling inward. Here the net inward transport of air causes a shrinking of the area occupied by the air mass, a process that is termed *horizontal convergence*. Whenever air converges horizontally, it must pile up, that is, increase in height to allow for the decreased area it now occupies. This generates a "taller" and therefore heavier air column. Yet a surface low can exist only as long as the column of air exerts less pressure than that occurring in surrounding regions. We seem to have encountered a paradox—low-pressure centers cause a net accumulation of air, which increases

their pressure. Consequently, a surface cyclone should quickly eradicate itself in a manner not unlike what happens when a vacuum-packed can is opened.

In light of the preceding discussion, it should be apparent that for a surface low to exist for any appreciable time, compensation must occur at some layer aloft. For example, surface convergence could be maintained if divergence (spreading out) aloft occurred at a rate equal to the inflow below. Figure 14.9A shows diagrammatically the relationship between surface convergence (inflow) and divergence (outflow) aloft that is needed to maintain a low-pressure center. Note that surface convergence about a cyclone causes a net upward movement. The rate of this vertical movement is quite slow, generally less than 1 kilometer per day. Nevertheless, since rising air cools adiabatically, cloudy conditions and precipitation are often associated with the passage of a low-pressure system (Figure 14.10A). On occasion, divergence aloft may even exceed surface convergence, resulting in intensified surface inflow and increased vertical motion. For that reason, divergence aloft can intensify these storm centers as well as maintain them.

Figure 14.9
*Cross-sectional views of the airflow associated with cyclones and anticyclones. **A.** Converging winds and rising air are associated with a low. **B.** Highs are associated with descending air and diverging winds.*

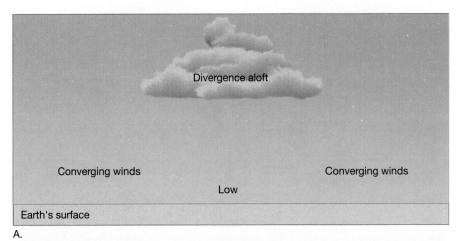

A.

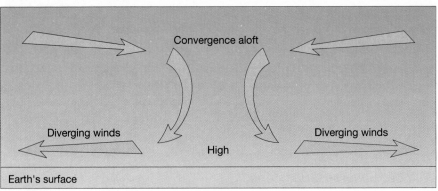

B.

Just as with cyclones, anticyclones, which are associated with surface divergence, must also be maintained from above. The mass outflow near the surface is accompanied by convergence aloft and general subsidence of the air column (Figure 14.9B). Since descending air is compressed and warmed, cloud formation and precipitation are unlikely in an anticyclone, and "fair" weather can usually be expected with the approach of a high (Figure 14.10B).

For reasons that should now be obvious, it has been common practice to print on barometers intended for household use the words "stormy" at the low end and "fair" on the high end. By noting whether the pressure is rising, falling, or steady, we have a good indication of what the forthcoming weather will be. Such a determination, called the **pressure**, or **barometric, tendency,** is a very useful aid in short-range weather prediction. The generalizations relating cyclones and anticyclones to the weather conditions just considered are the basis for several weather sayings and proverbs, including the one that follows. Note that *glass* refers to the barometer.

When the glass falls low,
Prepare for a blow;
When it rises high,
Let all your kites fly.

In conclusion, you should now be better able to understand why local television weather reporters emphasize the positions and projected paths of cyclones and anticyclones. The "villain" on these weather programs is always the cyclone, which produces "bad" weather in any season. Lows move in roughly a west-to-east direction across the United States and require a few days to more than a week for the journey. Their paths can be somewhat erratic; thus accurate prediction of their migration is difficult, although essential, for short-range forecasting. Meteorologists must also determine if the flow aloft will intensify an embryo storm or act to suppress its development. Due to the close tie between conditions at the surface and those aloft, a great deal of emphasis has been placed on the importance and understanding of the total atmospheric circulation, particularly in the mid-latitudes. Once we have

A.

B.

Figure 14.10
*These two photographs illustrate the basic weather generalizations associated with pressure centers. **A.** Centers of low pressure are frequently associated with cloudy conditions and precipitation. **B.** By contrast, clear skies and "fair" weather may be expected when an area is under the influence of high pressure. (Photos by E. J. Tarbuck)*

examined the workings of the general circulation, we will again consider the structure of the cyclone in light of these findings.

GENERAL CIRCULATION OF THE ATMOSPHERE

The underlying cause of wind is the unequal heating of the earth's surface. In tropical regions more solar radiation is received than is radiated back to space. In polar regions the opposite is true—less solar energy is received than is lost. Attempting to balance these differences, the atmosphere acts as a giant heat transfer system, moving warm air poleward and cool air equatorward. On a smaller scale, but for the same reason, ocean currents also contribute. The general circulation is very complex, and there is a great deal that has yet to be explained. We can, however, approximate its major components by first considering the circulation that would occur on a nonrotating earth having a uniform surface. We will then modify this system to fit observed patterns.

On a hypothetical nonrotating planet with a smooth surface of either all land or all water, two large thermally produced cells would form (Figure 14.11). The heated equatorial air would rise until it reached the tropopause, which, acting like a lid, would deflect the air poleward. Eventually, this upper-level airflow would reach the poles, sink, and spread out in all directions at the surface and move

back toward the equator. Once there, it would be reheated and start its journey over again. This hypothetical circulation system has upper-level air flowing poleward and surface air flowing equatorward.

If we add the effect of rotation, this simple convection system will break down into smaller cells. Figure 14.12 illustrates the three pairs of cells proposed

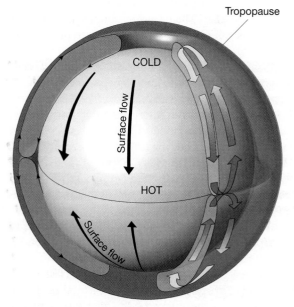

Figure 14.11
Global circulation on a nonrotating earth. A simple convection system is produced by unequal heating of the surface on a nonrotating planet.

Figure 14.12
Idealized global circulation.

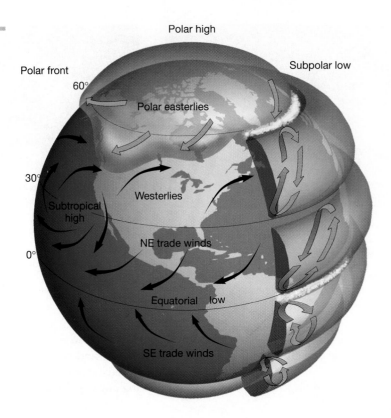

to carry on the task of heat redistribution on a rotating planet. The polar and tropical cells retain the characteristics of the thermally generated convection described earlier. The nature of the mid-latitude circulation is complex and will be discussed in more detail in a later section.

Near the equator, the rising air is associated with the pressure zone known as the **equatorial low**—a region marked by abundant precipitation. As the upper-level flow from the equatorial low reaches 20–30 degrees latitude, north or south, it will have cooled enough to sink toward the surface. This subsidence and associated adiabatic heating produce the hot, arid regions in this latitude range. The center of this zone of subsiding dry air is the **subtropical high**, which encircles the globe near 30 degrees latitude (Figure 14.12). Located here are extensive arid and semiarid regions. The great deserts of Australia, Arabia, and North Africa, for example, are dry primarily because of the stable conditions associated with the subtropical highs. At the surface, airflow is outward from the center of this high-pressure system. Some of the air travels equatorward and is deflected by the Coriolis effect, producing the reliable **trade winds**. The remainder travels poleward and

is also deflected, generating the prevailing **westerlies** of the mid-latitudes. As the westerlies move poleward, they encounter the cool **polar easterlies** in the region of the **subpolar low.** The interaction of these warm and cool winds produces the stormy belt known as the **polar front.** The source region for the variable polar easterlies is the **polar high.** Here, cold polar air is subsiding and spreading equatorward.

In summary, this simplified global circulation is dominated by four pressure zones. The subtropical and polar highs are areas of dry subsiding air that flows outward at the surface, producing the prevailing winds. The low-pressure zones of the equatorial and subpolar regions are associated with inward and upward airflow accompanied by precipitation.

Up to this point, we have described the surface pressure and associated winds as continuous belts around the earth. The only true zonal distribution of pressure exists in the region of the subpolar low in the Southern Hemisphere, where the ocean is continuous. At other latitudes, particularly in the Northern Hemisphere where landmasses are more prominent, large seasonal temperature differences disrupt this zonal pattern. Figure 14.13 shows the resulting pressure and wind patterns for July and January. The

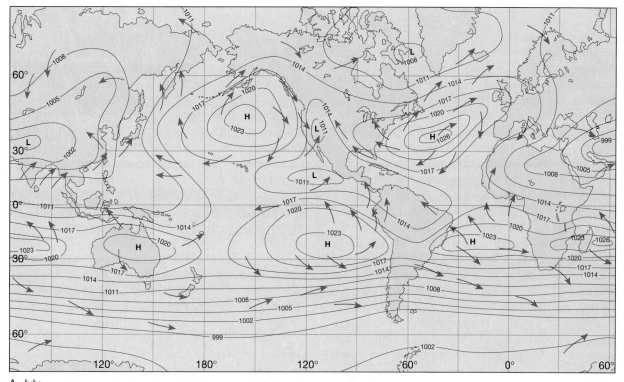

A. July

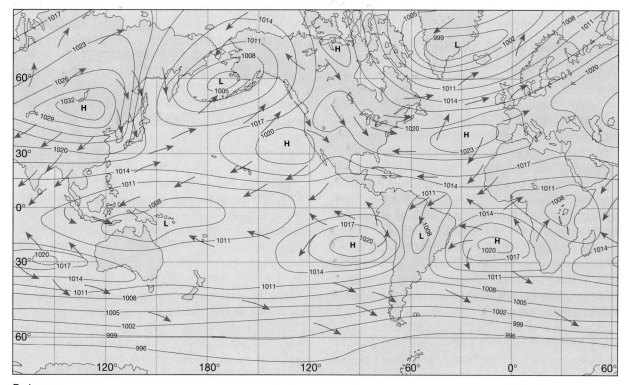

B. January

Figure 14.13
Average surface barometric pressure in millibars for **A***. July and* **B***. January, with associated winds.*

circulation over the oceans is dominated by semipermanent cells of high pressure in the subtropics and cells of low pressure over the subpolar regions. The subtropical highs are responsible for the trade winds and westerlies, as mentioned earlier. The large landmasses, on the other hand, particularly Asia, become cold in the winter and develop a seasonal high-pressure system from which surface flow is directed off the land (Figure 14.13). In the summer, the opposite occurs; the landmasses are heated and develop a low-pressure cell, which permits air to flow onto the land. These seasonal changes in wind direction are known as the **monsoons.** During warm months, areas such as India experience a flow of warm, water-laden air from the Indian Ocean, which produces the rainy summer monsoon. The winter monsoon is dominated by dry continental air. A similar situation exists, but to a lesser extent, over North America. In summary, the general circulation is produced by semipermanent cells of high and low pressure over the oceans and is complicated by seasonal pressure changes over land.

CIRCULATION IN THE MID-LATITUDES

The circulation in the mid-latitudes, the zone of the westerlies, is complex and does not fit the convection system proposed for the tropics. Between 30 and 60 degrees latitude, the general west-to-east flow is interrupted by the migration of cyclones—low-pressure systems often associated with precipitation—and anticyclones—high-pressure systems associated with clear skies. In the Northern Hemisphere these cells move from west to east around the globe, creating an anticyclonic (clockwise) flow or a cyclonic (counterclockwise) flow in their area of influence. A close correlation exists between the paths taken by these surface pressure systems and the position of the upper-level airflow, indicating that the upper air is responsible for directing the movement of cyclonic and anticyclonic systems.

Although much more is yet to be learned about the wavy upper-level flow of the westerlies, some of its basic features are understood with some degree of certainty. Among the most obvious features of the flow aloft are the seasonal changes. The change in wind speed is reflected on upper-air charts by more closely spaced contour lines in the cool season. The seasonal fluctuation of wind speeds is a consequence of the seasonal variation of the temperature gradient. The steep temperature gradient across the mid-

dle latitudes in the winter months corresponds to a stronger flow aloft. In addition, the polar jet stream fluctuates seasonally such that its mean position migrates southward with the approach of winter and northward as summer nears. By midwinter, the jet core may penetrate as far south as central Florida. Since the paths of cyclonic systems are guided by the flow aloft, we can expect the southern tier of states to experience most of their severe storms in the winter season. During the hot summer months, the storm track is across the northern states, and some cyclones never leave Canada. The northerly storm track associated with summer applies also to Pacific storms, which move toward Alaska during the warm months, thus producing an extended dry season for much of our west coast. The number of cyclones generated is seasonal as well, with the largest number occurring in the cooler months when the temperature gradients are greatest. This fact is in agreement with the role of cyclonic storms in the distribution of heat across the mid-latitudes.

Even in the cool season, the westerly flow goes through an irregular cyclic change. There may be periods of a week or more when the flow is nearly west to east. Under these conditions, relatively mild temperatures occur, and few disturbances are experienced in the region south of the jet stream (Figure 14.14A). Then, without warning, the upper flow begins to meander, producing large-amplitude waves and a general north-to-south flow (Figure 14.14B, C, and D). This change allows for an influx of cold air southward, which intensifies the temperature gradient and the flow aloft. During these periods, cyclonic activity dominates the weather picture. For a week or more, the cyclonic storms redistribute large quantities of heat across the mid-latitudes by moving cold air southward and warm air northward. This redistribution eventually results in a weakened temperature gradient and a return to a flatter flow aloft and less intense weather at the surface. These cycles, consisting of alternating periods of calm and stormy weather, can last from one to six weeks.

Because of the irregular and sometimes erratic behavior of the flow aloft, long-range weather prediction still remains essentially beyond the forecaster's reach. Nonetheless, attempts are being made to more accurately predict changes in the upper-level flow on a long-term basis. It is hoped that such research will eventually allow forecasters to answer such questions as, "Will next winter be colder than normal?" and, "Will the Great Plains experience a drought next year?"

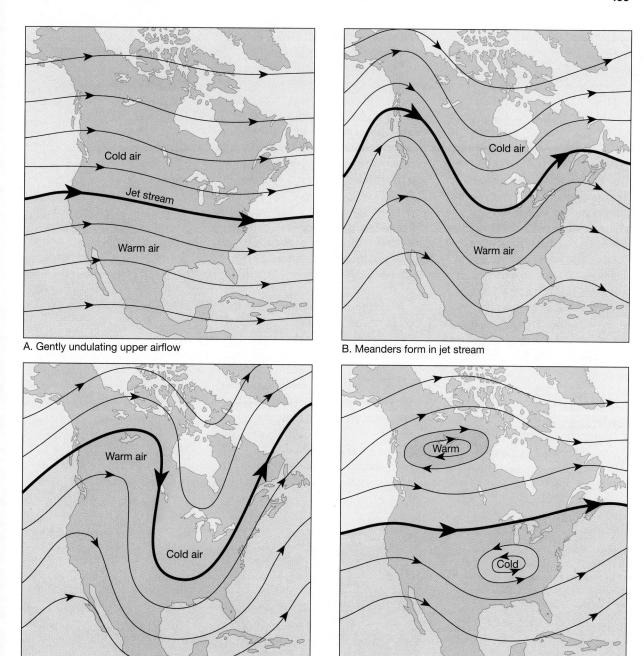

A. Gently undulating upper airflow

B. Meanders form in jet stream

C. Strong waves form in upper airflow

D. Cells of warm and cold air separate from main air mass

Figure 14.14
*Cyclic changes that occur in the upper-level airflow of the westerlies. The flow, which
has the jet stream as its axis, starts out nearly straight, then develops meanders,
which are eventually cut off.*

BOX 14.2
Scales of Atmospheric Motion

*T*he time and space scales commonly used to describe atmospheric motions are provided in Table 14.A. The largest-scale wind pattern is called a *macroscale wind system* and is exemplified by long waves in the westerlies. The planetary-scale flow patterns extend around the entire globe and often remain essentially unchanged for weeks at a time. A somewhat smaller type of macroscale circulation is commonly called the *synoptic scale*, or *weather map scale*, and consists mainly of individual traveling cyclones and anticyclones (Figure 14.A). The airflow in these less-than-global-sized macroscale circulations consists primarily of horizontal flow, with only modest amounts of vertical motion. In contrast, *mesoscale* winds influence smaller areas and exhibit extensive vertical flow, which can be rapid, as in a developing thunderstorm or tornado (Figure 14.B). The smallest scale of air motion is referred to as *microscale*. These small, often chaotic winds normally last seconds, or at most are measured in minutes. Examples include simple gusts, which hurl debris into the air, and well-developed vortices such as dust devils (Figure 14.C).

Although it is common practice to divide atmospheric motions according to scale, it is important to remember that this flow is complex—much like a turbulent river with smaller eddies within larger eddies within still larger eddies. Also, like the flow of a turbulent river, each scale of motion is related to the other. Let us examine the flow associated with the hurricanes

Figure 14.B
This tornado is one example of mesoscale circulation. (Photo by Sheila Beougher/Liaison International)

Figure 14.A
Traveling cyclonic storms, such as this hurricane, are examples of macroscale circulation. (Courtesy of the National Hurricane Center)

Table 14.A
Time and space scales for atmospheric motion.

Name of Scale	Time Scale	Size Scale	Examples
Macroscale			
Planetary Scale	Weeks to years	1000 to 40,000 km	Waves in the Westerlies
Synoptic Scale	Days to weeks	100 to 500 km	Cyclones, anticyclones, and hurricanes
Mesoscale	Minutes to days	½ to 100 km	Land-sea breezes, thunderstorms, and tornadoes
Microscale	Seconds to minutes	<½ km	Turbulence, dust devils, and gusts

Figure 14.C
Gusts are examples of microscale winds. (Photo by E. J. Tarbuck)

that form over the North Atlantic, for example. When we view one of these tropical cyclones on satellite images, the storm appears as a large whirling cloud migrating slowly across the ocean (see Figure 15.21). From this perspective (synoptic scale), the general counterclockwise rotation can be seen easily and the storm's path, which persists for a week or so, can be followed easily. When we average the winds of a hurricane, we find that they have net motion from east to west, thereby indicating that these large eddies are embedded in a still larger flow (the general circulation) that is moving westward across the tropical portion of the North Atlantic. When we examine a hurricane more closely by flying an airplane through it, some of the small-scale aspects of the storm become noticeable. As the plane approaches the outer edge of the system, it becomes evident that the large rotating cloud that we saw in the satellite images is made of numerous towering cumulonimbus clouds. Each of these mesoscale phenomena (thunderstorms) lasts for only a few hours and must be continually replaced by new ones if the hurricane is to persist. When we fly into these storms, we quickly realize that the individual clouds are also made up of even smaller-scale turbulences. The small thermals of rising air that compose these clouds can make for a rough trip. Thus a typical hurricane exhibits several different scales of motion, including many mesoscale thunderstorms, which, in turn, consist of numerous microscale turbulences, whereas the counterclockwise circulation of the hurricane is itself part of the larger general circulation in the tropics.

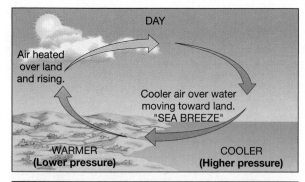

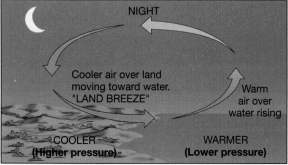

Figure 14.15
Sea and land breezes. During the daytime, land heats more intensely than water. A low is created over the land, and the air moves from the water (higher pressure) to the land (lower pressure). At night, the land cools more rapidly, resulting in higher pressure over land and a reversal of the wind.

LOCAL WINDS

Sea and Land Breezes

In coastal areas during the warm summer months, the land is heated more intensely during the daylight hours than the adjacent body of water (see Chapter 12). As a result, the air above the land surface heats and expands, creating an area of low pressure. A **sea breeze** then develops, blowing cooler air from the water (higher pressure) toward the land (lower pressure) (Figure 14.15). The sea breeze begins to develop shortly before noon and generally reaches its greatest intensity during the mid- to late afternoon. These cool winds can be a significant moderating influence on the temperatures in coastal areas. At night the reverse may take place; the land cools more rapidly than the sea and a **land breeze** develops (Figure 14.15). Small-scale sea breezes can also develop along the shores of large lakes. People who live in a city near the Great Lakes, such as Chicago, recognize the lake effect, especially in the summer. These people are reminded daily by weather reports of the cool temperatures near the lake as compared to warmer outlying areas.

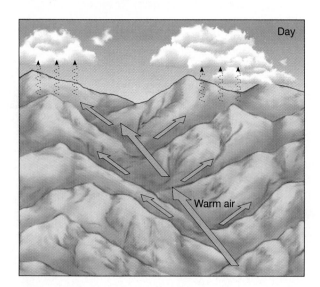

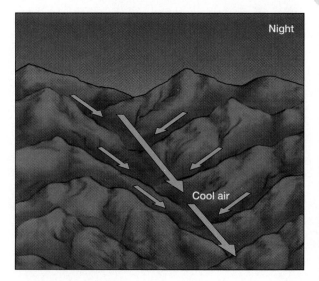

Figure 14.16
Mountain and valley breezes. Heating during the daylight hours warms the air along the mountain slopes. This warm air rises, generating a valley breeze. After sunset, cooling of the air near the mountain can result in cool air drainage into the valley, producing the mountain breeze.

Valley and Mountain Breezes

A daily wind similar to land and sea breezes occurs in many mountainous regions (Figure 14.16). Here during the daylight hours the air along the slopes of the mountains is heated more intensely than is the air at the same elevation over the valley floor. This warm air glides up along the slope and generates a **valley breeze.** The occurrence of these daytime upslope breezes can often be identified by the cumulus clouds that develop on adjacent mountain peaks. After sunset, the pattern may reverse. Rapid radiation cooling along the mountain slopes results in cool air drainage into the valley. This movement of air is called the **mountain breeze.** The same type of cool air drainage can occur in places that have very modest slopes. The result is that the coldest pockets of air are usually found in the lowest spots. Like many other winds, mountain and valley breezes have seasonal preferences. Although valley breezes are most common during the warm season when solar heating is most intense, mountain breezes tend to be more dominant in the cold season.

Chinook and Santa Ana Winds

Warm, dry winds are common on the eastern slopes of the Rockies, where they are called **chinooks.** Such winds are created when air descends the leeward side of a mountain and warms by compression. Since condensation may have occurred as the air ascended the windward side, releasing latent heat, the air descending the leeward slope will be warmer and drier than it was at a similar elevation on the windward side. Although the temperature of these winds is generally less than 10°C (50°F), which is not particularly warm, they generally occur in the winter and spring when the affected areas may be experiencing below-freezing temperatures. Thus, by comparison, these dry, warm winds often bring a drastic change. When the ground has a snow cover, these winds are known to melt it in short order. The word *chinook* literally means "snow-eater."

Another chinooklike wind that occurs in the United States is the **Santa Ana.** Found in southern California, these hot, desiccating winds greatly increase the threat of fire in this already dry area.

WIND MEASUREMENT

Two basic wind measurements, direction and speed, are particularly significant to the weather observer. Winds are always labeled by the direction *from* which they blow. A north wind blows from the north toward the south, an east wind from the east toward the west. The instrument most commonly used to determine wind direction is the **wind vane** (Figure 14.17). This instrument, which is a common sight on many buildings, always points into the wind. Often the wind direction is shown on a dial that is con-

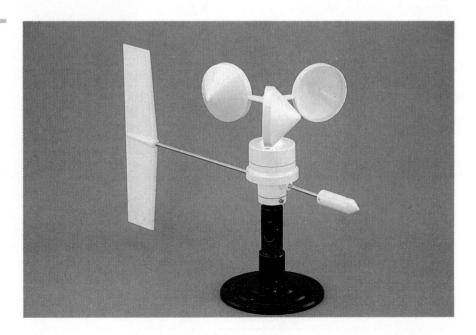

Figure 14.17
Wind vane and cup anemometer. The wind vane shows wind direction and the anemometer measures wind speed. (Courtesy of Qualimetrics, Inc.)

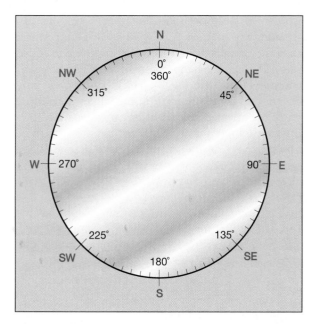

Figure 14.18
Wind direction. Wind direction may be expressed using the points of the compass or a scale of 0–360 degrees. Winds are always labeled according to the direction from which they are blowing.

nected to the wind vane (Figure 14.18). The dial will indicate the direction of the wind either by points of the compass, that is N, NE, E, SE, etc., or by a 0–360-degree scale. On the latter scale, 0 degrees or 360 degrees are both north, 90 degrees is east, 180 degrees is south, and 270 degrees is west. When the wind consistently blows more often from one direction than from any other, it is termed a **prevailing wind.**

Wind speed is commonly measured using a **cup anemometer** (Figure 14.17). The wind speed is read from a dial much like the speedometer of an automobile.

By knowing the locations of cyclones and anticyclones in relation to where you are, you can predict the changes in wind direction that will be experienced as the pressure center moves past. Since changes in wind direction often bring changes in temperature and moisture conditions, the ability to predict the winds can be very useful. In the Midwest, for example, a north wind may bring cool, dry air from Canada, while a south wind may bring warm, humid air from the Gulf of Mexico. Sir Francis Bacon summed it up nicely when he wrote, "Every wind has its weather."

WIND ENERGY

Wind has been used for centuries as an almost free and nonpolluting source of energy. Sailing ships and wind-powered grist mills represent two of the early ways that this renewable resource was harnessed. Further, as rural America was settled, there was a strong reliance on wind power to pump water and later to generate electricity. The world's largest wind generator was a two-bladed, 52-meter-diameter turbine developed and tested at Grandpa's Knob, Vermont, between 1941 and 1945. However, during much of the post–World War II period, abundant supplies of fossil fuels and the promise of cheap nuclear power caused a decline in the funding needed to support the development of large-scale wind turbines in the United States.

Figure 14.19
This wind farm near Palm Springs, California, consists of several hundred wind turbines. (Photo by Michael Collier)

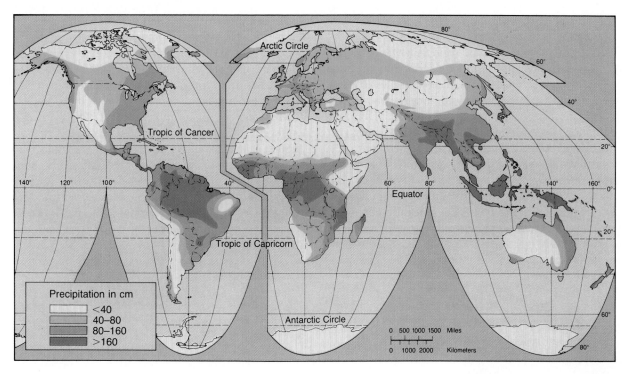

Figure 14.20
Average annual precipitation in centimeters.

Following the "energy crisis" that was precipitated by an oil embargo in the 1970s, interest in wind power once again increased. In 1980 the federal government initiated a program to develop wind-power systems. A project sponsored by the United States Department of Energy involved setting up experimental wind farms in mountain passes known to have strong persistent winds. One of these facilities, located at Altamont Pass near San Francisco, employs over 2000 wind turbines. Presently, California is planning to develop wind farms capable of supplying 8 percent of the state's electricity by the turn of the century (Figure 14.19).

Although the future for wind power appears promising, it is not without difficulties. There are many technical problems to overcome in building large, efficient turbines. In addition, noise pollution and the costs of large tracts of land in populated areas present other significant obstacles to development. One proposed solution is to erect very large wind turbines on offshore platforms. The wind energy would then be used to produce hydrogen by the electrolysis of water. This combustible gas would then be piped to land and used as a fuel.

GLOBAL DISTRIBUTION OF PRECIPITATION

Figure 14.20 shows the average annual precipitation for the world. Although this is a generalized map, an examination still shows a relatively complex pattern. Nevertheless, the major features of the earth's precipitation distribution can be explained by applying our knowledge of global wind and pressure systems. In general, regions influenced by high pressure, with its associated subsidence and divergent winds, experience relatively dry conditions. Conversely, regions under the influence of low pressure and its converging winds and ascending air receive ample precipitation. This pattern is illustrated by noting that the tropical region dominated by the equatorial low is the rainiest region on earth. It includes the rainforests of the Amazon basin in South America and the Congo basin in Africa. Here the warm, humid trade winds converge to yield abundant rainfall throughout the year. By way of contrast, the areas dominated by the subtropical high pressure cells clearly receive much smaller amounts of precipitation. These are regions of extensive deserts. In the Northern Hemisphere the largest is the Sahara. Examples in the Southern Hemisphere include the Kalahari in southern Africa and the dry lands of Australia.

If the earth's pressure and wind belts were the only factors controlling precipitation distribution, the pattern shown in Figure 14.20 would be simpler. The inherent nature of the air involved, however, is also an important factor in determining the potential for precipitation. Because cold air has a much lower capacity for moisture than warm air, we would expect a latitudinal variation in precipitation, with low latitudes receiving the greatest amounts of precipitation and high latitudes receiving the smallest amounts. An examination of Figure 14.20 does indeed reveal heavy rainfall in equatorial regions and meager precipitation in high-latitude areas. Recall that the dry region in the warm subtropics is explained by the presence of the subtropical high.

In addition to latitudinal variations in precipitation related to the global pattern of pressure and temperatures, the distribution of land and water complicates the precipitation pattern found over the earth. Large landmasses in the middle latitudes commonly experience decreased precipitation toward their interiors. For example, central North America and central Eurasia receive considerably less precipitation than coastal regions at the same latitude. Furthermore, the effects of mountain barriers also alter the precipitation patterns we would otherwise expect. Windward mountain slopes receive abundant rainfall resulting from orographic lifting, whereas leeward slopes and adjacent lowlands are usually deficient in moisture.

Review Questions

1. What is standard sea level pressure in millibars? In inches of mercury? In pounds per square inch?

2. Mercury is 13 times heavier than water. If you built a barometer using water rather than mercury, how tall would it have to be in order to record standard sea level pressure (in centimeters of water)?

3. Describe the principle of the aneroid barometer.

4. If a very flexible balloon rose to 31.2 kilometers, how many times larger would it be because of expansion than it was at sea level? (See Table 12.1.)

5. What force is responsible for generating wind?

6. Write a generalization relating the spacing of isobars to the speed of wind (see Figure 14.3).

7. How does the Coriolis effect modify air movement?

8. Contrast surface winds and upper-air winds in terms of speed and direction.

9. Describe the weather that usually accompanies a drop in barometric pressure and a rise in barometric pressure.

10. Sketch a diagram (isobars and wind arrows) showing the winds associated with surface cyclones and anticyclones in both the Northern and Southern hemispheres.

11. If you live in the Northern Hemisphere and are directly west of a cyclone, what most probably will be the wind direction? What will the wind direction be if you are west of an anticyclone?

12. The following questions relate to the global pattern of air pressure and winds.
    (a) The trade winds diverge from which pressure zone?
    (b) Which prevailing wind belts converge in the stormy region known as the polar front?
    (c) Which pressure belt is associated with the equator?

13. Describe the jet stream.

14. What influence does upper-level airflow seem to have on surface pressure systems?

15. Describe the monsoon circulation of India (Figure 14.13).

16. Why are sea breezes in the mid-latitudes most pronounced during the summer months?

17. A northeast wind is blowing *from* the _____ (direction) *toward* the _____ (direction).

18. The wind direction is 225 degrees. From what compass direction is the wind blowing (Figure 14.18)?

19. Which global pressure system is responsible for deserts such as the Sahara and Kalahari in Africa? With which global pressure belt are the tropical rainforests of the Amazon and Congo basins associated?

20. Other than the earth's pressure and wind belts, list two other factors that exert a significant influence on the global distribution of precipitation.

Key Terms

altimeter (p. 488)

aneroid barometer (p. 488)

anticyclone (high) (p. 491)

barograph (p. 488)

barometric tendency (p. 494)

chinook (p. 503)

convergence (p. 492)

Coriolis effect (p. 490)

cup anemometer (p. 504)

cyclone (low) (p. 491)

divergence (p. 492)

equatorial low (p. 496)

geostrophic wind (p. 491)

isobar (p. 489)

jet stream (p. 491)

land breeze (p. 502)

mercurial barometer (p. 486)

monsoon (p. 498)

mountain breeze (p. 503)

polar easterlies (p. 496)

polar front (p. 496)

polar high (p. 496)

pressure gradient (p. 489)

pressure tendency (p. 494)

prevailing wind (p. 504)

Santa Ana (p. 503)

sea breeze (p. 502)

subpolar low (p. 496)

subtropical high (p. 496)

trade winds (p. 496)

valley breeze (p. 503)

westerlies (p. 496)

wind (p. 488)

wind vane (p. 503)

Weather Patterns and Severe Storms

Opposite page: *Thunderstorm and lightning. (Photo
by Richard Kaylin/Allstock)*

Left: *Lightning. (Photo courtesy of National Center for
Atmospheric Research)*

Above: *Turbulent conditions surrounding a well-developed
thunderstorm. (Photo by Warren Faidley/Weatherstock)*

*T*ornadoes and hurricanes rank among nature's most destructive forces. Each spring, newspapers report the death and destruction left in the wake of a ``band'' of tornadoes. During late summer and fall we hear occasional reports about hurricanes in the news. Storms like Andrew, Camille, Gilbert, and Hugo can make front-page headlines. Thunderstorms, although less intense and more common than tornadoes and hurricanes, will also be part of our discussion in this chapter on the nature of severe weather disturbances. Before looking at violent weather, however, we shall study those atmospheric phenomena that most often affect our day-to-day weather: air masses, fronts, and traveling middle-latitude cyclones. Here we shall see the interplay of the elements of weather discussed earlier.

AIR MASSES

For many people who live in the middle latitudes, summer heat waves and winter cold spells are familiar experiences. In the first instance, several days of high temperatures and oppressive humidities may finally end when a series of thundershowers pass through the area, followed by a few days of relatively cool relief. By contrast, the clear skies that often accompany a span of frigid subzero days may be replaced by thick stratus clouds and a period of snow as temperatures rise to levels that seem mild when compared to those that existed just a day earlier. In both examples, what was experienced was a period of generally constant weather conditions followed by a relatively short period of change, and then the subsequent reestablishment of a new set of weather conditions that remained for perhaps several days before changing again.

The weather patterns just described are the result of the movements of large bodies of air, called air masses. An **air mass,** as the term implies, is an immense body of air, usually 1600 kilometers (1000 miles) or more across, and perhaps several kilometers thick, which is characterized by a homogeneity of temperature and moisture at any given altitude. When this air moves out of its region of origin, it will carry these temperatures and moisture conditions elsewhere, eventually affecting a large portion of a continent.

The horizontal uniformity of an air mass is not complete. Because air masses extend over large areas, small differences in temperature and humidity from place to place are to be expected. Nevertheless, the differences observed within an air mass are small in comparison to the rapid rates of change experienced across boundaries between air masses. Because it may take several days for an air mass to traverse an area, the region under its influence will probably experience fairly constant weather, a situation called **air-mass weather.** Certainly there may be some day-to-day variations, but the events will be quite unlike those in an adjacent air mass. For that reason the boundary between two adjoining air masses having contrasting characteristics, called a *front*, marks a change in weather.

Source Regions

When a portion of the lower troposphere moves slowly or stagnates over a relatively uniform surface, the air will assume the distinguishing features of that area, particularly with regard to temperature and moisture conditions (Figure 15.1).

The area where an air mass acquires its characteristic properties of temperature and moisture is called its **source region.** The source regions that produce air masses that influence North America are shown in Figure 15.2 on page 514.

Air masses are classified according to their source region. **Polar (P)** air masses originate in high latitudes, whereas those that form in low latitudes are called **tropical (T).** The designation *polar* or *tropical* gives an indication of the temperature characteristics of the air mass. *Polar* indicates cold, and *tropical* indicates warm. In addition, air masses are classified according to the nature of the surface in the source region. **Continental (c)** designates land, and **maritime (m)** indicates water. The designation *continental* or *maritime* thus suggests the moisture

Figure 15.1
Air masses form when air moves slowly or stagnates over a relatively uniform surface and assumes the distinguishing features of that region, especially with regard to temperature and moisture conditions. Air masses that form over the ocean are likely to be humid. (Photo by E. J. Tarbuck)

characteristics of the air mass. Continental air is likely to be dry and maritime air, humid. The four basic types of air masses according to this scheme of classification are continental polar (cP), continental tropical (cT), maritime polar (mP), and maritime tropical (mT).

Weather Associated with Air Masses

Continental polar and maritime tropical air masses influence the weather of North America most, especially east of the Rocky Mountains. Continental polar air masses originate in northern Canada, interior Alaska, and the Arctic—areas that are uniformly cold and dry in winter and cool and dry in summer. These air masses are usually associated with high pressure and typically have few clouds. In winter, an invasion of continental polar air brings the clear skies and cold temperatures we associate with a cold wave as it moves southward through Canada into the United States. In summer, this air mass may bring a few days of cooling relief.

Although cP air masses are not normally associated with heavy precipitation, those that cross the Great Lakes in winter sometimes bring snow to the leeward shores. These localized storms often form when the surface weather map indicates no apparent cause for a snowstorm to occur. Known as **lake-effect snows,** they are discussed in Box 15.1.

Maritime tropical air masses affecting North America most often originate in the Gulf of Mexico, the Caribbean Sea, or the adjacent Atlantic Ocean. As we might expect, these air masses are warm, moisture-laden, and usually unstable. Maritime tropical air is the source of much, if not most, of the precipitation received in the eastern two-thirds of the United States. In summer, when this air mass invades the central and eastern United States, and occasionally southern Canada, it brings the heat and oppressive humidity typically associated with its source region. The tropical Pacific is also a source region for maritime tropical air. However, this maritime tropical air seldom enters the continent. When it does, it may bring orographic rainfall to the southwestern United States and northern Mexico.

Of the two remaining air masses, maritime polar and continental tropical, the latter has the least influence upon the weather of North America. Hot, dry continental tropical air, originating in the Southwest and Mexico during the summer, seldom affects the weather outside its source region.

Every winter a unique and interesting weather phenomenon takes place along the downwind shores of the Great Lakes. Periodically brief, heavy snow showers issue from dark clouds that move onshore from the lakes. Seldom do the storms move more than about 40 kilometers inland from the shore before the snows come to an end. These highly localized storms occurring along the leeward shores of the Great Lakes create what are known as *lake-effect snows.*

Lake-effect storms account for a high percentage of the snowfall in many areas adjacent to the lakes. The strips of land that are most frequently affected, called *snowbelts,* are shown in Figure 15.A. When snowfall statistics for large cities were examined for a recent 10-year span, the two snowiest metropolitan areas in the United States turned out to be Buffalo and Rochester in upstate New York. Buffalo, on the eastern shore of Lake Erie, had more than 2700 centimeters (nearly 90 feet) during the period, while Rochester, situated on the south side of Lake Ontario, recorded slightly less (2635 centimeters). A comparison of average snowfall totals at Thunder Bay in Ontario, Canada, on the north shore of Lake Superior, and Marquette, Michigan, along the southern shore, provides another excellent example of lake-effect snow. Amounts are in centimeters.

Because Marquette is situated on the leeward shore of the lake, it receives substantial lake-effect snow and therefore has a much higher snowfall total than does Thunder Bay.

What causes lake-effect snow? The answer is closely linked to the differential heating of water and land (Chapter 12) and to the concept of atmospheric instability (Chapter 13). During the summer months water bodies, including the Great Lakes, absorb huge quantities of energy from the sun and from the warm air that passes over them. Although these water bodies do not reach particularly high temperatures, they nevertheless represent huge reservoirs of heat. The surround-

	October	November	December	January
Thunder Bay Ontario	3.0	14.9	19.0	22.6
Marquette Michigan	5.3	37.6	56.4	53.1

During the winter, maritime polar air masses coming from the North Pacific often originate as continental polar air masses in Siberia. As they move across the Pacific, they warm and gradually accumulate moisture. Upon entering North America, these air masses drop much of their moisture because of orographic lifting in the western mountains. Maritime polar air also originates off the coast of eastern Canada and occasionally influences the weather of the northeastern United States. When New England is on the northern or northwestern edge of a passing low, the cyclonic winds draw in maritime polar air. The result is a storm characterized by snow and cold temperatures, known locally as a *northeaster.*

FRONTS

Fronts are boundaries that separate air masses of different densities, one warmer and often higher in moisture content than the other. Ideally, fronts can form between any two contrasting air masses. Considering the vast size of the air masses involved, these 15- to 200-kilometer-wide bands of discontinuity are relatively narrow. On the scale of a weather map, they are generally narrow enough to be represented satisfactorily by a broad line.

Above the ground, the frontal surface slopes at a low angle so that warmer air overlies cooler air. In the ideal case, the air masses on both sides of the

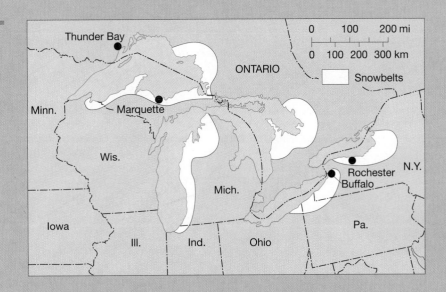

Figure 15.A
The snowbelts of the Great Lakes region are the zones that most frequently experience lake-effect snowstorms. As continental polar air crosses the lakes in winter, it acquires moisture and is made unstable because of warming from below. Snow showers on the lee side of the lakes are often the consequence of this air-mass modification.

ing land, on the other hand, cannot store heat nearly as effectively. Consequently, during autumn and winter, the temperature of the land drops quickly while water bodies lose their heat more gradually and cool slowly. From late November through late January, the contrasts in average temperatures between water and land range from about 8°C in the southern Great Lakes to 17°C farther north. However, the temperature differences can be much greater (perhaps 25°C) when a very cold cP air mass pushes southward across the lakes. When such a dramatic temperature contrast exists, the lakes interact with the air to produce major lake-effect storms. As a cP air mass moves across one of the Great Lakes, the air acquires large quantities of heat and moisture from the relatively warm lake surface. By the time it reaches the opposite shore, the air mass is humid and unstable and heavy snow showers are likely.

front move in the same direction and at the same speed. Under this condition, the front acts as a barrier with which the air masses must move, but through which they cannot penetrate. Generally, however, the pressure field across a front is such that air on one side is moving faster in the direction perpendicular to the front than the air mass on the other side of the front. Thus, one air mass actively advances into another and "clashes" with it. The boundaries were thus tagged *fronts* during World War I by Norwegian meteorologists, who visualized them as analogous to battle lines.

As one air mass moves into another, some mixing does occur along the frontal surface, but for the most part the air masses retain their identity as one air mass is displaced upward over the other. No matter which air mass is advancing, it is always the warmer, less dense air that is forced aloft, while the cooler, denser air acts as the wedge upon which lifting takes place.

Warm Fronts

When the surface position of a front moves so that warm air occupies territory formerly covered by cooler air, it is called a **warm front**. On a weather map, the surface position of a warm front is denoted by a line with semicircles extending into the cooler air. East of the Rockies, warm tropical air often enters the United States from the Gulf of Mexico and overruns receding cool air. As the cold wedge retreats, friction slows the advance of the surface po-

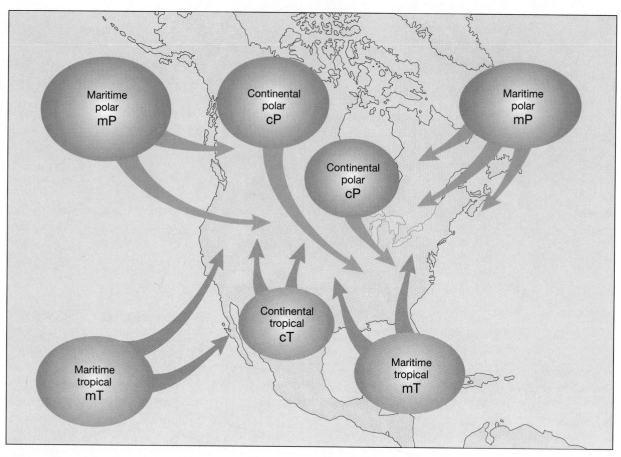

Figure 15.2
Air masses are classified on the basis of their source region. The designation continental (c) *or* maritime (m) *gives an indication of moisture content, whereas* polar (P) *and* tropical (T) *indicate temperature conditions.*

sition of the front more so than its position aloft; for this reason, the boundary separating these air masses acquires a small slope. The average slope of a warm front is about 1:200, which means that if you are 200 kilometers ahead of the surface location of a warm front, you will find the frontal surface at a height of 1 kilometer.

As warm air ascends the retreating wedge of cold air, it cools by adiabatic expansion to produce clouds and frequently, precipitation. The sequence of clouds shown in Figure 15.3A typically precedes a warm front. The first sign of the approach of a warm front is the appearance of cirrus clouds overhead (Figure 15.4). These high clouds form 1000 kilometers or more ahead of the surface front where the overrunning warm air has ascended high up the wedge of

cold air. As the front nears, cirrus clouds grade into cirrostratus, which blend into denser sheets of altostratus. About 300 kilometers ahead of the front, thicker stratus and nimbostratus clouds appear and rain or snow begins. Usually warm fronts produce several hours of moderate-to-gentle precipitation over a large region. This is in agreement with the relatively gentle slope of a warm front, which does not generally encourage convectional activity. However, on some occasions, warm fronts do produce cumulonimbus clouds and thunderstorm activity (Figure 15.3B). This occurs when the overrunning air is inherently unstable and the front is rather sharp. When these conditions exist, cirrus clouds are generally followed by cirrocumulus clouds, giving us the familiar "mackerel sky," which sailors saw as a warning of an impending storm, as indicated by the following proverb:

> Mackerel scales and mares' tails
> Make lofty ships carry low sails.

At the other extreme, a warm front associated with a dry air mass could pass practically unnoticed by those of us at the surface.

A gradual increase in temperature occurs with the passage of a warm front. As we would expect, the increase is most apparent when there is a large temperature difference between the adjacent air masses. Furthermore, a wind shift from the east to the southwest is often detectable. The reason for this shift will

be evident later. The moisture content and stability of the encroaching warm air mass largely determine when clear skies will return. During the summer, cumulus, and occasionally cumulonimbus, clouds are embedded in the warm unstable air mass that follows the front. Precipitation from these clouds is usually sporadic and not extensive.

Cold Fronts

When cold air is actively advancing into a region occupied by warmer air, the zone of discontinuity is called a **cold front** (Figure 15.5). As with warm fronts, friction tends to slow the surface position of

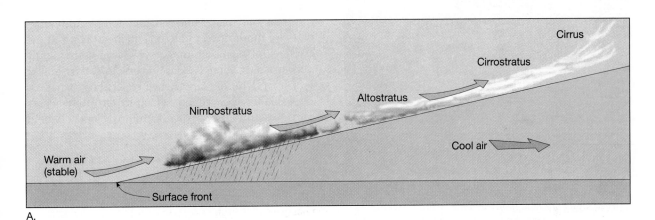

A.

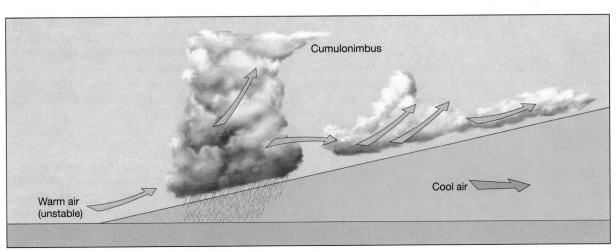

B.

Figure 15.3
A. Warm front with stable air and associated stratiform clouds. Precipitation is moderate and occurs within a few hundred kilometers of the surface front. **B.** Warm front with unstable air and cumuliform clouds. Precipitation is heavy near the surface front.

Figure 15.4
Cirrus clouds are high, thin clouds that are often the first sign that a warm front is approaching. As warm air is wedged aloft, clouds form at increasingly greater heights. (Photo by E. J. Tarbuck)

a cold front more so than its position aloft. However, because of the relative positions of the adjacent air masses, the cold front steepens as it moves. On the average, cold fronts are about twice as steep as warm fronts, having a slope of perhaps 1:100. In addition, cold fronts advance more rapidly than warm fronts. These two differences—rate of movement and steepness of slope—largely account for the more violent nature of cold-front weather. The displacement of air along a cold front is often rapid enough that the released latent heat appreciably increases the air's buoyancy. This frequently results in the sudden downpours and vigorous gusts of wind associated with mature cumulonimbus clouds. Since a cold front produces roughly the same amount of lifting as a warm front, but over a shorter distance, the intensity of precipitation is greater, but the duration is shorter.

A cold front is sometimes preceded by altocumulus clouds. As the front approaches, generally from the west or northwest, towering clouds are often seen in the distance. Near the front, a dark band of ominous clouds foretells the ensuing weather. Usually a marked temperature drop and a wind shift from the south to west or northwest accompany the passage of the front. On a weather map, the sometimes violent weather and sharp temperature contrast are indicated by a line with triangle-shaped points that extend into the warmer air mass.

The weather behind a cold front is dominated by a subsiding and relatively cold air mass. Hence, clearing conditions prevail after the front passes. Although general subsidence causes some adiabatic heating, this has a minor effect on surface temperatures. In the winter, the clear skies associated with these cold outbreaks further reduce surface temperatures because of more rapid radiation cooling at night. If the continental polar air mass, which most frequently accompanies a cold front, moves into a relatively warm and humid area, surface heating can produce shallow convection, which in turn may generate low cumulus or stratocumulus clouds behind the front.

Stationary Fronts

Occasionally the flow on both sides of a front is almost parallel to the position of the front. The surface position of the front does not move and it is therefore named a **stationary front**. On a weather map, stationary fronts are shown with triangular points on one side of the front and semicircles on the other. At times some overrunning occurs along a stationary front. This can result in an extended period of relatively widespread cloudiness, light rain, or light snow.

Occluded Fronts

Another commonly occurring front is the **occluded front**. Here an active cold front overtakes a warm front (Figure 15.6). As the advancing cold air wedges the warm front upward, a new front emerges between the advancing cold air and the air over which the warm front is gliding. The weather of an occluded front generally is quite complex. Most of the precipitation is associated with the warm air being forced aloft. However, when conditions are suitable, the newly formed front can produce precipitation of its own.

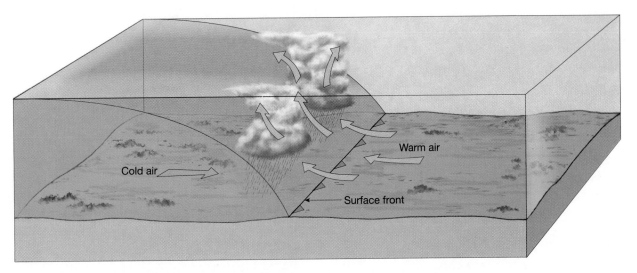

Figure 15.5
Fast-moving cold front and cumulonimbus clouds. Often thunderstorms occur if the warm air is unstable.

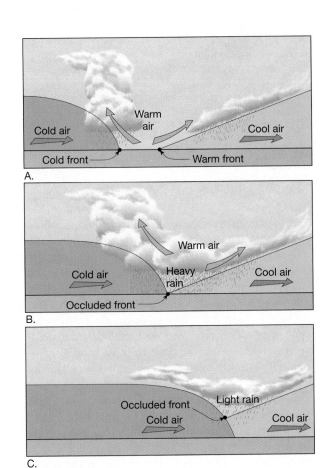

Figure 15.6
Stages in the formation and eventual dissipation of an occluded front.

THE MIDDLE-LATITUDE CYCLONE

The fronts that were just described continually influence the weather in the mid-latitudes. They are usually associated with a low-pressure system called a **middle-latitude,** or **wave, cyclone.**

As early as the 1800s it was known that cyclones were the bearers of precipitation and severe weather. But it was not until the early part of the twentieth century that a model of cyclone formation was developed (Figure 15.7). It was formulated by a group of Norwegian scientists and published in 1918. The Norwegian model of the middle-latitude or wave cyclone was created primarily from near-surface observations. Years later, as data from the middle and upper troposphere and from satellite images became available, some modifications were necessary. Yet this model is still an accepted working tool in interpreting the weather. It provides a visual picture of the dynamic atmosphere as it generates a storm. If you keep this model in mind when you consider changes in the weather, the observed changes will no longer come as a surprise. You should begin to see some order in what had appeared to be disorder, and you might even occasionally "predict" the impending weather.

Life Cycle of a Wave Cyclone

Wave cyclones form along fronts where they change in a somewhat predictable way. This life cycle can last for a few hours or for several days. Figure 15.7

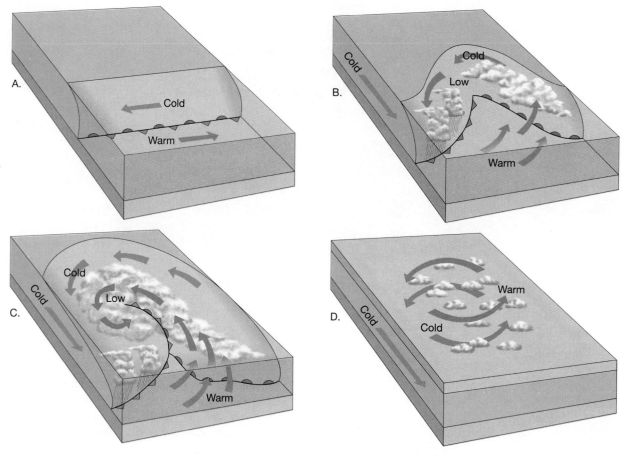

Figure 15.7
The life cycle of a hypothetical middle-latitude cyclone. **A.** *Front develops.* **B.** *Cyclonic circulation is developed.* **C.** *Occluded front is fully developed.* **D.** *Cyclone dissipates.*

is a schematic representation of the stages in the development of a "typical" wave cyclone. As the figure shows, cyclones originate along a front where air masses of different densities (temperatures) are moving parallel to the front in opposite directions. In the classic model, continental polar air associated with the polar easterlies would be north of the front, and maritime tropical air of the westerlies south of the front. The result of this opposing airflow is counterclockwise (cyclonic) rotation. (To better visualize this effect, place a pencil between the palms of your hands. Now move your right hand ahead of your left hand and notice that your pencil rotates in a counterclockwise fashion.) Under the correct conditions, the frontal surface will take on a wave shape. These waves are analogous to the waves produced on the surface of a water body by moving air, except the scale is different. The waves generated between two

contrasting air masses are usually several hundred kilometers long. Some waves tend to dampen out, whereas others become unstable and grow in amplitude. The latter ones change in shape with time much as a gentle ocean swell does as it moves into shallow water and becomes a tall breaking wave.

Once a small wave forms, warm air invades this weak spot along the front and extends itself poleward, while the surrounding cold air moves equatorward. This change causes a readjustment in the pressure field, which results in nearly circular isobars with the low pressure centered at the apex of the wave. The creation of the low encourages the inflow (convergence) of air and general vertical lifting, particularly where warm air is overrunning colder air. We can see in Figure 15.8 that the air in the warm sector is flowing from the southwest toward colder air flowing from the southeast. Since the warm air is

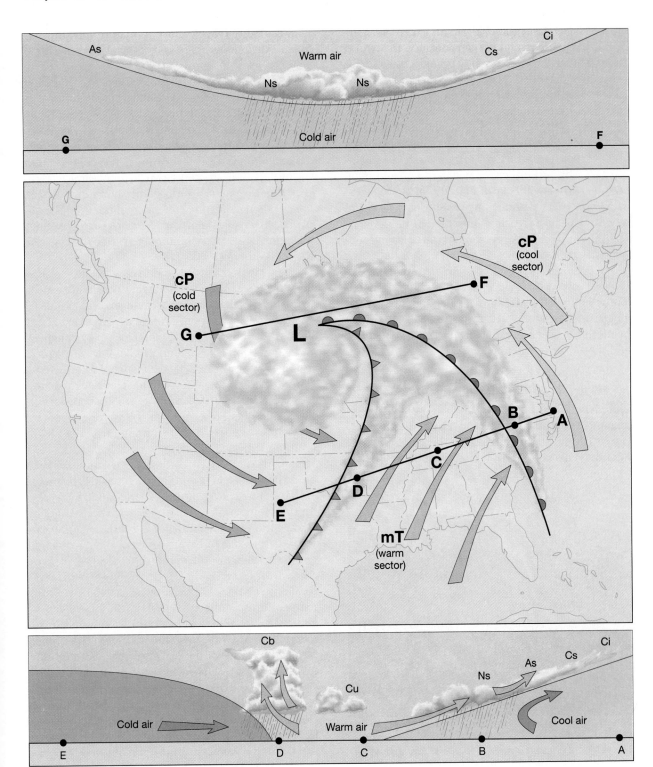

Figure 15.8
Distribution of clouds associated with an idealized middle-latitude cyclone.

moving faster than the cold air in a direction perpendicular to this front, we can conclude that warm air is invading a region formerly occupied by cold air; hence, this must be a warm front. Similar reasoning indicates that in the rear of the cyclonic disturbance, cold air is underrunning the air of the warm sector, generating a cold front there. Generally, the position of the cold front advances faster than the warm front and begins to close the warm sector as shown in Figure 15.7. This process, called **occlusion,** results in an occluded front with the displaced warm sector located aloft. The cyclone enters maturity (maximum intensity) when it reaches this stage in its development. A steep pressure gradient and strong winds develop as lifting continues. Eventually all of the warm sector is forced aloft and cold air surrounds the cyclone at low levels (Figure 15.7D). Once the sloping discontinuity (front) between the air masses no longer exists, the pressure gradient weakens. At this point, the cyclone has exhausted its source of energy, and the storm comes to an end.

Idealized Weather of a Wave Cyclone

The wave cyclone model provides a useful tool for examining the weather patterns of the middle latitudes. Figure 15.8 illustrates the distribution of clouds and thus the regions of possible precipitation associated

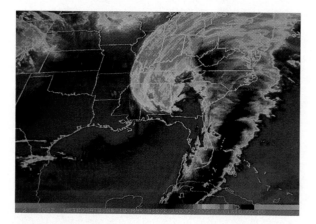

Figure 15.9
This false-color infrared satellite image shows a very strong cyclonic storm that struck March 13–15, 1993, and brought blizzard conditions to a region that stretched from Alabama to eastern Canada. It was responsible for at least 220 deaths and was called by many the "storm of the century." (Photo courtesy of NESDIS/NOAA)

with a mature wave cyclone. Compare this drawing to the satellite image of a cyclone shown in Figure 15.9.

Guided by the westerlies aloft, cyclones generally move eastward across the United States, so we can expect the first signs of their arrival in the west. However, often in the region of the Mississippi valley, cyclones begin a more northeasterly trajectory and occasionally move directly northward. A mid-latitude cyclone typically requires two to four days to pass over a given region. During that short time, rather abrupt changes in atmospheric conditions may be experienced. This is particularly true in the spring, when the largest temperature contrasts occur across the mid-latitudes.

Using Figure 15.8 as a guide, we will now consider these weather producers and what we should expect from them as they pass an area during the spring. To facilitate our discussion, profiles are provided along lines A–E and F–G. First, imagine the change in weather as you move along profile A–E. At point A the sighting of high cirrus clouds would be the first sign of the approaching cyclone. These high clouds can precede the surface front by 1000 kilometers or more and they generally will be accompanied by falling pressure. As the warm front advances, a lowering and thickening of the cloud deck is noticed. Usually within 12 to 24 hours after the first sighting of cirrus clouds, light precipitation begins (point B). As the front nears, the rate of precipitation increases, a rise in temperature is noticed, and winds begin to change from east or southeast to south or southwest. With the passage of the warm front, the area is under the influence of the maritime tropical air mass of the warm sector (point C). Generally the region affected by this sector of the cyclone experiences warm temperatures, south or southwest winds, and generally clear skies, although fair-weather cumulus or altocumulus are not uncommon here. The rather pleasant weather of the warm sector passes quickly and is replaced by gusty winds and precipitation generated along the cold front. The approach of a rapidly advancing cold front is marked by a wall of dark clouds (point D). Severe weather accompanied by heavy precipitation, hail, and an occasional tornado is a definite possibility at this time of year. The passage of the cold front is easily detected by a wind shift; the southerly flow is replaced by winds from the west to northwest and by a pronounced drop in temperature. Also, rising pressure hints of the subsiding cool, dry air behind the front. Once the front passes, the skies clear quickly as the cooler air invades the region (point *E*). Often a day or two of almost cloudless deep-blue skies can be expected unless another cyclone is edging into the region.

A very different set of weather conditions will prevail in those regions that encounter the portion of the cyclone containing the occluded front as shown along profile *F–G*. Here the temperatures remain cool during the passage of the storm; however, a continual drop in pressure and increasingly overcast conditions strongly hint at the approach of the low-pressure center. This sector of the cyclone most often generates snow or icing storms during the cool months. Further, the occluded front often moves more slowly than the other fronts; hence, the wishbone-shaped frontal structure shown in Figure 15.8 rotates in a counterclockwise manner, such that the occluded front appears to "bend over backwards." This effect adds to the misery of the region influenced by the occluded front since it remains over the area longer than the other fronts. The storm also reaches its greatest intensity during occlusion; consequently, the area affected by the developing occluded front receives the brunt of the storm's fury.

Cyclone Formation

When the earliest studies of cyclones were made, little was known about the nature of the airflow in the middle and upper troposphere. Since then a close relationship between surface disturbances and the flow aloft has been established. Whenever the flow aloft is relatively straight, that is, from west to east, very little cyclonic activity occurs at the surface. However, when the upper air begins to meander widely in a north-to-south direction, high-amplitude waves consisting of alternating troughs and ridges are produced, and surface cyclonic activity intensifies.

Before we discuss how cyclone formation is aided by the flow aloft, let us first review the nature of cyclonic and anticyclonic winds. Recall that the airflow about a surface low is inward, a fact that leads to mass convergence (coming together). The resulting accumulation of air must be accompanied by a corresponding increase in surface pressure. Consequently, we might expect a low-pressure system to "fill" rapidly and be eliminated, just as the vacuum in a coffee can is quickly dissipated when we open it. However, this does not occur. On the contrary, cyclones often exist for a week or longer. In order for this to happen, surface convergence must be offset by a mass outflow at some level aloft (see Figure 14.9). As long as divergence (spreading out) aloft is equal to, or greater than, the surface inflow, the low pressure and its accompanying convergence can be sustained.

Because cyclones are bearers of stormy weather, they have received far more attention than anticyclones. Nevertheless, a close relationship exists, which makes it difficult to separate any discussion of these two pressure systems. The surface air that feeds a cyclone, for example, generally originates as air flowing out of an anticyclone. Consequently, cyclones and anticyclones typically are found adjacent to one another. Like the cyclone, an anticyclone depends on the flow far above to maintain its circulation. In this instance, divergence at the surface is balanced by convergence aloft and general subsidence of the air column (see Figure 14.9).

We have seen that airflow aloft plays an important role in maintaining cyclonic and anticyclonic circulation. In fact, more often than not, these rotating surface wind systems are actually generated by upper-level flow (Figure 15.10). In cyclone development, the role of upper-level divergence is most significant. Frequently associated with the jet stream, upper-air divergence creates an environment analogous to a partial vacuum, which initiates upward flow. The fall in surface pressure that accompanies the outflow aloft will induce inward flow at the surface. The Coriolis effect will then come into play to produce the curved flow pattern associated with cyclonic flow.

Divergence aloft does not involve the outward, clockwise movement of air that occurs about a surface anticyclone. Instead, the flow aloft is nearly geostrophic. Its path is from west to east and generally has sweeping curves. One of the mechanisms responsible for the mass transport of air high above the earth's surface is a phenomenon known as **speed divergence.** It has been known for some time that the wind speeds along the axis of the jet stream are not constant. Some regions experience much higher wind speeds than others. Upon entering a zone of maximum wind velocity, air accelerates and therefore diverges. Conversely, when air leaves a zone of maximum wind velocity, a pile-up (convergence) results. An analogous situation occurs on a tollway in the region between toll stations. Upon exiting one toll booth and entering the zone of maximum speed, the automobiles diverge. As the automobiles slow to pay the next toll, they converge. Similarly, in the zone up-

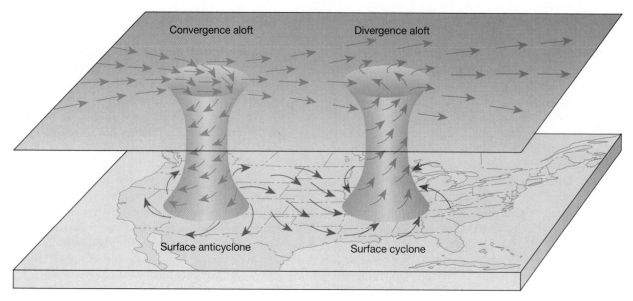

Figure 15.10
*Idealized diagram depicting the support that divergence and convergence aloft provide
to cyclonic and anticyclonic circulation at the surface.*

stream from a region of maximum speed, the air experiences speed divergence, while downstream, mass convergence occurs. Other factors, for example, the spreading out of the air stream, can also contribute to divergence aloft, while confluence of an air stream under correct conditions can cause mass convergence.

In conclusion, we find that divergence aloft initiates upward air movement, reduced surface pressure, and cyclonic flow. On the other hand, convergence along the jet stream results in general subsidence of the air column, increased surface pressure, and anticyclonic surface winds.

Figure 15.11
Cumulonimbus clouds can produce lightning, thunderstorms, and other forms of severe weather. (Warren Faidley/International Stock Photo)

BOX 15.2

What's in a Name?

So far we have examined the middle-latitude cyclones that play such an important role in causing day-to-day weather changes. Yet the use of the term *cyclone* is often confusing. To many people, the term implies only an intense storm, such as a hurricane or a tornado. When a hurricane unleashes its fury on India or Bangladesh, for example, it is usually reported in the media as a cyclone (the local term denoting a hurricane in that part of the word). Similarly, tornadoes are occasionally referred to as cyclones. This custom is particularly common in portions of the Great Plains of the United States. Recall that in the *Wizard of Oz,* Dorothy's house was carried from her Kansas farm to the land of Oz by a cyclone. Indeed, the nickname for the athletic teams at Iowa State University is the Cyclones. Although hurricanes and tornadoes are, in fact, cyclones, the vast majority of cyclones are not hurricanes or tornadoes. The term *cyclone* simply refers to the circulation around any low-pressure center, no matter how large or intense it is.

Tornadoes and hurricanes are both smaller and more violent than middle-latitude cyclones. Whereas middle-latitude cyclones may have a diameter of 1600 kilometers or more, hurricanes average only 600 kilometers across, and tornadoes, with a diameter of just ¼ kilometer, are much too small to show up on a weather map.

The thunderstorm, a much more familiar weather event, hardly needs to be distinguished from tornadoes, hurricanes, and mid-latitude cyclones. Unlike the flow of air about these latter storms, the circulation associated with thunderstorms is characterized by strong up-and-down movements. Winds in the vicinity of a thunderstorm do not follow the inward spiral of a cyclone, but they are typically variable and gusty.

Although thunderstorms form "on their own" away from cyclonic storms, they also form in conjunction with cyclones. For instance, thunderstorms are frequently spawned along the cold front of a mid-latitude cyclone, where on rare occasions a tornado may descend from the thunderstorm's cumulonimbus tower. Hurricanes also generate widespread thunderstorm activity. Thus, thunderstorms are related in some manner to all three types of cyclones mentioned here.

THUNDERSTORMS

This is the first of three discussions in this chapter that deal with severe weather. Sections on tornadoes and hurricanes follow this look at thunderstorms. Occurrences of severe weather have a fascination that ordinary weather phenomena cannot provide. The lightning display generated by a severe thunderstorm can be a spectacular event that elicits both awe and fear (Figure 15.11). Of course, hurricanes and tornadoes also attract a great deal of much-deserved attention. A single tornado outbreak or hurricane can cause billions of dollars in property damages as well as many deaths. During a recent 25-year span, the number of fatalities in the United States caused by these important meteorological hazards was nearly 5500 persons. It may surprise many that the number of deaths attributed to lightning (nearly 2400) was greater than those caused by either tornadoes (about 2250) or hurricanes (almost 820). However, this may not necessarily be the case in the future.

Thunderstorm Development

Most everyone has observed a small-scale phenomenon that is caused by the vertical motion of warm, unstable air. Perhaps on a hot day you have seen a small dust devil that formed over an open field and whirled its dusty load to great heights. Or maybe you have noticed a bird glide skyward effortlessly upon an invisible thermal of hot air. These examples illustrate the dynamic thermal instability that occurs during the development of a **thunderstorm.** Thunderstorm activity is associated with cumulonimbus clouds that generate heavy rainfall, thunder, lightning, and occasionally hail.

At any given time there are an estimated 2000 thunderstorms in progress over the face of the earth. As we would expect, the greatest proportion occur in the tropics where warmth, plentiful moisture, and instability are always present. About 45,000 thunderstorms take place each day and more than 16 million occur annually around the world. The lightning from these storms strikes the earth 100 times each second. Annually the United States experiences about 100,000 thunderstorms and millions of lightning strikes. A glance at Figure 15.12 shows that thunderstorms are most frequent in Florida and the eastern Gulf Coast region where activity is recorded on between 70 and 100 days each year. The region on the east side of the Rockies in Colorado and New Mexico is next, with thunderstorms occurring on 60 to 70 days each year. Most of the rest of the nation experiences thunderstorms on 30 to 50 days annually. Clearly the western margin of the United States has the least thunderstorm activity.

The greatest number of thunderstorms occur in association with relatively short-lived cumulonimbus clouds that produce local precipitation. In the United States these cells typically form within warm, humid (maritime tropical) air masses that originate over the Gulf of Mexico and migrate northward. Occasionally, thunderstorms grow very large and remain active for hours. These *severe thunderstorms* produce frequent lightning and are accompanied by locally damaging winds or hail. Most severe thunderstorms in the middle latitudes form along or ahead of cold fronts. Here, forceful lifting of unstable mT air masses triggers thunderstorm development.

All thunderstorms require warm, moist air, which, when lifted, will release sufficient latent heat to provide the buoyancy necessary to maintain its upward flight. Although this instability and associated buoyancy are triggered by a number of different processes, all thunderstorms have a similar life cycle. Because instability and buoyancy are enhanced by high sur-

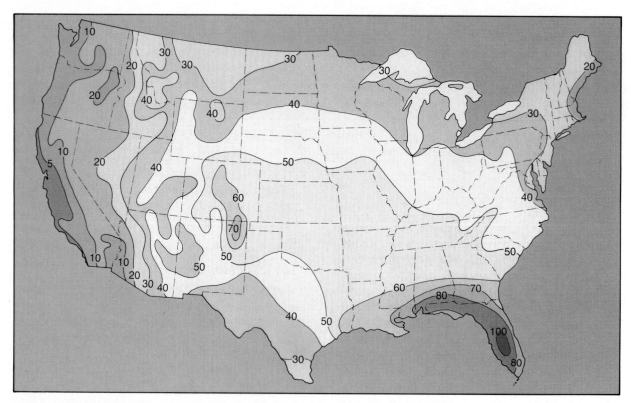

Figure 15.12
Average number of days per year with thunderstorms. Due to its close proximity to the source region for warm, humid, and unstable air masses, the Gulf Coast receives much of its precipitation from thunderstorms. (Source: Environmental Data Service, NOAA)

Figure 15.13
This developing cumulonimbus cloud became a towering August thunderstorm over the high plains near Boulder, Colorado. (Photo by Henry Lansford)

face temperatures, thunderstorms are most common in the afternoon and early evening. However, surface heating is not sufficient for the growth of towering cumulonimbus clouds. A solitary cell of rising hot air produced by surface heating could, at best, produce a small cumulus cloud, which would evaporate within 10–15 minutes. The development of 12,000-meter (or on rare occasions 18,000-meter) cumulonimbus towers requires a continual supply of moist air (Figure 15.13). Each new surge of warm air rises higher than the last, adding to the height of the cloud (Figure 15.14). These updrafts must occasionally reach speeds over 100 kilometers (62 miles) per hour to accommodate the size of hailstones they are capable of carrying upward. Usually within an hour the amount and size of precipitation that has accumulated is too much for the updrafts to support, and in one part of the cloud downdrafts develop, releasing

Figure 15.14
Stages in the development of a thunderstorm. During the cumulus stage, strong updrafts act to build the storm. The mature stage is marked by heavy precipitation and cool downdrafts in part of the storm. When the warm updrafts disappear completely, precipitation becomes light, and the cloud begins to evaporate.

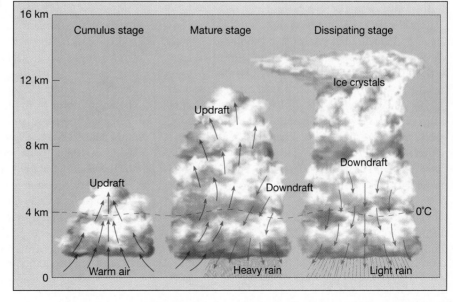

heavy precipitation. This represents the most active stage of the thunderstorm. Gusty winds, lightning, heavy precipitation, and sometimes hail are experienced. Eventually downdrafts dominate throughout the cloud. The cooling effect of falling precipitation coupled with the influx of colder air aloft mark the end of the thunderstorm activity. The life span of a cumulonimbus cell within a thunderstorm complex is only about an hour, but as the storm moves, fresh supplies of warm, water-laden air generate new cells to replace those that are dissipating.

Thunder and Lightning

One obvious feature of a thunderstorm is thunder. Due to the fact that thunder is produced by lightning, lightning must also be present (see Figure 15.11). **Lightning** is similar to the electrical shock you may have experienced on a very dry day upon touching a metal object. Only the intensities differ. During the development of a cumulonimbus cloud, a similar buildup of charge is generated. The reason for this, although not fully understood, hinges on the movement of precipitation within the cloud. The upper portion of the cloud acquires a positive charge, while the lower portion of the cloud maintains an overall negative charge while containing small positively charged pockets. These charges will build to millions, and even hundreds of millions, of volts before a lightning stroke acts to discharge the cloud by striking the earth, or possibly another cloud. The lightning we see as a single stroke is really several very rapid strokes between the cloud and the earth, which together last about only one-tenth of a second.

The electrical discharge of lightning rapidly heats the air, which, in turn, causes it to expand explosively. We hear this expansion as **thunder.** Since lightning and thunder occur simultaneously, it is possible to estimate the distance to the stroke. Lightning is seen instantaneously, whereas the rather slow sound waves, which travel approximately 330 meters per second, reach us some time later. Therefore, if thunder is heard 5 seconds after the lightning is seen, the lightning occurred about 1650 meters away (approximately 1 mile).

The thunder we hear as a rumble is produced along a rather long lightning path located at some distance from the observer. The sound that originates along the path nearest the observer arrives before the sound that originates farthest away. This lengthens the duration of the thunder. Reflection of the sound

waves further delays their arrival and adds to this effect. When lightning occurs more than 20 kilometers away, thunder is rarely heard. This type of lightning, popularly called *heat lightning*, is no different from that which we associate with thunder.

Since lightning moves easily through good electrical conductors, it is wise during thunderstorms to avoid touching or being close to any metallic or tall structure such as an antenna or a tree. Activities in water such as swimming and boating should also be avoided. Indeed, statistics gathered by the federal government show that lightning and the fires it may cause kill more people each year than tornadoes and hurricanes combined.

TORNADOES

Tornadoes are local storms of short duration that must be ranked high among nature's most destructive forces. Their sporadic occurrence and violent winds cause many deaths each year. Tornadoes are violent windstorms that take the form of a rotating column of air that extends downward from a cumulonimbus cloud (Figure 15.15). Pressures within some tornadoes have been estimated to be as much as 10 percent lower than immediately outside the storm. Drawn by the much lower pressure in the center of the storm, air near the ground rushes into the tornado from all directions. As the air streams inward, it is spiraled upward around the core until it eventually merges with the airflow of the parent thunderstorm deep in the cumulonimbus tower. Because of the tremendous pressure gradient associated with a strong tornado, maximum winds can sometimes approach 480 kilometers (300 miles) per hour.

An average of about 780 tornadoes are reported each year in the United States. Still, the actual numbers that occur from one year to the next vary greatly. During a recent 40-year span, for example, yearly totals ranged from a low of 421 in 1953 to a high of 1126 in 1990. Tornadoes occur during every month of the year. April through June is the period of greatest tornado frequency in the United States, while the number is lowest during December and January (Figure 15.16). Of the more than 30,000 confirmed tornadoes reported over the contiguous 48 states during the period, an average of almost five per day occurred during May. At the other extreme, a tornado was reported only every other day in January.

Figure 15.15
Tornado near Wichita, Kansas. A tornado is a violently rotating column of air in contact with the ground. The air column is visible when it contains condensation or when it contains dust and debris. Often the appearance is the result of both. When the column of air is aloft and does not produce damage, the visible portion is properly called a funnel cloud. *(Photo by Wade Balzer/SYGMA)*

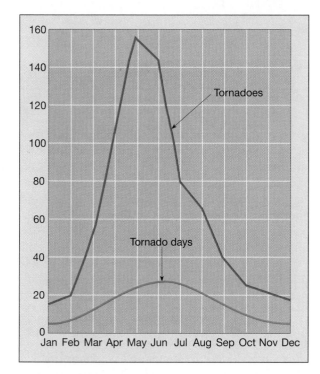

Figure 15.16
Average number of tornadoes and tornado days each month in the United States for a 27-year period. (After NOAA)

Tornadoes form in association with severe thunderstorms that produce high winds, heavy rainfall, and often damaging hail. Fortunately, less than 1 percent of all thunderstorms produce tornadoes. Although weather scientists are still not sure what triggers tornado formation, it has become apparent that they are products of the interaction between strong updrafts in the thunderstorm and winds in the troposphere. In spite of recent advances in modeling the many variables that eventually produce a strong tornado, our knowledge is still limited. Nevertheless, the general atmospheric conditions that are most likely to develop into tornado activity are known.

Severe thunderstorms—and hence tornadoes—are most often spawned along the cold front of a middle-latitude cyclone. Throughout spring, air masses associated with middle-latitude cyclones are most likely to have greatly contrasting conditions. Continental polar air from the Canadian Arctic may still be very cold and dry, whereas maritime tropical air from the Gulf of Mexico is warm, humid, and unstable. The greater the contrast, the more intense the storm. Because these two contrasting air masses are more likely to meet in the central United States, it is not surprising that this region generates more tornadoes than any other area of the country or, in fact, the world. Figure 15.17, which depicts tornado incidence in the United States based on data for a 27-year period, readily substantiates this fact.

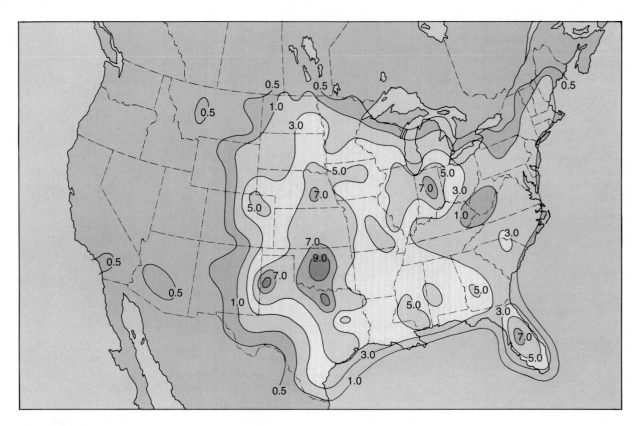

Figure 15.17
Average annual tornado incidence per 10,000 square miles (26,000 square kilometers) for a 27-year period.

The average tornado has a diameter of between 150 and 600 meters (500 and 2000 feet), travels across the landscape at approximately 45 kilometers (28 miles) per hour, and cuts a path about 10 kilometers (6 miles) long.* Since many tornadoes occur slightly ahead of a cold front, in the zone of southwest winds, most move toward the northeast. The Illinois example demonstrates this movement (Figure 15.18). What Figure 15.18 also shows is that many tornadoes do not fit the description of the "average" tornado. Many have had paths a great deal longer than 10 kilometers and have traveled not at 45 kilometers per hour, but at speeds in excess of 100 kilometers (62 miles) per hour. Furthermore, there have been tornadoes that have had diameters of up to 1.6

kilometers (1 mile)—more than four times the "average" size.

The potential for tornado destruction depends largely upon the strength of the winds generated by the storm. One commonly used guide to tornado intensity was developed by T. Theodore Fujita at the University of Chicago and is appropriately called the *Fujita intensity scale*, or simply the *F-scale* (Table 15.1). Since tornado winds cannot be measured directly, a rating on the F-scale is determined by assessing the worst damage produced by a storm.

Tornadoes take many lives each year, sometimes hundreds in a single day. When tornadoes struck an area stretching from Canada to Georgia on April 3, 1974, the death toll exceeded 300, the worst in half a century. Most tornadoes, however, do not result in a loss of life. In one statistical study that examined a 29-year period, there were 689 tornadoes that resulted in deaths. This figure represents slightly less than 4 percent of the total 19,312 reported storms. Although the percentage of tornadoes that result in

---

*   The 10-kilometer figure applies to documented tornadoes. Since many small tornadoes go undocumented, the real average path of all tornadoes is unknown, but shorter than 10 kilometers.

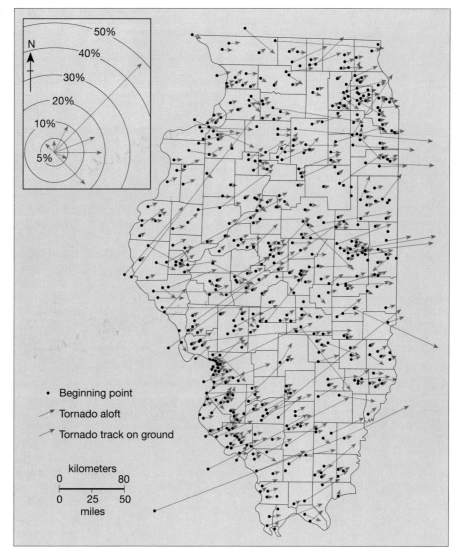

Figure 15.18
Paths of Illinois tornadoes (1916–1969). Since most tornadoes occur slightly ahead of a cold front, in the zone of southwest winds, they tend to move toward the northeast. Tornadoes in Illinois verify this. Over 80 percent exhibited directions of movement toward the northeast through east. (After J. W. Wilson and S. A. Changnon, Jr., Illinois Tornadoes, *Illinois State Water Survey Circular 103, 1971: 10, 24)*

• Beginning point

↗ Tornado aloft

↗ Tornado track on ground

kilometers
0   80
0   25   50
miles

death is small, each tornado is potentially lethal. When tornado fatalities and storm intensities are compared, the results are quite interesting: The majority (63 percent) of tornadoes are weak (F0 and F1) and the number of storms decreases as tornado intensity increases. The distribution of tornado fatalities, however, is just the opposite. Although only 2 percent of tornadoes are classified as violent (F4 and F5), they account for nearly 70 percent of the deaths. If there is some question as to the causes of tornadoes, there certainly is no question about the destructive effects of these violent storms.

Because tornadoes are small and relatively short-lived phenomena, they are among the most difficult weather features to forecast precisely. When conditions appear favorable for tornado formation, a **tornado watch** is issued for areas covering about 65,000 square kilometers. Between 35 and 40 percent of the predictions are correct; that is, one or more tornadoes occur somewhere in the specified region. The incorrect forecasts are about evenly divided between cases when no tornadoes are sighted and cases when tornadoes occur outside, but near, the watch area.

Whereas a tornado watch is designed to alert people to the possibility of tornadoes, a **tornado warning** is issued when a funnel cloud has actually been sighted or is indicated by radar. It warns of a high

Table 15.1
Fujita Intensity Scale.

| Scale | Wind Speed | | Expected Damage |
	Km/Hr	Mi/Hr	
F0	<116	<72	Light damage. Damage to chimneys and billboards; broken branches; shallow-rooted trees pushed over.
F1	116–180	72–112	Moderate damage. The lower limit is near the beginning of hurricane wind speed. Surfaces peeled off roofs; mobile homes pushed off foundations or overturned; moving autos pushed off the road.
F2	181–253	113–157	Considerable damage. Roofs torn off frame houses; mobile homes demolished; boxcars pushed over; large trees snapped or uprooted; light-object missiles generated.
F3	254–332	158–206	Severe damage. Roofs and some walls torn off well-constructed houses; trains overturned; most trees in forest uprooted; heavy cars lifted off ground and thrown.
F4	333–419	207–260	Devastating damage. Well-constructed houses leveled; structures with weak foundations blown some distance; cars thrown and large missiles generated.
F5	<419	>260	Incredible damage. Strong frame houses lifted off foundations and carried considerable distance to disintegrate; automobile-sized missiles fly through the air farther than 100 m; trees debarked; incredible phenomena occur.

probability of imminent danger. When severe weather threatens, the radar screens are monitored for very intense echoes, which in turn are associated with heavy precipitation and the greater likelihood of hail, strong winds, and tornadoes. In addition, the echo from a tornadic storm sometimes displays a hook-shaped appendage. If the direction and approximate speed of the storm are known, the storm's most probable path can be estimated. Because tornadoes often move erratically, the warning area is fan-shaped downwind from the point where the tornado has been spotted. Since the late 1960s, warnings have been given for most major tornadoes. It is believed that such warnings substantially reduce the number of deaths and serious injuries that might otherwise occur.

The tornado warning system that has been used throughout the United States for many years relies heavily on visual sightings by a few trained observers as well as the general public. Unfortunately, such a system is prone to incomplete coverage and mistakes. The errors are most likely to occur at night when tornadoes may go unnoticed or harmless clouds may be mistaken for funnel clouds. Hence, there may be a lack of adequate warning, on the one hand, or unnecessary warnings, on the other.

Many of the difficulties that have limited the accuracy of tornado warnings will be reduced or eliminated in the future when an advancement in radar technology, called **Doppler radar,** is put into general use. Doppler radar not only performs the same tasks as conventional radar, but also has the ability to detect motion directly. Doppler radar can detect the initial formation and subsequent development of a **mesocyclone,** an intense rotating wind system in the lower part of a thunderstorm that frequently precedes tornado development. Almost all mesocyclones produce damaging hail, severe winds, or tornadoes. Those that produce tornadoes (about 60 percent) can be distinguished by their stronger wind speeds and their sharper gradients of wind speeds.

In carefully planned tests conducted over several years, Doppler radar provided an average warning time of 21 minutes before tornado touchdown. By comparison, the average warning time provided by visual observation is less than two minutes. An additional advantage was a decreased false alarm rate. Predictions based on Doppler radar were correct 75 percent of the time—a considerable improvement over previous standards.

Doppler radar is not problem-free, however. Some operational concerns may arise when Doppler radar is deployed around the country. For example, studies indicate that nearly 40 percent of mesocyclones do not produce tornadoes. If this is a representative number, then a serious over-warning problem could result if warnings are issued every time a mesocyclone is detected. A second concern relates to weak tornadoes that rank at the bottom of the Fujita intensity scale (see Table 15.1). Doppler radar makes the detection of these weak tornadoes possible. Consequently, the potential exists for numerous warnings being issued for tornadoes that do little or no damage. This could desensitize the public to the dangers of the more rare, life-threatening tornadoes. Thus, a research goal should be the development of techniques that enable the forecasting of tornado intensity.

Although operational problems may exist, the benefits of Doppler radar are many. As a research tool, it not only provides data on the formation of tornadoes but is also helping meteorologists gain new insights into thunderstorm development, the structure and dynamics of hurricanes, and air turbulence hazards that plague aircraft. As a practical tool for tornado detection, it has significant advantages over a system that uses observers and conventional radar. Recognizing these advantages, the National Weather Service will be replacing its conventional weather radars with more advanced systems that use Doppler principles of direct wind speed and direction measurements. The new radars are part of the *Next Generation Weather Radar* (NEXRAD) program aimed at improving the forecasting of severe storms, tornadoes, and flash floods.

HURRICANES

The whirling tropical cyclones that on occasion have wind speeds reaching 300 kilometers (185 miles) per hour are known in the United States as **hurricanes** and are the greatest storms on earth. Out at sea, they can generate 15-meter (50-foot) waves capable of inflicting destruction hundreds of kilometers from their source. Should a hurricane smash into land, strong winds coupled with extensive flooding can impose billions of dollars in damages and great loss of life (Figure 15.19).

The vast majority of hurricane-related deaths and damage are caused by relatively infrequent, yet powerful, storms. This was clearly the case in the late 1980s and the early 1990s. The hurricane seasons of 1988 and 1989 produced five storms that were classified as very strong or devastating. This was the largest number of storms having such intensity since 1960 and 1961. Two of these storms were especially memorable—Hurricane Gilbert in 1988 and Hurricane Hugo in 1989. Gilbert was the most powerful hurricane yet recorded in the Western Hemisphere. Both storms were deadly and caused billions of dollars in property damages.

In 1992, yet another pair of memorable storms struck the United States. On August 24, Hurricane Andrew slammed into South Florida and then went on across the Gulf of Mexico to strike the Louisiana coast two days later. It was the costliest natural disaster in United States history (see Box 15.3). Just a little more than two weeks later, on September 11, Hurricane Iniki battered the Hawaiian island of Kauai. The direct hit devastated the 30-mile-wide island. It was the most powerful hurricane to hit the Hawaiian Islands this century.

Hurricanes are becoming a growing threat because more and more people are living and working

Figure 15.19
A. When Hurricane Hugo struck the South Carolina coast in September 1989, damages caused by the storm surge were extensive. (Photo by Tom Brunk) B. By contrast, when Hurricane Andrew struck South Florida in August 1992, the billions of dollars in property damages were primarily the result of the strong winds. (Photo by John Lopinot/Palm Beach Post)

A.

B.

along and near coasts. In 1990, 50 percent of the United States population lived within 75 kilometers (45 miles) of a coast. This number is projected to increase to 75 percent by the year 2010. The concentration of such large numbers of people near the shoreline means that hurricanes and other large storms place millions at risk. Moreover, the potential costs of property damage are incredible.

Hurricane Formation and Decay

Hurricanes form in all tropical waters (except those of the South Atlantic and eastern South Pacific) between the latitudes of 5 degrees and 20 degrees and are known by different names in various parts of the world. In the western Pacific, they are called *typhoons* and in the Indian Ocean, *cyclones*. The North Pacific has the greatest number of storms, averaging 20 per year. Fortunately for those living in the coastal regions of the southern and eastern United States, fewer than 5 hurricanes, on the average, develop each year in the warm sector of the North Atlantic.

Although many tropical disturbances develop each year, only a few reach hurricane status, which by international agreement requires wind speeds in excess of 119 kilometers (74 miles) per hour and a rotary circulation. Hurricanes average 600 kilometers (375 miles) in diameter and often extend 12,000 meters (40,000 feet) above the ocean surface. From the outer edge to the center, the barometric pressure has on occasion dropped 60 millibars, from 1010 millibars to 950 millibars. The lowest pressures ever recorded in the Western Hemisphere are associated with these storms. A steep pressure gradient generates the rapid, inward-spiraling winds of a hurricane.

As the inward rush nears the core of the storm, it turns upward and ascends in a ring of cumulonimbus towers. This doughnut-shaped wall of intense convective activity surrounding the center of the storm is called the **eye wall**. It is here that the greatest wind speeds and heaviest rainfall occur. Surrounding the eye wall are curved bands of clouds that trail away in a spiral fashion. Near the top of the hurricane the airflow is outward, carrying the rising air away from the storm center, thereby providing room for more inward flow at the surface.

At the very center of the storm is the **eye** of the hurricane (Figure 15.20). This well-known feature is a zone of about 20 kilometers (12.5 miles) in diameter where precipitation ceases and winds subside. It offers a brief but deceptive break from the extreme weather in the enormous curving wall clouds that surround it. The air within the eye gradually descends and heats by compression, making it the warmest part of the storm. Although many people believe that the eye is characterized by clear blue skies, such is usually not the case because the subsidence in the eye is seldom strong enough to produce cloudless conditions. Although the sky appears much brighter in this region, scattered clouds at various levels are common.

A hurricane can be described as a heat engine that is fueled by the latent heat liberated when huge quantities of water vapor condense. The amount of energy produced by a typical hurricane in just a single day is truly immense—roughly equivalent to the entire electrical energy production of the United States in a year. The release of latent heat warms the

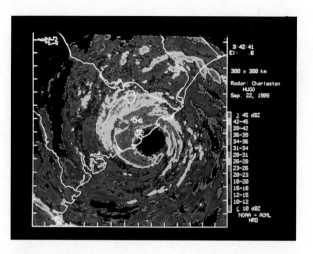

Figure 15.20
Radar image of Hurricane Hugo moving over the coast of South Carolina on September 22, 1989. Colors show rainfall intensity. Red is the most intense and blue is less intense. No rain is falling in the black areas. The rainless "hole" in the center is the eye. As expected, the heaviest rains are occurring in the eye wall. (Courtesy of Peter Dodge, NOAA, Hurricane Research Division)

air and provides buoyancy for its upward flight. The result is to reduce the pressure near the surface, which encourages a more rapid inward flow of air. To get this engine started, a large quantity of warm, moisture-laden air is required, and a continual supply is needed to keep it going.

Hurricanes develop most often in the late summer when water temperatures have reached 27°C (80°F) or higher and thus are able to provide the necessary heat and moisture to the air. This ocean-water temperature requirement is thought to account for the fact that hurricanes do not form over the relatively cool waters of the South Atlantic and the eastern South Pacific. For the same reason, few hurricanes form poleward of 20 degrees of latitude. Although water temperatures are sufficiently high, hurricanes do not form within 5 degrees of the equator, presumably because the Coriolis effect is too weak to initiate the necessary rotary motion.

The dynamics of the initial stage of a tropical cyclone's life cycle are not well understood. This lack of understanding is due in part to the complexity of these storms. An additional and very important factor is the lack of observations in regions where storms form. Although the exact mechanism of formation is not completely understood, we know that smaller tropical cyclones initiate the process. These initial disturbances are regions of low-level convergence and

BOX 15.3

Hurricane Andrew

*E*ver since 1965 when Hurricane Betsy struck, the residents of South Florida's vulnerable east coast have waited for the next big one. In the intervening years, many powerful storms skirted the region and caused destruction elsewhere. In August 1992, the inevitable became reality.

On August 16, 1992, a tropical depression formed out in the Atlantic, closer to the west coast of Africa than to the United States. The next day the National Hurricane Center declared that the depression had reached the status of a tropical storm and christened it *Andrew*. For the next week Andrew made its way westward, achieving hurricane status on August 22 (Figure 15.B). By Sunday, the 23rd, the spiraling storm was on a collision course with South Florida.

Hurricane Andrew made landfall south of Key Biscayne, Florida, during the dark, early morning hours of August 24 (Figure 15.C). Maximum sustained surface winds were about 230 kilometers (145 miles) per hour, with gusts exceeding 280 kilometers (175 miles) per hour. With a central pressure of 922 millibars (27.23 inches), Andrew was a category 4 storm on the Saffir-Simpson scale (see Table 15.2). It was close to being a rare category 5. Andrew had the third lowest pressure this century for a hurricane making landfall in the United States.

It took Hurricane Andrew just hours to cut its destructive path across Florida. Property damage from Andrew was primarily wind damage along a 40-kilometer-wide swath of destruction that was centered a few kilometers north of the town of Homestead. Fortunately, the highly developed coastline of Miami Beach was more than 27 kilometers

Figure 15.B
Positions for Hurricane Andrew, August 17–28, 1992. (National Hurricane Center)

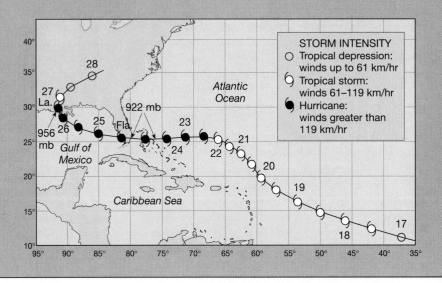

lifting. Many tropical disturbances of this type occur each year, but only a few develop into full-fledged hurricanes. By international agreement, lesser tropical cyclones are given different names based on the strength of their winds. When a cyclone's strongest winds do not exceed 61 kilometers (38 miles) per hour, it is called a **tropical depression;** when winds are between 61 and 119 kilometers (38 and 74 miles) per hour, the cyclone is termed a **tropical storm.**

Each year between 80 and 100 tropical storms develop, and of this total, half or more become hurricanes.

Hurricanes diminish in intensity whenever they (1) move over ocean waters that cannot supply warm, moist tropical air; (2) move onto land; or (3) reach a location where the large-scale flow aloft is unfavorable. Whenever a hurricane moves onto land, it loses its punch rapidly. The most important reason from the eye of the storm and so did not experience sustained hurricane-force winds. Two other aspects of the storm made it less damaging than it might otherwise have been: The storm surge was small and there was little rainwater flooding because the hurricane did not linger but advanced rapidly across the region.

The storm's destructive accomplishments were awesome. In one area, the wind carried 6-meter-long steel and concrete beams, with roofs still attached, more than 50 meters. Cars and boats were tossed about like toys, tens of thousands of homes and businesses were devastated, hundreds of acres of groves were uprooted, and Homestead Air Force Base was leveled. Government officials who toured the area said it looked like a "war zone." After crossing South Florida, Andrew went on to cross the Gulf of Mexico and lashed the Louisiana coastline on August 26 as a category 3 storm. In all, Hurricane Andrew was responsible for at least 62 deaths and caused $20–30 billion in damages. It was the costliest natural disaster in U.S. history.

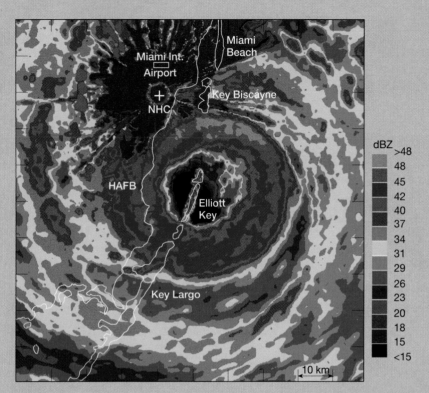

Figure 15.C
Color radar image of Hurricane Andrew at 4:35 A.M. EDT, August 24, 1992. The picture is from the last full sweep of the National Weather Service's Miami radar (located at the National Hurricane Center [NHC]) before the radar was destroyed by the storm. The digitized radar imagery shows the eye centered over Elliott Key just before landfall at Homestead Air Force Base (HAFB). As Andrew traveled due west, the heaviest damage occurred in those areas affected by the eye wall (doughnut-shaped region with echoes greater than 42 dBZ). The weather radar measures the power from the portion of the radar beam scattered back by raindrops and ice particles. The colors associated with higher dBZ (i.e., red) correspond to areas with larger amounts of rain, which typically are also regions of stronger winds. Areas with high dBZ in the center of the eye are because of ground clutter from islands. (Ground clutter is the reflection of the radar beam by terrain, large structures, and rough water.) Ground clutter in the vicinity of NHC has been removed and is shown in gray. (Courtesy of Hurricane Research Division/NOAA)

for this rapid demise is the fact that the storm's source of warm, moist air is cut off. When an adequate supply of water vapor does not exist, condensation and the release of latent heat must diminish. In addition, the increased surface roughness over land results in a rapid reduction in surface wind speeds. This factor causes the winds to move more directly into the center of the low, thus helping to eliminate the large pressure differences.

Hurricane Destruction

A location only a few hundred kilometers from a hurricane—just one day's striking distance away — may experience clear skies and virtually no wind. Prior to the age of weather satellites, such a situation made the job of warning people of impending storms very difficult.

The worst natural disaster in U.S. history resulted from a hurricane that struck an unprepared Galves-

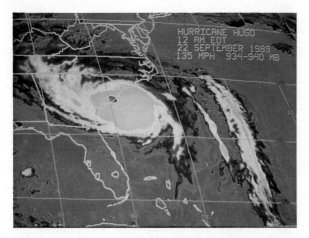

Figure 15.21
A satellite image of Hurricane Hugo as the storm moved onto the South Carolina coast. The eye is clearly visible in the center of the storm. The eye wall, the most intense part of the storm, is a doughnut-shaped wall of cumulonimbus development surrounding the eye. Satellites are essential tools for detecting and tracking hurricanes. (Courtesy of National Hurricane Center/NOAA)

ton, Texas, on September 8, 1900. The strength of the storm coupled with the lack of adequate warning caught the population by surprise and cost 6000 people in the city their lives. At least 2000 more were killed elsewhere. Fortunately, hurricanes are no longer the unheralded killers they once were. Once a storm develops cyclonic flow and the spiraling bands of clouds characteristic of a hurricane, it receives continuous monitoring. For example, when hurricanes such as Hugo (1989) and Andrew (1992) formed, satellites were able to identify and track the storms long before they made landfall (Figure 15.21). In the

United States, early warning systems have greatly reduced the number of deaths caused by hurricanes. At the same time, however, there has been an astronomical rise in the amount of property damage. The primary reason, of course, has been the rapid population growth in coastal areas.

Although the amount of damage caused by a hurricane depends on several factors, including the size and population density of the area affected and the nearshore bottom configuration, certainly the most significant factor is the strength of the storm itself. By studying the storms of the past, a scale has been established that is used to rank the relative intensity of hurricanes (Table 15.2). As Table 15.2 indicates, a *category 5* storm is the worst possible, whereas a *category 1* hurricane is least severe. During the hurricane season it is common to hear scientists and reporters alike use the numbers from the *Saffir-Simpson Hurricane Scale*. The famous Galveston hurricane just mentioned, with winds in excess of 209 kilometers (130 miles) per hour and a pressure of 931 millibars, would be placed in category 4. Storms that fall into category 5 are rare. Hurricane Camille, a 1969 storm that caused catastrophic damage along the coast of Mississippi, is one well-known example. More recently, Hurricane Gilbert was placed into category 5. Gilbert's central sea-level pressure was an extremely low 888 millibars, a new record minimum for Western Hemisphere storms.

Damage caused by hurricanes can be divided into three categories: (1) wind damage, (2) storm surge, and (3) inland freshwater flooding. Although wind damage is perhaps the most obvious of the categories, it is not necessarily responsible for the greatest amount of destruction. This is not to say, however, that wind damage cannot be significant. For some

Table 15.2
Saffir-Simpson Hurricane Scale.

Scale Number (category)	Central Pressure (millibars)	Winds (km/hr)	Storm Surge (meters)	Damage
1	≥980	119–153	1.2–1.5	Minimal
2	965–979	154–177	1.6–2.4	Moderate
3	945–964	178–209	2.5–3.6	Extensive
4	920–944	210–250	3.7–5.4	Extreme
5	<920	>250	>5.4	Catastrophic

structures, the force of the wind is sufficient to cause total destruction. This was demonstrated in South Florida in 1992. The billions of dollars in property damages from Hurricane Andrew were largely the result of strong winds (see Box 15.3).

The most devastating damage in the coastal zone is caused by the storm surge. It not only accounts for a large share of coastal property losses, but is also responsible for 90 percent of all hurricane-caused deaths. A **storm surge** is a dome of water 65 to 80 kilometers (40 to 50 miles) long that sweeps across the coast near the point where the eye makes landfall. If all of the wave activity were smoothed out, the storm surge would be the height of the water above normal tide level. In addition, superimposed upon the surge is tremendous wave activity. We can easily imagine the damage this surge of water could inflict on low-lying coastal areas. In the delta region of Bangladesh, for example, the land is mostly less than 2 meters above sea level. When a storm surge superimposed upon normal high tide inundated that area on November 13, 1970, the official death toll was 200,000; unofficial estimates ran to 500,000. This was one of the worst disasters of modern times.

The torrential rains that accompany most hurricanes represent a third significant threat—flooding. Although hurricanes weaken rapidly as they move in-land, the remnants of the storm can still yield 15 to 30 centimeters (6 to 12 inches) or more of rain as they move inland. A good example of such destruction is Hurricane Agnes. Although this was just a modest storm in terms of wind speed and pressure, it was one of the costliest hurricanes of the century, responsible for more than $2 billion in damages and 122 deaths. Most destruction was attributed to flooding caused by an inordinate amount of rainfall. Agnes dropped an estimated 105 cubic kilometers or more of rainwater over the eastern United States.

To summarize, extensive damage and loss of life in the coastal zone can result from storm surge, torrential rains, and strong winds. When loss of life occurs, it is commonly caused by the storm surge, which can devastate entire barrier islands or zones within a few blocks of the coast. Although wind damage is usually not as catastrophic as the storm surge, it affects a much larger area. Where building codes are inadequate, economic losses can be especially severe. Because hurricanes weaken as they move inland, most wind damage occurs within 200 kilometers of the coast. Far from the coast a weakening storm can produce extensive flooding long after the winds have diminished below hurricane levels. Sometimes the damage from inland flooding exceeds storm-surge destruction.

Review Questions

1. What are the characteristics of a maritime tropical air mass?

2. Describe the weather associated with a continental polar air mass in the winter and in the summer. When would this air mass be most welcome in the United States?

3. Where are the source regions for the maritime tropical air masses that affect North America? Where are the source regions for the maritime polar air masses?

4. Why are snowfall totals in the Great Lakes region highest on the leeward (downwind) shores of the lakes? What term is applied to these heavy snows? (See Box 15.1.)

5. Describe the weather along a cold front where very warm, moist air is being displaced.

6. Explain the basis for the following weather proverb:
   Rain long foretold, long last;
   Short notice, soon past.

7. The formation of an occluded front marks the beginning of the end of a wave cyclone. Why is this true?

8. For each of the weather elements that follow, describe the changes that an observer experiences when a wave cyclone passes with its center north of the observer: wind direction, pressure tendency, cloud type, cloud cover, precipitation, temperature.

9. Describe the weather conditions an observer would experience if the center of a wave cyclone passed to the south.

10. Briefly explain how the flow aloft aids the formation of cyclones at the surface.

11. Compare the wind speeds and sizes of middle-latitude cyclones, tornadoes, and hurricanes. How are thunderstorms related to each? (See Box 15.2.)

12. What is the primary requirement for the formation of thunderstorms?

13. Based on your answer to Question 12, where would you expect thunderstorms to be most common on the earth? Where specifically in the United States?

14. How is thunder produced? How far away is a lightning stroke if thunder is heard 15 seconds after the lightning is seen?

15. Why do tornadoes have such high wind speeds?

16. What general atmospheric conditions are most conducive to the formation of tornadoes?

17. When is the "tornado season"? That is, during what months is tornado activity most pronounced?

18. Distinguish between a tornado watch and a tornado warning.

19. What advantage does Doppler radar have over conventional radar? Describe some operational problems that may arise when Doppler radar is deployed.

20. Which has stronger winds, a tropical storm or a tropical depression?

21. Why does the intensity of a hurricane diminish rapidly when it moves onto land?

22. Hurricane damage can be divided into three broad categories. Name them. Which category is responsible for the highest percentage of hurricane-related deaths?

23. A hurricane has slower wind speeds than a tornado, yet it inflicts more total damage. How might this be explained?

Key Terms

air mass (p. 510)

air-mass weather (p. 510)

cold front (p. 515)

continental (c) air mass (p. 511)

Doppler radar (p. 531)

eye (p. 533)

eye wall (p. 533)

front (p. 512)

hurricane (p. 531)

lake-effect snow (p. 511)

lightning (p. 526)

maritime (m) air mass (p. 511)

mesocyclone (p. 531)

middle-latitude cyclone (p. 517)

occluded front (p. 516)

occlusion (p. 520)

polar (P) air mass (p. 511)

source region (p. 511)

speed divergence (p. 521)

stationary front (p. 516)

storm surge (p. 537)

thunder (p. 526)

thunderstorm (p. 523)

tornado (p. 526)

tornado warning (p. 529)

tornado watch (p. 529)

tropical (T) air mass (p. 511)

tropical depression (p. 534)

tropical storm (p. 534)

warm front (p. 513)

wave cyclone (p. 517)

16

Human Impact on the Atmosphere

Opposite page: Smog over the city of Los Angeles,
California. (Photo by John Mead/Photo Researchers,
Inc.)

Left: Industrial complex adding particulate matter and
gaseous emissions to the atmosphere. (Photo by E. J.
Tarbuck)

Above: Temperature inversion trapping smog near El
Cajon, California. (Photo by James E. Patterson)

Chapter 16 brings together meteorological principles and people. How people modify the atmospheric environment represents the major theme of this chapter. In addition to a close look at air pollution and the factors that affect the quality of the air we breathe, we will examine some of the ways that humans may inadvertently be changing the climate, both locally and globally. We will also explore some of the methods devised to intentionally alter weather processes. Can we make it rain? Can the fury of a hurricane be diminished? If so, how? These are some of the ideas that will be examined in the following pages.

The goal of this chapter is to survey the many ways by which humans modify their atmospheric environment and to understand the consequences of that modification. We will begin by examining inadvertent changes, that is, human-induced modifications of our atmosphere that are unintentional or accidental. First among these is the growing problem of air pollution. As we shall see, air pollution not only affects human health, it also threatens other constituents of our physical and biological environment. Following this, we will examine the most apparent human impact on climate—the building of cities. The construction of each factory, road, office building, and house destroys existing microclimates and creates new ones of great complexity. After looking at the climatic changes created by cities, we will shift our attention to a far greater scale and study the possible effects of human activities on the global climate. The chapter will conclude by examining the wide variety of ways by which people deliberately attempt to modify the weather. Intentional weather modification involves human intervention to influence and improve the atmospheric processes and events that comprise the weather, in other words, to "aim" weather at human purposes. Attempts to increase precipitation, disperse fog, reduce storm damage, and prevent frost damage to crops are all examples of intentional weather modification.

AIR POLLUTION

Air pollution is a continuing threat to our health and welfare. An average adult requires about 13.5 kilograms (30 pounds) of air each day, as compared to about 1.2 kilograms (2.65 pounds) of food and 2 kilograms (4.5 pounds) of water. The cleanliness of the air, therefore, should certainly be as important to us as the cleanliness of our food and water.

Air is never perfectly clean. There have always been many natural sources of air pollution (Figure 16.1). Ash from volcanic eruptions, salt particles from breaking waves, pollen and spores released by plants, smoke from forest and brush fires, and windblown dust are all examples of "natural" air pollution. From the time humans have been on the earth, however, they have added to the frequency and intensity of some of these natural pollutants, particularly smoke and dust (Figure 16.2). With the discovery of fire came an increased number of accidental as well as intentional burnings. Even today, in many parts of

Figure 16.1
Forest fires triggered by lightning are one of many natural sources of air pollution. (Warren Faidley/International Stock Photo)

the world fire is used to clear land for agricultural purposes (the so-called slash-and-burn method), filling the air with smoke and reducing visibility. When people clear the land of its natural vegetative cover for whatever purpose, soil is exposed and blown into the air. Be that as it may, when considering the air in a modern-day industrial city, these human-accentuated forms of pollution, although significant, may seem minor by comparison.

With the Industrial Revolution came "big-time" air pollution. Rather than simply accelerating natural sources, which of course continued to occur, people found many new ways to pollute the air (Figure 16.2) and many new agents to pollute it with.

This rapid increase in air pollution was not always viewed with great alarm. Instead, chimneys belching forth smoke and soot were considered a sign of growth and prosperity. For example, the following quotation from an 1880 speech by Robert Ingersoll,

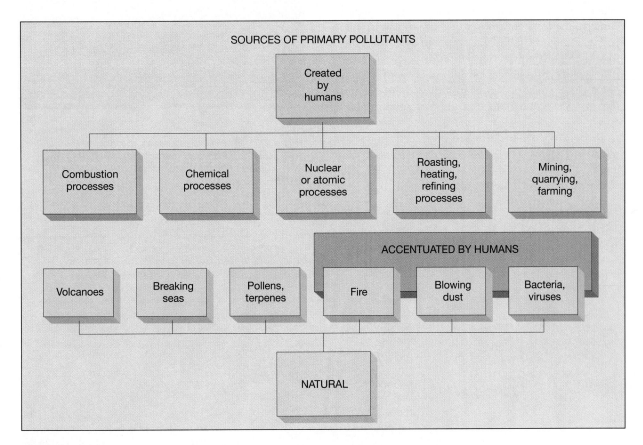

Figure 16.2
Sources of primary pollutants. [After Reid A. Bryson and John I. Kutzbach, Air Pollution. (Washington, D.C.: Association of American Geographers, 1968), fig. 3, p. 9]

a well-known lawyer and orator, is reported to have elicited great cheering from the audience:

> "I want the sky to be filled with the smoke of American industry and upon that cloud of smoke will rest forever the bow of perpetual promise. That is what I am for."

With the rapid growth of the world's population and accelerated industrialization, the quantities of atmospheric pollutants increased drastically.

Sources and Types of Air Pollution

Air pollutants are airborne particles and gases that occur in concentrations that endanger the health and well-being of organisms or disrupt the orderly functioning of the environment. Pollutants may be grouped into two categories: primary and secondary.

Primary pollutants are emitted directly from identifiable sources. They pollute the air immediately upon being emitted. Table 16.1 lists the major primary pollutants and the sources that produce them. When the various sources are examined, the significance of the transportation category is obvious. It accounts for nearly half of our air pollution (by weight). In addition to highway vehicles, this category includes trains, ships, and airplanes. Still, the tens of millions of cars and trucks on U.S. roads are, without a doubt, the greatest contributors in this category (Figure 16.3).

Secondary pollutants are produced in the atmosphere when certain chemical reactions take place among the primary pollutants. Sulfuric acid (H_2SO_4) is one example of a secondary pollutant. It is pro-

Figure 16.3
This traffic jam reminds us that the transportation category accounts for nearly half of the air pollution (by weight) in the United States. (Photo by D. I. MacDonald/The Stockhouse)

Table 16.1
Estimated nationwide emissions (millions of metric tons/year)

Source	Carbon Monoxide	Particulates	Sulfur Oxides	Volatile Organics	Nitrogen Oxides	Total
Transportation	43.5	1.6	1.0	5.1	7.3	58.5
Stationary Source Fuel Combustion	4.7	1.9	16.6	0.7	10.6	34.5
Industrial Processes	4.7	2.6	3.2	7.9	0.6	19.0
Solid Waste Disposal	2.1	0.3	0.0	0.7	0.1	3.2
Miscellaneous	7.2	1.2	0.0	2.8	0.2	11.4
Total	62.2	7.6	20.8	17.2	18.8	126.6

Source: National Air Pollutant Emission Estimates, 1900–1991, U.S. Environmental Protection Agency Publication No. EPA-454/R-92-013, 1992.

duced when sulfur dioxide combines with oxygen, yielding sulfur trioxide, which then combines with water to form an irritating and corrosive acid. A later section on acid precipitation will discuss the effects of this in some detail.

Air pollution in urban and industrial areas is often termed **smog.** The word was coined in 1905 by Dr. Harold A. Des Veaux, a London physician, and was created by combining the words *smoke* and *fog.* Dr. Des Veaux's term was indeed an apt description of London's principal air pollution threat, which was associated with the products of coal burning coupled with periods of high humidity. Today the term *smog* is used as a synonym for general air pollution and does not necessarily imply the smoke-fog combination. Therefore, when greater clarity is desired, we often find the term *smog* preceded by modifiers such as *London-type*, *classical*, *Los Angeles–type*, or *photochemical.* The first two refer to the original meaning of the term, whereas the last two refer to air quality problems created by secondary pollutants.

Many reactions that produce secondary pollutants are triggered by strong sunlight and are therefore termed **photochemical reactions.** One common example occurs when nitrogen oxides absorb solar radiation and initiate a chain of complex reactions. When certain volatile organic compounds are present, the result is the formation of a number of undesirable secondary products that are very reactive, irritating, and toxic. Collectively this noxious mixture of gases and particles is called *photochemical smog.* The major component in photochemical smog is ozone. Recall from Chapter 12 that ozone is also formed by natural processes in the stratosphere where it plays a vital role because of its ability to absorb damaging ultraviolet radiation. Nevertheless, because ozone is poisonous, it is considered a pollutant when it is produced near the earth's surface, where we breathe (see Box 16.1).

Meteorological Factors Affecting Air Pollution

There are two ways in which air pollution and meteorology are linked. One connection concerns the influence of weather conditions on the dilution of air pollutants. The second connection is the reverse of the first and has to do with the effect that air pollution has on weather and climate. The first of these associations will be examined in this section. The second, and equally important, relationship will be discussed in more detail later in this chapter.

In Chapter 13 the concept of atmospheric stability was introduced. We learned that the stability of air plays a significant role in controlling many aspects of daily weather. In this portion of Chapter 16 we will see that atmospheric stability is closely related to urban air pollution. Air quality is not just a function of the quantity and types of pollutants emitted into the air, but it is also closely linked to the atmosphere's ability to disperse these noxious substances. Dispersal, in turn, is related to the stability of the atmosphere.

Certainly the most obvious factor influencing air pollution is the quantity of contaminants emitted into the atmosphere. However, experience tells us that even when emissions remain relatively steady for extended periods, we often find wide variations in air quality from one day to the next. Indeed, when air pollution episodes occur, they are not generally the result of a drastic increase in the output of pollutants; instead, they occur because of changes in certain atmospheric conditions.

Perhaps you have heard the following well-known phrase: "The solution to pollution is dilution." To a significant degree this is true. If the air into which the pollution is released is not dispersed, the air will become more toxic. Two of the most important atmospheric conditions affecting the dispersion of pollutants are the strength of the wind and the stability of the air. These factors are critical because they determine how rapidly pollutants are diluted by mixing with the surrounding air after leaving the source.

The manner in which wind speed influences the concentration of pollutants is shown in Figure 16.4. When winds are weak, the concentration of pollutants is much higher than when winds are strong. For example, assume that a burst of pollution leaves a smoke stack every second. If the wind speed were 10 meters per second, the distance between each pollution "cloud" would be 10 meters. If the wind were reduced to 5 meters per second, the distance between "clouds" would be 5 meters. Consequently, because of the direct effect of wind speed, the concentration of pollutants is twice as great with the 5-meters-per-second wind as with the 10-meters-per-second wind. Therefore, it is easy to understand why air pollution problems are associated with periods when winds are weak or calm and seldom occur when winds are strong.

A second aspect of wind speed also influences air quality—the stronger the wind, the more turbulent the air. Thus, strong winds mix polluted air more rapidly with the surrounding air to cause the pollu-

BOX 16.1

Ozone: Good or Bad?

In any examination of atmospheric environmental issues, ozone is likely to be given a prominent place. Sometimes, however, there is confusion among the general public and media about the role of ozone in the atmosphere. The reason for this confusion is the fact that ozone is related to more than one environmental issue.

In Chapter 12 we learned that ozone occurs naturally in the region of the atmosphere known as the stratosphere. Here, 10 to 50 kilometers above the surface, ozone absorbs ultraviolet radiation from the sun. Since the wavelengths absorbed in the stratosphere are harmful to life at the earth's surface, ozone functions as a protective shield. Without this shield, potentially lethal intensities of ultraviolet radiation would reach the surface. We also learned that ozone-destroying chemicals, called chlorofluorocarbons (CFCs,) threaten the layer of stratospheric ozone. Therefore, in this context, ozone is considered a desirable atmospheric component that is critical to life on earth and must be preserved.

In contrast to the foregoing view of ozone as a beneficial atmospheric component, many people living in big cities view with alarm the presence of ozone in the atmosphere. Rather than wanting to "save the ozone," these individuals want to see ozone removed from the air. Are these urban dwellers referring to the *same* ozone that functions to shield life from damaging ultraviolet radiation? The answer is no.

Whereas ozone forms and exists naturally in the stratosphere, it is produced as a pollutant in the troposphere. Photochemical (Los Angeles–type) smog is a major environmental problem in big cities, especially during the summer months. Ozone is the major component of this noxious mixture of gases and particles that forms as a result of reactions among pollutants emitted by motor vehicles and industries. During air pollution episodes, ozone levels are frequently used as an index to the poor quality of urban air. The higher the ozone concentrations, the poorer the air quality. Therefore, in this context, ozone is an undesirable pollutant that threatens human health.

In conclusion, the answer to the question posed as the title to this box must be *yes*. Ozone is both good and bad. Whereas ozone in the upper atmosphere is beneficial to life by shielding the earth from harmful ultraviolet radiation, high concentrations at ground level are a significant health and environmental concern.

---

tion to be more dilute. Conversely, when winds are light, there is little turbulence and the concentration of pollutants remains high.

Whereas wind speed governs the amount of air into which pollutants are initially mixed, atmospheric stability determines the extent to which vertical motions will mix the pollution with cleaner air above. The vertical distance between the earth's surface and the height to which convectional movements extend is termed the **mixing depth.** Generally, the greater the mixing depth, the better air quality. When the mixing depth is several kilometers, pollutants are mixed through a large volume of cleaner air and dilute rapidly. When the mixing depth is shallow, pollutants are confined to a much smaller volume of air and concentrations can reach unhealthy levels. When air is stable, convectional motions are suppressed and mixing depths are small. Conversely, an unstable at-

mosphere promotes vertical air movements and greater mixing depths. Since heating of the earth's surface by the sun enhances convectional movements, mixing depths are usually greater during the afternoon hours. For the same reason, mixing depths during the summer months are typically greater than during the winter months.

Temperature inversion represents a situation in which the atmosphere is very stable and the mixing depth is significantly restricted. Warm air overlying cooler air acts as a lid and prevents upward movement, leaving the pollutants trapped in a relatively narrow zone near the ground. This effect is illustrated nicely by the photograph shown in Figure 16.5.

Inversions are generally classified into one of two categories—those that form near the ground and those that form aloft. The first type, called a *surface inversion*, develops close to the ground on clear and

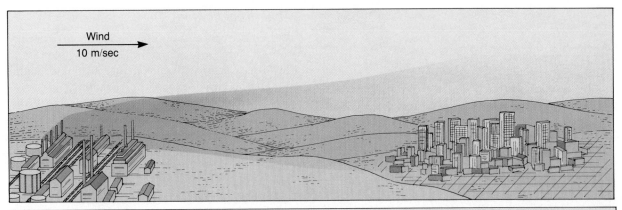

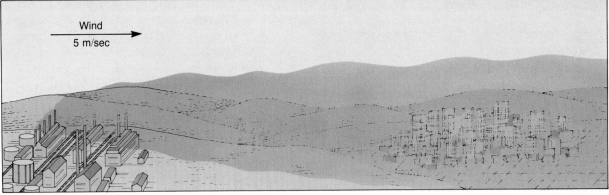

Figure 16.4
Effect of wind speed on the dilution of pollutants. The concentration of pollutants increases as wind speed decreases.

Figure 16.5
Temperature inversion in Los Angeles, California. Temperature inversions reduce mixing depth and inhibit the dispersal of pollutants. (Photo by Tom McHugh/Photo Researchers, Inc.)

relatively calm nights. It forms because the ground is a more effective radiator than the air above. This being the case, radiation from the ground to the clear night sky causes more rapid cooling at the surface than higher in the atmosphere. The result is that the air close to the ground is cooled more than the air above, yielding a temperature profile similar to the one shown in the upper portion of Figure 16.6. After sunrise the ground is heated and the inversion disappears. Although surface inversions are usually shallow, they may be quite thick in regions where the land surface is uneven. Due to the fact that cold air is denser (heavier) than warm air, the chilled air near the surface gradually drains from the uplands and slopes into adjacent lowlands and valleys. As might be expected, these thicker surface inversions will not dissipate as quickly after sunrise. Thus, although valleys are often preferred sites for manufacturing because they afford easy access to water transportation, they are also more likely to experience relatively thick surface inversions which, in turn, will have a negative effect upon air quality.

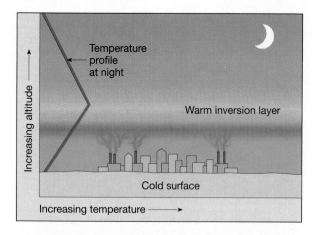

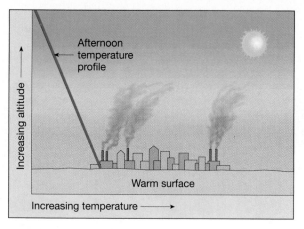

Figure 16.6
The upper portion of the figure shows a generalized temperature profile for a surface inversion. The lower portion shows how the temperature profile changes after the sun has heated the surface.

Many extensive and long-lived air pollution episodes are linked to temperature inversions that develop in association with the sinking air that characterizes slow-moving centers of high-pressure (Figure 16.7). As the air sinks to lower altitudes, it is compressed and so its temperature rises. Because turbulence is almost always present near the ground, this lower-most portion of the atmosphere is generally prevented from participating in the general subsidence. Thus an inversion develops aloft between the lower turbulent zone and the subsiding warmed layers above.

Acid Precipitation

As a consequence of burning tremendous quantities of fossil fuels, especially coal and oil, the United States discharges nearly 40 million metric tons of sulfur dioxide and nitrogen oxides into the atmosphere each year. Through a series of complex chemical reactions some of these pollutants are converted into acids, which then fall to the earth's surface as rain or snow. Another portion is deposited directly on surfaces and converted into acid after coming in contact with rain, dew, or fog.

In 1852, the English chemist Angus Smith coined the term *acid rain* to refer to the effect that industrial emissions had on precipitation in the British midlands. Well over a century later this phenomenon is the focus of research for thousands of environmental scientists as well as a topic having substantial international political importance. Although Smith clearly realized that acid rain caused environmental damage, large-scale effects were not recognized until the mid-twentieth century. Eventually, widespread public concern in the late 1970s led to significant government-sponsored studies of the problem. In the 1990s, such research activities continue to examine this still-unresolved environmental problem.

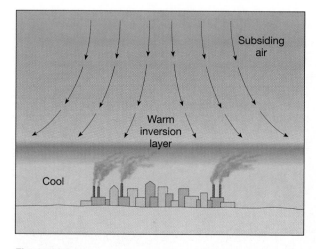

Figure 16.7
Inversions aloft frequently develop in association with slow-moving centers of high pressure where the air aloft subsides and warms by compression. The turbulent surface zone does not subside as much. Thus, an inversion often forms between the lower turbulent zone and the subsiding layers above.

BOX 16.2
The pH Scale

The *pH scale* is a common measure of the degree of acidity or alkalinity of a solution. As Figure 16.A illustrates, the scale ranges from 0 to 14, with a value of 7 denoting a solution that is neutral. Values below 7 indicate greater acidity, whereas numbers above 7 indicate greater alkalinity. The pH values of some familiar substances are shown on the diagram. Although distilled water is neutral (pH 7), rainwater is naturally acidic. It is important to note that the pH scale is logarithmic; that is, each whole number increment indicates a tenfold difference. Thus, pH 4 is 10 times more acidic than pH 5 and 100 times (10 × 10) more acidic than pH 6.

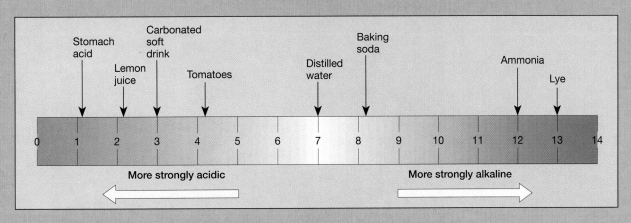

Figure 16.A
The pH scale.

Rain is somewhat naturally acidic because carbon dioxide dissolves in water in the atmosphere to produce carbonic acid. Until recently it was believed that unpolluted rain naturally had a mildly acidic pH of about 5.6. However, studies in uncontaminated remote areas have shown that precipitation usually has a pH closer to 5. Unfortunately, in most areas within several hundred kilometers of large centers of human activity, precipitation has much lower and hence more acidic pH values. This rain and snow is termed **acid precipitation.**

Many pollutants remain in the atmosphere for periods as long as five days, during which they may be transported great distances. Consequently, some of the acidity found in the northeastern United States originates hundreds of kilometers away in the industrialized regions of the Midwest and Ohio Valley. A similar situation exists in Europe where Swedish scientists suggest that as much as 75 percent of the human-generated sulfur in the atmosphere over southern Sweden originates from sources outside of their country.

Widespread acid rain has been known in northern Europe and eastern North America for decades. Recent studies have also shown that acid rain occurs in many other regions, including western North America, Japan, China, Russia, and South America. For many years, acid precipitation had not been directly linked to any adverse effects on human health. This is no longer true. There is a growing body of evidence that acid aerosols affect people's health. Exposure may impair the ability of the upper respiratory tract and the deep parts of the lungs to clear themselves of harmful particles. A link has also been made with the incidence of bronchitis among children. Beyond these possible impacts on health, the damaging effects of acid rain on the environment are believed to be already considerable in some areas and

Figure 16.8
The proportion of the world's population that is classified as urban continues to grow each year. The building of cities represents the most apparent human impact on climate. The changes produced by urban areas include all major elements. (Photo by Larry Hamill)

imminent in others. The best-known effect of acid precipitation is the lowering of pH in thousands of lakes in Scandinavia, the northeastern United States, and Canada. Accompanying this have been substantial increases in dissolved aluminum, which is leached from the soil by the acidic waters, and which in turn is toxic to fish and wildlife. As a consequence, some lakes are virtually devoid of fish while others are approaching this condition. Further, since ecosystems are characterized by many interactions at many levels of organization, evaluating the effects of acid precipitation on these complex systems is difficult and expensive, and therefore far from complete. It should also be pointed out that, even within small areas, the effects of acid precipitation can vary significantly from one lake to another. Much of this variation is related to the nature of the soil and rock materials in the area surrounding the lake. Because minerals such as calcite in some rocks and soils can neutral-

ize acid solutions, lakes surrounded by such materials are less likely to become acidic. On the other hand, lakes that lack this buffering material can be severely affected. Even so, over a period of time, the pH of lakes that have not yet been acidified may drop as the buffering material in the surrounding soil becomes depleted.

In addition to the thousands of lakes that can no longer support fish, research indicates that acid precipitation may also reduce agricultural crop yields and impair the productivity of forests. Acid rain not only harms the foliage but also damages roots and leaches nutrient minerals from the soil. Finally, acid precipitation is known to promote the corrosion of metals and contributes to the destruction of stone structures.

In summary, acid precipitation involves the delivery of acidic substances through the atmosphere to the earth's surface. These compounds are introduced into the air as by-products of combustion and industrial activity. The atmosphere is both the avenue by which offending compounds travel from sources to the sites where they are deposited and the medium in which the combustion products are transformed into acidic substances. In addition to its detrimental impact on aquatic systems, acid precipitation has a number of other harmful effects.

Acid precipitation is a complex and multifaceted issue. Because current knowledge regarding many aspects of the problem is incomplete, additional research is certainly needed. Yet it is clear that our understanding is sufficient to take corrective actions as well. One important step occurred with the passage of the Clean Air Act of 1990. It requires a substantial reduction in U.S. emissions of sulfur dioxide and modest cuts in emissions of nitrogen oxides by the year 2000. Canada has agreed to make comparable cuts in its emissions of these air pollutants.

THE CLIMATE OF CITIES

As far back as the early nineteenth century, Luke Howard, the Englishman who is best remembered for his cloud classification scheme, recognized that the weather in London differed from that in the surrounding rural countryside, at least in terms of reduced visibility and increased temperature. Indeed, with the coming of the industrial revolution in the 1800s, the trend toward urbanization accelerated, leading to significant changes in the climate in and near most cities. At the beginning of the last century, only about 2 percent of the world's population lived in cities of more than 100,000 people. Today, not only is the total world population dramatically larger, but also a far greater percentage resides in cities (Figure 16.8). World population has grown from about 1.6 billion in 1900 to 5.2 billion in 1990. During the decade of the 1990s, an estimated 960 million people will be added to the planet. In 1990 more than 2 billion people lived in urban areas. This figure represents abut 40 percent of the world's people. Projections show urban population growing rapidly, with most of this growth in the developing world.

As Table 16.2 illustrates, the climatic changes produced by urbanization involve all major surface conditions. Some of these changes are quite obvious and relatively easy to measure. Others are more subtle and sometimes difficult to measure. The amount of change in any of these elements, at any

time, depends on several variables, including the extent of the urban complex, the nature of industry, site factors such as topography and proximity to water bodies, time of day, season, and existing weather conditions.

The Urban Heat Island

The most studied and well-documented urban climatic effect is the **urban heat island**. The term simply refers to the fact that temperatures within cities are generally higher than in rural areas. The heat island is evident when temperature data such as appear in Table 16.3 are examined. As is typical, the data for Philadelphia show the heat island is most pronounced when minimum temperatures are examined. The magnitude of the temperature differences shown by this table is probably even greater

Table 16.2
Average climatic changes produced by cities.

Element	Comparison with Rural Environment
Particulate matter	10 times more
Temperature	
Annual mean	0.5–1.5°C higher
Winter	1–2°C higher
Solar radiation	15–30% less
Ultraviolet, winter	30% less
Ultraviolet, summer	5% less
Precipitation	5–15% more
Thunderstorm frequency	16% more
Winter	5% more
Summer	29% more
Relative humidity	6% lower
Winter	2% lower
Summer	8% lower
Cloudiness (frequency)	5–10% more
Fog (frequency)	60% more
Winter	100% more
Summer	30% more
Wind speed	25% lower
Calm winds	5–20% more

Source: After Landsberg, Changnon, and others.

than the figures indicate, because temperatures observed at suburban airports are usually higher than those in truly rural environments. Figure 16.9, which shows the distribution of average minimum temperatures in the Washington, D.C., metropolitan area for the three-month winter period over a five-year span, also illustrates a well-developed heat island. The warmest winter temperatures occurred in the heart of the city, while the suburbs and surrounding countryside experienced average minimum temperatures that were as much as 3.3°C lower. Remember that these temperatures are averages, because on many clear, calm nights the temperature difference between the city center and the countryside was considerably greater, often 11°C (20°F) or more. Conversely, on many overcast or windy nights the temperature differential approached zero degrees.

The radical change in the surface that results when rural areas are transformed into cities is a significant cause of the urban heat island. First, the tall buildings and the concrete and asphalt of the city absorb and store greater quantities of solar radiation than do the vegetation and soil typical of rural areas. In addition, because the city surface is impermeable, the runoff of water following a rain is rapid, resulting in a severe reduction in the evaporation rate. Hence, heat that once would have been used to convert liquid water to a gas now goes to further increase the surface temperature. At night, while both the city

and countryside cool by radiative losses, the stone-like surface of the city gradually releases the additional heat accumulated during the day, keeping the urban air warmer than that of the outlying areas.

A portion of the urban temperature rise must also be attributed to waste heat from sources such as home heating and air conditioning, power generation, industry, and transportation. Many studies have shown that the magnitude of human-generated energy in metropolitan areas is great when compared to the amount of energy received from the sun at the surface. For example, investigations in Sheffield, England, and Berlin showed that the annual heat production in those cities was equal to approximately one-third of that received from solar radiation. Another study of densely built-up Manhattan in New York City revealed that during the winter, the quantity of heat produced by combustion alone is 2½ times greater than the amount of solar energy reaching the ground. In summer, the figure dropped to ⅙.

There are other, somewhat less influential, causes of the heat island. For example, the "blanket" of pollutants over a city contributes to the heat island by absorbing a portion of the upward-directed long-wave radiation emitted at the surface and re-emitting some of it back to the ground. A somewhat similar effect results from the complex three-dimensional structure of the city. The vertical walls of office buildings, stores, and apartments do not allow radiation to escape as readily as in outlying rural areas where surfaces are relatively flat. As the sides of these structures emit their stored heat, a portion is reradiated between buildings rather than upward, and is therefore dissipated slowly.

In addition to retarding the heat loss from the city, tall buildings also alter the flow of air. Because of the greater surface roughness, wind speeds within an urban area are reduced. Estimates from available records suggest a decrease of the order of about 25 percent from rural values. The lower wind speeds decrease the city's ventilation by inhibiting the movement of cooler outside air, which, if allowed to penetrate, would reduce the higher temperatures of the city center.

Urban-Induced Precipitation

Most climatologists agree that cities influence the frequency and amount of precipitation in their vicinities. Several reasons have been proposed to explain

Table 16.3
Average temperatures (°C) for suburban Philadelphia Airport and downtown Philadelphia (ten-year averages).

	Airport	Down-town
Annual mean	12.8	13.6
Mean June maximum	27.8	28.2
Mean December maximum	6.4	6.7
Mean June minimum	16.5	17.7
Mean December minimum	−2.1	−0.4

Source: After H. Neuberger and J. Cahir, *Principles of Climatology* (New York: Holt, Rinehart and Winston, 1969), 128.

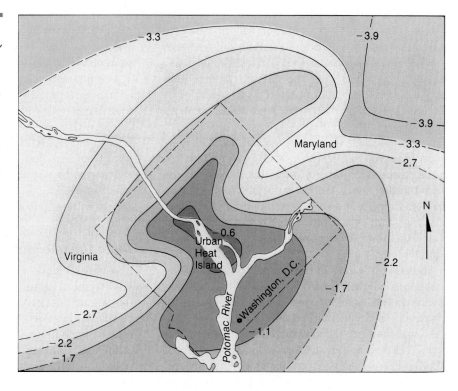

Figure 16.9
The heat island of Washington, D.C., as shown by average minimum temperatures (°C) during the winter season (December–February). The city center had an average minimum that was more than 3°C higher than in some outlying areas. [After Clarence A. Woollum, "Notes from a Study of the Microclimatology of the Washington, D.C., Area for the Winter and Spring Seasons," Weatherwise *17(1964): 264]*

why an urban complex might be expected to increase precipitation.

1. The urban heat island creates thermally induced upward motions which act to diminish the atmosphere's stability.
2. Clouds may be modified by the addition of condensation nuclei and freezing nuclei from industrial discharges.
3. The three-dimensional nature of urban areas creates an obstacle effect that impedes the passage of weather systems. Rain-producing processes may therefore linger over a city and increase precipitation.

Several studies comparing urban and rural precipitation have concluded that the amount of precipitation over a city is about 10 percent greater than over the nearby countryside. However, more recent investigations have shown that, although cities may indeed increase their own rainfall totals, the greatest effects may occur downwind of the city center.

A striking and controversial example of such a downwind effect was examined by Stanley Changnon in the late 1960s.* Records indicated that since 1925, LaPorte, Indiana, located 48 kilometers (30 miles)

downwind of the large complex of industries at Chicago, experienced a notable increase in total precipitation, number of rainy days, number of thunderstorm days, and number of days with hail. The magnitude of the changes and the absence of such changes in the surrounding area led to widespread public attention. Many questioned whether the anomaly was real or simply the result of such factors as observer errors and changes in the exposure of instruments. After completing his study, however, Changnon concluded that the observed differences were real and were likely produced by the large industrial complex west of LaPorte. Among the reasons for this conclusion was that the number of days with smoke and haze (a measure of atmospheric pollution) in Chicago after 1930 corresponded quite well with the LaPorte precipitation curve. An examination of Figure 16.10, for example, shows a marked increase in smoke-haze days after 1940, when the La-

---

* S. A. Changnon, Jr., "The LaPorte Weather Anomaly: Fact or Fiction?" *Bulletin of the American Meteorological Society* 49(1): 4–11. (NOTE: The term *anomaly* refers to a deviation in excess of normal variation.)

Porte graph begins its sharp rise. Further, the decrease in smoke-haze days after a peak in 1947 also generally matches a drop in the LaPorte curve. A second urban-related factor, steel production, was also found to correlate with the LaPorte precipitation curve. Records showed that seven peaks in steel production between 1923 and 1962 were associated with highs in the LaPorte curve.

The conclusions were accepted very cautiously by many and criticized by others who believed that because of the high natural rainfall variability of the area the data were neither sufficient nor accurate enough to make a sound determination. Nevertheless, this study was a pioneering effort, which illustrated the possible effect of human activity on local climates.

Subsequent studies of the complex problem of urban effects on precipitation have generally confirmed the view that cities increase rainfall in downwind areas. One study in the St. Louis area showed that the magnitude of the downwind precipitation increase climbed as industrial development expanded. Another line of evidence from the St. Louis study and other studies also supports the hypothesis that cities

are responsible for these increases. If cities are really modifying precipitation, the effects should be greater on weekdays, when urban activities are most intense, than on weekends, when much of the activity ceases. This is indeed the case. In the St. Louis study, analysis of precipitation on weekdays versus weekends revealed a significantly greater frequency of rain per weekday than per weekend day in the affected downwind area. Figure 16.11 shows similar findings that have been reported for Paris. Here a rise in precipitation can be noted from Monday through Friday, whereas the amounts for Saturday and Sunday are considerably smaller.

Other Urban Effects

In earlier sections, we saw that the greater air pollution in cities (1) contributes to the heat island by inhibiting the loss of long-wave radiation at night and (2) has a cloud-seeding effect that is believed to increase precipitation in and downwind of cities. These influences, however, are not the only ways in which pollutants influence urban climates. For example, the blanket of particulates over most large cities signifi-

Figure 16.10
Precipitation values at LaPorte, Indiana, and smoke-haze days at Chicago, both plotted as five-year moving totals. [Data from Stanley A. Changnon, Jr., "The LaPorte Weather Anomaly: Fact or Fiction?" Bulletin of the American Meteorological Society *49 (1968)]*

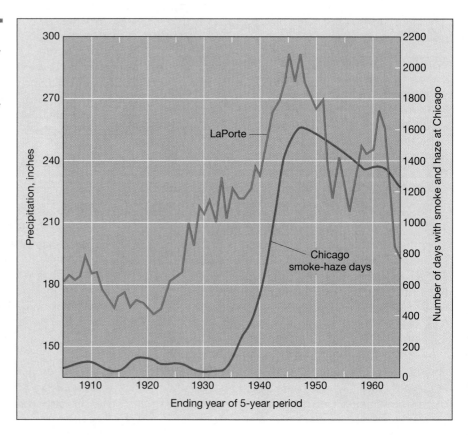

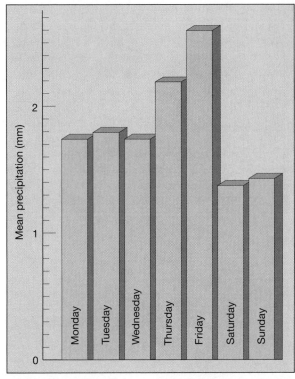

Figure 16.11
Average precipitation by day of the week in Paris for an eight-year period (1960–1967). Notice the gradual increase from Monday to Friday and the sharp drop in precipitation totals on Saturday and Sunday. The average for weekdays was 1.93 millimeters compared to only 1.47 millimeters on weekends, a difference of 24 percent.

cantly reduces the amount of solar radiation reaching the surface. As Table 16.2 indicates, the overall reduction in the receipt of solar energy is 15 percent, whereas short-wavelength ultraviolet is decreased by up to 30 percent. This weakening of solar radiation is, of course, variable. The decrease will be much greater during air pollution episodes than for periods when ventilation is good. Furthermore, particles are most effective in reducing solar radiation near the surface when the sun angle is low, because the length of the path through the dust increases as the sun angle drops. Hence, for a given quantity of particulate matter, solar energy will be reduced by the largest percentage at higher-latitude cities during the winter.

Relative humidities are generally lower in cities, because temperatures are higher and evaporation is reduced (see Table 16.2). Nevertheless, the frequency of fogs and the amount of cloudiness are greater. It is likely that the large quantities of condensation nuclei produced by human activities in urban areas lead to this increase in cloudiness and fog. When hygroscopic (water-seeking) nuclei are plentiful, water vapor readily condenses on them, even when the air is not yet saturated.

Finally, mention should be made of another urban-induced phenomenon, the **country breeze.** As the name implies, this circulation pattern is characterized by a light wind blowing into the city from the surrounding countryside. This breeze is best developed on relatively clear, calm nights, that is, on nights when the heat island is most pronounced. Higher city temperatures create upward air motion, which in turn initiates the country-to-city flow. One investigation in Toronto showed that the heat island created a rural-city pressure difference that was sufficient to cause an inward and counterclockwise circulation centered on the downtown area. When this circulation pattern exists, pollutants emitted near the urban perimeter tend to drift in and concentrate near the city's center.

HUMAN IMPACT ON GLOBAL CLIMATE

The proposals to explain global climatic change are many and varied. In Chapter 4 we examined some possible causes for ice-age climates. These hypotheses, which include the movement of lithospheric plates and variations in the earth's orbit, involved natural forcing mechanisms. Another natural forcing mechanism, discussed in Chapter 7, is the possible role of explosive volcanic eruptions in modifying the atmosphere. It is important to remember that these mechanisms, as well as others, not only have contributed to climatic changes in the geologic past, but will also be responsible for future shifts in climate. However, when relatively recent and future changes in our climate are considered, we must also examine the possible impact of human beings. In this section we shall examine the major way in which humans are believed to contribute to global climatic change. This impact results from the addition of carbon dioxide and other gases to the atmosphere (Figure 16.12).

It should be mentioned at this point, however, that human influence on regional and global climate probably did not just begin with the onset of the modern industrial period. There is good evidence that humans have been modifying the environment over extensive areas for thousands of years. The use of fire as well as the overgrazing of marginal lands by domesticated animals have negatively affected the abundance and distribution of vegetation. By altering ground cover, such important climatological factors as surface albedo, evaporation rates, and surface winds have been, and continue to be, modified. Commenting on this aspect of human-induced climatic modification, the authors of one study state,

> In contrast to the prevailing view that only modern humans are able to alter climate, we believe it is more likely that the human species has made a substantial and continuing impact on climate since the invention of fire.*

Finally, it should be pointed out that when any hypothesis of climatic change is examined, whether it depends upon natural or human-induced causes, a degree of caution must be exercised. It is safe to say that most, if not all, hypotheses are to some degree controversial and speculative. This is to be expected if we consider the fact that at present all of our models of the earth's climate are far from complete. Because planetary atmospheric processes are so large and complex, they cannot be physically reproduced in laboratory experiments. Instead, the climate must be simulated mathematically with the aid of computers. Although such models are sophisticated enough to serve as primary tools for climate research, they cannot yet approach the actual complexity of the atmosphere. Computer models are powerful and essential aids, but climate forecasts based on such simulations still contain many uncertainties. When the time comes that our atmosphere and its changes through time are more fully understood, we will no doubt find that many factors, both natural and human, influence climatic change.

Greenhouse Warming

In Chapter 12 we learned that, although carbon dioxide (CO_2) represents only about 0.03 percent of the gases composing clean, dry air, it is nevertheless a meteorologically significant component. The importance of CO_2 lies in the fact that it is transparent to

---

* Carl Sagan et al., "Anthropogenic Albedo Changes and the Earth's Climate," *Science* 206(4425): 1367.

incoming, short-wavelength solar radiation but is not transparent to some of the longer-wavelength outgoing terrestrial radiation. A portion of the energy leaving the ground is absorbed by CO_2 and subsequently re-emitted, part of it toward the surface, keeping the air near the ground warmer than it would be without CO_2. Thus, along with water vapor, CO_2 is largely responsible for the greenhouse effect of the atmosphere. Because CO_2 is an important heat absorber it follows that any change in the atmosphere's CO_2 content should alter temperatures in the lower atmosphere.

Paralleling the rapid growth of industrialization, which began in the nineteenth century, has been the

Figure 16.12
Paralleling the rapid growth of industrialization, which began in the nineteenth century, has been the combustion of fossil fuels, which has added great quantities of carbon dioxide to the atmosphere. (Wilson North/International Stock Photo)

consumption of fossil fuels. The combustion of these fuels has added great quantities of CO_2 to the atmosphere. Although the use of coal and other fuels is the most prominent means by which humans add CO_2 to the atmosphere, it is not the only way. The clearing of forests also contributes substantially. The carbon dioxide is released as vegetation is burned or decays. Deforestation is particularly pronounced in the tropics, where vast tracts are cleared for ranching and agriculture or subjected to inefficient commercial logging operations. Although some of the CO_2 is dissolved in the ocean and some is used by plants, between 45 and 50 percent remains in the atmosphere. Consequently, since 1860 there has been an increase of about 25 percent in the atmosphere's CO_2 content. Since 1958 scientists at Mauna Loa Observatory in Hawaii have continuously measured carbon dioxide concentrations (Figure 16.13). These measurements clearly show an upward trend from about 315 parts per million (ppm) to more than 350 ppm. Seasonal fluctuations occur because CO_2 is removed from the air by plants during the growing season and returned later when the plants decay. Naturally, we might expect that global temperatures have already increased as a result of growing carbon dioxide levels. Yet a steady temperature rise that can be attributed to carbon dioxide has not yet been detected. Most climatologists believe that the effects of the CO_2 increase are not yet large enough to show up clearly in the climatic record. However, the time is not far off when the effects will be evident.

If we assume that the use of fossil fuels will continue to increase at projected rates, current estimates indicate that the atmosphere's present carbon dioxide content of about 355 ppm will approach 400 parts per million by the year 2000 and will reach 600 ppm by some time in the second half of the next century. With such an increase in carbon dioxide, the enhancement of the greenhouse effect would be much more dramatic and measurable than in the past. When it is assumed that the atmosphere's carbon dioxide content will reach projected levels, the most realistic models predict a global surface temperature increase of between 1.5°C and 4.5°C. A change of this magnitude would be unprecedented in human history. Such an increase would come close to equaling the warming that has taken place since the peak of the most recent glacial stage 18,000 years ago, except it would occur much more rapidly.

Carbon dioxide is not the only gas contributing to a future global temperature increase. In recent years atmospheric scientists have come to realize that the industrial and agricultural activities of people are causing a buildup of certain trace gases that may also play a significant role. The substances are called *trace gases* because their concentrations are so much smaller than that of carbon dioxide. The trace gases that appear to be most important are methane (NH_4), nitrous oxide (N_2O), and certain types of chlorofluorocarbons. These gases absorb wavelengths of

Figure 16.13
Monitoring at Mauna Loa Observatory in Hawaii has revealed a rise in the concentration of carbon dioxide during the period shown here. The yearly oscillation is caused by the seasonal growth and decay of vegetation. (After NOAA)

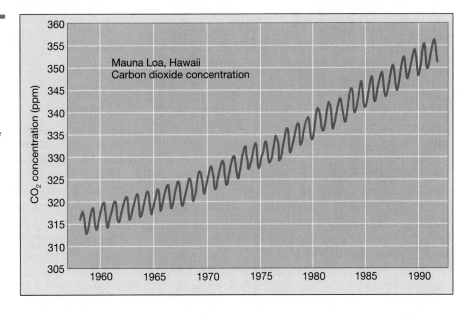

Methane is among the gases other than carbon dioxide that contribute to the greenhouse effect. It is present in much smaller amounts than CO_2, but its significance is greater than its relatively small concentration of about 1.7 ppm would indicate. The reason is that methane is 20 to 30 times more effective than CO_2 at absorbing infrared radiation emitted by the earth.

Methane is produced by certain bacteria in wet places where oxygen is scarce. Swamps, bogs, and other wetlands are important natural sources, as are the guts of termites. Methane is also produced in paddy fields used for growing rice ("artificial swamps") and in the intestinal tracts of cattle and sheep. Coal, oil, and natural gas exploitation is also a source. The concentration of methane in the atmosphere is believed to have about doubled since 1800, an increase that has been in step with the growth in human population. This relationship is a reflection of the close link between methane formation and agriculture. As the population numbers have risen, so have the number of cattle and rice paddies.

Nitrous oxide, sometimes called "laughing gas," is also building up in the atmosphere, although not as rapidly as methane. The increase is believed to be primarily the result of agricultural activity. When farmers use nitrogen fertilizers to boost crop yields, some of the nitrogen gets into the air as nitrous oxide. This gas is also produced as the result of high temperature combustion of fossil fuels. Although the annual amount released into the atmosphere is small, the lifetime of a nitrous oxide molecule is about 150 years. If the use of nitrogen fertilizers and fossil fuels grows at projected rates, nitrous oxide may make a contribution to greenhouse warming that approaches half that of methane.

Unlike methane and nitrous oxide, chlorofluorocarbons (CFCs) are not naturally present in the atmosphere. They are manufactured chemicals that have a number of uses, including a basic function in refrigeration and air conditioning equipment. CFCs have gained great notoriety because they are responsible for destroying ozone in the stratosphere. Their role in global warming generally is less well known; however, CFCs are very effective greenhouse gases. Although they were not developed until the 1920s and were not used in great quantities until the 1950s, they already contribute to the greenhouse effect at a level equal to methane.

Carbon dioxide from the burning of fossil fuels and deforestation is clearly the most important single cause for the projected global greenhouse warming. However, it is not the only contributor. When the effects of all human-generated greenhouse gases other than CO_2 are added together and projected into the future, their collective impact may double the impact of CO_2 alone.

outgoing earth radiation that would otherwise escape into space. Although individually their impact is modest, taken together the effects of these trace gases may be as great as CO_2 in warming the earth (see Box 16.3).

Sophisticated computer models of the atmosphere show that the warming of the lower atmosphere triggered by CO_2 and trace gases will not be the same everywhere. Rather, the temperature response in polar regions could be significantly greater than the global average. Part of the reason for such response is that the polar troposphere is very stable. This stability suppresses vertical mixing and thus limits the amount of surface heat that is transferred upward.

In addition, an expected reduction in sea ice would also contribute to the greater temperature increase.

Because the atmosphere is a very complex, interactive physical system, scientists must consider many possible outcomes when one of the system's elements is altered. These various possibilities, termed **climatic feedback mechanisms,** not only complicate climatic modeling efforts, but they also add greater uncertainty to climatic predictions.

The most important and obvious of the feedback effects related to increases in carbon dioxide arises from the fact that higher surface temperatures produce greater evaporation rates. This, in turn, increases the amount of water vapor in the atmosphere. Remember that water vapor is also a powerful absorber of terrestrial radiation. Therefore, with more water vapor in the air, the temperature increase caused by carbon dioxide alone is reinforced.

Recall that the temperature increase at high latitudes is expected to be as much as two or three times greater than the global average. This assumption is based in part on the likelihood that the area covered by floating pack ice will decrease as surface temperatures rise. Since ice reflects a much larger percentage of incoming solar radiation than does open water, the melting of the pack ice would replace a highly reflective surface with a relatively dark surface. The result would be a substantial increase in the amount of solar energy absorbed at the surface. This, in turn, would feed back to the atmosphere and magnify the initial temperature increase created by higher carbon dioxide levels.

Up to this point the climate feedback mechanisms we have discussed have magnified the temperature rise caused by the buildup of carbon dioxide. The effects that reinforce an initial change are termed **positive feedback mechanisms.** As we shall see, however, other effects must be classified as **negative feedback mechanisms** because they produce results that are just the opposite of the initial change and tend to offset it.

One probable result of a global rise in temperatures would be an accompanying increase in cloud cover due to the higher moisture content of the atmosphere. Most clouds are good reflectors of solar radiation. At the same time, however, they are also good absorbers and emitters of terrestrial radiation. For this reason, clouds produce two opposite effects. They act as a negative feedback mechanism because they increase albedo and thus diminish the amount of solar energy available to heat the atmosphere. On the other hand, clouds act as a positive feedback mechanism by absorbing and emitting terrestrial radiation that would otherwise be lost from the troposphere.

Which effect, if either, is stronger? Atmospheric modeling shows that the negative effect of a higher albedo is the more dominant. Therefore, the net result of an increase in cloudiness should be a decrease in air temperature. The magnitude of this negative feedback, however, is not believed to be as strong as the positive feedback caused by added moisture and decreased pack ice. Thus, although increases in cloud cover may partly offset a global temperature increase, climatic models show that the ultimate effect of a doubling of the atmosphere's carbon dioxide will still be a temperature increase.

The problem of global warming induced by increases in carbon dioxide and trace gases has been and continues to be one of the most studied and discussed aspects of climatic change. Although there are no models that can yet incorporate the full range of potential influences and feedbacks, the consensus in the scientific community is that altering atmospheric composition will eventually lead to a warmer planet with a different distribution of climatic regimes.

Possible Consequences of a Greenhouse Warming

What consequences can be expected if the carbon dioxide content of the atmosphere reaches a level twice that of the mid-nineteenth century? Although climatic models contain many uncertainties, they are providing us with some possible answers to this very basic question.

As has already been noted, the magnitude of the temperature increase will not be the same everywhere. The temperature rise will probably be smallest in the tropics and increase toward the poles. Further, the models indicate that some regions will experience a significant increase in precipitation and runoff, whereas other regions will experience a decrease in runoff either because of reduced precipitation or because of increased evaporation rates brought about by higher temperatures. Such changes could have a profound impact upon the distribution of the world's water resources and hence affect the productivity of agricultural regions that depend upon rivers for irrigation water. For example, a 2°C-temperature increase coupled with a 10-percent precipitation decrease in the region drained by the Colorado River could diminish the river's flow by 50 percent or more. Since the present flow of the river is barely enough to meet current demands for irrigated agriculture, the negative effect would be very serious (Figure 16.14). Many other rivers form the basis for extensive systems of irrigated agriculture and the projected reduction of their flow could have equally grave consequences. By contrast, large precipitation increases in other areas would increase the flow of some rivers and bring more frequent destructive floods.

In addition to affecting agricultural production, changes in temperature and precipitation could also

Figure 16.14
Decreased rainfall and increased evaporation rates could diminish the flow of the Colorado and other rivers, forcing the abandonment of much presently productive irrigated farmland. (Photo by James E. Patterson)

threaten natural ecosystems and cause changes in human settlement patterns. Not only might people engaged in agriculture be forced to relocate, but coastal residents also could face significant difficulties. For example, thermal expansion of the oceans and melting of glaciers are causing sea level to rise. Recent research indicates that sea level has risen between 10 and 15 centimeters over the past century and that the trend will continue at an accelerated rate in the years to come. It is believed that the rise could exceed 30 centimeters by the middle of the next century. Although such a change may seem modest, many scientists agree that any rise in sea level along a gently sloping shoreline, such as the Atlantic and Gulf coasts of the United States, will lead to significant erosion and shoreline retreat.* Concern is also

expressed that a warmer climate will cause glaciers to melt. In fact, about one-half of the 10- to 15-centimeter rise in sea level over the past century is attributed to the melting of small glaciers and ice caps. However, it should be emphasized that a significant melting of major ice sheets, although possible at some future date, is *not* expected during the next century.

Atmospheric scientists also expect that weather patterns will change as a result of the projected global warming. Potential weather changes include

1. a higher frequency and greater intensity of hurricanes because of warmer ocean temperatures,
2. shifts in the paths of large-scale cyclonic storms, which, in turn, would affect the distribution of precipitation and the occurrence of severe storms, including tornadoes, and
3. increases in the frequency and intensity of heat waves and droughts.

It should be emphasized that the impact upon global climate of an increase in the atmosphere's content of CO_2 and trace gases is obscured by many unknowns and uncertainties. The changes that occur will probably take the form of environmental shifts that will be imperceptible to most people from year to year. Nevertheless, although the changes may seem gradual, the effects will clearly have powerful economic, social, and political consequences. Stephen Schneider, a respected authority on climatic change, summarizes the situation as follows:

> There is little doubt that if human trends in population, industrial, and economic growth continue into the next century, there will be substantial increases in the greenhouse properties of the earth's atmosphere, which are virtually certain to create environmental change. The controversies begin when we try to put numbers on the amount of warming, its timing, and its implications for ecosystems and society. Most controversial, of course, is what to do about this prospect.*

INTENTIONAL WEATHER MODIFICATION

Certainly a desire to change the weather is far from modern. From earliest recorded times people have tried to use prayer, wizardry, dancing, and even magic to alter the weather. Until modern times, how-

---

* For a more extensive discussion of rising sea level and shoreline erosion problems, see page 407 in Chapter 11.

* Stephen H. Schneider, *Global Warming: Are We Entering the Greenhouse Century?* (San Francisco: Sierra Club Books, 1989), 23.

ever, attempts at weather modification remained largely in the realm of the mystic. By the nineteenth century such devices as smudge pots, sprinklers, and wind machines to fight frost were in use. During the Civil War, observations that rainfall apparently increased following some battles led to experiments in which cannons were fired at clouds to bring more rain. Unfortunately, these experiments, as well as many others, proved unsuccessful.

Weather modification strategies fall into three broad divisions. The first relies upon the injection of energy by "brute force." The use of powerful heat sources or the intense mechanical mixing of the air (such as by helicopters) are both examples of techniques undertaken at some airports to disperse fog. The second division involves the alteration of land and water surfaces in order to change their natural interactions with the lower atmosphere. One unproven technique is to blanket a land area with a dark substance. If this were done, the amount of heat absorbed by the surface would increase and lead to stronger upward air currents that in turn could aid cloud formation. Finally, the third division involves triggering, intensifying, and/or redirecting the atmosphere's natural energies. The seeding of clouds for many purposes, including precipitation enhance-

ment, is the primary example of this category. Because it offers a relatively inexpensive and easily used technique, **cloud seeding** has been and will likely continue to be the main focus of modern weather modification technology.

Cloud Seeding

Recall from the discussion of precipitation formation in Chapter 13 that at temperatures between $0°C$ and $-40°C$ small particles known as freezing nuclei act as catalysts to trigger the formation of ice crystals. Once ice crystals form, they grow larger at the expense of the remaining liquid droplets, and upon reaching sufficient size, they fall as precipitation. Freezing nuclei, however, are not abundant in the atmosphere. For example, nuclei capable of triggering ice formation at $-20°C$ have a typical concentration of only 0.1 to 1.0 per liter of air. By contrast, condensation nuclei, the particles on which cloud droplets form, often have concentrations of up to one million parts per liter of air.

Due to the natural rarity of freezing nuclei, the basis of **cloud seeding** efforts is that the addition of nuclei will spur the formation of additional precipitation. One of two agents is commonly used in cloud seeding. Dry ice may be introduced into a cloud to chill the air and cause ice crystals to form. Silver iodide crystals, however, are more commonly used. The similarity in the crystalline structures of silver iodide and ice accounts for silver iodide's ability to initiate the growth of ice crystals. Thus, unlike dry ice, silver iodide crystals act as freezing nuclei rather than as a cooling agent. For either substance to be successful, certain atmospheric conditions must exist. Clouds must be present, for seeding cannot generate them. In addition, at least the top portion of the cloud must be supercooled; that is, it must be composed of liquid droplets having a temperature below $0°C$ ($32°F$).

The most common method of seeding clouds is to fly through or below a cloud with silver iodide flares that release large numbers of freezing nuclei (Figure 16.15). This method relies on the naturally present turbulence to diffuse the material through large volumes of air. In some experiments, freezing nuclei have been introduced into clouds by firing rockets or artillery shells. However, conflicts with aviation and concerns for public safety have made this a seldom-used method. Burning silver iodide mixtures from burners on the ground is a third method of cloud seeding. Ground-based burners seem to be most effective in mountainous terrain where the upward-

Figure 16.15
Cloud seeding using silver iodide flares is one way that freezing nuclei are supplied to clouds. (Courtesy of National Center for Atmospheric Research/National Science Foundation)

BOX 16.4
Project STORMFURY

Reducing hurricane destruction cannot rely on tracking alone. If possible, methods that reduce the intensity of the storm must also be sought. This was the goal of the hurricane modification experiments between 1962 and 1983 that were part of the federal government's Project STORMFURY. Because the destructive potential of a hurricane increases rapidly as its maximum wind speed increases, a reduction as small as 10 percent was thought to be worthwhile.

As the inward rush of warm, moist air approaches the center of a hurricane, wind speeds increase. To explain why wind speeds increase near the storm center, we must understand the *law of conservation of angular momentum.* This law states that the product of the velocity of an object around a center of rotation (axis) and the distance squared of the object from the axis is constant. Therefore, when a parcel of air moves toward the center of a storm (the center of rotation), the product of its distance squared and velocity must remain unchanged. Consequently, if the distance decreases, the rotational velocity must increase. A common example of the conservation of angular momentum occurs when a figure skater starts whirling on the ice with both arms extended. Her arms are traveling in a circular path about an axis (her body). When the skater pulls her arms inward, she decreases the radius of the circular path of her arms. As a result, her arms go faster and the rest of her body must follow, thereby increasing her rate of spinning.

The proposed hurricane modification technique involved the artificial stimulation of cloud development outside the eye wall through seeding with silver iodide. This working hypothesis was basic to the experimental program and came to be known as the STORMFURY hypothesis. Recall that the energy for hurricanes comes from the release of latent heat in the ring of rising air, called the eye wall. The object of the hurricane modification experiments was to seed portions of the storm beyond this zone of maximum wind velocity. It was believed that doing so would cause supercooled droplets to freeze, thus releasing latent heat that would then stimulate cloud growth beyond the eye wall (Figure 16.B). In this way, the inward-spiraling flow of moist air would be intercepted and made to ascend at a greater distance from the storm center. The net effect would be an outward displacement of the eye wall and a reduction in maximum wind speeds. Recall from the discussion of the conservation of angular momentum that an increase in radius must be accompanied by a reduction in rotational speed.

Over the span of Project STORMFURY, modification was attempted on four different hurricanes on eight separate days. On four of these days the winds decreased between 10 and 30 percent. Lack of response on the other days was interpreted to be the result of faulty execution of the experiment or poorly selected subjects. These early re-

forced air helps the seeding agent into the clouds.

The seeding of winter clouds rising over some mountain barriers has been one of the most actively pursued examples of cloud seeding. Since only about 20 to 50 percent of the water that condenses from orographic clouds actually falls as precipitation, these clouds are good candidates for seeding to increase the percentage of water falling on mountain slopes. The purpose is to increase winter snowpack, which melts and runs off during warmer months. The runoff can be collected in reservoirs for use in irrigation and hydroelectric power generation. Although such projects have shown promise, their apparently positive results have not yet been confirmed to the satisfaction of most weather-modification scientists.

Seeding of nonorographic cumulus clouds is more difficult because of their great variability in time and space. Of the many experiments that appeared to have positive results, only a single project, one conducted in Israel, was able to show a statistically convincing confirmation of an increase in precipitation after cloud seeding. Results from two separate phases of testing indicated a 15 percent increase in rainfall. When the results became known, the Israeli government stopped experimenting and began a program of operational seeding at every opportunity.

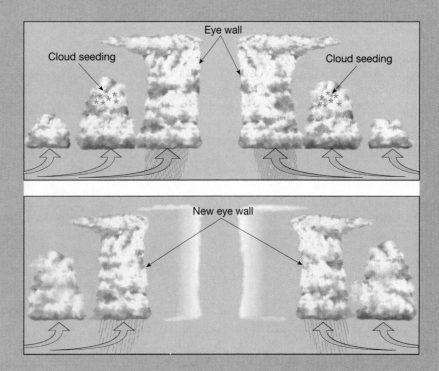

Figure 16.B
Schematic diagram showing the proposed effect of cloud seeding in a hurricane. Seeding clouds outside the zone of maximum winds stimulates cloud growth at a greater distance from the storm's center and leads to the decay of the old eye wall. [After Richard A. Anthes, Tropical Cyclones, Their Evolution, Structure and Effects, Meteorological Monographs, vol. 19 (Boston: American Meteorological Society, 1982), 41:121]

sults were encouraging and seemed to support the STORM-FURY hypothesis. However, as more and more data were gathered about hurricane structure and dynamics, it eventually became apparent that the seeding experiments were flawed. It also became clear that the positive results inferred from the seeding experiments could just as easily have been natural changes in hurricane behavior that were unrelated to human intervention.

A great deal about hurricanes was learned as a result of the experiments and observations made during Project STORM-FURY. Among the important lessons was that the STORM-FURY hypothesis of hurricane modification was not feasible. Although human control of hurricanes seemed attainable when the research began, successful intervention turned out to be an unattainable goal.

In light of the uncertainty associated with even the better-designed experiments, many scientists believe that a "return to basics" is necessary. Although there is little question that many weather events can be modified by human intervention, there is also no doubt that more basic knowledge of atmospheric processes is necessary before weather modification can be carried out with scientifically predictable results. Certainly a major result of the last four and a half decades of work in cloud seeding has been the sobering realization that even the simplest of weather events are exceedingly complex and not yet fully understood.

Fog and Cloud Dispersal

Perhaps the simplest of all cloud-seeding experiments involves spreading dry-ice pellets or silver iodide particles into layers of supercooled fog and low stratus clouds. These applications transform water droplets in clouds into ice crystals. The results are easily observed and may be dramatic. Initially the ice crystals are too small to fall to the ground, but as they grow at the expense of the remaining water droplets and as turbulent air motions disperse them through nearby air, they can grow large enough to produce a snow shower. The amount of snowfall produced in this manner is minor, but visibility almost always improves and often a large hole is opened in the stratus clouds or supercooled fog. The U.S. Air Force has practiced this technology for many years at various airbases, as have commercial airlines at selected airports in the northwestern United States. The possibility of using this technology to open large holes in winter clouds to increase the amount of solar radiation reaching the ground has been discussed but thus far has received little research attention.

Most fog is of the warm type, which means that it does not consist of supercooled water droplets. Unfortunately, these fogs are more difficult and expensive to combat because seeding will not diminish them. Successful attempts at dispersing warm fogs have involved mechanical mixing of the fog with drier, warmer air from above or by heating the air. When the layer of fog is very shallow, helicopters have sometimes been used to disperse it. By flying just above the fog, the helicopter creates a strong downdraft that pulls drier air toward the surface and mixes it with the saturated air near the ground. If the air aloft is dry enough, sufficient evaporation will occur to improve visibility significantly. At some airports where warm fogs are a common hazard, it has become more usual to heat and thus evaporate the fog. A sophisticated thermal fog dissipation system, called Turboclair, was installed at Orly Airport in Paris. This system consists of eight jet engines located in underground chambers alongside the upwind edge of the runway. Although expensive to install, the system is capable of improving visibility for a distance of about 900 meters (3000 feet) along the approach and touchdown zones.

Hail Suppression

Hail is believed to form by the collecting and freezing of supercooled droplets around a nucleus. Updrafts of moist air in a cumulonimbus cloud are often so strong that small hailstones may be recirculated in the upper part of the cloud and collect additional coatings of water that freeze to enlarge the stone. Normally only a small number of hail embryos exist because freezing nuclei are relatively scarce, and so the embryos grow freely until they fall from the cloud. Modern attempts at hail suppression have been based largely on the idea that introducing silver iodide crystals into appropriate clouds will interrupt the formation and growth of hailstones. It is as-

Figure 16.16
This wheat field in northeastern Colorado was battered to stubble by hail. (Photo by Henry Lansford)

sumed that each of these crystals, acting as a freezing nucleus, attracts a portion of the cloud's water supply and thereby increases the competition for the available supercooled water in the upper portion of the cloud. Without ample supplies of supercooled water, hailstones cannot grow large enough to be destructive when they fall.

Because of dramatic crop and property losses (Figure 16.16), hail suppression technology has been the focus of much interest and many field and laboratory studies. The results have, at best, been mixed. In France, a cloud-seeding project that used silver iodide burners to combat hail damage had encouraging results. In Russia, scientists using rockets and artillery shells to carry freezing nuclei to the clouds have claimed extraordinary success. On the other hand, the major effort in the United States, known as the National Hail Research Experiment, had disappointing results. This carefully planned and executed multiyear experiment found that no statistically significant reduction in hail was produced by cloud seeding. Commenting on current possibilities for successful hail suppression, one report states, "While certain positive results have been obtained, many important questions remain regarding the important chain of events leading to hail and to hypothesized hail sup-

A.

B.

Figure 16.17
Two types of freeze controls used in Florida citrus groves. **A.** *A wind machine mixes warmer air aloft with cooler surface air. (Photo by Dr. L. Parsons, University of Florida)* **B.** *Orchard heaters are often effective in preventing frost damage. (Courtesy of the Florida Citrus Commission)*

pression."* Clearly the current status of hail suppression must be described as one of uncertainty.

Frost Prevention

Frost, the fruit-growers' plight, occurs when the air temperature falls to 0°C (32°F) or below. It may be accompanied by deposits of ice crystals commonly called *white frost*. This, however, happens only if the air becomes saturated. White frost is not a requirement for crop damage.

Frost hazards exist when a cold air mass moves into a region, or when ample radiation cooling occurs on a clear night. The conditions accompanying the invasion of cold air are characterized by low daytime temperatures, long periods of effective frost, strong winds, and widespread damage. Frost induced by radiation loss is a nighttime phenomenon associated with a surface temperature inversion and is much easier to combat.

Several methods of frost prevention have been used with varying degrees of success (Figure 16.17). Generally these attempts are directed at reducing the amount of heat lost during the night or at adding heat to the lowermost layer of air. Heat conservation methods include covering plants with material having a low thermal conductivity, such as paper and cloth, and producing particles which, when suspended in air, reduce the radiation loss. Smudge fires have been used for particle production but have generally proven unsatisfactory. In addition to the pollution problems created by the dense clouds of black smoke, the carbon particles impede daytime warming by reducing the amount of solar radiation that can reach the surface. This reduction in daytime warming offsets the benefits gained during the night.

---

* Stanley A. Changnon et al., "Review of the Tenth Conference on Planned and Inadvertent Weather Modification," *Bulletin of the American Meteorological Society* 67, no. 12 (December 1986): 1502.

Methods of warming include sprinkling, air mixing, and the use of orchard heaters (Figure 16.17). Sprinklers distribute water to the plants and add heat in two ways, first from the warmth of the water, but more importantly from latent heat of fusion, which is released when the water freezes. As long as an ice-water mixture remains on the plant, the latent heat released will keep the temperature from dropping below 0°C (32°F). Air mixing is successful when the temperature at 15 meters (50 feet) above the ground is 5°C (9°F) higher than the surface temperature. By using a wind machine, the warmer air aloft is mixed with the colder surface air. Orchard heaters are probably the most successful. As many as 30 or 40 heaters per acre are required, and the fuel cost can be significant. Nevertheless, the effectiveness of the method seems to warrant the cost.

Review Questions

1. Consult Table 16.1 to answer the following.
   (a) Which source is responsible for the most pollution?
   (b) What is the single greatest air pollutant by weight?
   (c) What source produces the greatest quantity of sulfur oxides?

2. What is the difference between primary and secondary pollutants? What is a photochemical reaction?

3. What is the origin of the word *smog*? How has its meaning changed?

4. Why are air pollution episodes less likely when winds are strong?

5. How do temperature inversions form? Describe the role of inversions in air pollution episodes.

6. How much more acidic is a solution with a pH of 3 compared to a solution with a pH of 5? (See Box 16.2.)

7. List three effects of acid precipitation.

8. List two ways in which the difference between a "typical" rural surface and a "typical" city surface adds to the urban heat island.

9. How do each of the following factors contribute to the urban heat island: heat production, "blanket" of pollutants, three-dimensional city structure?

10. List the three factors that are likely causes of greater precipitation in and downwind of cities.

11. Describe the reasons for the following urban climatic characteristics.
    (a) Reduced solar radiation
    (b) Reduced relative humidity
    (c) Increased fog and cloud frequency
    (d) Reduced wind speeds and more days when calm winds prevail

12. What is a "country breeze"? Describe its cause.

13. Why has the atmosphere's carbon dioxide level been rising for more than 130 years?

14. How are temperatures in the lower atmosphere likely to change as carbon dioxide levels continue to increase? Why?

15. Is carbon dioxide the only gas contributing to a future global temperature change? (See Box 16.3.)

16. What are climatic feedback mechanisms? Give some examples.

17. List three possible consequences of a greenhouse warming.

18. What is the principle behind seeding supercooled clouds?

19. List two applications of cloud seeding aside from increasing precipitation.

20. How do frost and white frost differ?

21. Describe how sprinkling and mixing air are used in frost prevention.

Key Terms

acid precipitation (p. 549)

air pollutant (p. 544)

climatic feedback mechanism (p. 558)

cloud seeding (p. 561)

country breeze (p. 555)

frost (p. 565)

mixing depth (p. 546)

negative feedback mechanism (p. 559)

photochemical reaction (p. 545)

positive feedback mechanism (p. 559)

primary pollutant (p. 544)

secondary pollutant (p. 544)

smog (p. 545)

temperature inversion (p. 546)

urban heat island (p. 551)

PART 4

Astronomy

The Horsehead Nebula, a dark nebula in a region of glowing nebulosity in Orion. (Courtesy of Anglo-Australian Observatory, by David Malin)

17

The Earth's Place in the Universe

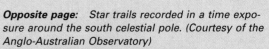

Opposite page: Star trails recorded in a time exposure around the south celestial pole. (Courtesy of the Anglo-Australian Observatory)

Left: Orion rising over Kitt Peak Observatory. (Courtesy of NOAO)

Above: Star trails over Kitt Peak Observatory. (Courtesy of NOAO)

*T*he earth is one of nine planets and numerous smaller bodies that orbit the sun. The sun is part of a much larger family of perhaps 100 billion stars that compose the Milky Way, which in turn is only one of billions of galaxies in an incomprehensibly large universe. This view of the earth's position in space is considerably different from that held only a few hundred years ago, when the earth was thought to occupy a privileged position as the center of the universe. This chapter unfolds with some events that led to modern astronomy. In addition, it examines the earth's place in time and space.

Long before recorded history, which began about 5,000 years ago, people were aware of the close relationship between events on earth and the positions of heavenly bodies, the sun in particular. People noted that changes in the seasons and floods of great rivers like the Nile in Egypt occurred when the celestial bodies, including the sun, moon, planets, and stars, reached a particular place in the heavens. Early agrarian cultures, which were dependent on the weather, believed that if the heavenly objects could control the seasons, they must also strongly influence all earthly events. This belief undoubtedly was the reason that early civilizations began keeping records of the positions of celestial objects. The Chinese, Egyptians, and Babylonians in particular are noted for this.

These cultures recorded the locations of the sun, moon, and the five planets visible to the unaided eye as these objects moved slowly against the background of "fixed" stars. In addition, the Chinese kept quite accurate records of comets and "guest stars" (Figure 17.1). Today we know that a "guest star" is really a normal star, usually too faint to be visible, that increases its brightness as it explosively ejects gases from its surface, a phenomenon we call a nova.

A study of Chinese archives shows that they recorded every appearance of the famous Halley's comet for at least ten centuries. However, because this comet appears only once in a lifetime—once every 76 years—they were unable to link these appearances to establish that what they saw was the same object each time. Thus, like most ancients, the Chinese considered comets to be mystical. Comets were seen as bad omens and were blamed for a variety of disasters, from wars to plagues (Figure 17.2).

ANCIENT ASTRONOMY

The "Golden Age" of early astronomy (600 B.C.–A.D. 150) was centered in Greece. The early Greeks have been criticized, and rightly so, for using philosophical arguments to explain natural phenomena. However, they did rely on observational data as well. The basics of geometry and trigonometry, which they had developed, were used to measure the sizes and distances of the largest-appearing bodies in the heavens—the sun and the moon.

Figure 17.1
Comet West. (Courtesy of Celestron International)

Early Greeks

Many astronomical discoveries have been credited to the Greeks. They held the **geocentric** ("earth-centered") view, believing that the earth was a sphere that stayed motionless at the center of the universe. Orbiting the earth were the moon, sun, and the known planets—Mercury, Venus, Mars, and Jupiter. Beyond the planets was a transparent, hollow sphere **(celestial sphere)** on which the stars traveled daily around the earth (this is how it looks, but, of course, the effect is actually caused by the earth's rotation

about its axis). Some early Greeks realized that the motion of the stars could be explained just as easily by a rotating earth, but they rejected that idea, because the earth exhibited no sense of motion and seemed too large to be movable. In fact, proof of the earth's rotation was not demonstrated until 1851, a topic we will consider later in Box 17.1

To the Greeks, all of the heavenly bodies, except seven, appeared to remain in the same relative position to one another. These seven wanderers (*planetai* in Greek) included the sun, the moon, Mercury, Venus, Mars, Jupiter, and Saturn. Each was thought to have a circular orbit around the earth. Although this system was incorrect, the Greeks refined it to the point that it explained the apparent movements of all celestial bodies.

As early as the fifth century B.C., the Greeks understood what causes the phases of the moon. Anaxagoras reasoned that the moon shines by reflected sunlight, and because it is a sphere, only half is illuminated at one time. As the moon orbits the earth, that portion of the illuminated half that is visible from the earth is always changing. Anaxagoras also realized that an eclipse of the moon occurs when it moves into the shadow of the earth.

The famous Greek philosopher Aristotle (384–322 B.C.) concluded that the earth is spherical because it always casts a curved shadow when it eclipses the moon. Although most of the teachings of Aristotle were passed along and were considered infallible by many, his belief in a spherical earth was abandoned during the Middle Ages.

The first Greek to profess a sun-centered, or **heliocentric,** universe was Aristarchus (312–230 B.C.).

Figure 17.2
The Bayeux Tapestry that hangs in Bayeux, France, shows the apprehension caused by Halley's comet in A.D. 1066. This event preceded the defeat of King Harold by William the Conqueror. (Courtesy of Wide World Photos)

Aristarchus also used simple geometric relations to calculate the relative distances from the earth to the sun and the moon. He later used these data to calculate their sizes. Due to observational errors beyond his control, he came up with measurements that were much too small. However, he did learn that the sun was many times more distant than the moon and many times larger than the earth. The latter fact may have prompted him to suggest a sun-centered universe. Nevertheless, because of the strong influence of Aristotle, the earth-centered view dominated Western thought for nearly 2000 years.

The first successful attempt to establish the size of the earth is credited to Eratosthenes (276–194 B.C.). Eratosthenes observed the angles of the noonday sun in two Egyptian cities that were roughly north and south of each other—Syene (presently Aswan) and Alexandria (Figure 17.3). Finding that the angles differed by 7 degrees, or 1/50 of a complete circle, he concluded that the circumference of the earth must be 50 times the distance between these two cities. The cities were 5000 *stadia* apart, giving him a measurement of 250,000 *stadia*. Many historians believe the *stadia* was 157.6 meters (517 feet), which would make Eratosthenes' calculation of the earth's circumference—39,400 kilometers (24,428 miles)—a measurement very close to the modern value of 40,075 kilometers (24,902 miles).

Probably the greatest of the early Greek astronomers was Hipparchus (second century B.C.), best known for his star catalog. Hipparchus determined the location of almost 850 stars, which he divided into six groups according to their brightness. He measured the length of the year to within minutes of the modern value and developed a method for predicting the times of lunar eclipses to within a few hours.

Although many of the Greek discoveries were lost during the Middle Ages, the earth-centered view that the Greeks proposed became established in Europe. Presented in its finest form by Claudius Ptolemy, this geocentric outlook became known as the **Ptolemaic system.**

The Ptolemaic System

Much of our knowledge of Greek astronomy comes from a thirteen-volume treatise, *Almagest* ("the great work"), which was compiled by Ptolemy in A.D. 141 and survived thanks to the work of Arab scholars. In this work, Ptolemy is credited with developing a model of the universe that accounted for the observable motions of the planets (Figure 17.4). The precision with which his model was able to predict planetary motion is attested to by the fact that it went virtually unchallenged, in principle if not in detail, for nearly thirteen centuries.

Figure 17.3
Orientation of the sun's rays at Syene (Aswan) and Alexandria in Egypt on June 21 when Eratosthenes calculated the earth's circumference.

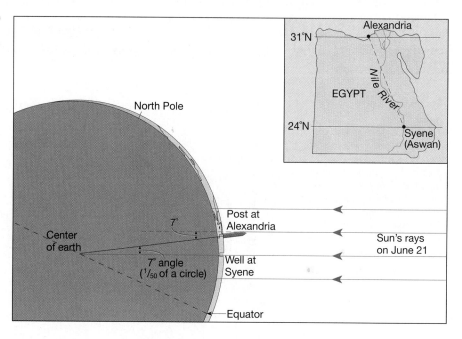

Figure 17.4
The universe according to Ptolemy, second century A.D.
A. *Ptolemy believed that the star-studded celestial sphere made a daily trip around a motionless earth. In addition he proposed that the sun, moon, and planets made trips of various lengths along individual orbits.*
B. *Retrograde motion as explained by Ptolemy.*

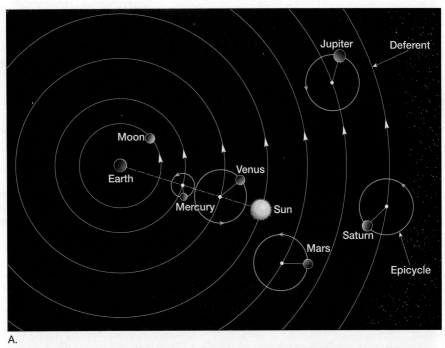

A.

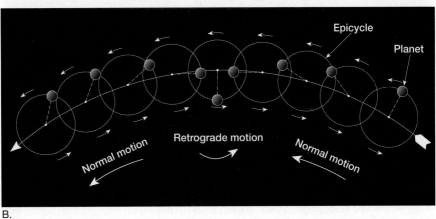

B.

In the Greek tradition, the Ptolemaic model had the planets moving in circular orbits around a motionless earth. (The circle was considered the pure and perfect shape by the Greeks.) However, the motion of the planets, as seen against the background of stars, is not so simple. Each planet, if watched night after night, moves slightly eastward among the stars. Periodically, each planet appears to stop, reverse direction for a period of time, and then resume an eastward motion. The apparent westward drift is called **retrograde motion.** This rather odd *apparent* motion results from the combination of the motion of the earth and the planet's own motion around the sun.

Figure 17.5 illustrates the retrograde motion of Mars. The earth has a faster orbital speed than Mars so it overtakes its neighbor. While doing so, Mars *appears* to be moving backward, in retrograde motion. This is analogous to what a race-car driver sees out the side window when passing a slower car. The slower planet, like the slower car, appears to be going backward, although its actual motion is in the same direction as the faster-moving body.

Figure 17.5 shows how retrograde motion works. It is much more difficult to accurately represent retrograde motion using the incorrect earth-centered model, but Ptolemy was able to do so (Figure 17.4B). Rather than using a simple circle for each planet's or-

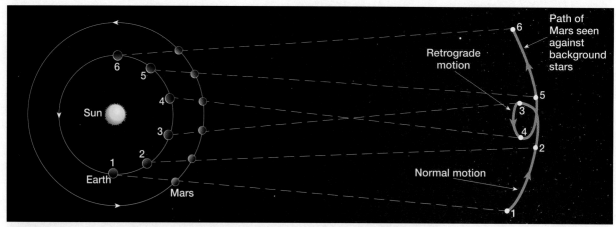

Figure 17.5
Retrograde (backward) motion of Mars as seen against the background of distant stars. When viewed from the earth, Mars moves eastward among the stars each day, then periodically appears to stop and reverse direction. This apparent westward drift is a result of the fact that the earth has a faster orbital speed than Mars and overtakes it. As this occurs, Mars appears to be moving backward; that is, it exhibits retrograde motion.

bit, he showed that the planets orbited on small circles *(epicycles)*, revolving along large circles *(deferents)*. By trial and error, he found the right combinations of circles to produce the amount of retrograde motion observed for each planet. (An interesting note is that almost any closed curve can be produced by the combination of two circular motions, a fact that can be verified by persons who have used the Spirograph® design-drawing toy.)

It is a tribute to Ptolemy's genius that he was able to account for the planets' motions as well as he did, considering that he used an incorrect model. Some suggest that he did not mean his model to represent reality, but only to be used for calculating the positions of the heavenly bodies. We probably will never know his intentions. However, the Roman Catholic Church, which dominated European thought for centuries, accepted Ptolemy's theory as the correct representation of the heavens, and this created problems for those who found fault with it.

THE BIRTH OF MODERN ASTRONOMY

Modern astronomy was not born overnight. Its development involved a break from deeply entrenched philosophical and religious views and, during the 1500s and 1600s, the founding of a "new and greater universe" governed by discernible laws. Let us now look at the work of five noted scientists involved in this transition: Nicolaus Copernicus, Tycho Brahe, Johannes Kepler, Galileo Galilei, and Sir Isaac Newton.

Nicolaus Copernicus

For almost thirteen centuries after the time of Ptolemy, very few astronomical advances were made in Europe. The first great astronomer to emerge after the Middle Ages was Nicolaus Copernicus (1473–1543) from Poland (Figure 17.6). Copernicus became convinced that the earth is a planet, just like the other five then-known planets. The daily motions of the heavens, he reasoned, could be better explained by a rotating earth. To counter the Ptolemaic objection that the earth would fly apart if it rotated, Copernicus suggested that the much-larger celestial sphere would be even more likely to fly apart if *it* rotated!

Having concluded that the earth is a planet, Copernicus reconstructed the solar system with the sun at the center and the planets Mercury, Venus, Earth, Mars, Jupiter, and Saturn orbiting around it. This was a major break from the ancient idea that a motionless earth lies at the center of all movement. However, Copernicus retained a link to the past and used circles, which were considered to be the perfect

Figure 17.6
Polish astronomer Nicolaus Copernicus (1473–1543) believed that the earth was just another planet. (Yerkes Observatory Photograph)

geometric shape, to represent the orbits of the planets. Although these circular orbits were close to reality, they didn't quite match what people saw. Unable to get satisfactory agreement between predicted locations of the planets and their observed positions, Copernicus found it necessary to add epicycles like those used by Ptolemy. The discovery that the planets have *elliptical* orbits would wait another century for the insights of Johannes Kepler.

Also, like his predecessors, Copernicus used philosophical justifications, such as the following, to support his point of view.

> . . . in the midst of all stands the sun. For who could in this most beautiful temple place this lamp in another or better place than that from which it can at the same time illuminate the whole?

Copernicus's monumental work, *De Revolutionibus, Orbium Coelestium (On the Revolution of the Heavenly Spheres)*, which set forth his controversial ideas, was published as he lay on his deathbed.

Hence, he never suffered the criticisms that fell on many of his followers.

The Copernican system challenged the primacy of the earth in the universe and was considered heretical by the Church. Expounding the idea cost at least one person his life. Giordano Bruno was seized by the Inquisition, a Church tribunal, in 1600 and, refusing to denounce the Copernican theory, was tied to a stake and burned alive.

Tycho Brahe

Tycho Brahe (1546–1601) was born of Danish nobility three years after the death of Copernicus. Reportedly, Tycho became interested in astronomy while viewing a solar eclipse that had been predicted by astronomers. He persuaded King Fredrich II to establish an observatory, which he headed, near Copenhagen. There he designed and built pointers (the telescope would not be invented for a few more decades), which he used for 20 years to systematically measure the locations of the heavenly bodies (Figure 17.7). These observations, particularly of Mars, were far more precise than any made previously and are his legacy to astronomy.

Tycho did not believe in the Copernican (sun-centered) system, because he was unable to observe an apparent shift in the position of stars that would be caused by the earth's motion. His argument went like this: If the earth does revolve along an orbit around the sun, the position of a nearby star, when observed from extreme points in the earth's orbit six months apart, should shift with respect to the more distant stars. His *idea* was correct, and this apparent shift of the stars is called *stellar parallax* (see Figure 20.1).

The principle of parallax is easy to visualize: Close one eye, and with your index finger vertical, use your eye to line up your finger with some distant object. Now, without moving your finger, view the object with your other eye and notice that the object's position appears to change. The farther away you hold your finger, the less the object's position seems to shift. Herein lay the flaw in Tycho's argument. He was right about parallax, but, because the distance to even the nearest stars is enormous compared to the width of the earth's orbit, the shift that occurs is too small to be noticed by using the first primitive telescopes, let alone the unaided eye.

With the death of his patron, the King of Denmark, Tycho was forced to leave his observatory. It was

Figure 17.7
Tycho Brahe (1546–1601) in his observatory, Uraniborg, on the Danish island of Hveen. Tycho (central figure) and the background are painted on the wall of the observatory within the arc of the sighting instrument called a quadrant. In the far right, Tycho can be seen "sighting" a celestial object through the "hole" in the wall. Tycho's accurate measurements of Mars enabled Johannes Kepler to formulate his three laws of planetary motion. (Courtesy of Thomas Clarke, McLaughlin Planetarium)

probably his arrogant and extravagant nature that caused a conflict with the next ruler, so Tycho moved to Prague in the Czech Republic. Here, in the last year of his life, he acquired an able assistant, Johannes Kepler. Kepler retained most of the observations made by Tycho and put them to exceptional use. Ironically, the data Tycho collected to refute the Copernican view would later be used by Kepler to support it.

Johannes Kepler

If Copernicus ushered out the old astronomy, Johannes Kepler (1571–1630) ushered in the new (Figure 17.8). Armed with Tycho's data, a good mathematical mind, and, of greater importance, a strong faith in the accuracy of Tycho's work, Kepler derived three basic laws of planetary motion. The first two laws resulted from his inability to fit Tycho's observations of Mars to a circular orbit. Unwilling to concede that the discrepancies were due to observational error, he searched for another solution. This endeavor led him to discover that the orbit of Mars is not a per-

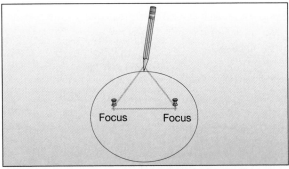

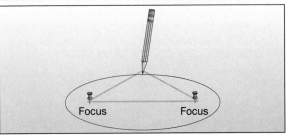

Figure 17.9
Drawing ellipses with various eccentricities. Using two straight pins for foci and a loop of string, trace out a curve while keeping the string taut, and you will have drawn an ellipse. The further the pins (the foci) are moved apart, the more flattened (more eccentric) is the resulting ellipse.

fect circle but is elliptical (Figure 17.9). About the same time, he realized that the orbital speed of Mars varies in a predictable way. As it approaches the sun, it speeds up, and as it moves away from the sun, it slows down.

In 1609, after almost a decade of work, Kepler proposed his first two laws of planetary motion:

1. The path of each planet around the sun is an ellipse with the sun at one focus (Figure 17.9). The other focus is symmetrically located at the opposite end of the ellipse.
2. Each planet revolves so that an imaginary line connecting it to the sun sweeps over equal areas in equal intervals of time (Figure 17.10). This law of equal areas expresses geometrically the variations in orbital speeds of the planets.

Figure 17.10 illustrates the second law. Note that in order for a planet to sweep equal areas in the same amount of time, it must travel more rapidly when it is nearer the sun and more slowly when it is farther from the sun.

Kepler was very religious and believed that the Creator made an orderly universe. The uniformity he

Figure 17.8
German astronomer Johannes Kepler (1571–1630) helped establish the era of modern astronomy by deriving three laws of planetary motion. (Smithsonian Institution Photo #56123)

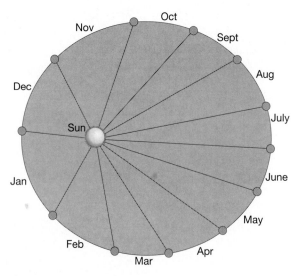

Figure 17.10
Kepler's law of equal areas. A line connecting a planet (earth) to the sun sweeps out an area in such a manner that equal areas are swept out in equal times. Thus, the earth revolves slower when it is farther from the sun (aphelion) and faster when it is closest (perihelion). The eccentricity of the earth's orbit is greatly exaggerated in this diagram.

tried to find eluded him for nearly a decade. Then, in 1619, he published his third law in *The Harmony of the Worlds*.

3. The orbital periods of the planets and their distances to the sun are proportional.

In its simplest form, the orbital period of revolution is measured in earth years, and the planet's distance to the sun is expressed in terms of the earth's mean distance to the sun. The latter "yardstick" is called the **astronomical unit** (AU) and averages about 150 million kilometers (93 million miles).

Using these units, Kepler's third law states that the planet's orbital period squared is equal to its mean solar distance cubed ($p^2 = d^3$). Consequently, the solar distances of the planets can be calculated when their periods of revolution are known. For example, Mars has a period of 1.88 years, which squared equals 3.54. The cube root of 3.54 is 1.52, and that is the distance to Mars in astronomical units (Table 17.1).

Kepler's laws assert that the planets revolve around the sun and therefore support the Copernican theory. Kepler, however, did fall short of determining the *forces* that act to produce the planetary motion he had so ably described. That task would remain for Galileo Galilei and Sir Isaac Newton.

Galileo Galilei

Galileo Galilei (1564–1642) was the greatest Italian scientist of the Renaissance (Figure 17.11). He was a contemporary of Kepler and, like Kepler, strongly supported the Copernican theory of a sun-centered solar system. Galileo's greatest contributions to science were his descriptions of the behavior of moving objects. These he derived from experimentation. The method of using experiments to determine natural laws had essentially been lost since the time of the early Greeks.

An example of such experimentation is the legend that Galileo climbed to the top of the Leaning Tower in Pisa, Italy, where he dropped two objects of differing masses. To the astonishment of the observers, they hit the ground at the same time. (Galileo probably did not attempt this experiment, although a colleague may have.) Galileo correctly concluded that variations in the rate of fall of very light objects, like feathers, are due to air resistance, not differences in mass. This same experiment was performed most dramatically on the airless moon nearly four centuries later when David Scott, an *Apollo 15* astronaut, showed that a feather and a hammer fall at the same rate.

All astronomical discoveries before Galileo's time were made without the aid of a telescope. In 1609, Galileo heard that a Dutch lens maker had devised a system of lenses that magnified objects. Apparently without ever seeing a telescope, Galileo constructed his own, which magnified distant objects to three

Table 17.1
Period of revolution and solar distances of planets.

Planet	Solar Distance (AU)*	Period (years)
Mercury	0.39	0.24
Venus	0.72	0.62
Earth	1.00	1.00
Mars	1.52	1.88
Jupiter	5.20	11.86
Saturn	9.54	29.46
Uranus	19.18	84.01
Neptune	30.06	164.80
Pluto	39.44	247.70

* AU = astronomical unit.

Figure 17.11
Italian scientist Galileo Galilei (1564–1642) used a new invention, the telescope, to observe the sun, moon, and planets in more detail than ever before. (Yerkes Observatory Photograph)

times the size seen by the unaided eye. He immediately made others, the best having a magnification of about 30.

With the telescope, Galileo was able to view the universe in a new way. He made many important discoveries that supported the Copernican view of the universe, such as:

1. The discovery of four satellites, or moons, orbiting Jupiter. Galileo accurately determined their periods of revolution, which range from 2 to 17 days (Figure 17.12). This find dispelled the old idea that the earth was the only center of motion in the universe; for here, plainly visible, was another center of motion—Jupiter. It also countered the argument, frequently used by those opposed to the sun-centered system, that the moon would be left behind if the earth really revolved around the sun.

2. The discovery, through observation, that the planets are circular disks rather than just points of light, as was previously thought. This indicated that the planets might be earthlike.

3. The discovery that Venus has phases just like the moon, demonstrating that Venus orbits its source of light—the sun. He saw that Venus appears smallest when it is in full phase, and thus is farthest from the earth (Figure 17.13). In the Ptolemaic system, as shown in Figure 17.4, the orbit of Venus lies between the earth and the sun, which means that only the crescent phase of Venus could be seen from earth.

4. The discovery that the moon's surface is not a smooth glass sphere, as the ancients had suspected and the Church had decreed. Rather, Galileo saw mountains, craters, and plains. He thought the plains might be bodies of water, and this idea was strongly promoted by others, as we can tell from the names given to these features (Sea of Tranquility, Sea of Storms, and so forth).

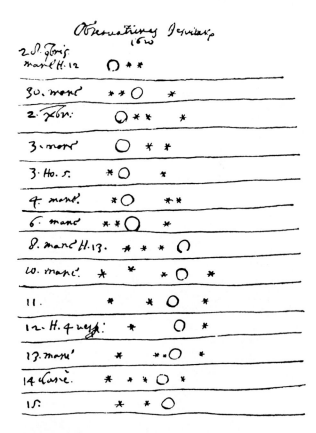

Figure 17.12
Sketch by Galileo of how he saw Jupiter and its four largest satellites through his telescope. The positions of Jupiter's four largest moons (drawn as stars) change nightly. You can observe these same changes with binoculars. (Yerkes Observatory Photograph)

Figure 17.13
Using a telescope, Galileo discovered that Venus has phases just like the moon. **A.** In the Ptolemaic (earth-centered) system, the orbit of Venus lies between the sun and the earth as shown in Figure 17.4A. Thus, in an earth-centered solar system, only the crescent phase of Venus would be visible from earth. **B.** In the Copernican (sun-centered) system, Venus orbits the sun and hence all of the phases of Venus should be visible from the earth. **C.** As Galileo observed, Venus goes through a series of moonlike phases. Further, Venus appears smallest during the full phase when it is farthest from the earth and largest in the crescent phase when it is closest to the earth. This verified Galileo's belief that the sun was the center of the solar system. (Photo courtesy of Lowell Observatory)

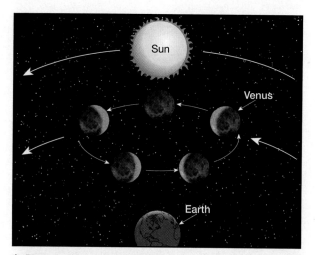

A. Phases of Venus as seen from the earth in the earth-centered model.

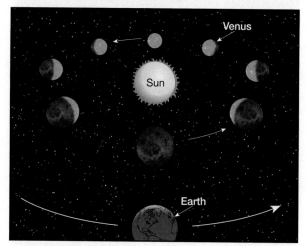

B. Phases of Venus as seen from the earth in the sun-centered model.

C.

5. The discovery that the sun (the viewing of which may have caused the eye damage that later blinded him) had sunspots (dark regions caused by slightly lower temperatures). He tracked the movement of these spots and estimated the rotational period of the sun as just under a month. Hence, another heavenly body was found to have both "blemishes" and rotational motion.

In 1616 the Church condemned the Copernican theory as contrary to Scripture, and Galileo was told to abandon it. Unwilling to accept this verdict, Galileo began writing his most famous work, *Dialogue of the Great World Systems*. Despite poor health, he completed the project and in 1630 went to Rome, seeking permission from Pope Urban VIII to publish. Because the book was a dialogue that expounded both the Ptolemaic and Copernican systems, publication was allowed. However, Galileo's enemies were quick to realize that he was promoting the Copernican view at the expense of the Ptolemaic system. Sale of the book was quickly halted, and Galileo was called before the Inquisition. Tried and convicted of proclaiming doctrines contrary to religious doctrine, he was sentenced to permanent house arrest, under which he remained for the last ten years of his life.

Despite this restriction, and his age, and his grief after the death of his eldest daughter, Galileo continued to work. In 1637 he became totally blind, yet during the next few years completed his finest scientific work, a book on the study of motion. When Galileo died in 1642, the Grand Duke of Tuscany wanted to erect a monument in his honor, but fear that it might offend the Church prevailed, and it was never built. It was not until 1992 that Galileo was finally exonerated by the Church.

Sir Isaac Newton

Sir Isaac Newton (1643–1727) (Figure 17.14) was born in the year of Galileo's death. His many accomplishments in mathematics and physics led a successor to say that "Newton was the greatest genius that ever existed . . . ."

Although Kepler and those who followed attempted to explain the forces involved in planetary motion, their explanations were less than satisfactory. Kepler believed that some force pushed the planets along in their orbits. Galileo, however, correctly reasoned that no force is required to keep an object in motion. Galileo proposed that the natural tendency for a moving object (that is unaffected by an outside

Figure 17.14
English scientist Sir Isaac Newton (1643–1727) explained gravity as the force that holds planets in orbit around the sun. (Yerkes Observatory Photograph)

force) is to continue moving at a uniform speed and in a straight line. This concept, *inertia*, was later formalized by Newton as his first law of motion.

The problem, then, was not to explain the force that keeps the planets moving but rather to determine the force that *keeps them from going in a straight line out into space*. It was to this end that Newton conceptualized the force of *gravity*. At the early age of 23, he envisioned a force that extends from the earth into space and holds the moon in orbit around the earth. Although others had theorized the existence of such a force, he was the first to formulate and test the *law of universal gravitation*. It states:

> Every body in the universe attracts every other body with a force that is directly proportional to their masses and inversely proportional to the square of the distance between them.

Thus, the gravitational force decreases with distance, so that two objects 3 kilometers apart have 3^2, or 9, times less gravitational attraction than if the same objects were 1 kilometer apart.

BOX 17.1
Proof for the Earth's Rotation

*E*very school child learns that the earth rotates on its axis once each day to produce periods of daylight and darkness. However, day and night and the apparent motions of the stars can be accounted for equally well by a sun and celestial sphere that revolve around a stationary Earth. Copernicus realized that a rotating Earth greatly simplified the existing model of the universe and strongly advocated this as the correct view. He was unable, however, to *prove* that the earth rotates. The first substantial proof was presented, 300 years

after his death, by the French physicist Jean Foucault.

In 1851 Foucault used a free-swinging pendulum to demonstrate that the earth does, in fact, turn on its axis. To envision Foucault's experiment, imagine a large pendulum swinging over the North Pole (Figure 17.A). Keep in mind that once a pendulum is put into motion, it continues swinging in the same plane unless acted upon by some outside force. Assume that a sharp stylus is attached to the bottom of this pendulum, marking the snow as it oscillates. When we observe the marks made by the stylus, we note that the pendulum is slowly but continually changing position. At the end of 24 hours it has returned to the starting position (Figure 17.A).

Because no outside force acted on the pendulum to change its position, what we ob-

served must have been the earth rotating beneath it. Foucault conducted a similar experiment when he suspended a long pendulum from the dome of the Pantheon in Paris. Today, Foucault pendulums can be found in a number of museums to recreate this famous scientific experiment (Figure 17.B).

Figure 17.B
Foucault pendulum housed in Museum of Science and Industry, Chicago. (Courtesy Museum of Science and Industry)

Figure 17.A
Apparent movement of a pendulum at the North Pole caused by the earth rotating beneath it.

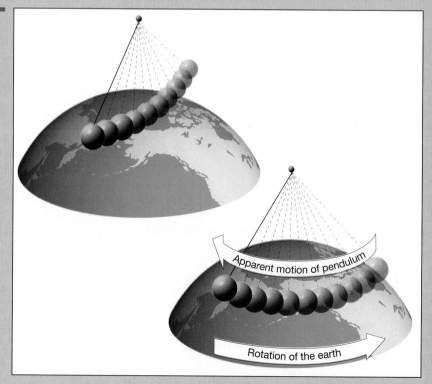

Apparent motion of pendulum

Rotation of the earth

The law of gravitation also states that the greater the mass of the object, the greater its gravitational force. For example, the large mass of the moon has a gravitational force strong enough to cause ocean tides on the earth, whereas the tiny mass of a communications satellite has no measurable effect on the earth. The mass of an object is a measure of the total amount of matter it contains. But more often mass is measured by determining the resistance an object exhibits in response to any effort made to change its state of motion.

Often we confuse the concept of *mass* with our notion of *weight*. Specifically, weight is the force of gravity acting upon an object. Therefore, weight varies when gravitational forces change. An object weighs less on the moon than on the earth because the moon is much less massive than the earth. However, unlike weight, the mass of an object does not change. For example, a person weighing 120 pounds on earth weighs ⅙ as much, or 20 pounds, on the moon, but the person's mass remains unchanged.

With his laws of motion, Newton proved that the force of gravity, combined with the tendency of a planet to remain in straight-line motion, results in the elliptical orbits discovered by Kepler. The earth, for example, moves forward in its orbit about 30 kilometers (18½ miles) each second, and during the same second, the force of gravity pulls it toward the sun about ½ centimeter (⅛ inch). Therefore, as Newton concluded, it is the combination of the earth's forward motion and its "falling" motion that defines its orbit (Figure 17.15). If gravity were somehow eliminated, the earth would move in a straight line out into space. On the other hand, if the earth's forward motion suddenly stopped, gravity would pull it directly toward the sun.

Up to this point, we have discussed the earth as if the only forces involved in its motion were caused by its gravitational relationship with the sun. However, all bodies in the solar system have gravitational effects on the earth and on each other. For this reason, the orbit of the earth is not the perfect ellipse determined by Kepler. Any variance in the orbit of a body from its predicted path is called **perturbation.** For example, Jupiter's gravitational pull on Saturn reduces Saturn's orbital period by nearly one week from the predicted period. As we shall see, the application of this concept led to the discovery of the planet Neptune, because of Neptune's gravitational effect on the orbit of Uranus.

Newton used the law of universal gravitation to redefine Kepler's third law, which states the relationship between the orbital periods of the planets and their solar distances. When restated, Kepler's third law takes into account the masses of the bodies involved and thereby provides a method for determining the mass of a body when the orbit of one of its satellites is known. For example, the mass of the sun is known from the orbit of the earth, and the mass of the earth has been determined from the orbit of the moon. In fact, the mass of any body with a satellite can be determined. The masses of bodies that do not have satellites can be determined only if the bodies noticeably affect the orbit of a neighboring body or of a nearby artificial satellite.

CONSTELLATIONS

As early as 5000 years ago, people became fascinated with the star-studded skies and began to name the patterns they saw. These configurations, called **constellations,** were named in honor of mythological characters or great heroes, such as Orion. It takes a good bit of imagination to make out the intended subjects, since most constellations were probably not thought of as likenesses in the first place. Although we inherited many of the constellations from Greek mythology, it is believed that Greek astronomers acquired most of theirs from the Babylonians, Egyptians, and Mesopotamians.

Although the stars that make up a constellation all appear to be the same distance from the earth,

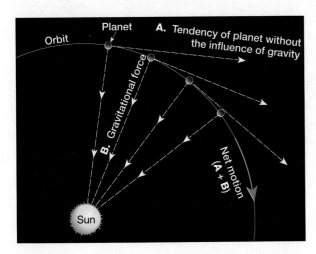

Figure 17.15
Orbital motion of the earth and other planets.

BOX 17.2
Astrology

Many people confuse astrology and astronomy to the point of believing these terms to be synonymous. Nothing can be farther from the truth. *Astronomy* is a scientific probing of the universe to derive the properties of celestial objects and the laws under which the universe operates. *Astrology,* on the other hand, is based on ancient super-stitions that hold that an individual's actions and personality are based on the positions of the planets and stars now, and at the person's birth. Scientists do not accept astrology, regarding it as a pseudoscience ("false science"). It is hoped that most people read horoscopes only as a pastime and do not let them influence daily living.

Apparently astrology had its origin more than 3000 years ago when the positions of the planets were plotted as they regularly migrated against the background of the "fixed" stars. Because the solar system is "flat" like a whirling frisbee, the planets orbit the sun along nearly the same plane. Therefore, the planets, sun, and moon all appear to move along a band around the sky known as the *zodiac.* Because the earth's moon cycles through its phases about twelve times each year, the Babylonians divided the zodiac into twelve constellations (Figure 17.C). Thus, each successive full moon can be seen against the backdrop of the next constellation.

The twelve constellations of the zodiac ("Zone of Animals," so named because some constellations represent animals) are Aries, Taurus, Gemini, Cancer, Leo, Virgo, Libra, Scorpio, Sagittarius, Capricorn, Aquarius, and

Figure 17.C
The twelve constellations of the zodiac. Earth is shown in its autumn (September) position in orbit, from which the sun is seen against the background of the constellation Virgo.

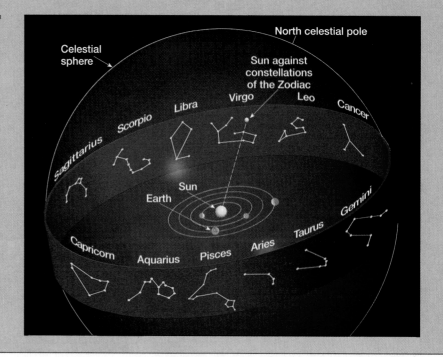

this is not the case. Some are many times farther away than others. Thus, the stars in a particular constellation are not associated with each other in any physical way. In addition, various cultural groups, including Native Americans, attached their own names, pictures, and stories to the constellations. For example, the constellation Orion, the Hunter, was known as the White Tiger to ancient Chinese astronomers.

Today, 88 constellations are recognized, and they are used to divide the sky into units, just as state boundaries divide the United States. Every star in the sky is in, but is not necessarily part of, one of these constellations. Constellations therefore enable astronomers to roughly identify the area of the heavens

Pisces. These names may be familiar to you as the astrological signs of the zodiac. When first established, the vernal equinox (first day of spring) occurred when the sun was viewed against the constellation Aries. However, during each succeeding vernal equinox, the position of the sun shifts very slightly against the background of stars. Now, over 2000 years later, the vernal equinox occurs when the sun is in Pisces (Figure 17.C). In several years, it will occur when the sun appears against Aquarius. (Hence, the "Age of Aquarius" is coming.)

Although astrology is not a science and has no basis in fact, it did contribute to the science of astronomy. The positions of the moon, sun, and planets at the time of a person's birth (sign of the zodiac) were considered to have great influence on that person's life.* Even the great astronomer Kepler was required to make horoscopes as part of his duties. In order to make horoscopes for the future, astrologers attempted to predict the future positions of the celestial bodies. Consequently, some of the improvements in astronomical instruments were probably due to the desire for more accurate predictions of events such as eclipses, which were considered highly significant in a person's life.

Even prehistoric people built observatories. The structure known as Stonehenge, in England, was undoubtedly an attempt at better solar prediction (Figure 17.D). At the time of midsummer in the Northern Hemisphere (June 21–22—the summer solstice), the rising sun emerges directly above the heel stone of Stonehenge. Besides keeping this calendar, Stonehenge may also have provided a method of determining eclipses. The remnants of other early observatories exist elsewhere in the Americas, Europe, Asia, and Africa.

---

* It is interesting to note that 2000 years ago a person born on July 28 was considered a Leo because the sun was in that constellation. During modern times the sun appears in the constellation Cancer on this date, but individuals born during this time frame are still dubbed Leos.

Figure 17.D
Stonehenge, an ancient solar observatory in England. On June 21–22 (summer solstice), the sun can be observed rising above the heel stone. (Photo by Robert Llewellyn)

they are observing. For the student, the constellations provide a good way to become familiar with the night sky (see Appendix E).

Some of the brightest stars were given proper names, such as Sirius, Arcturus, and Betelgeuse. In addition, the brightest stars in a constellation are generally named in order of their brightness by the letters of the Greek alphabet—alpha (α), beta (β), and so on—followed by the name of the parent constellation. For example, Sirius, the brightest star in the constellation Canis Major (Larger Dog), is called Alpha (α) Canis Majoris.

POSITIONS IN THE SKY

If you gaze away from the city lights on a clear night, you will get the distinct impression that the stars produce a spherical shell surrounding the earth. This impression seems so real that it is easy to understand why many early Greeks regarded the stars as being fixed to a crystalline celestial sphere. Although we realize that no such sphere exists, it is convenient to use this concept for locating stars.

One method for doing this, called the **equatorial system,** divides the celestial sphere into a coordinate system. It is very similar to the latitude-longitude system used for locations on the earth's surface (Figure 17.16). Because the celestial sphere appears to rotate around an imaginary line extending from the earth's axis, the north and south celestial poles are in line with the terrestrial North Pole and South Pole. The north celestial pole happens to be very near the bright star whose various names reflect its location: "pole star," Polaris, and North Star. To an observer in the Northern Hemisphere, the stars appear to circle Polaris, because it, like the North Pole, is in the center of motion (Figure 17.17). Figure 17.18

shows how to locate the North Star by using two stars in the very visible constellation called the Big Dipper.

Now, imagine a plane through the earth's equator, a plane that extends outward from the earth and intersects the celestial sphere. The intersection of this plane with the celestial sphere is the *celestial equator* (see Figure 17.16). In the equatorial system, the terms *declination*, which is analogous to latitude, and *right ascension*, which is analogous to longitude, are used (Figure 17.16). **Declination,** like latitude, is the angular distance north or south of the celestial equator. **Right ascension** is the angular distance measured eastward along the celestial equator from the position of the vernal equinox. (The *vernal equinox* is at the point in the sky where the sun crosses the celestial equator, at the onset of spring.) While declination is expressed in degrees, right ascension is usually expressed in hours, where each hour is equivalent to 15 degrees.

To visualize distances on the celestial sphere, it helps to remember that the moon and sun have an apparent width of about ½ degree. The declination and right ascension of some of the brightest stars (in

Figure 17.16
Astronomical coordinate system on the celestial sphere.

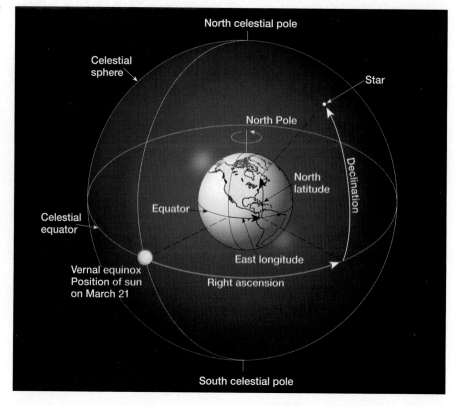

Figure 17.17
Star trails in the region of Polaris (north celestial pole) on a time exposure. (Courtesy of National Optical Astronomy Observatories)

other words, their locations) can be determined from the star charts in Appendix E.

MOTIONS OF THE EARTH

The two primary motions of the earth are rotation and revolution. **Rotation** is the turning, or spinning, of a body on its axis. **Revolution** is the motion of a body, such as a planet or moon, along a path around some point in space. For example, the earth *revolves* around the sun and the moon *revolves* around the earth. The earth also has another very slow motion known as **precession,** which is the slight movement, over a period of 26,000 years, of the earth's axis.

Rotation

The main consequences of the earth's rotation are day and night. The earth's rotation has become a standard method of measuring time because it is so dependable and easy to use. Each rotation equals about 24 hours. You may be surprised to learn that we can measure the earth's rotation in two ways, making two kinds of days. Most familiar is the **mean solar day,** the time interval from one noon to the next, which averages about 24 hours. Noon is when the sun has reached its zenith (highest point in the sky).

The **sidereal day,** on the other hand, is the time it takes for the earth to make one complete rotation (360 degrees) with respect to a star other than the

sun. The sidereal day is measured by the time required for a star to reappear at the identical position in the sky where it was observed the day before. The sidereal day has a period of 23 hours, 56 minutes, and 4 seconds (measured in solar time), which is almost 4 minutes shorter than the mean solar day. This difference results because the direction to distant stars changes only infinitesimally due to the earth's slow revolution along its orbit, whereas the direction to the sun changes by almost 1 degree each day. This difference is shown in Figure 17.19. If it is not apparent why we use the mean solar day rather than the sidereal day as a measure of our day, consider the fact that in sidereal time, "noon" occurs 4 minutes earlier each day. Therefore, after a span of six months, "noon" occurs at "midnight." Astronomers use sidereal time because the stars appear in the same position in the sky every 24 sidereal hours. Usually, an observatory will begin its sidereal day when the position of the vernal equinox is directly overhead, that is, over the meridian on which the observatory is located. Therefore, when the observatory's sidereal clock is the same as the star's right ascension, the star will be overhead, or at its highest point. For example, the brightest star in the heavens, Sirius, has a right ascension of 6 hours, 42 minutes, and 56 seconds and will be overhead when the clock at the observatory indicates that time.

Figure 17.18
Locating the North Star (Polaris) from the pointer stars in the Big Dipper, which is part of the constellation Ursa Major. The Big Dipper is shown soon after sunset in December (lower figure), April (upper figure), and August (left). Refer to the star charts in Appendix E to find the current position of the Big Dipper.

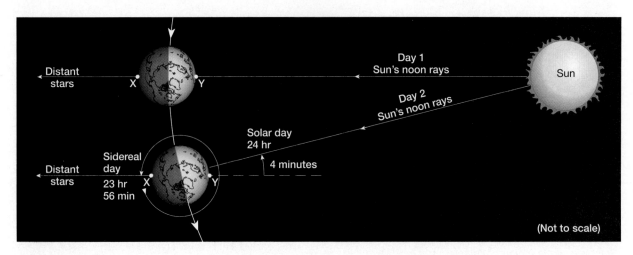

Figure 17.19
*The difference between a solar day and a sidereal day. Locations X and Y are directly
opposite each other. It takes the earth 23 hours and 56 minutes to make one rotation
with respect to the stars (sidereal day). However, notice that after the earth has rotated
once with respect to the stars, point Y is not yet returned to the "noon position" with
respect to the sun. The earth has to rotate another 4 minutes to complete the solar
day.*

Figure 17.20
*The earth's orbital motion
causes the apparent position
of the sun to shift about
1 degree each day on the
celestial sphere.*

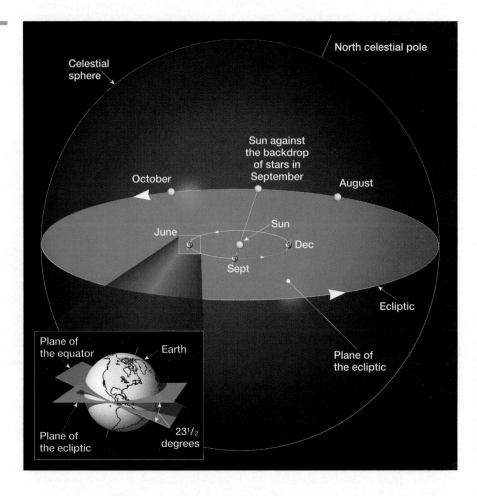

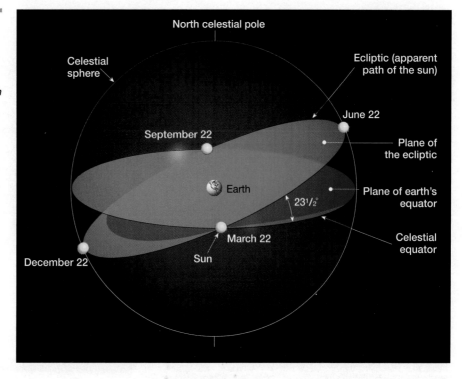

Figure 17.21
The apparent position of the sun plotted on the celestial sphere. The path of the sun (ecliptic) crosses the celestial equator on two occasions each year, March 20–21 and September 22–23. These are known as the equinox positions because the lengths of daylight and darkness on earth are equal.

Revolution

The earth revolves around the sun in an elliptical orbit at an average speed of 107,000 kilometers (66,000 miles) per hour. Its average distance from the sun is 150 million kilometers (93 million miles), but because its orbit is an ellipse, the earth's distance from the sun varies. At **perihelion** (closest to the sun), it is 147 million kilometers (91 million miles) distant, which occurs about January third each year. At **aphelion** (farthest from the sun), the earth is 152 kilometers (94.5 million miles) distant, which occurs about July fourth.

Due to the earth's annual movement around the sun, each day the sun appears to be displaced among the constellations at a distance equal to about twice its width, or 1 degree. The apparent annual path of the sun against the backdrop of the celestial sphere is called the **ecliptic** (Figure 17.20). Generally, the planets and the moon travel in nearly the same plane as the earth. Hence, their paths on the celestial sphere lie near the ecliptic. (The most notable exception is Pluto, which has an orbit that is tilted 17 degrees to the plane of the earth's orbit.)

The imaginary plane that connects the earth's orbit with the celestial sphere is called the **plane of the**

ecliptic. From this reference plane, the earth's axis of rotation is tilted about 23½ degrees. Because of the earth's tilt, the apparent path of the sun (ecliptic) and the celestial equator intersect each other at an angle of 23½ degrees (Figure 17.20). This angle is very important to the inhabitants of the earth. Because of the inclination of the earth's axis to the plane of the ecliptic, the earth exhibits its yearly cycle of seasons, a topic discussed in detail in Chapter 12.

When the *apparent* position of the sun is plotted on the celestial sphere over a period of a year's time, its path intersects the celestial equator at two points (Figure 17.21). From a Northern Hemisphere point of view, these intersections are called the vernal (spring) equinox (March 20–21) and autumnal equinox (September 22–23). On June 21–22, the date of the summer solstice, the sun appears 23½ degrees north of the celestial equator, and six months later, on December 21–22, the date of the winter solstice, the sun appears 23½ degrees south of the celestial equator.

Precession

A third and very slow movement of the earth is called *precession.* Although the earth's axis maintains ap-

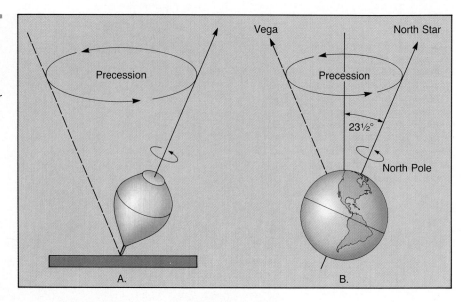

Figure 17.22
A. Precession illustrated by a spinning top. **B.** Precession of the earth causes the North Pole to point to different parts of the sky during a 26,000-year cycle. Today the North Pole points to Polaris (North Star). In 13,000 years, Vega will be the North Star.

proximately the same angle of tilt, the direction in which the axis points continually changes. As a result, the axis traces a circle on the sky. This movement is very similar to the movement (wobble) of a spinning top (Figure 17.22A). At the present time, the axis points toward Polaris. In A.D. 14,000, it will point toward the bright star Vega, which will then become the North Star for a few thousand years (Figure 17.22B). The period of precession is 26,000 years. By the year 28,000, Polaris will once again be the North Star.

Precession has only a minor effect on the seasons, because the angle of tilt changes only slightly. It does, however, cause the positions of the seasons (equinox and solstice) to move slightly each year among the constellations.

In addition to its own movements, the earth shares numerous motions with the sun. It accompanies the sun as the entire solar system speeds in the direction of the bright star Vega at 20 kilometers (12 miles) per second. Also, the sun, like other nearby stars, revolves around the galaxy, a trip that requires 230 million years to traverse at speeds approaching 250 kilometers (150 miles) per second. In addition, the galaxies themselves are in motion. We are presently approaching one of our nearest galactic neighbors, the Great Galaxy in Andromeda. In summary, the motions of the earth are many and complex, and its speed in space is very great.

MOTIONS OF THE EARTH-MOON SYSTEM

The earth has one natural satellite, the moon. In addition to accompanying the earth in its annual trek around the sun, our moon orbits the earth within a period of about one month. When viewed from above the North Pole, the direction of this motion is counterclockwise (eastward). Because the moon's orbit is elliptical, its distance to the earth varies by about 6 percent, averaging 384,401 km (238,329 mi).

The motions of the earth-moon system constantly change the relative positions of the sun, earth, and moon. The results are some of the most obvious of astronomical phenomena, namely, the *phases of the moon* and the occasional *eclipses of the sun and moon.*

Phases of the Moon

The first astronomical phenomenon to be understood was the regular cycle of the **phases of the moon.** On a monthly basis, we observe the phases as a systematic change in the amount of the moon that appears illuminated (Figure 17.23). We will choose the "new-moon" position in the cycle as a starting point. About two days after the new moon, a thin sliver (*crescent phase*) appears low in the western sky just after sunset. During the following week, the illuminated portion of the moon visible from earth increases (*waxing*) to a half circle (*first-quarter phase*)

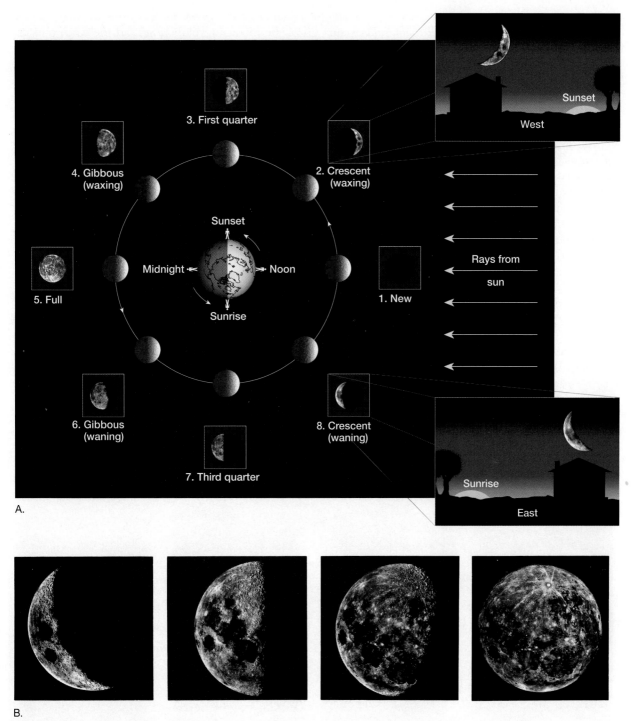

A.

B.

Figure 17.23
Phases of the moon. **A.** *The outer figures show the phases as seen from the earth.*
B. *Compare these photographs with the diagram. (Courtesy of Lick Observatory)*

and can be seen from about noon to midnight. In another week, the complete disk *(full-moon phase)* can be seen rising in the east as the sun is sinking in the west. During the next two weeks, the percentage of the moon that can be seen steadily declines *(waning)*, until the moon disappears altogether *(new-moon phase)*. The cycle soon begins anew with the reappearance of the crescent moon.

The lunar phases are a consequence of the motion of the moon and the sunlight that is reflected from its surface (Figure 17.23B). Half of the moon is illuminated at all times (note the inner group of moon sketches in Figure 17.23A). But to an earthbound observer, the percentage of the bright side that is visible depends on the location of the moon with respect to the sun and the earth. When the moon lies *between* the sun and the earth, none of its bright side faces the earth, so we see the new-moon ("no-moon") phase. Conversely, when the moon lies on the side of the earth opposite the sun, all of its lighted side faces the earth, so we see the full moon. At all positions between these extremes, an intermediate amount of the moon's illuminated side is visible from the earth.

Lunar Motions

The cycle of the moon through its phases requires 29½ days, a time span called the **synodic month.** Recall that this cycle was the basis for the first Ro-

man calendar. However, this is the *apparent* period of the moon's revolution around the earth and not the *true* period, which takes only 27⅓ days and is known as the **sidereal month.** The reason for the difference of nearly 2 days each cycle is shown in Figure 17.24. Note that as the moon orbits the earth, the earth-moon system also moves in an orbit around the sun. Consequently, even after the moon has made a complete revolution around the earth, it has not yet reached its starting position, which was directly between the sun and earth (new-moon phase). The additional motion to reach the starting point takes another 2 days.

An interesting fact concerning the motions of the moon is that its period of rotation about its axis and its revolution around the earth are the same—27⅓ days. Because of this, the same lunar hemisphere always faces the earth. All of the manned *Apollo* missions have been confined to the earth-facing side. Only orbiting satellites and astronauts have seen the "back" side of the moon.

Since the moon rotates on its axis only once every 27⅓ days, any location on its surface experiences periods of daylight and darkness lasting about two weeks. This, along with the absence of an atmosphere, accounts for the high surface temperature of 127°C (261°F) on the day side of the moon and the low surface temperature of −173°C (−280°F) on its night side.

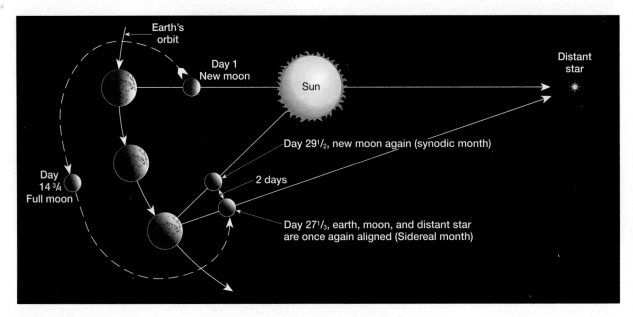

Figure 17.24
The difference between the sidereal month (27⅓ days) and the synodic month (29½ days). Distances and angles are not shown to scale.

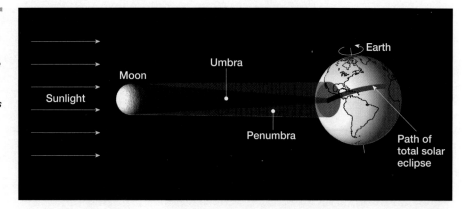

Figure 17.25
Solar eclipse. Observers in the zone of the umbral shadow see a total solar eclipse. Those in the penumbra see a partial eclipse. The path of the solar eclipse moves eastward across the globe.

Eclipses

Along with understanding the moon's phases, the early Greeks also realized that eclipses are simply shadow effects. When the moon moves in a line directly between the earth and the sun, which can occur only during the new-moon phase, it casts a dark shadow on the earth, producing a **solar eclipse** (Figure 17.25). On the other hand, the moon is eclipsed **(lunar eclipse)** when it moves within the shadow of the earth, a situation that is possible only during the full-moon phase (Figure 17.26).

Why then does a solar eclipse not occur with every new-moon phase and a lunar eclipse with every full-moon phase? They would, if the orbit of the moon lay exactly along the plane of the earth's orbit (the plane of the ecliptic). However, the moon's orbit is inclined about 5 degrees to the plane that contains the earth and the sun (Figure 17.27). Thus, during most new-moon phases, the shadow of the moon misses the earth (passes above or below); and during most full-moon phases, the shadow of the earth misses the moon. Only when a new- or full-moon phase occurs where the moon's orbit crosses the plane of the ecliptic can an eclipse take place. Since these conditions are normally met only twice a year, the usual number of eclipses is four. These occur as a set of one solar and one lunar eclipse, followed six months later with another set (Figure 17.27). Occasionally the alignment is such that three eclipses can be squeezed into a one-month period. These occur as a solar eclipse flanked by two lunar eclipses, or vice versa. Furthermore, it occasionally happens that the first set of eclipses occurs at the very beginning of a year, and a third set occurs before the year ends, resulting in six eclipses in that year. More rarely, if one of these sets is a three-eclipse kind, the total number of eclipses in a year can reach seven, which is the maximum.

During a total lunar eclipse, the earth's circular shadow can be seen moving slowly across the disk of the full moon. When totally eclipsed, the moon is

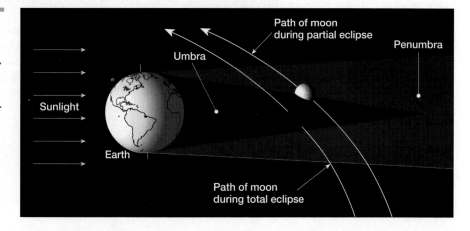

Figure 17.26
Lunar eclipse. During a total lunar eclipse the moon's orbit carries it into the dark shadow of the earth (umbra). During a partial eclipse only a portion of the moon enters the umbra.

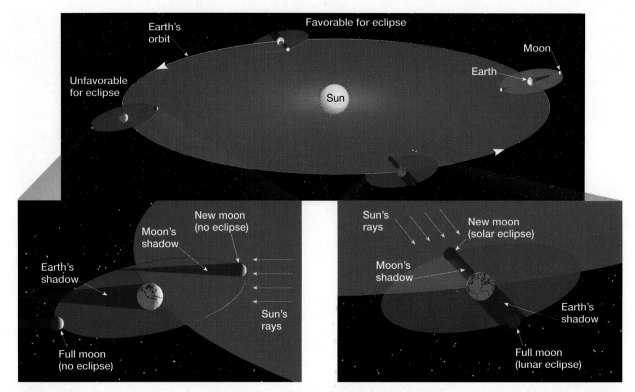

Figure 17.27
The moon's orbit is inclined about 5 degrees to the plane that contains the sun and the earth. Thus, during most new-moon phases, the shadow of the moon misses the earth (passes above or below), and during most full-moon phases, the shadow of the earth misses the moon. Only when a new- or full-moon phase occurs where the moon's orbit crosses the earth-sun plane can an eclipse take place. These conditions are met at roughly six-month intervals.

Figure 17.28
This sequence of photos starting from the upper left to the lower right shows the stages of a total solar eclipse. (From Foundations of Astronomy, *Third Edition, by Michael Seeds, reprinted by permission of the author and Wadsworth Publishing Company)*

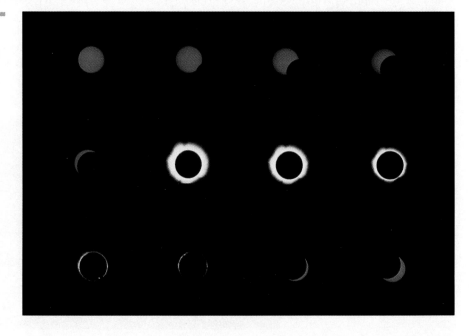

BOX 17.3
Calendars

One of the earliest responsibilities of astronomers was the keeping of the calendar. Although the oldest known calendars date from the eighth century B.C., others probably existed much earlier. These calendars had as their basis the movements of the heavenly bodies. Although our seven-day week has no astronomical significance, the seven days are clearly named after the seven moving celestial bodies known to the ancients: the sun, the moon, and the five planets (Mercury, Venus, Mars, Jupiter, and Saturn), which are easily seen with the unaided eye. For example, Sunday is "Sun's day," Monday is "Moon's day," and Saturday is "Saturn's day." The remaining days of the week were named in the Romance languages after Mars, Mercury, Jupiter, and Venus, in that order. For example, in Spanish, Tuesday is *Martes*.

Beyond the daily cycle of night and day, the most noticeable heavenly phenomenon is the 29.5-day cycle of the moon through its phases. This cycle became the *moonth,* now called the month. The first Western calendars based the year on the phases of the moon. Twelve months equaled a year. However, because there are actually 12.4 lunar cycles in a year, the calendar had to have a full month added every three years to keep the seasons in accord. Even with this correction, the calendar still fell slowly behind. When Julius Caesar gained power in Rome 2000 years ago, the calendar indicated spring, while the weather indicated the middle of winter. To correct this situation, he ordered that 80 days be added to the year of 46 B.C. That year, for obvious reasons, was called the "year of confusion." Also at the direction of an astronomer, Caesar ordered that the calendar be based on the tropical year of 365.25 days. One tropical year is the time required between two successive passages of the sun through the vernal equinox, or 365.2422 days. Calendar makers achieved this by having 365 days each normal year and adding an extra day every fourth year. So began the tradition of *leap year* and the *Julian calendar.* However, because the tropical year is slightly less than 365.25 days (about 10 minutes per year), the Julian calendar worked just like a fast clock. Both would require adjustment if used long enough. By the sixteenth century, the Julian calendar was ahead by 10 days.

In 1582 the *Gregorian calendar,* which we presently use, was developed. The extra 10 days were eliminated by making Friday, October 15, the day after Thursday, October 4. To slow the calendar, selected leap years were eliminated. A leap year is no longer added for centennial years except those divisible by 400. Hence, the years 1600, 2000, and 2400 are leap years, but all centennial years between them (e.g., 1900) are not. Our present calendar is accurate to within 1 day in 3000 years.

Not all countries adopted the new calendar at the same time. When George Washington was born, the calendar indicated February 11, 1732, but when the American colonies adopted the Gregorian calendar in 1752, his birth date became February 22, the day we now observe.

completely within the earth's shadow but still is visible as a coppery disk, because the earth's atmosphere bends and transmits some long-wavelength light (red) into its shadow. A total eclipse of the moon can last up to four hours and is visible to anyone on the side of the earth facing the moon.

During a total solar eclipse, the moon casts a circular shadow that is never wider than 275 kilometers (170 miles), about the size of South Carolina. Anyone observing in this region will see the moon slowly block the sun from view and the sky darken (Figure 17.28). Near totality, a sharp drop in temperature of a few degrees is experienced. The solar disk is completely blocked for at most seven minutes, because the moon's shadow is so small, and then one edge reappears.

At totality, the dark moon is seen covering the complete solar disk, and only the sun's brilliant white outer atmosphere is visible (see Figure 19.21). Total solar eclipses are visible only to people in the dark part of the moon's shadow *(umbra)*, while a partial eclipse is seen by those in the light portion *(penumbra)* (see Figure 17.25).

Partial solar eclipses are more common in the polar regions, because it is this zone that the penumbra covers when the dark umbra of the moon's shadow just misses the earth. A total solar eclipse is a rare event at any given location. The next one that will be visible from the contiguous United States will take place on August 21, 2017.

Review Questions

1. Why did the ancients think that celestial objects had some control over their lives?

2. Describe what produces the retrograde motion of Mars. What geometric arrangement did Ptolemy use to explain this motion?

3. What major change did Copernicus make in the Ptolemaic system? Why was this change philosophically significant?

4. What was Tycho Brahe's contribution to science?

5. Does the earth move faster in its orbit near perihelion (January) or near aphelion (July)? Keeping your answer to the previous question in mind, is the solar day longest in January or in July?

6. Use Kepler's third law ($p^2 = d^3$) to determine the period of a planet whose solar distance is
   (a) 10 AU.
   (b) 1 AU.
   (c) 0.2 AU.

7. Use Kepler's third law to determine the distance from the sun of a planet whose period is
   (a) 5 years.
   (b) 10 years.
   (c) 10 days.

8. Did Galileo invent the telescope?

9. Explain how Galileo's discovery of a rotating sun supported the Copernican view of a sun-centered universe.

10. Using a diagram, explain why the fact that Venus appears full when it is smallest supports the Copernican view and is inconsistent with the Ptolemaic system.

11. Newton learned that the orbits of the planets are the result of two actions. Explain these actions.

12. Of what value are constellations to modern-day astronomers?

13. Express the declination and right ascension of the star Arcturus (see Appendix E).

14. Explain the difference between the mean solar day and the sidereal day.

15. What is the approximate length of the cycle of the phases of the moon?

16. What is different about the crescent phase that precedes the new-moon phase and that which follows the new-moon phase?

17. What phase of the moon occurs approximately one week after the new moon? Two weeks?

18. When you observe the crescent phase early in the evening, is the visible moon waxing (growing) or waning (declining)? (See Figure 17.23.)

19. What phenomenon results from the fact that the moon's periods of rotation and revolution are the same?

20. The moon rotates very slowly (once in 27⅓ days) on its axis. How does this affect the lunar surface temperature?

21. Describe the locations of the sun, moon, and earth during a solar eclipse and during a lunar eclipse.

22. How many eclipses normally occur each year?

23. Solar eclipses are slightly more common than lunar eclipses. Why, then, is it more likely that your region of the country will experience a lunar eclipse?

24. How long can a total eclipse of the moon last? How about a total eclipse of the sun?

Key Terms

aphelion (p. 591)

astronomical unit (AU) (p. 580)

celestial sphere (p. 573)

constellation (p. 585)

declination (p. 588)

ecliptic (p. 591)

equatorial system (p. 588)

geocentric (p. 573)

heliocentric (p. 573)

lunar eclipse (p. 595)

mean solar day (p. 589)

perihelion (p. 591)

perturbation (p. 585)

phases of the moon (p. 592)

plane of the ecliptic (p. 591)

precession (p. 589)

Ptolemaic system (p. 574)

retrograde motion (p. 575)

revolution (p. 589)

right ascension (p. 588)

rotation (p. 589)

sidereal day (p. 589)

sidereal month (p. 594)

solar eclipse (p. 595)

synodic month (p. 594)

The Solar System

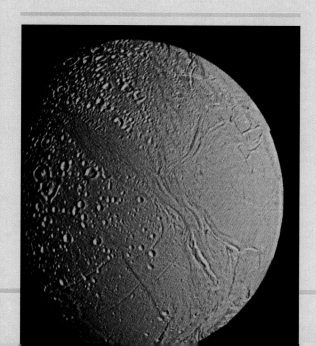

Opposite page: Saturn photographed from 18 million kilometers by Voyager 1. *(Courtesy of NASA).*

Left: Saturn's moon Enceladus. *(Courtesy of NASA)*

Above: Montage of Saturn and some of its satellites. *(Courtesy of NASA)*

601

When people first recognized that the planets were more similar to Earth than to the stars, a great deal of interest was generated. An important issue was, and still is, the possibility of intelligent life elsewhere in the universe. Space exploration has renewed this interest. However, no evidence for the existence of extraterrestrial life within our solar system has emerged. Nevertheless, since all the planets probably formed from the same primeval cloud of dust and gases, they provide valuable information concerning Earth's formation and early history. Recent space explorations have been organized with this goal in mind. To date, Mercury, Venus, Mars, Jupiter, Saturn, Uranus, Neptune, and the moon have been explored by space probes.

The sun is the hub of a huge rotating system consisting of nine planets, their satellites, and numerous small but interesting bodies, including asteroids, comets, and meteroids. An estimated 99.85 percent of the mass of our solar system is contained within the sun, while the planets collectively make up most of the remaining 0.15 percent. The planets, in order from the sun, are Mercury, Venus, Earth, Mars, Jupiter, Saturn, Uranus, Neptune, and Pluto (Figure 18.1).

Under the control of the sun's gravitational force, each planet maintains an elliptical orbit and each travels in the same direction. The nearest planet to the sun, Mercury, has the fastest orbital motion, 48 kilometers per second, and the shortest period of revolution, 88 days. By contrast, the most distant planet, Pluto, has an orbital speed of 5 kilometers per second and requires 248 years to complete one revolution.

Imagine a planet's orbit drawn on a flat sheet of paper. The paper represents the planet's *orbital plane*. The orbital planes of all nine planets lie within 3 degrees of the plane of the sun's equator, except for those of Mercury and Pluto, which are inclined 7 and 17 degrees, respectively.

THE PLANETS: AN OVERVIEW

Careful examination of Table 18.1 shows that the planets fall quite nicely into two groups: the **terrestrial** (Earthlike) **planets** (Mercury, Venus, Earth, and Mars) and the **Jovian** (Jupiterlike) **planets** (Jupiter, Saturn, Uranus, and Neptune). Pluto is not included in either category, because its great distance from Earth and its small size make this planet's true nature a mystery.

The most obvious difference between the terrestrial and the Jovian planets is their size (Figure 18.2). The largest terrestrial planet (Earth) has a diameter only one-quarter as great as the diameter of the smallest Jovian planet (Neptune), and its mass is only one-seventeenth as great. Hence, the Jovian planets are often called *giants*. Also, because of their relative locations, the four Jovian planets are referred to as the *outer planets*, while the terrestrial planets are called the *inner planets*. As we shall see, there appears to be a correlation between the positions of these planets and their sizes.

Other dimensions along which the two groups markedly differ include density, composition, and

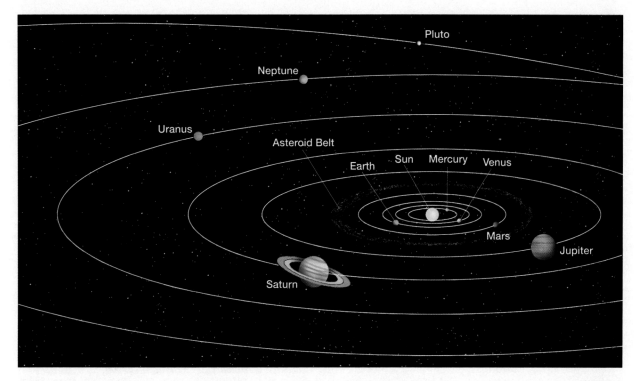

Figure 18.1
Orbits of the planets to scale.

rate of rotation. The densities of the terrestrial planets average about 5 times the density of water, whereas the Jovian planets have densities that average only 1.5 times that of water. One of the outer planets, Saturn, has a density only 0.7 that of water, which means that Saturn would float in water. Variations in the compositions of the planets are largely responsible for the density differences.

The substances that compose both groups of planets are divided into three groups—*gases, rocks,* and *ices*—based on their melting points.

1. The gases, hydrogen and helium, are those with melting points near absolute zero (0 Kelvin or −273°C), the lowest possible temperature.
2. The rocks are principally silicate minerals and metallic iron, which have melting points exceeding 700°C.
3. The ices have intermediate melting points (for example, H_2O has a melting point of 0°C) and include ammonia (NH_3), methane (CH_4), carbon dioxide (CO_2), and water (H_2O).

The terrestrial planets are mostly rocks: dense, rocky, and metallic material, with minor amounts of gases. The Jovian planets, on the other hand, contain a large percentage of gases (hydrogen and helium), with varying amounts of ices (mostly water, ammonia, and methane). This accounts for their low densities. (The outer planets may contain as much rocky and metallic material as the terrestrial planets, but this material would be concentrated in their small central cores.)

The Jovian planets have very thick atmospheres consisting of varying amounts of hydrogen, helium, methane, and ammonia. By comparison, the terrestrial planets have meager atmospheres at best. A planet's ability to retain an atmosphere depends on its temperature and mass. Simply stated, a gas molecule can "evaporate" from a planet if it reaches a speed known as the **escape velocity.** For Earth, this velocity is 11 kilometers (7 miles) per second. Any material, including a rocket, must reach this speed before it can leave Earth and go into space.

The Jovian planets, because of their greater surface gravities, have higher escape velocities (21–60 km/sec) than the terrestrial planets. Consequently, it is more difficult for gases to "evaporate" from them. Also, because the molecular motion of a gas is tem-

Table 18.1
Planetary data.

Planet	Symbol	Mean Distance from Sun: AU*	Mean Distance from Sun: Millions of Miles	Mean Distance from Sun: Millions of Kilometers	Period of Revolution	Inclination of Orbit	Orbital Velocity: mi/s	Orbital Velocity: km/s
Mercury	☿	0.387	36	58	88^{d}	7°00′	29.5	47.5
Venus	♀	0.723	67	108	224.7^{d}	3°24′	21.8	35.0
Earth	⊕	1.000	93	150	365.25^{d}	0°00′	18.5	29.8
Mars	♂	1.524	142	228	687^{d}	1°51′	14.9	24.1
Jupiter	♃	5.203	483	778	11.86^{yr}	1°18′	8.1	13.1
Saturn	♄	9.539	886	1427	29.46^{yr}	2°29′	6.0	9.6
Uranus	⛢	19.182	1783	2870	84^{yr}	0°46′	4.2	6.8
Neptune	♆	30.058	2794	4497	165^{yr}	1°46′	3.3	5.3
Pluto	♇	39.440	3666	5900	248^{yr}	17°12′	2.9	4.7

Planet	Period of Rotation	Diameter: Miles	Diameter: Kilometers	Relative Mass (Earth = 1)	Average Density (g/cm³)	Polar Flattening (%)	Eccentricity	Number of Known Satellites
Mercury	59^{d}	3015	4878	0.056	5.43	0.0	0.206	0
Venus	243^{d}	7526	12,104	0.82	5.24	0.0	0.007	0
Earth	$23^{h}56^{m}04^{s}$	7920	12,756	1.00	5.52	0.3	0.017	1
Mars	$24^{h}37^{m}23^{s}$	4216	6794	0.108	3.93	0.5	0.093	2
Jupiter	$9^{h}48^{m}$	88,700	143,884	317.869	1.32	6.7	0.048	16
Saturn	$10^{h}15^{m}$	75,000	120,536	95.143	0.70	10.4	0.056	21
Uranus	$17^{h}14^{m}$	29,000	51,118	14.560	1.25	2.3	0.047	15
Neptune	$16^{h}06^{m}$	28,900	50,530	17.207	1.77	1.8	.009	8
Pluto	6.4^{d}	~1500	2445	0.111	4.7	0.0	0.250	1

* AU = astronomical unit, Earth's mean distance from the sun.

perature-dependent, at the low temperatures of the Jovian planets even the lightest gases are unlikely to acquire the speed needed to escape.

On the other hand, a comparatively warm body with a small surface gravity, like our moon, is unable to hold even the heaviest gas and thus lacks an atmosphere. The slightly larger terrestrial planets of Earth, Venus, and Mars retain some heavy gases like carbon dioxide, but even their atmospheres make up only an infinitesimally small portion of their total mass.

It is hypothesized that the primordial cloud of dust and gas from which all the planets are thought to have condensed had a composition somewhat similar to that of Jupiter. However, unlike Jupiter, the terrestrial planets today are nearly void of light gases and ices. Were the terrestrial planets once much larger? Did they contain these materials but lose them because of their relative closeness to the sun? In the following section we will consider the evolutionary histories of these two diverse groups of planets in an attempt to answer these questions.

Figure 18.2
The planets drawn to scale.

ORIGIN OF THE SOLAR SYSTEM

The orderly nature of our solar system leads most astronomers to conclude that the planets formed at essentially the same time and from the same primordial material as the sun. This material formed a vast cloud of dust and gases called a *nebula*. The **nebular hypothesis** suggests that all bodies of the solar system formed from an enormous nebular cloud consisting of approximately 80 percent hydrogen, 15 percent helium, and a few percent of all the other heavier elements known to exist. The heavier substances in this frigid cloud of dust and gases consisted mostly of elements such as silicon, aluminum, iron, and calcium—the substances of today's common rocky materials. Also prevalent were other familiar elements, including oxygen, carbon, and nitrogen.

About 5 billion years ago, and for reasons not yet fully understood, this huge cloud of minute rocky fragments and gases began to contract under its own gravitational influence (Figure 18.3A). The contracting clump of material apparently had some rotational motion. As this slowly rotating cloud gravitationally contracted, it rotated faster and faster for the same reason ice skaters do when they draw their arms toward their bodies. This rotation caused the nebular cloud to assume a disk shape (Figure 18.3B). Within this rotating disk, relatively small contractions, like eddies in a stream, formed the nuclei from which the planets would eventually develop (Figure 18.3C). However, the greatest concentration of material was gravitationally pulled toward the center, forming the *protosun*.

As this gaseous cloud collapsed, the temperature of the central mass continued to increase. Nebular material near the protosun reached temperatures of several thousand degrees and was completely vaporized. However, at distances beyond the orbit of Mars, the temperatures probably remained very low, both then and now. Here, at −200°C, the dust fragments were most likely covered with a thick layer of ices made of water, carbon dioxide, ammonia, and methane. The disk-shaped cloud also contained appreciable amounts of the lighter gases, hydrogen and helium, which had not been consumed by the protosun.

In a relatively short time after the protosun formed, the temperature in the inner portion of the nebula dropped significantly. This temperature decrease caused those substances with high melting

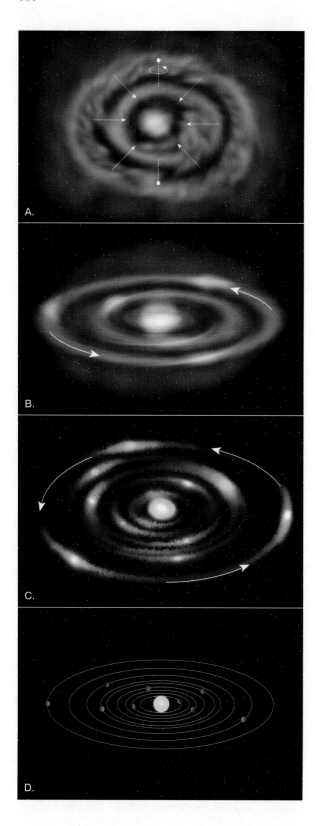

Figure 18.3
Nebular hypothesis. **A.** *A huge rotating cloud of dust and gases (nebula) begins to contract.* **B.** *Most of the material is gravitationally pulled toward the center, producing the sun. However, due to rotational motion, some dust and gases remain, orbiting the central body as a flattened disk.* **C.** *The planets begin to accrete from the material that is orbiting within the flattened disk.* **D.** *In time, most of the remaining debris collects into the nine planets and their moons, or is swept into space by the solar wind.*

points to condense into sand-sized particles. Materials such as iron and nickel solidified first. Next to condense were the elements of which the rock-forming minerals are composed—silicon, calcium, iron, and so forth. As these fragments collided, they joined into larger asteroid-sized objects which, in a few tens of millions of years, accreted into the four inner planets we call Mercury, Venus, Earth, and Mars (Figure 18.3D).

As more and more of the nebular debris was swept up by these *protoplanets*, the inner solar system began to clear, allowing solar radiation to pass through to heat the planets' surfaces. Due to their relatively high temperatures and weak gravitational fields, the inner planets were unable to accumulate much of the lighter components of this nebular cloud. These lighter components—hydrogen, ammonia, methane, and water—were eventually whisked from the inner solar system by the solar winds.

Shortly after the four terrestrial planets formed, the decay of radioactive isotopes within them, plus the heat from the colliding particles, produced at least some melting of the planets' interiors. Melting, in turn, allowed the heavier elements, principally iron and nickel, to sink toward the center, while the lighter silicate minerals floated upward.

During this period of *chemical differentiation*, gaseous materials escaped from the planets' interiors, much like what happens during a volcanic event on Earth. The hottest and second-smallest planet, Mercury, was unable to retain even the heaviest of these gases. Mars, on the other hand, being 40 percent larger and considerably cooler than Mercury, retained a thin layer of carbon dioxide and some water in the form of ice. The largest of the terrestrial planets, Venus and Earth, have surface gravitations strong enough to retain a substantial amount of the heavier gases. However, when compared to the four Jovian planets, even the atmospheres of these two terrestrial planets must be looked upon as meager at best.

At the same time that the terrestrial planets were forming, the larger Jovian planets, along with their extensive satellite systems, were also developing. However, because of the frigid temperatures existing far from the sun, the fragments from which these planets formed contained a high percentage of ices— water, carbon dioxide, ammonia, and methane. Perhaps by random chance, two of the outer planets, Jupiter and Saturn, grew many times larger (by mass) than Uranus and Neptune. (For comparison, Jupiter is 318 and Saturn is 95 times more massive than Earth. However, Uranus and Neptune have masses about 15 and 17 times greater, respectively, than Earth.)

When Jupiter and Saturn reached a certain size, estimated to be about 10 Earth masses, their surface gravitation was sufficient to attract and hold even the lightest materials—hydrogen and helium. It is thought that these gases gravitationally collapsed onto these large protoplanets as they swept through their region of the solar system. Thus, much of their size is attributable to the large envelope of light elements, which exists as a dense liquid below a thick hydrogen-rich atmosphere. Jupiter and Saturn therefore consist of a central core of ices and rock, and a much larger outer envelope containing mostly hydrogen and helium.

By contrast, the smaller Jovian planets, Uranus and Neptune, grew more slowly and contain proportionately much smaller amounts of hydrogen and helium. Nevertheless, hydrogen, methane, and ammonia are still the major constituents of their dense atmospheres. Perhaps a thin outer ocean of hydrogen exists on these planets as well. Thus, Uranus and Neptune are proposed to have a small rocky-iron core and a large mantle of water, ammonia, and methane surrounded by a thin ocean of liquid hydrogen. Consequently, these planets structurally resemble Jupiter and Saturn, but without their large hydrogen-helium envelopes.

In many respects, the development of the outer planets with their large satellite systems roughly parallels the events that formed the solar system as a whole. Like their parent planets, the satellites of the outer planets are composed primarily of icy materials with lesser amounts of rocky substances. However, because of their small size, they could not retain appreciable amounts of hydrogen and helium.

In the remainder of this chapter, we will consider each planet in more detail, as well as some minor members of the solar system. First, however, a discussion of our moon, Earth's companion in space, is appropriate.

THE MOON

Only one natural satellite, the moon, accompanies Earth on its annual journey around the sun. Although other planets have moons, our planet-satellite system is unique in the solar system, because Earth's moon is unusually large compared to its parent planet. The diameter of the moon is 3475 kilometers (2150 miles), about one-fourth of the Earth's 12,751 kilometers. From the calculation of the moon's mass, its density is 3.3 times that of water. This density is comparable to that of crustal rocks on Earth but is a fair amount less than the Earth's average density, which is 5.5 times that of water. Geologists have suggested that this difference can be accounted for if the moon's iron core is relatively small. The gravitational attraction at the lunar surface is one-sixth of that experienced on the Earth's surface (a 100-pound person on Earth weighs only 17 pounds on the moon). This difference allows an astronaut to lift a "heavy" life-support system with relative ease. If it were not necessary to carry such a load, an astronaut could jump six times higher than on earth.

The Lunar Surface

When Galileo first pointed his telescope toward the moon, he saw two different types of terrain (Figure 18.4). The dark areas he observed are now known to be fairly smooth lowlands, whereas the bright regions are densely cratered highlands (Figure 18.5). Because the dark regions resembled seas on Earth, they were later named **maria** (singular, *mare:* Latin for "sea"). This name is unfortunate because the moon's surface is totally void of water.

Today we know that the moon has no atmosphere and lacks water as well. Therefore, the processes of weathering and erosion that continually modify the Earth's surface was virtually lacking on the moon. In addition, tectonic forces are not active on the moon, so events such as earthquakes and volcanic eruptions no longer occur. However, because the moon is un-

Figure 18.4
Telescopic view from Earth of the lunar surface. (Courtesy of Lick Observatory)

protected by an atmosphere, a different kind of erosion occurs: tiny particles from space (micrometeorites) continually bombard its surface and ever so gradually smooth the landscape. Rocks, for example, can become slightly rounded on top if exposed at the lunar surface for a long enough period. Nevertheless, it is unlikely that the moon has changed appreciably in the last 3 billion years, except for a few craters created by large meteorites.

Craters. The most obvious features of the lunar surface are craters. They are so profuse that craters-within-craters-within-craters are the rule! The larger ones in the lower portion of Figure 18.4 are about 250 kilometers (155 miles) in diameter, roughly the width of Indiana. Most craters were produced by the impact of rapidly moving debris (meteoroids), a phenomenon that was considerably more common in the early history of the solar system than it is today.

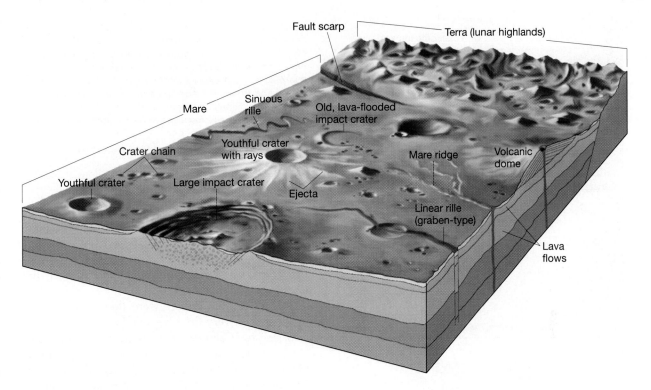

Figure 18.5
Block diagram illustrating major topographic features on the lunar surface.

By contrast, Earth has only about a dozen easily recognized impact craters. This difference can be attributed to the Earth's atmosphere. Friction with the air burns up small debris before it reaches the ground. In addition, evidence for most of the craters that formed early in the Earth's history has been obliterated by erosion or tectonic processes.

The formation of an impact crater is illustrated in Figure 18.6. Upon impact, the high-speed meteoroid compresses the material it strikes, then almost instantaneously the compressed rock rebounds, ejecting material from the crater. This process is analogous to the splash that occurs when a rock is dropped into water, and it often results in the formation of a central peak, as seen in the large crater in Figure 18.7. Most of the ejected material *(ejecta)* lands near the crater, building a rim around it. The heat generated by the impact is sufficient to melt some of the impacted rock. Astronauts have brought back samples of glass beads produced in this manner, as well as rock formed when angular fragments and dust were welded together by the impact. The latter material is called **lunar breccia.**

A meteoroid only 3 meters (10 feet) in diameter can blast out a 150-meter- (500-foot-) wide crater.

A few of the large craters such as Kepler and Copernicus, shown in Figure 18.4, formed from the impact of bodies 1 kilometer, or more, in diameter. These two large craters are thought to be relatively young because of the bright *rays* ("splash" marks) that radiate outward for hundreds of kilometers. These bright rays consist of fine debris ejected from the primary crater, including impact-generated glass beads, as well as material displaced during the formation of smaller, secondary craters.

Highlands. Densely pockmarked highland areas make up most of the lunar surface. In fact, all of the "back" side of the moon is characterized by such topography. Within the highland regions are mountain ranges that have been named for mountainous terrains on Earth. The highest lunar peaks reach elevations approaching 8 kilometers, only 1 kilometer lower than Mount Everest.

Maria. Although highlands predominate, the less-rugged maria have attracted a great deal of interest. The origin of maria basins as enormous impact craters produced by the violent impact of at least a dozen asteroid-sized bodies was hypothesized before

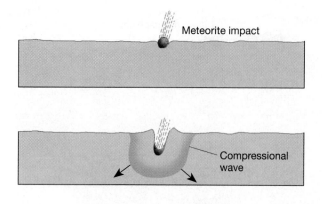

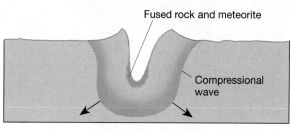

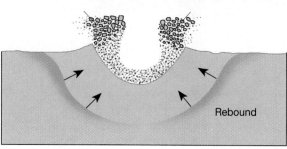

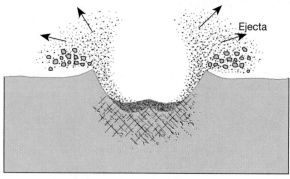

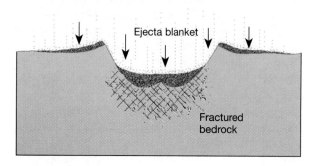

Figure 18.6
Formation of an impact crater. The energy of the rapidly moving meteoroid is transformed into heat energy and compressional waves. The rebound of the compressed rock causes debris to be ejected from the crater, and the heat melts some material, producing glass beads. Small secondary craters are formed by the material "splashed" from the impact crater. (After E. M. Shoemaker)

the turn of the century by the noted American geologist G. K. Gilbert (Figure 18.8A). However, it remained for the *Apollo* missions to determine what filled these depressions to produce the relatively flat topography (Figure 18.8B).

Apparently the craters were flooded with layer upon layer of very fluid basaltic lava somewhat resembling the Columbia Plateau in the northwestern United States (see Figure 7.20). The photograph in Figure 18.9 reveals the forward edge of one of the lava flows that filled the Imbrium basin. Astronauts have also viewed and photographed the layered nature of maria. The lava flows are often over 30 meters (100 feet) thick, and the total thickness of the material that fills the maria must approach thousands of meters.

On several occasions the lava overflowed the impact crater, engulfing the surrounding lowlands. Where the rim of a remnant crater can be seen above the lava, an estimate of the flow's thickness can be made. Examples of basins in which all of the disturbed regions were filled to overflowing include Mare Tranquillitatis (Sea of Tranquility), the site where astronaut Neil Armstrong was the first person to step onto the lunar surface, and Mare Imbrium (Sea of Rains). Some maria basalts fill only the central crater; these appear as dark, smooth crater floors in Figure 18.4.

Regolith.   All lunar terrains are mantled with a layer of gray, unconsolidated debris derived from a few billion years of meteoric bombardment (Figure 18.10). This soil-like layer, properly called **lunar regolith,** is composed of igneous rocks, breccia, glass beads, and fine particles commonly called *lunar dust.* As meteoroid after meteoroid collided with the lunar surface, the thickness of the lunar regolith increased, while the size of the bombarded debris diminished. In the maria that have been explored by *Apollo* as-

Crater ray

Secondary crater chain

Central peak

Continuous ejecta

Discontinuous ejecta

Figure 18.7
The 20-kilometer-wide lunar crater Euler in the southwestern part of Mare Imbrium. Clearly visible are the bright rays, central peak, secondary craters, and the large accumulation of ejecta near the crater rim. (Courtesy of NASA)

tronauts, the lunar regolith is apparently just over 3 meters (10 feet) thick, but it is believed to form a thicker mantle upon the older highlands.

Lunar History

Although the moon is our nearest planetary neighbor and astronauts have sampled its surface, much is still unknown about its origin. Until recently, most scientists argued that the moon, Earth, and the other planets formed together. That is, the moon consolidated from minute rock fragments and gases in a disk-shaped structure that orbited the protosun. Debris from this disk collided and accumulated by gravity into larger masses which, in turn, accreted into planetary-sized bodies.

A new hypothesis supported by many scientists suggests that a giant asteroid collided with Earth to produce the moon. The explosion caused by the impact of a Mars-sized body upon a semimolten Earth is thought to have ejected huge quantities of mantle rock from the primordial Earth. A portion of this ejecta remained in orbit around the Earth, gradually accumulating to form the moon. This giant-impact hypothesis is plausible, but raises many questions.

Despite the moon's uncertain origin, planetary geologists have worked out basic details of the moon's history. One method used in lunar study is to observe variations in crater density (quantity per unit area). Simply stated, the greater the crater density, the longer the topographic feature must have existed. From evidence such as this, scientists conclude that the moon evolved in three phases: formation of the original crust (highlands), maria basins, and rayed craters.

Figure 18.8
Formation of lunar maria.
A. *Impact of an asteroid-sized mass produced a huge crater hundreds of kilometers in diameter and disturbed the lunar crust far beyond the crater.* ***B.*** *Filling of the impact area with fluid basalts, perhaps derived from partial melting deep within the lunar mantle.*

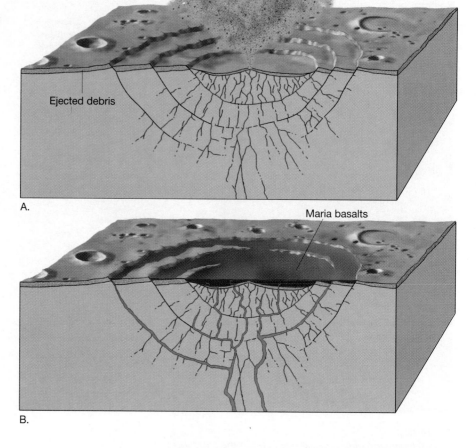

Ejected debris

A.

Maria basalts

B.

Original Crust (Highlands).

During its early history the moon was continually impacted as it swept up debris. This continuous bombardment, and perhaps radioactive decay, generated enough heat to melt the moon's outer shell, and quite possibly the interior as well.

When a large percentage of the debris had been gathered, the outer layer of the moon began to cool and form a crystalline crust. From samples obtained by *Apollo* astronauts, rocks of the primitive lunar crust are thought to be largely a calcium-rich feldspar (anorthite). This feldspar mineral crystallized early and, because it was less dense than the remaining melt, floated to the top and formed a surface scum. While this process was taking place, iron and other heavy metals probably sank to form a small central core. Even after the crust had solidified, the moon's surface was continually bombarded.

Remnants of the original crust occupy the densely cratered highlands, which have been estimated to be

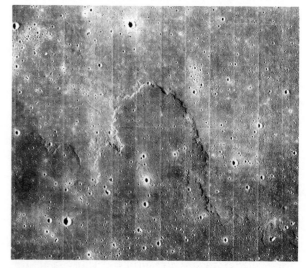

Figure 18.9
Margin of a lava flow on the surface of Mare Imbrium. Note how subsequent meteoroid impacts have cratered the basalt flows. (Courtesy of National Space Data Center)

Figure 18.10
Astronaut Harrison Schmitt sampling the lunar surface. Notice the footprints (insert) in the lunar "soil." (Courtesy of NASA)

as much as 4.5 billion years old—about the same age as Earth.

Maria Basins. The next major event in the moon's evolution was the formation of maria basins (see Figure 18.8). The meteoroids that produced these huge pits ejected mountainous quantities of lunar rock into piles rising 5 kilometers or more. The lunar Apennine mountain range, which typifies such an accumulation, was produced in conjunction with the formation of the Imbrium basin, the site explored by the *Apollo 15* astronauts. The crater density in the ejected material is greater than that of the surface of the associated mare. This confirms that an appreciable time span elapsed between the formation of these basins and their filling with lava. Radiometric dating of the maria basalts puts their age between 3.2 and 3.8 billion years, roughly a billion years younger than the initial crust. In places, the lava flows overlap the highlands, another testimonial to the younger age of the maria deposits.

Rayed Craters. The last prominent features to form on the lunar surface were the rayed craters, as exemplified by the crater Copernicus (see Figure 18.4). Material ejected from these "young" depressions is clearly seen blanketing the surface of the maria and many older rayless craters. By contrast, the older craters have rounded rims, and their rays have been erased by the impact of small debris. However, even a relatively young crater like Copernicus must be millions of years old. Had it formed on the Earth, erosional forces would have long since obliterated it.

Evidence carried back from the lunar landings indicates that most, if not all, of the moon's tectonic activity ceased about 3 billion years ago. The youngest maria lava flows are almost as old as the oldest rocks so far discovered on the Earth. If photos of the moon taken several hundreds of millions of years ago were available, they would reveal that the moon has changed little in the intervening years. By all standards of measure, the moon is a dead body wandering through space and time.

THE PLANETS: AN INVENTORY
Mercury: The Innermost Planet

Mercury, the innermost and smallest planet, has a diameter of 4878 kilometers. It is hardly larger than Earth's moon and is smaller than three other moons in the solar system. Also, like the moon, it absorbs most of the sunlight that strikes it, reflecting only 6 percent into space. This is characteristic of terrestrial bodies that have no atmosphere. (By way of comparison, the Earth reflects about 30 percent of the light that strikes it, most of it from clouds.)

Mercury's close proximity to the sun makes viewing from earthbound telescopes difficult at best. The first good glimpse of this planet came in the spring of 1974, when *Mariner 10* passed within 800 kilometers (500 miles) of its surface (Table 18.2). Its striking resemblance to the moon was immediately evident from the high-resolution images that were

Figure 18.11
Photomosaic of Mercury. This view of Mercury is remarkably similar to the "far side" of the moon. (Courtesy of NASA)

Table 18.2
Most significant space probes and planned space probes.

Mariner 2	1962	Fly-by of Venus (first to any planet)
Mariner 4, 6, 7	1965 1969 1969	Fly-by missions to Mars
Mariner 9	1971	Mars orbiter
Apollo 8	1968	Astronauts circled the moon and returned to the earth
Apollo 11	1969	First astronaut landed on the moon
Apollo 17	1972	Last of six manned *Apollo* missions to the moon
Mariner 10	1974	Orbited the sun, passing Mercury several times; fly-by of Venus
Pioneer 10, 11	1973 1974	First close-up views of Jupiter
Venera 8, 9, 10	1972 1975	Soviet landers on Venus (each survived about one hour)
Venera 13, 14	1982	First color images of Venus
Viking 1, 2	1976	Mars orbiters and landers
Voyager 1	1979 1980	Fly-by of Jupiter Fly-by of Saturn
Voyager 2	1979 1981 1986 1989	Fly-by of Jupiter Fly-by of Saturn Fly-by of Uranus Fly-by of Neptune
Giotto	1986	First photo of comet's nucleus
Galileo	1990 1992 1995	Fly-by of Earth, Venus Fly-by of Earth, moon Jupiter orbiter
Magellan	1990	Radar imaging orbiter of Venus
Mars Observer	1993	Mars orbiter

radioed back. In fact, the similarity was so great that a project scientist remarked that these photos could be substituted for ones of the back side of the moon and most nonscientists would not know the difference (Figure 18.11).

Mercury has cratered highlands, much like the moon, and vast smooth terrains which resemble maria. However, unlike the moon, Mercury is a very dense planet, which implies that it contains an iron core, perhaps larger than the Earth's. Also, Mercury has very long scarps that cut across the plains and craters alike. One proposal is that these scarps resulted from crustal shortening as the planet cooled and shrank.

Mercury revolves quickly but rotates slowly. The time from one sunrise to the next on Mercury is 179 Earth days. Thus, a night on Mercury lasts for about three months and is followed by three months of daylight. The night temperatures drop as low as $-173°C$ ($-280°F$) and the noontime temperatures exceed $427°C$ ($800°F$), hot enough to melt tin and lead. Mercury has the greatest temperature extremes of any planet. The odds of life as we know it existing on Mercury are nil.

Venus: The Veiled Planet

Venus, second only to the moon in brilliance in the night sky, is named for the goddess of love and beauty. It orbits the sun in a nearly perfect circle once every 225 days. Venus is similar to Earth in size, density, and mass, and is also located in the inner portion of the solar system. Thus, it has been referred to as "Earth's twin." Due to these similarities, it is hoped that a detailed study of Venus will provide geologists with a better understanding of Earth's evolutionary history.

Venus is shrouded in thick clouds impenetrable to visible light. Nevertheless, radar mapping by unmanned spacecraft, as well as by earthbound instruments, have revealed a varied topography with features somewhat between those of Earth and Mars. Simply, radar pulses in the microwave range are sent toward the Venusian surface, and the heights of features such as plateaus and mountains are measured by timing the return of the radar echo. In 1990, *Magellan* scientists began receiving crisp radar images with a resolution that permits structures the size of Mount St. Helens to be clearly defined. These new data have confirmed earlier proposals that basaltic volcanism and tectonic deformation are the dominant processes operating on Venus. Further, based on the low density of impact craters, researchers concluded that volcanism and tectonic deformation were also very active during the recent geologic past (Figure 18.12).

Over 80 percent of the Venusian surface consists of relatively subdued plains that are mantled by volcanic flows. Some lava channels extend for tens to

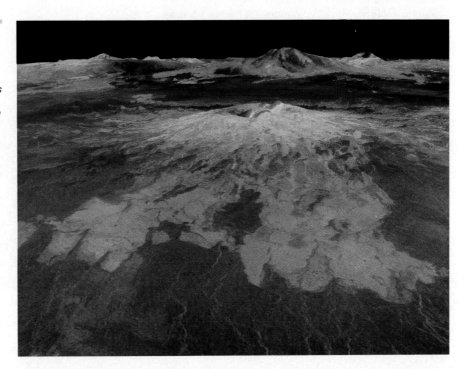

Figure 18.12
Sapas Mons, a volcanic cone, is seen in the center of this computer-generated view of Venus. Light-colored lava flows extending for hundreds of kilometers across the fractured plains are seen in the foreground, and Maat Mons, a large volcano, is located on the horizon. (Courtesy of NASA)

Figure 18.13
These domelike volcanic structures on Venus average 25 kilometers in diameter and are less than 1 kilometer high. They are interpreted as very thick lava flows. (Photo courtesy of Jet Propulsion Laboratory)

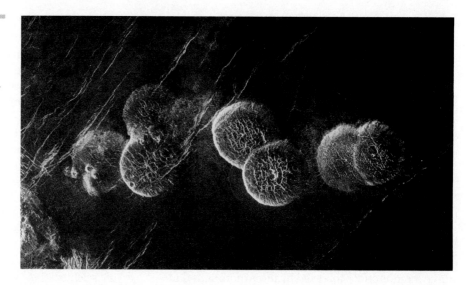

hundreds of kilometers and have narrow widths of from 1 to 2 kilometers. One of these lava channels meanders 6800 kilometers across the planet. Thousands of volcanic structures have already been identified. Most are small shield volcanoes a few hundred meters high with diameters of 2 to 8 kilometers. In addition, over 1500 volcanoes with diameters in excess of 20 kilometers have also been mapped. One of these, Sapas Mons, is approximately 400 kilometers (249 miles) across and 1.5 kilometers (0.9 mile) high. Many of the flows from this volcano erupted from the flanks rather than the summit. This eruptive style is shared by large shield volcanoes on Earth, such as those found in Hawaii. Also like the Hawaiian volcanoes, these large structures appear to be associated with the upwelling of hot plumes within the planet's interior. Other volcanic structures discovered on Venus are circular, pancake-shaped domes about 25 kilometers (15 miles) in diameter and nearly 1 kilometer high (Figure 18.13). These domes are thought to be the result of outpouring of very viscous lava, much like volcanic domes on Earth.

Only 8 percent of the Venusian surface consists of highlands that may be likened to continental areas on Earth. The highlands of Venus are dominated by large plateau-like structures; however, discrete mountain ranges, including Maxwell Montes, which rises 11 kilometers above the lowlands, also exist.

Tectonic activity on Venus seems to be driven by upwelling and downwelling of material in the planet's interior. Parallel sets of faults or fractures, as shown in Figure 18.14, are widespread on the plains. However, no evidence for appreciable horizontal displacement, as occurs along the San Andreas

fault, has yet to be detected. Some fracture zones are associated with numerous volcanic structures, suggesting that the region may have been uplifted and stretched as a result of convective upwelling. The greatest deformation on Venus is seen in the highlands, which reach elevations in excess of 10 kilometers. One hypothesis proposes that the mountain areas formed where convective currents crumpled and thickened the crust. Although mantle convection still operates on Venus, the process of plate tectonics

Figure 18.14
Two sets of parallel fractures indicating tectonic activity on Venus unlike any seen before in the solar system. (Photo courtesy of Jet Propulsion Laboratory)

which recycles rigid lithosphere does not appear to have contributed to the present Venusian topography.

Before the advent of space vehicles, Venus and Mars were considered the most hospitable sites in the solar system (Earth excluded) for living organisms. However, evidence from *Mariner* fly-by space probes and Russian *Venera* landings on Venus indicates differently. The surface of Venus reaches temperatures of 475°C (900°F), and the Venusian atmosphere is 97 percent carbon dioxide. Only minor amounts of water vapor and nitrogen have been detected. Its atmosphere contains an opaque cloud deck about 25 kilometers thick, which begins approximately 70 kilometers from the surface. Although the unmanned *Venera 8* survived less than one hour on the Venusian surface, it determined that the atmospheric pressure on that planet is 90 times that on Earth's surface. This hostile environment makes it unlikely that life as we know it exists on Venus and makes crewed space flights to Venus improbable (see Box 18.1).

Mars: The Red Planet

Mars has evoked greater interest than any other planet, for astronomers as well as for nonscientists. When we think of intelligent life on other worlds, "little green Martians" may come to mind. Interest in Mars stems mainly from this planet's accessibility to observation. All other planets within telescopic range

have their surfaces hidden by clouds, except for Mercury, whose nearness to the sun makes viewing difficult. Through the telescope, Mars appears as a reddish ball interrupted by some permanent dark regions that change intensity during the Martian year. The most prominent telescopic features of Mars are its brilliant white polar caps, resembling Earth's.

The Martian Atmosphere. The Martian atmosphere is only 1 percent as dense as that of Earth. It is primarily carbon dioxide with very small amounts of water vapor. Data radioed back to Earth from Mars probes confirmed speculations that the polar caps of Mars are made of water ice, covered by a relatively thin layer of frozen carbon dioxide. As the winter nears in either hemisphere, we see the equatorward growth of that hemisphere's ice cap as additional carbon dioxide is deposited. This finding is compatible with the temperatures observed at the polar caps. Here temperatures reach a frigid −125°C (−193°F), cold enough to solidify carbon dioxide.

Although the atmosphere of Mars is very thin, extensive dust storms do occur and may be responsible for the color changes observed from Earth-based telescopes. Winds up to 270 kilometers (170 miles) per hour can persist for weeks. Images radioed back by *Viking 1* and *2* revealed a Martian landscape remarkably similar to a rocky desert on Earth (Figure 18.15). Sand dunes are abundant, and many Mar-

Figure 18.15
This spectacular picture of the Martian landscape by the Viking 1 *lander shows a dune field with features remarkably similar to many seen in the deserts of Earth. The dune crests indicate that recent wind storms were capable of moving sand over the dunes in the direction from lower right to upper left. The large boulder at the left is about 10 meters from the spacecraft and measures 1 by 3 meters. (Photo courtesy of NASA)*

BOX 18.1

Venus and the Runaway Greenhouse Effect: A Lesson for Planet Earth?

Carl Sagan, one of the foremost experts on extraterrestrial life, called Earth "the Heaven of the solar system" and Venus "the Hell." Why should the environments of two planets that are nearly the same size and are located in close proximity to one another be so dramatically different? The primary reason is the blistering temperature of the Venusian atmosphere, which is caused by a runaway greenhouse effect.

The greenhouse effect occurs when a planet's atmosphere acts like the glass in a greenhouse and allows visible solar energy to penetrate to the surface. Also, like the glass in the greenhouse, certain gases in the planet's atmosphere are quite opaque to outgoing radiation. Thus, as in a greenhouse, the surface temperatures of these planets are warmer than expected because the heat is temporarily trapped.

Since both Earth and Venus experience a greenhouse effect, why do the surface temperatures of Venus reach 475°C (900°F) when Earth is very hospitable? The answer is that carbon dioxide, which is a main contributor to the greenhouse effect, makes up 97 percent of the Venusian at-

mosphere (Figure 18.A). The primary source of carbon dioxide is outgassing during volcanic eruptions. Why does Earth, which has numerous active volcanoes, have an atmosphere that consists of only a small fraction of 1 percent carbon dioxide? The answer can be found in Earth's abundant plant life.

Plants use carbon dioxide and water in photosynthesis to generate organic matter, while oxygen is released as a byproduct. Thus, over millions of years, plant life has altered our atmosphere, making it carbon dioxide-poor and oxygen-rich. In addition, carbon dioxide dissolved in seawater is used by a vast

Figure 18.A
Venus is shrouded in a hot, cloud-filled atmosphere composed mainly of carbon dioxide. (Photo courtesy of NASA)

number of organisms for the production of carbonate shells. These shells are eventually deposited as sediment on the ocean floor. Consequently, huge quantities of carbon dioxide from Earth's atmosphere are continually being converted into organic matter and carbonate sediments. Apparently, Venus does not possess living organisms or a chemical mechanism capable of removing atmospheric carbon dioxide. Hence, our neighboring planet has a carbon dioxide concentration that has reached extreme proportions.

The discovery of a hostile Venusian atmosphere provides a good example of the practical benefits that people derive from space exploration. These data clearly reveal what can happen to a planet's environment when the greenhouse effect runs rampant. Some climatologists believe that a similar situation may be in the making on Earth. Through the burning of fossil fuels, we are converting increasingly larger amounts of oxygen and organic materials into carbon dioxide and water. In this manner, we are altering Earth's atmosphere by rapidly undoing what plant life has taken millions of years to accomplish.

We expect that further studies of the Venusian atmosphere will provide insights into how Earth's atmosphere evolved and how it will respond to changes induced by human activity. Such insights could keep our planet from becoming, as some have suggested, "hothouse Earth."

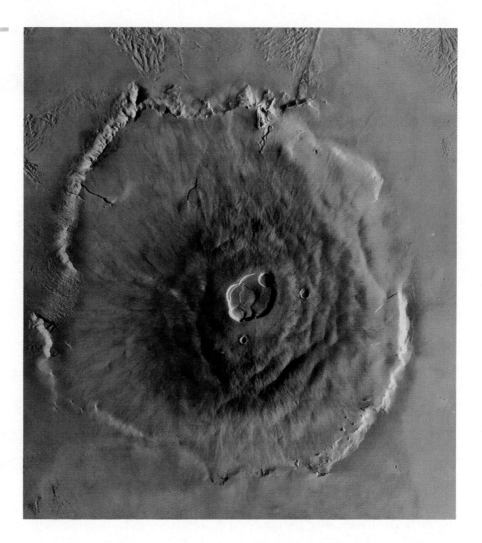

Figure 18.16
Image of Mons Olympus, an inactive shield volcano on Mars that covers an area about the size of the state of Ohio. (Courtesy of the U.S. Geological Survey)

tian impact craters have flat bottoms because they are partially filled with dust. Thus, unlike the craters of the moon, the oldest craters on Mars might be completely obscured by deposits of windblown material.

Mars's Dramatic Surface. *Mariner 9*, the first artificial satellite to orbit another planet, reached Mars in 1971. It witnessed a large dust storm raging, and only the ice caps were initially observable. It was spring in Mars's southern hemisphere, and that polar cap was receding rapidly. About mid-summer, when the rate of evaporation should have been greatest, the size of the polar cap showed little change. A residual cap remained through the summer, providing strong evidence that the residual polar cap is made of water ice. (Water ice has a much lower rate of evaporation than frozen carbon dioxide.)

When the dust cleared, images of Mars's northern hemisphere revealed numerous large volcanoes. The largest, Mons Olympus, covers an area the size of Ohio and is no less than 23 kilometers (75,000 feet) in elevation. This gigantic volcano and others resemble the shield volcanoes found on Earth, like those of Hawaii (Figure 18.16). Their extreme size is thought to result from the absence of plate movements on Mars. Therefore, rather than a chain of volcanoes forming as we find in Hawaii, single, larger cones developed.

Impact craters are notably less abundant in the region where the volcanic activity appears most prevalent. This indicates that at least some of the volcanic topography formed more recently in that planet's history, somewhat after the early period of heavy bombardment. Nevertheless, age determinations based on crater densities obtained from *Viking*

Figure 18.17
This image shows the entire Valles Marineris canyon system, over 2000 kilometers long and up to 8 kilometers deep. The dark red spots on the left edge of the image are huge volcanoes, each about 25 kilometers high. (Courtesy of U.S. Geological Survey)

images indicate that most Martian surface features are old by Earth standards. The highly cratered Martian southern hemisphere is probably similar in age to comparable lunar highlands (3.5–4.5 billion years old). The discovery of several highly cratered and weathered volcanoes on Mars further indicates that volcanic activity began early and had a long history. However, even the relatively fresh-appearing volcanic features of the northern hemisphere may be older than 1 billion years. This fact, coupled with the absence of "marsquake" recordings by *Viking* seismographs, points to a tectonically dead planet.

Another surprising find made by *Mariner 9* was the existence of several canyons that dwarf even Earth's Grand Canyon of the Colorado River. One of the largest, Valles Marineris, is roughly 6 kilometers deep, up to 160 kilometers wide, and extends for almost 5000 kilometers along the Martian equator (Figure 18.17). This vast chasm is thought to have formed by slippage of crustal material along huge faults in the crustal layer. In this respect, it would be comparable to the rift valleys of Africa.

Water on Mars? Not all valleys on Mars have a tectonic origin. Many Martian valleys have tributaries exhibiting drainage patterns similar to those of stream valleys found on Earth. In addition, *Viking* orbiter images have revealed streamlined features that are unmistakably ancient islands located in what is now a dry stream bed. When these streamlike channels were first discovered, some observers speculated that a thick water-laden atmosphere capable of generating torrential downpours once existed on Mars. What happened to this water? The present Martian atmosphere contains only traces of water. Moreover, the environment of Mars is far too harsh to allow water to exist as a liquid. Despite these difficulties, the work of flowing water still remains the most acceptable explanation for many of the Martian channels.

Many planetary geologists do not accept the premise that Mars once had an active water cycle similar to that on Earth. Rather, they believe that many of the large streamlike valleys were created by the collapse of surface material caused by the slow melting of subsurface ice (Figure 18.18). If this is the

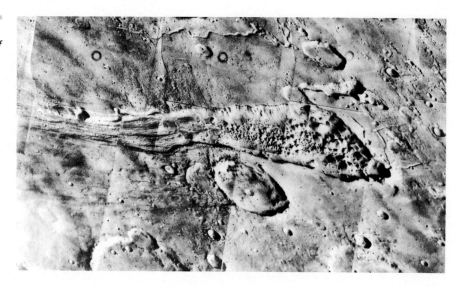

case, these large valleys would be more akin to features formed by mass wasting processes on Earth. Some of the smaller channels, they contend, have been cut by the gradual release of subsurface water, which flowed out of the ground like springs. Further, the heat from volcanic activity or meteoroid impact may have rapidly melted subsurface ice, which, in turn, caused local flooding. These floods may have carved some of the channels. An occasional flood would also account for the streamlined features found in what are presently dry river beds. The question of which phenomenon created the branching valleys of Mars—rainfall, gradual seepage of groundwater, or the collapse of surface material—will be debated for some time. The answer might not come until we obtain images of Mars that reveal greater detail than those presently available.

The Martian Satellites. Due to their small size, Phobos and Deimos, the two satellites of Mars, were not discovered until 1877. Phobos is nearer to its parent than any other natural satellite in the solar system. Only 5500 kilometers from the Martian surface, Phobos requires just 7 hours and 39 minutes for one revolution. Deimos, which is smaller and 20,000 kilometers away from Mars, revolves in 30 hours and 18 minutes. *Mariner 9* revealed that both satellites are irregularly shaped and have numerous impact craters, much like their parent (Figure 18.19).

The maximum diameter of Phobos is 24 kilometers and the maximum diameter of Deimos is only about 15 kilometers. Undoubtedly, these moons are asteroids which were captured by Mars. One of the most interesting coincidences in astronomy is the close resemblance between Phobos and Deimos and the two fictional satellites of Mars described by Jonathan Swift in *Gulliver's Travels*, written about 150 years before these satellites were actually discovered.

Jupiter: The Lord of the Heavens

Jupiter, the largest planet in our solar system, has a mass 2½ times greater than the combined mass of all the remaining planets, satellites, and asteroids. It is truly a giant among planets. In fact, had Jupiter been about ten times larger, it would have evolved into a small star. Despite its great size, however, it is only ⅟₈₀₀ as massive as the sun. Jupiter also rotates more rapidly than any other planet, completing one

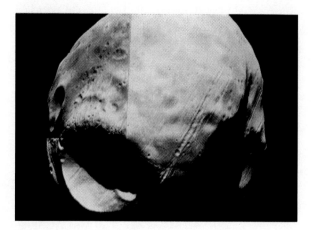

Figure 18.19
Phobos, the larger and innermost of the two Martian satellites. (Courtesy of NASA)

Figure 18.20
Jupiter with two of its largest satellites as seen from Voyager I. Io is seen against the Great Red Spot and Europa appears against a white oval. (Courtesy of NASA)

rotation in slightly less than ten hours. The effect of Jupiter's rapid rotation causes the equatorial region to be slightly bulged and the polar dimension to be flattened (see "Polar Flattening" column in Table 18.1).

When viewed through a telescope or binoculars, Jupiter appears to be covered with alternating bands of multicolored clouds aligned parallel to its equator (Figure 18.20). The most striking feature on the disk of Jupiter is the *Great Red Spot* in its southern hemisphere (Figure 18.21). Although its color varies greatly in intensity, the Great Red Spot has been a prominent feature since it was first discovered more than three centuries ago. When *Voyager 2* swept by Jupiter in July 1979, the size of the red spot was 11,000 kilometers by 22,000 kilometers, which equals two Earth-sized circles placed side by side. On occasion it has grown even larger. Although the Great Red Spot varies in size, it does remain the same distance from the Jovian equator.

The cause of the spot has been attributed to everything from volcanic activity to a large cyclonic storm. Images obtained by *Pioneer 11* as it moved to within 42,000 kilometers of Jupiter's cloud tops in December 1974, support the latter view. The Great Red Spot apparently is a counterclockwise-rotating storm caught between two jetstream-like bands of atmosphere flowing in opposite directions. This huge hurricane-like storm rotates once every 12 Earth days. Although several smaller storms have been observed in other regions of Jupiter's atmosphere, none has survived for more than a few days.

Structure of Jupiter. Jupiter's atmosphere is composed primarily of hydrogen and helium, with methane, ammonia, water, and sulfur compounds as minor constituents. The wind systems generate the light- and dark-colored bands that encircle this giant. Unlike the winds on Earth, which are driven by solar energy, Jupiter gives off nearly twice as much heat as it receives from the sun. Thus, it is the heat emanating from Jupiter's interior that produces huge convection currents in the atmosphere (Figure 18.22).

The areas of light-colored clouds *(zones)* are regions where gases are ascending and cooling. The light color is thought to be produced by ammonia crystallizing into "snowflakes" near the cloud tops, where temperatures of −120°C have been measured. The cold material from the light zones spills over onto the lower dark regions *(belts)*, where air is descending and heating. The reason for the rusty-brown color of the dark belts is unknown; they may be colored by sulfur compounds.

Atmospheric pressure at the top of the clouds is equal to sea-level pressure on Earth. Because of Jupiter's immense gravity, the pressure increases rapidly toward its surface. At 1000 kilometers below the clouds, the pressure is great enough to compress hydrogen gas into a liquid. Consequently, the surface of Jupiter is thought to be a gigantic ocean of liquid hydrogen. Less than halfway into Jupiter's interior,

Figure 18.21
The Great Red Spot. This Voyager *image shows the turbulent atmosphere and rotating storm that has raged on Jupiter for at least 300 years.*

Figure 18.22
The structure of Jupiter's atmosphere. The areas of light clouds (zones) are regions where gases are ascending and cooling. Sinking dominates the flow in the darker cloud layers (belts). This convective circulation, along with the rapid rotation of the planet, generates the high-speed winds observed between the belts and zones.

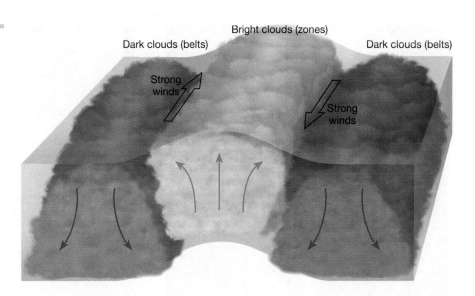

Bright clouds (zones)
Dark clouds (belts)   Dark clouds (belts)
Strong winds
Strong winds

pressures of unimaginable magnitude cause the liquid hydrogen to turn into *liquid metallic* hydrogen. Although scientists have never seen this unusual material, its properties are thought to be predictable.

Jupiter is also believed to contain as much rocky and metallic material as is found in the terrestrial planets, such as Earth. Whether this "earthly" material is scattered as small bits or makes up a central core is not known with certainty, but the latter idea is generally accepted.

Jupiter's Moons. Jupiter's satellite system, consisting of at least sixteen moons, resembles a miniature solar system. The four largest satellites were discovered by Galileo and travel in nearly circular orbits around the parent with periods of from two to seventeen days (see Figure 18.20). The two largest Galilean satellites, Callisto and Ganymede, surpass Mercury in size, while the two smaller ones, Europa and Io, are about the size of Earth's moon. These Galilean moons can be observed with a small telescope and are interesting in their own right. Because their orbits are along Jupiter's equatorial plane, and because they all have the same orbital direction, these moons most probably formed from "leftover" debris in much the same way as the planets did.

By contrast, Jupiter's four outermost satellites are very small (20 kilometers in diameter), revolve in a direction that is opposite the other moons, and have orbits that are steeply inclined to the Jovian equator. These satellites appear to be asteroids that passed near enough to be captured gravitationally by Jupiter.

The images obtained by *Voyagers 1* and *2* in 1979 revealed, to the surprise of almost everyone, that each of the four Galilean satellites has a character all its own. The entire surface of Callisto, the outermost of the Galilean satellites, is densely cratered, much like the surfaces of Mercury and the moon (Figure 18.23A). However, on Callisto, the impacts occurred in a crust that appears to be a dirty, frozen ocean of water ice. The largest features discovered are sets of bright concentric rings surrounding impact craters. These bright rings are thought to resemble the ripples produced when a pebble is dropped into calm water. They are probably composed of blocks of ice that were deformed and uplifted by the impact that generated the central crater.

Europa, smallest of the Galilean satellites, has an icy surface that is crisscrossed by many linear features (Figure 18.23B). Although some of these linear markings are thousands of kilometers in length, they have very little surface expression. Further, this satellite is notably void of large impact craters. Therefore, the present surface of Europa must have formed sometime after the early period of bombardment, when rocky chunks were far more numerous in the solar system. The crust of Europa may be a thick, frozen ice layer which caps a slushy ocean.

Ganymede, the largest Jovian satellite, contains the most diverse terrain. Like our moon, it has densely cratered regions and other very smooth areas where a younger icy layer covers the older cratered surface (Figure 18.23C). In addition, Ganymede has numerous parallel grooves. This suggests some type of tec-

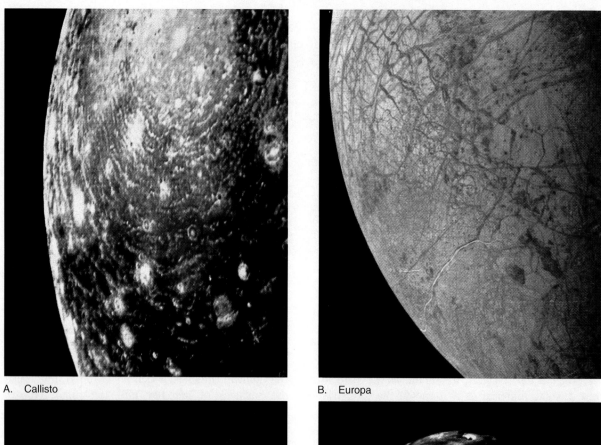

A.   Callisto

B.   Europa

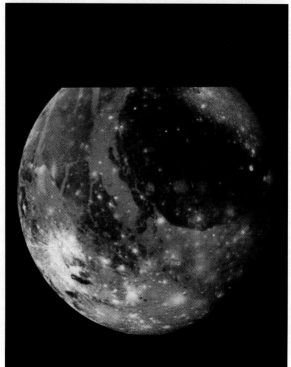

C.   Ganymede

D.   Io

Figure 18.23
Jupiter's four largest (Galilean) moons. These are the four moons discovered by Galileo. **A.** *Callisto, the outermost of the Galilean satellites, is densely cratered, much like the moon.* **B.** *Europa, smallest of Jupiter's moons, has an icy surface that is criss-crossed by many linear features.* **C.** *Ganymede, the largest Jovian satellite, contains cratered areas, smooth regions, and areas covered by numerous parallel grooves.* **D.** *The innermost moon, Io, is one of only three volcanically active bodies in the solar system. (Courtesy of NASA)*

tonic activity in the distant past. In some places, structural features show evidence of lateral displacement along strike-slip faults, a phenomenon previously found only on Earth.

The innermost of the Galilean moons, Io, is the only volcanically active body discovered in our solar system, other than Earth and Neptune's moon Triton (Figure 18.23D). To date, eight active sulfurous volcanic centers have been discovered. Umbrella-shaped plumes have been seen rising from the surface of Io to heights approaching 200 kilometers (Figure 18.24). The surface of Io is very colorful, a result of its sulfurous "rocks" that change colors from red to yellow and from black to white, depending upon temperature.

The heat source for Io's volcanic activity is thought to be tidal energy generated by a relentless "tug of war" between Jupiter and the Galilean satellites. Because Io is gravitationally locked to Jupiter, the same side always faces the giant planet. The gravitational influence of Jupiter and the other nearby satellites pulls and pushes on Io's tidal bulge as its slightly eccentric orbit takes it alternately closer to and farther from Jupiter. This gravitational flexing of Io is transformed into heat energy.

One of the most interesting discoveries made by *Voyager 1* is the ring system of Jupiter. Thought to be less than 30 kilometers thick, the apparently continuous ring may extend outward from the surface of Jupiter to a distance equal to twice the diameter of the planet. This ring system is believed to be different from that encircling Saturn. Rather than particles held in planetary-type orbits, the particles composing Jupiter's ring appear to be temporarily entrapped by the planet's intense magnetic field. The source of the ring material may be sulfur from the volcanoes of Io, ionized and trapped by Jupiter's magnetic field.

Saturn: The Elegant Planet

Requiring 29.46 years to make one revolution, Saturn is almost twice as far from the sun as Jupiter, yet its atmosphere, composition, and internal structure are thought to be remarkably similar to Jupiter's.

Saturn's Rings. The most prominent feature of Saturn is its system of rings (Figure 18.25). This spectacular characteristic was discovered by Galileo in 1610 A.D. Because he could not resolve them with his primitive telescope, they appeared to him as two smaller bodies adjacent to the planet. Their ring nature was revealed 50 years later by the Dutch astronomer Christian Huygens. Until the recent discovery that Jupiter, Uranus, and Neptune have very faint ring systems, this phenomenon was thought to be unique to Saturn.

When viewed from Earth, Saturn's rings appear as distinct bands, which classically have been called the A, B, and C rings (Figure 18.25). The A ring is the outermost of the bright rings and is separated from the brightest ring (B ring) by a large gap. This space, called the **Cassini division,** is easily seen in photographs (see chapter opening photograph). Having a width of 5000 kilometers, the Cassini division is large enough to accommodate Earth's moon.

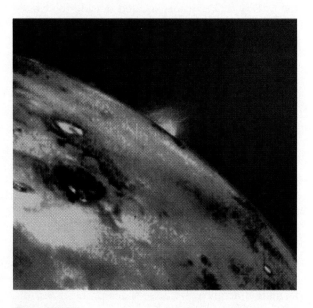

Figure 18.24
An eruption on Io. This umbrella-shaped plume is rising over 100 kilometers (60 miles) above Io's surface. (Courtesy of NASA)

Figure 18.25
A view of the dramatic ring system of Saturn.

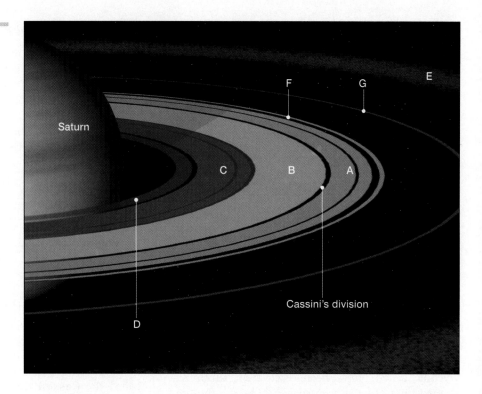

During the twentieth century, evidence has been gathered for the existence of four other very faint bands; the D ring is located closest to the planet and the E, F, and G rings are located outside the classically known rings (Figure 18.25). It turns out that even the Cassini division is not empty but contains a number of very faint bands.

From the Earth, our view of Saturn slowly changes as both planets proceed along their orbits, continually changing relative positions. This changes our angle of view of Saturn's rings. Once every 15 years, we see them edge-on, and they appear as an extremely fine line.

Saturn Close-Up. In 1980 and 1981, fly-by missions of the nuclear-powered *Voyagers 1* and *2* space vehicles came within 100,000 kilometers of the surface of Saturn. More information on Saturn and its satellite system was gained in a few days than had been acquired since Galileo first viewed this elegant planet telescopically (Figure 18.26). Some of the information acquired by these space probes is as follows:

1. Saturn's atmosphere is very dynamic with winds roaring at up to 1500 kilometers (930 miles) per hour.

2. Large cyclonic "storms" similar to Jupiter's Great Red Spot, although much smaller, occur in Saturn's atmosphere.
3. Eleven additional moons were discovered.
4. The icy rings of Saturn were discovered to be more complex than expected. Each of the seven rings is made of numerous ringlets, resembling the grooves on a phonograph record. The rings in the faint F ring are intertwined in stable, kinked, and braidlike configurations. Further, the B ring develops perplexing outwardly radiating spokes that survive for hours.
5. Satellite images reveal the thickness of the ring system to be no more than a few hundred meters, whereas its lateral extent exceeds 200,000 kilometers. We easily see the thin rings from more than a billion kilometers distance because they are highly reflective.

Although none of the images obtained so far has the resolution needed to "see" the fine structures of the rings, they undoubtedly are composed of relatively small particles (moonlets) that orbit the planet much like any other satellite. Radar observations indicate that most of the rings' particles are no larger than 10 meters, and the more abundant particles are perhaps as small as 10 centimeters (4 inches).

Further, it has been determined that the moonlets are good reflectors of visible light and poor reflectors of near-infrared wavelengths, both characteristics of water ice. Hence, at least the outer surfaces of these particles are probably coated with a highly reflective layer of water ice. Saturn's rings therefore are actually swarms of small ice-covered debris that orbit the planet in nearly circular paths, aligned with its equatorial plane. Most probably they are composed of material that failed to accrete into moons at the time the Saturnian system was forming.

The origin of the rings is believed to be related to their distance from the surface of Saturn. Since they are very close, the disruptive gravitational force of the parent prevented the individual particles from accreting into a larger satellite. Stated another way, objects cannot be held together by *self-gravitation* when they are within the influence of a stronger external gravitational field. In fact, should one of Saturn's large icy satellites approach closer than the outer edge of the bright rings, it would be destroyed and its remains distributed among the rings. The gravitational (tidal) force of Saturn pulling on the near side of such a satellite would be sufficiently greater than the force pulling on its far side, and it would tear the satellite apart.

Beyond the outermost bright ring (A ring), some moonlets have accreted to form very small satellites having diameters on the order of 100 kilometers (60 miles). Five of these asteroid-sized moons have been discovered orbiting within the faint outer rings, and others probably exist. The gravitational influence of these so-called "shepherding" satellites is believed to be responsible for keeping the particles confined within the rings, thereby producing the sharp edges of the rings.

Planetary geologists are very interested in the gravitational interactions among the objects that comprise Saturn's ring system. It is hoped that this information will reveal how material from the primordial cloud of dust and gases condensed to produce the planets.

Figure 18.26
Montage of the Saturnian system. Dione looms in the foreground; Tethys and Mimas are at lower right; Enceladus and Rhea are off ring's left; and Titan, the largest satellite but shown here at a distance, is at upper right. (Courtesy of NASA)

Saturn's Moons. The Saturnian satellite system consists of at least twenty-one bodies (Figure 18.27). All but two have nearly circular, counterclockwise orbits along Saturn's equatorial plane (see Figure 18.26). The largest Saturnian moon, Titan, is larger than Mercury and is the second-largest satellite in the solar system (after Jupiter's Ganymede). It is the only satellite in the solar system known to have a substantial atmosphere. Due to its dense gaseous cover, the atmospheric pressure at the surface of Titan is about one and one-half times that at Earth's surface.

Although Titan's atmosphere was predicted to be composed largely of methane, data from *Voyager 1* revealed that it is roughly 80 percent nitrogen, with methane probably accounting for less than 6 percent. The orange color of Titan's atmosphere may result from a photochemical "smog" of hydrocarbon molecules, including ethylene and hydrogen cyanide. This planet-sized moon appears to have polar ice caps that show seasonal variations in size. Its surface, if unfrozen, would be an ocean of liquid nitrogen.

Figure 18.28
This computer-assembled mosaic of images depicts the Uranian satellite Miranda (Voyager 2, 1986). In addition to crater-marked areas, note Miranda's regions of folded ridges that produce curving patterns unique to this satellite. (Courtesy of NASA)

Uranus and Neptune: The Twins

If any two planets in the solar system can be considered twins, Uranus and Neptune can. Only one percent different in diameter, both appear a pale greenish-blue, attributable to the methane in their atmospheres. Their structure and composition are believed to be similar. Neptune, however, is colder, because it is half again as distant from the sun's warmth as is Uranus.

Uranus, the Sideways Planet. A unique feature of Uranus is that it rotates "on its side"—its axis of rotation lies only 8 degrees from the plane of its orbit. Its rotational motion, therefore, has the appearance of rolling, rather than the toplike spinning of the other planets. Because the axis of Uranus is inclined almost 90 degrees, the sun is nearly overhead at one of its poles once each revolution, and then half a revolution later, it is overhead at the other.

A surprise discovery in 1977 revealed that Uranus is surrounded by rings, much like those encircling Jupiter. This find occurred as Uranus passed in front of a distant star and blocked its view, a process called **occultation** (the word *occult* means "hidden"). Observers saw the star "wink" briefly five times before the primary occultation and again five times afterward. Later studies have indicated that Uranus has at least nine distinct belts of debris orbiting its equatorial region.

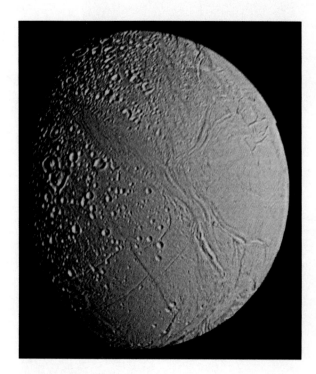

Figure 18.27
View of Enceladus, one of Saturn's moons, showing its cratered regions as well as areas that exhibit riftlike valleys. (Photo courtesy of NASA)

The first close-up views of Uranus and its satellites were transmitted to Earth by *Voyager 2* in January 1986. Studies of these images revealed ten previously unknown satellites that are much smaller than the five large moons known from Earth-based observations. Two additional rings also were discovered, along with evidence that others may encircle the planet.

Spectacular views of the five largest moons of Uranus showed quite varied terrains. Some contain long, deep canyons and linear scars, whereas others possess large, smooth areas on otherwise crater-riddled surfaces. A spokesperson for the Jet Propulsion Laboratory described Miranda, the innermost of the five largest moons, as having a greater variety of landforms than any body yet examined in the solar system (Figure 18.28). In particular, Miranda has three vast areas with families of grooves that encircle one another, much like the lanes of a gigantic racetrack.

Neptune. Even when the most powerful telescope is focused on Neptune, it appears as a bluish fuzzy disk. Until *Voyager 2*'s 1989 encounter with Neptune, astronomers knew very little about this planet, except that it had an atmosphere mostly of hydrogen, helium, and methane; that it had arcs or incomplete rings; and that it had two moons. However, *Voyager 2*'s 12-year, nearly 3-billion-mile journey has provided investigators with so much new information about Neptune and its satellites that years are needed to analyze it all.

Images from *Voyager 2* show that Neptune has a dynamic atmosphere, much like those of Jupiter and Saturn (Figure 18.29). Winds exceeding 1000 kilometers per hour (600 miles per hour) encircle the planet, making it one of the windiest places in the solar system. In addition, Neptune's dynamic atmosphere contains an Earth-sized blemish called the *Great Dark Spot* that is reminiscent of Jupiter's Great Red Spot. Like the Great Red Spot, this atmospheric disturbance is assumed to be a large rotating storm.

Perhaps the most surprising feature of the Neptunian atmosphere is the white, cirruslike clouds that occupy a layer about 50 kilometers above the main cloud deck. Although these clouds appear to change configurations, one group seems permanently attached to the Great Dark Spot. At the extremely low temperatures found on Neptune, frozen methane is the most likely constituent of these upper-level clouds.

Six new satellites, ranging from 50 to 400 kilometers in diameter, were discovered in the *Voyager* images. This brings the total number of known satel-

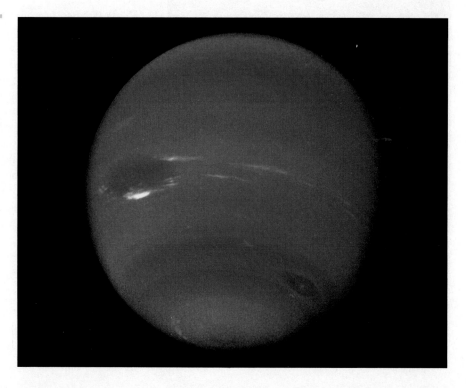

Figure 18.29
This image of Neptune shows the Great Dark Spot (left center). Also visible are bright cirruslike clouds that travel at high speed around the planet. A second oval spot is at 54° south latitude on the east limb of the planet. (Photo courtesy of the Jet Propulsion Laboratory)

lites for Neptune to eight. All of the newly discovered moons orbit the planet in a direction opposite that of the two larger satellites.

Voyager images also confirmed the existence of a ring system around Neptune's center. At least two narrow and two broad rings were discovered circling the planet. The outermost has three thicker arclike segments. These brighter regions should be spread relatively evenly around the ring, unless they are restrained by the gravitational influence of shepherding satellites like those in Saturn's rings. Although no such satellites were discovered in the *Voyager 2* images, objects smaller than 12 kilometers could not be detected by the cameras aboard that spacecraft.

Triton.   Triton, Neptune's largest moon, proved to be a most interesting object. Its diameter is roughly 2700 kilometers (nearly as large as Earth's moon).

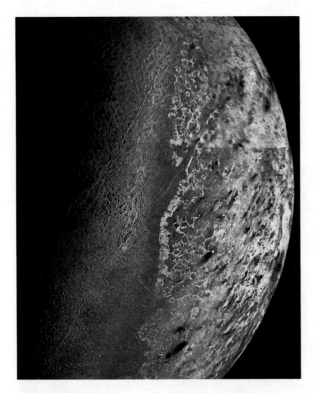

Figure 18.30
A photomosaic of Neptune's largest moon, Triton. The large south polar cap is on the right half of the image. Seasonal ice (probably nitrogen) covers the region. Because spring in Triton's southern hemisphere extends from 1960 to the year 2000, some of the polar cap has evaporated. (Photo courtesy of the Jet Propulsion Laboratory)

Triton is the only large moon in the solar system that has a highly inclined retrograde orbit, meaning that it orbits Neptune in the direction opposite to the direction in which all the planets travel. This indicates that Triton formed independently of Neptune and was gravitationally captured by the parent planet.

Much has been learned about Triton. It has the lowest surface temperature of any body in the solar system, 38 K ($-391°F$). Its very thin atmosphere is composed mostly of nitrogen, with a little methane. The surface of Triton apparently is largely water ice, covered with layers of solid nitrogen and methane.

Despite low surface temperatures, Triton displays a volcaniclike activity. Two active plumes were discovered that extended to an altitude of 8 kilometers and were blown downwind for more than 100 kilometers. Presumably, solar energy is absorbed more readily by the surface layers of darker methane ice. Such surface warming vaporizes some of the underlying nitrogen ice. As subsurface pressures increase, explosive eruptions eventually result.

Triton's surface appears to be geologically very young because it lacks heavily cratered areas. As you can see in Figure 18.30, parts of its surface have pits and dimples that are crisscrossed by ridges. Other areas seem to have frozen lakes that resemble lunar maria. Many of these lakes contain terraces, suggesting a succession of meltings and refreezings of water ice.

As *Voyager 2* advances toward interstellar space, our understanding of Neptune and its rings and satellites will continue to improve. In the meantime, both *Voyager* spacecraft will continue to explore the outer reaches of the solar system, sending back measurements of the intensity of the solar wind. If these spacecraft continue to operate as well in the future as they have in the past, they should be able to collect and transmit data for another twenty-five years. By that time both vehicles will be over 100 astronomical units from the sun and may cross the *heliopause*, where true interstellar space begins and where there is no longer any measurable influence from the sun.

Pluto: Planet X

Pluto lies on the fringe of the solar system, almost 40 times farther from the sun than the Earth. It is 10,000 times too small to be visible to the unaided eye. Because of its great distance and slow orbital speed, it takes Pluto 248 years to orbit the sun. Since its dis-

BOX 18.2

Discovery of Planet X

Du「uring the mid-1800s Neptune, the eighth planet in our solar system, was first sighted. After this discovery, unresolved discrepancies in the orbit of Uranus led some astronomers to suggest that the gravitational effect of a yet-unknown planet might be responsible. A vigorous search for this planet was begun in 1906 by Percival Lowell at the observatory in Flagstaff, Arizona, that now bears his name. He unsuccessfully searched for the object he called "Planet X" until his death 10 years later.

Some years afterward, a new photographic telescope was installed at the Lowell Observatory and the search was resumed. Two photographs of the same portion of the sky were taken at intervals of a week or so and then compared. In such short time intervals, stars appear stationary, but a planet shifts its position relative to its background. Twenty-five years had lapsed since Lowell's death, and 2 million stars were examined before Clyde Tombaugh discovered a body that had a shift that was about right for an object 1 billion miles beyond the orbit of Neptune (Figure 18.B).

This newly discovered planet was named Pluto, the Greek god of the underworld, because the first two letters are Percival Lowell's initials. Although the planet was located in the constellation Gemini as Lowell had calculated, its mass was much less than he had estimated. It seems unlikely that a planet of Pluto's size could have altered the orbit of Uranus measurably. Most probably, the discovery of Pluto can be credited to an extensive search of the correct portion of the sky for the wrong reasons.

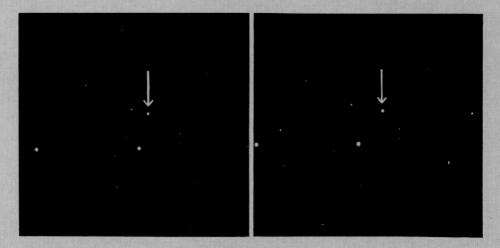

Figure 18.B
Pluto appears starlike on a photograph but is revealed by its motion to be a planet. (Courtesy of Hale Observatories)

covery in 1930, it has completed less than one-fifth of a revolution. Pluto's orbit is noticeably elongated (highly eccentric), causing it to occasionally travel inside the orbit of Neptune, where it currently resides. There is no chance that Pluto and Neptune will ever collide, because their orbits are inclined to each other and do not actually cross.

In June 1978, the moon Charon was discovered orbiting Pluto. Because of its close proximity to the planet, the best ground-based images of Charon show it only as an elongated bulge. In 1990, the Hubble Space Telescope produced an image that clearly resolves the separation between these two icy worlds (Figure 18.31). Charon orbits Pluto once every 6.4

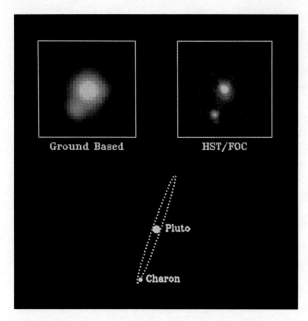

Figure 18.31
*Pluto and its moon Charon. The Hubble Space Tele-
scope produced the first image (upper right) that resolved these two icy worlds into separate objects. The
image in the upper left is the best ground-based photo
produced to date. (Courtesy of NASA)*

days at a distance of about 19,700 kilometers, or 20
times closer than our moon is to Earth.

The discovery of Charon greatly altered earlier estimates of Pluto's size. Current data indicate that
Pluto has a diameter of 2445 kilometers, about one-
fifth the size of the Earth, making it the smallest
planet in the solar system. Charon is about 1300 kilometers across.

The average temperature of Pluto is estimated at
−210°C, cold enough to solidify most gases that
might be present. Thus, Pluto probably has a meager atmosphere composed mainly of nitrogen. Pluto

might best be described as a large, dirty iceball made
up of a mixture of frozen gases with lesser amounts
of rocky substances.

A recent proposal suggests that Pluto once was a
satellite of Neptune and was displaced from its original orbit when it collided with a large foreign object.
The discovery of a satellite around Pluto is considered evidence that this event broke the Neptunian
satellite into two pieces and sent them into an elongated orbit around the sun.

MINOR MEMBERS OF THE SOLAR SYSTEM

Asteroids

Asteroids are relatively small bodies that have been
likened to "flying mountains." The largest, Ceres, is
about 1000 kilometers (620 miles) in diameter, but
most of the 50,000 that have been observed are only
about one kilometer across. The smallest asteroids
are assumed to be no larger than grains of sand. Most
asteroids lie between the orbits of Mars and Jupiter
and have periods ranging from three to six years (Figure 18.32). Some asteroids have very eccentric orbits
and travel very near the sun, and a few larger ones
regularly pass close to Earth and its moon. Many of
the most recent impact craters on the moon and
Earth were probably caused by collisions with asteroids (see Box 18.4).

Because many asteroids have irregular shapes,
planetary geologists first speculated that they might
have formed from the breakup of a planet that once
occupied an orbit between Mars and Jupiter (Figure
18.33). However, the total mass of the asteroids is estimated to be only one-thousandth that of Earth,
which itself is not a large planet. What, then might
have happened to the remainder of the original

Figure 18.32
*The orbits of most asteroids
lie between Mars and Jupiter.
Also shown are the orbits
of a few known near-Earth
asteroids. Perhaps a thousand
or more asteroids have near-
Earth orbits. Luckily, only a
few dozen are thought to be
larger than 1 kilometer in
diameter.*

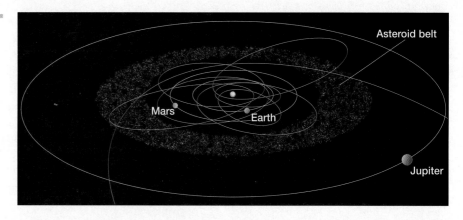

planet? Others have hypothesized that several larger bodies once coexisted in close proximity and that their collisions produced numerous smaller ones. The existence of several "families" of asteroids has been used to support this explanation. However, no conclusive evidence has been found for either hypothesis.

Comets

Comets are among the most spectacular and unpredictable bodies in the solar system. They have been compared to large, dirty snowballs, since they are made of frozen gases (water, ammonia, methane, carbon dioxide, and carbon monoxide) that hold together small pieces of rocky and metallic materials. Many comets travel along very elongated orbits that carry them beyond Pluto. On their return, these comets become visible again when they are within the orbit of Saturn.

When first observed, a comet appears very small; but as it approaches the sun, solar energy begins to vaporize the frozen gases, producing a glowing head called the **coma** (Figure 18.34). The size of the coma

Figure 18.33
Image of asteroid 951 (Gaspra) obtained by the Jupiter-bound Galileo spacecraft. Like other asteroids, Gaspra is probably a collision-produced fragment of a larger body. (Courtesy of NASA)

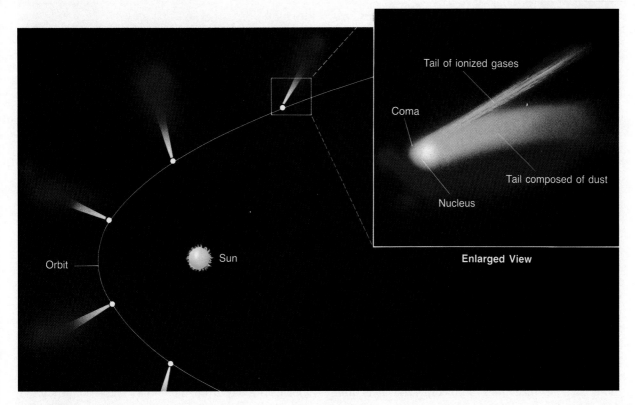

Figure 18.34
Orientation of a comet's tail as it orbits the sun.

BOX 18.3
Halley's Comet

The most famous short-period comet is Halley's comet (Figure 18.C). Its orbital period averages 76 years, and every one of its 29 appearances since 240 B.C. has been recorded by the Chinese. This record is a testimonial to their dedication as astronomical observers and the endurance of their culture. When seen in 1910, Halley's comet had developed a tail nearly 1.6 million kilometers (1 million miles) long and was visible during the daylight hours.

In 1986 the unspectacular showing of Halley's comet was a disappointment to many people in the Northern Hemisphere. Yet it was during this most recent visit to the inner solar system that a great deal of new information was learned about this most famous of comets. The new data were gathered by space probes that were sent to rendezvous with the comet. Most notably, the European probe *Giotto* approached to within 600 kilometers of the comet's nucleus and obtained the first images of this elusive structure. On the basis of these images, it is now clear that the nucleus of Halley's comet is not spherical but rather potato-shaped, with dimensions of roughly 16 kilometers by 8 kilometers.

Scientists were surprised to learn that the surface of Halley's comet is irregular and full of craterlike pits. Furthermore, the gases and dust that vaporize from the nucleus to form the coma and tail of the comet appeared to gush from its surface as bright jets or streams. Only about 10 percent of the comet's total surface was emitting these jets of dust and gas at the time of these observations. The remaining surface area of the comet appeared to be covered with a dark layer that may consist of organic material.

Figure 18.C
Photograph of Halley's comet showing two somewhat separate tails. The straight tail of ionized gases has a bluish cast, whereas the slightly curved dust tail exhibits a yellow coloration. (Photo by William Liller, NASA International Halley Watch, Easter Island)

varies greatly from one comet to another. Extremely rare ones exceed the size of the sun, but most approximate the size of Jupiter. Within the coma, a small glowing nucleus with a diameter of only a few kilometers can sometimes be detected. As comets approach the sun, some, but not all, develop a tail that extends for millions of kilometers. Despite the enormous size of their tails and comas, comets are thought to have an insignificant mass.

It seems strange that the tail of a comet points away from the sun in a slightly curved manner (Figure 18.34), but this fact led early astronomers to propose that the sun has a repulsive force that pushes the particles of the coma away, thus forming the tail. Today, two solar forces are known to contribute to this formation. One, *radiation pressure*, pushes dust particles away from the coma. The second, known as *solar wind*, is responsible for moving the ionized gases, particularly carbon monoxide (Figure 18.34). Usually a single tail composed of both types of materials is produced, but sometimes a dual tail is observed (see Box 18.3).

As a comet moves away from the sun, the gases begin to recondense, the tail disappears, and the comet once again returns to "cold storage." The material that was blown from the coma to form the tail is lost from the comet forever. Consequently, it is believed that most comets cannot survive more than a few hundred close orbits of the sun. Once all the gases are expended, the remaining material—a swarm of unconnected metallic and stony particles—continues the orbit without a coma or a tail.

Little is known about the origin of comets. The most widely accepted proposal considers them to be members of the solar system that formed at great distances from the sun. Accordingly, millions of comets are believed to orbit the sun beyond Pluto with periods measured in hundreds of years. It is proposed that the gravitational effect of stars passing nearby sends some of them into the highly eccentric orbits that carry them toward the center of our solar system. Here, the gravitation of the larger planets, particularly Jupiter, alters a comet's orbit and ac-

celerates its period of revolution. Many short-period comets of this type have been discovered. However, since they have a short life expectancy, we can be fairly certain that they are always being replaced by other long-period comets that are being deflected toward the sun.

Meteoroids

Nearly everyone has seen a **meteor,** popularly (but inaccurately) called a "shooting star." This streak of light lasts for, at most, a few seconds and occurs when a small solid particle, a **meteoroid,** enters Earth's atmosphere from interplanetary space. The friction between the meteoroid and the air heats both and produces the light we see. Although an occasional meteoroid is as large as an asteroid, most are about the size of sand grains and weigh less than $\frac{1}{100}$ gram. Consequently, they vaporize before reaching Earth's surface. Some, called *micrometeorites*, are so tiny that their rate of fall becomes too slow to cause them to burn up, so they drift down as "space dust." Each day, the number of meteoroids that enter Earth's atmosphere must reach into the thousands. After sunset, a half dozen or more are bright enough to be seen with the naked eye each hour from anywhere on Earth.

Occasionally, meteor sightings increase dramatically to 60 or more per hour. These spectacular displays, called **meteor showers,** result when Earth encounters a swarm of meteoroids traveling in the same direction and at nearly the same speed as Earth. The close association of these swarms to the orbits of some short-term comets strongly suggests that they represent material lost by these comets (Table 18.3). Some swarms not associated with orbits of known comets are probably the remains of the nucleus of a long-defunct comet. The meteor showers that occur regularly each year around August 12 are believed to be the remains of the Comet 1862 III, which has a period of 110 years.

Meteoroids associated with comets are small and not known to reach the ground. Most meteoroids large enough to survive the heated fall are thought to originate among the asteroids where chance collisions modify their orbits and send them toward Earth. Earth's gravitational force does the rest.

The remains of meteoroids, when found on Earth, are referred to as **meteorites.** A few very large meteoroids have blasted out craters on Earth's surface that are not much different in appearance from those

Table 18.3
Major meteor showers.

Shower	Approximate Dates	Associated Comet
Quandran-tids	January 4–6	—
Lyrids	April 20–23	Comet 1861 I
Eta Aquarids	May 3–5	Halley's comet
Delta Aquarids	July 30	—
Perseids	August 12	Comet 1862 III
Draconids	October 7–10	Comet Giaco-bini-Zinner
Orionids	October 20	Halley's comet
Taurids	November 3–13	Comet Encke
Androme-dids	November 14	Comet Biela
Leonids	November 18	Comet 1866 I
Geminids	December 4–16	—

BOX 18.4

Is Earth on a Collision Course?

T he solar system is cluttered with meteoroids, asteroids, and active, as well as extinct, comets. These fragments, which travel at supersonic speeds, can strike Earth with the explosive force of a very strong nuclear weapon.

In the last few decades, it has become increasingly clear that comets and asteroids have collided with Earth far more frequently than had previously been believed. The evidence for these ancient catastrophic events is in the form of giant impact structures called *astroblems*. More than 100 astroblems have been identified to date (Figure 18.D). Many were once considered to be the result of some poorly understood volcanic process. Although most astroblems are so old that they no longer resemble impact craters,

evidence of an intense shock remains (Figure 18.E). One notable exception is a very fresh-looking crater located near Winslow, Arizona, known as Meteor Crater (see Figure 18.35).

Evidence is mounting that about 65 million years ago a large asteroid about 10 kilometers (6 miles) in diameter collided with Earth. This impact may have caused the extinction of the dinosaurs, as well as nearly 50 percent of all plant and animal species (see Box 9.3).

More recently, a spectacular explosion has been attributed to

Figure 18.D
World map of major impact structures (astroblems). Others are being identified every year. (Data from Griffith Observatory)

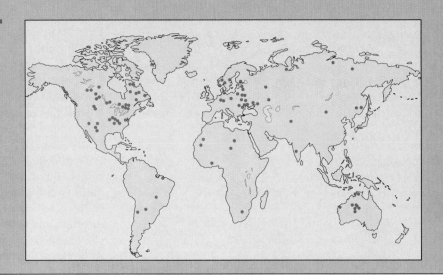

on the lunar surface. The most famous is Meteor Crater in Arizona (Figure 18.35). This huge cavity is about 1.2 kilometers (0.75 mile) across, 170 meters (560 feet) deep, and has an upturned rim that rises 50 meters (165 feet) above the surrounding countryside. Over 30 tons of iron fragments have been found in the immediate area, but attempts to locate the main body have been unsuccessful. Judging from the amount of erosion, it appears that the impact occurred within the last 20,000 years.

Prior to moon rocks brought back from lunar exploration, meteorites were the only samples of extraterrestrial material that could be directly ex-

amined (Figure 18.36). Meteorites are classified by their composition: (1) **irons**—mostly iron with 5–20 percent nickel, (2) **stony**—silicate minerals with inclusions of other minerals, and (3) **stony-irons**—mixtures. Although stony meteorites are probably more common, most meteorites found by people are irons. This is understandable, since irons tend to withstand the impact, weather more slowly, and are much easier for a lay person to distinguish from terrestrial rocks than are stony meteorites.

Iron meteorites are probably fragments of once-molten cores of large asteroids or small planets. One rare kind of meteorite, called a *carbonaceous chon-*

the collision of our planet with a comet or asteroid. In 1908, in a remote region of Siberia, a fireball that appeared to be more brilliant than the sun exploded with a violent force. The shock waves rattled windows and caused sounds that were heard up to 1000 kilometers away. The so-called Tunguska event scorched, delimbed, and flattened trees for a distance of 30 kilometers from the center of the explosion. Interestingly, expeditions to the area following World War I found no evidence of an impact crater nor any metallic fragments. Evidently, the explosion (which equalled at least a 10-megaton nuclear bomb) occurred a few kilometers above the surface. Most likely it was the demise of a comet or perhaps a stony asteroid.

The dangers of living with these relatively small but deadly objects was again brought to light in March 1989 when an asteroid nearly one kilometer across shot past Earth. It was a near miss, coming as close as about twice the distance to the moon. Traveling at 70,000 kilometers (44,000 miles) per hour, it could have produced a crater 10 kilometers (6 miles) in diameter and perhaps 2 kilometers (1.2 miles deep. As an observer noted, "Sooner or later it will be back." As it was, it crossed our orbit just 6 hours ahead of Earth. Statistics show that collisions of this tremendous magnitude should take place every few hundred million years and could have drastic consequences for life on Earth.

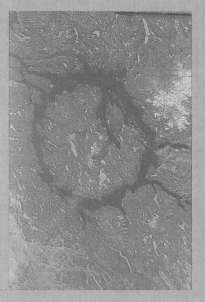

Figure 18.E
Manicouagan, Quebec, is a 200-million-year-old eroded impact crater. The lake, outlines the crater remnant, which is 70 kilometers (42 miles) across. Fractures related to this event extend outward for an additional 30 kilometers. (Courtesy of U.S. Geological Survey)

drite, was found to contain some simple amino acids and other organic compounds, which are the basic building blocks of life. This discovery confirms similar findings in observational astronomy which indicate that numerous organic compounds exist in the frigid realm of outer space.

If the composition of meteorites is representative of the material that makes up the Earthlike planets, as some planetary geologists believe, then Earth must contain a much larger percentage of iron than is indicated by the surface rock. This is one of the reasons why geologists suggest that the core of Earth may be mostly iron and nickel. In addition, the dating of meteorites has indicated that our solar system has an age which certainly exceeds 4.5 billion years. This "old age" has been confirmed by data obtained from lunar samples.

Figure 18.35
Meteor Crater, about 32 kilometers (20 miles) west of Winslow, Arizona. (Photo by Michael Collier)

Figure 18.36
Iron meteorite found near Meteor Crater, Arizona. (Courtesy of Meteor Crater, Northern Arizona, USA)

Review Questions

1. Compare the rotational periods of the terrestrial and Jovian planets.

2. How fast does a location on the equator of Jupiter rotate? Note that Jupiter's circumference equals approximately 452,000 kilometers (280,000 miles).

3. By what criteria are the planets placed into either the Jovian or terrestrial group?

4. What are the three types of materials thought to make up the planets? How are they different? How does their distribution account for the density differences between the terrestrial and Jovian planetary groups?

5. Briefly describe the events that are thought to have led to the formation of the solar system.

6. How is crater density used in the relative dating of features on the moon?

7. Why are large-rayed craters considered relatively young features on the lunar surface?

8. Briefly outline the history of the moon.

9. How are the maria of the moon thought to be similar to the Columbia Plateau?

10. Why has Mars been the planet most studied telescopically?

11. What surface features does Mars have that are also common on Earth?

12. Although Mars has valleys that appear to be products of stream erosion, what fact makes it unlikely that Mars has had a water cycle like that on Earth?

13. The two "moons" of Mars were once suggested to be artificial. What characteristics do they have that would cause such speculation?

14. What is thought to be the nature of Jupiter's Great Red Spot?

15. Why are the Galilean satellites of Jupiter so named?

16. What is distinctive about Jupiter's satellite Io?

17. Describe the nature of the surface of Callisto.

18. Why are the four outer satellites of Jupiter thought to have been captured?

19. What evidence indicates that Saturn's rings are composed of individual moonlets rather than consisting of solid disks?

20. Explain why a large satellite cannot exist closer to Saturn than the outer edge of its ring system.

21. What is unique about Saturn's satellite Titan?

22. Which of the five largest moons of Uranus has the most varied terrain?

23. What three bodies in the solar system exhibit volcaniclike activity?

24. What do you think would happen if Earth passed through the tail of a comet?

25. Describe the origin of comets according to the most widely accepted hypothesis.

26. Compare a meteoroid, meteor, and meteorite.

27. Why are meteorite craters more common on the moon than on Earth, even though the moon is much smaller?

28. It has been estimated that Halley's comet has a mass of 100 billion tons. Further, this comet is thought to lose 100 million tons of material during the few months that its orbit brings it close to the sun. With an orbital period of 76 years, what is the maximum remaining life span of Halley's comet?

Key Terms

asteroid (p. 632)

Cassini division (p. 625)

coma (p. 633)

comet (p. 633)

escape velocity (p. 603)

iron meteorite (p. 636)

Jovian planet (p. 602)

lunar breccia (p. 609)

lunar regolith (p. 610)

maria (p. 607)

meteor (p. 635)

meteorite (p. 635)

meteoroid (p. 635)

meteor shower (p. 635)

nebular hypothesis (p. 605)

occultation (p. 628)

stony-iron meteorite (p. 636)

stony meteorite (p. 636)

terrestrial planet (p. 602)

Light, Astronomical Observations, and the Sun

Opposite page: Twenty-seven radio telescopes operate together to form the Very Large Array. (Photo by Geoff Chester).

Left: Steerable radio telescope. (Courtesy of National Radio Astronomy Observatory)

Above: The Very Large Array, located near Socorro, New Mexico. (Photo by Geoff Chester)

*A*stronomers are in the business of gathering and studying light. Almost everything that is known about the universe beyond the earth comes by analyzing light from distant sources. Consequently, an understanding of the nature of light is basic to modern astronomy. This chapter deals with the study of light and the tools used by astronomers to gather light in order to probe the universe. In addition, we will examine the nearest source of light, our sun. By understanding the processes that operate on the sun, astronomers can better grasp the nature of more distant celestial objects.

THE STUDY OF LIGHT

The vast majority of our information about the universe is obtained from the study of the light emitted from celestial bodies (Figure 19.1). Although visible light is most familiar to us, it constitutes only a small part of an array of energy generally referred to as **electromagnetic radiation.** Included in this array are gamma rays, x rays, ultraviolet light, visible light, infrared light, and radio waves (see Figure 12.15). All radiant energy travels through the vacuum of space in a straight line at the rate of 300,000 kilometers (186,000 miles) per second.* Over a 24-hour day, this equals a staggering 26 billion kilometers.

Nature of Light

Experiments have demonstrated that light can be described in two ways. In some instances, light behaves like waves, and in others, like discrete particles. In the wave sense, light is analogous to swells in the ocean. This motion is characterized by the wavelength, which is the distance from one wave crest to the next. Wavelengths vary, from several kilometers for radio waves to less than a billionth of a centimeter for gamma rays. Most of these waves are either too long or too short for our eyes to detect.

The narrow band of electromagnetic radiation we can see is sometimes referred to as *white light.* However, white light consists of an array of waves having various wavelengths, a fact easily demonstrated with a prism (Figure 19.2). As white light passes through a prism, the color with the shortest wavelength, violet, is bent more than blue, which is bent more than green, and so forth (Table 19.1). Thus, white light can be separated into its component colors in the order of their wavelengths, producing the familiar rainbow of colors (Figure 19.2).

Wave theory, however, cannot explain some effects of light. In some cases, light acts like a stream of particles, analogous to infinitesimally small bullets fired from a machine gun. These particles, called **photons,** can exert a pressure ("push") on matter, which is called **radiation pressure.** Photons from the sun are responsible for "pushing" material away from a comet to produce its tail. Each photon has a specific

---

* Light rays are "bent" slightly when they pass nearby a very massive object such as the sun.

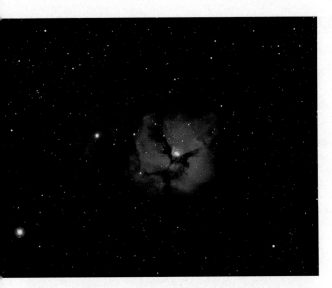

Figure 19.1
The Trifid Nebula, in the constellation Sagittarius. This colorful nebula is a cloud of dust, plus hydrogen and helium gases. These gases are excited by the radiation of the hot, young stars within and produce a pink glow. (Courtesy of National Optical Astronomy Observatories)

Table 19.1
Colors and corresponding wavelengths.

Color	Wavelength (nanometers*)
Violet	380–440
Blue	440–500
Green	500–560
Yellow	560–590
Orange	590–640
Red	640–750

* One nanometer is 10^{-9} meter.

amount of energy, which is related to its wavelength in a simple way: *Shorter wavelengths* correspond to *more energetic* photons. Thus, blue light has more energetic photons than red light.

Which theory of light—the wave theory or the particle theory—is correct? Both, since each will predict the behavior of light for certain phenomena. As George Abell, a prominent astronomer, stated about all scientific laws, "The mistake is only to apply them to situations that are outside their range of validity."

Spectroscopy

When Sir Isaac Newton used a prism to disperse white light into its component colors, he unknowingly initiated the field of **spectroscopy,** which is the study of properties of light that depend on wavelength. The rainbow of colors Newton produced is called a continuous spectrum, because all wavelengths of light are included. It was later learned that two other types of spectra exist, and that all three are generated under somewhat different conditions (Figure 19.3).

1. A **continuous spectrum** is produced by an incandescent solid, liquid, or gas under high pressure. It consists of an uninterrupted band of color (Figure 19.3A). One example would be light generated by a common light bulb. (*Incandescent* means "to emit light when hot.")

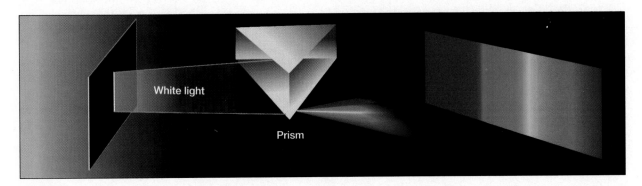

White light

Prism

Figure 19.2
A spectrum is produced when sunlight (white light) is passed through a prism, which bends each wavelength at different angles.

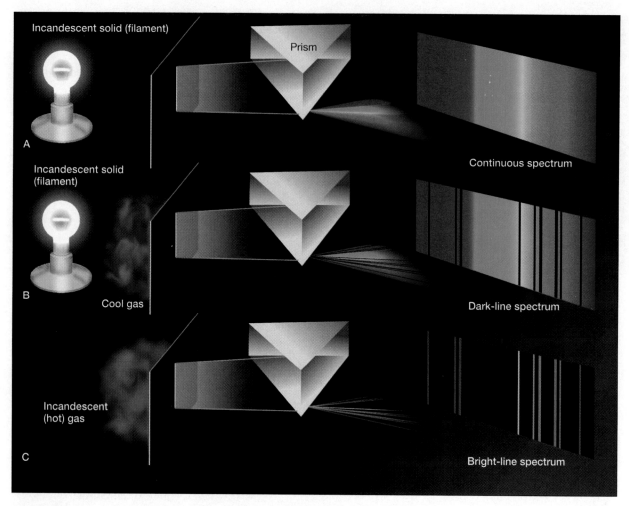

Figure 19.3
Formation of the three types of spectra. **A.** *Continuous spectrum.* **B.** *Dark-line spectrum.* **C.** *Bright-line spectrum.*

2. A **dark-line spectrum (absorption spectrum)** is produced when white light is passed through a comparatively cool gas under low pressure. The gas absorbs selected wavelengths of light, so the spectrum that is produced appears as a continuous spectrum, but with a series of dark lines running through it (Figure 19.3B).

3. A **bright-line spectrum (emission spectrum)** is produced by a hot (incandescent) gas under low pressure. It is a series of bright lines of particular wavelengths, depending on the gas that produces them (Figure 19.3C). These bright lines appear in the exact location as the dark lines that are produced by this gas in a dark-line (absorption) spectrum.

The spectra of most stars are of the dark-line type. The importance of these spectra is that each element or compound that is in gaseous form (in a star, material is usually in the gaseous form) produces a unique set of spectral lines. When the spectrum of a star is studied, the spectral lines act as "fingerprints," which identify the elements present.

In an admittedly oversimplified manner, we can imagine how the sun and other stars create a dark-line spectrum. A continuous spectrum is produced in the interior of the sun, where the gases are under very high pressure. When this light passes through the less dense gases of the solar atmosphere, they absorb selected wavelengths, which appear as dark lines in the spectrum.

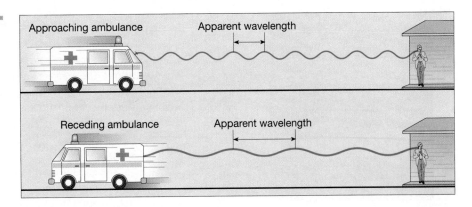

Figure 19.4
The Doppler effect, illustrating the apparent lengthening and shortening of wavelengths caused by the relative motion between a source and an observer.

When Newton studied solar light, he obtained a continuous spectrum. However, when a prism is used in conjunction with lenses, the solar spectrum can be dispersed even further. An instrument that does this is called a **spectroscope.** The spectrum of the sun contains thousands of dark lines. Over 60 elements have been identified by matching these lines with those of elements known on earth.

Two other facts concerning a radiating body are important. First, if the temperature of a radiating surface is increased, the total amount of energy emitted increases. The rate of increase is stated by the Stefan-Boltzmann law: The energy radiated by a body is directly proportional to the fourth power of its absolute temperature. For example, if the temperature of a star is doubled, the total radiation emitted increases by 2^4 ($2 \times 2 \times 2 \times 2$), or 16 times. Second, as the temperature of an object increases, a larger proportion of its energy is radiated at shorter wavelengths. To illustrate this, imagine a metal rod that is heated slowly. The rod first appears dull red (long wavelengths), and later bluish white (short wavelengths). From this, it follows that blue stars are hotter than yellow stars, which are hotter than red stars.

The Doppler Effect

You may have heard the change in pitch of a car horn or ambulance siren as it passes by. When it is approaching, the sound seems to have a higher-than-normal pitch, and when it is moving away, the pitch sounds lower than normal. This effect, which occurs for both sound and light waves, was first explained by Christian Doppler in 1842 and is called the **Doppler effect.** The reason for the difference in pitch is that it takes time for the wave to be emitted.

If the source is moving away, the beginning of the wave is emitted nearer to you than the end, which "stretches" the wave, that is, gives it a longer wavelength (Figure 19.4). The opposite is true for an approaching source.

In the case of light, when a source is moving away, its light appears redder than it actually is, because its waves appear lengthened. Objects approaching have their light waves shifted toward the blue (shorter wavelength). Thus, if a source of red light approached you at a very high speed (near the speed of light), it would actually appear blue. The same effect would be produced if you moved and the light were stationary.

Therefore, the Doppler effect reveals whether the earth is approaching or receding from a star or another celestial body. In addition, the amount of shift allows us to calculate the rate at which the relative movement is occurring. Larger Doppler shifts indicate higher velocities; smaller Doppler shifts indicate slower velocities. Doppler shifts are generally measured from the dark lines in the spectra of stars by comparing them with a standard spectrum produced in the laboratory (Figure 19.5).

ASTRONOMICAL TOOLS

Having examined the nature of light, we will now turn our attention to the tools astronomers use to intercept and study the energy emitted by distant objects in the universe (Figure 19.6). Because the basic principles of detecting radiation were originally developed through visual observations, we will consider optical telescopes first.

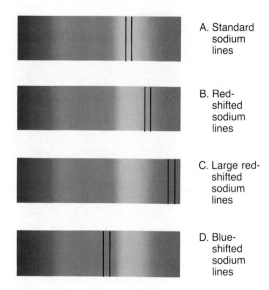

A. Standard sodium lines

B. Red-shifted sodium lines

C. Large red-shifted sodium lines

D. Blue-shifted sodium lines

Figure 19.5
A. *An illustration of standard sodium lines produced in a laboratory compared to sodium lines as they would appear when the source is receding (red shift,* **B.** *and* **C.**) *or approaching (blue shift,* **D.**).

Refracting Telescopes

Galileo is considered to be the first person to use telescopes for astronomical observations. Having learned about the newly invented instrument, Galileo built one of his own design that was capable of magnifying objects 30 times. Because this early instrument, as well as its modern counterparts, used a lens to bend or refract light, it is known as a **refracting telescope.**

The most important lens in a refracting telescope, the **objective lens,** produces the image by bending light from a distant object in such a way that the light converges at an area called the **focus** (Figure 19.7). For an object such as a star, the image appears as a point of light, but for a nearby object, it appears as an inverted replica of the original.

You can easily demonstrate the latter case by holding a lens in one hand and with the other hand placing a white card behind the lens. Now, vary the distance between them until an image appears on the card. The distance between the focus (where the image appears) and the lens is called the **focal length** of the lens.

Astronomers usually study an image from a telescope by first photographing the image. However, if a telescope is used to examine an image directly, a second lens, called an **eyepiece,** is required (Figure 19.7). The eyepiece magnifies the image produced by the objective lens. In this respect, it is similar to a magnifying glass. Thus, the objective lens produces a

Figure 19.6
Aerial view of Cerro Tololo Inter-American Observatory located near La Serena, Chile. (Courtesy of National Optical Astronomy Observatories)

Figure 19.7
Simple refracting telescope.

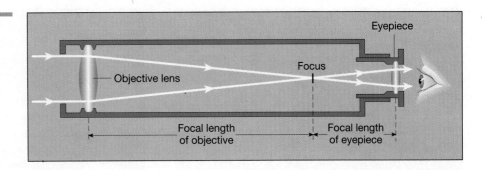

very small, bright image of an object, and the eyepiece enlarges the image so that details can be seen.

Although used extensively in the nineteenth century, refracting telescopes suffer a major optical defect. As light passes through any lens, the shorter wavelengths of light are bent more than the longer wavelengths. (Recall the effect of a prism in separating the colors of the spectrum.) Consequently, when a refracting telescope is in focus for red light, blue and violet light are out of focus. This troublesome effect, known as **chromatic** ("color") **aberration,** weakens the image and produces a halo of color around it. When blue light is in focus, a reddish halo appears, and vice versa. Although this effect cannot be eliminated completely, it is reduced by using a second lens made of a different type of glass.

Reflecting Telescopes

Newton was so bothered by chromatic aberration that he built and used telescopes that reflected light from a shiny surface (mirror). Because reflected light is not dispersed into its component colors, the problem is avoided. **Reflecting telescopes** use a concave mirror that focuses the light in front of the objective (the mirror), rather than behind it, like a lens (Figure 19.8). The mirror is generally made of glass that is finely ground. In the case of the 5-meter (200-inch) Hale Telescope, the grinding is accurate to about 1 millionth of a centimeter. The surface is then coated with a highly reflective material, usually an aluminum compound.

To focus parallel incoming light rays onto one spot, the mirror is ground to a special curved surface called a *paraboloid.* This is the same shape as that used for the reflector in the headlights of automobiles. In the case of the auto bulb, however, the light source is at

the focus, and the light goes out in parallel rays, rather than coming in.

Because the focus of a reflecting telescope is in front of the mirror, provisions have to be made to view the image without blocking too much of the incoming light. Figure 19.9 illustrates the most common arrangements. Most large telescopes employ more than one type. When using a very large reflecting telescope, the observer enters a viewing cage positioned at the focus to make the observations. The viewing cage blocks only about 10 percent of the total incoming light, and this is more than compensated for by the large objective (mirror) that is used.

In the "good old days," an astronomer would spend numerous long nights outdoors in the cold mountain air, perched in a viewing cage. But advances in photographic materials and computer-enhancing technologies now allow indoor work and reduce the time required to obtain an image.

Nearly all large optical telescopes built today are reflectors. Among the reasons is the monumental task of producing a large piece of high-quality, bubble-free glass for refracting telescopes. Because light does not pass through a mirror, the glass for a reflecting telescope does not have to be of optical quality, nor does the instrument suffer from chromatic aberration. In addition, a lens can be supported only around the edge, so it sags. Mirrors, on the other hand, can be supported fully from behind.

Large reflecting telescopes of about 4 meters (150 inches) are located at Kitt Peak, Arizona (Figure 19.10); Mauna Kea, Hawaii; Cerro Tololo, Chile; and Siding Spring, Australia (see Box 19.1). By comparison, the largest refractor in the world is the 1-meter (40-inch) telescope at Yerkes Observatory in Williams Bay, Wisconsin. This instrument was built before the turn of the century.

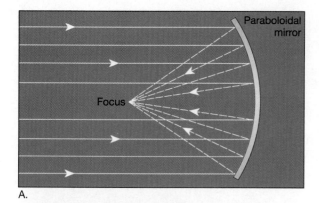

A.

B.

Figure 19.8
A. Diagram illustrating how concave mirrors, like those used in reflector telescopes, gather light. B. Preparation of the 2.4-meter mirror for the Hubble Space Telescope. (Courtesy of Space Telescope Science Institute)

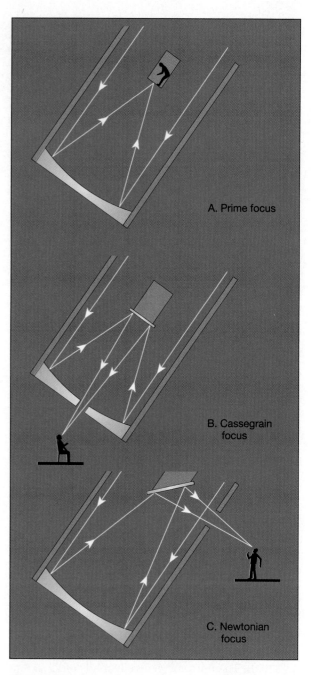

Figure 19.9
Viewing methods used with reflecting telescopes. A. Prime focus method (only used in very large telescopes). B. Cassegrain method (most common). C. Newtonian method.

BOX 19.1

The Largest Telescopes

The world's largest conventional telescope is the 6-meter (236-inch) reflector in the Caucasus Mountains in Russia. However, because of problems with location and optics, it has been less useful than many smaller telescopes.

Just behind it in size is the 5-meter (200-inch) Hale Telescope on Mount Palomar in California. The Hale Telescope is a 500-ton steel and glass marvel that floats into position on oil bearings. To ready the instrument for the night's vigil, the astronomer uses an elevator to reach the 2-meter-wide viewing cage, located inside the telescope almost 24 meters (80 feet) above the floor of the dome. From this perch, the astronomer can see the 5-meter mirror shimmering below.

Casting and grinding a large mirror is time-consuming and expensive. Thus, astronomers driven by the desire to obtain even greater light-gathering and resolving power have looked for innovative techniques for telescope building.

A new technology for casting an 8-meter mirror is currently being developed under the football stadium at the University of Arizona. This new oven turns like a merry-go-round, so that molten glass is forced outward in the mold, thereby producing a concaved outer surface. Once cooled, this glass casting has nearly the correct shape, which

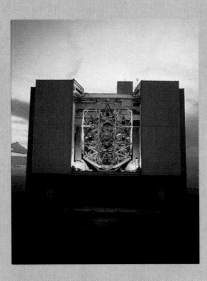

Figure 19.A
The Multiple Mirror Telescope Observatory. This unique telescope consists of six 1.8-meter mirrors that are brought to a common focus. (Courtesy of Fred Lawrence Whipple Observatory)

reduces the grinding time considerably. Twin 8-meter telescopes are planned. One will be located in the Northern Hemisphere, on Hawaii's Mauna Kea at an altitude of nearly 4200 meters (13,800 feet). The other is to be located in Chile, on the western slopes of the Andes Mountains.

Another effort is the Multiple Mirror Telescope (MMT), which uses six 1.8-meter mirrors that are linked together to focus at a single point (Figure 19.A). The result is a telescope with a light-gathering capability of a mirror 4.5 meters in diameter.

A different innovative design is the recently completed 10-meter Keck Telescope located on Mauna Kea. This instrument uses a mosaic of 36 six-sided mirrors carefully positioned by computer to give the optical effect of a 10-meter (400-inch) mirror.

Even more ambitious proposals are being considered. Plans are under consideration to employ four 8-meter mirrors in a "multiple mirror configuration" that would result in the equivalent of a single 16-meter instrument. If completed, this instrument would have 10 times the light-gathering capacity of the 5-meter Hale Telescope.

Properties of Optical Telescopes

Telescopes have three properties that aid astronomers in their work. They provide the observer with light-gathering power, resolving power, and magnifying power.

Since most celestial objects are very faint sources of light, astronomers are *most* interested in improving the *light-gathering* power of their instruments. As seen in Figure 19.11, a telescope with a large lens (or mirror) intercepts more light from distant objects and thereby produces brighter images. Since very distant stars appear very dim as well, a great deal of light must be collected before the image is bright enough to be seen. Consequently, telescopes with large objectives "see" farther into space than those with small objectives.

Another advantage of telescopes with large-diameter objectives is their greater *resolving power*

Figure 19.10
The 4-meter Mayall Telescope at Kitt Peak, south of Tucson, Arizona. This modern reflecting telescope is one of the world's largest. (Courtesy of National Optical Astronomy Observatories)

(Figure 19.12), which allows for sharper images and finer detail. For example, with the unaided eye, the Milky Way appears as a vague band of light in the night sky, but even a small telescope is capable of resolving (separating it into) individual stars. Even so, the condition of the earth's atmosphere (which is known as *seeing*) greatly limits the resolving power of earthbound telescopes. On a night when the stars twinkle, the seeing is poor, because the air is moving rapidly. This causes the image to move about, blurring it. Conversely, when the stars shine steadily, the seeing is described as good. Even under ideal conditions, however, some blurring occurs, eliminating the fine details. Thus, even the largest telescopes cannot photograph lunar features less than 0.5 kilometer (0.3 mile) in size.

To eliminate the problems of earthbound viewing, the United States built the Hubble Space Telescope, which was put into earth orbit in April 1990 (Figure 19.13). This 2.4-meter (94-inch) space telescope has 10 billion times more light-gathering power than the human eye. Although technical problems

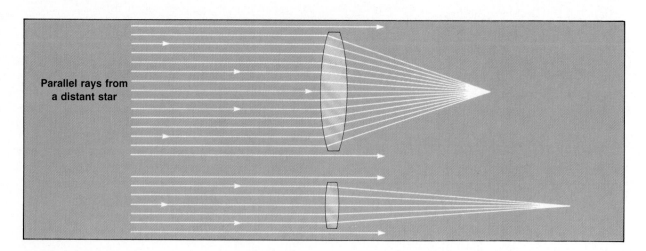

Parallel rays from a distant star

Figure 19.11
Comparison of the light-gathering ability of two lenses.

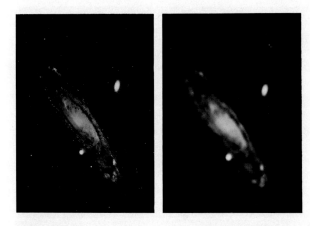

Figure 19.12
Appearance of the galaxy in the constellation Androm-
eda using telescopes of different resolution. (Courtesy
of Leiden Observatory)

have limited the usefulness of this important scientific tool, many spectacular images have been received. For example, as shown in Figure 18.31, the Hubble Space Telescope has provided images that clearly resolve the separation between Pluto and its moon, Charon. Pluto is at the center of the photo while Charon is the faint object in the lower left. For comparison, the best ground-based image taken to date is shown in the upper left of the figure.

When you think of the power of a telescope, you probably think of its *magnifying power*, the ability to make an object larger. Magnification is calculated by dividing the focal length of the objective by the focal length of the eyepiece. Thus, the magnification of a telescope can be changed by simply changing the eyepiece. However, increased magnification does not necessarily improve the clarity of the image. What can be viewed telescopically is limited by atmospheric conditions and the resolving power of the telescope. Any part of an image that is not clear at low magnification will appear only as a larger blur at higher magnification. Furthermore, increasing magnification spreads out the light and decreases the brightness of the object. Thus, astronomers describe telescopes not in terms of their magnification but by the diameter of the objective mirror or lens, because it is this factor that determines both the light-gathering power and the resolving power of a telescope.

Most modern telescopes have supplemental devices that enhance the image. A simple, but important, example is a photographic plate that can be exposed for long periods of time, thereby collecting enough light from a star to make an image that otherwise would be undetectable. One of the latest advances makes use of high-speed computers that adjust the optics to partially remove the distortion caused by the atmosphere. This process greatly enhances the image sharpness.

Detecting Invisible Radiation

As we said, sunlight is made up of more than just the radiation that is visible to our eyes. Gamma rays, x rays, ultraviolet radiation, infrared radiation, and radio waves are also produced by stars. Photographic film that is sensitive to ultraviolet and infrared radiation has been developed, thereby extending the limits of our vision. However, most of this radiation cannot penetrate our atmosphere, so balloons, rockets, and satellites must transport cameras "above" the atmosphere to record it.

Of great importance is a narrow band of radio radiation that does penetrate the atmosphere. One particular wavelength is the 21-centimeter (8-inch) line produced by neutral hydrogen (hydrogen atoms that lack an electrical charge). Measurement of this radiation has permitted us to map the galactic distribution of hydrogen—the material from which stars are made.

Figure 19.13
The deployment of the Hubble Space Telescope in earth
orbit, April 24, 1990, from Space Shuttle Discovery.
(Courtesy of Space Telescope Science Institute)

Figure 19.14
A. The 43-meter (140-foot) steerable radio telescope at Green Bank, West Virginia. The dish acts like the mirror of a reflector-type optical telescope to focus radio waves onto the antenna. The antenna is the small round object supported on four "legs" above the dish. B. Twenty-seven identical antennas operate together to form the Very Large Array near Socorro, New Mexico. (Courtesy of National Radio Astronomy Observatory)

A.

B.

The detection of radio waves is accomplished by "big dishes" called **radio telescopes** (Figure 19.14A). In principle, the dish of one of these telescopes operates in the same manner as the mirror of an optical telescope. It is parabolic in shape and focuses the radio waves on an antenna, which absorbs and transmits these waves to an amplifier, just like any radio antenna.

Because radio waves are about 100,000 times longer than visible radiation, the surface of the dish need not be as smooth as a mirror. In fact, except for the shortest radio waves, a wire mesh is a good re-

flector. On the other hand, because radio signals from celestial sources are very weak, large dishes are necessary to intercept an adequate signal. The largest radio telescope is a bowl-shaped antenna hung in a natural depression in Puerto Rico (Figure 19.15). It is 300 meters (1000 feet) in diameter and has some directional flexibility in its movable antenna. The largest steerable types have about 100-meter (330-foot) dishes like that currently being constructed at the National Radio Astronomy Observatory in Green Bank, West Virginia.

Radio telescopes also have rather poor resolution,

Figure 19.15
The 300-meter (1000-foot) radio telescope at Arecibo, Puerto Rico. (Courtesy of National Astronomy and Ionosphere Center's Arecibo Observatory, operated by Cornell University under contract with the National Science Foundation)

making it difficult to pinpoint the radio source. Pairs or groups of telescopes are used to reduce this problem. When several radio telescopes are wired together, the resulting network is called a **radio interferometer** (Figure 19.14B).

Radio telescopes have some advantages over optical telescopes. They are much less affected by turbulence in the atmosphere, clouds, and the weather in general. No protective dome is required, which reduces the cost of construction, and "viewing" is possible 24 hours a day. More importantly, radio telescopes can "see" through interstellar dust clouds that obscure our view at visible wavelengths. Radio signals from distant points in the universe pass unhindered through the dust, giving us an unobstructed view. Furthermore, radio telescopes can detect clouds of gases too cool to emit visible light. These cold gas clouds are important because they are the sites of star formation.

Radio telescopes are, however, hindered by human-made radio interference. Thus, while optical telescopes are placed on remote mountaintops to reduce interference from city lights, radio telescopes are often hidden in valleys to block human-made radio interference.

Radio telescopes have revealed such spectacular events as the collision of two galaxies. Of even greater interest was the discovery of *quasars* (quasi-stellar radio sources). These perplexing objects are the most distant things known in the universe and will be examined further in Chapter 20.

THE SUN

The sun is one of the 200 billion stars that make up the Milky Way galaxy. Although the sun is of no significance to the universe as a whole, to us on earth, it is the primary source of energy (Figure 19.16). Everything from the fossil fuels we burn in our au-

Figure 19.16
The sun, the source of over 99 percent of all energy on earth. This photo was taken during a solar eclipse on January 4, 1992. (Courtesy of National Optical Astronomy Observatories)

Figure 19.17
The unique Robert J. McMath Solar Telescope at Kitt Peak, near Tucson, Arizona. Movable mirrors at the top follow the sun, reflecting its light down the sloping tunnel. Note the glow of the lights of Tucson in this time exposure. (Courtesy of National Optical Astronomy Observatories)

tomobiles and power plants to the food we eat is ultimately derived from solar energy. The sun is also important to astronomers, since it is the only star close enough to permit study of the surface. Even with the largest telescopes, other stars appear only as points of light.

Due to the sun's brightness and its damaging radiation, it is not safe to observe it directly. However, a small telescope will project its image on a piece of cardboard held behind the telescope's eyepiece, and the sun may be studied safely in this manner. This basic method is used in several telescopes around the world, which keep a constant vigil of the sun. One of the finest is at the Kitt Peak National Observatory in southern Arizona (Figure 19.17). It consists of a 150-meter sloped enclosure that directs sunlight to a mirror situated below ground. From the mirror, an 85-centimeter (33-inch) image of the sun is projected to an observing room, where it is studied.

Compared to other stars of the universe, many of which are larger, smaller, hotter, cooler, more red, or more blue, the sun is an "average star." However, on the scale of our solar system, it is truly gigantic, having a diameter equal to 109 earth diameters (1.35 million kilometers) and a volume 1¼ million times as great as that of the earth. Yet, because of its gaseous nature, its density is only ¼ that of the solid earth, nearly the density of water.

Figure 19.18
Diagram of solar structure in cutaway view.

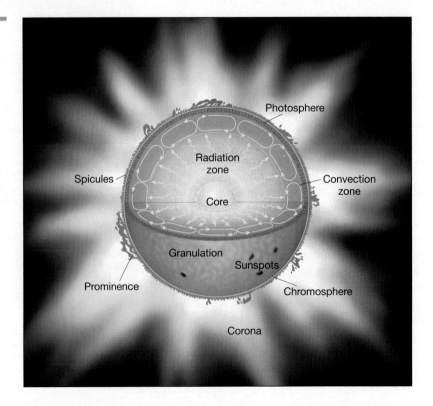

Structure of the Sun

For convenience of discussion, we divide the sun into four parts: the solar interior; the visible surface, or photosphere; and the two layers of its atmosphere, the chromosphere and the corona (Figure 19.18). Because the sun is gaseous throughout, no sharp boundaries exist between these layers. The sun's interior makes up all but a tiny fraction of the solar mass, and unlike the outer three layers, it is not accessible to direct observation. We shall discuss the visible layers first.

Photosphere. The **photosphere** ("sphere of light") is aptly named, for this radiates most of the sunlight we see and therefore appears as the bright disk of the sun. Although it is considered to be the sun's "surface," it is unlike most surfaces we are accustomed to. The photosphere consists of a layer of incandescent gas 300 kilometers (200 miles) thick

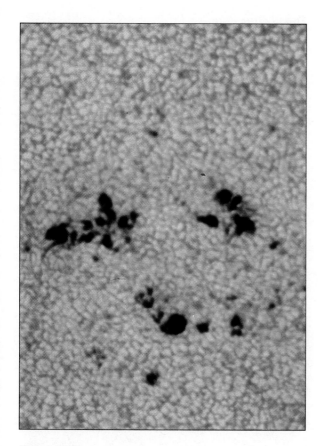

Figure 19.19
Granules of the solar photosphere. (Courtesy of National Optical Astronomy Observatories)

having a pressure less than 1/100 of our atmosphere. Furthermore, it is neither smooth nor uniformly bright as the ancients had imagined. It has numerous blemishes.

When viewed through a telescope under ideal conditions, the photosphere's grainy texture is apparent. This is the result of numerous, relatively small, bright markings called **granules,** which are surrounded by narrow, dark regions (Figure 19.19). Granules are typically 1000 kilometers (620 miles) in diameter and owe their brightness to hotter gases that are rising from below. As this gas spreads laterally, cooling causes it to darken and sink back into the interior. Although each granule survives only a few minutes, the combined motion of all granules gives the photosphere the appearance of boiling. This up-and-down movement of gas is called *convection*. Besides producing the grainy appearance of the photosphere, convection is believed to be responsible for the transfer of energy in the uppermost part of the sun's interior (Figure 19.18).

The composition of the photosphere is revealed by the dark lines of its absorption spectrum (see Figure 19.3). When these "fingerprints" are compared to the spectra of known elements, they indicate that most of the elements found on earth also occur on the sun. When the strengths of the absorption lines are analyzed, the relative abundance of the elements can be determined. These studies reveal that 90 percent of the sun's surface atoms are hydrogen, almost 10 percent are helium, and only minor amounts of the other detectable elements are present. Other stars also indicate similar disproportionate percentages of these two lightest elements, a fact we shall consider later.

Chromosphere. Just above the photosphere lies the **chromosphere** ("color sphere"), a relatively thin layer of hot, incandescent gases a few thousand kilometers thick. The chromosphere is observable for a few moments during a total solar eclipse or by using a special instrument that blocks out the light from the photosphere. Under such conditions, it appears as a thin red rim around the sun. Because the chromosphere consists of hot, incandescent gases under low pressure, it produces a bright-line spectrum that is nearly the reverse of the dark-line spectrum of the photosphere. One of the bright lines of hydrogen contributes a good portion of its total light and accounts for this sphere's red color.

A study of the chromospheric spectrum conducted in 1868 revealed the existence of an unknown ele-

Figure 19.20
Spicules of the chromosphere on the edge of the solar disk. (Courtesy of Sacramento Peak Observatory, Air Force Cambridge Research Laboratories)

ment. It was named helium, from *helios*, the Greek word for "sun." Originally, helium was thought to be an element unique to the stars, but 27 years later, it was discovered in natural-gas wells on earth.

The top of the chromosphere contains numerous **spicules,** which extend upward into the lower corona, almost like the trees which reach into our atmosphere (Figure 19.20). The spicules may be continuations of the turbulent motion of the granules below.

Corona. The outermost portion of the solar atmosphere, the **corona,** is very tenuous and, as with the chromosphere, is visible only when the brilliant photosphere is covered (Figure 19.21). This envelope of ionized gases normally extends a million kilometers from the sun and produces a glow about half as bright as the full moon.

At the outer fringe of the corona, the ionized gases have speeds great enough to escape the gravitational pull of the sun. The streams of protons and electrons that "boil" from the corona constitute the **solar wind.** They travel outward through the solar system at very high speeds (250–800 kilometers a second) and eventually are lost to interstellar space. During its journey, the solar wind interacts with the bodies

of the solar system, continually bombarding lunar rocks and altering their appearance. Although the earth's magnetic field prevents the solar winds from reaching our surface, these winds do affect our atmosphere, as we shall discuss later.

Studies of the energy emitted from the photosphere indicate that its temperature averages about 6000 K (10,000°F). Upward from the photosphere, the temperature unexpectedly increases, exceeding 1 million K at the top of the corona. It should be noted that, although the coronal temperature exceeds that of the photosphere by many times, it radiates much less energy overall because of its very low density. Surprisingly, the high temperature of the corona is probably caused by sound waves generated by the convective motion of the photosphere. Just as boiling water makes noise, the energetic sound waves generated in the photosphere are believed to be absorbed by the gases of the corona and thereby raise their temperatures.

The Active Sun

Sunspots. The most conspicuous features on the surface of the sun are the dark blemishes called **sunspots** (Figure 19.22A). Although large sunspots were occasionally observed before the advent of the telescope, they were generally regarded as opaque

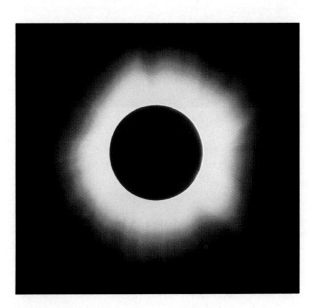

Figure 19.21
Solar corona photographed during a total eclipse. (Courtesy of National Optical Astronomy Observatories)

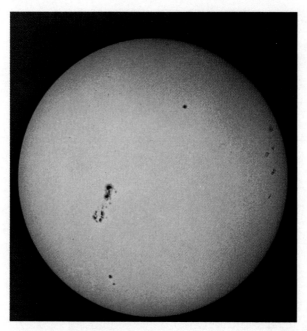
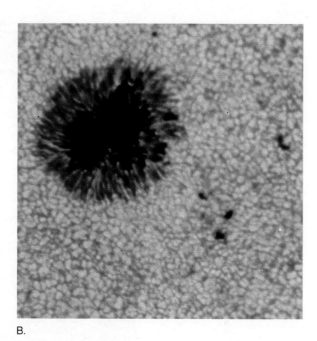

B.

Figure 19.22
A. Large sunspot group on the solar disk. (Celestron 8 photo courtesy of Celestron International) *B.* Sunspots having visible umbra (dark central area) and penumbra (lighter area surrounding umbra). (Courtesy of National Optical Astronomy Observatories)

objects located somewhere between the sun and the earth. In 1610 Galileo concluded that they were residents of the solar surface, and from their motion, he deduced that the sun rotates on its axis about once a month.

Observations made later indicated that not all parts of the sun rotate at the same speed. The sun's equator rotates once in 25 days, while a place located 70 degrees from the solar equator, either north or south, requires 33 days for one rotation. If the earth rotated in a similar disjointed manner, imagine the consequences! The sun's nonuniform rotation is a testimonial to its gaseous nature.

Sunspots begin as small dark pores about 1600 kilometers (1000 miles) in diameter. While most pores last for only a few hours, some grow into blemishes many times larger than the earth and last for a month or more. The largest spots often occur in pairs surrounded by several smaller spots. An individual spot contains a black center, the *umbra*, which is rimmed by a lighter region, the *penumbra* (Figure 19.22B). Sunspots appear dark only by contrast with the brilliant photosphere, a fact accounted for

by their temperature, which is about 1500 K less than that of the solar surface. If these dark spots could be observed away from the sun, they would appear many times brighter than the full moon.

During the early 1800s, it was believed that a tiny planet named Vulcan orbited between Mercury and the sun. In the search for Vulcan an accurate record of sunspot occurrences was kept. Although the planet was never found, the sunspot data collected did reveal that the number of sunspots observable on the solar disk varies in an 11-year cycle. First, the number of sunspots increases to a maximum, with perhaps a hundred or more visible at a given time. Then, over a period of 5–7 years, their numbers decline to a minimum, when only a few or even none are visible. At the beginning of each cycle, the first sunspots form about 30 degrees from the solar equator, but as the cycle progresses and their numbers increase, they form nearer the equator. During the period when sunspots are most abundant, the majority form about 15 degrees from the equator. They rarely occur more than 40 degrees away from the sun's equator, or within 5 degrees of it.

Figure 19.23
Solar disk photographed in hydrogen alpha light, showing the manifestations of the active sun. (This composite courtesy of Hale Observatories and Sacramento Peak Observatory, Air Force Cambridge Research Laboratories)

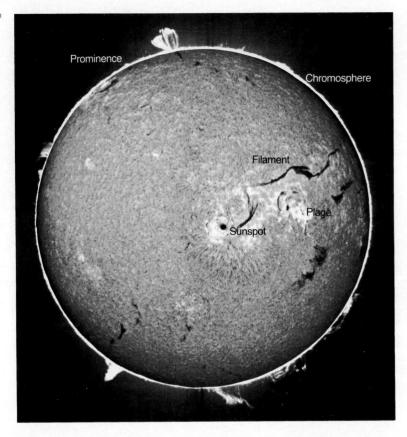

Another interesting characteristic of sunspots was discovered by astronomer George Hale, for whom the Hale Telescope is named. Hale deduced that the large spots are strongly magnetized, and when they occur in pairs, they have opposite magnetic poles. For instance, if one member of the pair is a north magnetic pole, then the other member is a south magnetic pole, as with the north and south poles of the earth's magnetic field. Also, every pair located in the same hemisphere is magnetized in the same manner. However, all pairs in the *other* hemisphere are magnetized in the opposite manner. At the beginning of each sunspot cycle, the situation reverses, and the polarity of these sunspot pairs is opposite those of the previous cycle. The cause of this change in polarity, in fact the cause of sunspots themselves, is not fully explained. However, other solar activity varies in the same cyclic manner as sunspots do, indicating a common origin.

Plages and Prominences. The brilliance of the photosphere makes viewing the activity occurring above the solar surface very difficult. To overcome this problem, numerous photographic methods are employed to filter solar radiation so that light of a particular spectral band (color) can be viewed. Using such techniques, large "clouds" can be seen in the chromosphere directly above sunspot clusters (Figure 19.23). These bright centers of solar activity are called **plages** and occasionally can even be viewed before or after the sunspot occurrences.

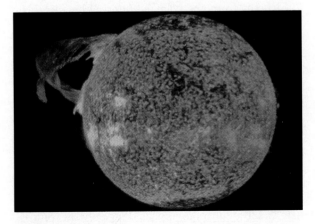

Figure 19.24
A huge solar prominence. (Courtesy of NASA)

Among the more spectacular features of the active sun are **prominences.** These huge cloudlike structures are best observed when they are on the edge, or limb, of the sun, where they often appear as great arches that extend well into the corona (Figure 19.24). Many prominences have the appearance of a fine tapestry and seem to hang motionless for days at a time, but motion pictures reveal that the material within them is continually falling like luminescent rain. Apparently these quiet prominences are condensations of coronal material that are gracefully "sliding down" the lines of magnetic force back into the chromosphere. More rarely, the material within a prominence rises almost explosively away from the sun. These active prominences reach velocities up to 1000 kilometers (620 miles) per second and may leave the sun entirely. Prominences can also be seen against the bright disk of the sun, in which case they appear as dark, thin streaks called *filaments* (Figure 19.23).

Figure 19.25
Aurora borealis (northern lights) as seen from Alaska. The same phenomenon occurs toward the South Pole, where it is called the Aurora australis (southern lights). (Courtesy of Gustav Lamprecht)

Solar Flares. These are the most explosive events associated with sunspots. **Solar flares** are brief outbursts that normally last an hour or so and appear as a sudden brightening of the region above a sunspot cluster. During their existence, enormous quantities of energy are released, much of it in the form of ultraviolet, radio, and x-ray radiation. Simultaneously, fast-moving atomic particles are ejected, causing the solar wind to intensify noticeably. Although a major flare could conceivably endanger a manned space flight, they are relatively rare. About a day after a large outburst, the ejected particles reach the earth and disturb the ionosphere,* affecting long-distance radio communications.

The most spectacular effect of solar flares, however, is the **auroras,** also called the northern and southern lights (Figure 19.25). Following a strong solar flare, the earth's upper atmosphere near its magnetic poles is set aglow for several nights. The auroras appear in a wide variety of forms. Sometimes the display consists of vertical streamers with considerable movement. At other times, the auroras appear as a series of luminous expanding arcs or as a quiet, almost foglike, glow. Auroral displays, like other solar activities, vary in intensity with the 11-year sunspot cycle.

The Solar Interior

The interior of the sun cannot be observed directly. For that reason, all we know about it is based on information acquired from the energy it radiates and from theoretical studies. The source of the sun's energy, **nuclear fusion,** was not discovered until the late 1930s.

Deep in its interior, a nuclear reaction called the **proton-proton chain** converts four hydrogen nuclei (protons) into the nucleus of a helium atom. The energy released from the proton-proton reaction results because some of the matter involved is actually converted to energy. This can be illustrated by noting that four hydrogen atoms have a combined atomic mass of 4.032 (4 × 1.008), whereas the atomic mass of helium is 4.003, or 0.029 less than the combined mass of the hydrogen. The tiny missing mass is emitted as energy according to Einstein's formula $E = mc^2$,

---

* The ionosphere is a complex atmospheric zone of ionized gases that extends between about 80 and 400 kilometers (50 and 250 miles) above the earth's surface.

BOX 19.2

Variable Sun and Climatic Change

Among the most persistent hypotheses of climatic change have been those based on the idea that the sun is a variable star and that its output of energy varies through time. The effect of such changes would seem direct and easily understood: Increases in solar output would cause the atmosphere to warm, and reductions would result in cooling. This notion is appealing because it can be used to explain climatic changes of any length or intensity. Still, there is at least one major drawback: No major long-term variations in the total intensity of solar radiation have yet been measured.

Several proposals for climatic change, based on a variable sun, relate to *sunspot cycles*. The most conspicuous and best-known features on the surface of the sun are the dark blemishes called *sunspots*. Although their origin is uncertain, it has been established that sunspots are huge magnetic storms that extend from the sun's surface deep into the interior. Moreover, these spots are associated with the sun's ejection of huge masses of particles that, on reaching the earth's atmosphere, interact with gases there to produce auroral displays.

Along with other solar activity, the number of sunspots increases and decreases on a regular basis, creating a cycle of about 11 years. A curve of the annual number of sunspots, beginning in the early 1700s, appears to be very regular (Figure 19.B).

Studies indicate that there have been prolonged periods when sunspots were absent, or nearly so. Moreover, it was found that these events corresponded closely with cold periods in Europe and North America. Conversely, periods characterized by plentiful sunspots were found to correlate well with warmer times in these regions. Referring to these excellent matches, the solar astronomer John Eddy stated: "These early results in comparing ing solar history with climate make it appear that changes on the sun are the dominant agent of climatic changes lasting between 50 and several hundred years."

But other scientists seriously questioned this conclusion. Their hesitation stems in part from subsequent investigations, using different climate records from around the world, that failed to find a significant correlation between variations in solar activity and climate. Even more troubling seems to be the fact that no testable physical mechanism exists to explain the purported effect.

Figure 19.B
Mean annual sunspot numbers.

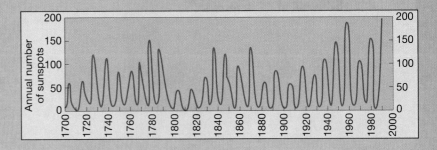

where E equals energy, m equals mass, and c equals the speed of light. Because the speed of light is very great, the amount of energy released from even a small amount of mass is enormous.

The conversion of just one pinhead's worth of hydrogen to helium generates more energy than burning thousands of tons of coal. Most of this energy is in the form of high energy photons that work their way toward the solar surface, being absorbed and re-emitted many times until they reach an opaque layer just below the photosphere. Here, convection currents serve to transport this energy to the solar surface, where it radiates through the transparent chromosphere and corona (see Figure 19.18).

Only a small percentage (0.7 percent) of the hydrogen in the proton-proton reaction is actually converted to energy. Nevertheless, the sun is consuming an estimated 600 million tons of hydrogen each second, with about 4 million tons of it being converted to energy. As hydrogen is consumed, the product of this reaction, helium, forms the solar core, which continually grows in size.

Just how long can the sun produce energy at its present rate before all of its fuel (hydrogen) is consumed? Even at the enormous rate of consumption, the sun has enough fuel to last easily another 100 billion years. However, evidence from other stars indicates that the sun will grow dramatically and engulf the earth long before all of its hydrogen is gone. It is thought that a star the size of the sun can exist in its present stable state for 10 billion years. Since the sun is already 5 billion years old, it is "middle-aged."

To initiate the proton-proton reaction, the sun's internal temperature must have reached several million degrees. What was the source of this heat? As previously noted, the solar system is believed to have formed from an enormous cloud of dust and gases (mostly hydrogen) that condensed gravitationally. The consequence of squeezing (compressing) a gas is to increase its temperature. Although all of the bodies in the solar system were compressed, the sun was the only one, because of its size, that became hot enough to trigger the proton-proton reaction. Astronomers currently estimate its internal temperature at 15 million K.

The planet Jupiter is basically a hydrogen-rich gas ball; if it were about ten times more massive, it too might have become a star. The idea of one star orbiting another may seem odd, but recent evidence indicates that about 50 percent of the stars in the universe probably occur in pairs or multiples!

Review Questions

1. What term is used to describe the collection that includes gamma rays, x rays, ultraviolet light, visible light, infrared light, and radio waves?

2. Which color has the longest wavelength? The shortest?

3. What is spectroscopy?

4. Describe a continuous spectrum. Give an example of a natural phenomenon that exhibits a continuous spectrum.

5. What produces emission lines (bright lines) in a spectrum?

6. What can be learned about a star (or other celestial objects) from a dark-line (absorption) spectrum?

7. How can astronomers determine whether a star is moving toward or away from the earth?

8. The primary objective of a refracting telescope is a mirror: True or false?

9. What three properties do telescopes have that aid astronomers?

10. Of the three properties listed in Question 9, which is the least important to astronomers?

11. What is the advantage of a space telescope over a similar earthbound instrument?

12. Why do astronomers seek to design telescopes with larger and larger objectives?

13. Why do all large optical telescopes use mirrors to collect light rather than lenses?

14. With reflecting telescopes, why are special viewing systems needed?

15. Explain the following statement: "Photography has extended the limits of our vision."

16. Why would the moon make a good site for an observatory?

17. Why are radio telescopes much larger than optical telescopes?

18. What are some of the advantages of radio telescopes over optical telescopes?

19. Compare the diameter of the sun to that of the earth.

20. Describe the photosphere, chromosphere, and corona.

21. List the features associated with the active sun and describe each.

22. Explain how a sunspot can be very hot and yet appear dark.

23. What is the solar wind?

24. What "fuel" does the sun consume?

25. What happens to the matter that is consumed in the proton-proton chain reaction?

Key Terms

aurora (p. 661)

bright-line (emission) spectrum (p. 646)

chromatic aberration (p. 649)

chromosphere (p. 657)

continuous spectrum (p. 645)

corona (p. 658)

dark-line (absorption) spectrum (p. 646)

Doppler effect (p. 647)

electromagnetic radiation (p. 644)

eyepiece (p. 648)

focal length (p. 648)

focus (p. 648)

granules (p. 657)

nuclear fusion (p. 661)

objective lens (p. 648)

photon (p. 644)

photosphere (p. 657)

plage (p. 660)

prominence (p. 661)

proton-proton chain (p. 661)

radiation pressure (p. 644)

radio interferometer (p. 655)

radio telescope (p. 654)

reflecting telescope (p. 649)

refracting telescope (p. 648)

solar flare (p. 661)

solar wind (p. 658)

spectroscope (p. 647)

spectroscopy (p. 645)

spicule (p. 658)

sunspot (p. 658)

20

Beyond Our Solar System

Opposite page: The Trifid Nebula in the constellation
Sagittarius. (Courtesy of NOAO)

Left: The Pleiades star cluster. (Courtesy of Hale
Observatories)

Above: Eagle Nebula in the constellation Serpens.
(Courtesy of NOAO)

The star Proxima Centauri is about 4.3 light-years away, roughly 100 million times farther than the moon. Yet, other than our own sun, it is the closest star to the earth. This fact suggests that the universe is incomprehensibly large. What is the nature of this vast cosmos beyond our solar system? Are the stars distributed randomly, or are they organized into distinct clusters? Do stars move, or are they permanently fixed features, like lights strung out against the black cloak of outer space? Does the universe extend infinitely in all directions, or is it bounded? To consider these questions, this chapter will examine the universe by taking a census of the stars—the most numerous objects in the night sky.

PROPERTIES OF STARS

Although the sun is the only star whose surface we can observe, a great deal is known about the universe beyond our solar system. In fact, more is known about the stars than about our outermost planet, Pluto. This knowledge hinges on the fact that stars, and even gases in the "empty" space between stars, radiate energy in all directions into space (Figure 20.1). The key to understanding the universe is to collect this radiation and unravel the secrets it holds. Astronomers have devised many ingenious methods to do just that. We will begin by examining stellar distances and some intrinsic properties of stars, including color, brightness, mass, temperature, and size.

Distances to the Stars

Measuring the distance to a star is difficult, at best. Obviously, we cannot journey to the star. However, astronomers have developed some indirect methods of measuring stellar distances. The most basic of these measurements is called stellar parallax.

Recall from Chapter 17 that **stellar parallax** is the extremely slight back-and-forth shifting in a nearby star's position due to the orbital motion of the earth. The principle of parallax is easy to visualize. Close one eye and with your index finger in a vertical position, use your eye to line your finger up with some distant object. Now, without moving your finger, view the object with your other eye and notice that its position appears to have changed. The farther away you hold your finger, the less its position seems to shift.

In practice, parallax is determined by photographing a nearby star against the background of distant stars. Then, six months later, when the earth has moved halfway around its orbit, a second photograph is taken. When these photographs are compared, the position of the nearby star appears to have shifted with respect to the background stars. Figure 20.2 illustrates this shift and the parallax angle determined from it. The nearest stars have the largest parallax angles, while those of distant stars are too slight to measure. Recall that the medieval astronomer Tycho Brahe was unable to detect stellar parallax, leading him to reject the idea that the earth orbits the sun.

It should be emphasized that parallax angles are *very* small. The parallax angle to the nearest star, Proxima Centauri, is less than 1 second of arc, which

Figure 20.1
Lagoon Nebula. It is in glowing clouds like these that gases and dust particles become concentrated into stars. (Courtesy of National Optical Astronomy Observatories)

equals 1/3600 of a degree. To put this in perspective, fully extend your arm and raise your little finger. Your finger is roughly 1 degree wide. Try doing this on a moonlit night, covering the moon with your finger. The moon is only about ½ degree wide. Now, imagine detecting a movement that is only 1/3600 as wide as your finger. It should be apparent why Tycho Brahe, without the aid of a telescope, was unable to observe stellar parallax.

The distances to stars are so large that the conventional units, such as kilometers or astronomical units, are often too cumbersome to use. Astronomers generally express stellar distances in units called **parsecs.** One parsec is 206,265 astronomical units, or 206,265 times the earth-sun distance. Our nearest stellar neighbor, Proxima Centauri, is 1.31 parsecs away, or nearly three hundred thousand times the distance from the earth to the sun.

Another unit used to express stellar distance is the **light-year,** which is the distance light travels in a year—about 9.5 trillion kilometers (5.8 trillion miles). For purposes of conversion, 1 parsec equals 3.26 light-years. The distances to some nearby stars are given in all three units in Table 20.1.

In principle, the method used to measure stellar distances is elementary and was known to the ancient Greeks. But in practice, the tiny angles involved and the fact that the sun, as well as the designated

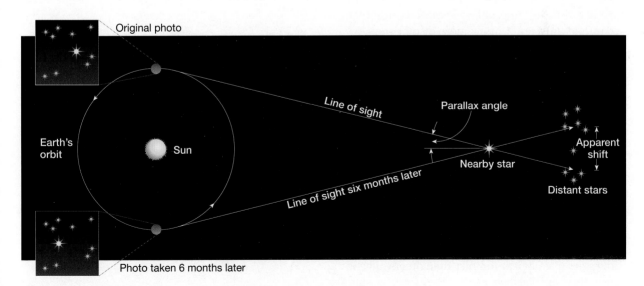

Figure 20.2
Geometry of stellar parallax. The parallax angle shown here is enormously exaggerated to illustrate the principle. Because the distance to even the nearest stars is thousands of times greater than the earth-sun distance, the triangles that astronomers work with are very long and narrow, making the angles that are measured very small.

Table 20.1
Distance to some stars.

Name	Par-secs	Light-years	Kilometers
Proxima Centauri	1.3	4.27	4.1×10^{13}
Sirius	2.7	8.70	8.3×10^{13}
Arcturus	11	36	3.4×10^{14}
Betelgeuse	159	520	4.9×10^{15}
Rigel	276	900	8.5×10^{15}
Deneb	491	1600	1.5×10^{16}

star, also has actual motion greatly complicate the measurements. The first accurate stellar parallax was not determined until 1838. Even today, parallaxes for only a few thousand of the nearest stars are known with certainty.

Stars more distant than 100 parsecs have such small shifts that accurate measurement is not possible. Fortunately, a few other methods have been derived for estimating distances to these stars. Also, the Hubble Space Telescope, which is not hindered by the earth's turbulent atmosphere, is expected to obtain accurate parallaxes of many more distant stars.

Stellar Brightness

Three factors control the apparent brightness of a star as seen from earth: *how big* it is, *how hot* it is, and *how far away* it is. The stars in the night sky are a grand assortment of sizes, temperatures, and distances, so their brightnesses vary widely.

Apparent Magnitude.　Stars have been classified according to their brightness since the second century B.C., when Hipparchus placed about 1000 of them into six categories. The measure of a star's brightness is called its **magnitude.** As mentioned, some stars may appear dimmer than others only because they are farther away. Therefore, a star's brightness, *as it appears when viewed from the earth*, has been termed its **apparent magnitude.**

When numbers are employed to designate relative brightness, the larger the magnitude number, the

dimmer the star. Stars that appear the brightest are of the first magnitude, while the faintest stars visible to the unaided eye are of the sixth magnitude. With the invention of the telescope, many stars fainter than the sixth magnitude were discovered.

In the middle of the nineteenth century, a method was developed whereby magnitudes could be used to quantitatively compare the brilliance of stars. So, just as we can compare the brightness of a 50-watt light bulb to that of a 100-watt bulb, we can compare the brightness of stars having different magnitudes. It was determined that a first-magnitude star was about 100 times brighter than a sixth-magnitude star. Therefore, on the scale that was devised, two stars that differ by 5 magnitudes have a ratio in brightness of 100 to 1. Hence, a seventh-magnitude star is 100 times brighter than a twelfth-magnitude star. It follows, then, that the brightness ratio of two stars differing by only one magnitude is about 2.5*. Thus, a star of the first magnitude is about 2.5 times brighter than a star of the second magnitude. Table 20.2 shows some magnitude differences and the corresponding brightness ratios.

Because some stars are brighter than first-magnitude stars, zero and negative magnitudes were introduced. The brightest star in the night sky, Sirius, has an apparent magnitude of −1.4, about 10 times brighter than a first-magnitude star. On this scale, the sun has an apparent magnitude of −26.7. At its

---

*　Note that 2.512 × 2.512 × 2.512 × 2.512 × 2.512 or 2.512 raised to the fifth power equals 100.

Table 20.2
Ratios of star brightness.

Difference in Magnitude	Brightness Ratio
0.5	1.6:1
1	2.5:1
2	6.3:1
3	16:1
4	40:1
5	100:1
10	10,000:1
20	100,000,000:1

brightest, Venus has a magnitude of −4.3. At the other end of the scale, the 5-meter (200-inch) Hale Telescope can view stars with an apparent magnitude of 23, approximately 100 million times dimmer than stars that are visible to the unaided eye.

Absolute Magnitudes. Astronomers are also interested in the "true" brightness of stars, called their **absolute magnitudes.** Stars of the same luminosity, or brightness, usually do not have the same apparent magnitude, because their distances from us are not equal. To compare their true, or intrinsic, brightness, astronomers determine what magnitude the star would have if it were at a standard distance of 10 parsecs, or about 32.6 light-years. For example, the sun, which has an apparent magnitude of −26.7, would, if located at a distance of 32.6 light-years, have an absolute magnitude of about 5. Thus, stars with absolute magnitudes greater than 5 (smaller numerical value) are intrinsically brighter than the sun, but because of their distance, they appear much dimmer.

Table 20.3 lists the absolute and apparent magnitudes of some stars as well as their distances from the earth. Most stars have an absolute magnitude between −5 and 15, which puts the sun near the middle of this range.

Table 20.3
Distance, apparent magnitude, and absolute magnitude of some stars.

Name	Distance (Parsecs)	Apparent Magnitude*	Absolute Magnitude*
Sun	NA	−26.7	5.0
Alpha Centauri	1.3	0.0	4.4
Sirius	2.7	−1.4	1.5
Epsilon Indi	3.4	4.7	7.0
Kruger 60B	3.9	11.2	13.2
Arcturus	11.0	−0.1	−0.3
Antares	120.0	1.0	−4.5
Betelgeuse	150.0	0.8	−5.5
Deneb	430.0	1.3	−6.9

* The more negative, the brighter; the more positive, the dimmer.

Figure 20.3
Time-lapse photograph of stars in the constellation Orion. These star trails show some of the various star colors. It is important to note that the eye sees color somewhat differently than does film. (Courtesy of National Optical Astronomy Observatory)

Stellar Color and Temperatures

The next time you are outdoors on a clear night, take a good look at the stars and note their color (Figure 20.3). Some that are quite colorful can be found in the constellation Orion (see Appendix E, winter chart). One of the two brightest stars in Orion, Betelgeuse (α Orionis), is definitely red, whereas the other, Rigel (β Orionis), appears blue.

Very hot stars with surface temperatures above 30,000 K emit most of their energy in the form of short-wavelength light and therefore appear blue. Red stars, on the other hand, are much cooler, generally less than 3000 K, and most of their energy is emitted as longer-wavelength red light. Stars with temperatures between 5000 and 6000 K appear yellow, like the sun. Because color is primarily a manifestation of a star's temperature, this characteristic can provide the astronomer with useful information about a star.

Binary Stars and Stellar Masses

Persons with good eyesight can resolve the second star in the handle of the Big Dipper (Mizar) as two stars. (For the sky location, see Appendix E, first star chart, top.) With their new tool, the telescope, astronomers discovered numerous such star pairs during the eighteenth century. One of the stars in the pair is usually fainter than the other, and for this rea-

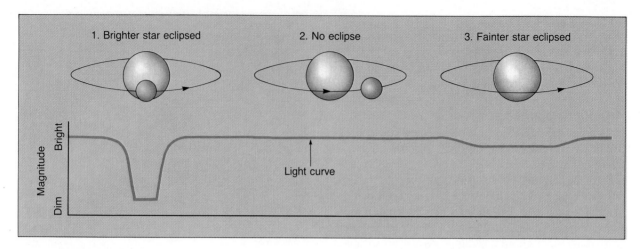

Figure 20.4
Idealized light curve and geometry of an eclipsing binary star.

son, was considered to be farther away. In other words, the stars were not considered true pairs but were thought only to lie in the same line of sight.

In the early nineteenth century, careful examination of numerous pairs by William Herschel revealed that many of these star pairs actually orbit one another. The two stars are in fact united by their mutual gravitation. Newton's law of gravitation had been extended to the stars. These pairs of stars, in which the members are far enough apart to be resolved telescopically, are called **visual binaries.** The idea of one star orbiting another may seem unusual, but evidence indicates that more than 50 percent of the stars in the universe occur in pairs or multiples.

One of the more interesting groups of non-visual binaries is detectable by changes in brightness. This change occurs when the plane of the orbits of two stars lies in our view, such that the stars continually eclipse one another (Figure 20.4). When the fainter of the two stars eclipses the brighter, a major drop in intensity is noted, and when the brighter star eclipses the fainter one, a lesser drop is observed.

Binary stars are used to determine the star property most difficult to calculate—its mass. Recall that the mass of a body can be established if it is gravitationally attached to a partner, which is the case for any binary star system. Using Kepler's third law as redefined by Newton, the sum of the masses of two stars equals the cube of their mean distance (measured in astronomical units, AU) divided by their pe-

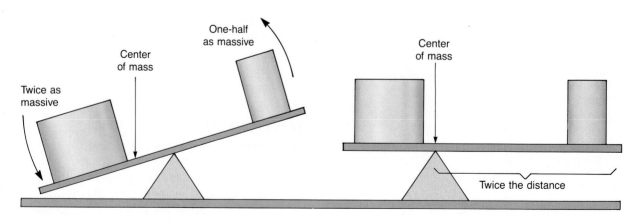

Figure 20.5
A seesaw used to illustrate the center of mass.

riod squared. However, this gives information only about the combined masses of the two stars, not about the individual masses.

Fortunately, binary stars orbit each other around a common point called the *center of mass* in a manner dependent on their individual masses. For stars of equal mass, the center of mass lies exactly halfway between them. If one star is more massive than its partner, their common center will be located closer to the more massive one. Thus, if the sizes of their orbits can be observed, a determination of their individual masses can be made. You can experience this relationship on a seesaw by trying to balance a person who has a much greater mass (Figure 20.5).

For illustration, when one star has an orbit half the size of its companion, it is twice as massive. If their combined masses, as determined by Kepler's law, are equal to 3 times the mass of the sun, then the larger will be twice as massive as the sun, and the smaller will have a mass equal to that of the sun. Most stars have masses that range between 1/10 and 50 times the mass of the sun.

STELLAR CLASSIFICATION

The vast majority of stars have continuous spectra on which a number of dark absorption lines are superimposed. Recall from our discussion of light that a given set of dark lines can be attributed to the presence of a particular element. Early studies of stellar spectra revealed that an element such as calcium or helium produces strong absorption lines in the spectra of some stars, but only weak lines in others, and no lines in still others. After the spectra of numerous stars were obtained, it was inevitable that someone would try to group stars according to their spectra.

The first classification scheme was based on the relative strengths of selected absorption lines. Thus, stars with similar-appearing spectra were placed in the same **spectral class.** This scheme assumed that stars with an abundance of a particular element, for instance, calcium, would produce strong calcium lines. Elements for which no absorption lines appeared were considered absent or rare.

Later it was determined that the spectral variations observed were primarily the result of *temperature* differences, not *compositional* differences. For example, very cool red stars (class M) have strong absorption lines of molecular titanium oxide (TiO). However, in the spectra of stars whose temperatures are above 5000 K, these lines do not appear. At these temperatures, the titanium oxide molecules are unstable and are split into atomic titanium and atomic oxygen. Hence, hotter stars probably contain as much titanium as cooler stars, but the absorption lines of the titanium oxide molecule cannot be produced because that molecule cannot exist in that environment. Similarly, the presence of most absorption lines was found to be dependent on the star's surface temperature, not its composition.

The original spectral classes had not been labeled with letters of the alphabet and were arranged by composition. They then were rearranged in order of decreasing temperatures but retained the old letter designations. The temperature order is O, B, A, F, G, K, M (Table 20.4). O-type stars are hot, blue stars with temperatures up to 50,000 K, and the coolest red stars (temperatures less than 3500 K) belong to the M spectrum class. The sun is a yellow G-type star, and the somewhat cooler K-type stars appear orange.

By now it should be obvious that once a star's spectrum has been classified, its surface temperature has also been established. As we shall see, a star's temperature is very useful in determining its size.

HERTZSPRUNG-RUSSELL DIAGRAM

Early in this century, Einar Hertzsprung and Henry Russell independently studied the relation between the brightness (absolute magnitude) and temperature (spectral class) of stars. From this research each developed a graph, now called a **Hertzsprung-Russell diagram,** or simply **H-R diagram,** that exhibits these intrinsic stellar properties. By studying H-R diagrams, we learn a great deal about the sizes of stars.

To produce an H-R diagram, astronomers survey a portion of the sky and plot each star according to its luminosity and temperature (Figure 20.6). Notice that the plots in Figure 20.6 are not uniformly distributed. Rather, about 90 percent of all stars fall along a band that runs from the upper-left corner to the lower-right corner of an H-R diagram. These "ordinary" stars are called **main-sequence stars.** As shown in Figure 20.6, the hottest main-sequence stars are intrinsically the brightest, and vice versa.

The luminosity of the main-sequence stars is also related to their mass. The hottest (blue) stars are about 50 times more massive than the sun, whereas the coolest (red) stars are only 1/10 as massive. Therefore, on the H-R diagram, the main-sequence

Table 20.4
Characteristics of spectral classes.

Spectral Class	Color	Approximate Temperature (K)	Principal Spectral Characteristics	Approximate Mass of Main-Sequence Star (Sun = 1)	Stellar Example
O	Blue	>30,000	Strong lines of ionized helium; weak hydrogen lines	40.0	10 Lacertae
B	Blue-white	10,500–30,000	Stronger hydrogen lines than O type; lines of neutral helium dominate	10.0	Rigel
A	White	7500–10,500	Hydrogen lines reach maximum strength; lines of ionized metals strong	3.0	Vega
F	Yellow-white	6000–7500	Hydrogen lines weaken; lines of neutral metals strengthen; lines of ionized metals weaken	1.5	Canopus
G	Yellow	5000–6000	Strong calcium lines; strong lines of neutral and ionized metals; weak hydrogen lines	1.0	Sun
K	Orange	3500–5000	Lines of neutral metals dominate; hydrogen lines weak but detectable	0.8	Arcturus
M	Red	<3500	Numerous lines of neutral metals; band TiO molecules	0.1	Antares

stars appear in a decreasing order, from *hotter, more massive* blue stars to *cooler, less massive* red stars.

Note the location of the sun in Figure 20.6. The sun is a G-type, main-sequence star with an absolute magnitude of about 5. Because the magnitudes of a vast majority of main-sequence stars lie between −5 and 15, and because the sun falls midway in this range, the sun is often considered an average star. However, more main-sequence stars are cooler and less massive than our sun.

Just as all humans do not fall into the normal size range, some stars do not fit in with the main-sequence stars. Above and to the right of the main sequence in the H-R diagram in Figure 20.6 lies a group of very luminous stars called **giants** or, on the basis of their color, **red giants.** The size of these giants can be estimated by comparing them with stars of known size that have the same surface temperature. We know that objects having equal surface temperatures radiate the same amount of energy per unit

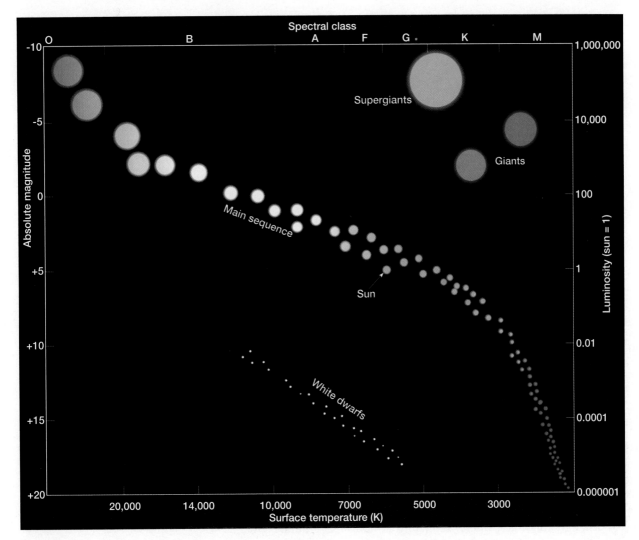

Figure 20.6
Idealized Hertzsprung-Russell diagram on which stars are plotted according to temperature and absolute magnitude.

area. Therefore, any difference in the brightness of two stars having the same surface temperature is attributable to their relative sizes.

As an example, let us compare the sun, which has a luminosity of 1, with another more luminous G-type star that has a luminosity of 100. Because both stars have the same surface temperature, they both radiate the same amount of energy per unit area. Therefore, in order for the more luminous star to be 100 times brighter than the sun, it must have 100 times more surface area. It should be clear why stars whose plots fall in the upper-right position of an H-R diagram are called *giants*.

Some stars are so large that they are called **supergiants.** Betelgeuse, a bright-red supergiant in the constellation Orion (Appendix E, winter chart), has a radius about 800 times that of the sun. If this star were at the center of the solar system, it would extend beyond the orbit of Mars, and the earth would find itself inside the star. Other red giants that are easy to locate in our sky are Arcturus in the constellation Bootes and Antares in Scorpius (Appendix E, summer chart).

In the lower-central portion of the H-R diagram, the opposite situation occurs. These stars are much fainter than main-sequence stars of the same tem-

Figure 20.7
Photographs of Nova Herculis (a nova in the constellation Hercules), taken about two
months apart, showing the decrease in brightness. (Courtesy of Lick Observatory)

perature, and by using the same reasoning as before, they must be much smaller. Some probably approximate the earth in size. This group has come to be called **white dwarfs,** although not all are white.

Soon after the first H-R diagrams were developed, astronomers realized their importance in interpreting stellar evolution. Just as with living things, a star is born, ages, and dies. Due to the fact that almost 90 percent of the stars lie on the main sequence, we can be relatively certain that stars spend most of their active years as main-sequence stars. Only a few percent are giants, and perhaps 10 percent are white dwarfs. After considering some variable stars, we will return to the topic of stellar evolution.

VARIABLE STARS

Stars that fluctuate in brightness are known as *variables.* Some, called **pulsating variables,** fluctuate regularly in brightness by expanding and contracting in size. The importance of one member of this group (cepheid variables) in determining stellar distances is discussed in Box 20.1. Other pulsating variables have irregular periods and are of no value for this purpose.

The most spectacular variables belong to a group known as **eruptive variables.** When one of these explosive events associated with these stars occurs, it appears as a sudden brightening of a star, called a **nova** (Figure 20.7). The term *nova*, meaning "new," was used by the ancients because these stars were unknown to them before their abrupt increase in luminosity.

During the outburst, the outer layer of the star is ejected at high speed. The "cloud" of ejected material occasionally can be captured photographically (Figure 20.8). A nova generally reaches maximum brightness in a couple of days, remains bright for only a few weeks, then slowly returns in a year or so to its original brightness. Because the star returns to its prenova brightness, we can assume that only a small amount of its mass is lost during the flare-up. Some stars have experienced more than one such event. In fact, the process probably occurs repeatedly.

The modern explanation of novae proposes that they occur in binary systems consisting of an expanding red giant and a hot white dwarf which are in close proximity. Hydrogen-rich gas from the oversized giant encroaches near enough to the white dwarf to be gravitationally transferred. Eventually,

BOX 20.1

**Determining Distance
from Magnitude**

For a star too distant for parallax measurements, knowing its absolute and apparent brightness provides astronomers with a tool for approximating its distance. The apparent magnitude is measured with a photometer (light meter) attached to a telescope. If we also know a star's true brightness, we can determine just how far away that star would have to be for it to have the brightness we observe.

You use this same principle when you drive at night and estimate the distance to an oncoming car from the brightness of its headlights. You can do this because you know the true brightness of an automobile headlight. But how do astronomers determine the intrinsic brightness of a star? Fortunately, some stars have characteristics that provide the necessary data.

One important group of such stars is called cepheid variables. These are pulsating stars that get brighter and fainter in a rhythmic fashion. The interval between two successive occurrences of maximum brightness of a pulsating variable is called its light period. Most cepheid variables pulsate with periods of between 2 and 50 days. For example, the North Pole Star (Polaris) varies about 10 percent in brightness over a period of 4 days. In general, the longer the light period of a cepheid, the greater is its absolute magnitude (Figure 20.A). So, by determining the light period of a cepheid, its absolute magnitude can be calculated. When this absolute magnitude is compared to the apparent magnitude, a good approximation of its distance can be made.

Figure 20.A
Relationship between the light period (two successive occurrences of maximum brightness) and absolute magnitude of pulsating stars (cepheid variables).

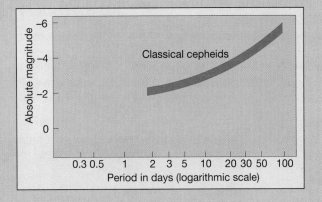

---

enough of the hydrogen-rich gas is added to the dwarf to cause it to ignite explosively. Such a thermonuclear reaction rapidly heats and expands the outer gaseous envelope of the hot dwarf to produce a nova event. In a relatively short time, the white dwarf returns to its prenova state, where it remains inactive until the next build-up occurs.

INTERSTELLAR MATTER

New stars are born out of accumulations of dust and gases that are scattered between existing stars. The name applied to these concentrations of interstellar matter is **nebula** (Latin for "cloud"). If this interstellar matter is close to a very hot (blue) star, usually an O- or B-type star, it will glow and is called a **bright nebula.** The two main types of bright nebulae are known as *emission nebulae* and *reflection nebulae.*

Emission nebulae are gaseous masses that consist largely of hydrogen and absorb *ultraviolet radiation* emitted by an embedded or nearby hot star. Because these gases are under very low pressure, they reradiate this energy as *visible light.* This conversion of ultraviolet light to visible light is known as *fluorescence,* an effect you observe daily in fluorescent lights. A well-known emission nebula easily seen with

Figure 20.8
Expanding "cloud" around Nova Persei. (Courtesy of Hale Observatories. Copyright by the California Institute of Technology and Carnegie Institution of Washington)

of rather dense clouds of large particles called **interstellar dust.** This view is supported by the fact that atomic gases with low densities could not reflect light sufficiently to produce the glow observed.

When a dense cloud of interstellar material is not close enough to a bright star to be illuminated, it is referred to as a **dark nebula.** Exemplified by the Horsehead nebula in Orion, dark nebulae appear as opaque objects silhouetted against a bright background (Figure 20.11). Dark nebulae can also easily be seen as starless regions—"holes in the heavens"—when viewing the Milky Way.

Although nebulae appear very dense, they actually consist of very thinly scattered matter. Because of their enormous size, however, the total mass of rarified particles and molecules may be many times that of the sun. Interstellar matter is of great interest to astronomers because it is from this material that stars and planets are formed.

binoculars is in the sword of the hunter in the constellation Orion (Figure 20.9).

Reflection nebulae, as the name implies, merely reflect the light of nearby stars (Figure 20.10). This fact was discovered when the spectra of these nebulae were found to closely match those of the nearby stars. Reflection nebulae are thought to be composed

STELLAR EVOLUTION

The idea of describing how a star is born, ages, and then dies may seem a bit presumptuous, for many of these objects have life spans that surely exceed billions of years. However, by studying stars of different ages, astronomers have been able to piece

Figure 20.9
Orion nebula is a well-known emission nebula. Bright enough to be seen by the naked eye, the Orion nebula is located in the sword of the hunter in the constellation of the same name. (Courtesy of National Optical Astronomy Observatories)

Figure 20.10
A faint reflection nebula, in the Pleiades star cluster, is caused by the reflection of starlight from dust in the nebula. This star cluster, just visible to the naked eye in the constellation Taurus, is spectacular when viewed through binoculars or a small telescope. (Hale Observatories photo. Copyright by the California Institute of Technology and Carnegie Institution of Washington)

squeezed to unimaginable pressures, its temperature rises, igniting its nuclear furnace, and a star is born. A star is a ball of very hot gases, caught between the opposing forces of gravity trying to contract it and thermal nuclear energy trying to expand it. Eventually, all of a star's nuclear fuel will be exhausted and gravity takes over, collapsing the stellar remnant into a small, dense body.

Star Birth

The birthplaces of stars are dark, cool, interstellar clouds, which are comparatively rich in dust and gases (Figure 20.12). In the neighborhood of the Milky Way, these gaseous clouds consist of 92 percent hydrogen, 7 percent helium, and less than 1 percent of the remaining heavier elements. By some mechanism not yet fully understood, these thin gaseous clouds become concentrated enough to begin to gravitationally contract. One proposal to explain the triggering of stellar formation is a shock wave traveling from a catastrophic explosion (supernova) of a nearby star. But, regardless of the force that initiates the concentration of interstellar matter, once it is accomplished, mutual gravitational attraction of the particles squeezes the cloud, pulling every particle toward the center. As the cloud shrinks, gravitational energy (potential energy) is converted into energy of motion, or heat energy.

The initial contraction spans a million years or so. With the passage of time, the temperature of this gaseous body slowly rises, eventually reaching a temperature sufficiently high to cause it to radiate en-

together a plausible model for stellar evolution. The method that was used to create this model is analogous to what an alien being, upon reaching the earth, might do to determine the evolution of human life. By examining a large number of humans, this stranger would be able to observe the birth of human life, the activities of children and adults, and the death of the elderly. From this information, the alien would attempt to put the stages of human development into their proper sequence. Based on the relative abundance of humans in each stage of development, it would even be possible to conclude that humans spend more of their lives as adults than as toddlers. In a similar fashion, astronomers have pieced together the story of the stars.

Simply, stars exist because of gravity. The mutual gravitational attraction of particles in a thin, gaseous cloud causes the cloud to collapse. As the cloud is

Figure 20.11
The Horsehead Nebula, a dark nebula in a region of glowing nebulosity in Orion. (Courtesy of Anglo-Australian Observatory, by David Malin)

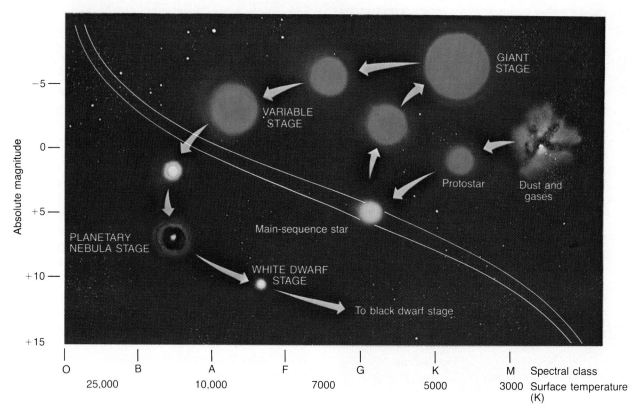

Figure 20.12
Diagram of stellar evolution on H-R diagram for a star about as massive as the sun.

ergy from its surface in the form of long-wavelength red light. Because this large red object is not hot enough to engage in nuclear fusion, it is not yet a star. The name **protostar** is applied to these bodies.

Protostar Stage

During the protostar phase, gravitational contraction continues, slowly at first, then much more rapidly (Figure 20.12). This collapse causes the core of the developing star to heat much more intensely than the outer envelope. When the core has reached at least 10 million K, the pressure within is so great that groups of four hydrogen nuclei are fused together into single helium nuclei. Astronomers refer to this nuclear reaction as **hydrogen burning** because an enormous amount of energy is released. However, keep in mind that thermonuclear "burning" is not burning in the usual chemical sense. (This reaction, also called the *proton-proton chain reaction*, is discussed in Chapter 19 in the subsection "The Solar Interior.")

Heat from hydrogen fusion causes the stellar gases to increase their motion. This in turn results in an increase in the outward gas pressure. At some point, this outward pressure exactly balances the inward force of gravity. When this balance is reached, the star becomes a stable *main-sequence star* (Figure 20.12). Stated another way, a stable main-sequence star is balanced between two forces: *gravity*, which is trying to squeeze it into a smaller sphere, and *gas pressure*, which is trying to expand it.

Main-Sequence Stage

From this point in the evolution of a main-sequence star until its death, the internal gas pressure struggles to offset the unrelenting force of gravity. Typically, hydrogen burning continues for a few billion years and provides the outward pressure required to support the star from gravitational collapse.

Different stars age at different rates. Hot, massive blue stars radiate energy at such an enormous rate that they substantially deplete their hydrogen fuel in only a few million years. By contrast, the very smallest (red) main-sequence stars may remain stable for hundreds of billions of years. A yellow star, such as the sun, remains a main-sequence star for about 10 billion years. Since the solar system is about 5 billion years old, it is comforting to know that the sun will remain stable for another 5 billion years.

An average star spends 90 percent of its life as a hydrogen-burning, main-sequence star. Once the hydrogen fuel in the star's core is depleted, it evolves rapidly and dies. However, with the exception of the least-massive (red) stars, a star can delay its death by burning other fuels and becoming a giant.

Red Giant Stage

The evolution to the red giant stage results because the zone of hydrogen burning continually migrates outward, leaving behind an inert helium core. Eventually, all of the hydrogen in the star's core is consumed. While hydrogen fusion is still progressing in the star's outer shell, no fusion is taking place in the core. Without a source of energy, the core no longer has enough pressure to support itself against the inward force of gravity. As a result, the core begins to contract. Although the core cannot generate nuclear energy, it does grow hotter by converting gravitational energy into heat energy. Some of this energy is radiated outward, initiating a more vigorous level of hydrogen fusion in the star's outer shell. This energy in turn heats and enormously expands the star's outer envelope, producing a giant body hundreds to thousands of times its main-sequence size (Figure 20.12). As the star expands, its surface cools, which explains the star's reddish appearance. Eventually the star's gravitational force will stop this outward expansion. Once again, the two opposing forces, gravity and gas pressure, will be in balance, and this gaseous mass will be a stable but much larger star. Some red giants overshoot the equilibrium point and rebound like an overextended spring. Such stars continue to oscillate in size, becoming variable stars. The cepheid variables discussed earlier are examples of variable giants.

While the envelope of a red giant expands, the core continues to collapse and heat until it reaches 100 million K. At this temperature, it is hot enough to initiate a nuclear reaction in which helium is converted to carbon. Thus, a red giant consumes both hydrogen and helium to produce energy. In stars more massive than the sun, still other thermonuclear re-

actions occur that generate all the elements on the periodic table up to number 26, iron. Nuclear burning of elements heavier than iron requires an additional source of energy to keep the reaction progressing. Hence, these elements are not produced in ordinary stars.

Eventually, all the usable nuclear fuel in these giants will be consumed. The sun, for example, will spend less than a billion years as a giant, and the more massive stars will pass through this stage even more rapidly. The force of gravity will again control the star's destiny as it squeezes the star into the smallest, most dense piece of matter possible.

Burnout and Death

Most of the events of stellar evolution discussed thus far are well-documented. What happens to a star after the red-giant phase is more speculative. We do know that a star, regardless of its size, must eventually exhaust all of its usable nuclear fuel and collapse in response to its immense gravitational force. Further, the evolution and death of each star is determined primarily by its initial mass. We should note, however, that during their long life span, stars may lose mass through a number of processes, including particles streaming away from their surfaces. Therefore, it is not currently possible to set exact mass limits when considering stellar evolution. With this in mind, we will now consider the final stage of stars in three different mass categories.

Death of Low Mass Stars. Stars less than one-half the mass of the sun (0.5 solar mass) consume their fule at a comparatively slow rate (Figure 20.13A). Consequently, these small, cool red stars may remain on the main sequence for up to 100 billion years. Because the interior of a low mass star never attains sufficiently high temperatures and pressures to fuse helium, its only energy source is hydrogen fusion. Thus, low mass stars never evolve to become bloated red giants. Rather, they remain as stable main-sequence stars until they consume their hydrogen fuel and collapse into a hot, dense *white dwarf*. As we shall see, white dwarfs are small, compact objects unable to support nuclear burning.

Death of Medium Mass (Sunlike) Stars. Main-sequence stars with masses between roughly 0.5 and 3 solar masses evolve in essentially the same way (Fig-

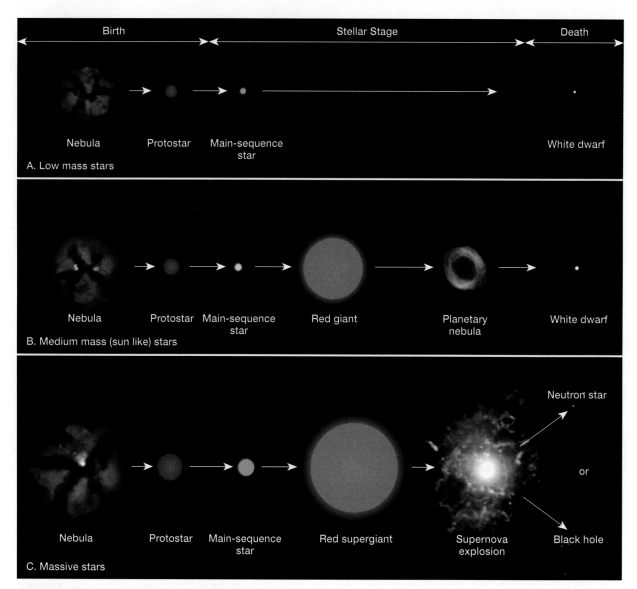

Figure 20.13
The evolutionary stages of stars having various masses.

ure 20.13B). During their giant phase, sunlike stars fuse hydrogen and helium fuel at an accelerated rate. Once this fuel is exhausted, these stars (like low mass stars) collapse into an earth-sized body of great density—a white dwarf. The density of a white dwarf is as great as physics will allow short of destroying protons and electrons.

The gravitational energy supplied to a collapsing white dwarf is reflected in its high surface temperature. However, without a source of nuclear energy, a white dwarf becomes cooler and dimmer as it continually radiates its remaining thermal energy into space.

During their collapse from red giants to white dwarfs, medium mass stars are thought to cast off their bloated outer atmosphere, creating an expanding spherical cloud of gas. The remaining hot, central white dwarf heats the gas cloud, causing it to glow. These often beautiful, gleaming spherical clouds are called **planetary nebulae.** A good example of a planetary nebula is the Helix nebula in the constellation Aquarius (Figure 20.14). This nebula appears as a ring because our line of sight through the center traverses less gaseous material than at the nebula's edge. It is, nevertheless, spherical in shape.

Figure 20.14
The Helix Nebula, the nearest planetary nebula to our solar system. A planetary nebula is the ejected outer envelope of a sunlike star that formed during the star's collapse from a red giant to a white dwarf. (Courtesy of Anglo-Australian Observatory, by David Malin)

Death of Massive Stars. In contrast to sunlike stars, which expire gracefully, stars over 3 solar masses have relatively short life spans and terminate in a brilliant explosion called a **supernova** (Figure 20.13C). During a supernova event, a star becomes millions of times brighter than its prenova stage (see Box 20.1). If one of the nearest stars to the earth produced such an outburst, its brilliance would surpass that of the sun. Supernovae are rare; none have been observed in our galaxy since the advent of the telescope, although Tycho Brahe and Galileo each recorded one about 30 years apart (see Box 20.2). What was an even larger supernova was recorded in A.D. 1054 by the Chinese. Today, the remnant of this great outburst is the Crab nebula, shown in Figure 20.15.

A supernova event is thought to be triggered when a massive star consumes most of its nuclear fuel. Without a heat engine to generate the gas pressure required to balance its immense gravitational field, it collapses. This implosion is of cataclysmic proportion, resulting in a shock wave that moves out from the star's interior. This energetic shock wave destroys the star and blasts its outer shell into space, generating the supernova event.

Figure 20.15
Crab Nebula in the constellation Taurus: the remains of the supernova of A.D. 1054. (Courtesy of Lick Observatory)

BOX 20.2
Supernova 1987A

*T*he first naked-eye supernova in 383 years was discovered in the southern sky in February 1987 (Figure 20.B). This stellar explosion was officially named SN 1987A (*SN* stands for "supernova" and 1987A indicates that it was the first supernova observed in 1987). Naked-eye supernovae are rare. Only a few have been recorded in historic times. Arabic observers saw one in 1006, and the Chinese recorded one in 1054 at the present location of the Crab Nebula. In addition, the astronomer Tycho Brahe observed a supernova in 1572 and Kepler saw one shortly thereafter in 1604.

Prior to this event, researchers could only test their hypotheses on dim supernovae seen in distant galaxies. Thus, when SN 1987A occurred, as-

tronomers quickly focused every available telescope in the Southern Hemisphere on this spectacular event. As one astronomer remarked, "This supernova is better studied by far than any supernova in history." More importantly, this event has allowed astronomers to use observational data to test their theoretical models of stellar evolution.

Supernova 1987A occurred about 170,000 light-years away in the Large Magellanic Cloud, a satellite galaxy to our own Milky Way. As expected, the supernova rapidly increased in brightness to a peak magnitude of 2.4, outshining all the other stars in the Large Magellanic Cloud. Also as predicted, within a few weeks it began to fade. However, SN 1987A did provide some surprises.

From old photographs taken of the area, researchers identified the exploded star as Sanduleak −69°202. Astronomers were surprised to find that the parent star was a hot, blue star about 15 times the mass of the sun. Recall that only cool red gi-

ants are thought to die as a supernova event. Further, the Hubble Space Telescope made another unexpected discovery. Its camera revealed a very large shell of gas that predates the supernova explosion by about 40,000 years.

Astronomers now think that Sanduleak −69°202 was once a red supergiant that had blown away its outer shell, exposing a hot, blue core. It is this ejected outer shell that appears in the image produced by the Hubble Space Telescope. Then, some 40,000 years later, the remaining hot core of the red supergiant collapsed, producing the supernova of 1987.

Despite these twists, the theory of stellar evolution has held up very well. Theory predicts that the expanding remnants of Supernova 1987A will be large enough to be observed by the turn of the century. Thus, astronomers continue to monitor SN 1987A to unravel its secrets and to confirm or refute their theories about the final stages of stellar evolution.

Figure 20.B
The great Supernova 1987A. The photo on the left was made prior to the supernova and the one on the right was made following the event. (Courtesy of Anglo-Australian Observatory, by David Malin)

Table 20.5
Summary of evolution for stars of various masses.

Initial Mass of Interstellar Cloud (Sun = 1)	Main-Sequence Stage	Giant Phase	Evolution After Giant Phase	Terminal State (Final Mass*)
0.001	None (Planet)	No	None	Planet (0.001)
0.1	Red (Type M)	No	None	White dwarf (0.1)
1	Yellow (Type G)	Yes	Planetary nebula	White dwarf (< 1.4)
6	White (Type A)	Yes	Supernova	Neutron star (1.4–3)
20	Blue (Type O)	Yes Supergiant	Supernova	Black hole (> 3.0)

* These mass numbers are estimates.

Theoretical work predicts that during a supernova, the star's interior condenses into a very hot object, possibly no larger than 20 kilometers in diameter. These incomprehensibly dense bodies have been named *neutron stars*. Some supernovae events are thought to produce even smaller, and most intriguing, objects called *black holes*. We will consider the nature of neutron stars and black holes in the Stellar Remnants section.

H-R Diagrams and Stellar Evolution

The Hertzsprung-Russell diagrams have been very helpful in formulating and testing models of stellar evolution. They are also useful for illustrating the changes that take place in an individual star during its life span. Figure 20.12 on page 680 shows on an H-R diagram the evolution of a star about the size of the sun. Keep in mind that the star does not physically move along this path, but rather that its position on the H-R diagram represents the color (temperature) and absolute magnitude (brightness) of the star at various stages in its evolution.

For example, on the H-R diagram, a protostar would be located to the right and above the main sequence (Figure 20.12). It is found to the right because of its relatively cool surface temperature (red color), and above because it would be more luminous than

a main-sequence star of the same color, a fact attributable to its large size. Careful examination of Figure 20.12 should help you visualize the evolutionary changes experienced by a star the size of the sun. In addition, Table 20.5 provides a summary of the evolutionary history of stars having various masses.

STELLAR REMNANTS

Eventually, all stars consume their nuclear fuel and collapse into one of three final states—white dwarf, neutron star, or black hole. Although different in some ways, these small, compact objects are all composed of incomprehensibly dense material and all have extreme surface gravity.

White Dwarfs

White dwarfs, shown at the bottom of Figure 20.6, page 675, are extremely small stars with densities greater than any known terrestrial material. It is believed that white dwarfs once were low or medium mass stars whose internal heat energy was able to keep these gaseous bodies from collapsing under their own gravitational force. Although some white dwarfs are no larger than the earth, the mass of such a dwarf can equal 1.4 times that of the sun. Thus,

BOX 20.3

From Stardust to You

*D*uring a supernova implosion, the internal temperature of a star may reach 1 billion K, a condition thought to produce very heavy elements such as gold and uranium. These heavy elements, plus the debris of novae and the planetary nebulae, are continually returned to interstellar space where they are available for the formation of other stars (Figure 20.C).

Astronomers believe that the earliest stars were made of nearly pure hydrogen. Fusion during the life and death of stars in turn produced heavier elements, some of which were returned to space. Because the sun contains some heavy elements but has not yet reached the stage in its evolution where it could have produced them, it must be at least a second-generation star. Thus, our sun, as well as the rest of the solar system, is believed to have formed from debris scattered from pre-existing stars. If this is the case, the atoms in your body were produced billions of years ago inside a star, and the gold in your jewelry formed during a supernova event that occurred trillions of kilometers away. Without these catastrophic events, the development of life on earth would not have been possible.

Figure 20.C
Eagle nebula in the constellation Serpens. This gaseous nebula is the site of recent star formation. (Courtesy of National Optical Astronomy Observatories)

their densities may be a million times greater than water. A spoonful of such matter would weigh several tons. Densities this great are possible only when electrons are displaced from their regular orbits and pushed closer to the nucleus, allowing the atoms to take up less than the "normal" amount of space. Material in this state is called **degenerate matter.**

To better visualize degenerate matter, consider that in ordinary atoms the electrons circle the nucleus at distances proportional to the orbits of the planets around the sun. Thus, like our solar system, atoms are made up mostly of empty space. You might envision an atom as having a very small nucleus surrounded by speedy electrons that produce a comparatively large electrically charged cloud.

In ordinary gas, such as that found in an automobile tire, the motion of these atoms causes them to collide with each other and, with the walls of the tire to provide the air pressure, keep the tire inflated. In a similar manner, gas pressure supports the outer envelope of a star and prevents gravitational collapse. When a gas cools, the atoms slow, thus reducing the pressure they exert. In the envelope of a star, this cooling results in a reduced volume until finally the electron clouds of the atoms are touching. However, if the gravitational force of the star is great enough, the atoms themselves will collapse, producing degenerate matter.

In degenerate matter, the atoms have been squeezed together so tightly that the electrons are displaced much nearer to the nucleus. Degenerate matter uses electrical repulsion rather than molecular motion to support itself from total collapse. Although atomic particles in degenerate matter are much closer together than in normal earth matter, they still are not packed as tightly as possible. Stars made of matter that has an even greater density are thought to exist.

As a star contracts into a white dwarf, its surface becomes very hot, sometimes exceeding 25,000 K. Even so, without a source of energy, it can only become cooler and dimmer. Although none have been observed, the terminal stage of a white dwarf must be a small, cold, nonluminous body called a *black dwarf.*

Neutron Stars

A study of white dwarfs produced what might at first appear to be a surprising conclusion. The smallest white dwarfs are the most massive, and the largest are the least massive. The explanation for this is that a more massive star, because of its greater gravitational force, is able to squeeze itself into a smaller, more densely packed object than can a less massive star. Thus, the smallest white dwarfs were produced from the collapse of larger, more massive stars than were the larger white dwarfs.

The previous data led to the prediction that stars smaller and more massive than white dwarfs must exist. Named **neutron stars,** these objects are thought to be the remnants of supernova events. Theoretical calculations indicate that if a supernova explosion leaves behind a remnant between 1.4 and 3 solar masses, gravity will be strong enough to cause it to collapse beyond the white dwarf stage into a smaller, yet more massive, neutron star. In a white dwarf, the electrons are pushed close to the nucleus, whereas in a neutron star, the electrons are forced to combine with protons to produce neutrons (hence the name). If the earth were to collapse to the density of a neutron star, it would have a diameter equivalent to the length of a football field. A pea-sized

sample of this matter would weigh 100 million tons. This is approximately the density of an atomic nucleus; thus neutron stars can be thought of as large atomic nuclei.

During a supernova implosion, the envelope of the star is ejected (Figure 20.16), while the core collapses into a very hot neutron star about 20 kilometers (12.4 miles) in diameter. Although neutron stars have high surface temperatures, their small size would greatly limit their luminosity. Consequently, locating one visually would be extremely difficult.

However, theory predicts that a neutron star would have a very strong magnetic field. Further, as a star collapses, it will rotate faster, for the same reason ice skaters rotate faster as they pull in their arms. If the sun were to collapse to the size of a neutron star, it would increase its rate of rotation from once every 25 days to nearly 1000 times per second. Radio waves generated by these rotating stars would be concentrated into two narrow zones that would align

Figure 20.16
Veil nebula in the constellation Cygnus is the remnant of an ancient supernova implosion. (Hale Observatories photo. Copyright by the California Institute of Technology and Carnegie Institution of Washington)

Figure 20.17
This illustration shows how astronomers believe a binary pair (red giant/black hole) might function.

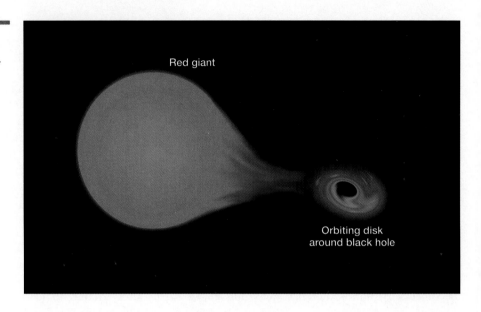

Red giant

Orbiting disk
around black hole

with the star's magnetic poles. Consequently, these stars would resemble a rapidly rotating beacon emitting strong radio waves. If the earth happened to be in the path of these beacons, the star would appear to blink on and off as the waves swept past.

In the early 1970s, a source that radiates short pulses of radio energy, called a **pulsar** (pulsating radio source), was discovered in the Crab nebula. Visual inspection of this radio source revealed it to be a small star centered in the nebula. The pulsar found in the Crab nebula is undoubtedly the remains of the supernova of A.D. 1054 (see Figure 20.15). Thus, the first neutron star had been discovered.

Black Holes

Are neutron stars made of the most dense materials possible? No. During a supernova event, remnants of stars greater than 3 solar masses apparently collapse into objects even smaller and denser than neutron stars. Even though these objects would be very hot, their surface gravity would be so immense that even light could not escape the surface. Consequently, they would literally disappear from sight. These incredible bodies have appropriately been named **black holes.** Anything that moved too near a black hole would be swept in by its irresistible gravity and devoured forever.

How can astronomers find an object whose gravitational field prevents the escape of all matter and energy? One strategy is to seek evidence of matter being rapidly swept into a region of apparent nothing-

ness. Theory predicts that as matter is pulled into a black hole, it should become very hot and emit a flood of x rays before being engulfed. Because isolated black holes would not have a source of matter to engulf, astronomers first looked at binary star systems.

A likely candidate for a black hole is Cygnus X-1, a strong x-ray source in the constellation Cygnus. In this case, the x-ray source can be observed orbiting a supergiant companion with a period of 5.6 days. It appears that gases are pulled from this companion and spiral into a disc-shaped structure around the black hole (Figure 20.17). The result is a stream of x rays. Because x rays cannot penetrate our atmosphere efficiently, the existence of black holes was not confirmed until recently. The first x-ray sources were discovered in 1971 by detectors on satellites. Cygnus X-1 is such a source.

STAR CLUSTERS

Like the members of ancient nomadic tribes, many stars travel in groups called **clusters.** Two common types are **open clusters,** like the Pleiades in the constellation Taurus (Figure 20.18) and **globular clusters** such as the one in the constellation Hercules (Figure 20.19). As their names suggest, the members of a globular cluster are arranged in a spherical shape, while the members of an open cluster are less well organized. Other differences between these groups include size, star density, and location within the galaxy.

A large globular cluster may contain 100,000 stars, which are packed near enough together to give the appearance of a solid when viewed from a great distance. However, there is plenty of space, so the chance of stellar collision is almost nonexistent. Globular clusters are located in all directions from the center of our galaxy, but appear in greater numbers nearest the center. The typical open clusters, on the other hand, contain tens to hundreds of stars and are almost exclusively confined to the plane of our galaxy.

It is assumed that the stars of each cluster have similar origins and compositions, and nearly the same "birthday." Consequently, they are very beneficial in testing concepts of stellar evolution. At birth, each star cluster presumably contained stars of every spectral type (color). Today, however, some clusters are notably missing the hot, massive O- and B-type main-sequence stars, and others are nearly depleted of all stars hotter than the G type. These differences can be explained if the clusters are of differing ages and the massive O-type stars do, in fact, use their fuel most rapidly, followed by B-type stars, and so on. Thus, studies of clusters support the proposal that the more massive (brighter) stars evolve to another stage, departing the main sequence before less massive (dimmer) stars do. Because the Pleiades, shown in Figure 20.18, contain numerous hot blue stars, we conclude

Figure 20.19
Globular star cluster in the constellation Hercules. (Courtesy of U.S. Naval Observatory)

that it is a young star cluster. In a similar manner, other aspects of stellar evolution have been tested using star clusters.

THE MILKY WAY

On a clear, moonless night away from city lights, you can see a truly marvelous sight—our own Milky Way galaxy (Figure 20.20). With his telescope, Galileo discovered that this band of light was produced by countless individual stars that the unaided eye is unable to resolve. Today, we realize that the sun is actually part of this vast system of stars, which number about one hundred billion. The "milky" appearance of our galaxy results because the solar system is located within the flat *galactic disk*. Thus, when it is viewed from the "inside," a higher concentration of stars appears in the direction of the galactic plane than in any other direction.

When astronomers began to telescopically survey the stars located along the plane of the Milky Way, it appeared that equal numbers lay in every direction. Could the earth actually be at the center of the galaxy? A better explanation was put forth. Imagine that the trees in an enormous forest represent the stars in the galaxy. After hiking into this forest a short distance, you look around. What you see is an equal number of trees in every direction. Are you really in the center of the forest? Not necessarily; anywhere in the forest, except at the very edge, you will seem to be in the middle.

Figure 20.18
The Pleiades, an open star cluster in the constellation Taurus, shown embedded in the interstellar clouds. The clouds reflect the light of the brighter cluster members. (Hale Observatories photo. Copyright by the California Institute of Technology and Carnegie Institution of Washington)

Figure 20.20
Panorama of our galaxy, the Milky Way. Notice the dark bands caused by the presence of interstellar dark nebulae. (Lund Observatory photograph)

Structure. Attempts to visually inspect the Milky Way are hindered by the large quantities of interstellar matter that lie in our line of sight. Nevertheless, with the aid of radio telescopes, the gross structure of our galaxy has been determined. The Milky Way is a rather large spiral galaxy whose disk is about 100,000 light-years wide and about 10,000 light-years thick at the nucleus (Figure 20.21). As viewed from the earth, the center of the galaxy lies beyond the constellation Sagittarius.

Radio telescopes reveal the existence of at least three distinct *spiral arms*, with some showing splintering (Figure 20.22). The sun is positioned in one of these arms about two-thirds of the way from the center, at a distance from the hub of about 30,000 light-years. The stars in the arms of the Milky Way rotate around the *galactic nucleus*, with the most outward ones moving the slowest, such that the ends of the

arms appear to trail. The sun and the arm it is in require about 200 million years for each orbit around the nucleus.

Surrounding the galactic disk is a nearly spherical *halo* made of very tenuous gas and numerous globular clusters. These star clusters do not participate in the rotating motion of the arms, but rather have their own orbits that carry them through the disk. Although some clusters are very dense, they pass among the stars of the arms with plenty of room to spare.

History. The Milky Way probably formed from an enormous hydrogen cloud, which began to condense some 10 billion years ago. As it contracted, its density increased, eventually becoming sufficient for stellar formation to begin. Initially, stars developed in scattered globular-type clusters. The globular clusters of our galaxy all lack hot O- or B-type stars,

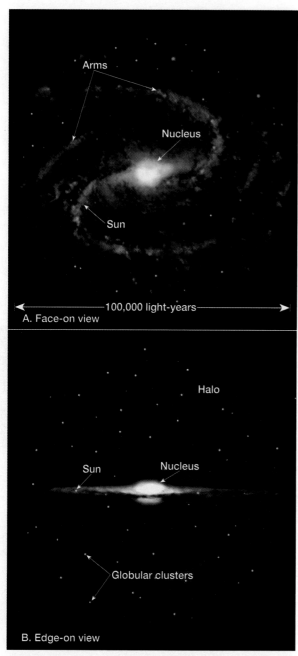

Figure 20.21
Structure of the visible portion of the Milky Way galaxy.

As this gaseous mass continued to condense, its rotational speed increased, causing it to flatten. While this was taking place, the original stars were generating heavy elements, and more and more of this matter was being added to the condensing gases. Therefore, the stars that formed later in the disk of the galaxy are richer in heavy elements. These younger stars are said to be **population I stars.** The sun is a population I star about 5 billion years old found in the galactic disk. The arms of the galaxy still contain large quantities of dust and gases, and stars are still forming there. These new stars will contain an even higher percentage of heavy elements.

GALAXIES

In the mid-1700s, German philosopher Immanuel Kant proposed that the telescopically visible fuzzy patches of light scattered among the stars were actually distant galaxies like the Milky Way. Kant described them as "island universes." Each galaxy, he felt, contained billions of stars and, as such, was a universe in itself. The weight of opinion, however, favored the hypothesis that they were dust and gas clouds (nebulae) within our galaxy. This matter was not resolved until the 1920s, when American astronomer Edwin Hubble (for whom the Hubble Space Telescope is named), was able to identify some cepheid variables in the Great Galaxy in Andromeda

Figure 20.22
If the Milky Way was photographed from a distance, it might appear like the spiral galaxy NGC 2997. (Courtesy of Anglo-Australian Observatory, by David Malin)

which testifies to their age. Astronomers call old stars like these, which formed early from nearly pure hydrogen, **population II stars.** At the time of the formation of population II stars, the galaxy had the spheroidal shape and size that is presently outlined by the most distant of the globular clusters (Figure 20.21).

(Figure 20.23). Hubble realized that because these intrinsically very bright stars had very faint apparent magnitudes (18), they must lie outside the Milky Way at a great distance.

Recall that cepheid variables can be used to determine stellar distances. When they were used to find the distance to the Great Galaxy in Andromeda, astronomers determined it to be over a million light-years away. Hubble had extended the universe far beyond the limits of our imagination, to include hundreds of billions of galaxies, each containing hundreds of billions of stars. It has been said that a million galaxies are found in that portion of the sky bounded by the cup of the Big Dipper. There are more stars in the heavens than grains of sand in all the beaches on earth.

Four Types of Galaxies

From the hundreds of billions of galaxies, four basic types have been identified: elliptical, spiral, barred spiral, and irregular.

Only 10 percent of the known galaxies lack symmetry and are classified as **irregular galaxies.** The best known irregular galaxies, the Large and Small Magellanic Clouds in the Southern Hemisphere, are easily visible with the unaided eye. Named after the explorer Ferdinand Magellan, who observed them when he circumnavigated the earth in 1520, they are our nearest galactic neighbors—only 150,000 light-years away.

The Milky Way and the famous Great Galaxy in Andromeda are examples of fairly large **spiral galaxies** (Figure 20.24). Andromeda can be seen with the unaided eye as a fuzzy fifth-magnitude object. Typically, spiral galaxies are disk-shaped with a somewhat greater concentration of stars near their centers, but there are numerous variations. Viewed broadside, arms are often seen extending from the central nucleus and sweeping gracefully away. The outermost stars of these arms rotate most slowly, giving the galaxy the appearance of a fireworks pinwheel.

One type of spiral galaxy, however, has the stars arranged in the shape of a bar, which rotates as a rigid system. This requires that the outer stars move faster than the inner ones, a fact not easy for astronomers to reconcile with the laws of motion. Attached to each end of these bars are curved spiral arms. These have become known as **barred spiral galaxies** (Figure 20.25). Spiral galaxies are generally quite large, ranging from 20,000 to about 125,000 light-years in diameter. About 10 percent of all galaxies are thought to be barred spirals and another 20 percent are regular spiral galaxies like the Milky Way.

The most abundant group, making up 60 percent of the total, is the **elliptical galaxies.** These are generally smaller than spiral galaxies. Some are so much smaller, in fact, that the term *dwarf* has been applied. Since these dwarf galaxies are not visible at great distances, a survey of the sky reveals more of the conspicuous large spiral galaxies. Although most elliptical galaxies are small, the very largest known galaxies (200,000 light-years in diameter) are also elliptical. As their name implies, elliptical galaxies have an ellipsoidal shape that ranges to nearly spherical, and they lack spiral arms.

The two dwarf companions of Andromeda shown in Figure 20.23 are elliptical galaxies. Some resemble

Figure 20.23
Great Galaxy, a spiral galaxy, in the constellation Andromeda. The two bright spots to the left and right are dwarf elliptical galaxies. (Hale Observatories photo. Copyright by the California Institute of Technology and Carnegie Institution of Washington)

Figure 20.24
Two views illustrating the idealized structure of spiral galaxies. (Courtesy of U.S. Naval Observatory)

the dense nucleus of a spiral galaxy, a fact that generated speculation that they may be the remains of a spiral galaxy whose arms wound around themselves. However, this has not been shown to be the case.

One of the major differences among the galactic types is the age of the stars that make them up. The irregular galaxies are composed mostly of young population I stars, whereas the elliptical galaxies contain old population II stars. The Milky Way and other spiral galaxies consist of both populations, with the youngest stars located in the arms.

Figure 20.25
Barred spiral galaxy. (Courtesy of Hale Observatories. Copyright by the California Institute of Technology and Carnegie Institution of Washington)

Galactic Clusters. Once astronomers discovered that stars were associated in groups, they set out to determine whether galaxies also were grouped or just randomly distributed throughout the universe. They found that, like stars, galaxies are grouped in **galactic clusters** (Figure 20.26). Some abundant clusters contain thousands of galaxies. Our own, called the **Local Group,** contains at least 28 galaxies. Of these, three are spirals, eleven are irregulars, and fourteen are ellipticals. Galactic clusters also reside in huge swarms called superclusters. From visual observations, it appears that superclusters may be the largest entities in the universe.

RED SHIFTS

One of the most important discoveries of modern astronomy was made in 1929 by Edwin Hubble. Observations completed several years earlier revealed that most galaxies have Doppler shifts in their spectral lines toward the red end of the spectrum. Recall that red shift occurs because the light waves are "stretched," indicating that the source is moving away (see Chapter 19). Hubble set out to explain the predominance of red shift which had been observed in the spectra of galaxies.

Recall that Hubble established a method of estimating the distance to galaxies by using stars called cepheid variables. He also realized that the dimmer galaxies were probably farther away. Thus, he tried

Figure 20.26
Numerous galaxies grouped in the constellation Hercules. (Courtesy of Hale Observatories. Copyright by the California Institute of Technology and Carnegie Institution of Washington)

to determine if there is a relationship between the distances to galaxies and their red shifts. Using these estimated distances and the observed Doppler red shifts, Hubble discovered that galaxies that exhibit the greatest red shifts are the most distant. A very important benefit of this discovery was that it gave astronomers another method of determining distances in the universe. Since many galaxies are too far away to permit observation of individual stars such as cepheid variables, the amount of red shift in their spectra can be used to estimate their distance.

Another consequence of the universal red shift is that it predicts that most galaxies (except for a few nearby) are receding from us. Recall that the amount of Doppler red shift is dependent on the velocity at which the object is moving away. Greater red shifts indicate faster recessional velocities. Because more distant galaxies have greater red shifts, Hubble concluded that they must be retreating from us at greater

velocities. This idea is currently called **Hubble's law** and states that galaxies are receding from us at a speed that is proportional to their distance.

An Expanding Universe. Hubble was surprised at this discovery because it implied that the most distant galaxies are moving away from us many times faster than those nearby. What type of cosmological theory can explain this fact? It was soon realized that an *expanding universe* can adequately account for the observed red shifts.*

To help visualize the nature of this expanding universe, we will employ a popularly used analogy. Imagine a loaf of raisin bread dough that has been set out to rise for a few hours. As the dough doubles in size, so does the distance between all of the raisins. However, the raisins that were originally farther apart traveled a greater distance in the same time span than those located closer together. We therefore conclude that, in an expanding universe, as in our analogy, those objects located farther apart recede at greater velocities.

Another feature of the expanding universe can be demonstrated using the raisin bread analogy. Let the raisins represent clusters of galaxies. During expansion, the size of the raisins, like the space occupied by the clusters, does not get larger, only the space separating them increases. It is believed that the galaxies within each cluster are held together by their mutual gravitational attraction. Within each cluster, galaxies have random motion in addition to the universal expansion. This explains why some of the galaxies in the Local Group exhibit a blue shift, indicating they are moving toward us. In addition, no matter which raisin you select, it will move away from all the other raisins. Likewise, no matter where one is located in the universe, every other galaxy (except those in the same cluster) will be receding. Edwin Hubble had indeed advanced our understanding of the universe.

---

* One other reason for a red shift is known. An immense gravitational field can stretch light waves. However, a gravitational red shift would produce other visible effects that are not observed.

THE BIG BANG

The universe—did it have a beginning? Will it have an end? Cosmologists are trying to answer these questions, and that makes them a rare breed.

First and foremost, any viable theory regarding the origin of the universe must account for the fact that all galaxies (except for the very nearest) are moving away from us. Because all galaxies appear to be moving away from the earth, is the earth in the center of the universe? If we are not in the center of the solar system and the solar system is not in the center of the galaxy, it seems unlikely that we should be in the center of the universe. A more probable explanation exists for this fact. Imagine a balloon with paper punch "dots" glued to its surface. When the balloon is inflated, each dot spreads apart from every other dot. Similarly, if the universe is expanding, every galaxy would be moving away from every other galaxy.

This belief in an expanding universe led to the widely accepted **Big Bang** theory. According to this theory, the entire universe was at one time confined to a dense, hot, supermassive ball. Then about 20 billion years ago, a cataclysmic explosion occurred, hurling this material in all directions. The big bang marks the inception of the universe; all matter and space was created at this instant. The ejected masses of gas cooled and condensed, forming the stellar systems we now observe fleeing from their birthplace.

Will the stars eventually dim from view and invisible galaxies travel on forever? It has been suggested that after a certain point, perhaps 20 billion years in the future, the galaxies will slow and eventually stop their outward flight. Gravitational contraction would follow. The galaxies would then collide and coalesce, and a new fireball would be born. For this event to occur, the universe must have an average density of one atom for every cubic meter of space. But present estimates indicate that the concentration of matter in the universe is far less than this amount, so this hypothesis remains unsubstantiated.

However, this subject does not end here. It has been proposed that heretofore undetected matter exists in large quantities in the universe. For example, numerous black holes may occupy many of the voids in the universe. If this is true, the galaxies could, in fact collapse upon themselves.

"Absence of evidence is not evidence of absence."
—Anonymous

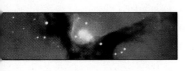

Review Questions

1. How far away in light-years is our nearest stellar neighbor, Proxima Centauri? Convert your answer to kilometers.

2. What is the most basic method of determining stellar distances?

3. Explain the difference between a star's apparent and absolute magnitudes. Which one is an intrinsic property of a star?

4. What is the ratio of brightness between a twelfth-magnitude and fifteenth-magnitude star?

5. What information about a star can be determined from its color?

6. What color stars have the highest and lowest surface temperatures?

7. Which property of a star can be determined from binary star systems?

8. Why were the original spectral classes of stars rearranged?

9. Make a generalization relating the mass and luminosity of main-sequence stars.

10. The disk of a star cannot be resolved telescopically. Explain the method that astronomers have used to estimate the size of stars.

11. Where on an H-R diagram does a star spend most of its lifetime?

12. How does the sun compare in size and brightness to other main-sequence stars?

13. Why is interstellar matter important to stellar evolution?

14. Compare a bright nebula and a dark nebula.

15. What element is the fuel for main-sequence stars? Red giants?

16. What causes a star to become a giant?

17. Why are less massive stars thought to age more slowly than more massive stars, even though they have much less "fuel"?

18. Enumerate the steps thought to be involved in the evolution of sunlike stars.

19. What is the final state of a low mass (red) main-sequence star?

20. What is the final state of a medium mass (sunlike) star?

21. How do the "lives" of the most massive stars end? What are the two possible products of this event?

22. Describe the general structure of the Milky Way.

23. Why is the sun considered a second-generation star?

24. Compare the three general types of galaxies.

25. Explain why astronomers consider elliptical galaxies more abundant than spiral galaxies, even though more spiral galaxies have been sighted.

26. How did Edwin Hubble determine that the Great Galaxy in Andromeda is located beyond our galaxy?

27. What evidence supports the Big Bang theory?

Key Terms

absolute magnitude (p. 671)

apparent magnitude (p. 670)

barred spiral galaxy (p. 692)

Big Bang (p. 695)

black hole (p. 688)

bright nebula (p. 677)

cluster (p. 688)

dark nebula (p. 678)

degenerate matter (p. 686)

elliptical galaxy (p. 692)

emission nebula (p. 677)

eruptive variable (p. 676)

galactic cluster (p. 693)

giant (p. 674)

globular cluster (p. 688)

Hertzsprung-Russell (H-R) diagram (p. 673)

Hubble's law (p. 694)

hydrogen burning (p. 680)

interstellar dust (p. 678)

irregular galaxy (p. 692)

light-year (p. 669)

Local Group (p. 693)

magnitude (p. 670)

main-sequence stars (p. 673)

nebula (p. 677)

neutron star (p. 687)

nova (p. 676)

open cluster (p. 688)

parsec (p. 669)

planetary nebula (p. 682)

population I stars (p. 691)

population II stars (p. 691)

protostar (p. 680)

pulsar (p. 688)

pulsating variable (p. 676)

red giant (p. 674)

reflection nebula (p. 678)

spectral class (p. 673)

spiral galaxy (p. 692)

stellar parallax (p. 668)

supergiant (p. 675)

supernova (p. 683)

visual binaries (p. 672)

white dwarf (p. 676)

APPENDIX A

Metric and English Units Compared

UNITS

1 kilometer (km)   = 1000 meters (m)
1 meter (m)   = 100 centimeters (cm)
1 centimeter (cm) = 0.39 inches (in.)
1 mile (mi)   = 5280 feet (ft)
1 foot (ft)   = 12 inches (in.)
1 inch (in.)   = 254 centimeters (cm)
1 square mile (mi^2) = 640 acres (a)
1 kilogram (kg)   = 1000 grams (g)
1 pound (lb)   = 16 ounces (oz)
1 fathom   = 6 feet (ft)

CONVERSIONS

When you want to convert:	Multiply by:	To find:
Length		
inches	2.54	centimeters
centimeters	0.39	inches
feet	0.30	meters
meters	3.28	feet
yards	0.91	meters
meters	1.09	yards
miles	1.61	kilometers
kilometers	0.62	miles
Area		
square inches	6.45	square centimeters
square centimeters	0.15	square inches
square feet	0.09	square meters
square meters	10.76	square feet
square miles	2.59	square kilometers
square kilometers	0.39	square miles

When you want to convert:	Multiply by:	To find:
Volume		
cubic inches	16.38	cubic centimeters
cubic centimeters	0.06	cubic inches
cubic feet	0.028	cubic meters
cubic meters	35.3	cubic feet
cubic miles	4.17	cubic kilometers
cubic kilometers	0.24	cubic miles
liters	1.06	quarts
liters	0.26	gallons
gallons	3.78	liters
Masses and Weights		
ounces	20.33	grams
grams	0.035	ounces
pounds	0.45	kilograms
kilograms	2.205	pounds

Temperature

When you want to convert degrees Fahrenheit (°F) to degrees Celsius (°C), subtract 32 degrees and divide by 1.8

When you want to convert degrees Celsius (°C) to degrees Fahrenheit (°F), multiply by 1.8 and add 32 degrees.

When you want to convert degrees Celsius (°C) to kelvins (K), delete the degree symbol and add 273.

When you want to convert kelvins (K) to degrees Celsius (°C), add the degree symbol and subtract 273.

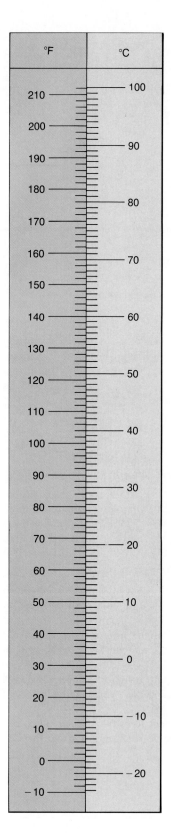

Figure A.1

APPENDIX B

Mineral Identification Key

Hardness	Streak	Other Diagnostic Properties	Name (Chemical Composition)
Harder than glass	Black streak	Black; magnetic; hardness = 6; specific gravity = 5.2; often granular	Magnetite (Fe_3O_4)
	Greenish-black streak	Brass yellow; hardness = 6; specific gravity = 5.2; generally an aggregate of cubic crystals	Pyrite (FeS_2)— fool's gold
	Red-brown streak	Gray or reddish brown; hardness = 5–6; specific gravity = 5; platy appearance	Hematite (Fe_2O_3)
Softer than glass	Greenish-black streak	Golden yellow; hardness = 4; specific gravity = 4.2; massive	Chalcopyrite ($CuFeS_2$)
	Gray-black streak	Silvery gray; hardness = 2.5; specific gravity = 7.6 (very heavy); good cubic cleavage	Galena (PbS)
	Yellow-brown streak	Yellow brown to dark brown; hardness variable (1–6); specific gravity = 3.5–4; often found in rounded masses; earthy appearance	Limonite ($Fe_2O_3 \cdot H_2O$)
	Gray-black streak	Black to bronze; tarnishes to purples and greens; hardness = 3; specific gravity = 5; massive	Bornite (Cu_5FeS_4)
Softer than your fingernail	Dark gray streak	Silvery gray; hardness = 1 (very soft); specific gravity = 2.2; massive to platy; writes on paper (pencil lead); feels greasy	Graphite (C)

Group II
Nonmetallic luster (dark colored).

Hardness	Cleavage	Other Diagnostic Properties	Name (Chemical Composition)
Harder than glass	Cleavage present	Black to greenish black; hardness = 5–6; specific gravity = 3.4; fair cleavage, two directions at nearly 90 degrees	Augite (Ca, Mg, Fe, Al silicate)
		Black to greenish black; hardness = 5–6; specific gravity = 3.2; fair cleavage, two directions at nearly 60 degrees and 120 degrees	Hornblende (Ca, Na, Mg, Fe, OH, Al silicate)
		Red to reddish brown; hardness = 6.5–7.5; conchoidal fracture; glassy luster	Garnet (Fe, Mg, Ca, Al silicate)
	Cleavage not prominent	Gray to brown; hardness = 9; specific gravity = 4; hexagonal crystals common	Corundum (Al_2O_3)
		Dark brown to black; hardness = 7; conchoidal fracture; glassy luster	Smoky quartz (SiO_2)
		Olive green; hardness = 6.5–7; small glassy grains	Olivine $(Mg, Fe)_2SiO_4$
Softer than glass	Cleavage present	Yellow brown to black; hardness = 4; good cleavage in six directions, light yellow streak that has the smell of sulfur	Sphalerite (ZnS)
		Dark brown to black; hardness = 2.5–3, excellent cleavage in one direction; elastic in thin sheets; black mica	Biotite (K, Mg, Fe, OH, Al silicate)
	Cleavage absent	Generally tarnished to brown or green; hardness = 2.5; specific gravity = 9; massive	Native copper (Cu)
Softer than your fingernail	Cleavage not prominent	Reddish brown; hardness = 1–5; specific gravity = 4–5; red streak; earthy appearance	Hematite (Fe_2O_3)
		Yellow brown; hardness = 1–3; specific gravity = 3.5; earthy appearance; powders easily	Limonite ($Fe_2O_3 \cdot H_2O$)

Group III
Nonmetallic luster (light colored).

Hardness	Cleavage	Other Diagnostic Properties	Name (Chemical Composition)
Harder than glass	Cleavage present	Flesh colored or white to gray; hardness = 6; specific gravity=2.6; two directions of cleavage at nearly right angles	Potassium feldspar ($KAlSi_3O_8$) Plagioclase feldspar ($NaAlSi_3O_8$ to $CaAl_2Si_2O_8$)
	Cleavage absent	Any color; hardness = 7; specific gravity = 2.65; conchoidal fracture; glassy appearance; varieties: milky, rose, smoky, amethyst (violet)	Quartz (SiO_2)
Softer than glass	Cleavage present	White, yellowish to colorless; hardness = 3; three directions of cleavage at 75 degrees (rhombohedral); effervesces in HCl; often transparent	Calcite ($CaCO_3$)
		White to colorless; hardness = 2.5; three directions of cleavage at 90 degrees (cubic); salty taste	Halite ($NaCl$)
		Yellow, purple, green, colorless; hardness = 4; white streak; translucent to transparent; four directions of cleavage	Fluorite (CaF_2)
Softer than your fingernail	Cleavage present	Colorless; hardness = 2–2.5; transparent and elastic in thin sheets; excellent cleavage in one direction; light mica	Muscovite (K, OH, Al silicate)
		White to transparent, hardness = 2; when in sheets, is flexible but not elastic; varieties: selenite (transparent, three directions of cleavage); satin spar (fibrous, silky luster); alabaster (aggregate of small crystals)	Gypsum ($CaSO_4 \cdot 2H_2O$)
	Cleavage not prominent	White, pink, green; hardness = 1–2; forms in thin plates; soapy feel; pearly luster	Talc (Mg silicate)
		Yellow; hardness = 1–2.5	Sulfur (S)
		White; hardness = 2; smooth feel; earthy odor; when moistened, has typical clay texture	Kaolinite (Hydrous Al silicate)
		Green; hardness = 2.5; fibrous; variety of serpentine	Asbestos (Mg, Al silicate)
		Pale to dark reddish brown; hardness = 1–3; dull luster; earthy; often contains spheroidal-shaped particles; not a true mineral	Bauxite (Hydrous Al oxide)

APPENDIX C

The Earth's Grid System

A glance at any globe reveals a series of north-south and east-west lines that together make up the earth's grid system, a universally used scheme for locating points on the earth's surface. The north-south lines of the grid are called **meridians** and extend from pole to pole (Figure C.1) All are halves of great circles. A **great circle** is the largest possible circle that may be drawn on a globe; if a globe were sliced along one of these circles, it would be divided into two equal parts called **hemispheres.** By viewing a globe or Figure C.1, it can be seen that meridians are spaced farthest apart at the equator and converge toward the poles. The east-west lines (circles) of the grid are known as **parallels.** As their name implies, these circles are parallel to one another (Figure C.1). While all meridians are parts of great circles, all parallels are not. In fact, only one parallel, the equator, is a great circle.

LATITUDE AND LONGITUDE

Latitude may be defined as distance, measured in degrees, *north* and *south* of the equator. Parallels are used to show latitude. Since all points that lie along the same parallel are an identical distance from the equator, they all have the same latitude designation. The latitude of the equator is 0 degrees, while the north and south poles lie 90 degrees N and 90 degrees S, respectively.

Longitude is defined as distance, measured in degrees, *east* and *west* of the zero or prime meridian. Since all meridians are identical, the choice of a zero line is obviously arbitrary. However, the meridian that passes through the Royal Observatory at Greenwich, England, is universally accepted as the reference meridian. Thus, the longitude for any place on the globe is measured east or west from this line. Longitude can vary from 0 degrees along the prime meridian to 180 degrees, halfway around the globe.

It is important to remember that when a location is specified, directions must be given, that is, north or south latitude and east or west longitude (Figure C.2) If this is not done, more than one point on the globe is being des-

ignated. The only exceptions, of course, are places that lie along the equator, the prime meridian, or the 180-degree meridian. It should also be noted that while it is not incorrect to use fractions, a degree of latitude or longitude is usually divided into minutes and seconds. A minute (') is $\frac{1}{60}$ of a degree, and a second ('') is $\frac{1}{60}$ of a minute. When locating a place on a map, the degree of exactness will depend upon the scale of the map. When using a small-scale world map or globe, it may be difficult to estimate latitude and longitude to the nearest whole degree or two. On the other hand, when a large-scale map of an area is used, it is often possible to estimate latitude and longitude to the nearest minute or second.

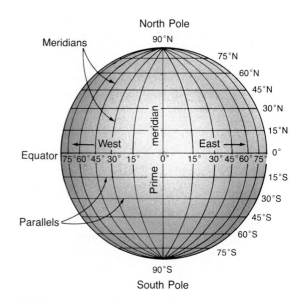

Figure C.1
The earth's grid system. (After J. B. Hoyt, Man and the Earth, *3rd ed.,* © *1966. Adapted by permission of Prentice-Hall, Inc., Englewood Cliffs, N.J.)*

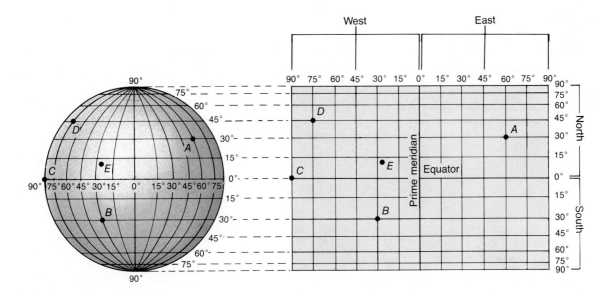

Figure C.2
Locating places using the grid system. For both diagrams: Point A is latitude 30 degrees N, longitude 60 degrees E; Point B is latitude 30 degrees S, longitude 30 degrees W; Point C is latitude 0 degrees, longitude 90 degrees W; Point D is latitude 45 degrees N, longitude 75 degrees W; Point E is approximately latitude 10 degrees N, longitude 25 degrees W. (After J. B. Hoyt, Man and the Earth, *3rd ed., © 1966. Adapted by permission of Prentice-Hall, Inc., Englewood Cliffs, N.J.)*

DISTANCE MEASUREMENT

The length of a degree of longitude depends upon where the measurement is taken. At the equator, which is a great circle, a degree of east-west distance is equal to approximately 111 kilometers (69 miles). This figure is found by dividing the earth's circumference—40,075 kilometers (24,900 miles)—by 360. However, with an increase in latitude, the parallels become smaller, and the length of a degree of longitude diminishes (see Table C.1). Thus, at about latitude 60 degrees N and S, a degree of longitude has a value equal to about half of what it was at the equator.

Since all meridians are halves of great circles, a degree of latitude is equal to about 111 kilometers (69 miles), just as a degree of longitude along the equator is. However, the earth is not a perfect sphere but is slightly flattened at the poles and bulges slightly at the equator. Because of this, there are small differences in the length of a degree of latitude.

Determining the shortest distance between two points on a globe can be done easily and fairly accurately using the "globe and string" method. It should be noted here that the arc of a great circle is the shortest distance between two points on a sphere. In order to determine the great circle distance (as well as observe the great circle route) between two places, stretch the string between the locations in question. Then, measure the length of the string along the equator (since it is a great circle with degrees marked on it) to determine the number of degrees between the two points. To calculate the distance in kilometers or miles, simply multiply the number of degrees by 111 or 69, respectively.

Table C.1
Longitude as distance.

°Lat.	Length of 1° Long.		°Lat.	Length of 1° Long.		°Lat.	Length of 1° Long.	
	km	miles		km	miles		km	miles
0	111.367	69.172	30	96.528	59.955	60	55.825	34.674
1	111.349	69.161	31	95.545	59.345	61	54.131	33.622
2	111.298	69.129	32	94.533	58.716	62	52.422	32.560
3	111.214	69.077	33	93.493	58.070	63	50.696	31.488
4	111.096	69.004	34	92.425	57.407	64	48.954	30.406
5	110.945	68.910	35	91.327	56.725	65	47.196	29.314
6	110.760	68.795	36	90.203	56.027	66	45.426	28.215
7	110.543	68.660	37	89.051	55.311	67	43.639	27.105
8	110.290	68.503	38	87.871	54.578	68	41.841	25.988
9	110.003	68.325	39	86.665	53.829	69	40.028	24.862
10	109.686	68.128	40	85.431	53.063	70	38.204	23.729
11	109.333	67.909	41	84.171	52.280	71	36.368	22.589
12	108.949	67.670	42	82.886	51.482	72	34.520	21.441
13	108.530	67.410	43	81.575	50.668	73	32.662	20.287
14	108.079	67.130	44	80.241	49.839	74	30.793	19.126
15	107.596	66.830	45	78.880	48.994	75	28.914	17.959
16	107.079	66.509	46	77.497	48.135	76	27.029	16.788
17	106.530	66.168	47	76.089	47.260	77	25.134	15.611
18	105.949	65.807	48	74.659	46.372	78	23.229	14.428
19	105.337	65.427	49	73.203	45.468	79	21.320	13.242
20	104.692	65.026	50	71.727	44.551	80	19.402	12.051
21	104.014	64.605	51	70.228	43.620	81	17.480	10.857
22	103.306	64.165	52	68.708	42.676	82	15.551	9.659
23	102.565	63.705	53	67.168	41.719	83	13.617	8.458
24	101.795	63.227	54	65.604	40.748	84	11.681	7.255
25	100.994	62.729	55	64.022	39.765	85	9.739	6.049
26	100.160	62.211	56	62.420	38.770	86	7.796	4.842
27	99.297	61.675	57	60.798	37.763	87	5.849	3.633
28	98.405	61.121	58	59.159	36.745	88	3.899	2.422
29	97.481	60.547	59	57.501	35.715	89	1.950	1.211
30	96.528	59.955	60	55.825	34.674	90	0.000	0.000

APPENDIX D

Topographic Maps

A map is a representation on a flat surface of all or a part of the earth's surface drawn to a specific scale. Maps are often the most effective means for showing the locations of both natural and cultural features, their sizes, and their relationships to one another. Like photographs, maps readily display information that would be impractical to express in words.

While most maps show only the two horizontal dimensions, geologists, as well as other map users, often require that the third dimension, elevation, be shown on maps. Maps that show the shape of the land are called **topographic maps.** Although various techniques may be used to depict elevations, the most accurate method involves the use of contour lines.

CONTOUR LINES

A **contour line** is a line on a map representing a corresponding imaginary line on the ground that has the same elevation above sea level along its entire length. While many map symbols are pictographs resembling the objects they represent, a contour line is an abstraction that has no counterpart in nature. It is, however, an accurate and effective device for representing the third dimension on paper.

Some useful facts and rules concerning contour lines are listed as follows. This information should be studied in conjunction with Figure D.1.

1. Contour lines bend upstream or upvalley. The contours form Vs that point upstream, and in the upstream direction the successive contours represent higher elevations. For example, if you were standing on a stream bank and wished to get to the point at the same elevation directly opposite you on the other bank, without stepping up or down, you would need to walk upstream along the contour at that elevation to where it crosses the stream bed, cross the stream, and then walk back downstream along the same contour.

2. Contours near the upper parts of hills form closures. The top of a hill is higher than the highest closed contour.
3. Hollows (depressions) without outlets are shown by closed, hatched contours. Hatched contours are contours with short lines on the inside pointing downslope.
4. Contours are widely spaced on gentle slopes.
5. Contours are closely spaced on steep slopes.
6. Evenly spaced contours indicate a uniform slope.
7. Contours usually do not cross or intersect each other, except in the rare case of an overhanging cliff.
8. All contours eventually close, either on a map or beyond its margins.
9. A single high contour never occurs between two lower ones, and vice versa. In other words, a change in slope direction is always determined by the repetition of the same elevation either as two different contours of the same value or as the same contour crossed twice.
10. Spot elevations between contours are given at many places, such as road intersections, hill summits, and lake surfaces. Spot elevations differ from control elevation stations, such as bench marks, in not being permanently established by permanent markers.

RELIEF

Relief refers to the difference in elevation between any two points. Maximum relief refers to the difference in elevation between the highest and lowest points in the area being considered. Relief determines the **contour interval,** which is the difference in elevation between succeeding contour lines that is used on topographic maps. Where relief is low, a small contour interval, such as 10 or 20 feet, may be used. In flat areas, such as wide river valleys or broad, flat uplands, a contour interval of 5 feet is often used. In rugged mountainous terrain, where relief is many hundreds of feet, contour intervals as large as 50 or 100 feet are used.

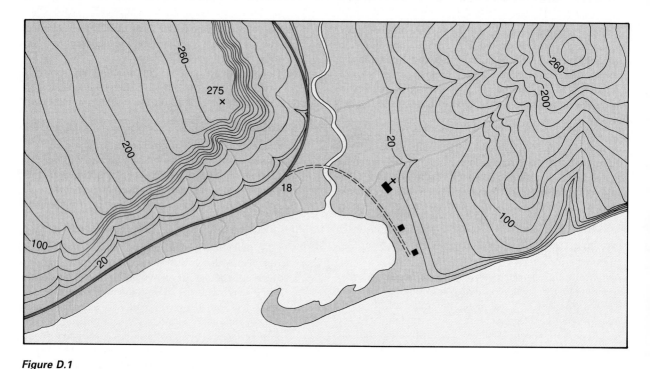

Figure D.1
Perspective view of an area and a contour map of the same area. These illustrations show how features are depicted on a topographic map. The upper illustration is a perspective view of a river valley and the adjoining hills. The river flows into a bay, which is partly enclosed by a hooked sandbar. On either side of the valley are terraces through which streams have cut gullies. The hill on the right has a smoothly eroded form and gradual slopes, whereas the one on the left rises abruptly in a sharp precipice, from which it slopes gently and forms an inclined plateau traversed by a few shallow gullies. A road provides access to a church and two houses situated across the river from a highway that follows the seacoast and curves up the river valley. The lower illustration shows the same features represented by symbols on a topographic map. The contour interval (vertical distance between adjacent contours) is 20 feet. (After U.S. Geological Survey)

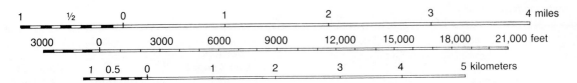

Figure D.2
Graphic scale.

SCALE

Map **scale** expresses the relationship between distance or area on the map to the true distance or area on the earth's surface. This is generally expressed as a ratio or fraction, such as 1:24,000 or 1/24,000. The numerator, usually 1, represents map distance, and the denominator, a large number, represents ground distance. Thus, 1:24,000 means that a distance of 1 unit on the map represents a distance of 24,000 such units on the surface of the earth. It does not matter what the units are.

Often, the graphic, or bar, scale is more useful than the fractional scale, because it is easier to use for measuring distances between points. The graphic scale (Figure D.2) consists of a bar that is divided into equal segments, which represent equal distances on the map. One segment on the left side of the bar is usually divided into smaller units to permit more accurate estimates of fractional units.

Topographic maps, which are also referred to as quadrangles, are generally classified according to publication scale. Each series is intended to fulfill a specific type of map

need. To select a map with the proper scale for a particular use, remember that large-scale maps show more detail and small-scale maps show less detail. The sizes and scales of topographic maps published by the U.S. Geological Survey are shown in Table D.1.

COLOR AND SYMBOL

Each color and symbol used on a U.S. Geological Survey topographic map has significance. Common topographic map symbols are shown in Figure D.3 on page 709. The meaning of each color is as follows:

Blue—water features
Black—construction works, such as homes, schools, churches, roads, and so forth
Brown—contour lines
Green—woodlands, orchards, and so forth
Red—urban areas, important roads, public land subdivision lines

Table D.1
National topographic maps.

Series	Scale	1-Inch Represents	Standard Quadrangle Size (latitude-longitude)	Quadrangle Area (square miles)	Paper Size E-W N-S Width Length (inches)
7½-minute	1:24,000	2000 feet	7½′ × 7½′	49–70	22 × 27*
Puerto Rico 7½-minute	1:20,000	about 1667 feet	7½′ × 7½′	71	29½ × 32½
15-minute	1:62,500	nearly 1 mile	15′ × 15′	197–282	17 × 21*
Alaska 1:63,360	1:63,360	1 mile	15′ × 20′–36′	207–281	18 × 21**
U.S. 1:250,000	1:250,000	nearly 4 miles	1° × 2°†	4580–8669	34 × 22††
U.S. 1:1,000,000	1:1,000,000	nearly 16 miles	4° × 6°†	73.734–102,759	27 × 27

Source: U.S. Geological Survey.

* South of latitude 31 degrees, 7½-minute sheets are 23 × 27 inches; 15-minute sheets are 18 × 21 inches.

** South of latitude 62 degrees, sheets are 17 × 21 inches.

† Maps of Alaska and Hawaii vary from these standards.

†† North of latitude 42 degrees, sheets are 29 × 22 inches; Alaska sheets are 30 × 23 inches.

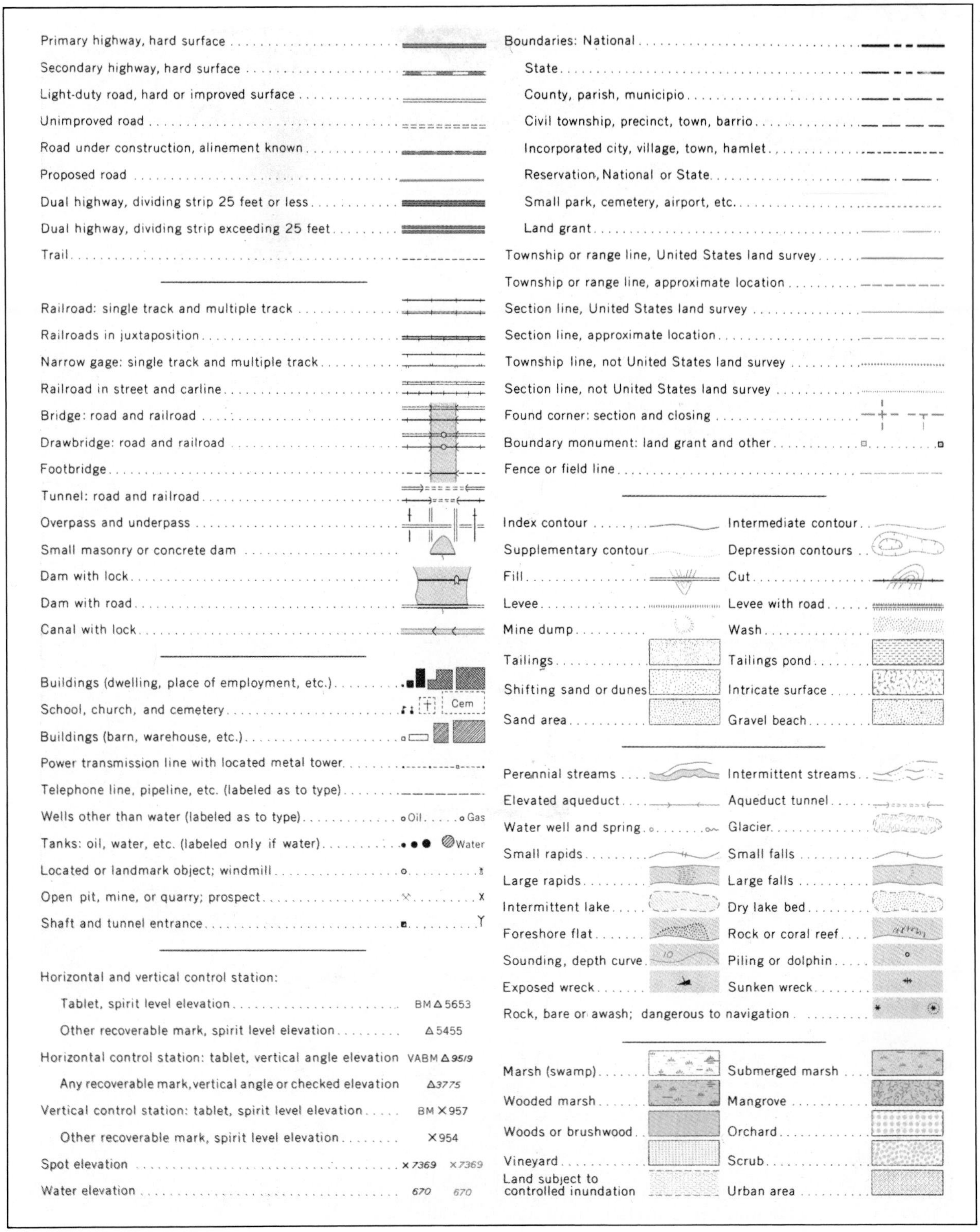

Primary highway, hard surface		Boundaries: National	
Secondary highway, hard surface		State	
Light-duty road, hard or improved surface		County, parish, municipio	
Unimproved road		Civil township, precinct, town, barrio	
Road under construction, alinement known		Incorporated city, village, town, hamlet	
Proposed road		Reservation, National or State	
Dual highway, dividing strip 25 feet or less		Small park, cemetery, airport, etc.	
Dual highway, dividing strip exceeding 25 feet		Land grant	
Trail		Township or range line, United States land survey	
		Township or range line, approximate location	
Railroad: single track and multiple track		Section line, United States land survey	
Railroads in juxtaposition		Section line, approximate location	
Narrow gage: single track and multiple track		Township line, not United States land survey	
Railroad in street and carline		Section line, not United States land survey	
Bridge: road and railroad		Found corner: section and closing	
Drawbridge: road and railroad		Boundary monument: land grant and other	
Footbridge		Fence or field line	
Tunnel: road and railroad			
Overpass and underpass		Index contour	Intermediate contour
Small masonry or concrete dam		Supplementary contour	Depression contours
Dam with lock		Fill	Cut
Dam with road		Levee	Levee with road
Canal with lock		Mine dump	Wash
		Tailings	Tailings pond
Buildings (dwelling, place of employment, etc.)		Shifting sand or dunes	Intricate surface
School, church, and cemetery	Cem	Sand area	Gravel beach
Buildings (barn, warehouse, etc.)			
Power transmission line with located metal tower		Perennial streams	Intermittent streams
Telephone line, pipeline, etc. (labeled as to type)		Elevated aqueduct	Aqueduct tunnel
Wells other than water (labeled as to type)	Oil Gas	Water well and spring	Glacier
Tanks: oil, water, etc. (labeled only if water)	Water	Small rapids	Small falls
Located or landmark object; windmill		Large rapids	Large falls
Open pit, mine, or quarry; prospect	X	Intermittent lake	Dry lake bed
Shaft and tunnel entrance	Y	Foreshore flat	Rock or coral reef
		Sounding, depth curve	Piling or dolphin
Horizontal and vertical control station:		Exposed wreck	Sunken wreck
Tablet, spirit level elevation	BM△5653	Rock, bare or awash; dangerous to navigation	
Other recoverable mark, spirit level elevation	△5455		
Horizontal control station: tablet, vertical angle elevation	VABM△95/9	Marsh (swamp)	Submerged marsh
Any recoverable mark, vertical angle or checked elevation	△3775	Wooded marsh	Mangrove
Vertical control station: tablet, spirit level elevation	BM×957	Woods or brushwood	Orchard
Other recoverable mark, spirit level elevation	×954	Vineyard	Scrub
Spot elevation	×7369 ×7369	Land subject to controlled inundation	Urban area
Water elevation	670 670		

Figure D.3

U.S. Geological Survey topographic map symbols. (Variations will be found on older maps.)

APPENDIX E

Star Charts[*]

The star charts on the next four pages can be used to locate the brighter stars and prominent constellations on the dates and times indicated on each chart. To use these charts, face southward and hold the chart overhead with the top toward the north.

---

[*] From Robert Dixon, *Dynamic Astronomy*, 6th Ed., Prentice-Hall, 1992.

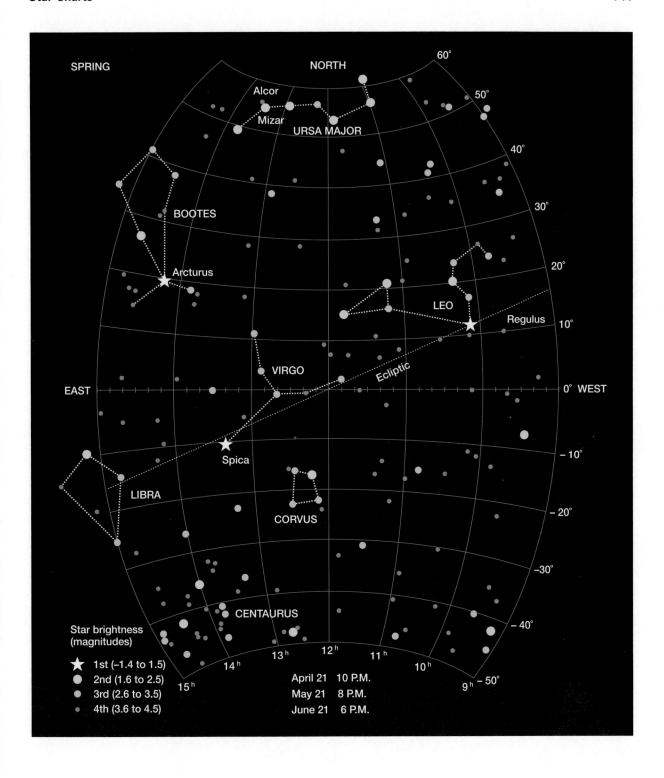

SPRING

NORTH

60°

50°

Alcor

Mizar

URSA MAJOR

40°

30°

BOOTES

20°

Arcturus

LEO

Regulus

10°

VIRGO

Ecliptic

EAST

0° WEST

Spica

−10°

LIBRA

CORVUS

−20°

−30°

CENTAURUS

−40°

Star brightness
(magnitudes)

★ 1st (−1.4 to 1.5)

● 2nd (1.6 to 2.5)

● 3rd (2.6 to 3.5)

· 4th (3.6 to 4.5)

15ʰ 14ʰ 13ʰ 12ʰ 11ʰ 10ʰ 9ʰ −50°

April 21 10 P.M.

May 21 8 P.M.

June 21 6 P.M.

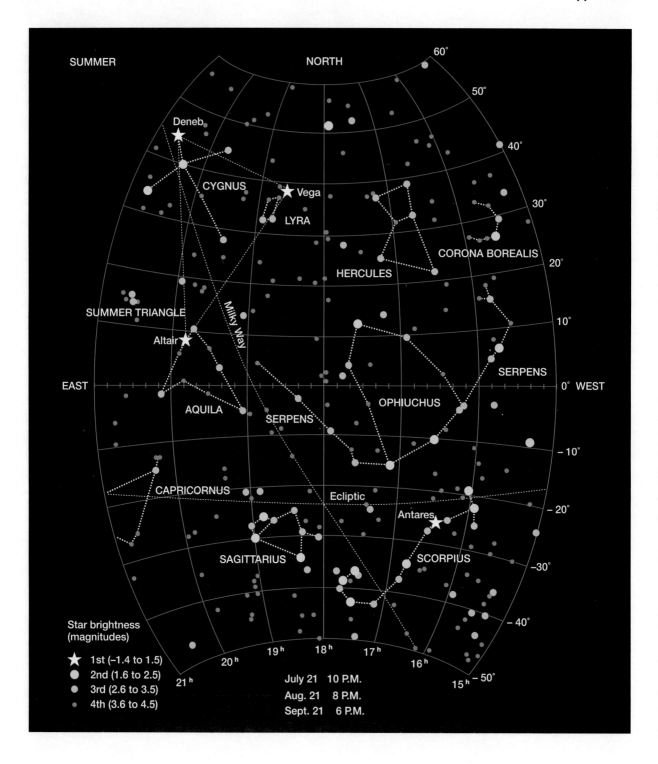

SUMMER

NORTH

60°

50°

40°

Deneb

30°

CYGNUS

Vega

20°

LYRA

CORONA BOREALIS

HERCULES

10°

SUMMER TRIANGLE

Milky Way

SERPENS

Altair

0° WEST

EAST

OPHIUCHUS

AQUILA

SERPENS

-10°

CAPRICORNUS

Ecliptic

-20°

Antares

SAGITTARIUS

SCORPIUS

-30°

-40°

Star brightness
(magnitudes)

20ʰ 19ʰ 18ʰ 17ʰ 16ʰ 15ʰ -50°

21ʰ

★ 1st (−1.4 to 1.5)

⬤ 2nd (1.6 to 2.5)

● 3rd (2.6 to 3.5)

· 4th (3.6 to 4.5)

July 21 10 P.M.
Aug. 21   8 P.M.
Sept. 21 6 P.M.

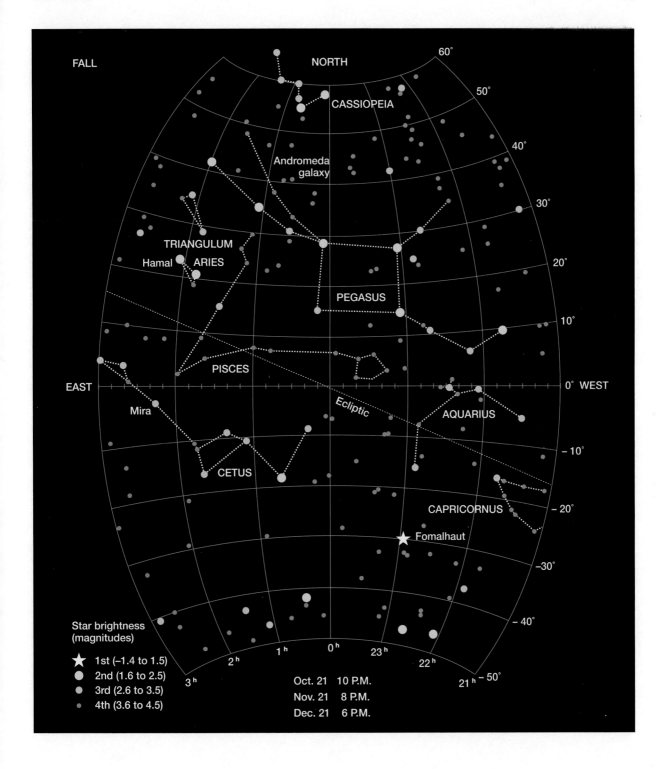

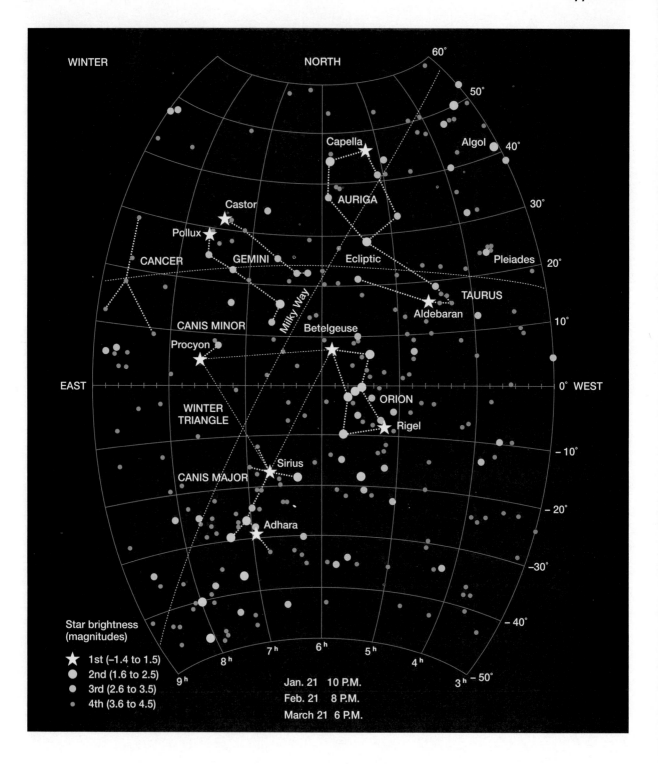

APPENDIX F

World Climates

The distribution of major atmospheric elements is very complex. Due to the many variations from place to place and from time to time at the same place, it is unlikely that any two sites on the earth's surface experience exactly the same weather. Since the number of places on earth is almost infinite, the number of different climates must also be extremely large. In order to cope with such variety, it is essential to devise a means of classifying this vast array of data. By establishing groups consisting of items that have certain important characteristics in common, order and simplicity are introduced. Organizing large amounts of information not only aids comprehension and understanding but also facilitates analysis and explanation.

We should remember that the classification of climates (or of anything else) is not a natural phenomenon, but the product of human ingenuity. The value of any particular classification is determined largely by its intended use. Here we shall use a system devised by the Russian-born German climatologist Wladimir Koeppen (1846–1940). As a tool for presenting the general world pattern of climates, it has been the best-known and most used system for more than fifty years. Koeppen believed that the distribution of natural vegetation was the best expression of the totality of climate. Therefore, the boundaries he chose were based largely on the limits of certain plant associations.

Five principal groups are recognized, with each group designated by a capital letter as follows:

A Humid tropical: winterless climates; all months have a mean temperature above 18°C.
B Dry: climates where evaporation exceeds precipitation; there is a constant water deficiency.
C Humid middle-latitude: mild winters; the average temperature of the coldest month is below 18°C but above −3°C.
D Humid middle-latitude: severe winters; the average temperature of the coldest month is below −3°C and the warmest monthly mean exceeds 10°C.
E Polar: summerless climates; the average temperature of the warmest month is below 10°C.

Notice that four of the major groups (A, C, D, E) are defined on the basis of temperature characteristics, and the fifth, the B group, has precipitation as its primary criterion. Each of the five groups is further subdivided by using the criteria and letter symbols presented in Table F.1.*

What follows is a brief summary of the major climatic types. While reading this summary, you should refer to Figure F.1 on the following two pages and Table F.2 on page 721, which shows data for representative places of each climatic type.

HUMID TROPICS (A CLIMATES)

Within the A group of climates, two main types are recognized—wet tropical climates (Af and Am) and tropical wet and dry (AW).

Wet Tropical Climates (Af, Am)

Since places with an Af or Am designation are lowland areas that lie near the equator, temperatures are consistently high. Consequently, not only is the annual mean high, but the annual range is very small. Total annual precipitation is high, often exceeding 200 centimeters; and although precipitation is not evenly distributed throughout the year, places with this climate are generally wet in all months. If a dry season exists, it is very short. In response to the constantly high temperatures and year-round rainfall, the wet tropics are dominated by the most luxuriant vegetation found in any climatic realm—the tropical rain forest.

---

* When classifying climatic data by using Table F.1, you should first determine whether or not the data meet the criteria for the E climates. If the station is not a polar climate, proceed to the criteria for B climates. If your data do not fit into either E or B groups, check the data against the criteria for A, C, and D climates, in that order.

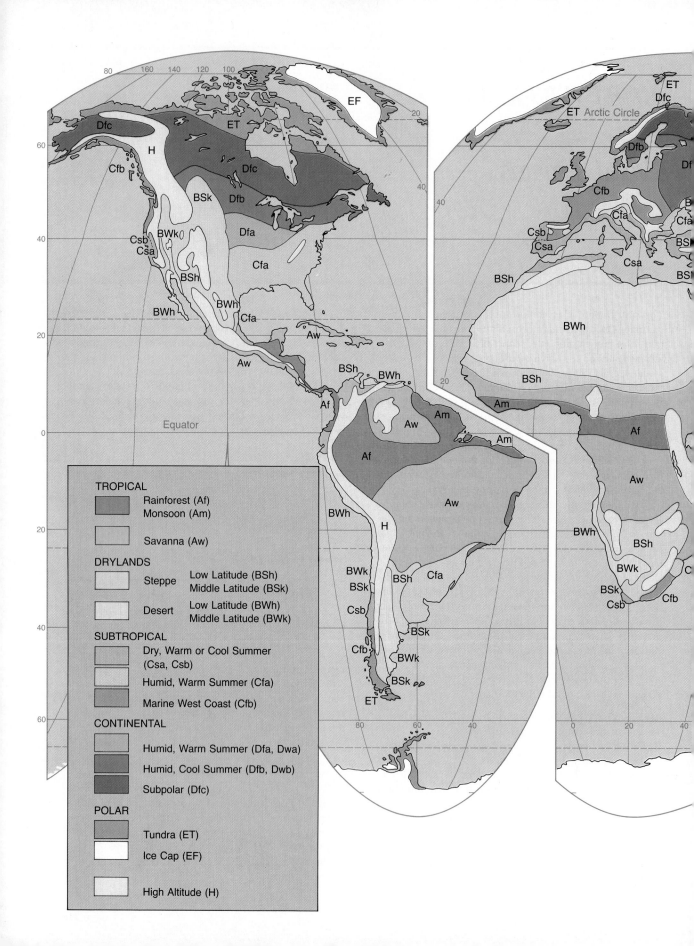

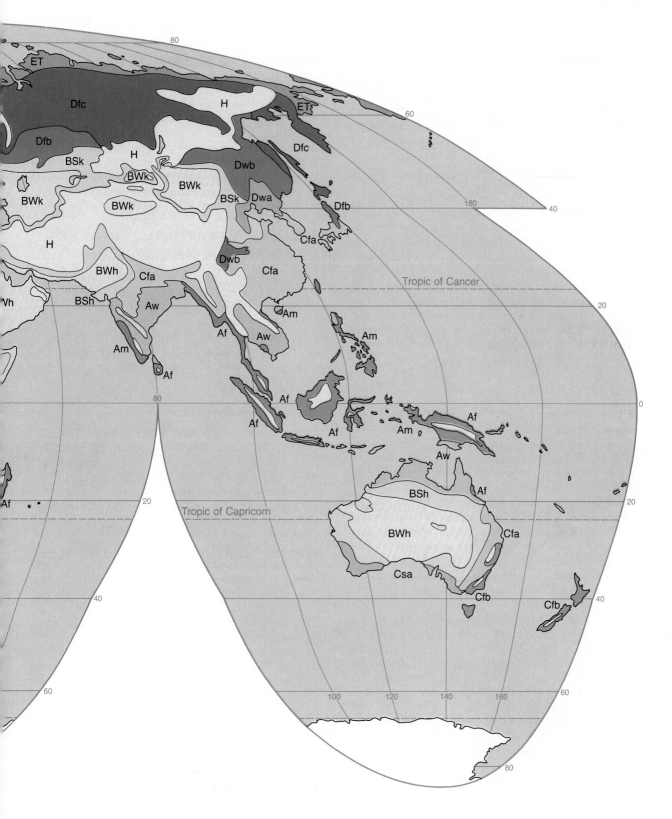

Figure F.1
Climates of the world (Koeppen). (Adapted from E. Willard Miller, Physical Geography,
Columbus, Ohio: Merrill, 1985. Plate 2)

Tropical Wet and Dry (Aw)

In the latitude zone poleward of the wet tropics and equatorward of the tropical deserts lies a transitional climatic region called tropical wet and dry. Here the rain forest gives way to the savanna, a tropical grassland with scattered drought-tolerant trees. Since temperature characteristics among all A climates are quite similar, the primary factor that distinguishes the Aw climate from Af and Am is precipitation. Although the overall amount of precipitation in the tropical wet and dry realm is often considerably less than in the wet tropics, the most distinctive feature of this climate is not the annual rainfall total but the markedly seasonal character of the rainfall. As the equatorial low advances poleward in summer, the rainy season commences and features weather patterns typical of the wet tropics. Later, with the retreat of the equatorial low, the subtropical high advances into the region and brings with it intense drought conditions. In some Aw regions such as India, Southeast Asia, and portions of Australia, the alternating periods of rainfall and drought are associated with a pronounced monsoon circulation (see Chapter 14).

DRY (B) CLIMATES

It is important to realize that the concept of dryness is a relative one and refers to any situation in which a water deficiency exists. Climatologists define a dry climate as one in which the yearly precipitation is not as great as the potential loss of water by evaporation. Thus, dryness is not only related to annual rainfall totals, but it is also a function of evaporation, which in turn is closely dependent upon temperature. To establish the boundary between dry and humid climates, the Koeppen classification uses formulas that involve three variables: average annual precipitation, average annual temperature, and seasonal distribution of precipitation. The use of average annual temperature reflects its importance as an index of evaporation. The amount of rainfall defining the humid-dry boundary increases as the annual mean temperature increases. The use of seasonal precipitation as a variable is also related to this idea. If rain is concentrated in the warmest months, loss to evaporation is greater than if the precipitation were concentrated in the cooler months.

Within the regions defined by a general water deficiency there are two climatic types: arid or desert (BW) and semiarid or steppe (BS). These two groups have many features in common; their differences are primarily a matter of degree. The semiarid is a marginal and more humid variant of the arid and represents a transition zone that surrounds the desert and separates it from the bordering humid climates.

The heart of the low-latitude dry climates (BWh and BSh) lies in the vicinity of the Tropic of Cancer and the Tropic of Capricorn. The existence and distribution of this rather extensive dry tropical realm is primarily a consequence of the subsidence and marked stability of the subtropical highs. Unlike their low-latitude counterparts, middle-latitude deserts and steppes are not controlled by the subsiding air masses of the subtropical anticyclones. Instead, these dry lands exist principally because of their positions in the deep interiors of large landmasses far removed from the oceans. In addition, the presence of high mountains across the paths of the prevailing winds further acts to separate these areas from water-bearing maritime air masses.

HUMID MIDDLE-LATITUDE CLIMATES WITH MILD WINTERS (C CLIMATES)

Although the term *subtropical* is often used for the C climates, it can be misleading. While many areas with C climates do indeed possess some near-tropical characteristics, other regions do not. For example, we would be stretching the use of the term *subtropical* to describe the climates of coastal Alaska and Norway, which belong to the C group. Within the C group of climates, several subgroups are recognized.

Humid Subtropics (Cfa)

Located on the eastern sides of the continents, in the 25- to 40-degree latitude range, this climatic type dominates the southeastern United States, as well as other similarly situated areas around the world. In the summer, the humid subtropics experience hot, sultry weather of the type one expects to find in the rainy tropics. Daytime temperatures are generally high, and since both specific and relative humidities are high, the night brings little relief. An afternoon or evening thunderstorm is also possible, for these areas experience such storms on an average of 40 to 100 days each year, the majority during the summer months. As summer turns to autumn, the humid subtropics lose their similarity to the rainy tropics. Although winters are mild, frosts are common in the higher-latitude Cfa areas and occasionally plague the tropical margins as well. The winter precipitation is also different in character from the summer. Some is in the form of snow, and most is generated along fronts of the frequent middle-latitude cyclones that sweep over these regions.

Marine West Coast Climate (Cfb, Cfc)

Situated on the western (windward) side of continents, from about 40 to 60 degrees north and south latitude, is a climatic region dominated by the onshore flow of oceanic air. The prevalence of maritime air masses means that mild winters and cool summers are the rule, as is an ample amount of rainfall throughout the year. Although there is no pronounced dry period, there is a drop in monthly pre-

Table F.1
Koeppen system of climatic classification.

1st	2nd	3rd	
	Letter Symbol		
A			Average temperature of the coldest month is 18°C or higher.
	f		Every month has 6 centimeters of precipitation or more.
	m		Short dry season; precipitation in driest month less than 6 centimeters, but equal to or greater than 10 − (R/25) (R is annual rainfall in centimeters).
	w		Well-defined winter dry season; precipitation in driest month less than 10 − (R/25).
	s		Well-defined summer dry season (rare).
B			Potential evaporation exceeds precipitation. The dry/humid boundary is defined by the following formulas: (Note: R is average annual precipitation in centimeters and T is average annual temperature in °C)
			$R < 2T + 28$ when 70% or more of rain falls in warmer 6 months.
			$R < 2T$ when 70% or more of rain falls in cooler 6 months.
			$R < 2T + 14$ when neither half year has 70% or more of total annual rain.
	S		Steppe ⎫ The BS/BW boundary is 1/2 the dry/humid boundary.
	W		Desert ⎭
		h	Average annual temperature is 18°C or greater.
		k	Average annual temperature is less than 18°C.
C			Average temperature of the coldest month is below 18°C and above −3°C.
	w		At least ten times as much precipitation in a summer month as in the driest winter month.
	s		At least three times as much precipitation in a winter month as in the driest summer month; precipitation in driest summer month less than 4 centimeters.
	f		Criteria for w and s cannot be met.
		a	Warmest month is over 22°C; at least 4 months over 10°C.
		b	No month above 22°C; at least 4 months over 10°C.
		c	One to 3 months above 10°C.
D			Average temperature of coldest month is −3°C or below; average temperature of warmest month is greater than 10°C.
	s		Same as under C.
	w		Same as under C.
	f		Same as under C.
		a	Same as under C.
		b	Same as under C.
		c	Same as under C.
		d	Average temperature of the coldest month is −38°C or below.
E			Average temperature of the warmest month is below 10°C.
	T		Average temperature of the warmest month is greater than 0°C and less than 10°C.
	F		Average temperature of the warmest month is 0°C or below.

cipitation totals during the summer. The reason for the reduced summer rainfall is the poleward migration of the oceanic subtropical highs. Although the areas of marine west coast climate are situated too far poleward to be dominated by these dry anticyclones, their influence is sufficient to cause a decrease in warm season rainfall.

Dry-Summer Subtropics (Csa, Csb)

This climate is typically located along the west sides of continents, between latitudes of 30 and 45 degrees. It is unique because it is the only humid climate that has a strong winter rainfall maximum, a feature that reflects its intermediate position between the marine west coast on the poleward side and the tropical steppes on the equatorward side. In summer the region is dominated by the stable eastern side of the oceanic subtropical highs. In winter, as the wind and pressure systems follow the sun equatorward, it is within range of the cyclonic storms of the polar frost. Thus, during the course of a year these areas alternate between being part of the dry tropics and being an extension of the humid middle latitudes. While middle-latitude changeability characterizes the winter, tropical constancy describes the summer.

HUMID MIDDLE-LATITUDE CLIMATES WITH SEVERE WINTERS (D CLIMATES)

The D climates are land-controlled climates, the result of broad continents in the middle latitudes. Because continentality is a basic feature, D climates are absent in the Southern Hemisphere where the middle-latitude zone is dominated by the oceans.

Humid Continental (Dfa, Dfb, Dwa, Dwb)

This climate is located in the central and eastern portions of North America and Eurasia in the latitude range between approximately 40 and 50 degrees north latitude. Both winter and summer temperatures may be characterized as relatively severe. Consequently, annual temperature ranges are high throughout the climate. Precipitation is characteristically greatest in summer. A portion of the winter precipitation is in the form of snow, the proportion increasing with latitude. Although the precipitation is generally considerably less during the cold season, it is usually much more conspicuous than the greater amounts that characterize the summer. An obvious reason is that snow remains on the ground, often for extended periods, and rain, of course, does not. Furthermore, while summer rains are often in the form of relatively short showers, winter snows generally occur over a prolonged period.

Subarctic Climate (Dfc, Dfd, Dwc, Dwd)

Situated north of the humid continental climate and south of the polar tundra is an extensive subarctic region. It is often referred to as the taiga climate because its extent closely corresponds to the northern coniferous forest of the same name. The outstanding feature in this realm is the dominance of winter. Not only is it long, but temperatures are also bitterly cold. By contrast, summers in the subarctic are remarkably warm, despite their short duration. However, when compared with regions farther south, this short season must be characterized as cool. The extremely cold winters and relatively warm summers combine to produce the highest annual temperature ranges on earth. Since these far northerly continental interiors are the source regions for cP air masses, there is very limited moisture available throughout the year. Precipitation totals are therefore small, with a maximum occurring during the warmer summer months.

POLAR (E) CLIMATES

Polar climates are those in which the mean temperature of the warmest month is below 10°C. Thus, just as the tropics are defined by their year-round warmth, the polar realm is known for its enduring cold. Since winters are periods of perpetual night, or nearly so, temperatures at most polar locations are understandably bitter. During the summer months temperatures remain cool despite the long days, because the sun is so low in the sky that its oblique rays are not effective in bringing about a genuine warming. Although polar climates are classified as humid, precipitation is generally meager. Evaporation, of course, is also limited. The scanty precipitation totals are easily understood in view of the temperature characteristics of the region. The amount of water vapor in the air is always small because low specific humidities must accompany low temperatures. Usually precipitation is most abundant during the warmer summer months when the moisture content of the air is highest.

Two types of polar climates are recognized. The tundra climate (ET) is a treeless climate found almost exclusively in the Northern Hemisphere. Due to the combination of high latitude and continentality, winters are severe, summers are cool, and annual temperature ranges are high. Further, yearly precipitation is small, with a modest summer maximum. The ice cap climate (EF) does not have a single monthly mean above 0°C. Consequently, since the average temperature for all months is below freezing, the growth of vegetation is prohibited and the landscape is one of permanent ice and snow. This climate of perpetual frost covers a surprisingly large area—more than 15.5 million square kilometers, or about 9 percent of the earth's land area. Aside from scattered occurrences in high mountain areas, it is confined to the ice caps of Greenland and Antarctica.

Table F.2
Climatic data for representative stations.

	J	F	M	A	M	J	J	A	S	O	N	D	Year
Iquitos, Peru (Af); lat. 3°39′S; 115 m												
Temp. (°C)	25.6	25.6	24.4	25.0	24.4	23.3	23.3	24.4	24.4	25.0	25.6	25.6	24.7
Precip. (mm)	259	249	310	165	254	188	168	117	221	183	213	292	2619
Rio de Janeiro, Brazil (Aw); lat. 22°50′S; 26 m												
Temp. (°C)	25.9	26.1	25.2	23.9	22.3	21.3	20.8	21.1	21.5	22.3	23.1	24.4	23.2
Precip. (mm)	137	137	143	116	73	43	43	43	53	74	97	127	1086
Faya, Chad (BWh); lat. 18°00′N; 251 m												
Temp. (°C)	20.4	22.7	27.0	30.6	33.8	34.2	33.6	32.7	32.6	30.5	25.5	21.3	28.7
Precip. (mm)	0	0	0	0	0	2	1	11	2	0	0	0	16
Salt Lake City, Utah (BSk); lat. 40°46′N; 1288 m												
Temp. (°C)	−2.1	0.9	4.7	9.9	14.7	19.4	24.7	23.6	18.3	11.5	3.4	−0.2	10.7
Precip. (mm)	34	30	40	45	36	25	15	22	13	29	33	31	353
Washington, D.C. (Cfa); lat. 38°50′N; 20 m												
Temp. (°C)	2.7	3.2	7.1	13.2	18.8	23.4	25.7	24.7	20.9	15.0	8.7	3.4	13.9
Precip. (mm)	77	63	82	80	105	82	105	124	97	78	72	71	1036
Brest, France (Cfb); lat. 48°24′N; 103 m												
Temp. (°C)	6.1	5.8	7.8	9.2	11.6	14.4	15.6	16.0	14.7	12.0	9.0	7.0	10.8
Precip. (mm)	133	96	83	69	68	56	62	80	87	104	138	150	1126
Rome, Italy (Csa); lat. 41°52′N; 3 m												
Temp. (°C)	8.0	9.0	10.9	13.7	17.5	21.6	24.4	24.2	21.5	17.2	12.7	9.5	15.9
Precip. (mm)	83	73	52	50	48	18	9	18	70	110	113	105	749
Peoria, Illinois (Dfa); lat. 40°45′N; 180 m												
Temp. (°C)	−4.4	−2.2	4.4	10.6	16.7	21.7	23.9	22.7	18.3	11.7	3.8	−2.2	10.4
Precip. (mm)	46	51	69	84	99	97	97	81	97	61	61	51	894
Verkhoyansk, Russia (Dfd); lat. 67°33′N; 137 m												
Temp. (°C)	−46.8	−43.1	−30.2	−13.5	2.7	12.9	15.7	11.4	2.7	−14.3	−35.7	−44.5	−15.2
Precip. (mm)	7	5	5	4	5	25	33	30	13	11	10	7	155
Ivigtut, Greenland (ET); lat. 61°12′N; 129 m												
Temp. (°C)	−7.2	−7.2	−4.4	−0.6	4.4	8.3	10.0	8.3	5.0	1.1	−3.3	−6.1	0.7
Precip. (mm)	84	66	86	62	89	81	79	94	150	145	117	79	1132
Mcmurdo Station, Antarctica (EF); lat. 77°53′S; 2 m												
Temp. (°C)	−4.4	−8.9	−15.5	−22.8	−23.9	−24.4	−26.1	−26.1	−24.4	−18.8	−10.0	−3.9	−17.4
Precip. (mm)	13	18	10	10	10	8	5	8	10	5	5	8	110

APPENDIX G

World Soils

Figure G.1, on pages 724–725, shows the generalized pattern of global soil orders according to the *Comprehensive Soil Classification System* (CSCS). It should be examined in conjunction with Table G.1, which briefly describes each of the soil orders depicted on the map. To avoid subjective decisions as to classification (a problem that plagued earlier systems), the CSCS defined its classes strictly in terms of soil characteristics. That is, it is based upon features that can be observed or inferred.

The CSCS uses a hierarchy of six categories, or levels. The system recognizes 10 major global *orders* that can be further subdivided into *suborders*, *great groups*, *subgroups*, *families*, and *series*. Note, however, that on the scale of a world map such as Figure G.1, only the largest units (soil orders) can be shown and then only in an extremely generalized way. Although the distribution pattern of major soil orders is more complex than can be shown in Figure G.1, the major distinguishing regional properties of world soils are depicted.

Table G.1
World soil orders.

Entisols	Youngest soils on the earth. Just beginning to develop in response to the weathering phenomena in the environment. Do not display natural horizons. Found in all climates. They weather slowly over thousands of years; consequently, volcanic ash deposits or sand deposits form the basis for entisols.
Vertisols	Soils containing large amounts of clay, which shrink upon drying and swell with the addition of water. Found in subhumid to arid climates, provided that adequate supplies of water are available to saturate the soil after periods of drought. Soil expansion and contraction exert stresses on human structures.
Inceptisols	Young soils that reveal developmental characteristics (horizons) in response to climate and vegetation. Exist from the Arctic to the tropics on young land surfaces. Common in alpine areas, on river floodplains, in stables and dune areas, and in areas once glaciated.
Aridisols	Soils that develop in dry places, such as the desert, where water —precipitation and groundwater—is insufficient to remove soluble minerals. Frequently irrigated for intensive agricultural production, although salt accumulation poses a problem.
Mollisols	Dark, soft soils that have developed under grass vegetation, generally found in prairie areas. Soil fertility is excellent because potential evaporation generally exceeds precipitation. Also found in hardwood forests with significant earthworm activity. Climatic range is boreal or alpine to tropical. Dry seasons are normal.
Spodosols	Soils found only in humid regions on sandy material. Range from the boreal coniferous forests into tropical forests. Beneath the dark upper horizon of weathered organic material lies a light-colored horizon of leached material, the distinctive property of this soil.
Alfisols	Mineral soils that form under boreal forests or broadleaf deciduous forests, rich in iron and aluminum. Clay particles accumulate in a subsurface layer in response to leaching in moist environments. Fertile, productive soils, because they are neither too wet nor too dry.
Ultisols	Soils that represent the products of long periods of weathering. Water percolating through the soil concentrates clay particles in the lower horizons (argillic horizons). Restricted to humid climates in the temperate regions and the tropics where the growing season is long. Abundant water and a long frost-free period contribute to extensive leaching, hence poorer soil quality.
Oxisols	Soils that occur on old land surfaces unless parent materials were strongly weathered before they were deposited. Generally found in the tropics and subtropical regions. Rich in iron and aluminum oxides, oxisols are heavily leached; hence are poor soils for agricultural activity. Few, if any, exist in the United States.
Histosols	Organic soils with little or no climatic implications. Can be found in any climate where organic debris can accumulate to form a "bog soil." Dark, partially decomposed organic material commonly referred to as *peat.*

Source: Robert E. Norris et al. *Geography: An Introductory Perspective,* Columbus, Ohio: Merrill, 1982.

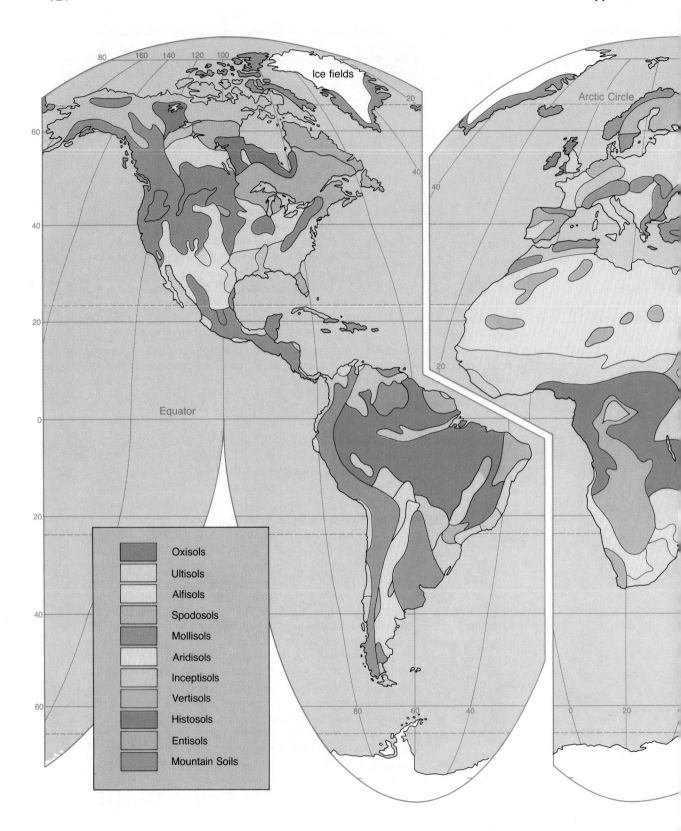

Ice fields

Arctic Circle

80 160 140 120 100

60

40

20

Equator

0

20

40

60

40

20

20

40

80 60 40 0 20

Oxisols
Ultisols
Alfisols
Spodosols
Mollisols
Aridisols
Inceptisols
Vertisols
Histosols
Entisols
Mountain Soils

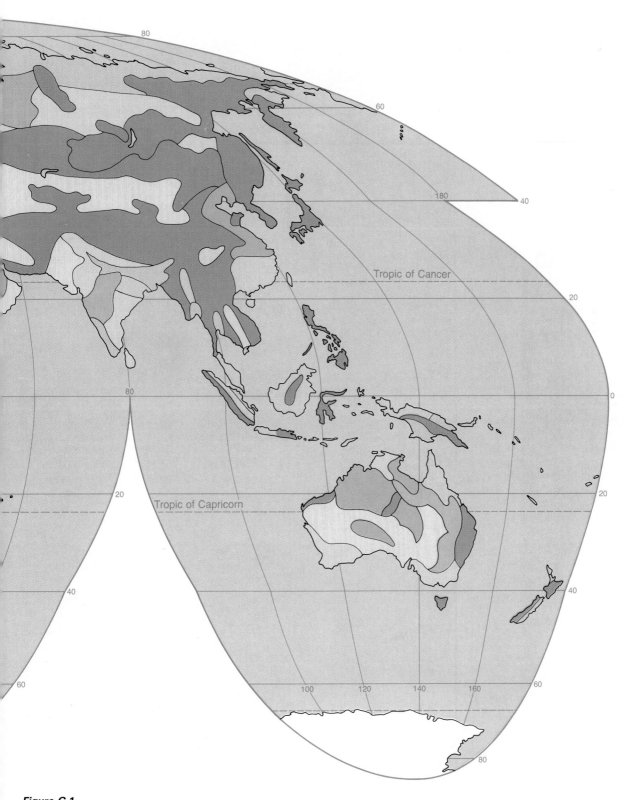

Figure G.1
Soil distribution. The pattern of global soil orders is remarkably similar to the pattern of major climates (see Figure F.1). Terminology employed is from the Comprehensive Soil Classification System. *(Adapted from E. Willard Miller,* Physical Geography, *Columbus, Ohio: Merrill, 1985, Plate 4)*

A P P E N D I X   H

Periodic Table of the Elements

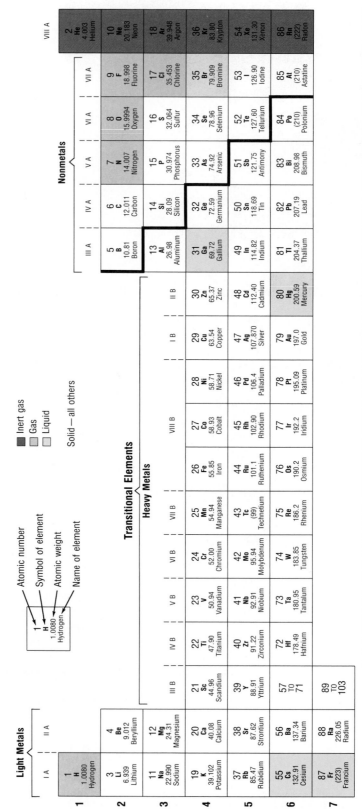

Glossary

Aa   A type of lava flow that has a jagged blocky surface.

Ablation   A general term for the loss of ice and snow from a glacier.

Abrasion   The grinding and scraping of a rock surface by the friction and impact of rock particles carried by water, wind, or ice.

Absolute dating   Determination of the number of years since the occurrence of a given geologic event.

Absolute humidity   The weight of water vapor in a given volume of air (usually expressed in grams/m^3).

Absolute instability   Air that has a lapse rate greater than the dry adiabatic rate.

Absolute magnitude   The apparent brightness of a star if it were viewed from a distance of 10 parsecs (32.6 lightyears). Used to compare the true brightness of stars.

Absolute stability   Air with a lapse rate less than the wet adiabatic rate.

Absorption spectrum   A continuous spectrum with dark lines superimposed.

Abyssal plain   Very level area of the deep-ocean floor, usually lying at the foot of the continental rise.

Accretionary wedge   A large wedge-shaped mass of sediment that accumulates in subduction zones. Here sediment is scraped from the subducting oceanic plate and accreted to the overriding crustal block.

Acid precipitation   Rain or snow with a pH value that is less than the pH of unpolluted precipitation.

Adiabatic temperature change   Cooling or warming of air caused when air is allowed to expand or is compressed, not because heat is added or subtracted.

Advection   Horizontal convective motion, such as wind.

Advection fog   A fog formed when warm, moist air is blown over a cool surface.

Aftershocks   Smaller earthquakes that follow the main earthquake.

Air mass   A large body of air that is characterized by a sameness of temperature and humidity.

Air-mass weather   The conditions experienced in an area as an air mass passes over it. Because air masses are large and fairly homogeneous, air-mass weather will be fairly constant and may last for several days.

Air pollutants   Airborne particles and gases that occur in concentrations that endanger the health and well-being of organisms or disrupt the orderly functioning of the environment.

Albedo   The reflectivity of a substance, usually expressed as a percentage of the incident radiation reflected.

Alluvial fan   A fan-shaped deposit of sediment formed when a stream's slope is abruptly reduced.

Alluvium   Unconsolidated sediment deposited by a stream.

Alpine glacier   A glacier confined to a mountain valley, which in most instances had previously been a stream valley.

Altimeter   An aneroid barometer calibrated to indicate altitude instead of pressure.

Altitude (of the sun)   The angle of the sun above the horizon.

Anemometer   An instrument used to determine wind speed.

Aneroid barometer   An instrument for measuring air pressure that consists of evacuated metal chambers that are very sensitive to variations in air pressure.

Angle of repose   The steepest angle at which loose material remains stationary without sliding downslope.

Angular unconformity   An unconformity in which the strata below dip at an angle different from that of the beds above.

Annual mean temperature   An average of the twelve monthly temperature means.

Annual temperature range   The difference between the highest and lowest monthly means.

Anthracite   A hard, metamorphic form of coal that burns clean and hot.

Anticline   A fold in sedimentary strata that resembles an arch.

Anticyclone   A high-pressure center characterized by a clockwise flow of air in the Northern Hemisphere.

Aphelion   The place in the orbit of a planet where the planet is farthest from the sun.

Apparent magnitude   The brightness of a star when viewed from the earth.

Aquicludes   Impermeable beds that hinder or prevent groundwater movement.

Aquifer   Rock or soil through which groundwater moves easily.

Archean eon   The second eon of Precambrian time, following the Hadean and preceding the Proterozoic. It extends between 3.8 and 2.5 billion years before the present.

Arête   A narrow knifelike ridge separating two adjacent glaciated valleys.

Arkose   A feldspar-rich sandstone.

Artesian well   A well in which the water rises above the level where it was initially encountered.

Asteroids   Thousands of small planetlike bodies, ranging in size from a few hundred kilometers to less than a kilometer, whose orbits lie mainly between those of Mars and Jupiter.

Asthenosphere   A subdivision of the mantle situated below the lithosphere. This zone of weak material exists below a depth of about 100 kilometers and in some regions extends as deep as 700 kilometers. The rock within this zone is easily deformed.

Astronomical theory   A theory of climatic change first developed by the Yugoslavian astronomer Milankovitch. It is based upon changes in the shape of the earth's orbit, variations in the obliquity of the earth's axis, and the wobbling of the earth's axis.

Astronomical Unit (AU)   Average distance from the earth to the sun; 1.5×10^8 km, or 93×10^6 miles.

Astronomy   The scientific study of the universe; it includes the observation and interpretation of celestial bodies and phenomena.

Atmosphere   The gaseous portion of a planet; the planet's envelope of air. One of the traditional subdivisions of the earth's physical environment.

Atoll   A continuous or broken ring of coral reef surrounding a central lagoon.

Atom   The smallest particle that exists as an element.

Atomic number   The number of protons in the nucleus of an atom.

Atomic weight   The average of the atomic masses of isotopes for a given element.

Aurora   A bright display of ever-changing light caused by solar radiation interacting with the upper atmosphere in the region of the poles.

Autumnal equinox   The equinox that occurs on September 21–23 in the Northern Hemisphere and on March 21–22 in the Southern Hemisphere.

Azoic zone   A well-known but incorrect theory formulated about 1850 by Edward Forbes stating that no life existed in the ocean below a depth of about 550 meters.

Back swamp   A poorly drained area on a floodplain that results when natural levees are present.

Barchan dune   Solitary sand dune shaped like a crescent with its tips pointing downwind.

Barchanoid dune   Dunes forming scalloped rows of sand oriented at right angles to the wind. This form is intermediate between isolated barchans and extensive waves of transverse dunes.

Barograph   A recording barometer.

Barometer   An instrument that measures atmospheric pressure.

Barometric tendency   *See* Pressure tendency.

Barred spiral   A galaxy having straight arms extending from its nucleus.

Barrier island   A low, elongate ridge of sand that parallels the coast.

Basalt   A fine-grained igneous rock of mafic composition.

Base level   The level below which a stream cannot erode.

Basin   A circular downfolded structure.

Batholith   A large mass of igneous rock that formed when magma was emplaced at depth, crystallized, and subsequently exposed by erosion.

Baymouth bar   A sandbar that completely crosses a bay, sealing it off from the open ocean.

Beach drift   The transport of sediment in a zigzag pattern along a beach caused by the uprush of water from obliquely breaking waves.

Beach nourishment   Large quantities of sand are added to the beach system to offset losses caused by wave erosion.

Bed load   Sediment that is carried by a stream along the bottom of its channel.

Benioff zone   Zone of inclined seismic activity that extends from a trench downward into the asthenosphere.

Bergeron process   A theory that relates the formation of precipitation to supercooled clouds, freezing nuclei, and the different saturation levels of ice and liquid water.

Big Bang theory   The theory that proposes that the universe originated as a single mass, which subsequently exploded.

Binary stars   Two stars revolving around a common center of mass under their mutual gravitational attraction.

Biogenous sediment   Sea-floor sediments consisting of material of marine-organic origin.

Biosphere   The totality of life on earth; the parts of the lithosphere, hydrosphere, and atmosphere in which living organisms can be found.

Bituminous   The most common form of coal, often called soft, black coal.

Black dwarf   A final state of evolution for a star, in which all of its energy sources are exhausted and it no longer emits radiation.

Black hole   A massive star that has collapsed to such a small volume that its gravity prevents the escape of all radiation.

Blowout (deflation hollow)   A depression excavated by the wind in easily eroded deposits.

Bode's law   A sequence of numbers that approximates the mean distances of the planets from the sun.

Body waves   Seismic waves that travel through the earth's interior.

Braided stream   A stream consisting of numerous intertwining channels.

Breakwater   A structure protecting a near-shore area from breaking waves.

Breccia   A sedimentary rock composed of angular fragments that were lithified.

Bright-line spectrum   The bright lines produced by an incandescent gas under low pressure.

Bright nebula   A cloud of glowing gas excited by ultraviolet radiation from hot stars.

Cactolith   A quasi-horizontal chonolith composed of anastomosing ductoliths, whose distal ends curl like a harpolith, thin like a sphenolith, or bulge discordantly like an akmolith or ethmolith.

Caldera   A large depression typically caused by collapse or ejection of the summit area of a volcano.

Calorie   The amount of heat required to raise the temperature of one gram of water 1°C.

Calving   Wastage of a glacier that occurs when large pieces of ice break off into water.

Capacity   The total amount of sediment a stream is able to transport.

Cassini division   A wide gap in the ring system of Saturn between the A ring and the B ring.

Catastrophism   The concept that the earth was shaped by catastrophic events of a short-term nature.

Cavern   A naturally formed underground chamber or series of chambers most commonly produced by solution activity in limestone.

Celestial sphere   An imaginary hollow sphere upon which the ancients believed the stars were hung and carried around the earth.

Cenozoic era   A time span on the geologic calendar beginning about 66 million years ago following the Mesozoic era.

Cepheid variable   A star whose brightness varies periodically because it expands and contracts. A type of pulsating star.

Chemical sedimentary rock   Sedimentary rock consisting of material that was precipitated from water by either inorganic or organic means.

Chemical weathering   The processes by which the internal structure of a mineral is altered by the removal and/or addition of elements.

Chinook   A wind blowing down the leeward side of a mountain and warming by compression.

Chromatic aberration   The property of a lens whereby light of different colors is focused at different places.

Chromosphere   The first layer of the solar atmosphere found directly above the photosphere.

Cinder cone   A rather small volcano built primarily of pyroclastics ejected from a single vent.

Circle of illumination   The great circle that separates daylight from darkness.

Cirque   An amphitheater-shaped basin at the head of a glaciated valley produced by frost wedging and plucking.

Cirrus   One of three basic cloud forms; also one of the three high cloud types. They are thin, delicate ice crystal clouds often appearing as veil-like patches or thin, wispy fibers.

Clastic rock   A sedimentary rock made of broken fragments of pre-existing rock.

Cleavage   The tendency of a mineral to break along planes of weak bonding.

Climate   A description of aggregate weather conditions; the sum of all statistical weather information that helps describe a place or region.

Climatic feedback mechanism   One of the several different outcomes that may result if one of the many elements in the atmosphere's extremely complex interactive system is altered.

Climatology   The scientific study of climate.

Cloud   A form of condensation best described as a dense concentration of suspended water droplets or tiny ice crystals.

Cloud seeding   The introduction into clouds of particles (most commonly dry ice or silver iodide) for the purpose of altering the cloud's natural development.

Clouds of vertical development   A cloud that has its base in the low height range but extends upward into the middle or high altitudes.

Cluster (star)   A large group of stars.

Coarse-grained texture   An igneous rock texture in which the crystals are roughly equal in size and large enough so that individual minerals can be identified with the unaided eye.

Col   A pass between mountain valleys where the headwalls of two cirques intersect.

Cold front   A front along which a cold air mass thrusts beneath a warmer air mass.

Collision-coalescence process   A theory of raindrop formation in warm clouds (above 0°C) in which large cloud droplets ("giants") collide and join together with smaller droplets to form a raindrop. Opposite electrical charges may bind the cloud droplets together.

Column   A feature found in caves that is formed when a stalactite and stalagmite join.

Columnar joints   A pattern of cracks that form during cooling of molten rock to generate columns that are generally six-sided.

Coma   The fuzzy, gaseous component of a comet's head.

Comet   A small body which generally revolves about the sun in an elongated orbit.

Competence   A measure of the largest particle a stream can transport; a factor dependent on velocity.

Composite cone   A volcano composed of both lava flows and pyroclastic material.

Compound   A substance formed by the chemical combination of two or more elements in definite proportions and usually having properties different than those of its constituent elements.

Condensation   The change of state from a gas to a liquid.

Condensation nuclei   Tiny bits of particulate matter that serve as surfaces on which water vapor condenses.

Conditional instability   Moist air with a lapse rate between the dry and wet adiabatic rates.

Conduction   The transfer of heat through matter by molecular activity. Energy is transferred through collisions from one molecule to another.

Cone of depression   A cone-shaped depression in the water table immediately surrounding a well.

Conformable   Layers of rock that were deposited without interruption.

Conglomerate   A sedimentary rock composed of rounded gravel-sized particles.

Constellation   An apparent group of stars originally named for mythical characters. The sky is presently divided into 88 constellations.

Contact metamorphism   Changes in rock caused by the heat from a nearby magma body.

Continental (c) air mass   An air mass that forms over land; it is normally relatively dry.

Continental drift theory   A theory that originally proposed that the continents are rafted about. It has essentially been replaced by the plate tectonics theory.

Continental margin   That portion of the sea floor adjacent to the continents. It may include the continental shelf, continental slope, and continental rise.

Continental rise   The gently sloping surface at the base of the continental slope.

Continental shelf   The gently sloping submerged portion of the continental margin extending from the shoreline to the continental slope.

Continental slope   The steep gradient that leads to the deep-ocean floor and marks the seaward edge of the continental shelf.

Continuous spectrum   An uninterrupted band of light emitted by an incandescent solid, liquid, or gas under pressure.

Convection   The transfer of heat by the movement of a mass or substance. It can take place only in fluids.

Convergence   The condition that exists when the distribution of winds within a given area results in a net horizontal inflow of air into the area. Since convergence at lower levels is associated with an upward movement of air, areas of convergent winds are regions favorable to cloud formation and precipitation.

Convergent boundary   A boundary in which two plates move together, causing one of the slabs of lithosphere to be consumed into the mantle as it descends beneath an overriding plate.

Coral reef   Structure formed in a warm, shallow, sunlit ocean environment that consists primarily of the calcite-rich remains of corals as well as the limy secretions of algae and the hard parts of many other small organisms.

Core   Located beneath the mantle, it is the innermost layer of the earth. The core is divided into an outer core and an inner core.

Coriolis force (effect)   The deflective force of the earth's rotation on all free-moving objects, including the atmosphere and oceans. Deflection is to the right in the Northern Hemisphere and to the left in the Southern Hemisphere.

Corona   The outer, tenuous layer of the solar atmosphere.

Correlation   Establishing the equivalence of rocks of similar age in different areas.

Country breeze   A circulation pattern characterized by a light wind blowing into a city from the surrounding countryside. It is best developed on clear and otherwise calm nights when the urban heat island is most pronounced.

Crater   The depression at the summit of a volcano, or that which is produced by a meteorite impact.

Creep   The slow downhill movement of soil and regolith.

Crevasse   A deep crack in the brittle surface of a glacier.

Cross-bedding   Structure in which relatively thin layers are inclined at an angle to the main bedding. Formed by currents of wind or water.

Cross-cutting   A principle of relative dating. A rock or fault is younger than any rock (or fault) through which it cuts.

Crust   The very thin outermost layer of the earth.

Crystal   An orderly arrangement of atoms.

Crystal form   The external appearance of a mineral as determined by its internal arrangement of atoms.

Crystallization   The formation and growth of a crystalline solid from a liquid or gas.

Cumulus   One of three basic cloud forms; also the name given one of the clouds of vertical development. Cumulus are billowy individual cloud masses that often have flat bases.

Cup anemometer   *See* Anemometer

Curie point   The temperature above which a material loses its magnetization.

Cutoff   A short channel segment created when a river erodes through the narrow neck of land between meanders.

Cyclone   A low-pressure center characterized by a counterclockwise flow of air in the Northern Hemisphere.

Daily mean The mean temperature for a day that is determined by averaging the 24 hourly readings or, more commonly, by averaging the maximum and minimum temperatures for a day.

Daily temperature range The difference between the maximum and minimum temperatures for a day.

Dark-line spectrum *See* Absorption spectrum.

Dark nebula A cloud of interstellar dust that obscures the light of more distant stars and appears as an opaque curtain.

Daughter product An isotope resulting from radioactive decay.

Declination (stellar) The angular distance north or south of the celestial equator denoting the position of a celestial body.

Deep-ocean trench A narrow, elongated depression on the floor of the ocean.

Deep-sea fan A cone-shaped deposit at the base of the continental slope. The sediment is transported to the fan by turbidity currents that follow submarine canyons.

Deflation The lifting and removal of loose material by wind.

Degassing The release of gases dissolved in molten rock.

Delta An accumulation of sediment formed where a stream enters a lake or ocean.

Dendritic pattern A stream system that resembles the pattern of a branching tree.

Density The weight per unit volume of a particular material.

Deposition The process by which water vapor is changed directly to a solid without passing through the liquid state.

Desalination The removal of salts and other chemicals from seawater.

Desert One of the two types of dry climate; the driest of the dry climates.

Desert pavement A layer of coarse pebbles and gravel created when wind removed the finer material.

Detrital sedimentary rock Rock formed from the accumulation of material that originated and was transported in the form of solid particles derived from both mechanical and chemical weathering.

Dew point The temperature to which air has to be cooled in order to reach saturation.

Differential weathering The variation in the rate and degree of weathering caused by such factors as mineral makeup, degree of jointing, and climate.

Diffused light Solar energy scattered and reflected in the atmosphere that reaches the earth's surface in the form of diffuse blue light from the sky.

Dike A tabular-shaped intrusive igneous feature that cuts through the surrounding rock.

Dip-slip fault A fault in which the movement is parallel to the dip of the fault.

Discharge The quantity of water in a stream that passes a given point in a period of time.

Disconformity A type of unconformity in which the beds above and below are parallel.

Disseminated deposit Any economic mineral deposit in which the desired mineral occurs as scattered particles in the rock but in sufficient quantity to make the deposit an ore.

Dissolved load That portion of a stream's load carried in solution.

Distributary A section of a stream that leaves the main flow.

Diurnal tide Tides characterized by a single high and low water height each tidal day.

Divergence The condition that exists when the distribution of winds within a given area results in a net horizontal outflow of air from the region. In divergence at lower levels the resulting deficit is compensated for by a downward movement of air from aloft; hence, areas of divergent winds are unfavorable to cloud formation and precipitation.

Divergent boundary A region where the rigid plates are moving apart, typified by the mid-oceanic ridges.

Divide An imaginary line that separates the drainage of two streams; often found along a ridge.

Dome A roughly circular upfolded structure similar to an anticline.

Doppler effect The apparent change in wavelength of radiation caused by the relative motions of the source and the observer.

Doppler radar In addition to the tasks performed by conventional radar, this new generation of weather radar can detect motion directly and hence greatly improve tornado and severe storm warnings.

Drainage basin The land area that contributes water to a stream.

Drawdown The difference in height between the bottom of a cone of depression and the original height of the water table.

Drift The general term for any glacial deposit.

Drumlin A streamlined asymmetrical hill composed of glacial till. The steep side of the hill faces the direction from which the ice advanced.

Dry adiabatic rate The rate of adiabatic cooling or warming in unsaturated air. The rate of temperature change is 1°C per 100 meters.

Dune A hill or ridge of wind-deposited sand.

Earthflow The downslope movement of water-saturated, clay-rich sediment. Most characteristic of humid regions.

Earthquake The vibration of the earth produced by the rapid release of energy.

Ebb current The movement of a tidal current away from the shore.

Eccentricity The variation of an ellipse from a circle.

Echo sounder An instrument used to determine the depth of water by measuring the time interval between emission of a sound signal and the return of its echo from the bottom.

Eclipse   The cutting off of the light of one celestial body by another passing in front of it.

Ecliptic   The yearly path of the sun plotted against the background of stars.

Elastic rebound   The sudden release of stored strain in rocks that results in movement along a fault.

Electromagnetic radiation   *See* Radiation.

Electromagnetic spectrum   The distribution of electromagnetic radiation by wavelength.

Electron   A negatively charged subatomic particle that has a negligible mass and is found outside an atom's nucleus.

Element   A substance that cannot be decomposed into simpler substances by ordinary chemical or physical means.

Elements of weather and climate   Those quantities or properties of the atmosphere that are measured regularly and that are used to express the nature of weather and climate.

Elliptical galaxy   A galaxy that is round or elliptical in outline. It contains little gas and dust, no disk or spiral arms, and few hot, bright stars.

Eluviation   The washing out of fine soil components from the A horizon by downward-percolating water.

Emergent coast   A coast where land that was formerly below sea level has been exposed either because of crustal uplift or a drop in sea level or both.

Emission nebula   A gaseous nebula that derives its visible light from the fluorescence of ultraviolet light from a star in or near the nebula.

End moraine   A ridge of till marking a former position of the front of a glacier.

Entrenched meander   A meander cut into bedrock when uplifting rejuvenated a meandering stream.

Environmental lapse rate   The rate of temperature decrease with increasing height in the troposphere.

Eon   The largest time unit on the geologic time scale, next in order of magnitude above era.

Ephemeral stream   A stream that is usually dry because it carries water only in response to specific episodes of rainfall. Most desert streams are of this type.

Epicenter   The location on the earth's surface that lies directly above the focus of an earthquake.

Epoch   A unit of the geologic calendar that is a subdivision of a period.

Equatorial low   A belt of low pressure lying near the equator and between the subtropical highs.

Equatorial system   A method of locating stellar objects much like the coordinate system used on the earth's surface.

Equinox   The time when the vertical rays of the sun are striking the equator. The length of daylight and darkness is equal at all latitudes at equinox.

Era   A major division on the geologic calendar; eras are divided into shorter units called periods.

Erosion   The incorporation and transportation of material by a mobile agent, such as water, wind, or ice.

Eruptive variable   A star that varies in brightness.

Escape velocity   The initial velocity an object needs to escape from the surface of a celestial body.

Esker   Sinuous ridge composed largely of sand and gravel deposited by a stream flowing in a tunnel beneath a glacier near its terminus.

Estuary   A funnel-shaped inlet of the sea that formed when a rise in sea level or subsidence of land caused the mouth of a river to be flooded.

Evaporation   The process of converting a liquid to a gas.

Evaporite   A sedimentary rock formed of material deposited from solution by evaporation of the water.

Evolution, (Theory of)   A fundamental theory in biology and paleontology that sets forth the process by which members of a population of organisms come to differ from their ancestors. Organisms evolve by means of mutations, natural selection, and genetic factors. Modern species are descended from related but different species that lived in earlier times.

Exfoliation dome   Large, dome-shaped structure, usually composed of granite, formed by sheeting.

Exotic stream   A permanent stream that traverses a desert and has its source in well-watered areas outside the desert.

Extrusive   Igneous activity that occurs outside the crust.

Eye   A zone of scattered clouds and calm averaging about 20 kilometers in diameter at the center of a hurricane.

Eyepiece   A short-focal-length lens used to enlarge the image in a telescope. The lens nearest the eye.

Eye wall   The doughnut-shaped area of intense cumulonimbus development and very strong winds that surrounds the eye of a hurricane.

Fall   A type of movement common to mass wasting processes that refers to the free falling of detached individual pieces of any size.

Fault   A break in a rock mass along which movement has occurred.

Fault-block mountain   A mountain formed by the displacement of rock along a fault.

Felsic   The group of igneous rocks composed primarily of feldspar and quartz.

Fetch   The distance that the wind has traveled across the open water.

Filaments   Dark, thin streaks that appear across the bright solar disk.

Fine-grained texture   A texture of igneous rocks in which the crystals are too small for individual minerals to be distinguished with the unaided eye.

Fiord A steep-sided inlet of the sea formed when a glacial trough was partially submerged.

Fissure eruption An eruption in which lava is extruded from narrow fractures or cracks in the crust.

Flare A sudden brightening of an area on the sun.

Flood basalts Flows of basaltic lava that issue from numerous cracks or fissures and commonly cover extensive areas to thicknesses of hundreds of meters.

Flood current The tidal current associated with the increase in the height of the tide.

Floodplain The flat, low-lying portion of a stream valley subject to periodic inundation.

Flow A type of movement common to mass wasting processes in which water-saturated material moves downslope as a viscous fluid.

Fluorescence The absorption of ultraviolet light, which is re-emitted as visible light.

Focal length The focal length of a lens is the distance from the lens to the point where it focuses parallel rays of light.

Focus (earthquake) The zone within the earth where rock displacement produces an earthquake.

Focus (light) The point where a lens or mirror causes light rays to converge.

Fog A cloud with its base at or very near the earth's surface.

Fold A bent rock layer or series of layers that were originally horizontal and subsequently deformed.

Foliated A texture of metamorphic rocks that gives the rock a layered appearance.

Foreshocks Small earthquakes that often precede a major earthquake.

Fossil fuel General term for any hydrocarbon that may be used as a fuel, including coal, oil, and natural gas.

Fossils The remains or traces of organisms preserved from the geologic past.

Fossil succession Fossil organisms succeed one another in a definite and determinable order, and any time period can be recognized by its fossil content.

Fracture Any break or rupture in rock along which no appreciable movement has taken place.

Freezing The change of state from a liquid to a solid.

Freezing nuclei Solid particles that serve as cores for the formation of ice crystals.

Front The boundary between two adjoining air masses having contrasting characteristics.

Frontal fog Fog formed when rain evaporates as it falls through a layer of cool air.

Frontal wedging Lifting of air resulting when cool air acts as a barrier over which warmer, lighter air will rise.

Frost Occurs when the temperature falls to 0°C or below. It may or may not be accompanied by deposits of ice crystals, commonly termed *white frost*. *See* White frost.

Frost wedging The mechanical breakup of rock caused by the expansion of freezing water in cracks and crevices.

Galactic cluster A system of galaxies containing from several to thousands of member galaxies.

Geocentric The concept of an earth-centered universe.

Geology The science that examines the earth, its form and composition, and the changes it has undergone and is undergoing.

Geostrophic wind A wind, usually above a height of 600 meters (2000 feet), that blows parallel to the isobars.

Geothermal energy Natural steam used for power generation.

Geyser A fountain of hot water ejected periodically.

Giant (star) A luminous star of large radius.

Glacial erratic An ice-transported boulder that was not derived from bedrock near its present site.

Glacial striations Scratches and grooves on bedrock caused by glacial abrasion.

Glacial trough A mountain valley that has been widened, deepened, and straightened by a glacier.

Glacier A thick mass of ice originating on land from the compaction and recrystallization of snow that shows evidence of past or present flow.

Glassy A term used to describe the texture of certain igneous rocks, such as obsidian, that contain no crystals.

Glaze A coating of ice on objects formed when supercooled rain freezes on contact.

Globular cluster A nearly spherically shaped group of densely packed stars.

Globule A dense, dark nebula thought to be the birthplace of stars.

Gondwanaland The southern portion of Pangaea consisting of South America, Africa, Australia, India, and Antarctica.

Graben A valley formed by the downward displacement of a fault-bounded block.

Graded bed A sediment layer that is characterized by a decrease in sediment size from bottom to top.

Graded stream A stream that has the correct channel characteristics to maintain exactly the velocity required to transport the material supplied to it.

Gradient The slope of a stream; generally measured in feet per mile.

Granules The fine structure visible on the solar surface caused by convective cells below.

Greenhouse effect The transmission of short-wave solar radiation by the atmosphere coupled with the selective absorption of longer-wavelength terrestrial radiation, especially by water vapor and carbon dioxide.

Groin A short wall built at a right angle to the shore to trap moving sand.

Ground moraine   An undulating layer of till deposited as the ice front retreats.

Groundwater   Water in the zone of saturation.

Guyot   A submerged flat-topped seamount.

Gyre   The large circular surface current pattern found in each ocean.

Hadean eon   The first eon on the geologic time scale; this eon ended 3.8 billion years ago and preceded the Archean eon.

Hail   Nearly spherical ice pellets having concentric layers and formed by the successive freezing of layers of water.

Half-life   The time required for one-half of the atoms of a radioactive substance to decay.

Halocline   A layer of water in which there is a high rate of change in salinity in the vertical dimension.

Hanging valley   A tributary valley that enters a glacial trough at a considerable height above its floor.

Hardness   The resistance a mineral offers to scratching.

Heliocentric   The view that the sun is at the center of the solar system.

Hertzsprung-Russell diagram   *See* H-R diagram.

High cloud   A cloud that normally has its base above 6000 meters; the base may be lower in winter and at high-latitude locations.

Hogback   A narrow, sharp-crested ridge formed by the upturned edge of a steeply dipping bed of resistant rock.

Horizon   A layer in a soil profile.

Horn   A pyramidlike peak formed by glacial action in three or more cirques surrounding a mountain summit.

Horst   An elongate, uplifted block of crust bounded by faults.

Hot spot   A concentration of heat in the mantle capable of producing magma which, in turn, extrudes onto the earth's surface. The intraplate volcanism that produced the Hawaiian Islands is one example.

Hot spring   A spring in which the water is 6°−9°C (10°−15°F) warmer than the mean annual air temperature of its locality.

H-R diagram   A plot of stars according to their absolute magnitudes and spectral types.

Hubble Law   Relates the distance to a galaxy and its velocity.

Humidity   A general term referring to water vapor in the air but not to liquid droplets of fog, cloud, or rain.

Humus   Organic matter in soil produced by the decomposition of plants and animals.

Hurricane   A tropical cyclonic storm having winds in excess of 119 kilometers (74 miles) per hour.

Hydroelectric power   Electricity generated by the energy of flowing or falling water.

Hydrogen burning   The conversion of hydrogen through fusion to form helium.

Hydrogenous sediment   Sea-floor sediments consisting of minerals that crystallize from seawater. An important example is manganese nodules.

Hydrosphere   The water portion of our planet; one of the traditional subdivisions of the earth's physical environment.

Hydrothermal solution   The hot, watery solution that escapes from a mass of magma during the later stages of crystallization. Such solutions may alter the surrounding country rock and are frequently the source of significant ore deposits.

Hygrometer   An instrument designed to measure relative humidity.

Hygroscopic nuclei   Condensation nuclei having a high affinity for water, such as salt particles.

Hypothesis   A tentative explanation that is tested to determine if it is valid.

Ice cap   A mass of glacial ice covering a high upland or plateau and spreading out radially.

Ice sheet   A very large, thick mass of glacial ice flowing outward in all directions from one or more accumulation centers.

Igneous rock   A rock formed by the crystallization of molten magma.

Immature soil   A soil lacking horizons.

Inclination of the axis   The tilt of the earth's axis from the perpendicular to the plane of the earth's orbit.

Inclusion   A piece of one rock unit contained within another. Inclusions are used in relative dating. The rock mass adjacent to the one containing the inclusion must have been there first in order to provide the fragment.

Index fossil   A fossil that is associated with a particular span of geologic time.

Inertia   A property of matter that resists a change in its motion.

Infiltration   The movement of surface water into rock or soil through cracks and pore spaces.

Infrared   Radiation with a wavelength from 0.7 to 200 micrometers.

Inner core   The solid innermost layer of the earth, about 1300 kilometers (800 miles) in radius.

Inselberg   An isolated mountain remnant characteristic of the late stage of erosion in an arid region.

Intensity (earthquake)   An indication of the destructive effects of an earthquake at a particular place. Intensity is affected by such factors as distance to the epicenter and the nature of the surface materials.

Interior drainage   A discontinuous pattern of intermittent streams that do not flow to the ocean.

Intermediate composition   The composition of igneous rocks lying between felsic and mafic.

Interstellar matter   Dust and gases found between stars.

Intrusive rock   Igneous rock that formed below the earth's surface.

Ion   An atom or molecule that possesses an electrical charge.

Ionosphere   A complex zone of ionized gases that coincides with the lower portion of the thermosphere.

Iron meteorite   One of the three main categories of meteorites. This group is composed largely of iron with varying amounts of nickel (5–20 percent). Most meteorite finds are irons.

Irregular galaxy   A galaxy that lacks symmetry.

Island arc   A group of volcanic islands formed by the subduction and partial melting of oceanic lithosphere.

Isobar   A line drawn on a map connecting points of equal atmospheric pressure, usually corrected to sea level.

Isostacy   The concept that the earth's crust is "floating" in gravitational balance upon the material of the mantle.

Isotherms   Lines connecting points of equal temperature.

Isotopes   Varieties of the same element that have different mass numbers; their nuclei contain the same number of protons but different numbers of neutrons.

Jet stream   Swift (120–240-kilometer per hour), high-altitude winds.

Jetties   A pair of structures extending into the ocean at the entrance to a harbor or river that are built for the purpose of protecting against storm waves and sediment deposition.

Joint   A fracture in rock along which there has been no movement.

Jovian planet   The Jupiter-like planets Jupiter, Saturn, Uranus, and Neptune. These planets have relatively low densities.

Kame   A steep-sided hill composed of sand and gravel originating when sediment collected in openings in stagnant glacial ice.

Karst   A topography consisting of numerous depressions called sinkholes.

Kettle holes   Depressions created when blocks of ice became lodged in glacial deposits and subsequently melted.

Laccolith   A massive igneous body intruded between pre-existing strata.

Lahar   Mudflows on the slopes of volcanoes that result when unstable layers of ash and debris become saturated and flow downslope, usually following stream channels.

Lake-effect snow   Snow showers associated with a cP air mass to which moisture and heat are added from below as the air mass traverses a large and relatively warm lake (such as one of the Great Lakes), rendering the air mass humid and unstable.

Land breeze   A local wind blowing from land toward the water during the night in coastal areas.

Lapse rate (normal)   The average drop in temperature (6.5°C per kilometer; 3.5°F per 1000 feet) with increased altitude in the troposphere.

Latent heat   The energy absorbed or released during a change in state.

Lateral moraine   A ridge of till along the sides of an alpine glacier composed primarily of debris that fell to the glacier from the valley walls.

Laterite   A red, highly leached soil type found in the tropics that is rich in oxides of iron and aluminum.

Laurasia   The northern portion of Pangaea consisting of North America and Eurasia.

Lava   Magma that reaches the earth's surface.

Law   A formal statement of the regular manner in which a natural phenomenon occurs under given conditions; e.g., the "law of superposition."

Law of conservation of angular momentum   The product of the velocity of an object around a center of rotation (axis) and the distance squared of the object from the axis is constant.

Leaching   The depletion of soluble materials from the upper soil by downward-percolating water.

Lightning   A sudden flash of light generated by the flow of electrons between oppositely charged parts of a cumulonimbus cloud or between the cloud and the ground.

Light-year   The distance light travels in a year; about 6 trillion miles.

Liquefaction   A phenomenon, sometimes associated with earthquakes, in which soils and other unconsolidated materials containing abundant water are turned into a fluidlike mass that is not capable of supporting buildings.

Lithification   The process, generally cementation and/or compaction, of converting sediments to solid rock.

Lithogenous sediment   Sea-floor sediments having their source as products of weathering on the continents.

Lithosphere   The rigid outer layer of the earth, including the crust and upper mantle.

Local group   The cluster of 20 or so galaxies to which our galaxy belongs.

Loess   Deposits of wind-blown silt, lacking visible layers, generally buff colored, and capable of maintaining a nearly vertical cliff.

Longitudinal (seif) dunes   Long ridges of sand oriented parallel to the prevailing wind; these dunes form where sand supplies are limited.

Longshore current   A near-shore current that flows parallel to the shore.

Low cloud   A cloud that forms below a height of 2000 meters.

Low-velocity zone   *See* Asthenosphere.

Luminosity   The brightness of a star. The amount of energy radiated by a star.

Lunar breccia   A lunar rock formed when angular fragments and dust are welded together by the heat generated by the impact of a meteoroid.

Lunar eclipse   An eclipse of the moon.

Lunar regolith   A thin, gray layer on the surface of the moon, consisting of loosely compacted, fragmented material believed to have been formed by repeated meteoritic impacts.

Luster   The appearance or quality of light reflected from the surface of a mineral.

Mafic   Igneous rocks with a low silica content and a high iron-magnesium content.

Magma   A body of molten rock found at depth, including any dissolved gases and crystals.

Magmatic differentiation   The process of generating more than one rock type from a single magma.

Magnitude (earthquake)   The total amount of energy released during an earthquake.

Magnitude (stellar)   A number given to a celestial object to express its relative brightness.

Main sequence   A sequence of stars on the Hertzsprung-Russell diagram, containing the majority of stars, that runs diagonally from the upper left to the lower right.

Manganese nodules   Rounded lumps of hydrogenous sediment scattered on the ocean floor consisting mainly of manganese and iron, and usually containing small amounts of copper, nickel, and cobalt.

Mantle   The 2900-kilometer (1800-mile)-thick layer of the earth located below the crust.

Maria   The Latin name for the smooth areas of the moon formerly thought to be seas.

Maritime (m) air mass   An air mass that originates over the ocean. These air masses are relatively humid.

Mass number   The number of neutrons and protons in the nucleus of an atom.

Mass wasting   The downslope movement of rock, regolith, and soil under the direct influence of gravity.

Meander   A looplike bend in the course of a stream.

Mean solar day   The average time between two passages of the sun across the local celestial meridian.

Mechanical weathering   The physical disintegration of rock, resulting in smaller fragments.

Medial moraine   A ridge of till formed when lateral moraines from two coalescing alpine glaciers join.

Melt   The liquid portion of magma excluding the solid crystals.

Melting   The change of state from a solid to a liquid.

Mercurial barometer   A mercury-filled glass tube in which the height of the mercury column is a measure of air pressure.

Mesocyclone   An intense, rotating wind system in the lower part of a thunderstorm that precedes tornado development.

Mesopause   The boundary between the mesosphere and the thermosphere.

Mesosphere   The layer of the atmosphere immediately above the stratosphere and characterized by decreasing temperatures with height.

Mesozoic era   A time span on the geologic calendar between the Paleozoic and Cenozoic eras—from about 245 to 66.4 million years ago.

Metamorphic rock   Rocks formed by the alteration of pre-existing rock deep within the earth (but still in the solid state) by heat, pressure, and/or chemically active fluids.

Metamorphism   The changes in mineral composition and texture of a rock subjected to high temperature and pressure within the earth.

Meteor   The luminous phenomenon observed when a meteoroid enters the earth's atmosphere and burns up; popularly called a "shooting star."

Meteorite   Any portion of a meteoroid that survives its traverse through the earth's atmosphere and strikes the earth's surface.

Meteoroid   Small solid particles that have orbits in the solar system.

Meteorology   The scientific study of the atmosphere and atmospheric phenomena; the study of weather and climate.

Meteor shower   Many meteors appearing in the sky caused when the earth intercepts a swarm of meteoritic particles.

Middle cloud   A cloud occupying the height range from 2000 to 6000 meters.

Middle-latitude cyclone   *See* Wave cyclone.

Mid-ocean ridge   A continuous mountainous ridge on the floor of all the major ocean basins and varying in width from 500 to 5000 kilometers (300 to 3000 miles). The rifts at the crests of these ridges represent divergent plate boundaries.

Mineral   A naturally occurring, inorganic crystalline material with a unique chemical composition.

Mineral resource   All discovered and undiscovered deposits of a useful mineral that can be extracted now or at some time in the future.

Mixed tide   A tide characterized by a large inequality in successive high water heights, low water heights, or both.

Mixing depth   The height to which convectional movements extend above the earth's surface. The greater the mixing depth, the better the air quality.

Mohorovičić discontinuity (Moho) The boundary separating the crust from the mantle, discernible by an increase in seismic velocity.

Mohs scale A series of ten minerals used as a standard in determining hardness.

Monsoon Seasonal reversal of wind direction associated with large continents, especially Asia. In winter, the wind blows from land to sea; in summer, from sea to land.

Monthly mean temperature The mean temperature for a month that is calculated by averaging the daily means.

Mountain breeze The nightly downslope winds commonly encountered in mountain valleys.

Mudflow The flow of debris containing a large amount of water; most characteristic of canyons and gullies in dry, mountainous regions.

Nansen bottle A device used by oceanographers to obtain subsurface seawater samples.

Natural levees The elevated landforms that parallel some streams and act to confine their waters, except during floodstage.

Neap tide Lowest tidal range, occurring near the times of the first- and third-quarter phases of the moon.

Nebula A cloud of interstellar gas and/or dust.

Nebular hypothesis The basic idea that the sun and planets formed from the same cloud of gas and dust in interstellar space.

Negative-feedback mechanism As used in climatic change, any effect that is opposite of the initial change and tends to offset it.

Neutron A subatomic particle found in the nucleus of an atom. The neutron is electrically neutral and has a mass approximately that of a proton.

Neutron star A star of extremely high density composed entirely of neutrons.

Nonconformity An unconformity in which older metamorphic or intrusive igneous rocks are overlain by younger sedimentary strata.

Nonfoliated Metamorphic rocks that do not exhibit foliation.

Nonmetallic mineral resource Mineral resource that is not a fuel or processed for the metals it contains.

Nonrenewable resource Resource that forms or accumulates over such long time spans that it must be considered as fixed in total quantity.

Normal fault A fault in which the rock above the fault plane has moved down relative to the rock below.

Normal polarity A magnetic field that is the same as that which exists at present.

Nova A star that explosively increases in brightness.

Nucleus The small heavy core of an atom that contains all of its positive charge and most of its mass.

Nuée ardente Incandescent volcanic debris buoyed up by hot gases that moves downslope in an avalanche fashion.

Objective lens In a refracting telescope, the long-focal-length lens that forms an image of the object viewed. The lens closest to the object.

Oblique fault A fault having both vertical and horizontal movement.

Obliquity The angle between the planes of the earth's equator and orbit.

Obsidian A volcanic glass of felsic composition.

Occluded front A front formed when a cold front overtakes a warm front. It marks the beginning of the end of a middle-latitude cyclone.

Occlusion The overtaking of one front by another.

Occultation An eclipse of a star or planet by the moon or a planet.

Oceanography The scientific study of the oceans and oceanic phenomena.

Open cluster A loosely formed group of stars of similar origin.

Orbit The path of a body in revolution around a center of mass.

Ore Usually a useful metallic mineral that can be mined at a profit. The term is also applied to certain nonmetallic minerals such as fluorite and sulfur.

Original horizontality Layers of sediments are generally deposited in a horizontal or nearly horizontal position.

Orogenesis The processes that collectively result in the formation of mountains.

Orographic lifting Mountains acting as barriers to the flow of air force the air to ascend. The air cools adiabatically, and clouds and precipitation may result.

Outer core A layer beneath the mantle about 2200 kilometers (1364 miles) thick that has the properties of a liquid.

Outwash plain A relatively flat, gently sloping plain consisting of materials deposited by meltwater streams in front of the margin of an ice sheet.

Oxbow lake A curved lake produced when a stream cuts off a meander.

Ozone A molecule of oxygen containing three oxygen atoms.

Pahoehoe A lava flow with a smooth-to-ropy surface.

Paleomagnetism The natural remnant magnetism in rock bodies. The permanent magnetization acquired by rock which can be used to determine the location of the magnetic poles and the latitude of the rock at the time it became magnetized.

Paleontology The systematic study of fossils and the history of life on earth.

Paleozoic era   A time span on the geologic calendar between the Precambrian and Mesozoic eras—from about 570 million to 245 million years ago.

Pangaea   The proposed supercontinent which 200 million years ago began to break apart and form the present land masses.

Parabolic dunes   The shape of these dunes resembles barchans except their tips point into the wind; they often form along coasts that have strong onshore winds, abundant sand, and vegetation that partly covers the sand.

Parallax   The apparent shift of an object when viewed from two different locations.

Parent material   The material upon which a soil develops.

Parsec   The distance at which an object would have a parallax angle of 1 second of arc (3.26 light-years).

Partial melting   The process by which most igneous rocks melt. Since individual minerals have different melting points, most igneous rocks melt over a temperature range of a few hundred degrees. If the liquid is squeezed out after some melting has occurred, a melt with a higher silica content results.

Pedalfer   Soil of humid regions characterized by the accumulation of iron oxides and aluminum-rich clays in the *B* horizon.

Pediment   A sloping bedrock surface fringing a mountain base in an arid region, formed when erosion causes the mountain front to retreat.

Pedocal   Soil associated with drier regions and characterized by an accumulation of calcium carbonate in the upper horizons.

Peneplain   In the idealized cycle of landscape evolution, an undulating plain near base level associated with old age.

Penumbra   The portion of a shadow from which only part of the light source is blocked by an opaque body.

Perched water table   A localized zone of saturation above the main water table created by an impermeable layer (aquiclude).

Peridotite   An igneous rock of ultramafic composition thought to be abundant in the upper mantle.

Perihelion   The point in the orbit of a planet where it is closest to the sun.

Period   A basic unit of the geologic calendar that is a subdivision of an era. Periods may be divided into smaller units called epochs.

Permeability   A measure of a material's ability to transmit water.

Perturbation   The gravitational disturbance of the orbit of one celestial body by another.

Phanerozoic eon   That part of geologic time represented by rocks containing abundant fossil evidence. The eon extending from the end of the Proterozoic eon (570 million years ago) to the present.

Phases of the moon   The progression of changes in the moon's appearance during the month.

Phenocryst   Conspicuously large crystals imbedded in a matrix of finer-grained crystals.

Photochemical reaction   A chemical reaction in the atmosphere that is triggered by sunlight, often yielding a secondary pollutant.

Photon   A discrete amount (quantum) of electromagnetic energy.

Photosphere   The region of the sun that radiates energy to space. The visible surface of the sun.

pH scale   A common measure of the degree of acidity or alkalinity of a solution, it is a logarithmic scale ranging from 0 to 14. A value of 7 denotes a neutral solution, values below 7 indicate greater acidity, and numbers above 7 indicate greater alkalinity.

Piedmont glacier   A glacier that forms when one or more valley glaciers emerge from the confining walls of mountain valleys and spread out to create a broad sheet in the lowlands at the base of the mountains.

Pipe   A vertical conduit through which magmatic materials have passed.

Placer   Deposit formed when heavy minerals are mechanically concentrated by currents, most commonly streams and waves. Placers are sources of gold, tin, platinum, diamonds, and other valuable minerals.

Plage   A bright region in the solar atmosphere located above a sunspot.

Plane of the ecliptic   The imaginary plane that connects the earth's orbit with the celestial sphere.

Planetary nebula   A shell of incandescent gas expanding from a star.

Plate   One of numerous rigid sections of the lithosphere that moves as a unit over the material of the asthenosphere.

Plate tectonics   The theory which proposes that the earth's outer shell consists of individual plates which interact in various ways and thereby produce earthquakes, volcanoes, mountains, and the crust itself.

Playa   A flat area on the floor of an undrained desert basin. Following heavy rain, the playa becomes a lake.

Playa lake   A temporary lake in a playa.

Pleistocene epoch   An epoch of the Quaternary period beginning about 1.6 million years ago and ending about 10,000 years ago. Best known as a time of extensive continental glaciation.

Plucking (quarrying)   The process by which pieces of bedrock are lifted out of place by a glacier.

Pluvial lake   A lake formed during a period of increased rainfall. During the Pleistocene epoch this occurred in some nonglaciated regions during periods of ice advance elsewhere.

Polar (P) air mass   A cold air mass that forms in a high-latitude source region.

Polar easterlies In the global pattern of prevailing winds, winds that blow from the polar high toward the subpolar low. These winds, however, should not be thought of as persistent winds, such as the trade winds.

Polar front The stormy frontal zone separating air masses of polar origin from air masses of tropical origin.

Polar high Anticyclones that are assumed to occupy the inner polar regions and are believed to be thermally induced, at least in part.

Polar wandering As the result of paleomagnetic studies in the 1950s, researchers proposed that either the magnetic poles migrated greatly through time or the continents had gradually shifted their positions.

Population I Stars rich in atoms heavier than helium. Nearly always relatively young stars found in the disk of the galaxy.

Population II Stars poor in atoms heavier than helium. Nearly always relatively old stars found in the halo, globular clusters, or nuclear bulge.

Porosity The volume of open spaces in rock or soil.

Porphyritic An igneous texture consisting of large crystals embedded in a matrix of much smaller crystals.

Positive-feedback mechanism As used in climatic change, any effect that acts to reinforce the initial change.

Pothole A depression formed in a stream channel by the abrasive action of the water's sediment load.

Precambrian All geologic time prior to the Paleozoic era.

Precession A slow motion of the earth's axis which traces out a cone over a period of 26,000 years.

Precipitation fog Fog formed when rain evaporates as it falls through a layer of cool air.

Pressure gradient The amount of pressure change occurring over a given distance.

Pressure tendency The nature of the change in atmospheric pressure over the past several hours. It can be a useful aid in short-range weather prediction.

Prevailing wind A wind that consistently blows from one direction more than from any other.

Primary pollutants Those pollutants emitted directly from identifiable sources.

Primary (P) wave A type of seismic wave that involves alternating compression and expansion of the material through which it passes.

Prominence A concentration of material above the solar surface that appears as a bright archlike structure.

Proterozoic eon The eon following the Archean and preceding the Phanerozoic. It extends between 2500 and 570 million years ago.

Proton A positively charged subatomic particle found in the nucleus of an atom.

Proton-proton chain A chain of thermonuclear reactions by which nuclei of hydrogen are built up into nuclei of helium.

Protostar A collapsing cloud of gas and dust destined to become a star.

Psychrometer A device consisting of two thermometers (wet bulb and dry bulb) that is rapidly whirled and, with the use of tables, yields the relative humidity and dew point.

Ptolemaic system An earth-centered system of the universe.

Pulsar A variable radio source of small size that emits radio pulses in very regular periods.

Pulsating variable A variable star that pulsates in size and luminosity.

Pyroclastic An igneous rock texture resulting from the consolidation of individual rock fragments that are ejected during a violent eruption.

Pyroclastic flow A highly heated mixture, largely of ash and pumice fragments, traveling down the flanks of a volcano or along the surface of the ground.

Pyroclastic material The volcanic rock ejected during an eruption, including ash, bombs, and blocks.

Radial pattern A system of streams running in all directions away from a central elevated structure, such as a volcano.

Radiation The transfer of energy (heat) through space by electromagnetic waves.

Radiation fog Fog resulting from radiation heat loss by the earth.

Radiation pressure The force exerted by electromagnetic radiation from an object such as the sun.

Radioactivity The spontaneous decay of certain unstable atomic nuclei.

Radiocarbon (carbon-14) The radioactive isotope of carbon, which is produced continuously in the atmosphere and is used in dating events as far back as 40,000 years.

Radio interferometer Two or more radio telescopes that combine their signals to achieve the resolving power of a larger telescope.

Radiometric dating The procedure of calculating the absolute ages of rocks and minerals that contain radioactive isotopes.

Radio telescope A telescope designed to make observations in radio wavelengths.

Rainshadow desert A dry area on the lee side of a mountain range. Many middle-latitude deserts are of this type.

Rapids A part of a stream channel in which the water suddenly begins flowing more swiftly and turbulently because of an abrupt steepening of the gradient.

Ray (lunar) Any of a system of bright elongated streaks, sometimes associated with a crater on the moon.

Recessional moraine An end moraine formed as the ice front stagnated during glacial retreat.

Rectangular pattern A drainage pattern characterized by numerous right angle bends that develops on jointed or fractured bedrock.

Red giant A large, cool star of high luminosity; a star occupying the upper right portion of the Hertzsprung-Russell diagram.

Reflecting telescope A telescope that concentrates light from distant objects by using a concave mirror.

Reflection nebula A relatively dense dust cloud in interstellar space that is illuminated by starlight.

Refracting telescope A telescope that employs a lens to bend and concentrate the light from distant objects.

Refraction The process by which the portion of a wave in shallow water slows, causing the wave to bend and tend to align itself with the underwater contours.

Regional metamorphism Metamorphism associated with the large-scale mountain-building processes.

Regolith The layer of rock and mineral fragments that nearly everywhere covers the earth's land surface.

Rejuvenation A change, often caused by regional uplift, that causes the forces of erosion to intensify.

Relative dating Rocks are placed in their proper sequence or order. Only the chronologic order of events is determined.

Relative humidity The ratio of the air's water vapor content to its water vapor capacity.

Renewable resource A resource that is virtually inexhaustible or that can be replenished over relatively short time spans.

Reserve Already identified deposits from which minerals can be extracted profitably.

Residual soil Soil developed directly from the weathering of the bedrock below.

Resolving power The ability of a telescope to separate objects that would otherwise appear as one.

Retrograde motion The apparent westward motion of the planets with respect to the stars.

Reverse fault A fault in which the material above the fault plane moves up in relation to the material below.

Reverse polarity A magnetic field opposite to that which exists at present.

Revolution The motion of one body about another, as the earth about the sun.

Richter scale A scale of earthquake magnitude based on the motion of a seismograph.

Rift zone A region of the earth's crust along which divergence is taking place.

Right ascension An angular distance measured eastward along the celestial equator from the vernal equinox. Used with declination in a coordinate system to describe the position of celestial bodies.

Rime A thin coating of ice on objects produced when supercooled fog droplets freeze on contact.

Rock A consolidated mixture of minerals.

Rock cycle A model that illustrates the origin of the three basic rock types and the interrelatedness of earth materials and processes.

Rock flour Ground-up rock produced by the grinding effect of a glacier.

Rockslide The rapid slide of a mass of rock downslope along planes of weakness.

Rotation The spinning of a body, such as the earth, about its axis.

Runoff Water that flows over the land rather than infiltrating into the ground.

Salinity The proportion of dissolved salts to pure water, usually expressed in parts per thousand $\left(\frac{0}{00}\right)$.

Saltation Transportation of sediment through a series of leaps or bounces.

Santa Ana The local name given a chinook wind in southern California.

Saturation The maximum possible quantity of water vapor that the air can hold at any given temperature and pressure.

Scoria Hardened lava which has retained the vesicles produced by escaping gases.

Sea arch An arch formed by wave erosion when caves on opposite sides of a headland unite.

Sea breeze A local wind blowing from the sea during the afternoon in coastal areas.

Sea-floor spreading The process of producing new sea floor between two diverging plates.

Seamount An isolated volcanic peak that rises at least 1000 meters (3000 feet) above the deep-ocean floor.

Sea stack An isolated mass of rock standing just offshore, produced by wave erosion of a headland.

Seawall A barrier constructed to prevent waves from reaching the area behind the wall. Its purpose is to defend property from the force of breaking waves.

Secondary enrichment The concentration of minor amounts of metals that are scattered through unweathered rock into economically valuable concentrations by weathering processes.

Secondary pollutants Pollutants that are produced in the atmosphere by chemical reactions that occur among primary pollutants.

Secondary (S) wave A seismic wave that involves oscillation perpendicular to the direction of propagation.

Sediment Unconsolidated particles created by the weathering and erosion of rock, by chemical precipitation from solution in water, or from the secretions of organisms and transported by water, wind, or glaciers.

Sedimentary rock Rock formed from the weathered products of pre-existing rocks that have been transported, deposited, and lithified.

Seismic sea wave A rapidly moving ocean wave generated by earthquake activity which is capable of inflicting heavy damage in coastal regions.

Seismogram   The record made by a seismograph.

Seismograph   An instrument that records earthquake waves.

Seismology   The study of earthquakes and seismic waves.

Semidiurnal tide   The predominant type of tide throughout the world, with two high waters and two low waters each tidal day.

Shadow zone   The zone between 104 and 143 degrees distance from an earthquake epicenter in which direct waves do not arrive because of refraction by the earth's core.

Sheeting   A mechanical weathering process characterized by the splitting off of slablike sheets of rock.

Shelf break   The point where a rapid steepening of the gradient occurs, marking the outer edge of the continental shelf and the beginning of the continental slope.

Shield volcano   A broad, gently sloping volcano built from fluid basaltic lavas.

Sidereal day   The period of rotation of the earth with respect to the stars.

Sidereal month   A time period based on the revolution of the moon around the earth with respect to the stars.

Silicate   Any one of numerous minerals that have the oxygen and silicon tetrahedron as their basic structure.

Silicon-oxygen tetrahedron   A structure composed of four oxygen atoms surrounding a silicon atom that constitutes the basic building block of silicate minerals.

Sill   A tabular igneous body that was intruded parallel to the layering of pre-existing rock.

Sink hole   A depression produced in a region where soluble rock has been removed by groundwater.

Sleet   Frozen or semifrozen rain formed when raindrops freeze as they pass through a layer of cold air.

Slide   A movement common to mass wasting processes in which the material moving downslope remains fairly coherent and moves along a well-defined surface.

Slip face   The steep, leeward slope of a sand dune; it maintains an angle of about 34 degrees.

Slump   The downward slipping of a mass of rock or unconsolidated material moving as a unit along a curved surface.

Smog   Originally used to describe a combination of smoke and fog, today the term is used as a synonym for general air pollution.

Snow   A solid form of precipitation produced by sublimation of water vapor.

Snowfield   An area where snow persists year-round.

Snowline   Lower limit of perennial snow.

Soil   A combination of mineral and organic matter, water, and air; that portion of the regolith that supports plant growth.

Soil horizon   A layer of soil that has identifiable characteristics produced by chemical weathering and other soil-forming processes.

Soil profile   A vertical section through a soil showing its succession of horizons and the underlying parent material.

Soil texture   The relative proportions of clay, silt, and sand in a soil. Texture strongly influences the soil's ability to retain and transmit water and air.

Solar constant   The rate at which solar radiation is received outside the earth's atmosphere on a surface perpendicular to the sun's rays when the earth is at an average distance from the sun.

Solar eclipse   An eclipse of the sun.

Solar flare   A sudden and tremendous eruption in the solar chromosphere.

Solar winds   Subatomic particles ejected at high speed from the solar corona.

Solifluction   Slow, downslope flow of water-saturated materials common to permafrost areas.

Solstice   The time when the vertical rays of the sun are striking either the Tropic of Cancer or the Tropic of Capricorn. Solstice represents the longest or shortest day (length of daylight) of the year.

Solum   The O, A, and B horizons in a soil profile. Living roots and other plant and animal life are largely confined to this zone.

Sorting   The process by which solid particles of various sizes are separated by moving water or wind. Also the degree of similarity in particle size in sediment or sedimentary rock.

Source region   The area where an air mass acquires its characteristic properties of temperature and moisture.

Specific gravity   The ratio of a substance's weight to the weight of an equal volume of water.

Specific humidity   The weight of water vapor compared with the total weight of the air, including the water vapor.

Spectral class   A classification of a star according to the characteristics of its spectrum.

Spectroscope   An instrument for directly viewing the spectrum of a light source.

Spectroscopy   The study of spectra.

Speed divergence   The divergence of air aloft that results from the variations in velocity that occur along the axis of a jet stream. On passing from a zone of slower wind speed to one of faster speed, air accelerates and therefore experiences divergence.

Spheroidal weathering   Any weathering process that tends to produce a spherical shape from an initially blocky shape.

Spicule   A narrow jet of rising material in the solar chromosphere.

Spiral galaxy   A flattened, rotating galaxy with pinwheel-like arms of interstellar material and young stars winding out from its nucleus.

Spit   An elongate ridge of sand that projects from the land into the mouth of an adjacent bay.

Spring   A flow of groundwater that emerges naturally at the ground surface.

Spring equinox   The equinox that occurs on March 21–22 in the Northern Hemisphere and on September 21–23 in the Southern Hemisphere.

Spring tide   Highest tidal range that occurs near the times of the new and full moons.

Stalactite   The iciclelike structure that hangs from the ceiling of a cavern.

Stalagmite   The columnlike form that grows upward from the floor of a cavern.

Star dune   Isolated hill of sand that exhibits a complex form and develops where wind directions are variable.

Stationary front   A situation in which the surface position of a front does not move; the flow on either side of such a boundary is nearly parallel to the position of the front.

Steam fog   Fog having the appearance of steam; produced by evaporation from a warm water surface into the cool air above.

Stellar parallax   A measure of stellar distance.

Steppe   One of the two types of dry climate. A marginal and more humid variant of the desert that separates it from bordering humid climates.

Stony-iron meteorite   One of the three main categories of meteorites. This group, as the name implies, is a mixture of iron and silicate minerals.

Stony meteorite   One of the three main categories of meteorites. Such meteorites are composed largely of silicate minerals with inclusions of other minerals.

Storm surge   The abnormal rise of the sea along a shore as a result of strong winds.

Strata   Parallel layers of sedimentary rock.

Stratified drift   Sediments deposited by glacial meltwater.

Stratopause   The boundary between the stratosphere and the mesosphere.

Stratosphere   The layer of the atmosphere immediately above the troposphere, characterized by increasing temperatures with height due to the concentration of ozone.

Stratovolcano   *See* Composite cone.

Stratus   One of three basic cloud forms; also the name given one of the low clouds. They are sheets or layers that cover much or all of the sky.

Streak   The color of a mineral in powdered form.

Striations (glacial)   Scratches or grooves in a bedrock surface caused by the grinding action of a glacier and its load of sediment.

Strike-slip fault   A fault along which the movement is horizontal.

Subduction   The process of thrusting oceanic lithosphere into the mantle along a convergent boundary.

Subduction zone   A long, narrow zone where one lithospheric plate descends beneath another.

Sublimation   The conversion of a solid directly to a gas without passing through the liquid state.

Submarine canyon   A seaward extension of a valley that was cut on the continental shelf during a time when sea level was lower, or a canyon carved into the outer continental shelf, slope, and rise by turbidity currents.

Submergent coast   A coast with a form that is largely the result of the partial drowning of a former land surface either because of a rise of sea level or subsidence of the crust or both.

Subpolar low   Low pressure located at about the latitudes of the Arctic and Antarctic circles. In the Northern Hemisphere the low takes the form of individual oceanic cells; in the Southern Hemisphere there is a deep and continuous trough of low pressure.

Subsoil   A term applied to the B horizon of a soil profile.

Subtropical high   Not a continuous belt of high pressure but rather several semipermanent, anticyclonic centers characterized by subsidence and divergence located roughly between latitudes 25 and 35 degrees.

Summer solstice   The solstice that occurs on June 21–22 in the Northern Hemisphere and on December 21–22 in the Southern Hemisphere.

Sunspot   A dark spot on the sun, which is cool by contrast to the surrounding photosphere.

Supercooled   The condition of water droplets that remain in the liquid state at temperatures well below 0°C.

Supergiant   A very large star of high luminosity.

Supernova   An exploding star that increases in brightness many thousands of times.

Superposition   In any undeformed sequence of sedimentary rocks, each bed is older than the layers above and younger than the layers below.

Surf   A collective term for breakers; also the wave activity in the area between the shoreline and the outer limit of breakers.

Surface soil   The uppermost layer in a soil profile: the A horizon.

Surface waves   Seismic waves that travel along the outer layer of the earth.

Suspended load   The fine sediment carried within the body of flowing water.

Swells   Wind-generated waves that have moved into an area of weaker winds or calm.

Syncline   A linear downfold in sedimentary strata; the opposite of anticline.

Synodic month   The period of revolution of the moon with respect to the sun, or its cycle of phases.

Talus   An accumulation of rock debris at the base of a cliff.

Tarn   A small lake in a cirque.

Tectonics The study of the large-scale processes that collectively deform the earth's crust.

Temperature inversion A layer in the atmosphere of limited depth where the temperature increases rather than decreases with height.

Temporary (local) base level The level of a lake, resistant rock layer, or any other base level that stands above sea level.

Terminal moraine The end moraine marking the farthest advance of a glacier.

Terrace A flat, benchlike structure produced by a stream, which was left elevated as the stream cut downward.

Terrane A crustal block bounded by faults, whose geologic history is distinct from the histories of adjoining crustal blocks.

Terrestrial planet Any of the Earth-like planets including Mercury, Venus, Mars, and Earth.

Texture The size, shape, and distribution of the particles that collectively constitute a rock.

Theory A well-tested and widely accepted view that explains certain observable facts.

Thermal gradient The increase in temperature with depth. It average 1°C per 30 meters (1−2°F per 100 feet) in the crust.

Thermocline A layer of water in which there is a rapid change in temperature in the vertical dimension.

Thermohaline circulation Movements of ocean water caused by density differences brought about by variations in temperature and salinity.

Thermosphere The region of the atmosphere immediately above the mesosphere and characterized by increasing temperatures due to absorption of very short-wave solar energy by oxygen.

Thrust fault A low-angle reverse fault.

Thunder The sound emitted by rapidly expanding gases along the channel of lightning discharge.

Thunderstorm A storm produced by a cumulonimbus cloud and always accompanied by lightning and thunder. It is of relatively short duration and usually accompanied by strong wind gusts, heavy rain, and sometimes hail.

Tidal current The alternating horizontal movement of water associated with the rise and fall of the tide.

Tidal delta A deltalike feature created when a rapidly moving tidal current emerges from a narrow inlet and slows, depositing its load of sediment.

Tidal flat A marshy or muddy area that is covered and uncovered by the rise and fall of the tide.

Tide Periodic change in the elevation of the ocean surface.

Till Unsorted sediment deposited directly by a glacier.

Tombolo A ridge of sand that connects an island to the mainland or to another island.

Tornado A small, very intense cyclonic storm with exceedingly high winds, most often produced along cold fronts in conjunction with severe thunderstorms.

Tornado warning A warning issued when a tornado has actually been sighted in an area or is indicated by radar.

Tornado watch A forecast issued for areas of about 65,000 square kilometers (25,000 square miles) indicating that conditions are such that tornadoes may develop; it is intended to alert people to the possibility of tornadoes.

Trade winds Two belts of winds that blow almost constantly from easterly directions and are located on the equatorward sides of the subtropical highs.

Transform boundary A boundary in which two plates slide past one another without creating or destroying lithosphere.

Transpiration The release of water vapor to the atmosphere by plants.

Transported soil Soils that form on unconsolidated deposits.

Transverse dunes A series of long ridges oriented at right angles to the prevailing wind; these dunes form where vegetation is sparse and sand is very plentiful.

Travertine A form of limestone ($CaCO_3$) that is deposited by hot springs or as a cave deposit.

Trellis pattern A system of streams in which nearly parallel tributaries occupy valleys cut in folded strata.

Trench An elongate depression in the sea floor produced by bending of oceanic crust during subduction.

Tropical depression By international agreement, a tropical cyclone with maximum winds that do not exceed 61 kilometers (38 miles) per hour.

Tropical storm By international agreement, a tropical cyclone with maximum winds between 61 and 119 kilometers (38 and 74 miles) per hour.

Tropic of Cancer The parallel of latitude, 23½ degrees north latitude, marking the northern limit of the sun's vertical rays.

Tropic of Capricorn The parallel of latitude, 23½ degrees south latitude, marking the southern limit of the sun's vertical rays.

Tropopause The boundary between the troposphere and the stratosphere.

Troposphere The lowermost layer of the atmosphere. It is generally characterized by a decrease in temperature with height.

Tsunami The Japanese word for a seismic sea wave.

Turbidite Turbidity current deposit characterized by graded bedding.

Turbidity current A downslope movement of dense, sediment-laden water created when sand and mud on the continental shelf and slope are dislodged and thrown into suspension.

Ultimate base level   Sea level; the lowest level to which stream erosion could lower the land.

Ultramafic   Igneous rocks composed mainly of iron and magnesium-rich minerals.

Ultraviolet   Radiation with a wavelength from 0.2 to 0.4 micrometer.

Umbra   The central, completely dark part of a shadow produced during an eclipse.

Unconformity   A surface that represents a break in the rock record, caused by erosion or nondeposition.

Uniformitarianism   The concept that the processes that have shaped the earth in the geologic past are essentially the same as those operating today.

Upslope fog   Fog created when air moves up a slope and cools adiabatically.

Upwelling   The rising of cold water from deeper layers to replace warmer surface water that has been moved away.

Urban heat island   Refers to the fact that temperatures within a city are generally higher than in surrounding rural areas.

Valley breeze   The daily upslope winds commonly encountered in a mountain valley.

Valley glacier   *See* Alpine glacier.

Valley train   A relatively narrow body of stratified drift deposited on a valley floor by meltwater streams that issue from a valley glacier.

Vapor pressure   That part of the total atmospheric pressure attributable to water vapor content.

Vein deposit   A mineral filling a fracture or fault in a host rock. Such deposits have a sheetlike, or tabular, form.

Ventifact   A cobble or pebble polished and shaped by the sandblasting effect of wind.

Vernal equinox   *See* Spring equinox.

Vesicular   A term applied to igneous rocks that contain small cavities called vesicles, which are formed when gases escape from lava.

Viscosity   A measure of a fluid's resistance to flow.

Visible light   Radiation with a wavelength from 0.4 to 0.7 micrometer.

Volcanic bomb   A streamlined pyroclastic fragment ejected from a volcano while molten.

Volcanic neck   An isolated, steep-sided, erosional remnant consisting of lava that once occupied the vent of a volcano.

Volcano   A mountain formed of lava and/or pyroclastics.

Warm front   A front along which a warm air mass overrides a retreating mass of cooler air.

Wash   A common term for a desert stream course which is typically dry except for brief periods immediately following a rain.

Water table   The upper level of the saturated zone of groundwater.

Wave-cut cliff   A seaward-facing cliff along a steep shoreline formed by wave erosion at its base and mass wasting.

Wave-cut platform   A bench or shelf in the bedrock at sea level, cut by wave erosion.

Wave cyclone   A cyclone that forms and moves along a front. The circulation around the cyclone tends to produce the wavelike deformation of the front.

Wave height   The vertical distance between the trough and crest of a wave.

Wavelength   The horizontal distance separating successive crests or troughs.

Wave of oscillation   A water wave in which the wave form advances as the water particles move in circular orbits.

Wave of translation   The turbulent advance of water created by breaking waves.

Wave period   The time interval between the passage of successive crests at a stationary point.

Wave refraction   *See* Refraction.

Weather   The state of the atmosphere at any given time.

Weathering   The disintegration and decomposition of rock at or near the surface of the earth.

Welded tuff   A pyroclastic rock composed of particles that have been fused together by the combination of heat still contained in the deposit after it has come to rest and the weight of overlying material.

Well   An opening bored into the zone of saturation.

Westerlies   The dominant west-to-east motion of the atmosphere that characterizes the regions on the poleward side of the subtropical highs.

Wet adiabatic rate   The rate of adiabatic temperature change in saturated air. The rate of temperature change is variable, but it is always less than the dry adiabatic rate.

White dwarf   A star that has exhausted most or all of its nuclear fuel and has collapsed to a very small size; believed to be near its final stage of evolution.

White frost   Ice crystals instead of dew that form on surfaces when the dew point is below freezing.

Wind   Air flowing horizontally with respect to the earth's surface.

Wind vane   An instrument used to determine wind direction.

Winter solstice   The solstice that occurs on December 21–22 in the Northern Hemisphere and on June 21–22 in the Southern Hemisphere.

Yazoo tributary   A tributary that flows parallel to the main stream because a natural levee is present.

Zodiac   A band along the ecliptic containing the twelve constellations of the zodiac.

Zone of accumulation   The part of a glacier characterized by snow accumulation and ice formation. Its outer limit is the snowline.

Zone of aeration   Area above the water table where openings in soil, sediment, and rock are not saturated but filled mainly with air.

Zone of fracture   The upper portion of a glacier consisting of brittle ice.

Zone of saturation   Zone where all open spaces in sediment and rock are completely filled with water.

Zone of wastage   The part of a glacier beyond the zone of accumulation where all of the snow from the previous winter melts, as does some of the glacial ice.

Index

ISBN 0-02-419025-X

9 780024 190253